Statistics for Experimenters

WILEY SERIES IN PROBABILITY AND STATISTICS

Established by WALTER A. SHEWHART and SAMUEL S. WILKS

A complete list of the titles in this series appears at the end of this volume.

Statistics for Experimenters

Design, Innovation, and Discovery

Second Edition

GEORGE E. P. BOX

J. STUART HUNTER

WILLIAM G. HUNTER

WILEY-INTERSCIENCE

A JOHN WILEY & SONS, INC., PUBLICATION

Published by John Wiley & Sons, Inc., Hoboken, New Jersey.
Published simultaneously in Canada.

Library of Congress Cataloging-in-Publication Data:

Box, George E. P.
Statistics for experimenters : design, discovery, and innovation / George E.P. Box, J. Stuart Hunter, William G. Hunter.– 2nd ed.
p. cm.
Includes bibliographical references and index.
ISBN 0-471-71813-0 (acid-free paper)
1. Experimental design 2. Analysis of variance. 3. Mathematical statistics. I. Hunter, J. Stuart, 1923– II. Hunter, William Gordon, 1937– III.Title.

QA279.B69 2005
519.5—dc22

2004063780

Printed in the United States of America.

10 9

To the memory of
WILLIAM GORDON HUNTER

WILLIAM G. HUNTER
1937–1986

Experiment!
Make it your motto day and night.
Experiment,
And it will lead you to the light.

The apple on the top of the tree
Is never too high to achieve,
So take an example from Eve...
Experiment!

Be curious,
Though interfering friends may frown.
Get furious
At each attempt to hold you down.

If this advice you only employ,
The future can offer you infinite joy
And merriment ...
Experiment
And you'll see!

COLE PORTER*

When the Lord created the world and people to live in it—an enterprise which, according to modern science, took a very long time—I could well imagine that He reasoned with Himself as follows: "If I make everything predictable, these human beings, whom I have endowed with pretty good brains, will undoubtedly learn to predict everything, and they will thereupon have no motive to do anything at all, because they will recognize that the future is totally determined and cannot be influenced by any human action. On the other hand, if I make everything unpredictable, they will gradually discover that there is no rational basis for any decision whatsoever and, as in the first case, they will thereupon have no motive to do anything at all. Neither scheme would make sense. I must therefore create a mixture of the two. Let some things be predictable and let others be unpredictable. They will then, amongst many other things, have the very important task of finding out which is which."

E. F. SCHUMACHER*

*From *Small Is Beautiful.* Used by permission

Greek Alphabet

Α α	alpha	Ν ν	nu
Β β	beta	Ξ ξ	xi
Γ γ	gamma	Ο o	omicron
Δ δ	delta	Π π	pi
Ε ε	epsilon	Ρ ρ	rho
Ζ ζ	zeta	Σ σ	sigma
Η η	eta	Τ τ	tau
Θ θ	theta	Υ υ	upsilon
Ι ι	iota	Φ ϕ	phi
Κ κ	kappa	Χ χ	chi
Λ λ	lambda	Ψ ψ	psi
Μ μ	mu	Ω ω	omega

Contents

Preface to the Second Edition

In rewriting this book, we have deeply felt the loss of our dear friend and colleague Bill Hunter (William G. Hunter, 1937–1986). We miss his ideas, his counsel, and his cheerful encouragement. We believe, however, that his spirit has been with us and that he would be pleased with what we have done.

The objectives for this revised edition of *Statistics for Experimenters* remain the same as those for the first:

1. To make available to experimenters scientific and statistical tools that can greatly catalyze innovation, problem solving, and discovery.
2. To illustrate how these tools may be used by and with subject matter specialists as their investigations proceed.

Developments that would have delighted Bill are the receptive atmosphere these techniques now encounter in industry and the present universal availability of very fast computers, allowing* where necessary, the ready use of computationally intensive methods.

Under such banners as "Six Sigma," management has realized the importance of training its work forces in the arts of economic investigation. With this democratization of the scientific method, many more people are being found with creative ability and unrealized aptitude for problem solving and discovery. Also, the "team idea" not only accelerates improvement but identifies such natural leaders of innovation and can allow them to lead. To make such initiatives possible the modern philosophy and methods of process improvement must be taught at all levels of an organization. We believe both trainers and trainees engaged in such efforts will find this book helpful. Also based on a long experience, we

*All the computations in this book can be done with the statistical language R (R Development Core Team, 2004), available at CRAN (http://cran.R-project.org). Functions for displaying anova and lambda plots, for Bayesian screening and model building are included in the BHH2 and BsMD R-packages and available at CRAN under contributed packages. There is as well commercial software, such as the SCA Statistical System, which some readers will find easier to use. All the data for the exercises, examples, and problems are also available at ftp://ftp.wiley.com/public/sci_tech_med/statistics_experimenters

believe the material in this book provides appropriate training for scientists and engineers at universities whose needs have in the past been frequently neglected.

The material of the original book has been rearranged and largely rewritten with the object of ensuring even greater accessibility to its users. In addition, a number of new things have been introduced or emphasized:

The many issues, in addition to statistical considerations, that must be attended to in a successful investigation.

The need to work closely with subject matter specialists.

The importance of choosing a relevant reference set.

The value of graphical techniques as an adjunct to more formal methods.

The use of inactive factor and canonical spaces to solve multiple response problems.

The greatly increased efficiency sometimes achievable by data transformation using "lambda plots."

The understanding of error transmission and its use in designing products and processess that are least affected by (robust to) variation in system components.

The value of split plot and similar arrangements in industrial experimentation particularly in the design of environmentally robust products and processes.

The value of a sequential approach to problem solving and in particular the sequential assembly of experimental designs.

How to choose the best follow-up runs.

The acquiring of investigational technique by hands-on design of some simple device (e.g., a paper helicopter).

How empiricism can lead to mechanism.

The use of randomization and blocking to allow analysis "as if" standard assumptions were true.

The use of complex experimental arrangements, in particular Plackett Burman designs and their analysis.

Use of the *design information function.*

An introduction to process control, to forecasting, and to time series analysis.

A discussion of evolutionary process operation.

USING THE QUESTIONS AND PROBLEMS

The questions and problems at the end of each chapter can be used in two ways. You can consider them for review *before* reading the chapter to help identify key points and to guide your reading and you can consider them for practice and exercise *after* reading this chapter. You may also find helpful the collection of quotes on the inside covers of this book.

As you apply these ideas and especially if you meet with unusual success or failure, we shall be interested to learn of your experiences. We have tried to write a book that is useful and clear. If you have any suggestions as to how it can be improved, please write us.

ACKNOWLEDGMENTS

For help with the preparation of this book we are especially grateful for the generous assistance of Ernesto Barrios Zamudio. We also wish to thank René Valverde Ventura, Carla Vivacqua and Carmen Paniagua Quiñones. We are indebted to David Bacon, Mac Berthouex, Søren Bisgaard, Bob DeBaun, David Steinberg, Dan Meyer, Bill Hill, Merv Muller and Daniel Peña for their critiques and valuable suggestions on earlier versions of the manuscript and to Claire Box and Tady Hunter who have provided patient encouragement throughout this long endeavor. We are particularly grateful to our editor Lisa Van Horn for cheerful and wise counsel throughout the book's preparation.

George E. P. Box
J. Stuart Hunter

Madison, Wisconsin
Princeton, New Jersey
April 2005

CHAPTER 1

Catalyzing the Generation of Knowledge

1.1. THE LEARNING PROCESS

Knowledge is power. It is the key to innovation and profit. But the getting of new knowledge can be complex, time consuming, and costly. To be successful in such an enterprise, you must learn about learning. Such a concept is not esoteric. It is the key to idea generation, to process improvement, to the development of new and robust products and processes. By using this book you can greatly simplify and accelerate the generation, testing, and development of new ideas. You will find that statistical methods and particularly experimental design catalyze scientific method and greatly increase its efficiency.

Learning is advanced by the iteration illustrated in Figure 1.1. An initial idea (or model or hypothesis or theory or conjecture) leads by a process of *deduction* to certain necessary consequences that may be compared with data. When consequences and data fail to agree, the discrepancy can lead, by a process called *induction,* to modification of the model. A second cycle in the iteration may thus be initiated. The consequences of the modified model are worked out and again compared with data (old or newly acquired), which in turn can lead to further modification and gain of knowledge. The data acquiring process may be scientific experimentation, but it could be a walk to the library or a browse on the Internet.

Inductive–Deductive Learning: An Everyday Experience

The iterative inductive–deductive process, which is geared to the structure of the human brain and has been known since the time of Aristotle, is part of one's everyday experience. For example, a chemical engineer Peter Minerex* parks

* Can you guess why he's called Peter Minerex?

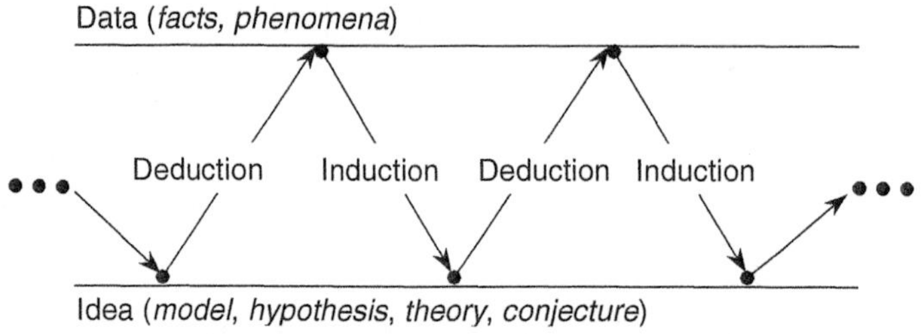

Figure 1.1. Iterative learning process.

his car every morning in an allocated parking space. One afternoon after leaving work he is led to follow the following deductive–inductive learning sequence:

Model: Today is like every day.
Deduction: My car will be in its parking place.
Data: It isn't.
Induction: Someone must have taken it.

Model: My car has been stolen.
Deduction: My car will not be in the parking lot
Data: No. It's over there!
Induction: Someone took it and brought it back.

Model: A thief took it and brought it back.
Deduction: My car will have been broken into.
Data: It's unharmed and unlocked.
Induction: Someone who had a key took it.

Model: My wife used my car.
Deduction: She probably left a note.
Data: Yes. Here it is.

Suppose you want to solve a particular problem and initial speculation produces some relevant idea. You will then seek data to further support or refute this theory. This could consist of some of the following: a search of your files and of the Web, a walk to the library, a brainstorming meeting with co-workers and executives, passive observation of a process, or active experimentation. In any case, the facts and data gathered sometimes confirm your conjecture, in which case you may have solved your problem. Often, however, it appears that your initial idea is only partly right or perhaps totally wrong. In the latter two cases, the difference between deduction and actuality causes you to keep digging. This can point to a modified or totally different idea and to the reanalysis of your present data or to the generation of new data.

Humans have a two-sided brain specifically designed to carry out such continuing deductive–inductive conversations. While this iterative process can lead to a solution of a problem, you should not expect the nature of the solution, or the route by which it is reached, to be unique.

Chemical Example

Rita Stoveing,* a chemist, had the following idea:

Model Because of certain properties of a newly discovered catalyst, its presence in a particular reaction mixture would probably cause chemical A to combine with chemical B to form, in high yield, a valuable product C.

Deduction Rita has a tentative hypothesis and deduces its consequences but has no data to verify or deny its truth. So far as she can tell from conversations with colleagues, careful examination of the literature, and further searches on the computer, no one has ever performed the operation in question. She therefore decides she should run some appropriate experiments. Using her knowledge of chemistry, she makes an experimental run at carefully selected reaction conditions. In particular, she guesses that a temperature of 600°C would be worth trying.

Data The result of the first experiment is disappointing. The desired product C is a colorless, odorless liquid, but what is obtained is a black tarry product containing less than 1% of the desired substance C.

Induction At this point the initial model and data do not agree. The problem worries Rita and that evening she is somewhat short with her husband, Peter Minerex, but the next morning in the shower she begins to think along the following lines. Product C might first have been formed in high yield, but it could then have been decomposed.

Model Theory suggests the reaction conditions were too severe.

Deduction A lower temperature might produce a satisfactory yield of C. Two further runs are made with the reaction temperature first reduced to 550°C and then to 500°C.

Data The product obtained from both runs is less tarry and not so black. The run at 550°C yields 4% of the desired product C, and the run at 500°C yields 17%

Induction Given these data and her knowledge of the theory of such reactions, she decides she should experiment further not only with temperature but also with a number of other factors (e.g., concentration, reaction time, catalyst charge) and to study other characteristics of the product (e.g., the levels of various impurities, viscosity).

*Can you guess why she's called Rita Stoveing?

To evaluate such complex systems economically, she will need to employ designed experiments and statistical analysis. Later in this book you will see how subsequent investigation might proceed using statistical tools.

Exercise 1.1. Describe a real or imagined example of iterative learning from your own field—engineering, agriculture, biology, genomics, education, medicine, psychology, and so on.

A Feedback Loop

In Figure 1.2 the deductive–inductive iteration is shown as a process of feedback. An initial idea (hypothesis, model) is represented by M_1 on the left of the diagram. By *deduction* you consider the expected consequences of M_1—what might happen if M_1 is true and what might happen if M_1 is false. You also *deduce* what data you will need to explore M_1. The experimental plan (design) you decide on is represented here by a frame through which some aspects of the true state of nature are seen. Remember that when you run an experiment the frame is at your choice* (that's *your* hand holding the frame). The data produced represent some aspect (though not always a relevant one) of the true state of nature obscured to a greater or lesser extent by "noise," that is, by experimental error. The analyzed data may be compared with the expected (deduced) consequences of M_1. If they agree, your problem may be solved. If they disagree, the way they disagree can allow you to see how your initial idea M_1 may need to be

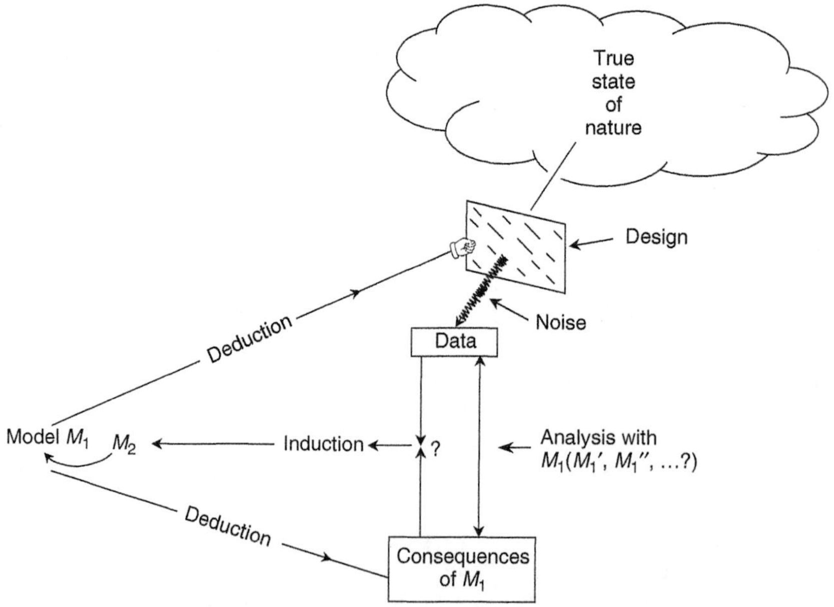

Figure 1.2. Iterative problem solving seen as a feedback loop.

*This is not true, of course, for happenstance data over which you have no control.

modified. Using the same data, you may consider alternative analyses and also possible modifications of the original model M_1', M_1'', ... It may become clear that your original idea is wrong or at least needs to be considerably modified. A new model M_2 may now be postulated. This may require you to choose a new or augmented experimental design to illuminate additional and possibly different aspects of the state of nature. This could lead to a satisfactory solution of the problem or, alternatively, provide clues indicating how best to proceed.

1.2. IMPORTANT CONSIDERATIONS

Subject Matter Knowledge

Notice the importance of subject matter knowledge to perceive and explore tentative models and to know where to look for help.

The Route to Problem Solving Is Not Unique

When he first noticed that his car was missing, Peter Minerex might easily have behaved differently. For example, he might immediately have phoned the police and thus initiated different (but perhaps not equally effective) routes to discovery. Similarly, in the chemical investigation a different experimenter, after studying the disappointing results, might have decided to explore an entirely different chemical route to obtain the desired product. The object is to *converge* to a satisfactory solution—the starting point and the route (and sometimes the nature of the solution) will be different for different investigators.

The game of 20 Questions illustrates these points. In the game the object is to identify an unknown object using no more than 20 questions, each of which has only one of two distinct answers. Suppose that the object to be guessed is *Abraham Lincoln's stove pipe hat.* The initial clue is *vegetable with animal associations.* For a competent team of players presented with this initial clue the game might go as follows:

TEAM A

Question	Answer
Are the animal associations human?	Yes
Male or female?	Male
Famous or not?	Famous
Connected with the arts	No
Politician?	Yes
USA or other?	USA
This century or not?	Not
Nineteenth or eighteenth century?	Nineteenth
Connected with the Civil War?	Yes
Lincoln?	Yes
Is the object Lincoln's hat?	Yes

But for a different team of competent players the game would almost certainly follow a different route. For example:

TEAM B

Question	Answer
Is the object useful?	Yes
Is it an item of dress?	Yes
Male of female?	Male
Worn above or below the belt?	Above
Worn on the head or not?	Head
Is it a famous hat?	Yes
Winston Churchill's hat?	No
Abraham Lincoln's hat?	Yes

The game follows the iterative pattern of Figures 1.1 and 1.2, where the "design" is the choice of question. At each stage conjecture, progressively refined, leads to an appropriate choice of a question that elicits the new data, which leads in turn to appropriate modification of the conjecture. Teams *A* and *B* followed different routes, but each was led to the correct answer because the data were generated by the truth.

The qualities needed to play this game are (a) subject matter knowledge and (b) knowledge of strategy. Concerning strategy, it is well known that at each stage a question should be asked that divides the objects not previously eliminated into approximately equiprobable halves. In these examples the players usually try to do this with questions such as "male or female?" or "worn above or below the belt?"*

Knowledge of strategy parallels knowledge of statistical methods in scientific investigation. Notice that without knowledge of strategy you can always play the game, although perhaps not very well, whereas without subject matter knowledge it cannot be played at all. However, it is by far best to use *both* subject matter knowledge and strategy. It is possible to conduct an investigation without statistics but impossible to do so without subject matter knowledge. However, by using statistical methods convergence to a solution is speeded and a good investigator becomes an even better one.

1.3. THE EXPERIMENTER'S PROBLEM AND STATISTICAL METHODS

Three problems that confront the investigator are *complexity, experimental error*, and the difference between *correlation and causation.*

*From a dictionary containing a million words a single word can be found playing 20 Questions. The questions begin, "Is it in the front half or the back half of the dictionary?" If the answer is, say, "The front half," then the next question is, "Is it in the front half or the back half of that half?" And so on. Note that $2^{20} \gg 10^6$.

Complexity

In experimentation for process improvement* and discovery it is usually necessary to consider simultaneously the influence of a number of "input variables" such as temperature, feed rate, concentration, and catalyst on a collection of output variables such as yield, impurity, and cost. We call controllable input variables *factors* and output variables *responses*. In studying the question as to how to improve a process the prime question is,

"What does what to what?"

With k factors and p responses there are $k \times p$ entities to consider. Further, while a certain subset of factors (e.g., temperature and pressure) might be available to change one response (e.g., yield), a quite different or likely overlapping subset (e.g., temperature and concentration) might influence a different response (e.g., purity). Compromises may be necessary to reach a satisfactory high yield and adequate purity. Also, some factors will interact in their influence on a particular response. For example, the change in yield induced by a particular change in temperature might itself change at different concentrations. To take account of all these matters simultaneously faces the experimenter with a daunting challenge. Pick and try guesses and the use of the "change one factor at a time" philosophy of experimentation is unlikely to produce a good result quickly and economically.

The use of statistical experimental design makes it possible while minimizing the influence of experimental error, to experiment with numbers of factors simultaneously. In this way you can get a clear picture of how they behave separately and together. Such understanding can lead to empirical solutions to problems, but it can do much more. A subject matter specialist provided with the results from a well-run experiment may reason along the following lines: "When I see what x_3 does to y_1 and y_2 and how x_1 and x_2 interact in their effect on y_3, this suggests to me that what is going on is thus and so and I think what we ought to do now is this." Theoretical understanding can spring from empirical representation.

Experimental Error

Variability not explained by known influences is called *experimental error*. Since some experimental error is inevitable, knowing how to cope with it is essential. Frequently, only a small part of experimental error is attributable to errors in measurement. Variations in raw materials, in sampling, and in the settings of the experimental factors often provide larger components. Good experimental design helps to protect real effects from being obscured by experimental error and conversely having the investigator mistakenly believe in effects that do not exist.

* The word *process* is used here in its general sense. Thus a process might be an analytical method, the manufacture of a product, or some medical procedure.

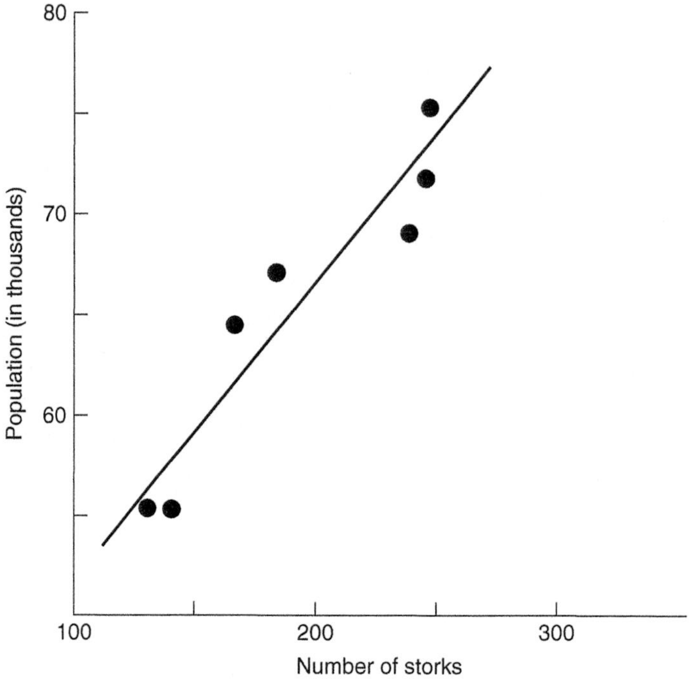

Figure 1.3. Number of storks versus population of Oldenburg.

The confusing influence of experimental error is greatly reduced by the wise use of statistical experimental design.* In addition, statistical analysis provides *measures of precision* of estimated quantities under study (such as differences in means or rates of change). This makes it possible to judge whether there is solid evidence for the existence of real effects and greatly increases the probability that the investigator will be led along a true rather than a false path.

Confusion of Correlation with Causation

Figure 1.3 shows the civilian population of Oldenburg plotted against the number of storks observed at the end of each of 7 years.† Although in this example few would be led to hypothesize that the increase in the number of storks *caused* the observed increase in population, investigators are sometimes guilty of this kind of mistake in other contexts. Correlation between two variables Y

*Another way to state this is to say that designed experiments can greatly increase the signal-to-noise ratio.

†These data cover the years 1930–1936. See *Ornithologische Monatsberichte*, **44**, No. 2, Jahrgang, 1936, Berlin, and **48**, No. 1, Jahrgang, 1940, Berlin, and *Statistiches, Jahrbuch Deutscher Gemeinden*, 27–22, Jahrgang, 1932–1938, Gustas Fischer, Jena. We are grateful to Lars Pallesen for these references.

and X often occurs because they are *both* associated with a third factor W. In the stork example, since the human population Y and the number of storks X both increased over this 7-year period, the common factor W was in this case *time*.

Exercise 1.2. Give other examples where correlation exists but causation does not.

1.4. A TYPICAL INVESTIGATION

To illustrate the process of iterative learning and to provide simultaneously a preview of what is discussed in this book, we employ a description of an imaginary investigation aimed at improving the quality of drinking water. Our investigators are the chemist Rita Stoveing and the chemical engineer Peter Minerex. As you read what follows, consider how identical *statistical* problems could confront investigators in any other experimental science.

The Problem

There is of course only a limited amount of usable water on this planet and what we have must be used and reused. The following investigation was necessary because a particular water supply contained an unacceptably high level of nitrate. Minerex and Stoveing had developed a new ion exchange resin that absorbed the offending nitrate. The attractive feature of this new resin was that it was specific for nitrate and potentially much cheaper to use and to regenerate than currently available resins. Unfortunately, it could only be made under laboratory conditions in experimental quantities. It was realized that a good deal of further experimentation might be necessary before a commercially viable product could be produced. The following outline shows how, as their investigation proceeded, they would be led to consider different questions and employ different statistical techniques* of varying degrees of sophistication.

Iterative Cycles of Investigation

Rita and Peter knew that observations recorded under apparently similar conditions could vary considerably. Thus, before beginning their investigation every effort was made to stabilize procedures and to reduce process variation. Furthermore, they expected to have to experiment with a number of different factors and to look at several different responses. To make such a study efficient and economical, they would need to make extensive use of statistical methods.

*This imaginary investigation has the property that in successive iterations it uses most of the techniques discussed in this book approximately in the order in which they appear. This is, of course, merely a pedagogical device. However, many investigations do go through various phases of statistical sophistication in somewhat the manner described here.

Commentary	The two investigators had each taken a statistics course but were currently a bit rusty.

ITERATION I

Question	Where can we find a quick summary of elementary statistical principles?
Design and Analysis	Chapter 2: Basics
Findings	Studying this chapter provided the needed review and prepared them for what followed.

ITERATION II

Commentary	Minerex believed, and Stoveing did not, that using a very pure (and more expensive) version of their resin would improve its performance.
Question	How does their "ordinary" resin compare with the more expensive high-purity version?
Design and Analysis	Chapter 3: Comparing Two Entities
Findings	Their expensive high-purity resin was about equal in performance to the ordinary resin. (She was right!)

ITERATION III

Commentary	Minerex was wrong about the high-purity version, but their ordinary resin still looked promising. They now decided to compare laboratory samples of their new resin with five standard commercially available resins.
Question	How does their new ordinary resin compare with the five commercially available resins?
Design and Analysis	Chapter 4: Comparing a Number of Entities
Findings	The laboratory samples of the new resin were as good as any of the commercially available alternatives and perhaps somewhat superior.

ITERATION IV

Commentary	The new resin had been shown to be capable of doing as well as its competitors. However, under the conditions contemplated for economic manufacture the removal of nitrate was insufficient to achieve the standard required for drinking water.
Question	What are the most important factors influencing nitrate removal? Can modifications in the present manufacturing equipment affecting such factors as flow rate, bed depth and regeneration time lead to improved nitrate removal?
Design and Analysis	Chapters 5, 6, 7, and 8: Studies using factorial and fractional factorial designs

Findings	With suitable modifications of the equipment, sufficiently low nitrate levels could be achieved.

ITERATION V

Commentary	The company now concluded that manufacture of this new resin was possible and could be profitable. To learn more, a pilot plant was built.
Question	How do the settings of the process factors affect the quality and cost of the new resin? What are the best settings?
Design and Analysis	Chapters 10, 11, and 12: The method of least squares, multidimensional modeling, and response surfaces
Findings	The pilot plant investigation indicated that with appropriate settings of the process factors a product could be produced with satisfactory quality at a reasonable cost.

ITERATION VI

Commentary	Before the process could be recommended for production the problems of sampling and testing had to be solved.
Question	How can sampling and testing methods be refined to give reliable determinations of the characteristics of the new resin?
Design and Analysis	Chapter 9: Multiple Sources of Variation
Findings	Components of variability in the sampling and chemical analysis of the product were identified and measured. Using this information, a sampling and testing protocol was devised to minimize the variance of the determinations at least cost.

ITERATION VII

Commentary	Before the new resin could be recommended commercially its behavior had to be considered under a variety of environmental conditions that might be encountered at different locations and at different times. It was necessary to design the resin process system so that it was insensitive to variations due to different environments in which the resin would be required to operate.
Question	How can the nitrate adsorption product be designed so as to be insensitive to factors likely to be different at different locations, such as pH and hardness of the water supply and the presence of trace amounts of likely impurities?

Design and Analysis	Chapter 13: Designing Robust Products and Processes
Findings	A process design was possible that ensured that the manufactured new resin was insensitive to changes in pH, water hardness, and moderate amounts of likely impurities.

ITERATION VIII

Commentary	The regeneration process for the resin was done automatically by a system containing a number of electronic and mechanical components. It was known that manufacturing variations transmitted by these components could affect system performance.
Question	How can the regeneration system be designed so that small manufacturing changes in the characteristics of its components do not greatly affect its performance?
Design and Analysis	Chapter 13: Designing Robust Products and Processes
Findings	It was found that in some instances it would be necessary to use expensive components with tightened specifications and that in other instances less expensive components with wider specifications could be substituted. A system that gave high performance at low cost was developed.

ITERATION IX

Commentary	It was found that the full-scale plant was not easy to control.
Question	How can better process control be achieved?
Design and Analysis	Chapter 14: Process Control, Forecasting, and Time Series
Findings	By the process of debugging using monitoring techniques and simple feedback adjustment, adequate control was obtained.

ITERATION X

Commentary	The initial process conditions for the full-scale process were arrived at from pilot plant experiments.
Question	How can improved conditions from the full-scale process be achieved?
Design and Analysis	Chapter 15: Evolutionary Process Operation
Findings	Operation in the evolutionary mode provided a steadily improving process.

The above by no means exhausts the application of statistical methods that might have been necessary to produce a marketable and profitable product. For example, it would be necessary to determine the extent of the possible market and to judge the performance of the new resin relative to competitive products. Also, production scheduling and inventory controls would need to be organized and monitored.

1.5. HOW TO USE STATISTICAL TECHNIQUES

All real problems have their idiosyncrasies that must be appreciated before effective methods of tackling them can be adopted. Consequently each new problem should be treated on its own merits and with respect. Being too hasty causes mistakes. It is easy to obtain the right answer to the wrong problem.

Find Out as Much as You Can About the Problem

Ask questions until you are satisfied that you fully understand the problem and are aware of the resources available to study it. Here are some of the questions you should ask and get answers too. What is the object of this investigation? Who is responsible? I am going to describe your problem; am I correct? Do you have any past data? How were these data collected? In what order? On what days? By whom? How? May I see them? How were the responses measured? Have the necessary devices been recently checked? Do you have other data like these? How does the equipment work? What does it look like? May I see it? May I see it work? How much physical theory is known about the phenomenon? If the process concerns a manufacturing process, what are the sampling, measurement, and adjustment protocols?

Don't Forget Nonstatistical Knowledge

When you are doing "statistics" do not neglect what you and your colleagues know about the subject matter field. Statistical techniques are useless unless combined with appropriate subject matter knowledge and experience. They are an adjunct to, not a replacement for, subject matter expertise.

Define Objectives

It is of utmost importance to (1) *define clearly the objectives* of the study; (2) be sure that all interested parties concur in these objectives; (3) make sure that the necessary equipment, facilities, scientific personnel, time, money, and adequate management support are available to perform the proposed investigation; (4) agree on the criteria that will determine when the objectives have been met; and (5) arrange that if the objectives have to be changed all interested parties are made aware of the new objectives and criteria. Not giving these matters sufficient attention can produce serious difficulties and sometimes disaster.

Learn from Each Other: The Interplay Between Theory and Practice

While experimenters can greatly gain from the use of statistical methods, the converse is even more true. A specialist in statistics can learn and benefit enormously from his or her discussions with engineers, chemists, biologists, and other subject matter specialists. The generation of really new ideas in statistics and good statistical tools seem to result from a genuine interest in practical problems. Sir Ronald Fisher, who was the originator of most of the ideas in this book, was a scientist and experimenter who liked to work closely with other experimenters. For him there was no greater pleasure than discussing their problems over a glass of beer. The same was true of his friend William S. Gosset (better known as "Student"), of whom a colleague* commented, "To many in the statistical world 'Student' was regarded as a statistical advisor to Guinness's brewery; to others he appeared to be a brewer who devoted his spare time to statistics .... Though there is some truth in both these ideas they miss the central point, which was the intimate connection between his statistical research and the practical problems on which he was engaged." The work of Gosset and Fisher reflects the hallmark of good science, the interplay between theory and practice. Their success as scientists and their ability to develop useful statistical techniques were highly dependent on their deep involvement in experimental work.

REFERENCES AND FURTHER READING

Box, G. E. P. (1987) In memoriam: William G. Hunter, 1937–1986, *Technometrics*, **29** 251–252.

Two very important books about the use of statistical methods in scientific investigations are:

Fisher, R. A. (1925) *Statistical Methods for Research Workers*, Edinburg and London, Oliver and Boyd.

Fisher, R. A. (1935) *The Design of Experiments*, Edinburg and London, Oliver and Boyd.

For some further information on statistical communication and consulting see the following articles and the references listed therein:

Derr, J. (2000) *Statistical Consulting: A Guide to Effective Communication*, Australia, Duxbury.

Chatfield, C. (1995) *Problem Solving: A Statistician's Guide*, London, Chapman and Hall.

Bajaria, H. J., and Copp, R. P. (1991) *Statistical Problem Solving*, Garden City, MI, Multiface Publishing.

Hoadley, A., and Kettenring J. (1990) Communications between statisticians and engineers/physical scientists, *Technometrics*, **32**, 243–274.

*L. McMullen in the foreword to "*Student's*" *Collected Papers,* edited by E. S. Pearson and J. Wishart, University Press Cambridge, London, 1942, issued by the Biometrika Office, University College, London.

Kimball, A. W. (1957) Errors of the third kind in statistical consulting, *J. Am. Statist. Assoc.*, **57**, 133.

Daniel C. (1969) Some general remarks on consultancy in statistics, *Technometrics*, **11**, 241.

Examples of "nonsense correlation" similar to the one in this chapter about storks and the birth rate are contained in:

Hooke, R. (1983) *How to Tell the Liars from the Statisticians*, Dekker, New York.

Huff, D. (1954) *How to Lie with Statistics*, Norton, New York.

Tufte, R. (1983) *The Visual Display of Quantitative Information*, Graphics Press, New York.

For discussion and illustration of iterative experimentation see, for example:

Box, G. E. P. (1976) Science and statistics, *J. Am. Statist. Assoc.*, **71**, 791.

Crouch, T. (2003) *The Bishop's Boys*, W. W. Norton, New York. (The Wright brothers' sequential experiments in developing and designing the first airplane.)

Deming, W. E. (1953) On the distinction between enumerative and analytic surveys, *J. Am. Stat. Assoc.*, **48**, 244–255.

Medawar, P. B. (1990) *Advice to a Young Scientist*, Sloan Science Series, Basic Books, New York.

Watson, J. D. (1968) *The Double Helix: A Personal Account of the Discovery of the Structure of DNA*, Touchstone: Simon & Schuster, New York.

Some useful references on statistical computing are:

Ross, I., and Gentleman, R. (1996). R, A language for data analysis and graphics, *J. Computat. Graphical Stat.*, **5**(3), 299–314.

R Development Core Team (2005). *R: A Language and Environment for Statistical Computing*, R Foundation for Statistical Computing, Vienna, Austria. http://www.R-project.org

NIST/SEMATECH e-Handbook of Statistical Methods, URL: http://www.itl.nist.gov/div898/handbook, 2005.

Some suppliers of statistical software useful for readers of this book are:

NIST/Dataplot URL: http://www.itl.nist.gov/div898/software.htm, 2005.

SCA: Scientific Computing Associates URL: http://www.scausa.com

JMP: The Statistical Discovery Software URL: http://www.jmp.com

MINITAB: Statistical Software URL: http://www.minitab.com

R: A Language and Environment for Statistical Computing URL: http://www.R-project.org

S-PLUS: Insightful S-PLUS URL: http://www.insightful.com/products/splus

QUESTIONS FOR CHAPTER 1

1. What is meant by the iterative nature of learning?

2. In what ways can statistics be useful to experimenters?

3. What is achieved by good statistical analysis? By good statistical design? Which do you believe is more important?

4. What are three common difficulties encountered in experimental investigations?

5. Can you give examples (if possible from your own field) of real confusion (perhaps controversy) that has arisen because of one or more of these difficulties?

6. Which techniques in this book do you expect to be most useful to you?

7. How should you use the techniques in this book?

8. Can you think of an experimental investigation that (a) was and (b) was not iterative?

9. Read accounts of the development of a particular field of science over a period of time (e.g., read the books *The Double Helix* by J. D. Watson and *The Bishops Boys* (the Wright brothers) by T. D. Crouch, 1989). How do these developments relate to the discussion in this chapter of a scientific investigation as an iterative process? Can you trace how the confrontation of hypotheses and data led to new hypotheses?

CHAPTER 2

Basics (Probability, Parameters, and Statistics)

Did you take a course in elementary statistics? If yes, and you remember it, please go on to Chapter 3. If no, read on. If you need to brush up on a few things, you will find this chapter helpful and useful as a reference.

2.1. EXPERIMENTAL ERROR

When an operation or an experiment is repeated under what are as nearly as possible the same conditions, the observed results are never quite identical. The fluctuation that occurs from one repetition to another is called noise, experimental variation, *experimental error*, or merely *error*. In a statistical context the word error is used in a technical and emotionally neutral sense. It refers to variation that is often unavoidable. It is not associated with blame.

Many sources contribute to experimental error in addition to errors of measurement, analysis, and sampling. For example, variables such as ambient temperature, skill or alertness of personnel, age and purity of reagents, and the efficiency or condition of equipment can all contribute. Experimental error must be distinguished from mistakes such as misplacing a decimal point when recording an observation or using the wrong chemical reagent.

It has been a considerable handicap to many experimenters that their formal training has left then unequipped to deal with the common situation in which experimental error cannot be safely ignored. Not only is awareness of the possible effects of experimental error essential in the analysis of data, but also its influence is a paramount consideration in *planning* the generation of data, that is, in the *design* of experiments. Therefore, to have a sound basis on which to build practical techniques for the design and analysis of experiments, some elementary understanding of experimental error and of associated probability theory is essential.

Statistics for Experimenters, Second Edition. By G. E. P. Box, J. S. Hunter, and W. G. Hunter

Experimental Run

We shall say that an experimental *run* has been performed when an apparatus has been set up and allowed to function under a specific set of experimental conditions. For example, in a chemical experiment a run might be made by bringing together in a reactor specific amounts of chemical reactants, adjusting temperature and pressure to the desired levels, and allowing the reaction to proceed for a particular time. In engineering a run might consist of machining a part under specific manufacturing conditions. In a psychological experiment a run might consist of subjecting a human subject to some controlled stress.

Experimental Data or Results

An experimental result or datum describes the outcome of the experimental run and is usually a numerical measurement. Ten successive runs made at what were believed to be identical conditions might produce the following *data*:

66.7 64.3 67.1 66.1 65.5 69.1 67.2 68.1 65.7 66.4

In a chemical experiment the data might be percentage yield, in an engineering experiment the data could be the amount of material removed in a machining operation, and in a psychological experiment the data could be the times taken by 10 stressed subjects to perform a specific task.

2.2. DISTRIBUTIONS

The Dot Diagram

Figure 2.1 is a dot diagram showing the scatter of the above values. You will find that the dot diagram is a valuable device for displaying the distribution of a small body of data. In particular, it shows:

1. The *location* of the observations (in this example you will see that they are clustered near the value 67 rather than, say, 85 or 35).
2. The *spread* of the observations (in this example they extend over about 5 units).
3. Whether there are some data points considerably more extreme that the rest. Such "outliers" might result from mistakes either in recording the data or in the conduct of a particular test.

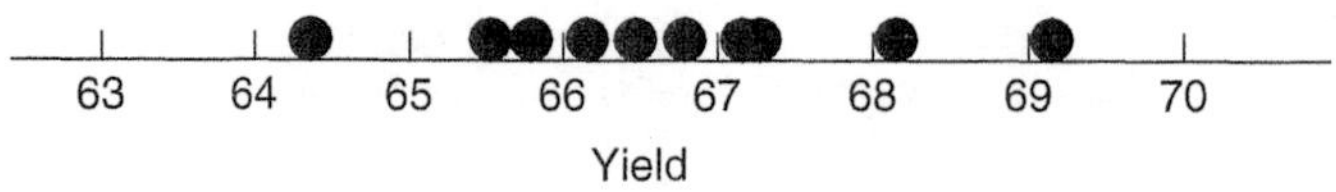

Figure 2.1. Dot diagram for a sample of 10 observations.

The Frequency Distribution

When you have a large number of results, you will be better able to appreciate the data by constructing a *frequency distribution*, also called a *histogram* or *frequency diagram*. This is accomplished by dividing the horizontal axis into intervals of appropriate size and constructing a rectangle over the ith interval with its *area* proportional to n_i, the number (frequency) of observations in that interval. Figure 2.2 shows a frequency distribution for $N = 500$ observations of yield from a production process.* In this example each observation was recorded to one decimal place. The smallest of the observations was between 56 and 57 and the largest between 73 and 74. It was convenient therefore to classify the observations into 18 intervals, each covering a range of one unit. There were two observations in the first interval, thus $n_1 = 2$. Since in this example all the intervals are of width 1, the frequency of observations n_i in the ith interval, $i = 1, 2, \ldots, 18$, is directly proportional to the height (ordinate) on the vertical axis.

Figure 2.2 gives a vivid impression of the 500 observations. In particular, it shows their location and spread. But other characteristics are brought to your

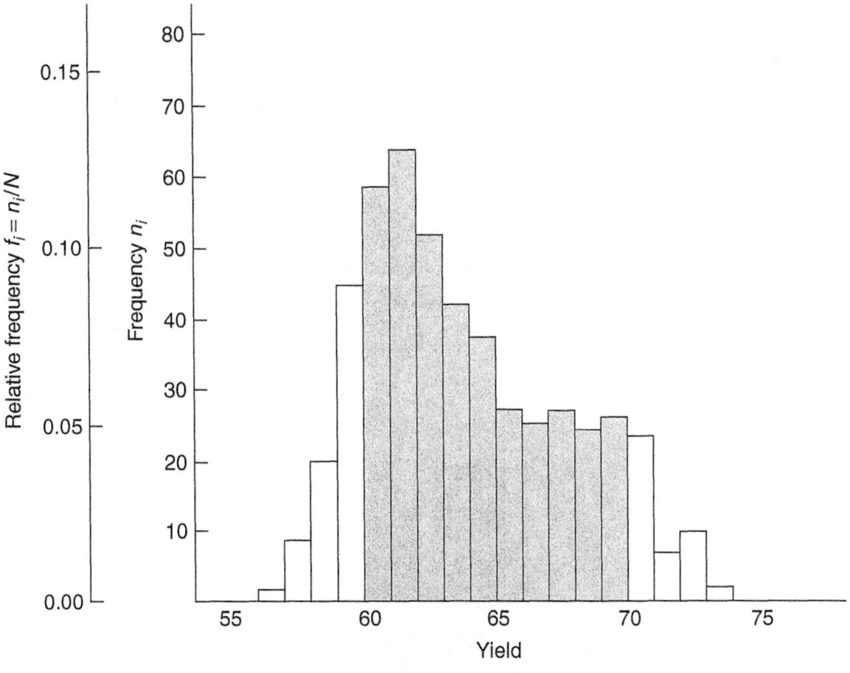

Figure 2.2. Frequency diagram (histogram, frequency distribution) for a sample of 500 observations.

*Many computer software programs provide such histograms. In this example the histogram (frequency diagram) has intervals of equal width and hence height is also proportional to n_i. However, histograms are sometimes constructed for data in which the intervals are of different width. In that case the *area* of the rectangle constructed over each interval must be proportional to n_i, the frequency of observations within that interval.

attention, for instance, that about $\frac{4}{5}$ of the observations lie between 60 and 70. This fraction, more precisely $\frac{382}{500}$, is represented by the shaded region under the frequency diagram. Notice too that the histogram is not symmetric but tails off slowly for observations having high values.

Exercise 2.1. Construct a dot diagram with data given in miles per gallon for five test cars: 17.8, 14.3, 15.8, 18.0, 20.2.

Exercise 2.2. Construct a histogram for these air pollution data given in parts per million of ozone: 6.5, 2.1, 4.4, 4.7, 5.3, 2.6, 4.7, 3.0, 4.9, 4.7, 8.6, 5.0, 4.9, 4.0, 3.4, 5.6, 4.7, 2.7, 2.4, 2.7, 2.2, 5.2, 5.3, 4.7, 6.8, 4.1, 5.3, 7.6, 2.4, 2.1, 4.6, 4.3, 3.0, 4.1, 6.1, 4.2.

Hypothetical Population of Results Represented by a Distribution

The total aggregate of observations that conceptually *might* occur as the result of repeatedly performing a particular operation is referred to as the *population* of observations. Theoretically, this population is usually assumed to be infinite, but for the purposes of this book you can think of it as having size N where N is large. The (usually few) observations that actually have occurred are thought of as some kind of *sample* from this conceptual population. With a large number of observations the bumps in the frequency diagram often disappear and you obtain a histogram with an appearance like that in Figure 2.3. (Until further notice you should concentrate on this histogram and ignore for the moment the smooth curve that is superimposed.) If you make the area of the rectangle erected on the ith interval of this histogram equal to the *relative frequency* n_i/N of values occurring in that interval, this is equivalent to choosing the vertical scale so that the area under the whole histogram is equal to unity.

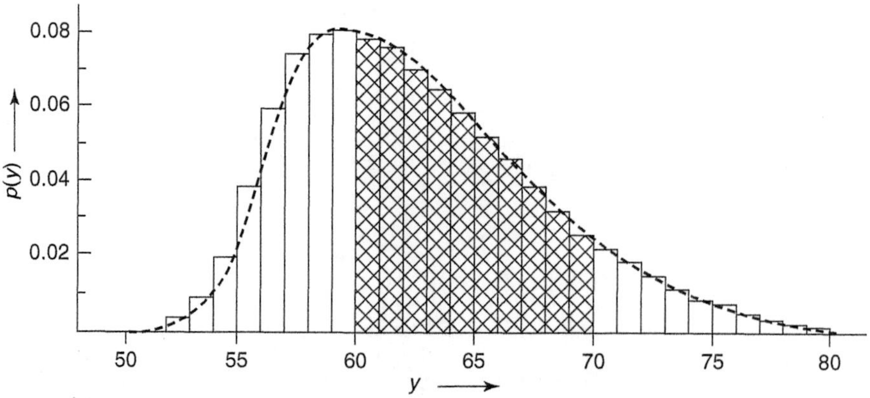

Figure 2.3. Hypothetical probability distribution for a conceptual population of observations.

Randomness and Probability

A *random* drawing is one where each member of the population has an equal chance of being chosen. Imagine you have written each of the N observations from the population on a ticket and put all these into a lottery drum and in a random drawing you obtain some value y. Then:

1. The probability that y is less than some value y_0, $\Pr(y < y_0)$, will be equal to the area under the histogram to the left of y_0 [for illustration, in Figure 2.3, if y_0 was equal to 60, the probability that y was less than 60 would be 0.361. i.e., $\Pr(y < y_0) = 0.361$].
2. The probability $\Pr(y > y_0)$ that y is greater than y_0 will be equal to the area under the histogram to the right of y_0.
3. The probability $\Pr(y_0 < y < y_1)$ that y is greater than y_0 but less than y_1 will be equal to the area under the histogram between y_0 and y_1. For example, in Figure 2.3 the shaded portion has an area of 0.545 so that $\Pr(60 < y < 70) = 0.545$.

To the accuracy of the grouping interval the diagram of relative frequencies for the whole population tells you everything you can know about the probability of a randomly chosen member of the population falling within any given range. It is therefore called a *probability distribution.*

Probability Density

In this example the width of the grouping interval happens to be one unit of yield, but suppose it is h units of yield. Suppose also that for a particular interval of size h the height of the constructed rectangle is $p(y)$ and its area is P. (Recall that this area $P = n/N$ is the probability of the interval containing a randomly chosen y.) Then $P = p(y) \times h$ and $p(y) = P/h$. The probability *density,* that is the ordinate $p(y)$, is thus obtained by dividing the probability, the area associated with a given interval, by the width of the interval. Notice that it is always an *area* under the probability distribution that represents probability. The ordinate $p(y)$ of the distribution, the density at the point y, is not itself a probability and only becomes one when it is multiplied by the appropriate interval width. Probability density has the same status as physical density. Knowing the density of a metal does not tell you whether a given piece of the metal will be heavy or light. To find this out, you must multiply its density by its volume to obtain its mass. Thus probability = probability density × interval size, just as mass = density × volume.

Representing a Probability Distribution by a Continuous Curve

Now if you imagine the interval h taken to be very small, the probability P associated with the interval becomes proportionally small also, but no matter how far you carry the process, the probability density $p(y) = P/h$ can still be finite. In the limit, as the interval becomes infinitesimally short, you can conceive of the population being represented by a *continuous* probability distribution like the continuous dashed line in Figure 2.3.

When a mathematical function is used to represent this continuous probability distribution, it is sometimes called the *(probability) density function.* One such theoretical function is the normal distribution, which is discussed later in this chapter. Just as before, the total area under a continuous probability distribution must be unity, and the area between any two values y_0 and y_1 is equal to $\Pr(y_0 < y < y_1)$.

A question sometimes asked is: Given that the theoretical population is represented by a *continuous* curve, what is the probability of getting a particular value of y, say $y = 66.3$? If the question referred to the probability of y being *exactly* 66.300000000 ... corresponding to a *point* on the horizontal axis, the answer would be zero, because $P = p(y) \times h$ and h is zero. However, what is usually meant is: Given that the observations are made, say, to the nearest 0.1, what is the probability of getting the value $y = 66.3$? Obviously, the answer to this question is not zero. If we suppose that all values between $y = 66.25$ and $y = 66.35$ are recorded as $y = 66.3$, then we require $\Pr(66.25 < y < 66.35)$. The required probability is given by the area under the curve between the two limits and is adequately approximated by $p(66.3) \times 0.1$. □

Sample Average and Population Mean

One important feature of a sample of n observations is its *average* value, denoted by $\bar{y}$(read as "y-bar"). For the sample of 10 observations plotted in Figure 2.1

$$\bar{y} = \frac{66.7 + 64.3 + \cdots + 66.4}{10} = 66.62$$

In general, for a sample of n observations, you can write

$$\bar{y} = \frac{y_1 + y_2 + \cdots + y_n}{n} = \frac{\sum y}{n}$$

where the symbol $\sum$, a capital sigma, before the y's means add up all the y's.* Since $\bar{y}$ tells you where the scatter of points is centered along the y axis, more precisely where it will be balanced, it is a measure of *location* for the sample. If you imagine a hypothetical population as containing some *very* large number N of observations, it is convenient to note the corresponding measure of location

*In some instances where you want to indicate a particular sequence of y's to be added together, you need a somewhat more elaborate notation. Suppose that $y_1, y_2, y_3, \ldots$ refer to the first, second, third observations, and so on, so that y_j means the jth observation. Then, for example, the sum of the observations beginning with the third observation y_3 and ending with the eighth y_8 is written as $\sum_{j=3}^{8} y_j$ and means $y_3 + y_4 + y_5 + y_6 + y_7 + y_8$.

by the Greek letter η (eta) and to call it the *population mean*, so that

$$\eta = \frac{\sum y}{N}$$

2.3. STATISTICS AND PARAMETERS

To distinguish between the sample and population quantities, η is called the population *mean* and $\overline{y}$ the sample *average*. A parameter like the mean η is a quantity directly associated with a *population*. A statistic like the average $\overline{y}$ is a quantity calculated from a set of data often thought of as some kind of *sample* taken from the population. Parameters are usually designated by Greek letters; statistics by Roman letters. In summary:

Population: a very large set of N observations from which your sample of observations can be imagined to come.

Sample: the small group of n observations actually available.

Parameter: population mean $\eta = \sum y/N$.

Statistic: sample average $\overline{y} = \sum y/n$.

The mean of the population is also called the *expected value* of y or the *mathematical expectation* of y and is often denoted as $E(y)$. Thus $\eta = E(y)$.

The Hypothesis of Random Sampling?

Since the hypothetical population conceptually contains all values that can occur from a given operation, any set of observations we may collect is *some* kind of sample from the population. An important statistical idea is that in certain circumstances a set of observations may be regarded as a *random* sample.

This *hypothesis* of random sampling however will often *not* apply to actual data. For example, consider daily temperature data. Warm days tend to follow one another, and consequently high values are often followed by other high values. Such data are said to be *autocorrelated* and are thus not directly representable by *random* drawings.* In both the analysis and the design of scientific experiments much hangs on the applicability of this hypothesis of random sampling. The importance of randomness can be appreciated, for example, by considering public opinion polls. A poll conducted on election night at the headquarters of one political party might give an entirely false picture of the standing of its candidate in the voting population. Similarly, information generated from a sample of apples taken from the top of the barrel can prove misleading.

It is unfortunate that the hypothesis of random sampling is treated in much statistical writing as if it were a natural phenomenon. In fact, for real data it is a property that can never be relied upon, although suitable precautions in the design of an experiment can make the assumption relevant.

* In Chapter 14 you will see how autocorrelated data can be *indirectly* represented by random drawings.

2.4. MEASURES OF LOCATION AND SPREAD

The Population Mean, Variance, Standard Deviation, and Coefficient of Variation

As you have seen, an important characteristic of a population is its *mean* value $\eta = \sum y/N$. This mean value η is called a *parameter* of the distribution. It is also called the *mathematical expectation of* y and denoted by $E(Y)$. Thus $E(Y) = \eta$. It is the first moment of the distribution of y, and it defines the point of balance along the horizontal axis, as illustrated in Figure 2.4. It thus supplies a measure of the *location* of the distribution, whether, for instance, the distribution is balanced at 0.6, 60, or 60,000. Knowledge of location supplies useful but incomplete information about a population. If you told visitors from another world that adult males in the United States had a mean height of approximately 70 inches, they could still believe that some members of the population were 1 inch and others 1000 inches tall. A measure of the *spread* of the distribution would help give them a better perspective.

The most useful such measure for our purposes is the *variance* of the population, denoted by σ^2 (sigma squared). A measure of how far any particular observation y is from the mean η is the deviation $y - \eta$; the variance σ^2 is the mean value of the squares of such deviations taken over the whole population. Thus

$$\sigma^2 = E(y - \eta)^2 = \frac{\sum(y - \eta)^2}{N}$$

Just as the special symbol $E(y)$ may be used to denote the mean value, $E(y) = \eta$, so too the special symbol $V(y)$ is used to denote the variance, and thus $V(y) = \sigma^2$. A measure of spread which has the same units as the original observations is σ, the positive square root of the variance—the root mean square of the deviations. This is called the *standard deviation*

$$\sigma = +\sqrt{\sigma^2} = +\sqrt{V(y)} = +\sqrt{E(y - \eta)^2} = +\sqrt{\frac{\sum(y - \eta)^2}{N}}$$

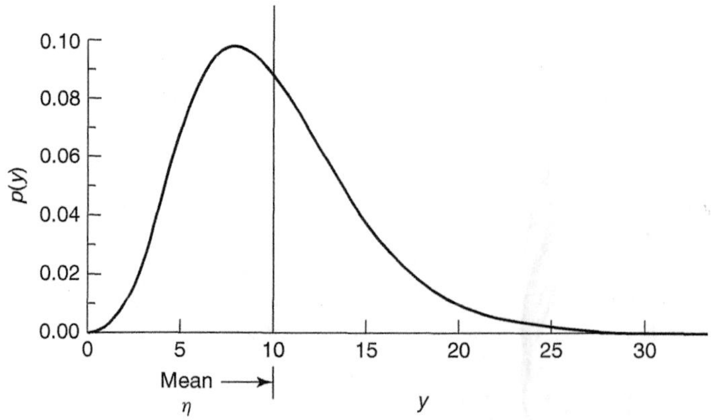

Figure 2.4. Mean $\eta = E(y)$ as the point of balance for the distribution.

Occasionally a subscript will accompany σ. Thus the symbol σ_y leaves no doubt that you are talking about the standard deviation of the population of observations y and not, for example, about some other population of observations z.

Average, Variance, and Standard Deviation of a Sample

The data available to an investigator may be thought of as a small *sample* of n observations from the hypothetical larger set N represented by the population. For the data in the dot diagram of Figure 2.1 the sample average $\bar{y} = \sum y/n = 666.2/10 = 66.62$ supplies a measure of the location of the sample. Similarly, the sample variance supplies a measure of the spread of the sample. The sample variance is calculated as

$$s^2 = \frac{\sum(y - \bar{y})^2}{n - 1} = \frac{\sum y^2 - n\bar{y}^2}{n - 1}$$

The positive square root of the sample variance gives the sample standard deviation,

$$s = +\sqrt{\frac{\sum(y - \bar{y})^2}{n - 1}}$$

which has the same units as the observations. Thus for the sample of 10 observations in Figure 2.1 the sample variance is

$$s^2 = \frac{(66.7 - 66.62)^2 + (64.3 - 66.62)^2 + \cdots + (66.4 - 66.62)^2}{9}$$

$$= \frac{44398.96 - 10(66.62)^2}{9} = 1.86$$

and the sample standard deviation is $s = 1.36$.

Once again, as was true for the mean η and the average $\bar{y}$, a Greek letter is used for the population parameter and a Roman letter for the corresponding sample statistic. Thus σ^2 and σ are parameters that denote the population variance and population standard deviation, and the statistics s^2 and s denote the sample variance and sample standard deviation. A summary is given in Table 2.1.

The Coefficient of Variation

Suppose you need to know how big is the standard deviation σ relative to the mean η. The ratio σ/η is called the *coefficient of variation.* When written as a percentage $100\,\sigma/\eta$, it is sometimes called the *percentage error.* A coefficient of variation of 3% would imply that σ was 3% of the mean η. As you will see later, it is very closely related to the standard deviation of $\log y$. The sample coefficient of variation is $s/\bar{y}$. Its inverse, $\bar{y}/s$, is sometimes referred to as a *signal-to-noise*

Table 2.1. Population and Sample Quantities

	Parameters	Statistics
Definition	*Population*: a hypothetical set of N observations from which the sample of observations actually obtained can be imagined to come (typically N is very large)	*Sample*: a set of n available observations (typically n is small)
Measure of location	Population mean $\eta = \sum y/N$	Sample average $\bar{y} = \sum y/n$
Measure of spread	Population variance $\sigma^2 = \sum(y-\eta)^2/N$	Sample variance $s^2 = \sum(y-\bar{y})^2/(n-1)$
	Population standard deviation $\sigma = +\sqrt{\sum(y-\eta)^2/N}$	Sample standard deviation $s = +\sqrt{\sum(y-\bar{y})^2/(n-1)}$

ratio. For the dot diagram data of Figure 2.1 the sample coefficient of variation $s/\bar{y}$ is $1.36/66.62 = 0.020 = 2\%$.

The Median

Another measure of a sample that is occasionally useful is the *median*. It may be obtained by listing the n data values in order of magnitude. The median is the middle value if n is odd and the average of the two middle values if n is even.

Residuals and Degrees of Freedom

The n deviations of observations from their sample average are called *residuals*. These residuals always sum to zero. Thus $\sum(y-\bar{y}) = 0$ constitutes a *linear constraint* on the residuals $y_1 - \bar{y}, y_2 - \bar{y}, \ldots, y_n - \bar{y}$ because any $n-1$ of them completely determine the other. The n residuals $y - \bar{y}$ (and hence the sum of squares $\sum(y-\bar{y})^2$ and the sample variance $s^2 = \sum(y-\bar{y})^2/(n-1)$) are therefore said to have $n-1$ *degrees of freedom*. In this book the number of degrees of freedom is denoted by the Greek letter ν (nu). For the dot diagram data the sample variance is $s^2 = 1.86$, the sample standard deviation is $s = \sqrt{1.86} = 1.36$, and the number of degrees of freedom is $\upsilon = n - 1 = 10 - 1 = 9$. The loss of one degree of freedom is associated with the need to replace the unknown population parameter η by the estimate $\bar{y}$ derived from the sample data. It can be shown that because of this constraint the best estimate of σ^2 is obtained by dividing the sum of squares of the residuals not by n but by $\upsilon = n - 1$.

In later applications you will encounter examples where, because of the need to calculate several sample quantities to replace unknown population parameters, several constraints are necessarily placed on the residuals. When there are

p independent linear constraints on n residuals, their sum of squares and the resulting sample variance and standard deviation are all said to have $\upsilon = n - p$ degrees of freedom.

Sample Variance If the Population Mean Were Known: The "Natural" Variance and Standard Deviation

If the population mean η were known, the sample variance would be calculated as the ordinary average of the squared deviations from this known mean,

$$\dot{s}^2 = \frac{\sum(y - \eta)^2}{n}$$

The statistic is designated by a dot to distinguish it from s^2. The sum of squares $\sum(y - \eta)^2$ and the associated statistic $\dot{s}^2$ would then have n degrees of freedom because all n quantities $y - \eta$ are free to vary; knowing $n - 1$ of the deviations does not determine the nth. In this book we will call $\dot{s}^2$ the *natural* variance and $\dot{s}$ the *natural* standard deviation.

Exercise 2.3. Calculate the average and standard deviation for the following data on epitaxial layer thickness in micrometers: 16.8, 13.3, 11.8, 15.0, 13.2. Confirm that the sum of the residuals $y - \overline{y}$ is zero. Illustrate how you would use this fact to calculate the fifth residual knowing only the values of the other four.

Answer: $\overline{y} = 14.02$, $s = 1.924$ with $\upsilon = 4$ degrees of freedom.

Exercise 2.4. A psychologist measured (in seconds) the following times required for 10 rats to complete a maze: 24, 37, 38, 43, 33, 35, 48, 29, 30, 38. Determine the average, sample variance, and sample standard deviation for these data.

Answer: $\overline{y} = 35.5$, $s^2 = 48.72$, $s = 6.98$ with $\upsilon = 9$ degrees of freedom

Exercise 2.5. The following observations on the lift of an airfoil (in kilograms) were obtained in successive trials in a wind tunnel: 9072, 9148, 9103, 9084, 9077, 9111, 9096. Calculate the average, sample variance, and sample standard deviation for these observations.

Answer: $\overline{y} = 9098.71$, $s^2 = 667.90$, $s = 25.84$ with $\upsilon = 6$ degrees of freedom

Exercise 2.6. Given the following liters of a reagent required to titrate Q grams of a substance: 0.00173, 0.00158, 0.00164, 0.00169, 0.00157, 0.00180. Calculate the average, the sample variance, and sample standard deviation for these observations.

Answer: $\overline{y} = 0.00167$, $s^2 = 0.798 \times 10^{-8}$, $s = 0.893 \times 10^{-4}$ with $\upsilon = 5$ degrees of freedom

2.5. THE NORMAL DISTRIBUTION

Repeated observations that differ because of experimental error often vary about some central value in a roughly symmetric distribution in which small deviations

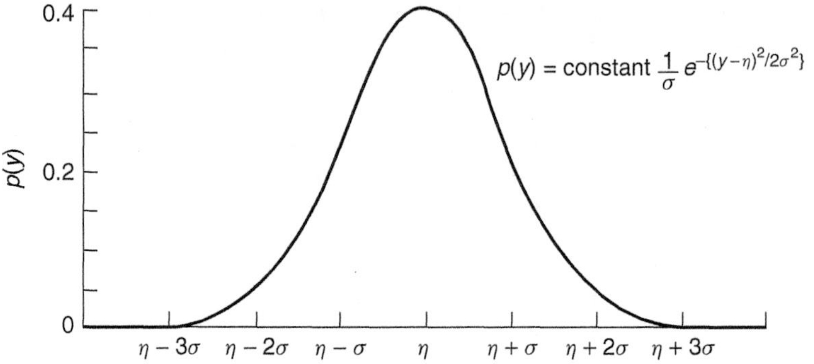

Figure 2.5. Normal (Gaussian) distribution.

occur much more frequently than large ones. For representing this situation a valuable theoretical distribution that occupies an important position in the theory of statistics is the *Gaussian*, or *normal, distribution.* The appearance of this distribution and its mathematical formula are shown in Figure 2.5. It is a symmetric curve with its highest ordinate at its center, tailing off to zero in both directions in a way intuitively expected of experimental error. It has the property that the logarithm of its density function is a quadratic function of the standardized residual $(y - \eta)/\sigma$.

Reasons for the Importance of the Normal Distribution

Two things explain the importance of the normal distribution:

1. The *central limit* effect that produces a tendency for real error distributions to be "normal like."
2. The *robustness* to nonnormality of some common statistical procedures, where "robustness" means insensitivity to deviations from theoretical normality.

The Central Limit Effect

Usually the "overall" error $y - \eta = e$ is an aggregate of a number of component errors. Thus a measurement of yield obtained for a particular experimental run may be subject to analytical error, chemical sampling error and process error produced by failures in meeting the exact settings of the experimental conditions, errors due to variation in raw materials, and so on. Thus e will be some function of a number of component errors $e_1, e_2, \ldots, e_n$. If each individual error is fairly small, it would usually be possible to approximate the overall error as a *linear* function of the component errors

$$e = a_1e_1 + a_2e_2 + \cdots + a_ne_n$$

where the a's are constants. The central limit theorem says that, under conditions almost always satisfied in the real world of experimentation, the distribution of such a linear function of errors will tend to normality as the number of its components becomes large. The tendency to normality occurs almost regardless of the individual distributions of the component errors. An important proviso is that *several* sources of error must make important contributions to the overall error and that no particular source of error dominate the rest.

Illustration: Central Limit Tendency for Averages

Figure 2.6*a* shows the distribution for throws of a *single* true six-sided die. The probabilities are equal to $\frac{1}{6}$ for each of the scores 1, 2, 3, 4, 5, or 6. The mean score is $\eta = 3.5$. Figure 2.6*b* shows the distribution of the *average* score obtained from throwing two dice, and Figures 2.6*c*, *d*, and *e* show the distributions of *average* scores calculated for throwing 3, 5, and 10 dice. To see how the central limit effect applies to averages, suppose $e_1, e_2, \ldots, e_n$ denote the deviations of the scores of the individual dice from the mean value of $\eta = 3.5$. Let e be the corresponding deviation of the average, that is, $e = \sum e_i/n$. Then e will satisfy the equation $e = a_1e_1 + a_2e_2 + \cdots + a_ne_n$ with all the a's set equal to $1/n$. Whereas the original, or *parent*, distribution of the individual observations (the scores from single throws) is far from normal shape, the ordinates of the distribution for the averages are remarkably similar to ordinates of the normal distribution even for n as small as 5.

In summary:

1. When, as is usual, an experimental error is an aggregate of a number of component errors, its distribution tends to normal form, even though the distribution of the components may be markedly nonnormal.
2. A sample average tends to be normally distributed, even though the individual observations on which it is based are not. Consequently, statistical methods that depend, not directly on the distribution of individual observations, but on the distribution of averages tend to be insensitive or robust to nonnormality.
3. Procedures that compare means are usually robust to nonnormality. However, this is not generally true for procedures for the comparison of variances.

because means tend to be normally distributed.

Robustness of Procedures to the Assumption of Normality

It is important to remember that all mathematical–statistical models are approximations. In particular, there never was, nor ever will be, an *exactly* straight line or observations that *exactly* follow the normal distribution. Thus, although many of the techniques described here are derived on the assumption of normality, approximate normality will usually be all that is required for them to be *useful*. In particular, techniques for the comparison of means are usually *robust* to *nonnormality*. Unless specifically warned, you should not be unduly worried about

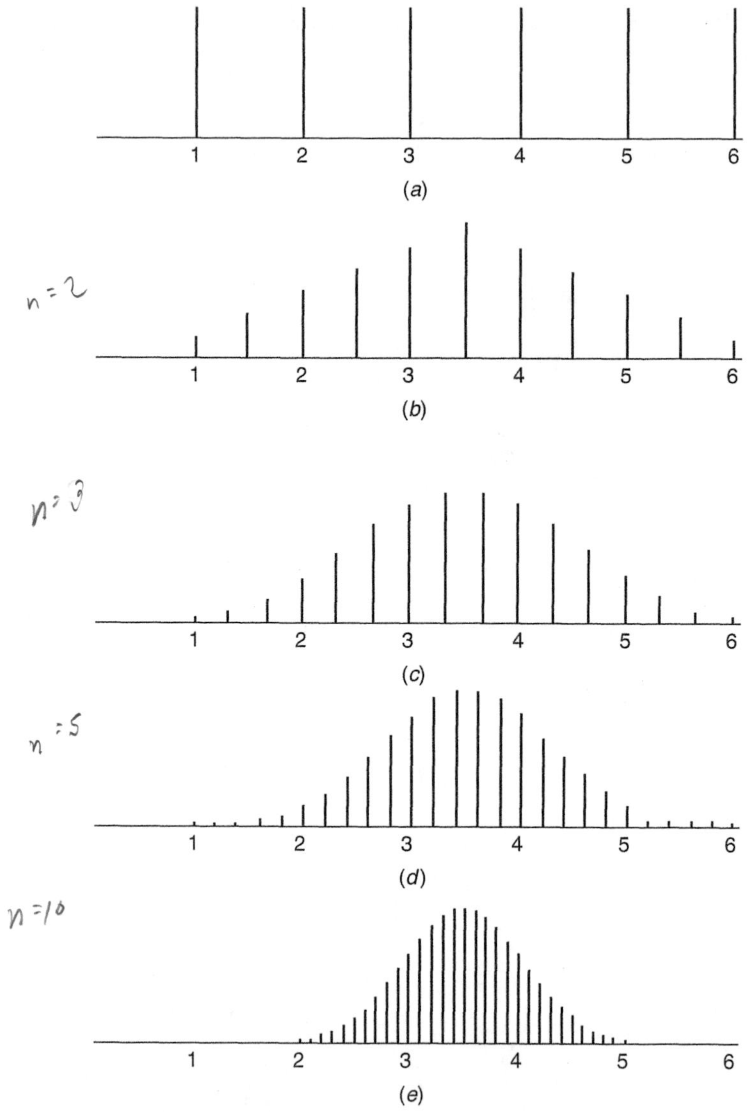

Figure 2.6. Distribution of averages scores from throwing various numbers of dice: (*a*) 1, (*b*) 2, (*c*) 3, (*d*) 5, and (*e*) 10 dice.

normality. You should of course be continually on the look-out and check for *gross* violations of this and all other assumptions.

Characterizing the Normal Distribution

Once the mean η and the variance σ^2 of a normal distribution are given, the entire distribution is characterized. The notation $N(\eta, \sigma^2)$ is often used to indicate

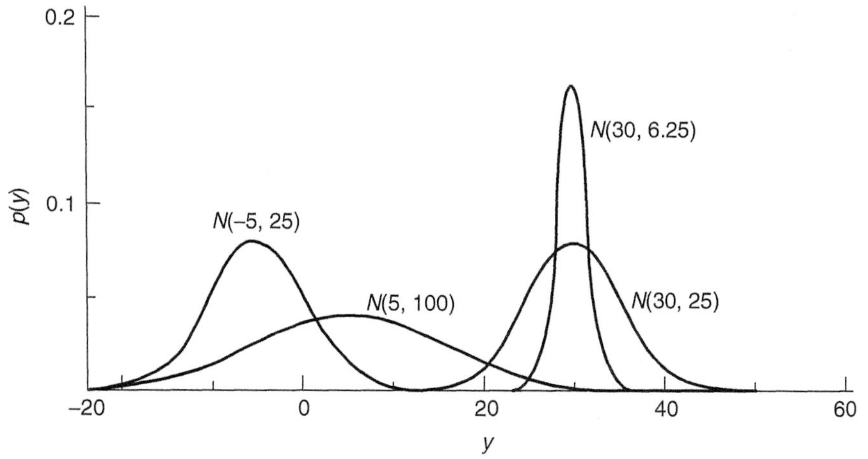

Figure 2.7. Normal distributions with different means and variances.

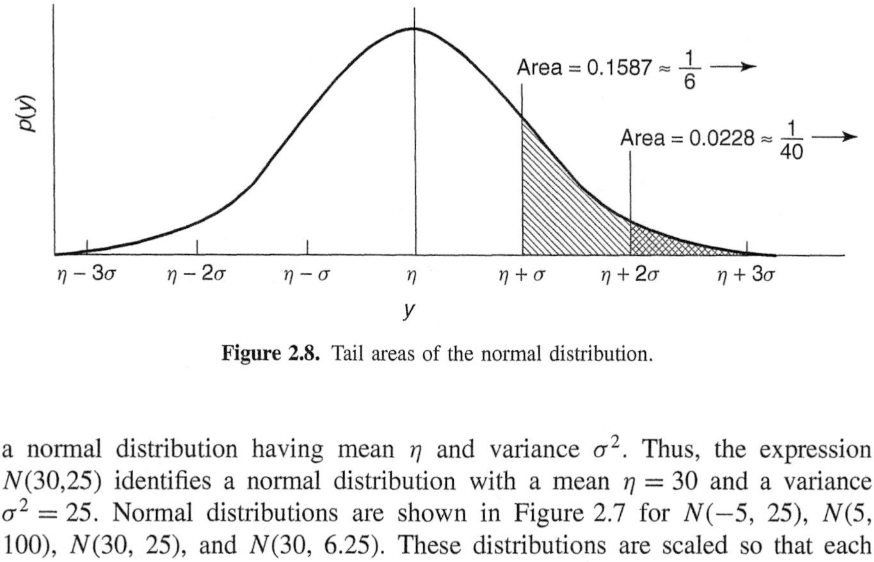

Figure 2.8. Tail areas of the normal distribution.

a normal distribution having mean η and variance σ^2. Thus, the expression $N(30{,}25)$ identifies a normal distribution with a mean $\eta = 30$ and a variance $\sigma^2 = 25$. Normal distributions are shown in Figure 2.7 for $N(-5,\ 25)$, $N(5,\ 100)$, $N(30,\ 25)$, and $N(30,\ 6.25)$. These distributions are scaled so that each area is equal to unity.

For a normal distribution the standard deviation σ measures the distance from its mean η to the point of inflection on the curve. The point of inflection (see Fig. 2.8) is the point at which the slope stops increasing and starts to decrease (or vice versa). The following should help you gain a fuller appreciation of the normal distribution:

1. The probability that a positive deviation from the mean will exceed one standard deviation is 0.1587 (roughly $\frac{1}{6}$). This is represented by the shaded "tail" area shown in Figure 2.8.

2. Because of symmetry, this probability is exactly equal to the chance that a negative deviation from the mean will exceed one standard deviation.
3. From these two statements it is clear that the probability that a deviation *in either direction* will exceed one standard deviation is $2 \times 0.1587 = 0.3174$ (roughly $\frac{1}{3}$), and consequently the probability of a deviation of less than one standard deviation is 0.6826 (roughly $\frac{2}{3}$).
4. The chance that a positive deviation from the mean will exceed two standard deviations is 0.0228 (roughly $\frac{1}{40}$). This is represented by the heavily shaded region in Figure 2.8.
5. Again, this is exactly equal to a chance that a negative deviation from the mean will exceed two standard deviations.
6. From these two statements the chance that a deviation in either direction will exceed two standard deviations is 0.0456 (roughly $\frac{1}{20}$ or 0.05)

A probability statement concerning some normally distributed quantity y is often best expressed in terms of a *standardized normal deviate* or *unit normal deviate,*

$$z = \frac{y - \eta}{\sigma}$$

The quantity z is said to have a distribution that is $N(0,1)$, that is, z has a normal distribution with a mean $\eta = 0$ and a variance $\sigma^2 = 1$. We can therefore rewrite the previous statements as follows:

1. $\Pr(y > \eta + \sigma) = \Pr[(y - \eta) > \sigma] = \Pr\left[\left(\frac{y - \eta}{\sigma}\right) > 1\right] = \Pr\ (z > 1) =$ 0.1587
2. $\Pr(z < -1) = 0.1587$
3. $\Pr(|z| > 1) = 0.3174$
4. $\Pr(z > 2) = 0.0228$
5. $\Pr(z < -2) = 0.0228$
6. $\Pr(|z| > 2) = 0.0455$

Using Tables of the Normal Distribution

In general, to determine the probability of an event y exceeding some value y_0, that is, $\Pr(y > y_0)$, you compute the normal deviate $z_0 = (y_0 - \eta)/\sigma$ and obtain $\Pr(z > z_0)$ from a computer program or from Table A at the end of this book. (The probabilities associated with the normal deviate are also available on many hand-held calculators.) For example, given a normal population with mean $\eta = 39$ and variance $\sigma^2 = 16$, what is the probability of obtaining an observation greater than 42, that is, $\Pr(y > 42)$? The normal deviate $z = (42 - 39)/4 = 0.75$ and the required probability is $\Pr(z > 0.75) = 0.2266$. As a second example, given that the distribution is $N(39, 16)$, what is the probability that an observation will lie in the interval $(38 < y < 42)$? The answer is 0.3721. To visualize this problem (and similar ones), it is helpful to sketch a normal distribution located with mean $\eta = 39$ with a standard deviation $\sigma = 4$ and shade in the area in question. Table A at the end of this book gives the area in the upper tail of the standardized distribution, that is, $\Pr(z > z_0)$.*

Exercise 2.7. The percent methanol in batches of product has an upper specification limit of 0.15%. Recorded data suggest the methanol observations may be characterized by a normal distribution with mean $\eta = 0.10\%$ and standard deviation $\sigma = 0.02\%$. What is the probability of exceeding the specification?

Exercise 2.8. The upper and lower specification limits on capacitance of elements within a circuit are 2.00 to 2.08 μF. Manufactured capacitors can be purchased that are approximately normally distributed with mean $\eta = 2.50$ μF and a standard deviation of 0.02 μF. What portion of the manufactured product will lie inside the specifications?

2.6. NORMAL PROBABILITY PLOTS

A normal distribution is shown in Figure 2.9*a*. Suppose the probability of an occurrence of some value *less* than x be given by a shaded area P. If you now plot P against x, you will obtain the sigmoid cumulative normal curve as in Figure 2.9*b*. Normal probability plots adjust the vertical scale (see Fig. 2.9*c*) so that P versus x is displayed as a straight line.

Now consider the dots in Figure 2.9*a* representing a random *sample* of 10 observations drawn from this normal distribution. Since the sample size is 10, the observation at the extreme left can be taken to represent the first 10% of the cumulative distribution so you would plot this first observation on Figure 2.9*b* midway between zero and 10%, that is, at 5%. Similarly, the second observation from the left can be taken as representative of the second 10% of the cumulative distribution, between 10 and 20%, and it is plotted at the intermediate value 15%. As expected, these sample values approximately trace out the sigmoid curve as in Figure 2.9*b*. Thus, when the same points are plotted on a normal probability scale, as in Figure 2.9*c*, they plot roughly as a straight line. Most computer software programs provide normal probability plots. Many offer comments on the adequacy of fitted lines and identify points either on or off the line. We believe that such commentary should be treated with some reserve and what to do about it is best left to the investigator. Scales for making your own normal probability plots are given in Table E at the end of this book. Intercepts such that $P_i = 100(i - 0.5)/m$ are given for the frequently needed values $m = 15, 31, 63, 16, 32, 64$. Instead of identifying the ordinates by percentages, it is common to

*Other normal tables may consider other areas, for example, that $\Pr(z < z_0)$ or $\Pr(0 < z < z_0)$. When you use a normal table, be sure to note which area (probability) is associated with z.

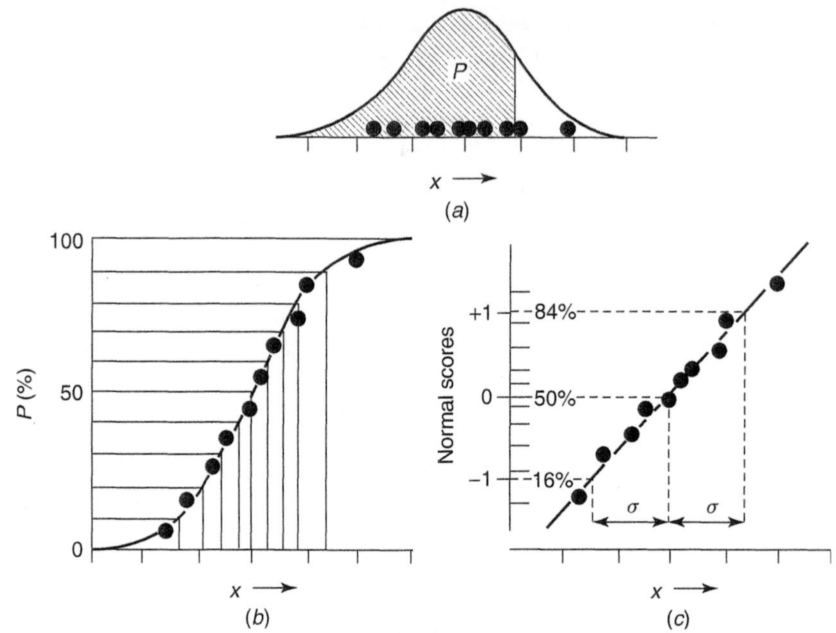

Figure 2.9. Normal distribution and the normal probability plot.

use as an alternative the *normal scores* (the number of standard deviations from the mean), also shown in Figure 2.9*c*.

Using Normal Probability Plots

Normal plots are used not so much for the purpose of checking distributional normality; a very large sample would be needed to do that. They can however point to suspect values that may have occurred. A further important application of normal probability plots is in the analysis of experiments using factorial and fractional factorial statistical designs. These applications are discussed in Chapters 5 and 6.

An Estimate of σ from the Normal Plot

The slope of the normal plot can provide an approximate estimate of the standard deviation σ. As you will see from Figure 2.9*c*, if you read off the value of x for probabilities 16%, 50%, and 84%, then an estimate of σ is equal to the distance on the x axis between the 16% and 50% points, or between the 50% and 84% points, or half the distance between the 16% and 84% values.

2.7. RANDOMNESS AND RANDOM VARIABLES

Suppose you *knew* the distribution $p(y)$ of the population of heights of recruits in the Patagonian army and suppose that a recruit was selected randomly. Without seeing him, what do you know of his height? Certainly you do not know it exactly. But it would be incorrect to say you know nothing about it because, knowing the distribution of heights of recruits, you can make statements of the following kind: The probability that the randomly chosen recruit is shorter than y_0 inches is P_0; the probability that he is taller than y_0 inches but shorter than y_1 inches is P_1. A quantity y, such as the randomly drawn recruit's height, which is not known exactly but for which you know the probability distribution is called a *random variable.*

Statistical Dependence

Suppose you were considering two characteristics, for example, the height y_1 in inches and the weight y_2 in pounds of the population of recruits. There will be a distribution of heights $p(y_1)$ and of weights $p(y_2)$ for these soldiers; thus the height and weight of a randomly drawn recruit have probability distributions and are random variables. But now consider the probability distribution of the *weights* of all recruits who were *75 inches tall.* This distribution is written as $p(y_2 \mid y_1 = 75)$. It is called the *conditional* distribution of weight y_2 given that y_1 is 75 inches (the vertical slash stands for the word "given"). Clearly, you would expect the conditional distribution $p(y_2 \mid y_1 = 75)$ for the weights of recruits who were 75 inches tall to be quite different from, say, the conditional distribution of weights $p(y_2 \mid y_1 = 65)$ of recruits 65 inches tall. The random variables height y_1 and weight y_2 would be said therefore to be *statistically dependent.*

Now suppose that y_3 was a measure of the IQ of the recruit. It might well be true that the conditional distribution of IQ would be the same for 75-inch recruits as for a 65-inch recruit. That is,

$$p(y_3|y_1 = 75) = p(y_3|y_1 = 65)$$

If the conditional distribution was the same *whatever* the height of the recruit, then

$$p(y_3|y_1) = p(y_3)$$

and y_1 and y_3 would be *statistically independent.*

The Joint Distribution of Two Random Variables

With height y_1 measured to the nearest inch and weight y_2 to the nearest pound, consider the joint probability of obtaining a recruit with height 65 inches and weight 140 pounds. This probability is denoted by $\Pr(y_1 = 65, y_2 = 140)$. One way of finding recruits of this kind is (1) to pick out the special class of recruits weighing 140 pounds and (2) to select *from this special class* those recruits who are 65 inches tall. The required joint probability is then the probability of finding a recruit who weighs 140 pounds multiplied by the probability of finding a recruit who is 65 inches tall *given* that he weighs 140 pounds. In symbols we have

$$\Pr(y_1 = 65, y_2 = 140) = \Pr(y_2 = 140) \times \Pr(y_1 = 65 \mid y_2 = 140)$$

You could have made the selection equally well by first picking the class with recruits with the desired height and then selecting from this class those with the required weight. Thus, it is equally true that

$$\Pr(y_1 = 65, y_2 = 140) = \Pr(y_1 = 65) \times \Pr(y_2 = 140 \mid y_1 = 65)$$

Parallel relationships apply to probability densities. Thus, if $p(y_1, y_2)$ is the joint density associated with specific (exact) values of two random variables y_1 and y_2, that density can always be factored as follows:

$$p(y_1, y_2) = p(y_1) \times p(y_2 \mid y_1) = \mathrm{p}(y_2) \times p(y_1 \mid y_2)$$

A Special Form for the Joint Distribution of Independent Random Variables

If y_1 and y_2 are *statistically independe*nt, then $p(y_2 \mid y_1) = p(y_2)$. Substitution of this expression into $p(y_1, y_2) = p(y_1) \times p(y_2 \mid y_1)$ shows that in the *special circumstance* of statistical independence the joint probability density may be obtained by *multiplying individual densities:*

$$p(y_1, y_2) = p(y_1) \times p(y_2)$$

A corresponding product formula is applied to probabilities of events. Thus, if y_3 was a measure of IQ independent of height y_1, the probability of obtaining a recruit with an IQ greater than 130 *and* a height greater than 72 inches would be given by the product of the two probabilities:

$$\Pr(y_1 > 72, y_3 > 130) = \Pr(y_1 > 72) \times \Pr(y_3 > 130)$$

The product formula does not apply to variables that are statistically dependent. For instance, the probability that a randomly chosen recruit weighs over 200 pounds ($y_2 > 200$) *and* measures over 72 inches in height ($y_1 > 72$) cannot be found by multiplying the individual probabilities. The product formula does not hold in this instance because $p(y_2 \mid y_1)$ is not equal to $p(y_2)$ but rather depends on y_1.

These arguments generalize to any number of variables. For example, *if* $y_1, y_2, \ldots, y_n$ are *independently* distributed random variables and $y_{10}, y_{20}, \ldots, y_{n0}$ are particular values of these random variables, then

$$\Pr(y_1 > y_{10}, y_2 > y_{20}, \ldots, y_n > y_{n0}) = \Pr(y_1 > y_{10}) \times \Pr(y_2 > y_{20}) \\ \times \cdots \times \Pr(y_n > y_{n0})$$

Application to Repeated Scientific Observations

In the examples above $y_1, y_2, \ldots, y_n$ represented different *kinds* of variables—height, weight, IQ. These formulas also apply when the variables are *observations* of the *same* phenomena. Suppose that $y_1, y_2, \ldots, y_n$ are measurements of specific gravity recorded in the order in which they were made. Consider the first observation y_1. Suppose it can be treated as a random variable, that is, it can be characterized as a drawing from some population of first observations typified by its density function $p(y_1)$. Suppose the subsequent observations $y_2, y_3, \ldots, y_n$ can be similarly treated and they have density functions $p(y_1), p(y_2), p(y_3), \ldots, p(y_n)$, respectively. If $y_1, y_2, \ldots, y_n$ are statistically independent, then $p(y_1, y_2, \ldots, y_n) = p(y_1) \times p(y_2) \times \cdots \times p(y_n)$. And if $p(y_1), p(y_2), p(y_3), \ldots, p(y_n)$ are not only independent but also identical in form (had the same location, spread, and shape), then these n repeated observations would be said to be *independently* and *identically* distributed (often shortened to IID). In that case the sample of observations $y_1, y_2, \ldots, y_n$ is *as if* it had been generated by random drawings from some fixed population typified by a single probability density function $p(y)$. If in addition the common population distribution was normal, they would be normally, identically, and independently distributed (often shortened to NIID). The nomenclature IID and NIID is used extensively in what follows.

2.8. COVARIANCE AND CORRELATION AS MEASURES OF LINEAR DEPENDENCE

A measure of linear dependence between, for example, height y_1 and weight y_2 is their *covariance*. Just as their variances are the mean values of the population of squared deviations of observations from their means, so their covariance (denoted by Cov) is the mean value of the population of the *products* of these deviations $y_1 - \eta_1$ with $y_2 - \eta_2$. Thus,

$$V(y_1) = E(y_1 - \eta_1)^2 = \sigma_1^2, \qquad V(y_2) = E(y_2 - \eta_2)^2 = \sigma_2^2$$

and

$$\text{Cov}(y_1, y_2) = E(y_1 - \eta_1)(y_2 - \eta) = \sum(y_1 - \eta_1)(y_2 - \eta_2)/N$$

In particular, if y_1 and y_2 were independent, $\text{Cov}(y_1, y_2)$ would be zero.*

In practice, recruits that deviated positively (negatively) from their mean height would tend to deviate positively (negatively) from their mean weight. Thus positive (negative) values of $y_1 - \eta_1$ would tend to be accompanied by positive (negative) values of $y_2 - \eta_2$ and the covariance between height and weight would be positive. Conversely, the covariance between speed of driver reaction and alcohol consumption would likely be negative; a decrease in reaction speed is associated with an increase in alcohol consumption, and vice versa.

The covariance is dependent upon the scales chosen. If, for example, height was measured in feet instead of inches, the covariance would be changed. A

*The converse is not true. For example, suppose, apart from error, that y_1 was a quadratic function of y_2 that plotted like the letter U. Then, although y_1 and y_2 would be statistically dependent, their covariance could be zero.

"scaleless covariance" called the *correlation coefficient* identified by the symbol $\rho(y_1, y_2)$ or simply ρ (the Greek letter rho) is obtained by dividing the cross product of the deviations $y_1 - \eta_1$ and $y_2 - \eta_2$ by σ_1 and σ_2, respectively. Thus

$$\rho = \rho(y_1, y_2) = E\left[\frac{(y_1 - \eta_1)}{\sigma_1}\frac{(y_2 - \eta_2)}{\sigma_2}\right] = \frac{\mathrm{Cov}(y_1, y_2)}{\sigma_1\sigma_2} = \frac{\mathrm{Cov}(y_1, y_2)}{\sqrt{V(y_1)V(y_2)}}$$

Equivalently, we can write $\mathrm{Cov}(y_1, y_2) = \rho\sigma_1\sigma_2$.

The *sample correlation coefficient* between y_1 and y_2 is then defined as

$$r = \frac{\sum(y_1 - \bar{y}_1)(y_2 - \bar{y}_2)/(n-1)}{s_1 s_2}$$

The numerator in this expression is called the *sample covariance*, and s_1 and s_2 are the sample standard deviations for y_1 and y_2.

Exercise 2.9. Compute the sample correlation coefficient for these data:

y_1 (height in inches)	65	68	67	70	75
y_2 (weight in pounds)	150	130	170	180	220

Answer: 0.83.

Serial Dependence Measured by Autocorrelation

When data are taken in sequence, there is usually a tendency for observations made close together in time (or space) to be more alike that those taken farther apart. This can occur because disturbances, such as high impurity in a chemical feedstock or high respiration rate in an experimental animal, can persist. There are occasional instances where consecutive observations are less alike than those taken farther apart. Suppose, for example, that the response measured is the apparent monthly *increase* in the weight of an animal calculated from the differences of monthly weighings. An abnormally high increase in weight recorded for October (perhaps because of water retention) can result in an unusually low increase being attributed to November.

If sufficient data are available, serial correlation can be seen by plotting each observation against the immediately proceeding one (y_t vs. y_{t-1}). Similar plots can be made for data two intervals apart (y_t vs. y_{t-2}), three units apart, and so on. The corresponding correlation coefficients are called *autocorrelation coefficients*. The distance between the observations that are so correlated is called the *lag*. The lag k *sample autocorrelation coefficient* is defined by

$$r_k = \frac{\sum(y_t - \bar{y})(y_{t-k} - \bar{y})}{\sum(y_t - \bar{y})^2}$$

Exercise 2.10. Compute r_1 for these data: 3, 6, 9, 8, 7, 5, 4.

Answer: 0.32.

Note: There is little point in calculating autocorrelation coefficients for such small samples. The above exercise is included only to help gain an appreciation of the meaning of the formula for r. Standard software programs can quickly provide all sample autocorrelations up to any desired lag k. The plot of r_k versus k is called the *sample autocorrelation function.*

2.9. STUDENT'S t DISTRIBUTION

You saw earlier that if you assume normality and you know the standard deviation σ you can determine the probability of y exceeding some value y_0 by calculating the normal deviate $z_0 = (y_0 - \eta)/\sigma$ and thus $\Pr(z > z_0) = \Pr(y > y_0)$ from the tabled normal distribution. For example, suppose the level of impurity in a reactor is approximately normally distributed with a mean $\eta = 4.0$ and a standard deviation $\sigma = 0.3$. What is the probability that the impurity level on a *randomly chosen* day will exceed 4.4? Here $y_0 = 4.4$, $\eta = 4.0$, and $\sigma = 0.3$, so that

$$z_0 = \frac{y_0 - \eta}{\sigma} = \frac{4.4 - 4.0}{0.3} = 1.33$$

From Table A at the end of this book you can determine that $\Pr(z > 1.33) = 0.0918$. There is thus about a 9% chance that the impurity level on a randomly chosen day will exceed 4.4.

In practice σ is almost always unknown. Suppose, then, that you substitute in the equation for the normal deviate an estimate s of σ obtained from n observations themselves. Then the ratio

$$t = \frac{y_0 - \eta}{s}$$

has a known distribution called *Student's* distribution.* Obviously a probability calculated using the t distribution must depend on how reliable is the substituted value s. It is not surprising therefore that the distribution of t depends on the number of degrees of freedom $\upsilon = n - 1$ of s. The t distributions for $\upsilon = 1$, 9, and ∞ are shown in Figure 2.10.

Thus, suppose that the *estimated* standard deviation s was 0.3 and was based on seven observations. Then the number of degrees of freedom would be 6 and the probability of a randomly drawn observation exceeding 4.4 would be obtained from

$$t_0 = \frac{y_0 - \eta}{s} = \frac{4.4 - 4.0}{0.3} = 1.33$$

*The t distribution was first discovered in 1908 by the chemist W. S. Gosset, who worked for the Guiness brewery in Dublin and wrote under the pseudonym "Student."

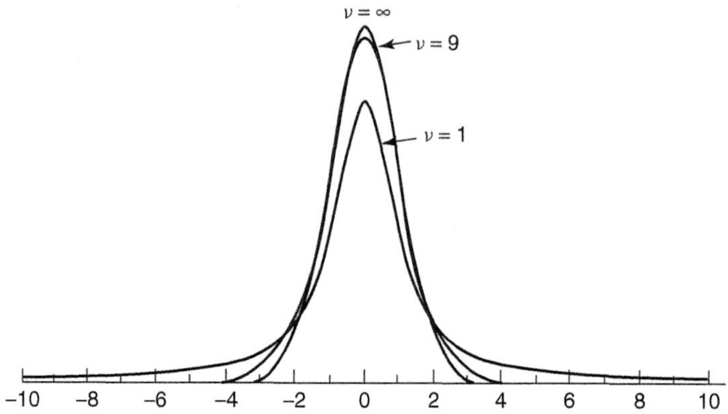

Figure 2.10. The t distribution for $\nu = 1, 9, \infty$.

The corresponding probability $\Pr(t > 1.33)$ when t has six degrees of freedom can be calculated by your computer or from Table B at the end of this volume. The value of $\Pr(t > 1.33)$ is found to be about 12%, which, as might be expected, is somewhat larger than the 9% found assuming that σ was known.

As illustrated in Figure 2.10, when the number of degrees of freedom ν is small, uncertainty about s results in a greater probability of extreme deviations and hence in a "heavier" tailed t distribution. The following values for $\Pr(t > 2)$ illustrate the point:

$\upsilon = \infty$ (normal distribution)	$\Pr(t > 2) = 2.3\%$
$\upsilon = 9$	$\Pr(t > 2) = 3.8\%$
$\upsilon = 1$	$\Pr(t > 2) = 14.8\%$

Except in the extreme tails of the distribution, the normal distribution provides a rough approximation of the t distribution when υ is greater than about 15.

Random Sampling, Independence, and Assumptions

Random *sampling* is a way of physically inducing statistical independence. Imagine a very large population of data with values written on tickets placed in a lottery drum. Suppose that after the tickets were thoroughly mixed two tickets were drawn at random. We would expect the number on the first ticket to be statistically independent of that on the second. Thus, random sampling would ensure the validity of what was called the IID assumption. The observations would then be identically and independently distributed. This means that the individual distributions are the same and that knowing one tells you nothing about the others.

It would be nice if you could always believe the random sampling model because this would ensure the validity of the IID assumption and produce certain dramatic simplifications. In particular, it would endow sample statistics such as $\bar{y}$ with very special properties.

The Mean and Variance of the Average $\bar{y}$ for Independent Identically Distributed Observations

If the random sampling model is appropriate so that the errors are IID, we have the simple rule that $\bar{y}$ varies about the population mean η with variance σ^2/n. Thus

$$E(\bar{y}) = \eta, \qquad V(\bar{y}) = \frac{\sigma^2}{n}$$

[However, when observation errors are not independent but correlated, the expression for the variance of $\bar{y}$ contains a factor C that depends on the degree of their correlation, that is, $V(\bar{y}) = C \times (\sigma^2/n)$. For independent data $C = 1$ but for autocorrelated data it can deviate greatly from this value. If, for example, $n = 10$ and only the *immediately adjacent* observations were autocorrelated, the factor C could be as large as 1.9 for positively autocorrelated observations or as small as 0.1 for negatively correlated observations. (See Appendix 2A.) Thus different degrees of lag 1 autocorrelation can change $V(\bar{y})$ by a factor of 19! It would be disastrous to ignore important facts of this kind.]

Random Sampling Distribution of the Average $\bar{y}$

To visualize what is implied by the expressions $E(\bar{y}) = \eta$ and $V(\bar{y}) = \sigma^2/n$, suppose that a very large number of white tickets in a white lottery drum represent the population of individual observations y and that this population has mean η and variance σ^2. Now suppose you randomly draw out a sample of $n = 10$ white tickets, calculate their average $\bar{y}$, write it down on a blue ticket, put the blue ticket in a second blue lottery drum, and repeat the whole operation many many times. Then the numbers in the blue drum, which now form the population of averages $\bar{y}$, will have the same mean η as the original distribution of white tickets but with a variance σ^2/n, which is only one nth as large as that of the original observations.

The original distribution of the individual observations represented by the drum of white tickets is often called the *parent* distribution. Any distribution derived from this parent by random sampling is called a *sampling* distribution. Thus the distribution of numbers shown on the blue tickets in the blue lottery drum is called the sampling distribution of $\bar{y}$.

Approach to Normality of the Sampling Distribution of $\bar{y}$

To justify the formulas for the mean and variance of $\bar{y}$, we must assume the random sampling model is appropriate, but these formulas would remain true no matter what was the shape of the parent distribution. In particular, the parent distribution would not need to be normal. If it is not normal, then, as is illustrated in Figure 2.11, on the random sampling model averaging not only reduces the standard deviation by a factor of $1/\sqrt{n}$ but also simultaneously produces a distribution for $\bar{y}$ that is more nearly normal. This is because of the central limit

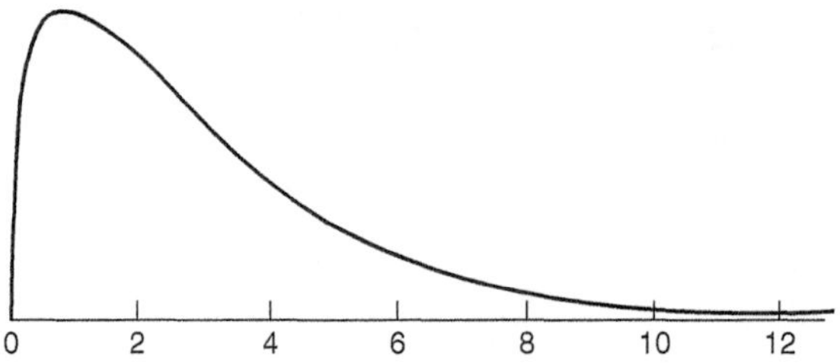

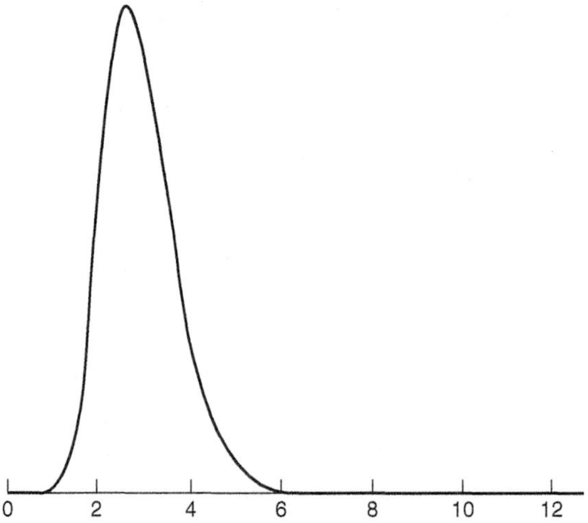

Figure 2.11. Distribution (b) is for the average of $n = 10$ observations randomly sampled from the skewed distribution (a).

effect discussed earlier. In summary, then, *on the IID hypothesis appropriate for random sampling:*

	Parent Distribution for Observations y	Sampling Distribution for Averages $\overline{y}$
Mean	η	η
Variance	σ^2	σ^2/n
Standard deviation	σ	$\sigma/\sqrt{n}$
Form of parent distribution	Any *	More nearly normal than the parent distribution

*This statement applies to all parent distributions commonly met in practice. It is not true for certain mathematical toys (e.g., the Cauchy distribution), which need not concern us here.

2.10. ESTIMATES OF PARAMETERS

The *quantity* $\bar{y}$ is the average of the particular sample of observations that happens to be available. Without further assumption there is little more you can say about it. If, however, the observations can be regarded as a *random* sample from some population with mean η and variance σ^2, then

1. $\bar{y}$ has for its mean the mean value η of the parent population and
2. $\bar{y}$ varies about η with standard deviation $\sigma/\sqrt{n}$.

Thus, as you imagine taking a larger and larger sample, $\bar{y}$ tends to lie closer and closer to η. In these circumstances it is reasonable to regard $\bar{y}$ as an estimate of η. Similarly, it may be shown that s^2 has a mean value σ^2 and varies about that value also with a standard deviation proportional to $1/\sqrt{n-1}$. On the statistical assumption of random sampling, therefore, you can regard s^2 as an estimate of σ^2.

The problem of choosing statistics to estimate parameters in the best way possible is complicated, depending heavily, as you might guess, on the definition of "best." The important point to remember is that it is the IID assumption, appropriate to the hypothesis of random sampling, that endows estimates with special distributional properties.

Sampling Distributions of a Sum and a Difference

Interest often centers on the distribution of the sum Y of two *independently* distributed random variables y_A and y_B. Let us suppose that y_A has a distribution with mean η_A and variance σ_A^2 and y_B has a distribution with mean η_B and variance σ_B^2. What can we say about the distribution of $Y = y_A + y_B$?

Again you could illustrate this question by considering lottery drums each with appropriate populations of A and B tickets. Imagine that after each random drawing from drum A to obtain y_A another random drawing is made from drum B to obtain y_B, and the sum $Y = y_A + y_B$ is written on a red ticket and put in a third lottery drum. After many such drawings what could you say about the distribution of the sums written of the red tickets in the third lottery drum?

It turns out that the mean value of sum Y is the sum of the mean values of y_A and y_B:

$$E(Y) = E(y_A + y_B) = E(y_A) + E(y_B) = \eta_A + \eta_B$$

and it can be shown for *independent* drawings that the variance of Y is the sum of the variances of y_A and y_B:

$$V(Y) = V(y_A + y_B) = \sigma_A^2 + \sigma_B^2$$

Correspondingly, for the difference of two random variables

$$E(Y) = E(y_A - y_B) = E(y_A) - E(y_B) = \eta_A - \eta_B$$

and for *independent* drawings

$$V(Y) = V(y_A - y_B) = \sigma_A^2 + \sigma_B^2$$

The result for the difference follows from that of the sum, for if you write $-y_B = y'_B$, then, of course, the variance does not change, $V(y_B) = V(y'_B)$, and thus

$$V(y_A - y_B) = V(y_A + y'_B) = \sigma_A^2 + \sigma_B^2$$

These results for the sum and the difference do not depend on the *form* of the parent distributions; and, because of the central limit effect, the distributions of both sum and difference tend to be closer to the normal than are the original parent distributions. If, then, y_A and y_B are *independent* drawings from the same population with variance σ^2 or, alternatively, from different populations having the same variance σ^2, the variance of the sum and the variance of the difference are identical, that is,

$$V(y_A + y_B) = 2\sigma^2, \qquad V(y_A - y_B) = 2\sigma^2$$

2.11. RANDOM SAMPLING FROM A NORMAL POPULATION

The following important results are illustrated in Figure 2.12 on the assumption of *random* sampling from a *normal* parent population or equivalently on the NIID assumptions. If a random sampling of n observations is drawn from a normal distribution with mean η and variance σ^2, then:

1. The distribution of $\bar{y}$ is also normal with mean η and variance σ^2/n.
2. The sample variance s^2 is distributed independently of $\bar{y}$ and is a scaled χ^2 (chi-square) distribution having $\upsilon = n - 1$ degrees of freedom. This is a skewed distribution whose properties are discussed later.
3. The quantity $(\bar{y} - \eta)/(s/\sqrt{n})$ is distributed in a t distribution with $\upsilon = n - 1$ degrees of freedom.

In the illustration in Figure 2.12 $n = 5$ so that $\upsilon = 4$. Result 3 is particularly remarkable because it allows the deviation $\bar{y} - \eta$ to be judged against an *estimate* of the standard deviation $s/\sqrt{n}$ obtained from *internal evidence from the sample itself.* Thus, no external data set is needed to obtain the necessary Student's t reference distribution *provided* the assumption of random sampling from a normal distribution can be made.

Standard Errors

The standard deviation of the average, $s/\sqrt{n}$, is the positive square root of the variance of the average and is often referred to as the "standard error" of the average. The square root of the variance of any statistic constructed from a sample of observations is commonly called that statistic's standard error.

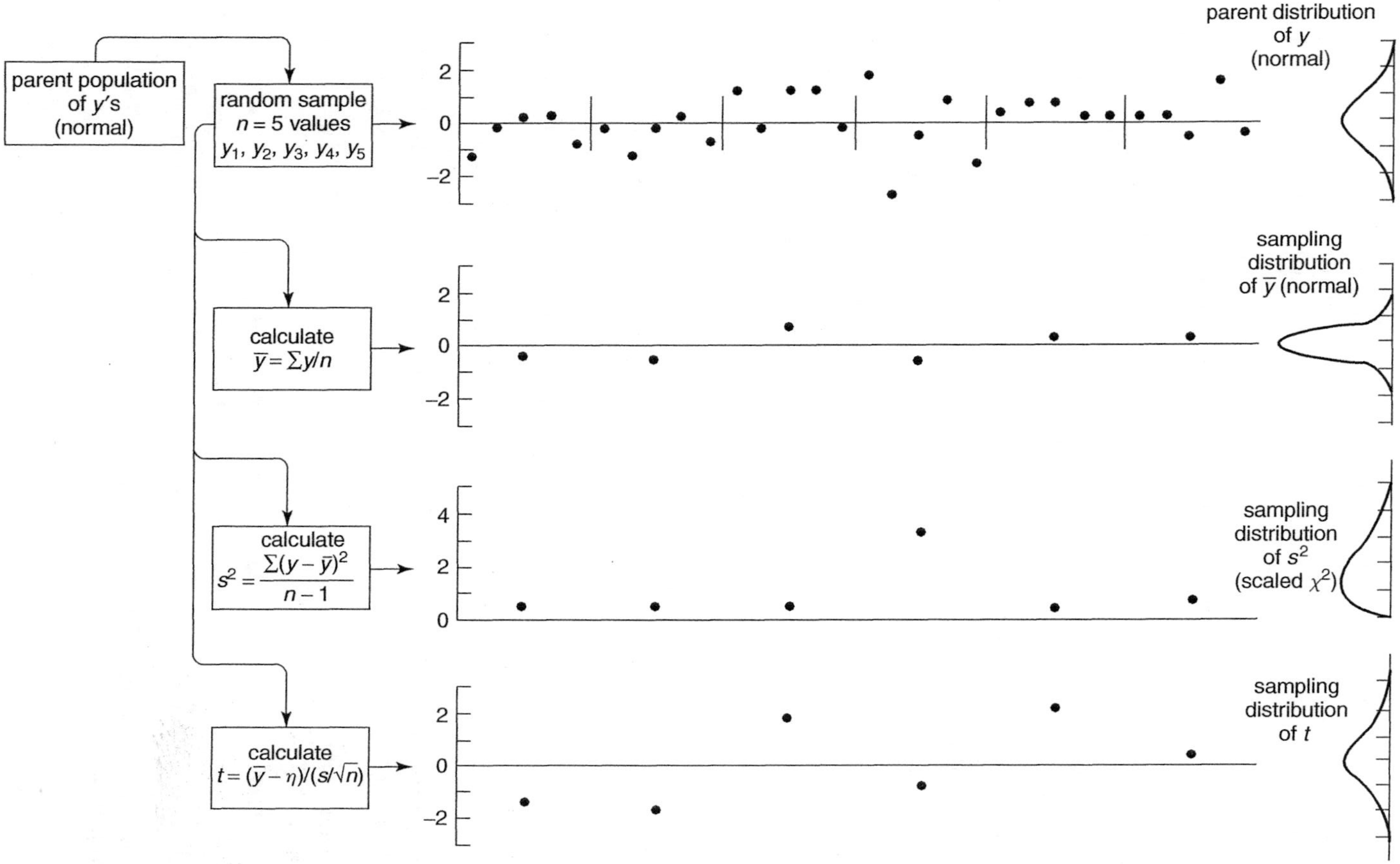

Figure 2.12. Random sampling from a normal distribution to produce sampling distributions for $\bar{y}$, s^2, and t.

Sufficiency of $\bar{y}$ and s^2

If the assumptions of random sampling from a normal distribution are exactly satisfied, Fisher showed that *all* the information in the sample $y_1, y_2, \ldots, y_n$ about η and σ^2 is contained in the two quantities $\bar{y}$ and s^2. These statistics are then said to be *jointly sufficient* estimates for η and σ^2.

2.12. THE CHI-SQUARE AND *F* DISTRIBUTIONS

Two important distributions, again derived on the assumption of random sampling from normal distributions, are the χ^2 (chi-square) distribution (from which you can obtain the distribution of the sample variance s^2) and also the F distribution (from which you can obtain the distribution of the ratio of two sample variances).

The Chi-Square Distribution

Suppose that $z_1, z_2, \ldots z_\nu$ are a set of ν (nu) independently distributed unit normal deviates, each having a mean zero and a variance unity. Then their sum of squares $\sum_{u=1}^{\nu} z_u^2 = z_1^2 + z_2^2 + \cdots + z_\nu^2$ has a distribution of special importance, called a χ^2 *(chi-square) distribution.* The number of *independent* squared normal variables determines the important parameter of the distribution—the number of degrees of freedom υ. We use the symbol $\sim$ to mean "is distributed as" so that

$$\sum_{u=1}^{\nu} z_u^2 \sim \chi_\nu^2$$

means that $\sum_{u=1}^{\nu} z_u^2$ has a chi-square distribution (is distributed as χ^2) with υ degrees of freedom. *Thus, the sum of squares of υ independent unit normal deviates has a chi-square distribution with υ degrees of freedom.*

The χ_ν^2 distribution has mean υ and variance 2υ. It is skewed to the right, but the skewness becomes less as υ increases, and for υ greater than 50 the chi-square distribution is approximately normal. You can determine probabilities associated with this distribution using an appropriate computer program or from the percentage points of the χ_ν^2 distribution given in Table *C* at the end of this book. The curve in Figure 2.11*a* is that of a χ^2 distribution having $\upsilon = 3$ degrees of freedom.

Distribution of Sample Variances Calculated from Normally Distributed Data

The following results are true on the NIID assumptions; specifically it is supposed that $y_1, y_2, \ldots, y_n$ are normally, independent, and identically distributed random variables having mean η and variance σ^2. Since $z_u = (y_u - \eta)/\sigma$ is normally distributed with mean zero and variance unity (a unit normal deviate), the sum of squares $\sum z_u^2$ of deviations from the population mean has a chi-square distribution with $\upsilon = n$ degrees of freedom, that is,

$$\frac{\sum (y_u - \eta)^2}{\sigma^2} \sim \chi_n^2$$

For the "natural" variance estimate $\dot{s}^2$, appropriate when the population mean *is known*, $\dot{s}^2 = \sum(y_u - \eta)^2/n$, it follows therefore that

$$\frac{n\dot{s}^2}{\sigma^2} \sim \chi_n^2$$

or, equivalently,

$$\dot{s}^2 \sim \frac{\sigma^2}{n}\chi_n^2$$

The much more common circumstance* is that in which η is unknown and the sample average $\bar{y}$ must be used instead of the population mean. The standardized sum of squares of deviations from $\bar{y}$ has a chi-square distribution with $\upsilon = n - 1$ degrees of freedom. Thus

$$\frac{\sum(y_u - \bar{y})^2}{\sigma^2} \sim \chi_{n-1}^2$$

and since $s^2 = \sum(y_u - \bar{y})^2/(n-1)$, it follows that $(n-1)s^2/\sigma^2 \sim \chi_{n-1}^2$ or, equivalently, $s^2 \sim [\sigma^2/(n-1)]\chi_{n-1}^2$. The distribution of s^2 is thus a scaled chi-square distribution with scale factor $\sigma^2/(n-1)$.

The F Distribution of the Ratio of Two Independent Sample Variances

Suppose that a sample of n_1 observations is randomly drawn from a normal distribution having variance σ_1^2, a second sample of n_2 observations is randomly drawn from a second normal distribution having variance σ_2^2, so that the estimates s_1^2 and s_2^2 of the two population variances have $\upsilon_1 = n_1 - 1$ and $\upsilon_2 = n_2 - 1$ degrees of freedom, respectively. Then s_1^2/σ_1^2 is distributed as $\chi_{\upsilon_1}^2/\upsilon_1$ and s_2^2/σ_2^2 is distributed as $\chi_{\upsilon_2}^2/\upsilon_2$, and the ratio $(\chi_{\upsilon_1}^2/\upsilon_1)/(\chi_{\upsilon_2}^2/\upsilon_2)$ has an F distribution with υ_1 and υ_2 degrees of freedom. The probability points for the F distribution are given in Table D at the end of this book. Thus

$$\frac{s_1^2/\sigma_1^2}{s_2^2/\sigma_2^2} \sim F_{\upsilon_1,\upsilon_2}$$

or, equivalently,

$$\frac{s_1^2}{s_2^2} \sim \frac{\sigma_1^2}{\sigma_2^2}F_{\upsilon_1,\upsilon_2}$$

Residuals

When η and σ^2 are known, they completely define the normal distribution, and the standardized residuals

$$\frac{(y_1 - \bar{y})}{s}, \frac{(y_2 - \bar{y})}{s}, \ldots, \frac{(y_n - \bar{y})}{s}$$

* The natural standard deviation has an important application discussed later.

or equivalently the *relative* magnitudes of the residuals

$$y_1 - \bar{y}, y_2 - \bar{y}, \ldots, y_n - \bar{y}$$

from such a distribution are *informationless*. The other side of the coin is that, when a *departure* from the hypothesis of random sampling from a normal population occurs, these residuals can provide hints about the nature of the departure.

A First Look at Robustness

The NIID assumptions, or equivalently the assumption of random sampling from a normal distribution, are never exactly true. However, for many important procedures, and in particular comparisons of means, the results obtained on NIID assumptions can frequently provide adequate approximations even though the assumptions are to some extent violated. Methods that are insensitive to specific assumptions are said to be *robust* to those assumptions. Thus methods of comparing means are usually robust to moderate nonnormality and inequality of variance. Most common statistical procedures are, however, highly *non*robust to autocorrelation of errors. The important question of the robustness of statistical procedures is discussed in greater detail in later chapters.

Exercise 2.11. The following are 10 measurements of the specific gravity y of the same sample of an alloy:

Date	October 8		October 9		October 10		October 11		October 12	
Time	AM	PM	AM	PM	AM	PM	AM	PM	AM	PM
y	0.36721	0.36473	0.36680	0.36487	0.36802	0.36396	0.36758	0.36425	0.36719	0.36333

(a) Calculate the average $\bar{y}$ and standard deviation s for this sample of $n = 10$ observations.

(b) Plot the residuals $y - \bar{y}$ in time order.

(c) It is proposed to make an analysis on the assumption that the 10 observations are a random sample from a normal distribution. Do you have any doubts about the validity of this assumption?

2.13. THE BINOMIAL DISTRIBUTION

Data sometimes occur as the proportion of times a certain event happens. On specific assumptions, the *binomial distribution* describes the frequency with which such proportions occur. For illustration, consider the strategy of a dishonest gambler Denis Bloodnok, who makes money betting with a biased penny that he knows comes up heads, on the average, 8 times out of 10. For his penny the *probability* p of a head is 0.8 and the *probability* $q = 1 - p$ of a tail is 0.2. Suppose he bets at even money that of $n = 5$ tosses of his penny at least four will come up heads. To make his bets appropriately, Bloodnok calculates the probability, with five tosses, of no heads, one head, two heads, and so on. With y representing the number of heads, what he needs are the $n + 1 =$ six values:

$$\Pr(y = 0), \Pr(y = 1), \ldots, \Pr(y = 5)$$

Call the tossing of the penny five times a *trial* and denote the *outcome* by listing the heads (H) and tails (T) in the order in which they occur. The outcome $y = 0$ of getting no heads can occur in only one way, so

$$\Pr(y = 0) = \Pr(\mathrm{T\ T\ T\ T\ T}) = q \times q \times q \times q \times q = q^5 = 0.2^5 = 0.00032$$

The outcome of just one head can happen in five different ways depending on the order in which the heads and tails occur. Thus

$$\Pr(y = 1) = \Pr(\mathrm{H\ T\ T\ T\ T}) + \Pr(\mathrm{T\ H\ T\ T\ T}) + \Pr(\mathrm{T\ T\ H\ T\ T}) + \Pr(\mathrm{T\ T\ T\ H\ T}) + \Pr(\mathrm{T\ T\ T\ T\ H})$$

and $\Pr(y = 1) = 5pq^4 = 5 \times 0.8 \times 0.2^4 = 5 \times 0.000128 = 0.00640$. The outcome of just two heads can happen in 10 different ways (orderings):

(H H T T T)	(H T H T T)	(H T T H T)	(H T T T H)	(T H H T T)
(T H T H T)	(T H T T H)	(T T H H T)	(T T H T H)	(T T T H H)

Thus $\Pr(y = 2) = 10p^2q^3 = 10 \times 0.8^2 \times 0.2^3 = 0.05120$, and so on. Evidently, to obtain $\Pr(y)$ for any y, you must

1. calculate the probability $p^y q^{n-y}$ of y heads and $n - y$ tails occurring in some specific order and
2. multiply by the number of different orderings in which y heads can occur in n throws: This is called the *binomial coefficient* and is given by

$$\binom{n}{y} = \frac{n!}{y!(n-y)!}$$

where $n!$ is read as n *factorial* and is equal to $n \times (n-1) \times (n-2) \times \cdots \times 2 \times 1$.

Thus the binomial distribution showing the probability of obtaining y heads in some order or other when there are n throws is

$$\Pr(y) = \binom{n}{y} p^y q^{n-y} = \frac{n!}{y!(n-y)!} p^y q^{n-y}$$

Table 2.2. Binomial Distribution for y Heads in n Trials When $p = 0.8$ and $n = 5$

Number of Heads y	$\binom{5}{y} = \frac{5!}{y!(5-y)!}$	$p^y q^{n-y}$	$\Pr(y)$
0	1	0.00032	0.00032
1	5	0.00128	0.00640
2	10	0.00512	0.05120
3	10	0.02048	0.20480
4	5	0.08192	0.40960
5	1	0.32768	0.32768
			1.00000

For Bloodnok's problem the distribution of the number of heads is calculated in Table 2.2 and plotted in Figure 2.13*a*.

In this way Bloodnok calculates that using his biased penny the probability of *at least* four heads [given by $\Pr(4) + \Pr(5)$] is about 0.74. For the fair penny $\Pr(4) + \Pr(5)$ yields a probability of only 0.19 (see Fig. 2.13*b*). Thus a wager at even money that he can throw at least four heads appears unfavorable to him. But by using his biased penny he can make an average of 48 cents for every dollar bet. (If he bets a single dollar on this basis 100 times, in 74 cases he will make a dollar and in 26 cases he will lose a dollar. His overall net gain is thus $74 - 26 = 48$ dollars per 100 bet.)

Exercise 2.12. Obtain $\Pr(y)$ for $y = 0, 1, 2, \ldots, 5$ for five fair pennies and confirm that $\Pr(4) + \Pr(5) = 0.19$. What would Bloodnok's average profit or loss be if we accepted his wager but made him use fair pennies?

Answer: A loss of 62 cents for every dollar bet.

Exercise 2.13. Using a standard deck of 52 cards, you are dealt 13. What is the probability of getting six spades? *Answer*: 0.13.

General Properties of the Binomial Distribution

The binomial distribution might be applied to describe the proportion of animals surviving a certain dosage of drug in a toxicity trial or the proportion of manufactured items passing a test in routine quality inspection. In all such trials there are just two outcomes: head or tail, survived or died, passed or failed. In general, it is convenient to call one outcome a *success* and the other a *failure*. Thus we can say the binomial distribution gives the probability of y successes out of a possible total of n in a trial where the fixed probability of an individual success is p and of a failure is $q = 1 - p$.

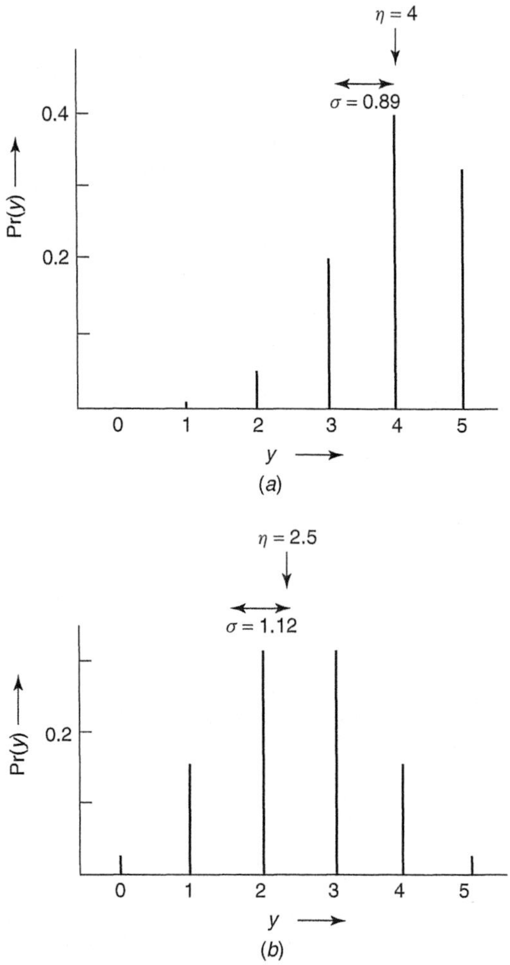

Figure 2.13. Binomial distributions for different choices of p: (*a*) $p = 0.8$ and $n = 5$ (Bloodnok's distribution); (*b*) $p = 0.5$ and $n = 5$ (distribution for fair penny).

Mean and Variance of the Binomial Distribution

The mean and variance of the binomial distribution are

$$\eta = np, \qquad \sigma^2 = npq$$

Thus the standard deviation is $\sigma = \sqrt{npq}$. For the gambler's trial of five tosses of the biased penny $\eta = 5 \times 0.8 = 4.0$ and $\sigma = \sqrt{5 \times 0.8 \times 0.2} = 0.89$, as illustrated in Figure 2.13*a*.

Distribution of the Observed Proportion of Successes y/n

In some cases you may wish to consider not the number of successes but the *proportion of successes* y/n. The probability of obtaining some value y/n is the same as the probability of obtaining y, so

$$\Pr\left(\frac{y}{n}\right) = \binom{n}{y} p^y q^{n-y}$$

The distribution of y/n is obtained simply by rescaling the horizontal axis of the binomial distribution $\Pr(y)$. Thus Figure 2.14a shows the distribution of y

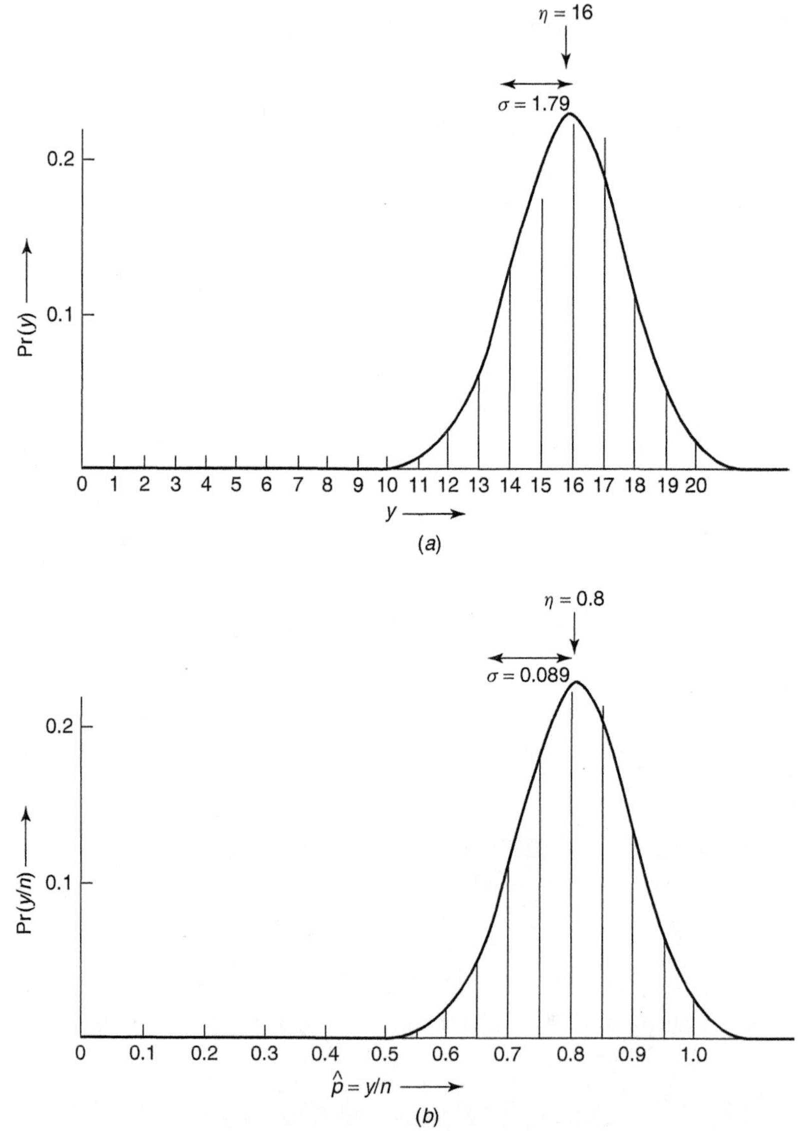

Figure 2.14. Normal approximation to binomial: (a) number of successes; (b) proportion of successes; (c) normal approximation showing tail area with Yates adjustment.

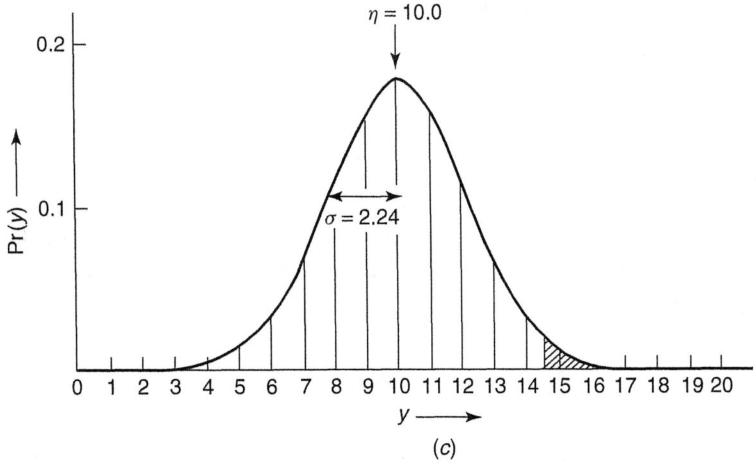

Figure 2.14. *(continued)*

successes for $p = 0.8$ and $n = 20$. Figure 2.14*b* shows the distribution of the proportion $\hat{p} = y/n$ of successes using the same ordinates. The observed proportion y/n is $\hat{p}$, an *estimate* of the probability p.*

Approach to Normality of the Binomial Distribution

In Figures 2.14*a,b,c* the continuous line shows a normal distribution having the same mean and standard deviation as the respective binomial distribution. In general, the normal approximation to the ordinates of the binomial distribution steadily improves as n increases and is better when p is not near the extremes of zero or unity. A rough rule is that for $n > 5$ the normal approximation will be adequate if the absolute value of $(1/\sqrt{n})(\sqrt{q/p} - \sqrt{p/q})$ is less than 0.3.

Yates Adjustment

In the above a continuous normal distribution is being used to approximate a discrete binomial distribution. A better normal approximation for $\Pr(y > y_0)$ is obtained by adding $\frac{1}{2}$ to the value of y_0. Thus, for a binomial distribution with $p = 0.5$ and $n = 20$, $\Pr(y > 14) = 2.06\%$. The approximately normal distribution using the same mean $\eta = 10.0$ and standard deviation 2.24 gives $P_0(y > 14) = 3.18\%$. A much better approximation is provided with Yates adjustment $P(y > 14.5) = 2.21\%$. If you imagine Figure 2.14*c* drawn as a histogram, you can easily see why.

* A caret over a symbol or group of symbols means "an estimate of." Thus $\hat{p}$ is read "an estimate of p."

2.14. THE POISSON DISTRIBUTION

If in the expression for the binomial distribution

$$\Pr(y) = \left\{ \frac{n!}{y!(n-y)!} \right\} p^y q^{n-y}$$

the quantity p is made smaller and smaller *at the same* time as n is made larger and larger so that $\eta = np$ remains finite and fixed, it turns out that $\Pr(y)$ tends closer and closer to the limit,

$$\Pr(y) = \frac{e^{-\eta}\eta^y}{y!}$$

This limiting distribution is called the Poisson distribution after its discoverer. The distribution has the property that the probability of a particular value depends *only* on the mean frequency η. For the Poisson distribution η is also the variance of the distribution. Thus η is the only parameter of the distribution. This limiting distribution is of importance because there are many everyday situations, such as the occurrence of accidents, where the probability p of an event at a particular time or place is small but the number of opportunities n for it to happen are large.

In general let y be the intensity of an event, that is, the count of the number of times a random event occurs within a certain unit (breaks per length of wire, accidents per month, defects per assembly). Then, so long as this intensity is not influenced by the size of the unit or the number of units, the distribution of y is that of the Poisson distribution with η the mean number of counts per unit.

To illustrate, when Henry Crun is driving his car, the probability that he will have an accident in any given minute is small. Suppose it is equal to $p = 0.00042$. Now suppose that in the course of one year he is at risk for a total of $n = 5000$ minutes. Then, although there is only a small probability of an accident in a given minute, his mean frequency of accidents *per year* is $np = 2.1$ and is not small. If the appropriate conditions* apply his frequency of accidents per year will be distributed as a binomial distribution with $p = 0.00042$ and $n = 5000$, shown in the first row of Table 2.3.

Thus you see that in 12.2% of the years (about 1 in 7) Henry would have no accidents. In 25.7% of years (about 1 in 4) he would have one accident, and so on. Occasionally, specifically in $(4.2 + 1.5 + 0.4 + 0.1)\% = 6.2\%$ of the years he would have five accidents or more.

Now consider the record of his girlfriend Minnie Bannister, who is a better driver than Henry. Minnie's probability p' of having an accident in any given minute is $p' = 0.00021$ (half that of Henry's), but since she drives a total of 10,000 minutes a year (twice that of Henry), her mean number of accidents per year is $\eta = n'p' = 2.1$, the same as his. If you study the individual frequencies

* An important assumption unlikely to be true would be that the probability of an accident in any given minute remains constant, irrespective of traffic density.

Table 2.3. Distribution of Accidents per Year Assuming a Binomial Distribution

	Probability That y Accidents Will Occur in a Given Year								
	$y=0$	$y=1$	$y=2$	$y=3$	$y=4$	$y=5$	$y=6$	$y=7$	$y=8$
(a) Henry Crun (exact binomial, $p=0.00042$, $n=5000$, $\eta=2.1$, $\sigma^2=2.0991$)	0.12240	0.25715	0.27007	0.18905	0.09924	0.04166	0.01457	0.00437	0.00115
(b) Minnie Bannister (exact binomial, $p=0.00021$, $n=10,000$, $\eta=2.1$, $\sigma^2=2.0996$)	0.12243	0.25716	0.27004	0.18903	0.09923	0.04167	0.01458	0.00437	0.00115
(c) Limiting Poisson distribution ($p\to 0$, $n\to\infty$ $\eta=2.1$, $\sigma^2=2.1$)	0.12246	0.25716	0.27002	0.18901	0.09923	0.04168	0.01459	0.00438	0.00115

of accidents in Table 2.3 you will see that not only Minnie's mean but also Minnie's distribution of accidents is virtually the same as Henry's. Her distribution of accidents per year calculated from the binomial formula with $n' = 10,000$ and $p' = 0.00021$ is shown in the second row (b) of Table 2.3.

For both Minnie and Henry the mean frequency of accidents per year is $\eta =$ 2.1 Now consider the same probabilities calculated from the Poisson distribution. They are

$$\Pr(0) = e^{-2.1} = 0.12246, \qquad \Pr(1) = e^{-2.1} \times 2.1 = 0.25716$$
$$\Pr(2) = e^{-2.1} \times (2.1^2)/2 = 0.27002, \text{ etc.}$$

and are given in Table 2.3(c) and plotted in Figure 2.15*a*. You will see that they approximate very closely the values calculated from the binomial distributions. The limiting Poisson distribution is important because it can approximate many everyday situations where the probability p of an event at a particular time or place is small but the number of opportunities n for it to happen is large.

For example, consider the problem faced by a manufacturer of raisin bran cereal who wants to ensure that on the average 90% of spoonfuls will each contain *at least* one raisin. On the assumption that the raisins are randomly distributed in the cereal, to what value should the mean η of raisins per spoonful be adjusted to ensure this? To solve this problem, imagine a spoonful divided up into raisin-sized cells, the number n of cells will be large, but the probability of any given cell being occupied by a raisin will be small. Hence, the Poisson distribution

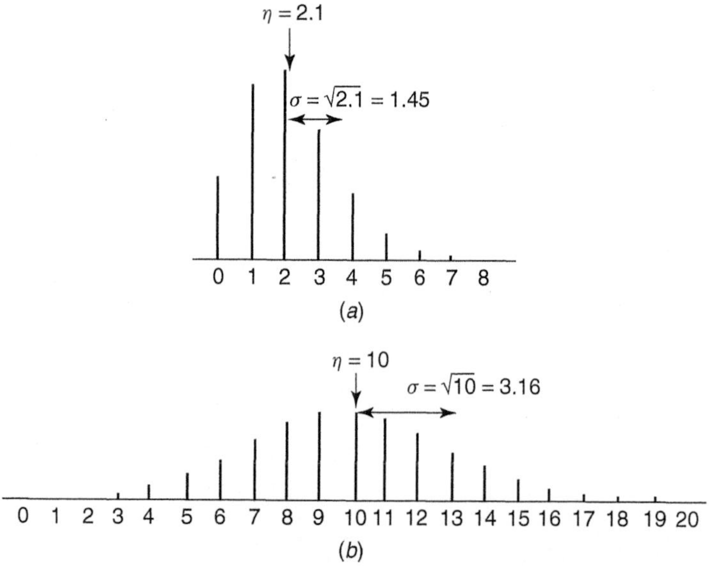

Figure 2.15. Poisson distributions for (a) $\eta = 2.1$ and (b) $\eta = 10$.

should supply a good approximation. The required probability is 1 − probability of no raisins. Thus you require that $1 - e^{-\eta} = 0.90$, that is, $\eta = \log_e 10 = 2.3$. Hence the manufacturer should ensure that on the average there are 2.3 raisins in every spoonful.

When η Is Not Too Small, the Poisson Distribution Can Be Approximated by the Normal Distribution

As the mean frequency (count) increases, the Poisson distribution approaches the normal distribution quite quickly. Thus the Poisson distribution with $\eta = \sigma^2 = 10$ in Figure 2.15*b* has ordinates well approximated by the normal distribution. When employing the normal approximation to compute the tail areas of a Poisson distribution Yates adjustment should be used, that is, you use $y_0 + 0.5$ instead of y_0 in the normal approximation. For example, $\Pr(y > 13)$ is approximated by the area under the normal curve to the right of $y = 13.5$.

Additive Property of the Poisson Distribution

The Poisson distribution has an important additive property. If $y_1, y_2, \ldots, y_k$ are observations from independent Poisson distributions with means $\eta_1, \eta_2, \ldots, \eta_k$, then $Y = y_1 + y_2 + \cdots + y_k$ also has a Poisson distribution with mean $\eta_1 + \eta_2 + \cdots + \eta_k$.

Sum of Normalized Squared Deviations Approximately Distributed as Chi Square

If y follows the Poisson distribution with mean η and η is of moderate size (say greater than 5), the quantity $(y - \eta)/\sqrt{\eta}$ is very approximately a standard normal deviate and its square $(y - \eta)^2/\eta$ is distributed approximately as χ^2 with one degree of freedom. It follows that if $y_1, y_2, \ldots, y_k$ are k observations from independent Poisson distributions with means $\eta_1, \eta_2, \ldots, \eta_k$ then

$$\sum_{j=1}^{k} \frac{(y_j - \eta_j)^2}{\eta_j} \approx \chi_k^2$$

where the notation $\approx$ means 'is approximately distributed as'.

In words, the sum of the *observed minus the expected frequencies squared divided by the expected frequency* is approximately distributed as chi square with k degrees of freedom.

APPENDIX 2A. MEAN AND VARIANCE OF LINEAR COMBINATIONS OF OBSERVATIONS

Suppose y_1, y_2, y_3 are *random variables* (not necessarily normal) with means η_1, η_2, η_3, variances σ_1^2, σ_2^2, σ_3^2, and correlation coefficients ρ_{12}, ρ_{13}, ρ_{23}. The mean and the variance of the linear combination $Y = a_1 y_1 + a_2 y_2 + a_3 y_3$, where a_1, a_2, a_3 are any positive or negative constants, are, respectively,

$$E(Y) = a_1\eta_1 + a_2\eta_2 + a_3\eta_3$$

and

$$V(Y) = a_1^2\sigma_1^2 + a_2^2\sigma_2^2 + a_3^2\sigma_3^2 + 2a_1a_2\sigma_1\sigma_2\rho_{12} + 2a_1a_3\sigma_1\sigma_3\rho_{13} + 2a_2a_3\sigma_2\sigma_3\rho_{23}$$

These formulas generalize in an obvious way. For a linear combination $Y = \sum_{i=1}^{n} a_i y_i$ of n random variables

$$E(Y) = \sum_{i=1}^{n} a_i\eta_i$$

and, $V(Y)$ will have n "squared" terms like $a_i^2\sigma_i^2$ and $n(n-1)/2$ cross-product terms like $2a_ia_j\sigma_i\sigma_j\rho_{ij}$. Specifically

$$V(Y) = \sum_{i=1}^{n} a_i^2\sigma_i^2 + 2\sum_{i=1}^{n}\sum_{j=i+1}^{n} a_ia_j\sigma_i\sigma_j\rho_{ij}$$

Notice that $\sigma_i\sigma_j\rho_{ij}$ is the $\text{Cov}(y_i, y_j)$, so that you can also write

$$V(Y) = \sum_{i=1}^{n} a_i^2 V(y_i) + 2\sum_{i=1}^{n}\sum_{j=i1}^{n} a_ia_j \,\text{Cov}(y_i, y_j)$$

Variance of a Sum and a Difference of Two Correlated Random Variables

Since a sum $y_1 + y_2$ can be written as $(+1)y_1 + (+1)y_2$ and a difference $y_1 - y_2$ as $(+1)y_1 + (-1)y_2$,

$$V(y_1 + y_2) = \sigma_1^2 + \sigma_2^2 + 2\rho_{12}\sigma_1\sigma_2$$
$$V(y_1 - y_2) = \sigma_1^2 + \sigma_2^2 - 2\rho_{12}\sigma_1\sigma_2$$

You will see that, if the correlation between y_1 and y_2 is zero, the variance of the sum is the same as that of the difference. If the correlation is positive, the variance of the sum is greater; if the correlation is negative, the opposite is true.

No Correlation

Consider the statistic Y,

$$Y = a_1y_1 + a_2y_{2\partial} + \cdots + a_ny_n$$

a linear combination of n random variables $y_1, y_2, \ldots, y_n$ each of which is assumed uncorrelated with any other; then

$$V(Y) = a_1^2\sigma_1^2 + a_2^2\sigma_2^2 + \cdots + a_n^2\sigma_n^2$$

No Correlation, Equal Variances

If, in addition, all the variances are equal, $E(Y)$ remains as before and

$$V(Y) = (a_1^2 + a_2^2 + \cdots + a_n^2)\sigma^2$$

Variance of an Average of n Random Variables with All Means Equal to η and All Variances Equal to σ^2

Since

$$\bar{y} = \frac{y_1 + y_2 + \cdots + y_n}{n} = \frac{1}{n}y_1 + \frac{1}{n}y_2 + \cdots + \frac{1}{n}y_n$$

the average is a linear combination of observations with all the a's equal to $1/n$. Thus, given the assumptions, $E(\bar{y}) = \eta$ and for the variance of the average $\bar{y}$,

$$V(\bar{y}) = \left(\frac{1}{n^2} + \frac{1}{n^2} + \cdots + \frac{1}{n^2}\right)\sigma^2 = \frac{n}{n^2}\sigma^2 = \frac{\sigma^2}{n}$$

the formula for the variance of the average used earlier.

Variance of the Average $\bar{y}$ for Observations That Are Autocorrelated at Lag 1

As illustrated earlier, the linear statistic $Y = \sum_{i=1}^{n} a_i y_i$ has expectation (mean)

$$E(Y) = \sum_{i=1}^{n} a_i \eta_i$$

and variance

$$V(Y) = \sum_{i=1}^{n} a_i^2\sigma_i^2 + 2\sum_{i=1}^{n}\sum_{j=i+1}^{n} a_i a_j \sigma_i \sigma_j \rho_{ij}$$

Suppose now that all the observations $y_1, y_2, \ldots, y_n$ have a constant variance σ^2 and the same lag 1 autocorrelation $\rho_{i,i+1} = \rho_1$. Suppose further that all correlations at greater lags are zero. Then

$$Y = n\bar{y} = y_1 + y_2 + \cdots + y_n$$

and making the necessary substitutions, you get

$$V(\bar{y}) = C \times \frac{\sigma^2}{n}$$

where

$$C = \left[1 + \frac{2(n-1)}{n}\rho_1\right]$$

It may be shown that for the special case being considered ρ_1 must lie between -0.5 and $+0.5$. Consequently C lies between $(2n-1)/n$ and $1/n$. Thus, for $n = 10$, the constant C lies between 1.9 and 0.1. (A range of nineteen!) Now for observations taken in sequence serial dependence is almost certain, so the consequence of ignoring it can be disastrous. (see e.g., Box and Newbold, (1971)).

REFERENCES AND FURTHER READING

Bhattacharya, G. K., and Johnson, R. A. (1977) *Statistical Concepts and Methods*, Wiley, New York.

Box, G. E. P., and Newbold, P. (1971) Some comments on a paper of Coen, Gomme and Kendall. *J. Roy. Stat. Soc. Series A*, **134**, 229–240.

Fisher, R. A., and Yates, F. (1938, 1963) *Statistical Tables for Biological, Agricultural and Medical Research*, Hafner, New York.

Pearson, E. S., and Hartley, H. O. (1969) *Biometrika Tables for Statisticians*, Cambridge University Press, Cambridge, UK.

Student (W. S. Gosset) (1908) The probable error of the mean, *Biometrika*, **6**, 1–25.

For the biographies of famous statisticians see:

Box, J. F. (1978) *R. A. Fisher, the Life of a Scientist*, Wiley, New York.

Johnson, N. L., and Kotz, S. (1997) *Leading Personalities in Statistical Sciences*, Wiley, New York.

Pearson, E. S. (1990) *Student: A Statistical Biography of William Seely Gosset*, Clarendon, Oxford, UK.

Stigler, S. (1988) *A History of Statistics*, Harvard, Cambridge, MA.

Tiao, G. C. (Ed) (2000) *Box on Quality and Discovery*, Wiley, New York.

QUESTIONS FOR CHAPTER 2

1. What are a dot diagram, a histogram, and a probability distribution?

2. What is meant by a *population* of observations, a *sample* of observations, *a random* drawing from the population, and a *random sample* of observations? Define *population mean, population variance, sample average*, and *sample variance.*

3. Distinguish between a standard deviation and a standard error.

4. What is the central limit effect and why is it important?

5. What does a normal distribution look like and why is it important?

6. What two parameters characterize the normal distribution?

7. What is meant by the IID assumption and the NIID assumption?

8. When must a t distribution be used instead of a normal distribution?

9. What are statistical dependence, randomness, correlation, covariance, autocorrelation, random sampling, and a sampling distribution?

10. What are the properties of the sampling distribution for the average $\overline{y}$ from a nonnormal distribution? Discuss and state your assumptions?

11. What are the sampling distributions of $\overline{y}$, s^2, and $t = (\overline{y} - \eta)/(s/\sqrt{n})$ from a normal distribution. Discuss and state your assumptions.

12. What are the formulas for the expectation and variance for the sum and for the differences of two random variables? How might these formulas be used to check for lag1 autocorrelation.

13. Consider the second difference $Y = y_i - 2y_{i-1} + y_{i-2}$. What is the variance of Y if successive observations y_i, $i = 1, 2, \ldots, n$, all have the same variance and are (a) independent and (b) autocorrelated with $\rho = 0.2$ and all other autocorrelations 0?

14. Two instruments used to measure a certain phenomenon give separate responses y_1 and y_2 with variances $\sigma^2{}_1$ and $\sigma^2{}_2$. What are two ways to determine whether these instrumental responses are correlated? Answer: (a) Estimate ρ from repeated pairs of y_1 and y_2 and (b) use the sums and difference of the responses.

15. What is meant by the *natural* variance and *natural* standard deviation of a sample of observations?

PROBLEMS FOR CHAPTER 2

Whether or not specifically asked, the reader should always (1) plot the data in any potentially useful way, (2) state the assumptions made, (3) comment on the appropriateness of the assumptions, and (4) consider alternative analyses.

1. Given the following observations: 5, 4, 8, 6, 7, 8. Calculate (a) the sample average, (b) the sample variance, and (c) the sample standard deviation.

2. The following are recorded shear strengths of spot welds (in pounds): 12,560, 12,900, 12,850, 12,710. Calculate the average, sample variance, and sample standard deviation.

3. Four ball bearings were taken randomly from a production line and their diameters measured. The results (in centimeters) were as follows: 1.0250, 1.0252, 1.0249, 1.0249. Calculate the average, sample variance, and sample standard deviation.

4. **(a)** The following are results of fuel economy tests (miles per gallon) obtained from a sample of 50 automobiles in 1975:

17.74	12.17	12.22	13.89	16.47
15.88	16.10	16.74	17.54	17.43
14.57	12.90	12.81	14.95	16.25
17.13	14.46	14.20	16.90	11.34
12.57	13.15	16.53	13.60	13.34
13.67	14.23	15.81	16.63	11.40
14.94	13.66	9.79	13.08	14.57
14.93	14.01	14.43	16.35	15.65
11.52	17.46	14.67	15.92	16.02
13.46	13.70	14.98	14.57	15.72

Plot the histogram of these data and calculate the sample average and sample standard deviation.

(b) The following 50 results were obtained in 1985:

24.57	24.79	22.21	25.84	25.35
22.19	24.37	21.32	22.74	23.38
25.10	28.03	29.09	29.34	24.41
25.12	25.27	27.46	27.65	27.95
21.67	22.15	24.36	26.32	24.05
28.27	26.57	26.10	24.35	30.04
25.18	27.42	24.50	23.21	25.10
23.59	26.98	22.64	25.27	25.84
27.18	24.69	26.35	23.05	23.37
25.46	28.84	22.14	25.42	21.76

Plot the histogram of these data and calculate the sample average and sample standard deviation. Considering both data sets, comment on any interesting aspects you find. Make a list of questions you would like answered that would allow you to investigate further.

5. If a random variable has a normal distribution with mean 80 and standard deviation 20, what is the probability that it assumes the following values?

(a) Less than 77.4

(b) Between 61.4 and 72.9

(c) Greater than 90.0

(d) Less than 67.6 or greater than 88.8

(e) Between 92.1 and 95.4

(f) Between 75.0 and 84.2

(g) Exactly 81.7

6. Suppose the daily mill yields at a certain plant are normally distributed with a mean of 10 tons and a standard deviation of 1 ton. What is the probability that tomorrow the value will be greater than 9.5? To obtain your answer, assume that the observations behave as statistically independent random variables.

7. Four samples are taken and averaged every hour from a production line and the hourly measurement of a particular impurity is recorded. Approximately one out of six of these *averages* exceeds 1.5% when the mean value is approximately 1.4%. State assumptions that would enable you to determine the proportion of the *individual* readings exceeding 1.6%. Make the assumptions and do the calculations. Are these assumptions likely to be true? If not, how might they be violated?

8. Suppose that 95% of the bags of a certain fertilizer mix weigh between 49 and 53 pounds. *Averages of three* successive bags were plotted, and 47.5% of these were observed to lie between 51 and X pounds. Estimate the value of X. State the assumptions you make and say whether these assumptions are likely to be true for this example.

9. For each sample of 16 boxes of cereal an *average box weight* is recorded. Past records indicate that 1 out of 40 of these *averages* is less than 8.1 ounces. What is the probability that an *individual* box of cereal selected at random will weigh less than 8.0 ounces? State any assumptions you make.

10. The lengths of bolts produced in a factory may be taken to be normally distributed. The bolts are checked on two "go–no go" gauges and those shorter than 2.9 or longer than 3.1 inches are rejected.

(a) A random sample of 397 bolts are checked on the gauges. If the (true but unknown) mean length of the bolts produced at that time was 3.06 inches and the (true but unknown) standard deviation was 0.03 inches, what values would you expect for n_1, the number of bolts found to be too short, and n_2, the number of bolts found to be too long?

(b) A random sample of 50 bolts from another factory are also checked. If for these $n_1 = 12$ and $n_2 = 12$, estimate the mean and the standard deviation for these bolts. State your assumptions.

11. The mean weight of items coming from a production line is 83 pounds. On the average, 19 out of 20 of these individual weights are between 81 and 85 pounds. The *average weights* of six randomly selected items are plotted. What proportion of these points will fall between the limits 82 and 84 pounds? What is the precise nature of the random sampling hypothesis that you need to adopt to answer this question? Consider how it might be violated and what the effect of such a violation might be on your answer.

12. After examination of some thousands of observations a Six Sigma Black Belt quality engineer determined that the observed thrusts for a certain aircraft engine were approximately normally distributed with mean 1000 pounds and standard deviation 10 pounds. What percentage of such measured thrusts would you expect to be

(a) Less than 985 pounds?

(b) Greater than 1020 pounds?

(c) Between 985 and 1020 pounds?

13. On each die in a set of six dice the pips on each face are diamonds. It is suggested that the weight of the stones will cause the "five" or "six" faces to fall downward, and hence the "one" and "two" faces to fall upward, more frequently than they would with fair dice. To test this conjecture, a trial is conducted as follows: The throwing of a one or two is called a success. The six dice are thrown together 64 times and the frequencies of throws with $0, 1, 2, \ldots, 6$ successes summed over all six pairs are as follows:

Successes out of six (number of dice showing a one or two)	0	1	2	3	4	5	6
Frequency	0	4	19	15	17	7	2

(a) What would be the theoretical probability of success and the mean and variance of the above frequency distribution if all the dice were fair?

(b) What is the empirical probability of success calculated from the data and what is the sample average and variance?

(c) Test the hypothesis that the mean and variance have their theoretical values.

(d) Calculate the expected frequencies in the seven "cells" of the table on the assumption that the probability of a success is exactly $\frac{1}{3}$.

(e) Can you think of a better design for this trial?

14. The level of radiation in the control room of a nuclear reactor is to be automatically monitored by a Geiger counter. The monitoring device works as follows: Every tenth minute the number (frequency) of "clicks" occurring in t seconds is counted automatically. A scheme is required such that if the frequency exceeds a number c an alarm will sound. The scheme should have the following properties: If the number of *clicks per second* is less than 4, there should be only 1 chance in 500 that the alarm will sound, but if the number reaches 16, there should be only about 1 change in 500 that the alarm will not sound. What values should be chosen for t and c? *Hint*: Recall that the square root of the Poisson distribution is roughly normally distributed with standard deviation 0.5. [*Answer*: $t \cong 2.25$, $c \cong 20$ (closest integer to 20.25).]

15. Check the approximate answer given for problem 14 by actually evaluating the Poisson frequencies.

CHAPTER 3

Comparing Two Entities: Reference Distributions, Tests, and Confidence Intervals

In this chapter we discuss the problem of comparing two entities experimentally and deciding whether differences that are found are likely to be genuine or merely due to chance. This raises important questions concerning the choice of *relevant* reference distributions with which to compare apparent differences and also the uses of randomization, blocking, and graphical tests.

3.1. RELEVANT REFERENCE SETS AND DISTRIBUTIONS

In the course of their scientific careers Peter Minerex and his wife Rita Stoveing had to move themselves and their family more than once from one part of the country to another. They found that the prices of suitable houses varied considerably in different locations. They soon determined the following strategy for choosing a new house. Having found temporary accommodation, they spent some time merely looking at houses that were available in the area. In this way they built up in their minds a "reference set" or "reference distribution" of available values for that particular part of the country. Once this distribution was established, they could judge whether a house being considered for purchase was modestly priced, exceptionally (significantly) expensive, or exceptionally (significantly) inexpensive.

The method of statistical inference called *significance testing* (equivalently *hypothesis testing*) parallels this process. Suppose an investigator is considering a particular result apparently produced by making some experimental modification to a system. He or she needs to know whether the result is easily explained by

Statistics for Experimenters, Second Edition. By G. E. P. Box, J. S. Hunter, and W. G. Hunter

mere chance variation or whether it is exceptional, pointing to the effectiveness of the modification. To make this decision, the investigator must in some way produce a relevant reference distribution that represents a characteristic set of outcomes which could occur if the modification *was entirely without effect*. The *actual* outcome may then be compared with this reference set. If it is found to be exceptional, the result is called statistically significant. In this way a *valid* significance test may be produced.

Validity

The analogy with pricing a house points to a very important consideration: Peter and Rita must be sure that the reference set of house prices that they are using is *relevant* to their present situation. They should not, for instance, use a reference set appropriate to their small country hometown if they are looking for a house in a metropolis. The experimenter must be equally discerning because many of the questions that come up in this and subsequent chapters turn on this issue.

In general, what is meant by a valid significance test is that an observed result can be compared with a *relevant* reference set of possible outcomes. If the reference set is not relevant, the test will be invalid.

The following example concerns the assessment of a modification in a manufacturing plant. The principles involved are of course not limited to this application and you may wish to relate them to an appropriate example in your own field.

An Industrial Experiment: Is the Modified Method (B) Better Than the Standard Method (A)?

An experiment was performed on a manufacturing plant by making *in sequence* 10 batches of a chemical using a standard production method (A) followed by 10 batches using a modified method (B). The results are given in Table 3.1. What evidence do the data provide that method B gives higher yields than method A?

To answer this question, the experimenters began very properly by plotting the data as shown in Figure 3.1 and calculating the averages obtained for methods A and B. They found that

$$\bar{y}_A = 84.24, \qquad \bar{y}_B = 85.54$$

where you will recall the bar over the symbol y is used to denote an arithmetic average. The modified method thus gave an average that was 1.3 units higher—a change that although small could represent significant savings. However, because of the considerable variability in batch to batch yields, the experimenters worried whether they could reasonably claim that new process B was better or whether the observed difference in the averages could just be a chance event. To answer this question, they looked at the yields of the process for the previous 210 batches. These are plotted in Figure 3.2 and recorded in Table 3B.1 at the end of this

Table 3.1. Yield Data from an Industrial Experiment

Time Order	Method	Yield
1	*A*	89.7
2	*A*	81.4
3	*A*	84.5
4	*A*	84.8
5	*A*	87.3
6	*A*	79.7
7	*A*	85.1
8	*A*	81.7
9	*A*	83.7
10	*A*	84.5
11	*B*	84.7
12	*B*	86.1
13	*B*	83.2
14	*B*	91.9
15	*B*	86.3
16	*B*	79.3
17	*B*	82.6
18	*B*	89.1
19	*B*	83.7
20	*B*	88.5

$$\bar{y}_A = 84.24 \qquad \bar{y}_B = 85.54$$
$$\bar{y}_B - \bar{y}_A = 1.30$$

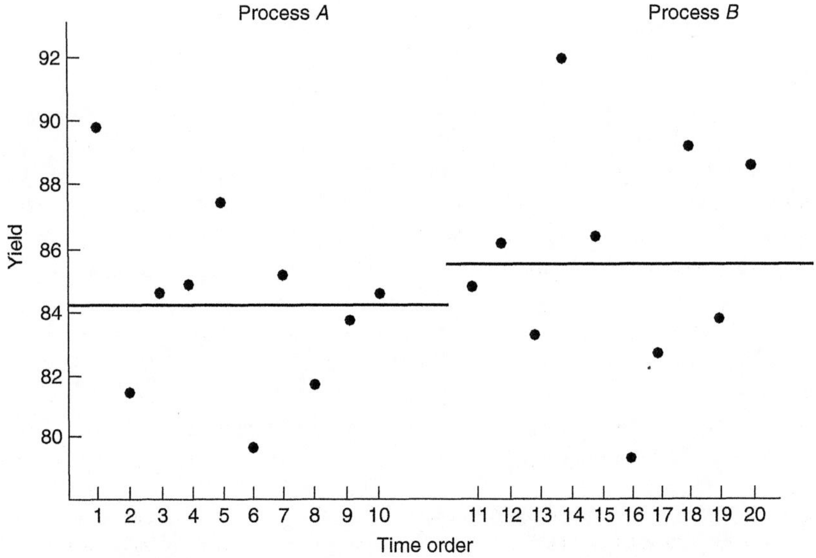

Figure 3.1. Yield values plotted in time order for comparative experiment.

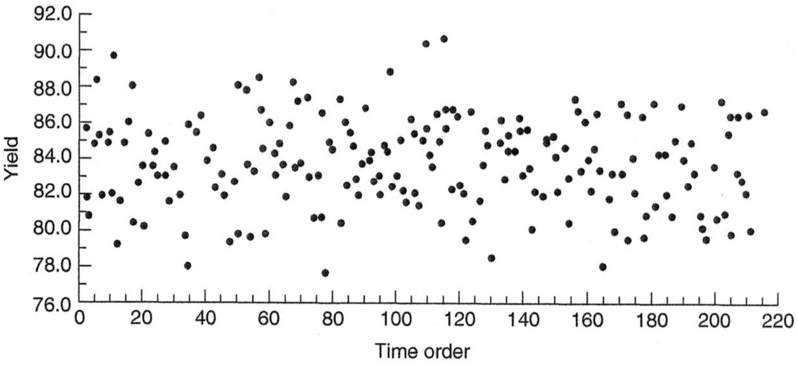

Figure 3.2. Plot of 210 past observations of yield from industrial process.

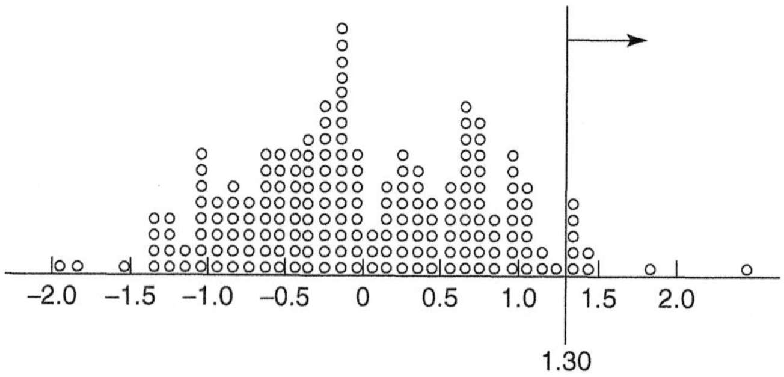

Figure 3.3. Reference distribution of 191 differences between averages of adjacent sets of 10 observations.

chapter. They used these 210 yield values to determine how often in the past had differences occurred between averages of *successive* groups of 10 observations that were at least as great as 1.30. If the answer was "frequently," the investigators concluded that the observed difference could be readily explained by the ordinary chance variations in the process. But if the answer was "rarely," a more tenable explanation would be that the modification has indeed produced a real increase in the process mean.

The 191 differences between averages of two successive groups of 10 observations* are plotted in Figure 3.3 and recorded in Table 3B.2. They provide a *relevant reference* set with which the observed difference of 1.30 may be compared.

It is seen that rather rarely, specifically in only nine instances, do the differences in the reference set exceed +1.30, the difference observed in the actual trial.

*These 191 differences were computed as follows. The first, −0.43, was obtained by subtracting the average 83.94 of batches 1 to 10 *from* average 83.51 of batches 11 to 20. This calculation was repeated using the averages calculated from batches 2 to 11 and 12 to 21 and so on through the data of Table 3B.1.

The "null" hypothesis, which would make the observed difference a member of the reference set, is thus somewhat discredited. In statistical parlance the investigators could say that, in relation to this reference set, the observed difference was *statistically significant* at the $9/191 = 0.047$ level of probability or, more simply stated, at the 4.7% significance level.

This test using the external reference distribution has appeal because it assumes very little. In particular, it does not assume normality or that the data are independent (and in particular not autocorrelated). It supposes only that whatever mechanisms gave rise to the past observations were operating during the plant trial. To use an external reference distribution, however, you must have a fairly extensive set of past data that can reasonably be regarded as typifying standard process operation.

A Recognizable Subset?

One other important consideration in seeking an appropriate reference set is that it does not contain one or more recognizable subsets. Suppose that after more careful study of the 210 previous data in the industrial experiment it was found, for example, the first 50 values were obtained when the process was run with a different raw material; then this would constitute a recognizable subset which should be omitted from the relevant reference set.

Similarly, if Peter and Rita were to be relocated, say, in Florida, they might decide that houses that did not have air conditioning were a recognizable subset that should be excluded from their reference set. Later in this chapter you will see how you might tackle the problem when you have no external reference distribution. But there you will also find that the importance of using a relevant reference set which does not contain recognizable subsets still applies.

Exercise 3.1. Suppose that the industrial situation was the same as that above except that only 10 experimental observations were obtained, the first 5 with method A (89.7, 81.4, 84.5, 84.8, 87.3) and the next 5 with B (80.7, 86.1, 82.7, 84.7, 85.5). Using the data in Table 3B.1, construct an appropriate reference distribution and determine the significance associated with $\bar{y}_B - \bar{y}_A$ for this test.

Answer: 0.085.

Exercise 3.2. Consider a trial designed to compare cell growth rate using a standard way of preparing a biological specimen (method A) and a new way (method B). Suppose that four trials done on successive days in the order A, A, B, B gave the results 23, 28, 37, 33 and also that immediately before this trial a series of preparations on successive days with method A gave the results 25, 23, 27, 31, 32, 35, 40, 38, 38, 33, 27, 21, 19, 24, 17, 15, 14, 19, 23, 22. Using an external reference distribution, compute the significance level for the null hypothesis $\eta_A = \eta_B$ when the alternative is $\eta_B > \eta_A$. Is there evidence that B gives higher rates than A.

Answer: The observed significance level is $0/17 = 0$. Evidence is that B gives higher growth rates.

Random Samples from Two Populations

A customary assumption is that a set of data, such as the 10 observations from method A in the plant trial, may be thought of as a *random sample* from a conceptual *population* of observations for method A. To visualize what is being assumed, imagine that tens of thousands of batch yields obtained from standard operation with method A are written on cards and placed in a large lottery drum, as shown diagrammatically at the top of Figure 3.4. The characteristics of this very large population of cards are represented by the distribution shown in the lower part of the same figure. *On the hypothesis of random sampling* the 10 observations with method A would be thought of as 10 *independent random* drawings from lottery drum A and hence from distribution A. In a similar way the observations from modified method B would be thought of as 10 random drawings from lottery drum B and hence from population B.

A null hypothesis that you might wish to test is that the modification is entirely without effect. If this hypothesis is true, the complete set of 20 observations can be explained as a random sample from a single *common* population. The alternative hypothesis is that the distributions from which the two random samples are drawn,

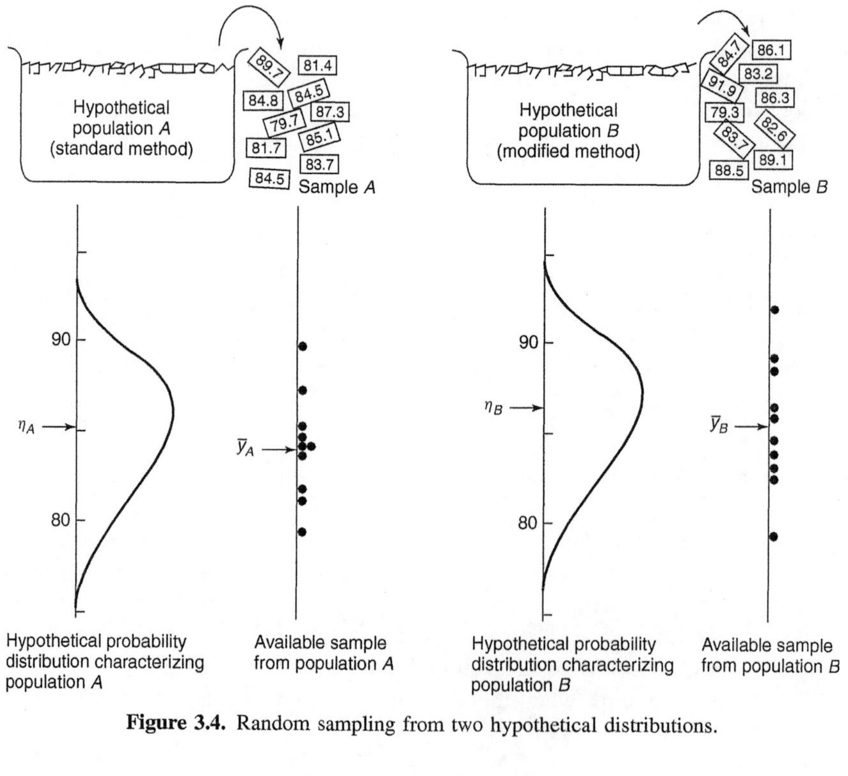

Figure 3.4. Random sampling from two hypothetical distributions.

though otherwise similar, have different means η_A and η_B. (You should note that neither distribution A nor distribution B is necessarily normal.)

Why It Would Be Nice If We Could Believe the Random Sampling Model

The assumption that the data can be represented as *random* samples implies the y's are *independently* distributed about their respective means. Thus the errors $y_1 - \eta_A, y_2 - \eta_A, \ldots, y_{10} - \eta_A, y_{11} - \eta_B, y_{12} - \eta_B, \ldots, y_{20} - \eta_B$ could be supposed to vary independently of one another and in particular the *order* in which the runs were made would have no influence on the results.

Industrial Experiment: Reference Distribution Based on the Random Sampling Model

Suppose for the industrial example that you had no external reference distribution and the only data available were the two sets of 10 yields obtained from methods A and B given in Table 3.1. Suppose you assumed that these were *random samples from populations* of approximately the same form (and in particular the same variance σ^2) but with possible different means η_A and η_B. Then the following calculations could be made.

Calculation of the Standard Error of the Differences in Averages

If the data were random samples with $n_A = 10$ observation from the first populations and $n_B = 10$ observations from the second, the variances of the calculated averages would be

$$V(\bar{y}_A) = \frac{\sigma^2}{n_A}, \qquad V(\bar{y}_B) = \frac{\sigma^2}{n_B}$$

Also, $\bar{y}_A$ and $\bar{y}_B$ would be distributed independently, so that

$$V(\bar{y}_B - \bar{y}_A) = \frac{\sigma^2}{n_A} + \frac{\sigma^2}{n_B} = \sigma^2\left(\frac{1}{n_A} + \frac{1}{n_B}\right)$$

Thus the standard deviation of the conceptual distribution of the difference of averages would be

$$\sqrt{V(\bar{y}_A - \bar{y}_B)} = \sigma\sqrt{\frac{1}{n_A} + \frac{1}{n_B}}$$

You will also remember that the square root of the variance of a statistic is often called the *standard error* of that statistic.

Even if the distributions of the original observations had been moderately nonnormal, the distribution of the difference $\bar{y}_B - \bar{y}_A$ between sample averages with $n_B = n_A = 10$ would be expected to be nearly normal because of the central limit effect. On this assumption of random sampling and denoting the true population difference by $\delta = \eta_B - \eta_A$,

$$z = \frac{(\bar{y}_B - \bar{y}_A) - \delta}{\sigma\sqrt{1/n_B + 1/n_A}}$$

would be approximately a normal deviate.

Estimating the Variance σ^2 from the Samples

Suppose now that the only evidence about the variance σ^2 is from the $n_A = 10$ observations from method A and the $n_B = 10$ observations from method B. These sample variances are

$$s_A^2 = \frac{\sum(y_A - \bar{y}_A)^2}{n_A - 1} = \frac{75.784}{9} = 8.42$$

$$s_B^2 = \frac{\sum(y_B - \bar{y}_B)^2}{n_B - 1} = \frac{119.924}{9} = 13.32$$

On the assumption that the two population variances for methods A and B are to an adequate approximation equal, these estimates may be combined to provide a *pooled* estimate s^2 of this common σ^2. This is done by adding the sums of squares of the errors and dividing by the sum of the degrees of freedom,

$$s^2 = \frac{\sum(y_A - \bar{y}_A)^2 + \sum(y_B - \bar{y}_B)^2}{n_A + n_B - 2} = \frac{75.784 + 119.924}{18} = 10.87$$

On the assumption of random *normal* sampling, s^2 provides an estimate of σ^2 with $n_A + n_B - 2 = 18$ degrees of freedom distributed independently of $\bar{y}_B - \bar{y}_A$. Substituting the estimate s for the unknown σ, you obtain

$$t = \frac{(\bar{y}_B - \bar{y}_A) - \delta}{s\sqrt{1/n_B + 1/n_A}}$$

in which the discrepancy $[(\bar{y}_B - \bar{y}_A) - \delta]$ is compared with the *estimated* standard error of $\bar{y}_B - \bar{y}_A$. On the above assumptions, then, this quantity follows a t distribution with $\nu = n_B + n_A - 2 = 18$ degrees of freedom.

For the present example $\bar{y}_B - \bar{y}_A = 1.30$ and $s\sqrt{1/n_A + 1/n_B} = 1.47$. The significance level associated with any hypothesized difference δ_0 is obtained by referring the statistic

$$t_0 = \frac{1.30 - \delta_0}{1.47}$$

to the t table with 18 degrees of freedom. In particular (see Table 3.2), for the null hypothesis $\delta_0 = 0$, $t_0 = 1.30/1.47 = 0.88$ and $\Pr(t > 0.88) = 19.5\%$. This differs substantially from the value of 4.7% obtained from the external reference distribution. Why?

Table 3.2. Calculation of Significance Level Based on the Random Sampling Hypothesis

Method A	Method B	$n_A = 10$	$n_B = 10$
89.7	84.7	Sum = 842.4	Sum = 855.4
81.4	86.1	Average $\overline{y}_A = 84.24$	Average $\overline{y}_B = 85.54$
84.5	83.3		
84.8	91.9	Difference $\overline{y}_B - \overline{y}_A = 1.30$	
87.3	86.3		
79.7	79.3	$\sum y_A^2 - (\sum y_A)^2/n_A = 75.784$	$\sum y_B^2 - (\sum y_B)^2/n_B = 119.924$
85.1	82.6		
81.7	89.1		
83.7	83.7		
84.5	88.5		

Pooled estimate of σ^2:
$$s^2 = \frac{75.784 + 119.924}{10 + 10 - 2} = \frac{195.708}{18} = 10.8727$$
with $\nu = 18$ degrees of freedom

Estimated variance of $\overline{y}_B - \overline{y}_A$: $s^2\left(\frac{1}{n_A} + \frac{1}{n_B}\right) = \frac{2s^2}{10}$

Estimated standard error of $\overline{y}_B - \overline{y}_A$: $\sqrt{\frac{s^2}{5}} = \sqrt{\frac{10.8727}{5}} = 1.47$

$$t_0 = \frac{(\overline{y}_B - \overline{y}_A) - \delta_0}{s\sqrt{1/n_B + 1/n_A}}$$

For $\delta_0 = 0$, $t_0 = \frac{1.30}{1.47} = 0.88$ with $\nu = 18$ degrees of freedom

$$\Pr(t \geq 0.88) = 19.5\%$$

Exercise 3.3. Remove the last two observations from method B (i.e., 83.7 and 88.5) and, again assuming the random sampling hypothesis, test the hypothesis that $\eta_B - \eta_A = 0$.

Answer: s^2pooled $= 11.4358$ with 16 degrees of freedom; $t = 1.17/1.60 = 0.73$, $\Pr(t \geq 0.73) = 0.238$.

Comparing the External and Internal Reference Distributions

Figure 3.5a (similar to Fig. 3.3) displays the reference distribution appropriate for the statistic $\overline{y}_B - \overline{y}_A$ with $n_A = n_B = 10$ constructed from an *external* set of 210 successively recorded observations. As you have seen, for this reference set $\Pr[(\overline{y}_B - \overline{y}_A) \geq 1.30] = 4.7\%$ and the observed difference $\overline{y}_B - \overline{y}_A = 1.30$ has a significance probability of 4.7% so that the null hypothesis is somewhat discredited, and it is likely that $\eta_B > \eta_A$.

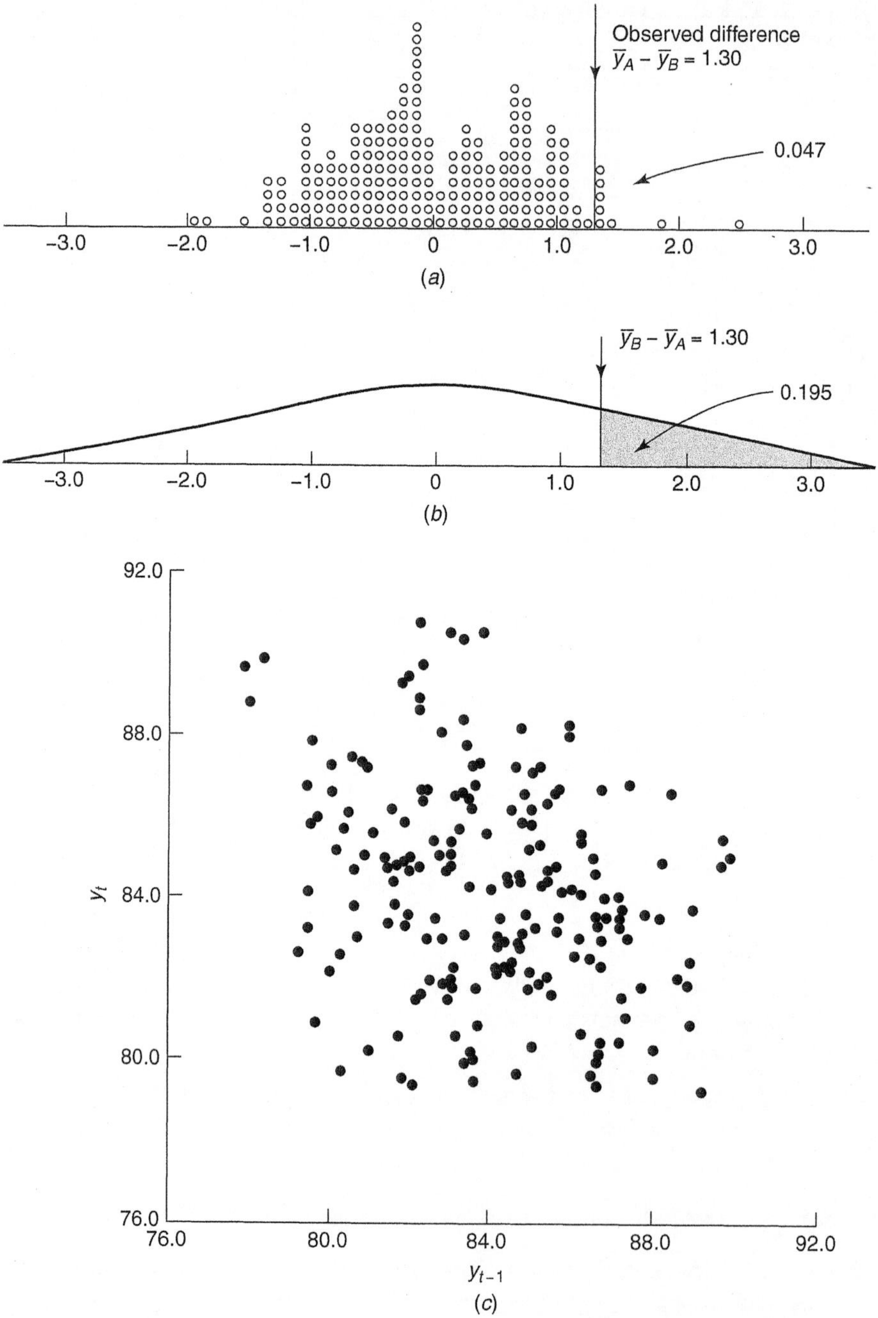

Figure 3.5. (*a*) Reference distribution based on external data. (*b*) Reference distribution based on random sampling model using *internal data.* (*c*) Plot of y_t versus y_{t-1} using the 210 observations of Table 3B.1 showing an estimated correlation (lag-on a autocorrelation) of $\hat{\rho}_1 = -0.29$.

In Figure 3.5*b*, however, you see displayed an appropriately scaled t reference distribution ($\nu = 18$ degrees of freedom) based solely on the *internal* evidence contained in the $n_A + n_B = 20$ observations listed in Table 3.1. This reference distribution assumes the random sampling model and the observed difference then has a significance probability of 19.5% which would provide little evidence against the hypothesis that $\eta_B - \eta_A = 0$.

Why the large disagreement in the probability statements provided by these two methods of analysis? It was caused by a rather small serial correlation (autocorrelation) of adjacent observations. The physical reason was almost certainly as follows. In this particular process the yield was determined by measuring the volume of liquid product in each successive batch, but because of the design of the system, a certain amount of product was left in the pipes and pumps at each determination. Consequently, if slightly less was pumped out on a particular day, the recorded yield would be low but on the following day it would tend to be high because the available product would be from both the current batch and any residual from the previous batch. Such an effect would produce a negative correlation between successive yields such as that found. More frequently in serial data positive autocorrelation occurs and can produce much larger discrepancies (see e.g. Box and Newbold, 1971).

Now Student's t is appropriate on the random sampling hypothesis that the errors are *not* correlated. In which case the variance of an average is $V(\bar{y}) = \sigma^2/n$. However, as noted in Appendix 2A, for autocorrelated data $V(\bar{y})$ can be very different. In particular, for a series of observations with a single autocorrelation of ρ_1 at lag 1 (i.e., where each observation is correlated with its immediate neighbors), it was shown there that the variance of $\bar{y}$ is not σ^2/n but

$$C \times \frac{\sigma^2}{n} \qquad \text{with} \qquad C = 1 + \frac{2(n-1)}{n}\rho_1$$

For the 210 successively recorded data that produced the external reference distribution the estimated lag 1 autocorrelation coefficient is $\hat{\rho}_1 = -0.29$. Using the above formula for averages of $n = 10$ observations, this gives an approximate value $\hat{C} = 0.48$. The negative autocorrelation in these data produces a reduction in the standard deviation by a factor of about $\sqrt{0.48} = 0.7$. Thus the reference distribution obtained from past data in Figure 3.5*a* has a *smaller* spread than the corresponding scaled t distribution in Figure 3.5*b*. Student's t test in this case gives the wrong answer because it assumes that the errors are independently distributed, thus producing an inappropriate reference distribution.

You are thus faced with a dilemma: most frequently you will not have past data from which to generate an external reference distribution, but because of influences such as serial correlation the t test based on NIID assumptions will not be valid. Fisher (1935) pointed out that for a randomized experiment the dilemma might be resolved by using a randomization distribution to supply a relevant reference set.

A Randomized Design Used in the Comparison of Standard and Modified Fertilizer Mixtures for Tomato Plants

The following data were obtained in an experiment conducted by a gardener whose object was to discover whether a change in fertilizer mixture applied to her tomato plants would result in improved yield. She had 11 plants set out in a single row; 5 were given the standard fertilizer mixture A, and the remaining 6 were fed a supposedly improved mixture B. The A's and B's were randomly applied to positions along the row to give the experimental design shown in Table 3.3. She arrived at this random arrangement by taking 11 playing cards, 5 marked A and 6 marked B. After thoroughly shuffling them she obtained the arrangement shown in the table. The first plant would receive fertilizer A as would the second, the third would receive fertilizer B, and so on.

Fisher argued that physical randomization such as was employed in this experiment would make it possible to conduct a valid significance test without making any other assumptions about the form of the distribution.

The reasoning was as follows: the experimental arrangement that was actually used is one of the $\frac{11!}{5!6!} = 462$ possible ways of allocating 5 A's and 6 B's to the 11 trials and any one of the allocations could equally well have been chosen. You can thus obtain a valid reference set appropriate to the hypothesis that the modification is without effect by computing the 462 differences in averages obtained from the 462 rearrangements of the "labels." The histogram in Figure 3.6 shows this randomization distribution together with the difference in the averages $\bar{y}_B - \bar{y}_A = 1.69$ actually observed. Since 154 of the possible 462 arrangements provide differences greater than 1.69, the significance probability is

Table 3.3. Results from a Randomized Experiment (Tomato Yields in Pounds)

Position in row	1	2	3	4	5	6	7	8	9	10	11
Fertilizer	A	A	B	B	A	B	B	B	A	A	B
Pounds of tomatoes	29.2	11.4	26.6	23.7	25.3	28.5	14.2	17.9	16.5	21.1	24.3

Standard Fertilizer A	Modified Fertilizer B
29.9	26.6
11.4	23.7
25.3	28.5
16.5	14.2
21.1	17.9
	24.3
$n_A = 5$	$n_B = 6$
$\bar{y}_A = 20.84$	$\bar{y}_B = 22.53$

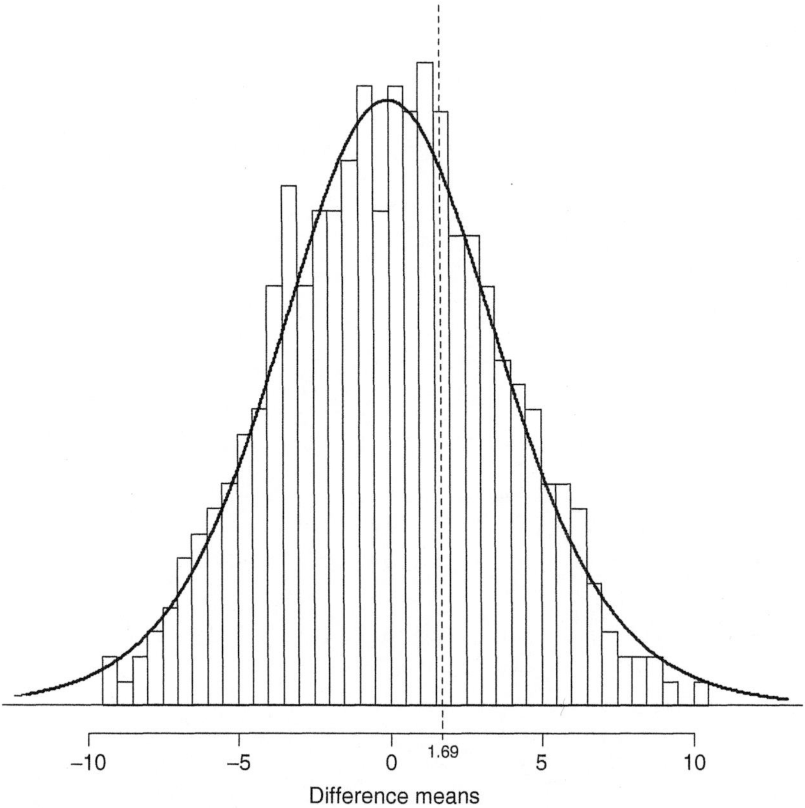

Figure 3.6. Randomization distribution with scaled t distribution: tomato plant example.

$154/462 = 33\%$. There was therefore little reason to doubt the null hypothesis that one fertilizer was as good as the other.

Comparison with the t Test

Writing δ for the difference in means $\eta_B - \eta_A$ on the NIID assumption of random sampling from a normal population, the quantity (see Table 3.4 for details)

$$\frac{(\overline{y}_B - \overline{y}_A) - \delta}{s\sqrt{1/n_B + 1/n_A}}$$

would be distributed as t with $\nu = n_B + n_A - 2$ *degrees of freedom*. For this example $\overline{y}_B - \overline{y}_A = 1.69$ so the null hypothesis that $\delta = 0$ may be tested by calculating

$$t = \frac{1.69 - 0}{3.82} = 0.44$$

Table 3.4. Calculation of t for the Tomato Experiment

$$S_A^2 = \frac{\sum(y_A - \bar{y}_A)^2}{n_A - 1}$$

$$\bar{y}_B - \bar{y}_A = 22.53 - 20.84 = 1.69$$

$$s_A^2 = \frac{\sum y_A^2 - \left(\sum y_A\right)^2 / n_A}{n_A - 1} = \frac{S_A}{\nu_A} = \frac{209.9920}{4} = 52.50$$

$$s_B^2 = \frac{\sum y_B^2 - \left(\sum y_B\right)^2 / n_B}{n_B - 1} = \frac{S_B}{\nu_B} = \frac{147.5333}{5} = 29.51$$

The pooled variance estimate is

$\nu_A = n_A - 1 = 5 - 1 = 4$

$\nu_B = n_B - 1 = 6 - 1 = 5$

$$s^2 = \frac{S_A + S_B}{\nu_A + \nu_B} = \frac{\nu_A s_A^2 + \nu_B s_B^2}{\nu_A + \nu_B} = \frac{4(52.50) + 5(29.51)}{4 + 5} = 39.73$$

with $\nu = n_A + n_B - 2 = \nu_A + \nu_B = 9$ degrees of freedom

The estimated variance of $\bar{y}_B - \bar{y}_A$ is $s^2(1/n_B + 1/n_A) = 39.73(1/6 + 1/5) = 14.57$. The standard error of $\bar{y}_B - \bar{y}_A$ is $\sqrt{14.57} = 3.82$,

$$t_0 = \frac{(\bar{y}_B - \bar{y}_S) - \delta_0}{\sqrt{s^2(1/n_B + 1/n_A)}} = \frac{(22.53 - 20.84) - \delta_0}{\sqrt{39.73(1/6 + 1/5)}} = \frac{1.69 - \delta_0}{3.82}$$

where δ_0 is the hypothesized value of δ. If $\delta_0 = 0$, then

$$t_0 = 0.44 \quad \text{with} \quad \nu = 9 \text{ degrees of freedom}$$

$$\Pr(t \geq t_0) = \Pr(t \geq 0.44) = 0.34$$

and referencing this value to a t table with $\nu = 9$ degrees of freedom, giving $\Pr(t \geq 0.44) = 34\%$, as compared with 33% given by the randomization test. So that in this example the result obtained from NIID theory is very close to that found from the randomization distribution. A *scaled* t distribution with the standard error of $\bar{y}_B - \bar{y}_A$, that is, 3.82, as the scale factor is shown together with the randomization distribution in Figure 3.6. For this example the distributions are in remarkably good agreement. As Fisher pointed out, the randomization test frees you from the *assumption* of independent errors as well as from the assumption of normality. At the time of its discovery calculation was slow, but with modern computers calculation of the randomization distribution provide a practical method for significance testing and for the calculation of confidence intervals.

Exercise 3.4. Given the following data from a randomized experiment, construct the randomization reference distribution and the approximate scaled t distribution. What is the significance level in each case?

A	B	B	A	B
3	5	5	1	8

Answer: 0.05, 0.04.

Exercise 3.5. Repeat the above exercise with these data:

B	A	B	A	A	A	B	B
32	30	31	29	30	29	31	30

Answer: 0.02, 0.01.

3.2. RANDOMIZED PAIRED COMPARISON DESIGN: BOYS' SHOES EXAMPLE

Often you can greatly increase precision by making comparisons *within matched pairs of experimental material.* Randomization is then particularly easy to carry out and can ensure the validity of such experiments. We illustrate with a trial on boys' shoes.

An Experiment on Boys' Shoes

The data in Table 3.5 are measurements of the amount of wear of the soles of shoes worn by 10 boys. The shoe soles were made of two different synthetic materials, a standard material A and a cheaper material B. The data are displayed in Figure 3.7 using black dots for material A and open dots for material B. If you consider only the overall dot plot in the right margin, the black dots and open dots overlap extensively and there is no clear indication that one material is better than the other.

But these tests were run in *pairs*—each boy wore a special pair of shoes, the sole of one shoe made with material A and the sole of the other with B. The decision as to whether the left or right sole was made with A or B was

Table 3.5. Boys' Shoes Example: Data on the Wear of Shoe Soles Made of Two Different Materials A and B

Boy	Material A	Material B	Difference $d = B - A$
1	13.2(L)	14.0(R)	0.8
2	8.2(L)	8.8(R)	0.6
3	10.9(R)	11.2(L)	0.3
4	14.3(L)	14.2(R)	−0.1
5	10.7(R)	11.8(L)	1.1
6	6.6(L)	6.4(R)	−0.2
7	9.5(L)	9.8(R)	0.3
8	10.8(L)	11.3(R)	0.5
9	8.8(R)	9.3(L)	0.5
10	13.3(L)	13.6(R)	0.3

Average difference $\bar{d} = 0.41$

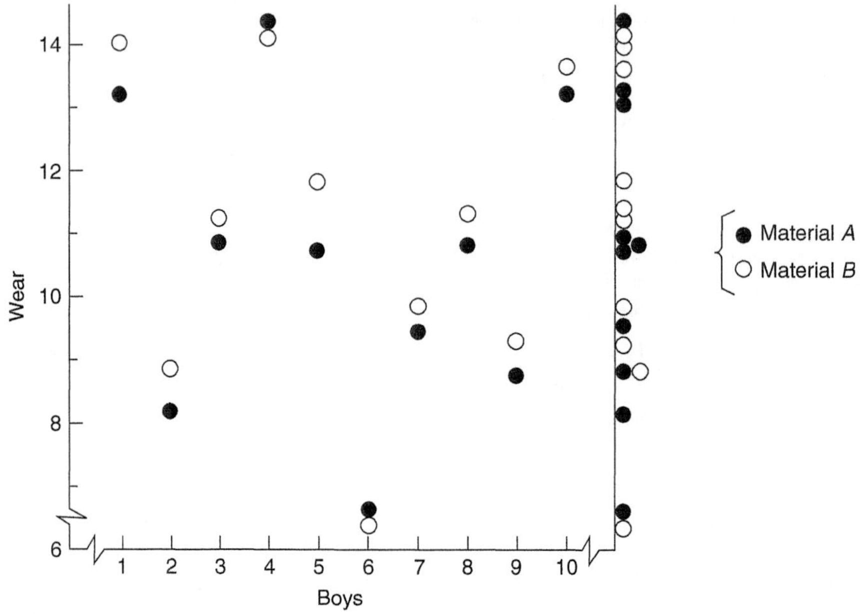

Figure 3.7. Data on two different materials *A* and *B*, used for making soles of boys' shoes.

determined by the flip of a coin. During the test some boys skuffed their shoes more than others, however for each boy his two shoes were subjected to the same treatment. Thus by working with the 10 differences $B - A$ most of the boy-to-boy variation could be eliminated. An experimental design of this kind is called a *randomized paired comparison* design. (You will see in the next chapter how this idea can be extended to compare more than two entities using "randomized block" designs.)

Notice that while it is desirable that the experimental error *within* pairs of shoes is made small it is not necessary or even desirable that the variation *between* pairs be made small. In particular, while you might at first think that it would be better to make all the tests with the same boy, reflection will show that this is not so. If this was done, conclusions might apply only to that particular boy. By introducing 10 different boys, you give the experiment what Fisher called a "wider inductive base." Conclusions drawn now depend much less on the choice of boys. This idea is a particular case of experimenting to produce robust products, as discussed later. You might say that the conclusions are *robust* to the choice of boys.

Statistical Analysis of the Data from the Paired Comparison Design

Material *A* was standard and *B* was a cheaper substitute that, it was feared, might result in an increased amount of wear. The immediate purpose of the experiment was to test the hypothesis that no change in wear resulted when switching from

A to B against the alternative that there was increased wear with material B. The data for this experiment are given in Table 3.5.

A Test of Significance Based on the Randomization Reference Distribution

Randomization for the paired design was accomplished by tossing a coin. A head meant that material A was used on the right foot. The sequence of tosses obtained was

T T H T H T T T H T

leading to the allocation of treatments shown in Table 3.5.

Consider the null hypothesis that the amount of wear associated with A and B are the same. On this hypothesis, the labeling given to a pair of results merely affects the sign associated with the difference. The sequence of 10 coin tosses is one of $2^{10} = 1024$ equiprobable outcomes. To test the null hypotheses, therefore, the average difference 0.41 actually observed may be compared with the other 1023 averages obtained by calculating $\bar{d}$ for each of the 1024 arrangements of signs in:

$$\bar{d} = \frac{\pm 0.8 \pm 0.6 \pm \cdots \pm 0.3}{10}$$

The resulting randomization distribution together with the appropriately scaled t distribution is shown in Figure 3.8.

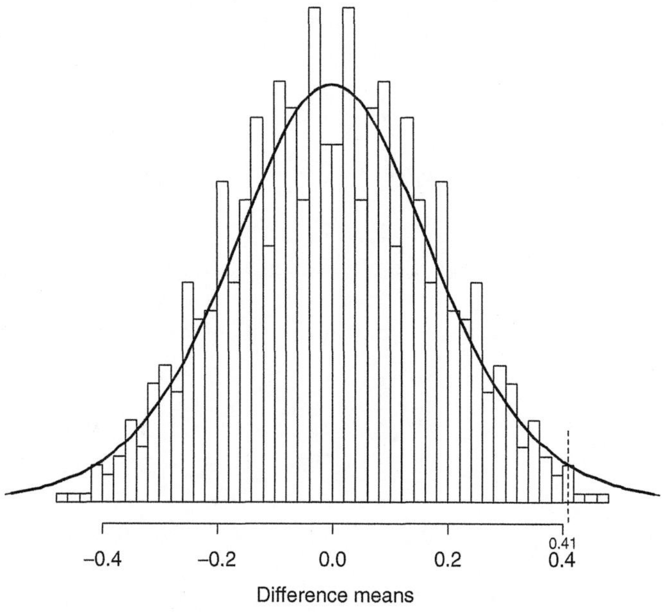

Figure 3.8. Randomization distribution and scaled t distribution: boys' shoes example.

The value $\bar{d} = 0.41$ that actually occurred is quite unusual. Only 3 of the 1024 differences produced by the randomization process give values of $\bar{d}$ greater than 0.41. Using Yates's continuity correction and including half of the four differences in which $\bar{d} = 0.41$, you get a significance level of $5/1024 = 0.5\%$. Thus the conclusion from this randomization test is that a highly statistical significant increase in the amount of wear is associated with the cheaper material B.

Comparison and Test

Consider again the differences 0.8, 0.6, ..., 0.3. If you accepted the hypothesis of random sampling of the d's from a normal population with mean δ, you could use the t distribution to compare $\bar{d}$ with any chosen value δ_0 of δ. If there were n differences $\delta - \delta_0$ then

$$s_d^2 = \frac{\sum(d - \bar{d})^2}{n - 1} = \frac{\sum d^2 - \left(\sum d\right)^2 / n_d}{n - 1} = \frac{3.030 - 1.681}{9} = 0.149$$

$$s_d = \sqrt{0.149} = 0.386 \qquad \text{and} \qquad s_{\bar{d}} = \frac{s_d}{\sqrt{n}} = \frac{0.386}{\sqrt{10}} = 0.122$$

The value of t_0 associated with the null hypothesis $d = \delta_0$ is

$$t_0 = \frac{\bar{d} - \delta_0}{s_{\bar{d}}} = \frac{0.41 - 0}{0.12} = 3.4$$

By referring this to the t table with nine degrees of freedom, you find

$$\Pr(t \geq 3.4) \cong 0.4\%$$

which agrees closely with the result of 0.5% given by the randomization distribution.

Exercise 3.6. Given these data from a randomization paired comparison design and using the usual null hypothesis, construct the randomization distribution and the corresponding scaled t distribution and calculate the significance levels based on each.

B	A	A	B	A	B	B	A	A	B
7	9	3	5	8	12	11	4	4	6

Exercise 3.7. Repeat the above exercise with the following data

B	A	B	A	A	B	A	B	B	A	B	A	A	B	B	A
29	29	32	30	23	25	36	37	39	38	31	31	27	26	30	27

A Complementary Graphical Display

As an adjunct to the t test and randomization test, a graphical procedure is introduced. In Figure 3.9a a dot plot of the residuals $d - \bar{d}$ supplies a reference distribution for d. The open dot on the same diagram indicates the scaled average $\bar{d} \times \sqrt{\nu} = 0.41 \times 3 = 1.23$. The scale factor $\sqrt{\nu}$ is such that if the true difference was zero the open circle could be plausibly considered as a member of the residual reference plot.* In Figure 3.9b a normal plot of the residuals is also appended. These graphical checks thus lead to the same conclusion as the formal t test and the corresponding randomization test.

Why the Graphical Procedure?

Since an "exact" t test is available supported by the randomization test, why bother with the graphical plot? A famous example will explain.

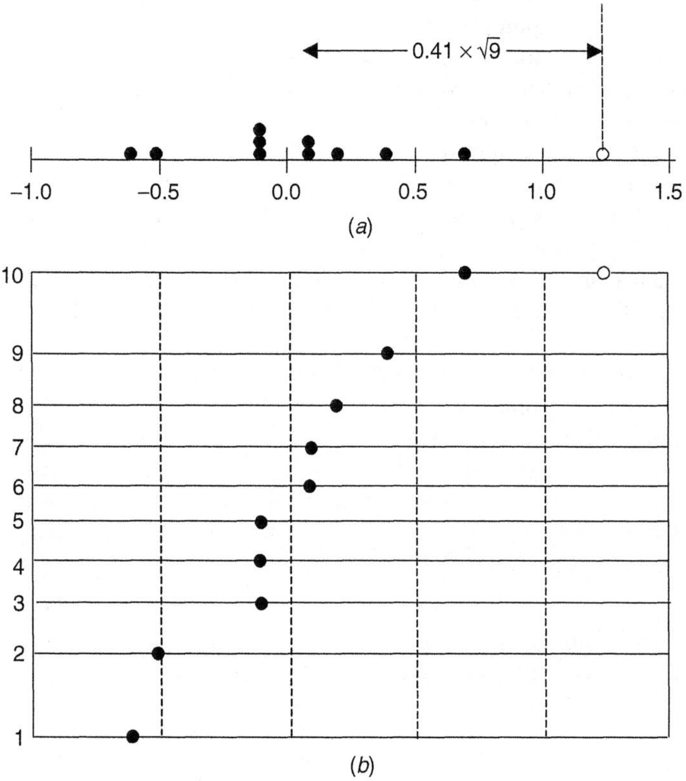

Figure 3.9. Boys' shoes example: (a) reference dot plot of residuals $d - \bar{d}$ and in relation to the scaled mean; (b) normal plot of the residuals.

* Specifically the natural variance of the open dot would be the same as that of the residual dots (see Appendix 4A for details).

Darwin's Data on the Heights of Plants

In his book *The Design of Experiments* Fisher (1935, p. 36) illustrated the paired t test with data due to Charles Darwin. An experiment was conducted in which the heights of 15 pairs of plants were measured. Of each pair one was self-fertilized and the other cross-fertilized and each pair was grown in the same pot. Darwin wanted to know if there was evidence of a difference in mean heights for the two types of plants. The 15 differences d in eighths of an inch are given in Table 3.6. Also shown are the average $\bar{d}$, its estimated standard error $s_{\bar{d}}$, and the value of $t = 2.15$ which has 14 degrees of freedom.

The 5% point for t is 2.145, and Fisher observed, "the result may be judged significant, though barely so." He noted that the flaw in Darwin's procedure was the absence of randomization and he went on to show that the result obtained from the t test was almost exactly duplicated by the randomization test using the Yates adjustment for continuity. He did this by evaluating only the extreme differences of the randomization distribution. The complete evaluation here is shown in Figure 3.11. As in the case of the two previous examples, the randomization distribution and the appropriately scaled t distribution are quite similar.

Thus there was some evidence that cross-fertilized plants tended to be taller than comparable self-fertilized plants and this reasoning seems to be supported by the dot plots in Figures 3.10*a and b* since the scaled average represented by the open dot does appear somewhat discrepant when judged against the residual reference set. But even more discrepant are the two outlying deviations shown on the left. (These are observations 2 and 15 in Fisher's table.) Associated discrepancies are displayed even more clearly in the normal plots of the original unpaired data in Figures 3.10*c,d*, from which it seems that two of the cross-fertilized plants and possibly one of the self-fertilized plants are suspect.

Thus, if around 1860 or so you were advising Darwin, you might say, "Charles, you will see there is some evidence that cross-fertilized plants may be taller than the self-fertilized; however results from these three plants seem questionable." To which Darwin might reply, "Yes. I see! Before we do anything else let's walk over to the greenhouse and look at the plants and also at my notebook. I may have made copying errors or we may find that the deviant plants have been affected in some way, possibly by disease."

Table 3.6. Differences in Eighths of a Inch between Cross- and Self-Fertilized Plants in the Same Pair

	49	23	56
	−67	28	24
	8	41	75
	16	14	60
	6	29	−48
$\bar{d} = 20.93$	$s_{\bar{d}} = 9.75$	$t = 2.15$	$\nu = 14$

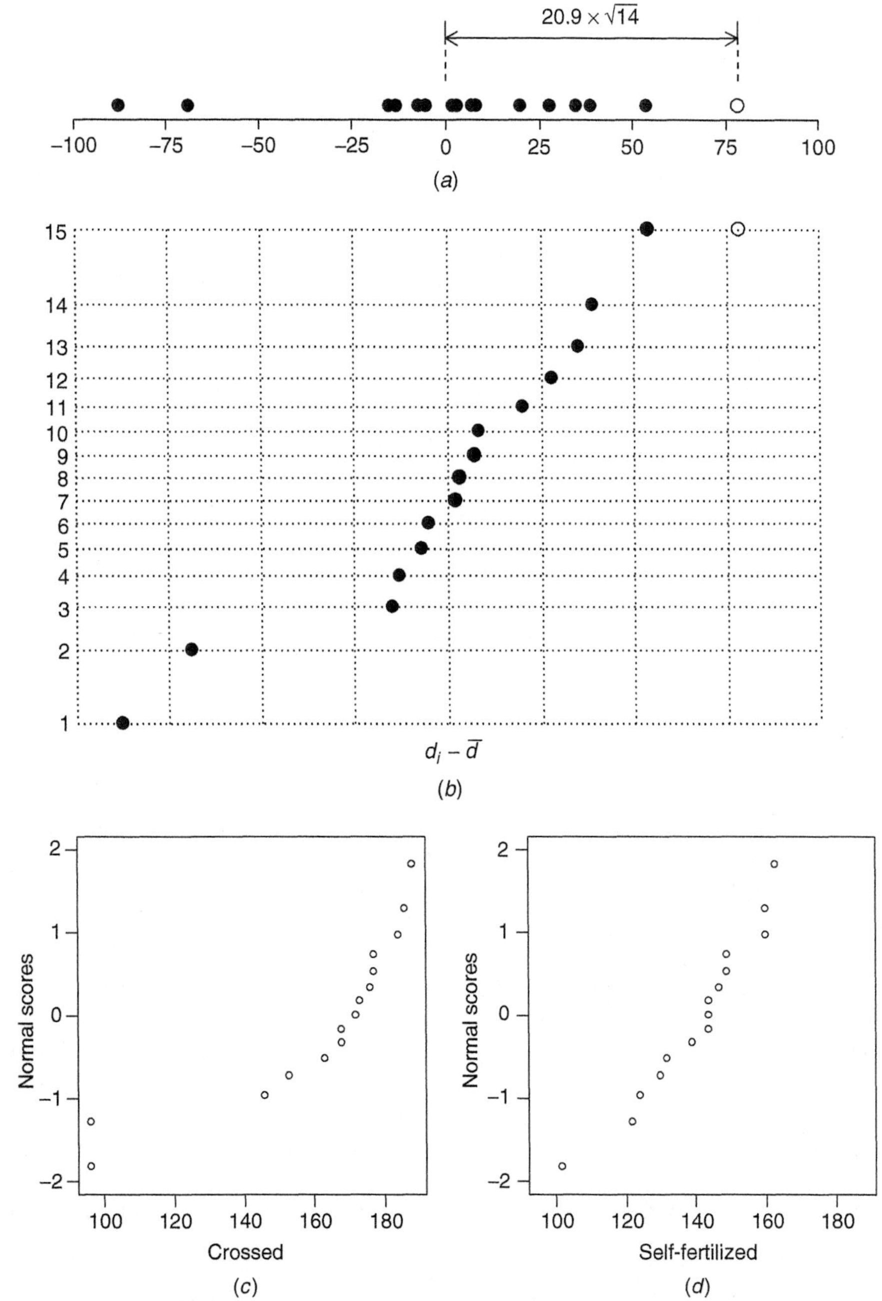

Figure 3.10. Darwin's data: (*a*) reference dot plot of the residuals $d - \bar{d}$ in relation to the scaled mean; (*b*) normal plot of the residuals; (*c*) normal plot for cross-fertilized plants; (*d*) normal plot for self-fertilized plants.

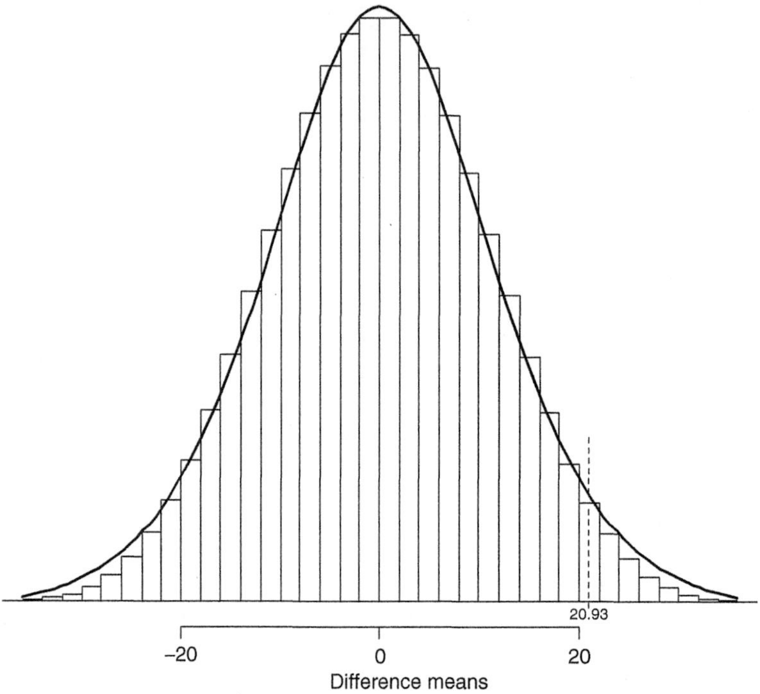

Figure 3.11. Reference distribution and scaled t distribution: Darwin's paired data.

Nothing other than looking carefully at the physical evidence and the data itself and derived values such as residuals can reveal discrepancies of this kind. Neither the randomization test nor any of the other so-called nonparametric distribution free tests will help you. In many cases, as in this example, such a test may merely confirm the result obtained by the parametric test (in this case a t test). See, for example, Box and Andersen (1955). To see why this is so, notice that the randomization test is based on Fisher's concept that after randomization, if the null hypothesis is true, the two results obtained from each particular pot will be *exchangeable*. The randomization test tells you what you could say if *that assumption were true*. But notice that inspite of randomization the basic hypothesis of exchangeability would be false if, for example, the discrepant plants could have been *seen* to be different (e.g., diseased). The data from such plants would then constitute a recognizable subset and there would be three distributions: self-fertilized, cross-fertilized, and diseased. So randomization as envisaged would not produce a relevant reference distribution for the comparison of interest. Such discrepancies can only be found by looking.

Discussion of a Constructed Example

From these three examples of randomization tests and associated t tests—the tomato plant data, the boys' shoes, and Darwin's data—you might be left with an impression that the two types of tests, randomization and t, will always coincide. The following example illustrates that this is not so. This appeared in an article urging the supposed superiority of randomization tests and was used to show that for the comparison of two means the randomization distribution may *not* necessarily be approximated by the t distribution. The authors Ludbrook and Dudley, (1998) used the following constructed example.

It was supposed that 12 men from a fitness center were asked to participate "in an experiment to establish whether eating fish (but not meat) resulted in lower plasma cholesterol concentrations than eating meat (but not fish)." The men were supposedly randomly allocated to the fish eating and meat eating groups and it was further supposed that at the end of one year their measured cholesterol concentrations were as follows:

Fish eaters: 5.42, 5.86, 6.156, 6.55, 6.80, 7.00, 7.11.

Meat eaters: 6.51, 7.56, 7.61, 7.84, 11.50.

The authors remarked that most investigators would note the outlying value (11.50) in the meat eating group (see normal plot in Fig. 3.12*a*) and having confirmed that "it did not result from an error of measurement or transcription" use some different test procedure. By contrast, they calculated the randomization distribution in Fig. 3.12*b* for the difference in the averages ($\bar{y}_{\text{meat}} - \bar{y}_{\text{fish}}$). As they found, the distribution is bimodal and bears no resemblance to the symmetric t distribution.

However, in a real investigation, having determined that the outlying result is genuine (e.g., not from a copying error), a very important question would be, "Was there something different about this particular subject or this particular test that produced the outlying result?" Were all the cholesterol determinations made in the same laboratory? What were the cholesterol readings for the subjects at the beginning of the study? Did the aberrant subject come from a family in which high cholesterol levels were endemic? Did the subject stop taking the fitness classes? Did he or she follow a different exercise routine? Did a physical examination and appropriate laboratory tests support other abnormalities? In a real scientific investigation the answers to such questions would greatly affect the direction of subsequent investigation.

The statistical question posed was whether or not the null hypothesis is sensible that meat eaters and fish eaters came from the same population or from two populations with different means. However, if any of the above explanations was found to be correct, the aberrant observation would have come from a third population (e.g., meat eaters who were members of a high-cholesterol family) and the outlying observation would be a member from a recognizable subset and the two-sample randomization reference distribution would not be relevant.

In Figure 3.12*c* we have added a third diagram which shows that the bimodality of the randomization distribution of $\bar{y}_{\text{meat}} - \bar{y}_{\text{fish}}$ is entirely produced by the aberrant data value 11.50. The shading indicates that almost all of the samples in the left hand hump as those where the aberrant value was allocated to $\bar{y}_{\text{fish}}$

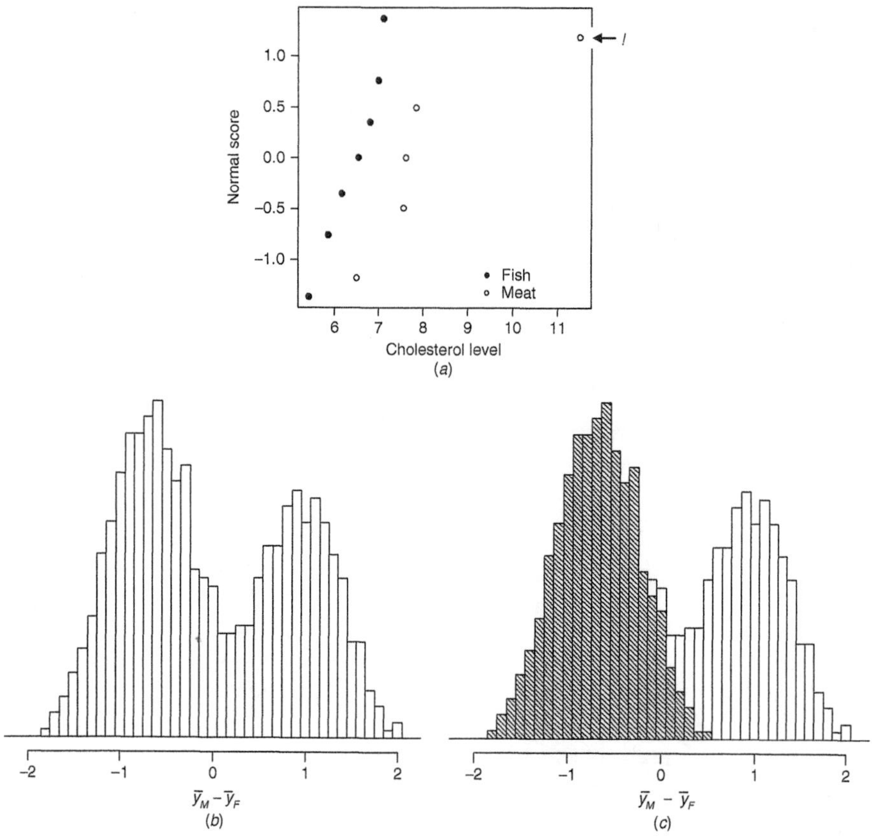

Figure 3.12. (*a*) Normal plots for fish and meat eaters. (*b*) Bimodal randomization distribution for $(\bar{y}_{\text{meat}} - \bar{y}_{\text{fish}})$. (*c*) Distribution with shaded portion indicating samples in which the outlier occurred in $\bar{y}_{\text{fish}}$.

and those in the right hand hump to $\bar{y}_{\text{meat}}$. Thus in an ongoing investigation reaction to these data should not be to use this bimodal distribution to supply a significance test or to look for some alternative test but to try to determine the reason for the discrepant observation. If this could be found it might constitute the most important discovery from this experiment.

What Does This Tell Us?

Whatever your field of expertise, there are important lessons to be learned from these examples. The statistics in textbooks and in many published papers concern themselves mostly with the analysis of data that are long dead. Observations that are apparently discrepant can be dealt with only in terms of various tests. Thus you will find a particular suspect "data set" analyzed and reanalyzed; one expert will find an outlier, another will argue there are probably two outliers, another will say that if an appropriate transformation on the data is made there may be none, and so on. But such postmortem analyses of data obtained from some experimental system that no longer exists are necessarily reactive and cannot lead to any certain conclusions. The question "So if you got these results, what would you *do* now?" is rarely asked, and if you are content only to be a data analyst, you can do nothing about it anyway. The reason for discordant runs that prove not to be copying mistakes can be decided only by looking and checking the system that generated the data and finding out, possibly by making one or two additional runs, if something exceptional was done that might explain suspect runs. (Experiment and you'll see.) This is only possible if you are part of the team carrying out the ongoing investigation.

Thus, just as you might in 1860 have walked with Charles Darwin down to his greenhouse to try to find out what was going on, when you are involved in an investigation that shows aberrant values, you must do likewise, that is, walk down* to the laboratory, to the pilot plant, or to the test bed where the data were generated and talk to those responsible for the experiment and the environment surrounding the process of data taking. This is one of the many reasons why the designer and analyst of the experiment must be part of the investigation.

When you visit the laboratory, process, or test bed where the data were obtained and talk to those running the experiments, the first thing to look for is whether the data you analyzed are the same as the data they originally recorded. If there were no copying errors, you should try to find out whether something unusual happened during a suspect run. Good experimenters, like Darwin, keep notebooks on the experimental runs they make. (You should make sure that your experimenters do too.) An experimenter might say, "Yes, in order to run this particular combination of specified conditions, we had to change certain other factors" (e.g., use a higher pressure or an increased flow rate). A deviation from the protocol of this kind may be the reason for unusually high values (perhaps good) or low values (perhaps bad). If the former, you may have stumbled on an unsuspected factor that could be used to improve the process! If the latter, then you may have found conditions to avoid. In either case, this search for an assignable cause for the unusual event is likely to teach you something and point the way where further experimentation is necessary.

Remember that your engineers and technologists, and your competitor's engineers and technologists have had similar training and possess similar beliefs. Therefore, finding out things that are *not* expected is doubly important because they would be unknown to your competitors. It is said that the hole in the ozone layer would have been found much earlier if the data continuously transmitted from the orbiting satellite had not been screened by a computer program that automatically removed or down-weighted outliers. If you just throw away "bad values," you may be throwing away important information.

*If the experiment is being conducted at a remote site that you cannot visit, much can be done by telephone or email.

3.3. BLOCKING AND RANDOMIZATION

Pairing observations is a special case of what is called *blocking*. You will find that blocking has important applications in many kinds of experiments. A block is a portion of the experimental material (the two shoes of one boy, two seeds in the same pot) that is expected to be more homogeneous than the aggregate (the shoes of all the boys, all the seeds not in the same pot). By confining comparisons to those within blocks (boys, pots), greater precision is usually obtained because the differences associated between the blocks are eliminated.

In the paired design the "block size" was 2, and we compared two treatments A and B. But if you had been interested in comparing four types of horseshoes, then, taking advantage of the happy circumstance that a horse has four legs, you could run this experiment with blocks of size 4. Such a trial would employ a number of horses, with each horse having the four different treatments applied randomly to the horseshoes on its four hooves. We discuss the use of such larger blocks in later chapters.

Pairs (Blocks) in Time and Space

Runs made close together in time or space are likely to be more similar than runs made further apart* and hence can often provide a basis for blocking. For example, suppose that in the comparison of two treatments A and B two runs could be made each day. If there was reason to believe that runs made on the same day would, on the whole, be more alike than those on different days, it would be advantageous to run the trial as a paired test in which a block would be an individual day and the order of running the two treatments within that day would be decided at random.

In the comparison of methods for treating leather specimens, pieces of leather close together on the hide would be expected to be more alike than pieces further apart. Thus, in a comparative experiment 6-inch squares might be cut from several hides, each square cut in two, and treatments A and B applied randomly to the halves.

A Possible Improvement in the Design of the Experiment on Tomato Plants

Earlier in this chapter we considered a randomized experiment carried out by a gardener to compare the effects of two different fertilizers on tomato plants. The fully randomized arrangement that was used was a valid one. On the assumption of exchangeability the difference in means was referred to a relevant reference set. However, an equally valid arrangement using randomized pairs might have been *more sensitive* for detecting real differences between the fertilizers. Plots close together could be used as a basis for pairing, and with six plots of two plants each the arrangement of treatments might have looked like that shown below:

$$(B \quad A) \quad (B \quad A) \quad (A \quad B) \quad (B \quad A) \quad (A \quad B) \quad (B \quad A)$$

*Notice, however, that the 210 consecutive yields for the production data were *negatively* serially correlated. This does not often happen, but it can!

Because the relevant error would now arise only from the differences between *adjacent* plants, this arrangement would usually be better than the unpaired design.

Occasionally, however, for either of two reasons the paired design could be less sensitive. For example:

(a) In an experiment using 12 plots the reference t distribution for the unpaired design would have 10 degrees of freedom. For the design arranged in six pairs it would have only 5 degrees of freedom. Thus you would gain from the paired design only if the reduction in variance from pairing outweighed the effects of the decrease in the number of degrees of freedom of the t distribution.

(b) It can occasionally happen (as in the industrial data of Table 3.1) that the errors associated with adjacent observations are negatively correlated so that comparisons within pairs are *less* alike because of the noise structure.

Block What You Can and Randomize What You Cannot

The lack of independence in experimental material provides both a challenge and an opportunity. Positive correlation between adjacent agricultural plot yields could be exploited to obtain greater precision. On the exchangeability assumption randomization could approximately validate the statistical tests.

Although blocking and randomization are valuable devices for dealing with *unavoidable* sources of variability, hard thinking may be required when faced with sources of variability that are avoidable. Extraneous factors likely to affect comparisons within blocks should be eliminated ahead of time, but realistic variation between blocks should be encouraged. Thus, in the boys' shoes example it would be advantageous to include boys of different habits and perhaps shoes of different styles. But choosing only boys from the football team would clearly reduce the scope of the inferences that could be drawn.

About the Nonparametric and Distribution Free Tests

The randomization tests mentioned in this chapter introduced in 1935 were the first examples of what were later called, "nonparametric" or "distribution free" tests.

Although something of a mouthful to say, *exchangeability theory* tests would be a more appropriate name for these procedures. The name would then point out the essential assumption involved. You replace the NIID assumption for the assumption of exchangeability. This assumption becomes much more plausible for an experiment that has been appropriately randomized. But as you have seen, you must always be on the lookout for "bad" data.

Sampling experiments that demonstrate the effects of nonnormality, serial dependence, and randomization on both kinds of tests are described in Appendix 3A. In particular, you will see that when the distribution assumption of serial independence is violated both kinds of tests are equally seriously affected.

3.4. REPRISE: COMPARISON, REPLICATION, RANDOMIZATION, AND BLOCKING IN SIMPLE EXPERIMENTS

What have you learned about the conduct of experiments to assess the possible difference between two means for treatments A and B?

1. Whenever possible, experiments should be comparative. For example, if you are testing a modification, the modified *and* unmodified procedures should be run side by side in the same experiment.
2. There should be genuine replication. Both A and B runs should be carried out several times. Furthermore, this should be done in such a way that variation among replicates can provide an accurate measure of errors that affect any comparison made between A runs and B runs.
3. Whenever appropriate, blocking (pairing) should be used to reduce error. Similarity of basic conditions for pairs of runs usually provides a basis for blocking, for example, runs made on the same day, from the same blend of raw materials, with animals from the same litter, or on shoes from the same boy.
4. Randomization should be planned as an integral part of experimentation. Having eliminated "known" sources of discrepancy, either by holding them constant during the experiment or by blocking, unknown discrepancies should be forced by randomization to contribute homogeneously to the errors of both treatments A and B. Given the speed of the modern computer a full randomization test might as well be made. When the exchangeability hypothesis makes sense, the randomization test is likely to agree closely with the parametric t test.
5. None of the above will necessarily alert you to or protect your experiment from the influence of bad values. Only by looking at the original data and at the appropriate checks on the original data can you be made aware of these.
6. In spite of their name, distribution free tests are not distribution free and are equally sensitive to violation of distributional IID assumptions other than that of normality.
7. Even with randomization the assumption of exchangeability can be violated.

3.5. MORE ON SIGNIFICANCE TESTS

One- and Two-Sided Significance Tests

The objective of the experiment comparing the two types of boys' shoe soles was to test the null hypothesis H_0 that no change in wear resulted when switching from material A to the cheaper material B. Thus it can be said that the null hypothesis $H_0 : \delta_0 = \eta_B - \eta_A = 0$ is being tested against the alternative hypothesis H_1 that there was *increased* wear with material B, that is, $H_1 : \eta_B - \eta_A > 0$. The data for this experiment were given in Table 3.5. The value of t_0 associated with this one-sided test of the null hypothesis is

$$t_0 = \frac{\bar{d} - \delta_0}{s_{\bar{d}}} = \frac{0.41 - 0}{0.12} = 3.4$$

Referring this to the t table Appendix B with nine degrees of freedom gives

$$\Pr(t \geq 3.4) = 0.4\%$$

If the modification under test could have affected the wear equally well in *either* direction, you would want to test the hypothesis that the true difference δ was zero against the alternative that it could be greater *or* less than zero, that is,

$$H_0 : \delta = \delta_0 = 0, \qquad H_1 : \delta \neq \delta_0$$

You would now ask how often t would *exceed* 3.4 *or fall short* of -3.4. Because the t distribution is symmetric, the required probability is obtained by doubling the previously obtained probability, that is,

$$\Pr(|t| > |t_0|) = 2 \times \Pr(\mathrm{t} > 3.4) = 0.8\%$$

If the true difference were zero, a deviation *in either direction* as large as that experienced or larger would occur by chance only about 8 times in 1000.

Exercise 3.8. Given the data below from a randomized paired design, calculate the t statistic for testing the hypothesis $\delta = 0$ and the probability associated with the two-sided significance test.

A	B	A	B	A	B	A	B	A	B
3	5	8	12	11	4	2	10	9	6

Answer: $t = 0.30$, p value $= 0.776$.

Conventional Significance Levels

A number of conventional significance levels are in common use. These levels are somewhat arbitrary but have been used as "critical" probabilities representing various degrees of skepticism that a discrepancy as large or larger than that observed might occur by chance. A discrepancy between observation and hypothesis producing a probability less than the "critical" level is said to be significant* at that level. A conventional view would be that you should be somewhat convinced of the reality of the discrepancy at the 5% level and fairly confident at the 1% level.

*The choice of the word *significant* in this context is unfortunate since it refers not to the *importance* of a hypothesized effect but to its *plausibility* in the light of the data. In other words, a particular difference may be statistically significant but scientifically unimportant and vice versa.

It is always best to state the probability itself. The statement that a particular deviation is "not significant at the 5% level" on closer examination is sometimes found to mean that the actual "significance probability" is 6%. The difference in mental attitude to be associated with a probability of 5% and one of 6% is of course negligible. In practice your *prior* belief in the possibility of a particular type of discrepancy, subject matter knowledge and the consequences (if any) of being wrong must affect your attitude. Complementary graphical analysis such as we show in Figures 3.9 and 3.10 and in later chapters can make it less likely that you will be misled.

Significance testing in general has been a greatly overworked procedure. Also in most cases where significance statements have been made it would have been better to provide an interval within which the value of the parameter in question might be expected to lie. Such intervals are called *confidence intervals.*

Confidence Intervals for a Difference in Means: Paired Comparison Design

The hypothesis of interest is not always the null hypothesis of "no difference." Suppose that in the boys' shoes example the true mean increase in wear of the cheaper material B compared to material A had some value δ. Then a $1-\alpha$ confidence interval for δ would be such that, using a two-sided significance test, all values of δ within the confidence interval did not produce a significant discrepancy with the data at the chosen value of probability α but all the values of δ outside the interval did show a significant discrepancy. The quantity $1-\alpha$ is sometimes called the *confidence coefficient.* For the boys' shoes example the average difference in wear was $\bar{d} = 0.41$, its standard error was 0.12, and there were nine degrees of freedom in the estimate of variance. The 5% level for the t distribution is such that $\Pr(|t| > 2.262) = 5\%$. Thus all values of δ such that

$$\left|\frac{0.41-\delta}{0.12}\right| < 2.262$$

would not be discredited by a two-sided significance test made at the 5% level and would therefore define a 95% confidence interval. Thus the 95% confidence *limits* for the mean δ are

$$0.41 \pm 2.262 \times 0.12 \qquad \text{or} \qquad 0.41 \pm 0.27$$

This interval thus extends from 0.14 to 0.68. The two values $\delta_- = 0.14$ and $\delta_+ = 0.68$ are called confidence *limits* and as required the *confidence interval* spans the values of δ for which the observed difference $\bar{d} = 0.41$ just achieves a significant difference.* See Figure 3.13.

Exercise 3.9. For the data given in the previous exercise compute the 80% confidence limits for the mean difference δ. *Answer*: 3.0, 6.6.

*It has the theoretical property that in "repeated sampling from the same population" a proportion $1-\alpha$ of intervals so constructed would contain δ.

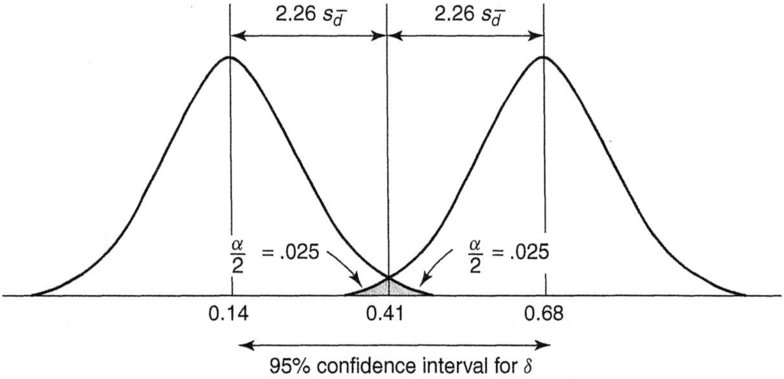

Figure 3.13. Generation of 95% confidence interval for δ: boys' shoes example.

In general, the $1 - \alpha$ confidence limits for δ would be

$$\bar{d} \pm t_{\nu, \frac{1}{2}\alpha} \cdot s_{\bar{d}}$$

where n differences $d_1, d_2, \ldots, d_n$ have an average value $\bar{d}$ and $s_{\bar{d}}$ is the standard error of $\bar{d}$. Thus

$$s_{\bar{d}}^2 = \frac{s_d^2}{n} = \frac{\sum_{u=1}^{n}(d_u - \bar{d})^2}{(n-1)n} \qquad \text{and} \qquad s_{\bar{d}} = \sqrt{s_{\bar{d}}^2}$$

The value of $t_{\nu,\alpha/2}$ denotes the t deviate with ν degrees of freedom corresponding to the single tail area $\alpha/2$. See the t table (Table B1 at the end of this book).

Confirmation Using Randomization Theory

Randomization theory can be used to verify the 95% confidence limits of $\delta_- = 0.14$ and $\delta_+ = 0.68$, which just produce a significant discrepancy with the data and so delineate the confidence interval. The necessary data for making the calculations are given in Table 3.7.

The hypothesis that the parent populations have a difference in mean equal to the upper limit $\delta_+ = 0.68$ implies that after subtracting 0.68 from each of the differences the signs (plus or minus) carried by the resulting quantities are a matter of random labeling. The resulting quantities given in column 2 of Table 3.7 have an average of $0.41 - 0.68 = -0.27$. Sign switching shows that just 25 out of the possible 1024 possible averages (i.e., 2.4%) have equal or smaller values. Similarly, after subtracting $\delta_- = 0.14$ the third column of the table is obtained, whose average value is $0.41 - 0.14 = 0.27$. Sign switching then produces just 19 out of 1024 possible averages (i.e., 1.9%) with equal or larger values. These values approximate* the probability of 2.5% provided by the t statistic.

*The adjustment for continuity due to Yates (see Fisher, 1935, p. 47) allowing for the fact that the randomization distribution is discrete produces even closer agreement.

Table 3.7. Calculations for Confirmation Via Randomization Theory of Confidence Interval

	Original Difference (1)	Difference − 0.68 (2)	Difference − 0.14 (3)
	0.80	0.12	0.66
	0.60	−0.08	0.46
	0.30	−0.38	0.16
	−0.10	−0.78	−0.24
	1.10	0.42	0.96
	−0.20	−0.88	−0.34
	0.30	−0.38	0.16
	0.50	−0.18	0.36
	0.50	−0.18	0.36
	0.30	−0.38	0.16
Average	0.41	−0.27	0.27

Note: Original Differences from Table 3.5, boys' shoes example.

In the analysis of randomized designs to compare means, we shall proceed from now on to calculate confidence intervals employing normal sampling theory on the basis that these procedures could usually be verified to an adequate approximation with randomization theory.[†]

Exercise 3.10. Using the data in the previous example, determine the approximate 80% confidence interval of δ using the randomization distribution.

Answer: 2.8 to 6.5 approximately.

Sets of Confidence Intervals

A better understanding of the uncertainty associated with an estimate is provided by a *set* of confidence intervals. For instance, using $\bar{d} \pm t_{\frac{1}{2}\alpha} \cdot s_{\bar{d}}$ and the t table (Table B1 at the back of the book), one can compute the set of confidence intervals given in Table 3.8. These intervals are shown diagrammatically in Figure 3.14.

A Confidence Distribution

A comprehensive summary of all confidence interval statements is supplied by a *confidence distribution.* This is a curve having the property that the area under the curve between any two limits is equal to the confidence coefficient for that pair

[†] Studies have shown (Welch, 1937, Pitman; 1937, Box and Andersen; 1955) that these properties extend to analysis of variance problems that we later discuss. Specifically Box and Andersen show that randomization tests can be closely approximated by corresponding parametric tests with slightly modified degrees of freedom.

Table 3.8. Confidence Intervals: Boys' Shoes Example

Significance Level	Confidence Coefficient	Confidence Interval δ_-	δ_+
0.001	0.999	−0.16	0.98
0.01	0.99	0.01	0.81
0.05	0.95	0.14	0.68
0.10	0.90	0.19	0.63
0.20	0.80	0.24	0.58

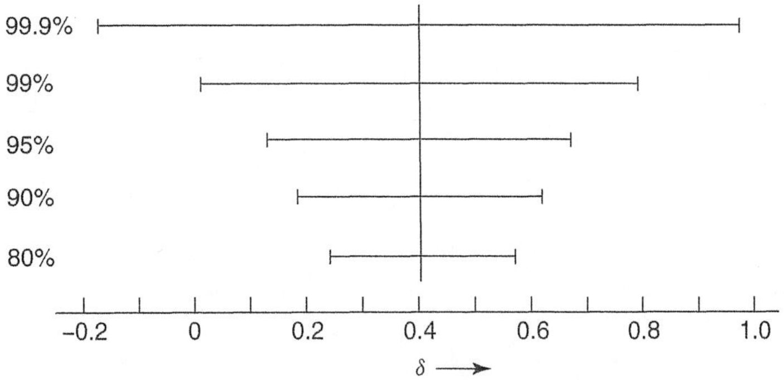

Figure 3.14. Set of confidence intervals: boys' shoes example.

of confidence limits. Such a curve is shown for the shoe data in Figure 3.15. For example, the area under the confidence distribution between the 90% limits 0.19 and 0.63 is exactly 90% of the total area under the curve. The curve is a scaled t distribution centered at $\bar{d} = 0.41$ and having scale factor $s_{\bar{d}} = 0.12$. Note that, on the argument we have given here, this confidence distribution merely provides a convenient way of summarizing all possible confidence interval statements. It is *not* the probability distribution for δ, which according to the frequency theory of statistical inference is a fixed constant and does not have a distribution. It is interesting, however, that according to the alternative statistical theory of *Bayesian inference* the same distribution can, on reasonable assumptions, be interpreted as a probability distribution. We do not pursue this tempting topic further here, but to us, at least, to do this would make more sense.

Confidence Intervals Are More Useful Than Single Significance Tests

The information given by a set of confidence intervals subsumes that supplied by a significance test and provides more besides. For example, consider the

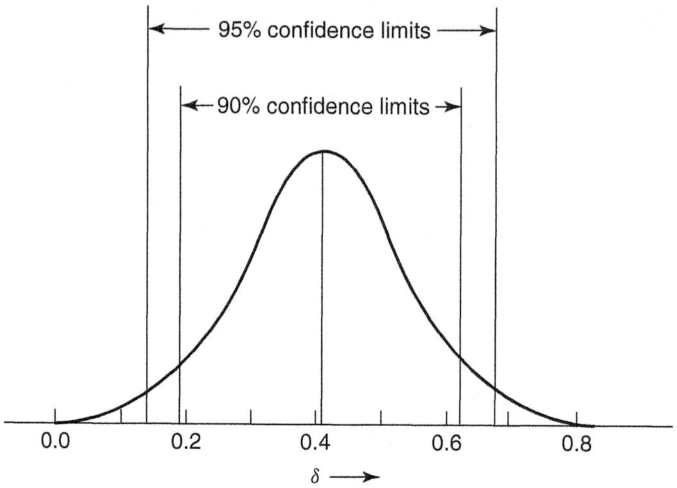

Figure 3.15. Confidence distribution for the difference in wear produced by two materials: boys' shoes example.

statement that the 95% interval for δ extends from $\delta_- = 0.14$ to $\delta_+ = 0.68$. That the observed difference $\bar{d} = 0.41$ is "significant" for $\delta = 0$ at the 5% level is evident in the confidence interval statement since the postulated value of $\delta = 0$ lies outside the interval. The interval statement, however, provides you with important additional facts:

1. At the 5% level the data contradict any assertion that the modification causes a change in the amount of wear *greater* than 0.68. Such a statement could be very important if the possible extent of a deleterious effect was under study.
2. Contemplation of the whole 95% interval (0.14, 0.68) makes it clear that, although we have demonstrated a "highly significant" difference between the amounts of wear obtained with materials A and B, since the mean wear is about 11, the *percentage change in wear* is quite small.
3. The width of the interval $0.68 - 0.14 = 0.54$ is large compared with the size of the average difference 0.41. If this difference had to be estimated more precisely, the information about the standard error of the difference would be useful in deciding roughly *how many* further experiments would be needed to reduce the confidence interval to any desired extent. For example, a further test with 30 more pairs of shoes would be expected to halve the interval. (The length of the interval would be inversely proportional to the *square root* of the total number of pairs tested.)
4. In addition to these calculations, a graphical analysis should of course be made that allows you to see all the data, to be made aware of suspect observations, and to more fully appreciate what the data do and do not tell you

Confidence Intervals for a Difference in Means: Unpaired Design

For a fully randomized (unpaired) design such as was used in the tomato plant example a confidence interval for the mean difference $\eta_B - \eta_A$ may be obtained by an argument similar to that used above. The two treatments had $n_B = 6$ and $n_A = 5$ observations and the difference in the averages $\bar{y}_B - \bar{y}_A = 1.69$. The hypothesis that the mean difference $\delta = \eta_B - \eta_A$ has some value δ_0 can therefore be tested by referring

$$t_0 = \frac{(\bar{y}_B - \bar{y}_A) - \delta_0}{s\sqrt{1/n_B + 1/n_A}} = \frac{1.69 - \delta_0}{3.82}$$

to a table of the t distribution with $(n_B - 1) + (n_A - 1) = 9$ degrees of freedom. For the two-tailed test $\Pr(|t| > 2.262) = 0.05$. Thus all the values of $\delta = \eta_B - \eta_A$ for which

$$\left|\frac{1.69 - \delta}{3.82}\right| < 2.262$$

are not discredited at the 5% level by a two-sided significance test. The 95% confidence limits are therefore

$$1.69 \pm 3.82 \times 2.262 \qquad \text{or} \qquad 1.69 \pm 8.64$$

and the 95% interval extends from -6.95 to 10.33. These limits can be verified approximately by a randomization test as before.

In general, the $1 - \alpha$ limits for $\delta = \eta_B - \eta_A$ are

$$(\bar{y}_B - \bar{y}_A) \pm t_{\alpha/2} s\sqrt{1/n_B + 1/n_A}$$

where

$$s^2 = \frac{(n_B - 1)s_B^2 + (n_A - 1)s_A^2}{(n_B - 1) + (n_A - 1)}$$

Exercise 3.11. Given the following data from a randomized experiment, compute the probability associated with the two-sided significance test of the hypothesis that $\eta_B - \eta_A = 0$:

A	*B*	*B*	*A*	*B*
3	5	5	1	8

Answer: 0.08.

Exercise 3.12. Repeat the above exercise with these data:

B	*A*	*B*	*A*	*A*	*A*	*B*	*B*
32	30	31	29	30	29	31	30

Answer: 0.02.

Exercise 3.13. Using the data in Exercise 3.11, compute the 95, 90, and 80% confidence interval for $\eta_B - \eta_A$. *Answer*: (−0.7,8.7), (0.5,7.5) (1.6,6.4).

Exercise 3.14. Using the data in Exercise 3.12, compute the 95, 90, and 80% confidence interval for $\eta_B - \eta_A$.

Answer: (−0.4,3.4), (0.3,2.7), (0.5,2.5), (0.8,2.2).

Exercise 3.15. Following the approach described in this chapter, derive the formula for a $1 - \alpha$ confidence interval for the mean η given a random sample of n observations $y_1, y_2, \ldots, y_n$ from a normal population.

Answer: $\bar{y} \pm t_{\alpha/2} s\sqrt{n}$, where $s^2 = \sum(y - \bar{y})^2/(n-1)$ with $\nu = (n-1)$ degrees of freedom.

Exercise 3.16. Using your answer to the above exercise, compute the 90% confidence interval for η given the following mileage readings: 20.4, 19.3, 22.0, 17.5, 14.3 miles per gallon. List all your assumptions. *Answer*: 18.7 ± 2.8.

In this chapter we have obtained formulas for $1 - \alpha$ confidence intervals of differences in means for paired and unpaired experiments. Notice that both these important formulas are of the form

$$\text{Statistic} \pm t_{\alpha/2} \times \text{standard error of the statistic}$$

where the standard error is the square root of the estimated variance of the statistic. For regression statistics, which we discuss later, as indeed for any statistic that is a linear function of approximately normally distributed data, similar intervals can be constructed.

Inferences about Variances of Normally Distributed Data

Sometimes it is the degree of *variation* that is of interest. For example, you might want the dye uptake of nylon thread, the potency of an antityphus vaccine, or the speed of color film to vary as little as possible. Process modifications that reduce variance can be of great importance. Again, it may be of interest to compare, for example, the *variation* of two analytical methods. The following describes how tests of significance and confidence intervals may be obtained for variances. The performance of these methods for comparing variances are, however, much more dependent on approximate normality of the parent distribution than are the corresponding methods for comparing means.

You saw in Chapter 2 that on the hypothesis of random sampling from a normal population the standardized sum of squares of deviations from $\bar{y}$ has a chi-square distribution with $\nu = n - 1$ degrees of freedom. Thus

$$\frac{\sum(y_u - \bar{y})^2}{\sigma^2} = \frac{(n-1)s^2}{\sigma^2} \sim \chi^2_{n-1}$$

A Significance Test

It is claimed that measurements of the diameter of a rod in thousandths of an inch have a variance σ^2 no larger than 10. Suppose you make six measurements and obtain as estimate of variance $s^2 = 13$. On NIID assumptions, is there any substantial evidence that the variance is higher than what is claimed? If you refer $(n-1)s^2/\sigma^2 = (5 \times 13)/10 = 6.5$ to a χ^2 table with five degrees of freedom, (see Table C at the end of this book), you will see that such a discrepancy happens more than 25% of the time so there is little reason to dispute the claim.

A Confidence Interval

The lower and upper $1 - \alpha$ confidence limits for σ^2 are of the form

$$\frac{(n-1)s^2}{B} \quad \text{and} \quad \frac{(n-1)s^2}{A}$$

where, as before, the confidence limits are values of the variance σ^2 that will just make the sample value significant at the stated level of probability.

Suppose, as before, you had obtained as estimate of variance $s^2 = 13$ based on five degrees of freedom. A 95% confidence interval for the population variance σ^2 could be obtained as follows. The χ^2 table shows that 0.975 and 0.025 probability points of χ^2 with $\nu = 5$ degrees of freedom are $A = 0.831$ and $B = 12.83$. Thus the required 95% confidence limits are

$$\frac{5 \times 13}{12.83} = 5.07 \quad \text{and} \quad \frac{5 \times 13}{0.831} = 78.22$$

Notice that, unless the number of degrees of freedom is high, the variance and its square root the standard deviation cannot be estimated very precisely. As a rough guide, the standard deviation of an estimate s expressed as a percentage of σ is $100/\sqrt{2\nu}$. Thus, for example, if we wanted an estimate of σ with no more than a 5% standard deviation, we would need a sample of about 200 independent observations!

Exercise 3.17. Using the data of Exercise 3.16, obtain a 90% confidence interval for the variance of the mileage readings. Carefully state your assumptions.

Testing the Ratio of Two Variances

Suppose that a sample of n_1 observations is randomly drawn from a normal distribution having variance σ_1^2, a second sample of n_2 observations is randomly drawn from a second normal distribution having variance σ_2^2, and estimates s_1^2 and s_2^2 of the two population variances are calculated, having ν_1 and ν_2 degrees of freedom. On the standard NIID assumptions, since s_1^2/σ_1^2 is distributed as $\chi_{\nu_1}^2/\nu_1$ and s_2^2/σ_2^2 is distributed as $\chi_{\nu_2}^2/\nu_2$, the ratio $(\chi_{\nu_1}^2/\nu_1)/(\chi_{\nu_2}^2/\nu_2)$ has an F distribution with ν_1 and ν_2 degrees of freedom. The probability points of the F distribution are given in Table D at the end of this book. Thus,

$$\frac{s_1^2/\sigma_1^2}{s_2^2/\sigma_2^2} \sim F_{\nu_1,\nu_2} \qquad \text{or equivalently} \qquad \frac{s_1^2}{s_2^2} \sim \frac{\sigma_1^2}{\sigma_2^2} F_{\nu_1,\nu_2}$$

For example, the sample variances for replicate analyses performed by an inexperienced chemist 1 and an experienced chemist 2 are $s_1^2 = 0.183$ ($\nu_1 = 12$) and $s_2^2 = 0.062$ ($\nu_2 = 9$). Assuming the chemists' results can be treated as normally and independently distributed random variables having variances σ_1^2 and σ_2^2, suppose you wish to test the null hypothesis that $\sigma_1^2 = \sigma_2^2$ against the alternative that $\sigma_1^2 > \sigma_2^2$. On the null hypothesis $\sigma_1^2 = \sigma_2^2$ the ratio s_1^2/s_2^2 is distributed as $F_{12,9}$. Referring the ratio $0.183/0.062 = 2.95$ to the table of F with $\nu_1 = 12$ and $\nu_2 = 9$, you will find that on the null hypothesis the probability of a ratio as large or larger than 2.95 is about 6%. There is some suggestion therefore, though not a very strong one, that the inexperienced chemist has a larger variance.

Confidence Limits for σ_1^2/σ_2^2 for Normally Distributed Data

Arguing exactly as before, on NIID assumptions, A and B are the $1 - \alpha/2$ and $\alpha/2$ probability points of the F distribution with ν_1 and ν_2 degrees of freedom, the lower and upper $1 - \alpha$ confidence limits for σ_1^2/σ_2^2 are

$$\frac{s_1^2/s_2^2}{B} \qquad \text{and} \qquad \frac{s_1^2/s_2^2}{A}$$

The $1 - \alpha/2$ probability points of the F distribution are not given but may be obtained from the tabulated $\alpha/2$ points. This is done by interchanging ν_1 and ν_2 and taking the reciprocal of the tabled value.

For instance, suppose you needed to calculate the 90% limits for the variance ratio σ_1^2/σ_2^2 where, as before, $s_1^2/s_2^2 = 2.95$ and the estimates had respectively 12 and 9 degrees of freedom. Entering the F table with $\nu_1 = 12$ and $\nu_2 = 9$, we find the 5% probability point $B = 3.07$. To obtain the 95% probability point, we enter the F table with $\nu_1 = 9$ and $\nu_2 = 12$ and take the reciprocal of the value 2.80 to obtain $A = 1/2.80 = 0.357$. Thus the required confidence limits are

$$\frac{2.95}{3.07} = 0.96 \qquad \text{and} \qquad \frac{2.95}{0.357} = 8.26$$

Lack of Robustness of Tests on Variances

Whereas tests to compare means are insensitive to the normality assumption, this is not true for the tests on variances. (See, e.g., comparisons made from the sampling experiments in Appendix 3A.) It is sometimes possible to avoid this difficulty by converting a test on variances to a test on means (see Bartlett and

Kendall, 1946). The logarithm of the sample variance s^2 is much more nearly normally distributed than is s^2 itself. Also on the normal assumption the variance of $\log s^2$ is independent of the population variance σ^2. If you have a number of variances to compare, you could carry out an approximate test that is insensitive to nonnormality by taking the logarithms of the sample variances and performing a t test on these "observations."

Consider the following example. Each week two analysts perform five tests on identical samples from sources that were changed each week. These special samples were included at random in the sequence of routine analyses and were not identifiable by the analysts. The variances calculated for the results and subsequent analyses are as follows.

	Analyst 1		Analyst 2		
Week	s_1^2	$\log 100s_1^2$	s_2^2	$\log 100s_2^2$	$d = \log 100s_1^2 - \log 100s_2^2$
1	0.142	1.15	0.043	0.63	0.52
2	0.09	0.96	0.079	0.90	0.06
3	0.214	1.33	0.107	1.03	0.30
4	0.113	1.05	0.037	0.43	0.62
5	0.082	0.91	0.045	0.65	0.26

Using the paired t test for means, you obtain

$$\overline{d} = 0.352, \qquad s_d = 0.226, \qquad s_{\overline{d}} = s_d/\sqrt{n} = 0.101$$

Thus

$$t = \frac{\overline{d} - 0}{s_{\overline{d}}} = \frac{0.352}{0.101} = 3.49$$

To allow for changes from week to week in the nature of the samples analyzed, we have in this instance performed a *paired t* test. The value of 3.49 with four degrees of freedom is significant at the 5% level. Hence, analyst 2 is probably the more precise of the two. In this example there is a natural division of the data into groups. When this is not so, the observations might be divided randomly into small groups and the same device used.

3.6. INFERENCES ABOUT DATA THAT ARE DISCRETE: BINOMIAL DISTRIBUTION

A discussion of the properties of the binomial distribution was given in Chapter 2. This would be of value to you if you had steadily lost money to the gambler Denis Bloodnok by betting how many times his penny would produce four or more heads in five throws. Now suppose you manage to get hold of his penny and you toss it 20 times and 15 turn up heads. Should this confirm your suspicions that his coin is biased?

You can answer this question by applying the binomial distribution whereby $\Pr(y = y_0) = \binom{n}{y} p^{y_0} q^{n-y_0}$ (see Chapter 2). Giving Bloodnok the benefit of the doubt, you might entertain the null hypothesis that his penny was fair ($p = 0.5$) against the alternative that it was biased in his favor ($p > 0.5$). You will find that

$$\begin{aligned}\Pr(y > 14) &= \Pr(y = 15) + \Pr(y = 16) + \Pr(y = 17) + \Pr(y = 18) + \cdots \\ &= 0.015 + 0.005 + 0.001 + 0.000 + 0.000 \\ &= 0.021\end{aligned}$$

Thus the observed event is significant at the 2.1% level. You might therefore regard your suspicion as justified. When the normal approximation to the binomial is used, as was seen in Chapter 2, a much closer approximation is established by using Yates adjustment and continuity.

A Confidence Interval for *p*

Suppose your null hypothesis was that $p = 0.8$. Then, after obtaining $y = 15$ heads in $n = 20$ trials you could again evaluate $\Pr(y > 14)$ by adding the binomial probabilities with $p = 0.8$. In this way you would obtain, for $p = 0.8$, $n = 20$,

$$\Pr(y > 14) = 0.175 + 0.218 + 0.205 + 0.137 + 0.058 + 0.012 = 0.805$$

Thus, whereas 0.5 is an implausible value for p, 0.8 is not. Now imagine a series of values for p confronting the data $y = 15$, $n = 20$. As p was increased from zero, there would be some value p_- less than y/n that *just* produced significance at, say, the 2.5% level for the null hypothesis $p = p_-$ tested against the alternative $p > p_-$. Similarly, there would be some other value p_+ greater than y/n that *just* produced significance at the 2.5% level for the null hypothesis that $p = p_+$ against the alternative that $p < p_+$. The values that do this are $p_- = 0.51$ and $p_+ = 0.94$. They are the limits of a 95% confidence interval for p. In repeated sampling 95% of the intervals calculated this way will include the true value of p. Confidence limits for p based on the estimated probability $\hat{p} = y/n$ may be read from the charts give in Table F at the end of this book or by using an appropriate computer software program.

Exercise 3.18. Recalculate the confidence limits for p using the normal approximation.

Some Uses of the Binomial Distribution

The following exercises illustrate the use of significance levels and confidence intervals for the binomial distribution.

Exercise 3.19. An IQ test is standardized so that 50% of male students in the general population obtain scores of over 100. Among 43 male applicants at an army recruiting office only 10 obtained scores higher than 100. Is there significant evidence that the applicants are not a random sample from the population used to standardize the test?

Answer: Yes. For $\hat{p} = \frac{10}{43}$ the significance level is 0.0003 on the hypothesis that $p = 0.5$ against the alternative $p < 0.5$. The normal approximation gives 0.0004.

Exercise 3.20. Using a table of random numbers, police in a certain state randomly stopped 100 cars passing along a highway and found that 37 of the drivers were not wearing seat belts. Obtain confidence limits for the probability of not wearing seat belts on the highway in question. Explain your assumptions.

Partial Answer: The 95% confidence limits are 0.275 and 0.475.

Exercise 3.21. In a plant that makes ball bearings, samples are routinely subjected to a stringent crushing test. A sampling inspection scheme is required such that if out of a random sample of n bearings more than y fail the batch is rejected. Otherwise it is accepted. Denote by p the proportion of bearings in a large batch that will fail the test. It is required that there be a 95% chance of accepting a batch for which p is as low as 0.3 and a 95% chance of rejecting a batch for which p is as high as 0.5. Using the normal approximation, find values of n and y that will be satisfactory. *Hint:* Show that to satisfy the requirements it is necessary that

$$\frac{(y_0 - 0.5) - 0.3n}{\sqrt{n \times 0.3 \times 0.7}} = \frac{0.5n - (y_0 - 0.5)}{\sqrt{n \times 0.5 \times 0.5}} = 1.645.$$

Answer: $n = 62$, $y = 25$.

Comparing Different Proportions

The effectiveness of mothproofing agents was determined by placing 20 moth larvae in contact with treated wool samples and noting the number that died in a given period of time. The experimenters wanted to compare two different methods A and B for applying the agent to the wool, but it was known that tests done in different laboratories gave widely different results. A cooperative experiment involving seven different laboratories resulted in the data shown in Table 3.9. Randomization was employed in selecting the particular larvae and wool samples for each trial.

If the number of larvae dying (or equivalently the percentage dying, $100y/n$) could be assumed to be approximately normally distributed with constant variance, a paired t test could be used to assess the null hypothesis that, irrespective of which application method was used, the mean proportion dying was the same.

Table 3.9. Comparison of Two Methods of Application of Mothproofing Agent in Seven Laboratories: Tests with Twenty Moth Larvae

Method of Application	Different Ways of Measuring Results	Laboratory						
		1	2	3	4	5	6	7
A	Number dead	8	7	1	16	10	19	9
	Percentage dead	40	35	5	80	50	95	45
	Score	43	40	14	71	50	86	47
B	Number dead	12	6	3	19	15	20	11
	Percentage dead	60	30	15	95	75	100	55
	Score	57	37	25	86	67	100	53
Difference in percentages		20	−5	10	15	25	5	10
Difference in scores		14	−3	11	15	17	14	6

	Average Difference	Standard Error of Average Difference	*t* Value	Significance Level (%) (Two-Sided Test)
Percentage	11.43	3.73	3.06	2.2
Score	10.57	2.63	4.03	0.7

As you can see from Table 3.9, the test yields the value $t = 3.06$ with six degrees of freedom. This is significant at the 2.2% level, suggesting that method B caused greater mortality of the larvae.

Variance Stabilizing Transformation for the Binomial

The proportion dying in the above experiment varied greatly from laboratory to laboratory. Indeed, it looks from the data as if p might easily vary from 0.05 to 0.95. If this were so, the variances npq could differ by a factor of 5. Before treating binomial proportions $y/n = \hat{p}$ by standard NIID theory techniques using t statistics (and the analysis of variance and regression methods discussed later), it is desirable to make a variance stabilizing transformation. Fisher showed that this could be done by analyzing not $\hat{p}$ but the "score" x given by

$$\sin x = \sqrt{\hat{p}} \qquad \text{where} \qquad \hat{p} = \frac{y}{n}$$

A graph of the score x* needed to stabilize the variance versus the proportion $\hat{p}$ (measured as a percentage) is shown in Figure 3.16. Whereas the variance of p is very different for various values of p, the variance of x is approximately constant. It also turns out that x is much more nearly normally distributed than $\hat{p}$. You can see from the graph that this stabilization is achieved by stretching out the scale at the ends of the range of x.

* Measured in grads (100 grads = 90°)

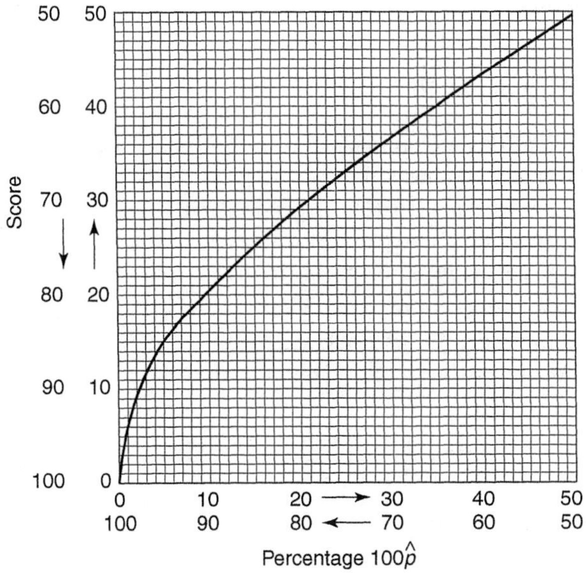

Figure 3.16. Variance stabilizing transformation for the binomial: score x as a function of the percentage $100\hat{p}$.

To carry out the modified test, you first transform the percentages to scores, for example by using the graph in Figure 3.16. Thus the entry 40% in the first column of the table transforms into a score of 43, and so on. A paired t test applied to the scores gives a value $t = 4.03$, which is even more significant than that obtained from the raw percentages. Because the assumptions on which the t test is derived are more nearly met, the transformed scores will on average produce tests of greater sensitivity. The reason the transformation produces a marked difference in the significance level for this example is that the observed values y/n cover a wide range. If this were not so, the transformation would have had less effect on the significance level.

The variance of the transformed score x is very nearly independent of p and has a theoretical value of $1013/n$. Even though this theoretical variance is available, it makes the assumption that the value of the parameter p does not change from trial to trial. Whenever possible, it is better to employ a variance calculated from the data. More explicitly, in this example it could be argued that each transformed value has a theoretical variance of $1013/20 = 50.7$ and therefore the test should be based on this value. If you use this route, however, you must make a direct assumption of the exact applicability of exact binomial sampling. It is always best to avoid assumptions you do not need to make.

Does the Probability Vary? Sampling Inspection of Ball Bearings

Each day 20 ball bearings randomly sampled from routine production are subjected to a crushing test. The number of bearings failing the test on 10 successive days are as follows:

Day	1	2	3	4	5	6	7	8	9	10
Number failing out of 20, y	8	4	10	6	2	3	7	12	5	7

Is there evidence of day-to-day variation in the probability p of failure?

Consider the null hypothesis that p (the probability that a bearing fails) does not vary from day to day. On this hypothesis the distribution of $(y - \eta)/\sigma = (y - np)/\sqrt{npq}$ may be approximated by a standard normal distribution. We do not know p, but an estimate is provided by $\hat{p} = \bar{y}/n = 6.4/20 = 0.32$. Now for a sample of k observations from a fixed binomial distribution, $\sum_{j=1}^{k}(y_j - n\hat{p})^2/n\hat{p}\hat{q}$ is approximately distributed as a χ^2 distribution with $k - 1$ degrees of freedom. Thus, to test the null hypothesis, the value

$$\frac{\sum (y_j - 6.4)^2}{20 \times 0.32 \times 0.68} = \frac{86.40}{4.352} = 19.85$$

may be referred to a χ^2 distribution with $k - 1 = 9$ degrees of freedom. This value yields significance at the 1.88% level. There is evidence, therefore, of real variation from day to day in failure rates as measured by the crushing test.

Suppose now that the value of p *differs* from day to day about its average value with variance σ_p^2. Then it turns out that σ_y^2 will not be given by npq but by $\sigma_y^2 = npq + n(n - 1)\sigma_p^2$, that is, $9.600 = 4.352 + 380\sigma_p^2$. A (very rough) estimate of σ_p^2 is therefore $(9.600 - 4.352)/380 = 0.014$. This yields an estimate standard deviation $\hat{\sigma}_p = \sqrt{0.014} = 0.12$, which is relatively large in comparison with the estimated probability $\hat{p} = \bar{y}/n = 0.32$. You would want to find out why such large variation in p occurred. The variation could arise, for example, from inadequacies in the test procedure or in the sampling method or because of inadequate process control. Further iterations in the investigation would take up these questions.

3.7. INFERENCES ABOUT FREQUENCIES (COUNTS PER UNIT): THE POISSON DISTRIBUTION

As explained in Chapter 2, the Poisson distribution is the limiting distribution to which the binomial distribution tends as p is made smaller and smaller and n is made larger and larger so that $\eta = np$ remains finite. It is often used to represent the frequency y in time or space of rare events such as the number of blemishes on a painted panel, the number of bacterial colonies on a petri dish, and the number of accidents occurring in a given time period. Its distribution function is

$$\Pr(y) = \frac{e^{-\eta}\eta^y}{y!}$$

where η is the mean number of events per unit. The variance σ^2 of the Poisson distribution also equals η.

A Significance Test

A supervisor of a raisin bran production unit suspected that raisins were not being added to the cereal in accordance with process specifications. In a test the quality control supervisor randomly selected one box per hour over a 12-hour period, thoroughly shook the box, and counted the raisins in one standard-sized scoopful. The specification required a mean of 36 raisins per scoopful. Random sampling variation would thus produce Poisson distributed frequencies (counts) with mean 36 and standard deviation $\sigma = \sqrt{36} = 6$. The inspector's data were as follows:

Hour	1	2	3	4	5	6	7	8	9	10	11	12	Total
Number of raisins in a scoopful	43	46	50	40	38	29	31	35	41	52	48	37	490

Average frequency: 40.83

Comparing the observed frequencies y_j with the hypothesized mean $\eta = 36$ provides a quantity approximately distributed as chi square. Thus

$$\frac{\sum (y_j - \eta)^2}{\eta} = \frac{\sum (y_j - 36)^2}{36} = \frac{866}{36} = 24.06 \sim \chi^2_{12}$$

Entering the chi-square tables with 12 degrees of freedom, you will find this value to be significant at about the 2% point. Two possible reasons for the discrepancies between observed and expected frequencies would be (a) the mean frequency of raisins η is greater than 36 and (b) the variance from box to box is greater than can be accounted for by random sampling variation.

To test the first possibility, you could use the additive property of the Poisson distribution. The sum of the frequencies 490 would be compared with the value $432 = 12 \times 36$ expected if the mean frequency per scoopful were really $\eta = 36$. If you refer the value $(490 - 432)^2/432 = 7.79$ to a chi-square table with one degree of freedom, you will find it is significant at about the 0.5% level. It appears therefore that the mean frequency is almost certainly greater than 36, the value expected if the process was operating correctly.

For the second possibility, for samples drawn from a Poisson distribution of *unknown* mean the quantity $\sum_{j=1}^{k} (y_j - \bar{y})^2/\bar{y}$ will be approximately distributed as χ^2 with $k - 1$ degrees of freedom. For this example, then, $\bar{y} = 490/12 = 40.83$ and

$$\frac{\sum (y_j - 40.83)^2}{40.83} = 14.34$$

Entering in the chi-square table with $k - 1 = 12 - 1 = 11$ degrees of freedom produces a significance level of about 20%.

This analysis indicates therefore that there are discrepancies from the specification frequency and that they are likely due to an excessively high mean level rather than to variation from hour to hour.

Exercise 3.22. The first step that should be taken is, as always, to plot the data. Make such a plot for the above example and suggest what additional enquiries the supervisor might make in furtherance of this interactive investigation.

Confidence Limits for a Poisson Frequency

Suppose you have an observed value y_0 for a Poisson distributed frequency. What are the $1 - \alpha$ confidence limits η_- and η_+ for the mean frequency η? As before, η_- is some value less than y_0 that just produces a result significant at the $\alpha/2$ level against the alternative $\eta > \eta_-$. Correspondingly, η_+ is some value greater than y_0 that just produces significance at the $\alpha/2$ level against the alternative $\eta < \eta_+$. It is tedious to compute these values directly from the distribution. The computer can supply these limits or alternatively Table G at the back of the book gives the 95 and 99% limits for $y_0 = 1, 2, \ldots, 50$.

Approach to Normality of the Poisson Distribution When η Is Not Too Small

As the mean frequency (count) increases, the Poisson distribution approaches normality quite quickly, as was illustrated in Figure 2.15*a*, which showed the Poisson distribution with mean $\eta = \sigma^2 = 2.1$ (representing, from Chapter 2, Minnie's and Henry's accident distributions). Figure 2.15*b* (on page 56) showed a Poisson distribution with $\eta = \sigma^2 = 10$. The former was quite skewed, but the latter could be fairly well approximated by a normal distribution with mean and variance equal to 10 using Yates's adjustment as with the binomial.

Variance Stabilizing Transformation for the Poisson Distribution

You can often analyze frequency data to an adequate approximation by using standard normal theory procedures involving t statistics and the analysis of variance and regression techniques that are discussed later. However, because the variance of a Poisson variable is equal to its mean, contrary to the assumptions of normal theory, the variance alters as the mean changes. If the frequencies cover wide ranges, it is best to work with scores obtained by an appropriate transformation. The appropriate variance stabilizing transformation is in this instance the square root. Thus, if y is distributed as a Poisson variable, the score $\sqrt{y}$ has an approximately constant variance equal to 0.25.

3.8. CONTINGENCY TABLES AND TESTS OF ASSOCIATION

Since for the Poisson distribution the mean η is equal to the variance, for sufficiently large η the quantity $(y - \eta)/\sqrt{\eta}$ is distributed approximately as a unit normal deviate and $(y - \eta)^2/\eta$ is distributed as χ^2 with one degree of freedom. Also, from the additive property of χ^2 for n cells $\sum_1^n (y_i - \eta_i)^2/\eta_i$ is distributed approximately as χ^2 with n degrees of freedom. These facts may be used for the analysis of "contingency" tables as is illustrated by the following example.

The data in Table 3.10 show the results from five hospitals of a surgical procedure designed to improve the functioning of certain joints impaired by disease. In this study the meanings to be attached to "no improvement," "partial functional restoration," and "complete functional restoration" were carefully defined in terms of measurable phenomena.

These data, when expressed as percentages, suggest differences in success at the five hospitals. However, percentages can be very misleading. For example, 28.6% can, on closer examination, turn out to represent 2/7 or 286/1000. To make an analysis, therefore, you must consider the *original data* on which these percentages were based. These are the frequencies shown in large type in Table 3.11a.

An important question was whether the data were explicable on the hypothesis that the mean frequencies in the various categories were distributed *in the*

Table 3.10. Results (Percentages) from a Surgical Procedure in Five Hospitals

	Hospital					
	A	*B*	*C*	*D*	*E*	Overall Percentage
No improvement	27.7	16.1	10.1	16.4	52.4	24.5
Partial restoration	38.3	32.3	45.6	43.8	35.4	40.6
Complete restoration	34.0	51.6	44.3	39.8	12.2	34.9

Table 3.11a. Results of Surgical Procedures: Original Frequencies Shown in Bold Type, Expected Frequencies in Upper Right Corner, Contribution to χ^2 in Top Left Corner of Each Cell

	Hospital					Totals
	A	B	C	D	E	
No improvement	0.19 11.53 **13**	0.89 7.60 **5**	6.67 19.37 **8**	3.44 31.39 **21**	26.05 20.11 **43**	**90**
Partial improvement	0.06 19.08 **18**	0.53 12.59 **10**	0.48 32.07 **36**	0.31 51.97 **56**	0.55 33.29 **29**	**149**
Complete restoration	0.01 16.39 **16**	2.49 10.81 **16**	2.01 27.55 **35**	0.91 44.64 **51**	12.10 28.60 **10**	**128**
Totals	**47**	**31**	**79**	**128**	**82**	**367**

χ^2 value 56.7, degrees of freedom 8, significance level < 0.001%

same proportions from hospital to hospital. If the mean (expected) frequencies were exactly known for each cell in the table, the χ^2 approximation mentioned above could be used to test the hypothesis. However, the only evidence of what the expected frequencies should be on this hypothesis comes from the marginal totals of the data. For instance, from the right-hand margin, 90 out of 367, or 24.5%, of all patients showed no improvement. Since there was a total of 47 patients from hospital A, 24.5% of 47, or 11.53, would be the expected frequency of patients showing no improvement in hospital A if the null hypothesis were true. In a similar way the other empirical expected frequencies can be computed from the marginal totals. In general, the expected frequency $\hat{\eta}_{i,j}$ to be entered in the ith row and jth column is $F_i F_j / F$, where F_i is the marginal frequency in the ith row and F_j is the marginal frequency in the jth column and F is the total frequency. These values are shown in the top right-hand corner of each cell. The value of chi square contributed by the cell in the ith row and jth column, $(y_{ij} - \hat{\eta}_{ij})^2/\hat{\eta}_{ij}$, is shown in the top left-hand corner of each cell. For the first entry the contribution is $(13 - 11.53)^2/11.53 = 0.19$. Adding together the 15 contributions gives a total chi-square value of 56.7.

Fisher showed that, when empirical expected frequencies were calculated from marginal totals, the chi-square approximation* could still be used but the correct number of degrees of freedom for a table with r rows and c columns is $(r - 1)(c - 1)$. This corresponds to the number of cells in the table less the number of known relationships (constraints) that the method of calculation has imposed on the expected values. Thus the calculated value of 56.7 should be referred to a chi-square table with $2 \times 4 = 8$ degrees of freedom. The value is highly significant. Thus, there is little doubt that the results from hospital to hospital are different.

Now hospital E was a referral hospital. It seemed relevant therefore to ask two questions:

1. How much of the discrepancy is associated with differences between the referral hospital on the one hand and the nonreferral hospitals *taken together* on the other?
2. How much of the discrepancy is associated with differences among the four nonreferral hospitals?

Tables 3.11b and 3.11c, which have the same format as Table 3.11a, permit the appropriate comparisons to be made. It is evident that the differences are mainly between the referral hospital E and the nonreferral hospitals. Differences in Table 3.11c, among nonreferral hospitals are readily explained by sampling variation.

You should notice that the discrepancies mainly arise because the results from the referral hospital are *not as good* as those from hospitals A, B, C, and D. This

*The χ^2 approximation is reasonably good provided the expected frequency in each cell is not less than 5.

Table 3.11b. Hospital *E* (Referral) Compared to Hospitals A, B, C and D (Nonreferral)

	Hospital		
Restoration	A + B + C + D	E	Total
None	7.50 69.89 **47**	26.06 20.11 **43**	**90**
Partial	0.16 115.71 **120**	0.55 33.29 **29**	**149**
Complete	3.48 99.40 **118**	12.10 28.60 **10**	**128**
Totals	**285**	**82**	**367**

χ^2 value 49.8, degrees of freedom 2, significance level < 1%

Table 3.11c. Comparison of Hospitals A, B, C and D (Nonreferral)

Hospital				
A	B	C	D	Total
3.55 7.75 **13**	0.00 5.11 **5**	1.94 13.03 **8**	0.00 21.11 **21**	**47**
0.16 19.79 **18**	0.71 13.05 **10**	0.23 33.26 **36**	0.08 53.89 **56**	**120**
0.61 19.46 **16**	0.78 12.84 **16**	0.16 32.71 **35**	0.18 53.00 **51**	**118**
47	**31**	**79**	**128**	**285**

χ^2 value 8.3, degrees of freedom 6, significance level about 20%

might be expected because typically the more difficult cases are sent to the referral hospital.

The 2 x 2 Table: Comparison of Repair Records for Television Sets

A special case of some importance occurs for the 2×2 contingency table when there are only two rows and two columns, and consequently the resulting chi square has only one degree of freedom. For instance, the data in Table 3.12 were obtained by a consumer organization from 412 buyers of new color television sets made by manufacturers A and B. The frequencies in the column headed "required service" denote the number of customers whose sets needed service at least once during a 10-year period.

A null hypothesis of interest is that the proportions requiring service are the same for both manufacturers, observed discrepancies being due to sampling variation. To test the hypothesis, expected frequencies may be calculated from the marginal totals as before. These are shown in the top right-hand corner of the cells in the table. By adding the contributions from the four cells, we obtain $\chi^2 = 15.51$, which is referred to the chi-square table with one degree of freedom. This gives a significance probability of less than 0.1%. Thus, if these samples can be regarded as *random* samples from the two manufacturers, there is little doubt that sets made by manufacturer A have a better repair record than those by manufacturer B.

Yates's Adjustment

For the 2×2 contingency table, where the chi-square approximation is most strained, the test tends to exaggerate significance. A considerably improved approximation is obtained by applying Yates's adjustment, discussed earlier. In the present context this adjustment consists of changing the observed frequencies by half a unit to give *smaller* deviations from expected values. For the color television data, for example, you should change the observed frequencies from 111, 162, 85, 54 to 111.5, 161.5, 84.5, 54.5, which yields a slightly smaller value for χ^2 of 14.70 again with a significance probability of less than 0.1%.

Table 3.12. Comparison of Two Brands of Color Television Sets

Brand	Required Service	Did Not Require Service	Total
A	2.74 129.87	2.49 143.13	
	111	**162**	**273**
B	5.38 66.13	4.89 72.87	
	85	**54**	**139**
Total	**196**	**216**	**412**

APPENDIX 3A. COMPARISON OF THE ROBUSTNESS OF TESTS TO COMPARE TWO ENTITIES

In this chapter, as a means of introducing some elementary principles of statistical analysis and design, you were deliberately confronted with a dilemma. The comparison of two means requires either (1) a long sequence of previous records that may not be available or (2) a random sampling assumption that may not be tenable. As was said earlier, practitioners might be forgiven for believing that a solution to these difficulties would be provided by what have been misleadingly called "distribution free" tests, also referred to as "nonparametric tests." For illustration, we compare under nonstandard conditions the performance of the t test and that of a widely used nonparametric test due to Wilcoxon.* We denote this test by W and apply to the analysis of the two samples of 10 observations in Table 3A.1.

You make the Wilcoxon test as follows:

1. Rank the combined samples in order of size from smallest to largest (here a -13 lies to the left and hence is "smaller" than a -10). Ties are scored with their average rank. The ranking of the observations is shown in Table 3A.1.
2. Calculate the sum of the ranks for methods A and B:

   For A: $2 + 3 + 4 + 7.5 + 9.5 + 9.5 + \cdots + 19 = 95.5$.
   For B: $1 + 5 + 6 + 7.5 + 11 + 14 + \cdots + 20 = 114.5$.
3. Refer one of the sums to the appropriate table. Alternatively, for equal groups of size n, if S is the larger of two sums, an approximate significance level is obtained by referring $z_0 = \sqrt{12}[S - 0.5n(2n + 1)]/[n\sqrt{2n + 1}]$ to

Table 3A.1. Wilcoxon Example: Comparing the Means of Two samples of Ten Observations Each

Rank	1	2	3	4	5	6	7.5	7.5	9.5	9.5
Observation	79.3	79.7	81.4	81.7	82.6	83.2	83.7	83.7	84.5	84.5
Method	*B*	*A*	*A*	*A*	*B*	*B*	*A*	*B*	*A*	*A*
Rank	11	12	13	14	15	16	17	18	19	20
Observation	84.7	84.8	85.1	86.1	86.3	87.3	88.5	89.1	89.7	91.9
Method	*B*	*A*	*A*	*B*	*B*	*A*	*B*	*B*	*A*	*B*

* Frank Wilcoxon was a scientist of the first rank who had many years of practical experience working with experimenters on the design and analysis of experiments at Boyce Thompson Institute and later at Lederle Laboratories. He was well aware of the essential role that randomization played in the strategy of investigation and its importance for validation of significance tests which later writers called "distribution free." His motivation for introducing this test was not to obtain a procedure that was insensitive to nonnormality. At a time when aids to rapid computing were not available, he often needed to make very large numbers of significance tests every day. He developed his test because it could be *done quickly*. For this purpose it was extremely valuable.

the tables of the unit normal distribution. For this example $z_0 = 0.718$ and $\Pr(z > 0.718) = 0.236$. This probability of 23.6% is comparable to the level 19.5% given by the t test, which also assumes random sampling.

As we explained, this test is not really distribution free but makes the distributional assumption of exchangeability. It gives erroneous results when that assumption is inappropriate. In particular the distribution free test assumes, as does the parametric test, that the observations are not autocorrelated. However, observations taken in sequence are unlikely to be independently distributed. For example, records from successive plots in an agricultural field trial are likely to be serially correlated* and so are successive observations in time.

Table 3A.2. Percentage of 20,000 Results Significant at the 5% Level When the Mean Difference $\delta = 0$ Using the t Test (t) and Wilcoxon Test (W)

Parent Distribution

ρ = *Autocorrelation between Successive Observations*

ρ		Rectangular	Normal	Skew[a]	Contaminated normal[b]
A. WITHOUT RANDOMIZATION					
0.0	t	5.1	5.0	4.7	5.0
	W	4.4	4.3	4.5	4.5
−0.4	t	0.8	0.6	0.5	1.8
	W	0.7	0.5	0.4	1.3
+0.4	t	19.5	20	19.6	13.0
	W	17.3	17.6	18.2	12.9
B. WITH RANDOMIZATION					
0.0	t	5.0	5.0	4.8	4.9
	W	4.4	4.4	4.4	4.3
−0.4	t	4.7	5.2	4.8	5.1
	W	4.0	4.5	4.4	4.5
+0.4	t	4.9	4.9	4.9	5.0
	W	4.1	4.1	4.4	4.4

[a] A chi square with four degrees of freedom.
[b] A normal distribution randomly contaminated with 5% of data at $+3\sigma$ and 5% at -3σ. The symbol ρ indicates the serial correlation of adjacent observations.

*This is of course the reason for arranging the trials within blocks and for randomization.

Table 3A.2 shows the effects of violations of the assumption of normality and independence of errors on the performance of the unpaired t test and the corresponding nonparametric W test. The t test was very little affected by severe nonnormality but both tests were greatly affected by serial correlation of the errors. So the use of the "nonparametric" procedure in place of t would protect you from circumstances that were not a threat while leaving you equally vulnerable to circumstances that were.

The results were obtained by taking two samples of 10 observations and making a t test (t) and a Wilcoxon test (W) for a difference in means. The sampling was repeated 20,000 times and the percentage of results significant at the 5% level was recorded. Thus for these tests made where there was no difference in the means the percentage should be close to 5%.

To determine the effect of major violations of the assumption of normality, data were generated from four sets of independently distributed random variables u_i having respectively a *rectangular* distribution, a *normal* distribution, a *highly skewed* distribution (a χ^2 distribution with four degrees of freedom), and a *"contaminated" normal* distribution with 10% of the observations randomly occurring at either $+3\sigma$ or -3σ.

From the first row of the table where the errors are uncorrelated you will see that the 5% level for the t test is affected very little by the gross violation of the normality assumption. In the first row of Table 3A.2, marked *without randomization,* the observations had independent errors $e_i = u_i$. In the second and third rows the data were not independent but were generated respectively so that successive errors had autocorrelations equal to $\rho_1 = -0.4$ and $\rho_1 = +0.4$.* You can see that these departures from assumptions have dramatic consequences that are much the same for both the t test and the W test but neither are much affected after randomization.

Comparison of variances

Although tests for comparisons of means are not affected very much by the distributional nonnormality, this happy state of affairs does not extend to the comparison of variances. Thus, for the four distributions used above, the percent probabilities of exceeding the normal theory 5% for an F test based on two samples of 10 observations each are

Rectangular	Normal	Skew	Contaminated Normal
1.1	5.0	14.4	9.2

Each result is based, as before, on 20,000 repetitions.

*The errors e_i were generated by a "first-order autoregressive process" with $e_i = u_i + \rho u_{i-1}$ with the ρ's appropriately chosen and the u's NIID random variables. For this process ρ is the first-order autocorrelation between adjacent observations.

APPENDIX 3B. CALCULATION OF REFERENCE DISTRIBUTION FROM PAST DATA

Table 3B.1. Production Records of 210 Consecutive Batch Yields

Observed	Average of 10 Observations	Observed	Average of 10 Observations	Observed	Average of 10 Observations	Observed	Average of 10 Observations	Observed	Average of 10 Observations	Observed	Average of 10 Observations
85.5	—	84.5	84.42	80.5	84.53	79.5	83.72	84.8	84.36	81.1	83.68
81.7	—	82.4	84.70	86.1	84.09	86.7	83.89	86.6	84.54	85.6	83.91
80.6	—	86.7	84.79	82.6	84.28	80.5	83.90	83.5	84.58	86.6	83.53
84.7	—	83.0	85.30	85.4	84.51	91.7	84.50	78.1	84.33	80.0	83.43
88.2	—	81.8	84.51	84.7	84.33	81.6	83.99	88.8	84.47	86.6	84.06
84.9	—	89.3	84.90	82.8	83.61	83.9	83.71	81.9	83.98	83.3	84.41
81.8	—	79.3	84.20	81.9	84.05	85.6	84.04	83.3	83.96	83.1	83.82
84.9	—	82.7	84.40	83.6	83.94	84.8	83.88	80.0	83.34	82.3	83.68
85.2	—	88.0	84.82	86.8	84.16	78.4	83.47	87.2	83.65	86.7	84.26
81.9	83.94	79.6	83.73	84.0	83.84	89.9	84.26	83.3	83.75	80.2	83.55
89.4	84.33	87.8	84.06	84.2	84.21	85.0	84.81	86.6	83.93		
79.0	84.06	83.6	84.18	82.8	83.88	86.2	84.76	79.5	83.22		
81.4	84.14	79.5	83.46	83.0	83.92	83.0	85.01	84.1	83.28		
84.8	84.15	83.3	83.49	82.0	83.58	85.4	84.38	82.2	83.69		
85.9	83.92	88.4	84.15	84.7	83.58	84.4	84.66	90.8	83.89		
88.0	84.23	86.6	83.88	84.4	83.74	84.5	84.72	86.5	84.35		
80.3	84.08	84.6	84.41	88.9	84.44	86.2	84.78	79.7	83.99		
82.6	83.85	79.7	84.11	82.4	84.32	85.6	84.86	81.0	84.09		
83.5	83.68	86.0	83.91	83.0	83.94	83.2	85.34	87.2	84.09		
80.2	83.51	84.2	84.37	85.0	84.04	85.7	84.92	81.6	83.92		
85.2	83.09	83.0	83.89	82.2	83.84	83.5	84.77	84.4	83.70		

87.2	83.91	84.8	84.01	81.6	83.72	80.1	84.16	84.4	84.19
83.5	84.12	83.6	84.42	86.2	84.04	82.2	84.08	82.2	84.00
84.3	84.07	81.8	84.27	85.4	84.38	88.6	84.40	88.9	84.67
82.9	83.77	85.9	84.02	82.1	84.12	82.0	84.16	80.9	83.68
84.7	83.44	88.2	84.18	81.4	83.82	85.0	84.21	85.1	83.54
82.9	83.70	83.5	84.07	85.0	83.43	85.2	84.11	87.1	84.28
81.5	83.59	87.2	84.82	85.8	83.77	85.3	84.08	84.0	84.58
83.4	83.58	83.7	84.59	84.2	83.89	84.3	84.19	76.5	83.51
87.7	84.33	87.3	84.90	83.5	83.74	82.3	83.85	82.7	83.62
81.8	83.99	83.0	84.90	86.5	84.17	89.7	84.47	85.1	83.69
79.6	83.23	90.5	85.47	85.0	84.51	84.8	84.94	83.3	83.58
85.8	83.46	80.7	85.18	80.4	83.93	83.1	85.03	90.4	84.40
77.9	82.82	83.1	85.31	85.7	83.96	80.6	84.23	81.0	83.61
89.7	83.50	86.5	85.37	86.7	84.42	87.4	84.77	80.3	83.55
85.4	83.57	90.0	85.55	86.7	84.95	86.8	84.95	79.8	83.02
86.3	83.91	77.5	84.95	82.3	84.68	83.5	84.78	89.0	83.21
80.7	83.83	84.7	84.70	86.4	84.74	86.2	84.87	83.7	83.18
83.8	83.87	84.6	84.79	82.5	84.57	84.1	84.85	80.9	83.62
90.5	84.15	87.2	84.78	82.0	84.42	82.3	84.85	87.3	84.08

Table 3B.2. Reference Set of Differences between Averages of Two Adjacent Sets of 10 Successive Batches[a]

—	−0.36	−0.32	1.09	−0.43
—	−0.52	−0.21	0.87	−1.32
—	−1.33	−0.36	1.11	−1.30
—	−1.81	−0.93	−0.12	−0.64
—	−0.36	−0.75	0.67	−0.58
—	−1.02	0.13	1.01	0.37
—	0.21	0.39	0.74	0.03
—	−0.29	0.38	0.98	0.75
—	−0.91	−0.22	**1.87**	0.44
−0.43	0.64	0.20	0.66	0.17
−1.24	−0.17	−0.37	−0.04	−0.23
−0.15	−0.17	−0.16	−0.60	0.97
−0.02	0.96	0.12	−0.93	0.72
−0.08	0.78	0.80	0.02	0.98
−0.15	−0.13	0.54	−0.50	−0.21
−0.79	0.30	0.08	−0.51	−0.81
−0.38	−0.34	−1.01	−0.67	0.29
−0.26	0.71	−0.55	−0.78	0.49
−0.10	0.68	−0.05	−1.15	−0.58
0.82	0.53	−0.30	−1.07	−0.30
0.90	1.01	0.33	−0.30	−0.01
−0.68	**1.46**	0.79	0.78	−0.61
−0.66	0.76	−0.11	0.95	0.40
−1.25	1.04	−0.42	−0.17	−1.06
−0.27	**1.35**	0.30	0.61	−0.13
0.13	1.37	**1.13**	0.74	−0.52
0.21	0.88	1.25	0.67	−1.07
0.24	−0.12	0.97	0.79	−1.40
0.29	0.20	0.68	0.66	0.11
−0.18	−0.12	0.68	1.00	0.46
0.43	−0.37	−0.45	−0.11	−0.01
1.47	−1.38	−0.62	−0.40	0.33
1.33	−0.90	−0.03	−0.45	−0.87
2.48	−0.80	0.54	0.10	−0.18
1.01	−1.04	−0.43	−0.30	0.51
1.33	−1.94	−1.24	−0.97	**1.39**
0.29	−0.90	−0.64	−0.82	0.61
0.57	−0.76	−0.86	−1.53	0.50
0.95	−0.63	−1.10	−1.20	0.64
−0.42	−0.94	−0.16	−1.10	−0.53

[a] Differences that exceed +1.30 are in bold type.

REFERENCES AND FURTHER READING

Bartlett, M. S., and Kendall, D. G. (1946) The Statistical Analysis of Variance Heterogeneity and the Logarithmic Transformation, *Supplement to the Journal of the Royal Statistical Society*, **8**, 128–138.

Box, G. E. P. (1990) Must we randomize our experiment, *Quality Eng.*, **2**(4) 497–502.

Box, G. E. P. (2001) Comparisons, absolute values, and how I got to go to the Folies Bergères, *Quality Eng.*, **14**(1), 167–169.

Box, G. E. P., and Andersen, S. L. (1955) Permutation theory in derivation of robust criteria and the study of departures from assumptions, *J. Roy. Statist. Soc., Series B*, **17**, 1–26.

Box, G. E. P., and Newbold, P. (1971) Some comments on a paper of Coen, Gomme and Kendall, *J. Roy. Statist. Soc., Series A*, **134**(2), 229–240.

Browne, J. (2000) *Charles Darwin, Vol. II: The Power of Place*, Knopf, New York.

Cochran, W. G. (1950), The comparison of percentages in matched sampoles, *Biometrika*, **37**, 256–266.

Cochran, W. G. (1954) Some methods for strengthening the common χ^2 tests, *Biometrics*, **10**, 417–451.

Darwin, C. (1876) *The Effects of Cross and Self Fertilization in the Vegetable Kingdom*, John Murray, London.

Fisher, R. A. (1935) *The Design of Experiments*, Oliver and Boyd, London.

Ludbrook, J., and Dudley, H (1998) Why permutation tests are superior to t and F tests in biomedical research, *Amer. Statist.*, **52** (2), 127–132.

Pitman, E. J. G. (1937) Significance tests which may be applied to samples from any populations: III The analysis of variance test, *Biometrika*, **29**(3/4), 322–335.

Welch, B. L. (1937), On the z-test in randomized blocks in Latin squares, *Biometrika*, **29**(1/2), 21–52.

Wilcoxon, F. (1945) Individual comparisons by ranking methods, *Biometrics*, **1**, 80–83.

Yates, F. (1934) Contingency tables involving small numbers and the χ^2 test, *Supplement to the J. Roy. Statist. Soc.* **1**, 217–235.

QUESTIONS FOR CHAPTER 3

1. What is meant by a valid significance test?
2. What is a reference distribution? What are the possible advantages derived from (a) an external reference set or (b) a randomized reference set?
3. What is randomization and what is its value?
4. What is meant by exchangeability?
5. What is meant by a "distribution free" test? Comment.
6. How can experiments be randomized in practice? Be specific.
7. Why can probabilities calculated on NIID assumptions be seriously affected by autocorrelation of observations?

8. What is a randomization distribution?

9. What is the value of pairing and blocking? In practice, how can a design be blocked?

10. Can you find (or imagine) an example of an experiment in your own field in which both randomization and blocking were (or could be) used?

11. Can you describe an actual situation from your own field in which trouble was encountered because the data were not obtained from a randomized experiment? If not, can you imagine one?

12. Will conducting (a) a nonparametric test or (b) a randomization test protect you from the effects of "bad values"? Comment.

13. How might you discover and what would you do about suspected "bad values"? Can you imagine a situation where wrong inferences could be made because of bad values?

14. Given that you have performed a significance test for a particular δ, how would you construct a 95% confidence interval for δ?

15. If $\eta_A = \eta_B$, is the 95% confidence interval for δ calculated using paired data always shorter than that using unpaired data? Explain how you would decide which experimental strategy to employ, paired or unpaired. How would you verify each interval using a randomization test?

16. Are tests on variances more or less sensitive to departures from assumptions? Comment.

PROBLEMS FOR CHAPTER 3

1. A civil engineer tested two different types (A and B) of a special reinforced concrete beam. He made nine test beams (five A's and 4 B's) and measured the strength of each. From the following strength data (in coded units) he wants to decide whether there is any real difference between the two types. What assumptions does he need to draw conclusions? What might he conclude? Give reasons for your answers.

Type A	67	80	106	83	89
Type B	45	71	87	53	

2. Every 90 minutes routine readings are made of the level of asbestos fiber present in the air of an industrial plant. A salesperson claimed that spraying with chemical S-142 could be beneficial. Arrangements were made for a comparative trial in the plant itself. Four consecutive readings, the first without S-142 and the second with S-142 were as follows: 8, 6, 3, and 4. In light of past data collected without S-142, given below, do you think S-142 works? Explain your answer.

Asbestos Data, 112 Consecutive Readings

November 2	9	10	9	8	9	8	8	8	7	6	9	10	11	9	10	11
November 3	11	11	11	10	11	12	13	12	13	12	14	15	14	12	13	13
November 4	12	13	13	13	13	13	10	8	9	8	6	7	7	6	5	6
November 5	5	6	4	5	4	4	2	4	5	4	5	6	5	5	6	5
November 6	6	7	8	8	8	7	9	10	9	10	9	8	9	8	7	7
November 7	8	7	7	7	8	8	8	8	7	6	5	6	5	6	7	6
November 8	6	5	6	6	5	4	3	4	5	5	6	5	6	7	6	5

3. Nine samples were taken from two streams, four from one and five from the other, and the following data on the level of a pollutant (in ppm) obtained:

Stream 1:	16	12	14	11	
Stream 2:	9	10	8	6	5

It is claimed that the data prove that stream 2 is cleaner than stream 1. An experimenter asked the following questions. When were the data taken? All in one day? On different days? Were the data taken during the same time period on both streams? Where the stream temperatures and flows the same? Where in the streams were the data taken? Why were these locations chosen? Are they representative? Are they comparable?
Why do you think she asked these questions? Are there other questions she should have asked? Is there any set of answers to these questions (and others you invent) that would justify the use of a t test to draw conclusions? What conclusions?

4. Fifteen judges rated two randomly allocated brands of beer, A and B, according to taste (scale: 1 to 10) as follows:

Brand A:	2	4	2	1	9	9	2	2
Brand B:	8	3	5	3	7	7	4	

Stating assumptions, test the hypothesis that $\eta_A = \eta_B$ against the alternative that $\eta_A \neq \eta_B$. Can you think of a better design for this experiment? Write *precise* instructions for the conduct of your preferred experiment.

✓**5.** These data were obtained in a comparison of two different methods for determining dissolved oxygen concentration (in milligrams per liter):

Sample	1	2	3	4	5	6
Method A (amperometric)	2.62	2.65	2.79	2.83	2.91	3.57
Method B (visual)	2.73	2.80	2.87	2.95	2.99	3.67

Estimate the difference between the methods. Give a confidence interval for the mean difference. What assumptions have you made? Do you think this experiment provides an adequate basis for choosing one method other the other? In not, what further experiments and data analysis would you recommend? Imagine different outcomes that could result from your analysis and say what each might show.

6. The following are smoke readings taken on two different camshafts:

Engine	Camshaft *A*	Camshaft *B*
1	2.7	2.6
2	2.9	2.6
3	3.2	2.9
4	3.5	3.3

Suppose your supervisor brings you these data and says that one group claims there is essentially no difference in the camshafts and the other says there is a difference. What question would you ask about this experiment? What answers would make a further analysis useful? What further analysis? Assuming any answers to your questions that you like, write a report on the experiment and the data. Somewhere in your report explain the difference between "statistically significant" and "technically important."

7. Two different designs for a solar energy collector were tested with the following results. The measured response was power (in watts).

design *A*	$1.8^{(1)}$	$1.9^{(2)}$	$1.1^{(5)}$	$1.4^{(7)}$
design *B*	$1.9^{(3)}$	$2.1^{(4)}$	$1.5^{(6)}$	$1.5^{(8)}$

The data were collected at eight different comparable time periods. The random order of the tests is indicated by the superscripts in parentheses. Is there evidence that a statistically significant difference exists between the mean values for the power attainable from these two designs? Do you need to assume normality to make a test?

8. In a cathode interference resistance experiment, with filament voltage constant, the plate current was measured on seven twin triodes. Analyze the data given below. Is there any evidence of a systematic difference between the plate current readings of compartments *A* and *B*?

Plate Current Readings

Tube	*A*	*B*
1	1.176	1.279
2	1.230	1.000
3	1.146	1.146
4	1.672	1.176
5	0.954	0.699
6	1.079	1.114
7	1.204	1.114

Explain your assumptions.

9. An agricultural engineer obtained the following data on two methods of drying corn and asked for statistical analysis:

Drying Rate

With Preheater	Without Preheater
16	20
12	10
22	21
14	10
19	12

What questions would you ask him? Under what circumstances would you be justified in analyzing the data using (a) a paired t test, (b) an unpaired t test, or (c) something else? Analyze the data for options (a) and (b).

10. An agricultural engineer is trying to develop a piece of equipment for processing a crop immediately after harvesting. Two configurations of this equipment are to be compared. Twenty runs are to be performed over a period of 5 days. Each run involves setting up the equipment in either configuration I or configuration II, harvesting a given amount of crop, processing it on the equipment, obtaining a quantitative measure of performance, and then cleaning the equipment. Since there is only one piece of equipment, the tests must be done one at a time. The engineer has asked you to consult on this problem.

(a) What questions would you ask her?

(b) What advice might you give about planning the experiment?

(c) The engineer believes the most accurate experiment requires that an equal number of runs be conducted with each configuration. What assumptions would make it possible to demonstrate in a quantitative way that this is true.

11. Assuming a standard deviation $\sigma = 0.4$, calculate a 90% confidence interval for the mean reaction time using the following data (in seconds):

1.4 1.2 1.2 1.3 1.5 1.0 2.1 1.4 1.1

Carefully state and criticize any assumptions you make. Repeat this problem assuming σ is unknown.

12. Five student groups in a surveying class obtain the following measurements of distance (in meters) between two points:

420.6 421.0 421.0 420.7 420.8

Stating precisely your assumptions, find an approximate 95% confidence interval of the mean measured distance.

13. Given the following data on egg production from 12 hens randomly allocated to two different diets, estimate the mean difference produced by the diets and obtain a 95% confidence interval for this mean difference. Explain your answer and the assumptions you make.

Diet *A*	166	174	150	166	165	178
Diet *B*	158	159	142	163	161	157

14. Two species of trees were planted on 20 randomly selected plots, 10 for *A* and 10 for *B*. The average height per plot was measured after 6 years. The results (in meters) were as follows:

A	3.2	2.7	3.0	2.7	1.7	3.3	2.7	2.6	2.9	3.3
B	2.8	2.7	2.0	3.0	2.1	4.0	1.5	2.2	2.7	2.5

Obtain a 95% confidence interval for the difference in mean height.

15. A chemical reaction was studied by making 10 runs with a new supposedly improved method (*B*) and 10 runs with the standard method (*A*). The following yield results were obtained:

Method *A*	**Method *B***
$54.6^{(16)}$	$74.9^{(12)}$
$45.8^{(10)}$	$78.3^{(19)}$
$57.4^{(11)}$	$80.4^{(14)}$
$40.1^{(2)}$	$58.7^{(6)}$
$56.3^{(20)}$	$68.1^{(8)}$
$51.5^{(18)}$	$64.7^{(3)}$
$50.7^{(9)}$	$66.5^{(7)}$
$64.5^{(15)}$	$73.5^{(4)}$
$52.6^{(1)}$	$81.0^{(17)}$
$48.6^{(5)}$	$73.7^{(13)}$

The superscripts in parentheses denote the time order in which the runs were made. Comment on the data and analyze them.

16. (These data are taken from U.S. Patent 3,505,079 for a process preparing a dry free-flowing baking mix.) Five recipes were used for making a number of cakes, starting from two different types of premix, *A* and *B*. The difference between the two premixes was that *A* was aerated and *B* was not. The volumes of the cakes were measured with the following results:

	Recipe *A*	**Recipe *B***
1	43	65
2	90	82
3	96	90
4	83	65
5	90	82

The recipes differed somewhat in the amount of water added, beating time, baking temperature, and baking time. The patent claims that significantly greater volume was obtained with A. Do these data support this claim? Make any relevant comments you think are appropriate. Include in your answer a calculated 95% confidence interval for the true difference between the volumes obtained with A and B. State any assumptions you make.

17. The following are results from a larger study on the pharmacological effects of nalbuphine. The measured response obtained from 11 subjects was the change in pupil diameter (in millimeters) after 28 doses of nalbuphine (B) or morphine (A):

A	+2.4	+0.08	+0.8	+2.0	+1.9	+1.0
B	+0.4	+0.2	−0.3	+0.8	0.0	

Assume that the subjects were randomly allocated to the drugs. What is the 95% confidence interval for $\eta_B - \eta_A$? What is the 90% confidence interval? State your assumptions. [These data are from H. W. Elliott, G. Navarro, and N. Nomof (1970), *J. Med.*, **1**, 77.]

18. In a set of six very expensive dice the pips on each face are diamonds. It is suggested that the weights of the stones will cause "five" and "six" faces to fall downward and hence the "one" and "two" faces to fall upward more frequently than they would with technically fair dice. To test the conjecture a trial is conducted as follows: The observance of a one or two on a die is called a success. The whole set of six dice is then thrown 64 times and the frequencies of throws with 0, 1, 2, . . ., 6 successes are as follows:

Number of dice showing a "success"	0	1	2	3	4	5	6
Observed frequency	0	4	19	15	17	7	2

(a) What would be the theoretical probability of success and the mean and variance of the frequency distribution if all the dice were fair?

(b) What is the empirical probability of success calculated from the data, and what is the sample average and variance?

(c) Test the hypotheses that the mean and variance have their theoretical values.

(d) Calculate the expected frequencies in the seven cells of the table on the assumption that the probability of success is exactly $\frac{1}{3}$.

(e) Make a chi-square test of agreement of the observed with the expected frequencies (combining the frequencies for zero and one successes and also for five and six successes so that the expected frequency within each cell is greater than 5). The number of degrees of freedom for the test is the number of frequencies compared *less* 1, since the total of the expected frequencies has been chosen to match the actual total number of trials.

(f) Calculate the expected frequencies based on the empirical estimate of probability, and make a chi-square test of agreement with the observed frequencies. The number of degrees of freedom is now the number of frequencies compared *less* 2, since expected frequencies have been chosen so that the total frequency and the mean both match the data.

(g) Can you think of a better design for this trial?

19. The level of radiation in the control room of a nuclear reactor is to be automatically monitored by a Geiger counter. The monitoring device works as follows: Every tenth minute the number (frequency) of "clicks" occurring in t seconds is counted automatically. A scheme is required such that, if this frequency exceeds some number c, an alarm will sound. The scheme should have the following properties: If the number of *clicks per second* is less than 4, there should be only about 1 chance in 500 that the alarm will sound, but if the number of clicks reaches 16, there should be only about 1 chance in 500 that the alarm will not sound. What values should be chosen for t and c?
Hint: Recall that the square root of a Poisson frequency is roughly normally distributed with a standard deviation 0.5. (*Answer*: $t \approx 2.25$, $c \approx 20$, closest integer to 20.25.)

20. Check the approximate answer given for the above problem by actually evaluating the Poisson frequencies.

21. Consider the following data on recidivism as measured by the rearrest of persons who have committed at least one previous offense. There are a number of points on which you would wish to be reassured before drawing conclusions from such data. What are they? Assuming that your questions received satisfactory answers, what conclusions could you draw from the following data?

	Treatment Received		
	No Therapy	**Group Therapy**	**Individual Therapy**
Back in prison within 1 year	24	10	41
Back in prison in 1–10 years	25	13	32
Never back in prison	12	20	9
Total	61	43	82

22. Why is a set of confidence intervals more useful than a significance test?

23. What is the formula for the confidence interval for the difference between two means for a paired design? For an unpaired (fully randomized) design? How are these formulas similar? How are they different? How would you verify each interval using a randomization distribution?

24. Suppose you are given data from a simple comparative experiment in which n_A observations have been made with treatment A and n_B observations with treatment B. It is necessary that $n_A = n_B$? Suppose that $n_A = n_B$ and the data are paired. Will the equation for the paired data always give a shorter interval? Why not?

25. Think of a specific problem of comparing two treatments preferably from your own field. How would you organize an experiment to find out which one was better? What are some of the problems that might arise? How would you analyze the results? How could past data be used to assist in choosing the design and the number of runs to be made?

26. Let y be the number of sixes recorded on the throw of three dice. Find $\Pr(y = 0)$, $\Pr(y = 1)$, $\Pr(y = 2)$, $\Pr(y = 3)$, $\Pr(y > 1)$, $\Pr(y < 3)$. Suppose that in 16 throws of a single die a six appears on 8 occasions. Does this discredit the hypothesis that the die is fair? Use these data to obtain a 95% confidence interval for p.

27. Possibly meaningful signals have been obtained from outer space. The data take the form of the number of pulses y received in each sequence of 127 minutes. A skeptic suggests that the variation in the frequencies observed, $y_1, y_2, \ldots, y_{127}$, might be ascribed to chance causes alone. Describe any test you might make to test the skeptic's hypothesis.

28. You are assigned to a research program to assess the effectiveness of anti-cancer drugs in animals and human beings. Write a report for the physician in charge of the program (who has no statistical knowledge) describing how methods using (a) the binomial distribution, (b) the Poisson distribution, and (c) contingency tables might be of value is assessing the results of the program. Illustrate your discussion with hypothetical experiments and data and give appropriate analyses.

29. Illustrate how the binomial and Poisson distributions may be used to design sampling inspection schemes.

CHAPTER 4

Comparing a Number of Entities, Randomized Blocks, and Latin Squares

4.1. COMPARING k TREATMENTS IN A FULLY RANDOMIZED DESIGN

Frequently you will want to compare more than two entities—treatments, processes, operators, or machines. This chapter is about how to do it. The first example is one in which there are $k = 4$ treatments randomly applied to $n = 24$ subjects.

Blood Coagulation Time Example

Table 4.1 gives coagulation times for samples of blood drawn from 24 animals receiving four different diets A, B, C, and D. (To help the reader concentrate on essentials, in this book we have adjusted the data so that the averages come out to be whole numbers.) These data are plotted in Figure 4.1. The animals were randomly allocated to the diets, and the blood samples were taken and tested in the random order indicated by the bracketed superscripts in the Table.

Consider the question, "Is there evidence to indicate real difference between the mean coagulation times for the four different diets?" The necessary calculations are frequently set out in an *analysis of variance* table, a valuable device due to Fisher. The idea is to determine whether the discrepancies *between* the treatment averages are greater than could be reasonably expected from the variation that occurs *within* the treatment classifications. For example your computer will produce an analysis of variance (ANOVA) table that looks like that in Table 4.2.

Statistics for Experimenters, Second Edition. By G. E. P. Box, J. S. Hunter, and W. G. Hunter

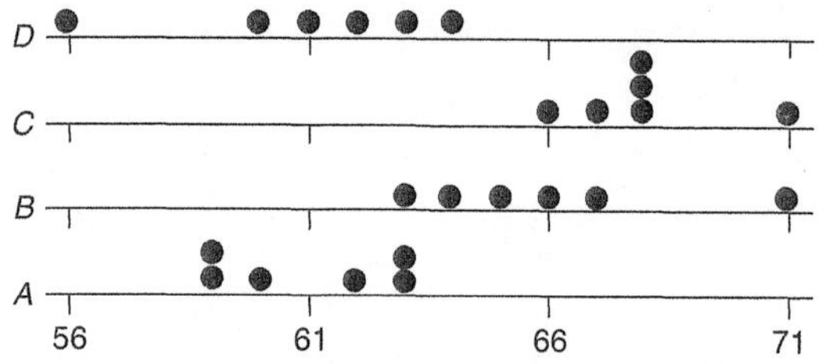

Figure 4.1. Dot plot of original data by diets.

Table 4.1. Coagulation Time for Blood Drawn from 24 Animals Randomly Allocated to Four Diets

	Diets (Treatments)			
	A	*B*	*C*	*D*
	$62^{(20)}$	$63^{(12)}$	$68^{(16)}$	$56^{(23)}$
	$60^{(2)}$	$67^{(9)}$	$66^{(7)}$	$62^{(3)}$
	$63^{(11)}$	$71^{(15)}$	$71^{(1)}$	$60^{(6)}$
	$59^{(10)}$	$64^{(14)}$	$67^{(17)}$	$61^{(18)}$
	$63^{(5)}$	$65^{(4)}$	$68^{(13)}$	$63^{(22)}$
	$59^{(24)}$	$66^{(8)}$	$68^{(21)}$	$64^{(19)}$
Treatment average	61	66	68	61
Grand average	64	64	64	64
Difference	−3	+2	+4	−3

Table 4.2. The Analysis of Variance (ANOVA) Table: Blood Coagulation Example

Source of Variation	Sum of Squares	Degrees of Freedom	Mean Square	
Between treatments	$S_T = 228$	$\nu_T = 3$	$m_T = 76.0$	$F_{3,20} = 13.63$
Within treatments	$S_R = 112$	$\nu_R = 20$	$m_R = 5.6$	
Total about the grand average	$S_D = 340$	$\nu_D = 23$		

To better understand this analysis look at Table 4.3. On the left you will see a table of the original observations **Y** and a table **D** of deviations from the grand average of 64. Thus, in the first row of **D** are the entries $-2 = 62 - 64$, $-1 = 63 - 64$, and so on. This table of deviations is now further decomposed into tables **T** and **R**. Table **T** represents the part due to the treatment deviations $(-3, +2, +4, -3)$ from the grand average. Table **R** represents the residual part

Table 4.3. Arithmetic Breakup of Deviations from the Grand Average $\bar{y} = 64$

Observations	Deviations from Grand Average of 64		Treatment Deviations		Residuals within-Treatment Deviations
y_{ti}	$y_{ti} - \bar{y}$		$\bar{y}_t - \bar{y}$		$y_{ti} - \bar{y}_t$
62 63 68 56	−2 −1 4 −8		−3 2 4 −3		1 −3 0 −5
60 67 66 62	−4 3 2 −2		−3 2 4 −3		−1 1 −2 1
63 71 71 60	−1 7 7 −4		−3 2 4 −3		2 5 3 −1
59 64 67 61	−5 0 3 −3		−3 2 4 −3		−2 −2 −1 0
63 65 68 63	−1 1 4 −1		−3 2 4 −3		2 −1 0 2
59 66 68 64	5 2 4 0		−3 2 4 −3		−2 0 0 3
Y	**D = Y − 64**	=	**T**	+	**R**
Sum of squares	340	=	228	+	112
degrees of freedom	23	=	3	+	20

that is left due to experimental error and model inadequacy. The individual items in this table are called *residuals*.

Entries in the ANOVA Table: Sums of Squares

The sums of squares S_D, S_T, and S_R in the analysis of variance (ANOVA) in Table 4.2 are the sums of the 24 entries in each table **D, T**, and **R**. Thus

$$S_D = (-2)^2 + (-1)^2 + (4)^2 + \cdots + (0)^2 = 340$$
$$S_T = (-3)^2 + (2)^2 + (4)^2 + \cdots + (-3)^2 = 228$$
$$S_R = (1)^2 + (-3)^2 + (0)^2 + \cdots + (3)^2 = 112$$

You will find that $S_D = S_T + S_R$ (for this example $340 = 228 + 112$). The additivity property of these sums of squares is true for *any* similar table of numbers split up in this way.

Entries in the ANOVA Table: Degrees of Freedom

The number of degrees of freedom is the number of elements in each of the decomposition tables that can be arbitrarily assigned. For example, **D** has 23 degrees of freedom because if you fill this table with 23 arbitrarily chosen numbers the 24th will be determined since the deviations of any set of numbers from their average must always sum to zero. On the same basis the elements of **T** have three degrees of freedom. The elements of **R** are constrained in two different ways—the elements in each column must add to zero and the sum of all of the elements must also sum to zero, and thus the number of residual degrees of freedom is $24 - 1 - 3 = 20$. Note that for *any data of this kind*, not only are the sums of squares additive, but also are the degrees of freedom.

Entries in the ANOVA Table: Mean Squares

The mean squares m_T and m_R are obtained by dividing S_T and S_R by their degrees of freedom ν_T and ν_R. On assumptions we discuss later, if there were *no differences* due to treatments (diets), the mean squares m_T and m_R would provide independent estimates of the error variance σ^2 and their ratio would have an F distribution with ν_T and ν_R degrees of freedom.

Computer calculations, or reference to the tables at the back of this book, show that the probability of a value of $F_{3,20} \geq 13.6$ is less than 0.001. You see that the result is highly supportive of the inference that the null hypothesis should be rejected and hence that the diets really do produce different coagulation times.

Graphical ANOVA

Walter Shewhart (1939, p. 88) once said, "Original data should be presented in a way that will preserve the evidence in the original data." The ANOVA table alone does not do this. But as you saw in Chapter 3 you can supplement more formal analyses with graphical methods and, as Yogi Berra says, "You can see a lot by just looking."

A graphical ANOVA is shown in Figure 4.2, which compares a suitably scaled dot diagram of the treatment deviations directly with a reference dot diagram of the residuals themselves. (Note: Since these are deviations from averages, these both have *known* means of zero.) Notice that this is a supplement to the standard ANOVA table. It would be deceptive if used alone because it takes no account of the individual degrees of freedom that determine the significance probabilities. But as commented by F. J. Anscombe (1973, p. 17), "A computer should make both calculations and graphs. Both kinds of output should be studied; each will contribute to understanding."

The scale factor for treatments is such that if there were no difference between the treatment means the *natural* variance of the *dots* in the dot diagram for treatments would be directly comparable to that for residuals. By natural variance is

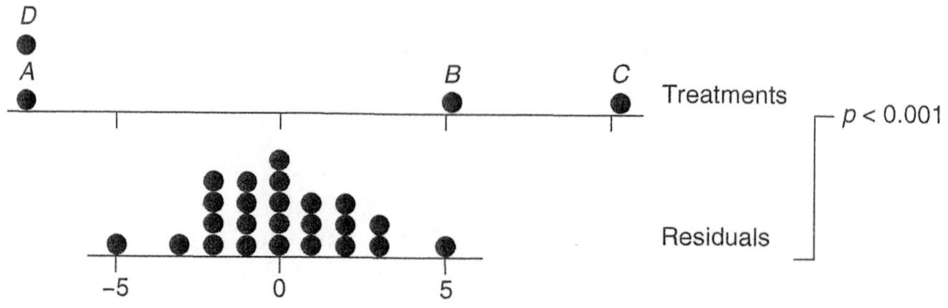

Figure 4.2. Dot diagram residuals and scaled treatment deviations.

meant the sum of squares of the deviations of the dot deviations divided by the *number* of dots (not the degrees of freedom). This measure of spread is appropriate because it shows the spread of the dots that the eye actually sees. The analysis asks the question, "Might the scaled treatment deviations just as well be part of the noise?" In Appendix 4A it is shown that the appropriate scale factor is $\sqrt{\nu_R/\nu_T} = \sqrt{20/3} = 2.6$. The scaled treatment deviations -7.8, 5.2, 10.4, and -7.8 are obtained therefore by multiplying the treatment deviations -3, $+2$, $+4$, -3 by 2.6. This graphic analysis thus obtained is shown in Figure 4.2. It visually supports the finding that the differences between treatments are unlikely to be due to chance. The ratio of the natural variances of the dot plots produces the usual F value. (See Appendix 4A.) It ensures that you appreciate the *nature* of the differences and similarities produced by the treatments, something the ANOVA table does not do. It also directs your attention to the individual residuals* that produce m_R and makes you aware of any large deviations that might call for further study. For instance, Figure 4.2 immediately makes clear that there is nothing suspicious about the distribution of the residuals. Also that treatments A and D are alike in their effects but C is markedly different and B produces an intermediate effect. Experimenters sometimes believe that a high level of significance necessarily implies that the treatment effects are accurately determined and separated. The graphical analysis discourages overreaction to high significance levels and avoids underreaction to "very nearly" significant differences.

In the first edition of this book the treatment deviations were referred to a reference t distribution. On NIID assumptions the t distribution may be regarded as a reference distribution that could be fitted to the residuals. Rather than take this additional theoretical step it seems preferable to use the residuals themselves as the reference distribution in the graphical analysis.

Geometry and the ANOVA Table

Look again at Table 4.3 and now think of the 24 numbers in each of the tables **D**, **T**, and **R** as constituting the elements of *vectors* **D**, **T**, and **R**. From geometry (whatever the number of dimensions), if the sum of products of the 24 elements in each of two vectors (sometimes called the inner product) is zero, the vectors are at right angles, that is, orthogonal. You can confirm, for example, that the vectors **T** and **R**, whose elements are set out in Table 4.3, are orthogonal by noticing that the inner product of their twenty four elements, $(-3)(1) + (2)(-3) + (4)(0) + \cdots + (-3)(3)$, equals zero. Indeed, for *any* series of numbers set out in a table of this kind, because of the constraints placed upon their elements, the vectors **T** and **R** will always be orthogonal. Also, since the vector **D** is the hypotenuse of a right triangle with sides **T** and **R** with S_T and S_R, the squared lengths of the vectors, the additive property of the sums of squares $S_D = S_T + S_R$ follows by extension of Pythagoras' theorem to n dimensions. Also, geometrically the

*If desired, a normal plot of the residuals may be appended.

degrees of freedom are the number of dimensions in which the vectors are free to move given the constraints. These results are shown geometrically in Figure 4.3 for just three observations.

Exercise 4.1. Each of 21 student athletes, grouped into three teams A, B, and C, attempts to successfully toss a basketball through a hoop within a fixed time period. The number of successes is given in the following table. Are there real differences between the three teams? Construct an ANOVA for these data and comment.

A	B	C
$21^{(14)}$	$13^{(9)}$	$15^{(17)}$
$19^{(6)}$	$16^{(7)}$	$16^{(8)}$
$17^{(1)}$	$15^{(11)}$	$14^{(14)}$
$21^{(13)}$	$12^{(2)}$	$15^{(5)}$
$22^{(21)}$	$19^{(16)}$	$16^{(12)}$
$23^{(18)}$	$19^{(19)}$	$12^{(15)}$
$17^{(3)}$	$18^{(20)}$	$17^{(10)}$

Assumptions

For the production of the ANOVA table, no assumptions are needed. You could have written any 24 numbers for the "observations" in Table 4.1 and completed an "analysis of variance" table like Table 4.2 and all the properties discussed so far would apply. However, the *relevance* of such an ANOVA table for solving the problem of comparing treatment means would depend on certain assumptions.

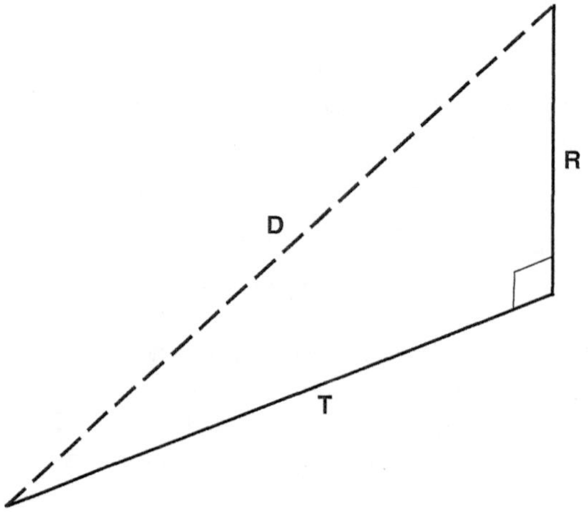

Figure 4.3. Right triangle of **D**, **T**, and **R**.

An Additive Model?

The analysis of the data in Table 4.1 implies tentative acceptance of the underlying additive model

$$y_{ti} = \eta + \tau_t + \varepsilon_{ti}$$

where y_{ti} is the ith observation in the tth column of the table, η is the overall mean, τ_t is the deviation produced by treatment t, and ε_{ti} is the associated error.

Errors Independently and Identically Distributed?

On the IID assumption that each error ε_{ti} varies *independently* of the others and has an *identical* distribution (and in particular the same variance), the expected (mean) values of m_T and m_R would be

$$E(m_T) = \sum \tau_t^2 + \sigma^2, \qquad E(m_R) = \sigma^2$$

Thus, if there were no differences in the four treatments so that $\tau_1 = \tau_2 = \tau_3 = \tau_4 = 0$ and $\sum \tau^2 = 0$, then both m_T and m_R, the mean squares in the ANOVA table, would be estimates of σ^2.

Normally Distributed?

If it could be further assumed that the ε_{ti} were normally distributed (that they were NIID), then m_T and m_R would be distributed independently, and on the null hypothesis that $\sum \tau^2 = 0$ the ratio $F = m_T/m_R$ would be the ratio of two independent estimates of σ^2 and so would be distributed in an $F_{3,20}$ distribution with 3 and 20 degrees of freedom. For the blood coagulation example Figure 4.4

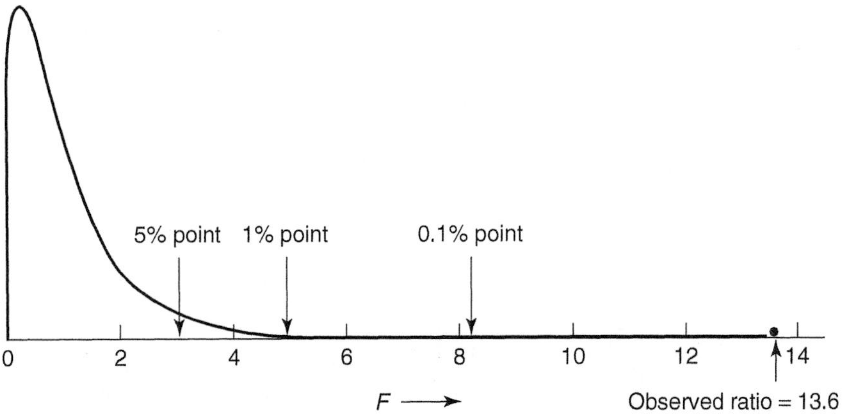

Figure 4.4. Observed value of the ratio $m_T/m_R = 13.6$ in relation to an F distribution with 3 and 20 degrees of freedom: blood coagulation example.

shows the appropriate $F_{3,20}$ distribution in relation to the observed value of 13.6. When the treatment effects are *not* all equal, the mean value of F is equal to $(\sum \tau^2 + \sigma^2)/\sigma^2$. The numerator in the F ratio is a measure of signal plus noise and the denominator a measure of noise alone. Thus, like many other classical statistical criteria (such as the t statistic), the F statistic measures the signal-to-noise ratio familiar to engineers, and the significance test tells you if the apparent signal could or could not easily be explained by the noise.

The above additive model would not always provide a good approximation for the original data, but as you will see in Chapter 7, a transformation of the data can sometimes remedy this and also produce a more efficient analysis.

Graphical Checks

The assumptions (additivity, IID errors, normality, constant variance) sound formidable, but they are not so limiting as might be thought. The ANOVA is quite robust (insensitive) to moderate nonnormality and to moderate inequality of group variances. Unfortunately, as you saw in Chapter 2, much more serious is the assumption of *independence* between errors for an unrandomized design. You must *expect* that data collected in sequence will *not* be independent but be serially correlated. It is well known that serial correlation can lead to very serious errors if it is ignored (Box and Newbold, 1971). A further concern mentioned earlier is the possibility that there are "bad values" or "outliers" among the data due to copying errors or mismanagement of particular experimental runs. The fact that the t and F distributions may not be greatly affected by outliers is, in this context, almost a disadvantage since* the associated nonparametric randomization tests can produce reference distributions very closely approximating their parametric counterparts even when, as in Darwin's data in Chapter 3, there are pronounced outliers. Graphical inspection of the data is therefore of considerable importance.

Exercise 4.2. Perform a graphical ANOVA on the data of Exercise 4.1.

Outliers?

By plotting residuals, as was done at the bottom of Figure 4.2, it may be possible to detect the presence of serious outliers. If their cause can be determined, they may provide important and unanticipated information.

Serial Correlation

Randomization can nullify the potentially serious effect of autocorrelation.

Are the Variances the Same for Different Treatments?

Figure 4.5*a* shows plots of the residuals for all four diets separately. A plot of this kind is useful not only as a check on the assumption of variance homogeneity

* See the analysis of Darwin's data in Table 3.6.

but to enable you to see whether some diets might be associated with greater variability than others. For this example there seems to be little evidence for such differences.

Does the Spread of Residuals Increase as the Mean Increases?

In Figure 4.5*b* the residuals $y_{ti} - \overline{y}_t$ are plotted against treatment averages $\overline{y}_t$. A tendency for the spread of errors to increase as the averages increase points to a

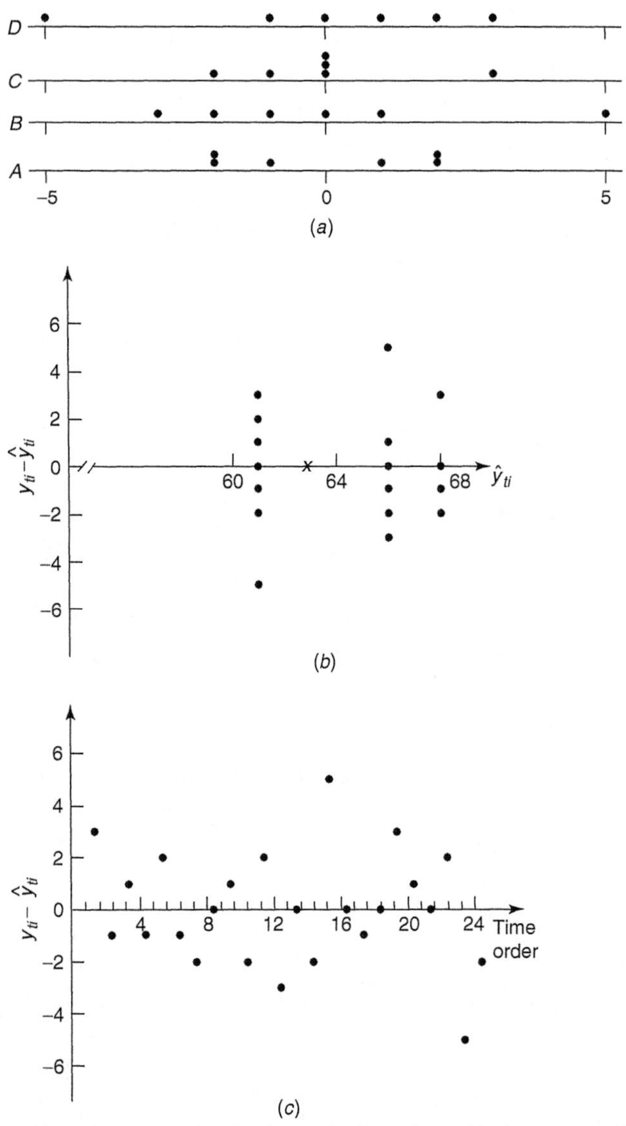

Figure 4.5. Dot diagrams: (*a*) residuals for each diet; (*b*) residuals versus estimated values; (*c*) residuals in time sequence.

possible need for data transformation. No such tendency appears here, but later you will see an example where this phenomenon does occur and data transformation has a profound influence on the conclusions.

Are Sizes of the Residuals Related to Their Time Order?

A plot of the residuals in time sequence like that in Figure 4.5*c* can detect a systematic drift occurring during the experiments. Because of randomization, such a drift will not invalidate your experiment. However, it might suggest your experimental procedure is sensitive to previously unsuspected environmental changes, for example, in the analytical laboratory. Correcting such a deficiency could produce a smaller variance in future experiments.

Exercise 4.3. The players in Exercise 4.1 were randomly assigned to the 21 time trials. The randomization sequence is given as a superscript attending each number of successes. (Thus, the first player to try out was the third member of team *A* who scored 17.) Comment.

A Conclusion Instead of an Argument—Pitfalls in Comparative Experiments

To better understand the rationale for randomization and other matters, it will help to dramatize things a bit. Suppose that the data in Table 4.1 and Figure 4.2 had come, not from a randomized animal experiment, but from an industrial trial on a pilot plant where the treatments *A*, *B*, *C*, and *D* were different process operating conditions with *A* the standard process. Suppose also that the data were measures of some criterion of efficiency that it is desired to increase. Further suppose that the arrangement of the experiment has been inadequately considered and in particular there had been no attempt to randomize.

The scene opens with seven people sitting around a table at a meeting to discuss the results. They are the plant manager, the process superintendent responsible for making the runs on the pilot plant, a design engineer who proposed modifications *B* and *C*, a chemical engineer who suggested modification *D*, a plant operator who took the samples of product for analysis, an analytical chemist who was responsible for the tests made on the samples, and a part-time data analyst who made the statistical calculations. After some preliminaries the dialogue might go something like this:

Plant manager (who would be happy if no changes were shown to be necessary)—I am not convinced that the modifications *B* and *C* are any better than the present plant process *A*. I accept that the differences are highly statistically significant and that, almost certainly, genuine differences did occur—but *I believe* the differences were not due to the process changes that we instituted. Have you considered *when* the runs were made? I find that all the runs with process *A* were made on a weekend and that the people responsible for operating the pilot plant at that time were new to the job. During the week, when modifications B, C, and D were made, I see that different operators were involved in making the runs.

Design engineer—There may have been some effects of that kind but I am *almost certain* they could not have produced differences as large as we see here.

Pilot plant superintendent—Also you should know that I went to some considerable trouble to supervise every one of these treatment runs. Although there were different operators, I'm *fairly sure* that correct operating procedures were used for all the runs. I am, however, *somewhat doubtful* as to the reliability of the method of the chemical testing which I understand has recently been changed. Furthermore *I believe* that not all the testing was done by the same person.

Analytical chemist—It is true that we recently switched to a new method of testing, but only after very careful calibration trials. Yes, the treatment samples came in at different times and consequently different people were responsible for the testing, but they are all excellent technicians and I am *fully confident* there could be no problem there. However, *I think* there is a question about the validity of the samples. As we know, getting a representative sample of this product is not easy.

Plant operator (sampler)—It used to be difficult to get a representative sample of the product, but you will remember that because of such difficulties a new set of stringent rules for taking samples was adopted some time ago. *I think* we can accept that during these trials these rules were exactly followed by the various operators who took the samples.

Chemical engineer (proposer of method D)—Before we go any further, are we sure that the statistical analysis is right? Does anyone here really *understand* the Analysis of Variance? Shouldn't the experiment have been randomized in some way?

Data analyst—I attended a special two-day short course on statistics and can assure the group that the correct *software program* was used for analyzing the data.

There were clearly many things to argue about and many uncertainties.* The plant manager commented "I believe," the design engineer was "almost certain," the plant superintendent was "somewhat doubtful," the analytical chemist "fully confident," and so on. Have you ever been so unlucky as to have to sit through a postmortem discussion like the above? The questions raised were about:

What was done?—operating procedures, sampling testing.

When was it done?—samples taken, samples tested.

Who and how many did it?—operators, samplers, testers, data analysts.

* There would be other questions that could have been raised but that no one had thought of at the time. Some of these might return to haunt the participants long after the investigation was over.

The points raised at the meeting all concerned matters that could cast doubt on any conclusions drawn. The way these questions were to be answered should have been settled before the experiment was performed. R. A. Fisher once said you cannot make an analysis of a poorly designed experiment—you can only carry out a postmortem to find out what it died of.

Preparation

The preparation for an investigation calls for much more than the choice of a statistical design. You must first consider the problems raised by such questions as:

Is the system of measurement and testing of sufficient accuracy and in proper control?

Is the system for sampling adequate?

Is it reasonably likely that all the factor combinations required by the proposed design can actually be run?

Do the operators and those responsible for sampling and testing really feel part of the team? Have they been involved in planning how the experimental runs can actually be made? Do we have their input?

Now that the plan has been finally agreed on, does everyone understand what they are supposed to do?

Have you tried to arrange, where possible, that the effects of known sources of inevitable variability are reduced by "blocking"? (See the boys' shoes example in the previous chapter and later examples.)

After you have done your best to deal with such problems, how can you protect the experiment from the many "lurking variables" of which you are currently unaware?

Fisher once said that designing an experiment was like playing a game of chance with the devil (aka Murphy). You cannot predict what ingenious schemes for invalidating your efforts he might produce. Think of a game of roulette in which you are the croupier. Gamblers can invent all sorts of systems that they imagine can beat the bank, but if the bank adopted any systematic strategy, as soon as this was suspected, the gambler **could** adopt a betting method to beat the bank.

Only a random strategy can defeat every betting system. Similarly, if experimental runs have been properly randomized, the known hazards and biases (and those not mentioned or even thought of) can be forced to occur randomly and so will not prejudice the conclusions.

Practical Considerations

In experimentation randomization of the *environment* in which each run is made is the objective. The features of the treatments themselves are not randomized away. For example, it may be that treatment B gives a much more variable result than some other process modification. However, this would be a characteristic of the treatment and not its environment so that this information would be preserved in a randomized experiment. In particular, graphical checks are not obscured by randomization.

Concerns were expressed at the meeting about such issues as the way in which sampling and testing of the product were carried out. These were important considerations. The fact that biases due to such factors can be made to act randomly does not mean that such issues can be ignored. Unless you can get these procedures under proper control, you will unnecessarily increase variation and make it more difficult to find the real treatment differences. You could produce a valid but very insensitive experiment. The study and improvement of sampling and testing methods are discussed in a special section of Chapter 9.

In animal experiments such as that set out in Table 4.1 it is easy to allocate animals randomly to different experimental conditions and run the experiments in random order. But in an industrial environment full-scale randomization would in most cases be difficult and in some impossible. Consequently a fully randomized arrangement is seldom used in industry because this is almost never the most *sensitive* arrangement or the easiest to carry out. Instead "randomized block" designs and "split-plot" designs, discussed later, would most often be used. Usually these designs are much easier to carry out and can provide more accurate results.

Extrapolation of Conclusions and Scaleup

In this pilot plant experiment one matter that was not mentioned at the meeting of the committee but in practice would almost certainly come up is the question of scaleup. Someone would have said, "Even if we accept that processes B and C are better on the *pilot plant*, it doesn't follow that they will be better on the full-scale plant." Scaleup necessarily calls on the subject matter expertise of engineers, chemists, and other technologists. Robustness studies discussed in Chapter 12 can help, but as Deming (1975) has pointed out, extrapolation of results from one environment to another must ultimately rest on a "leap of faith" based on subject matter knowledge. Good experiments can however make that leap less hazardous. (It is easier to leap over a canyon 2 feet across than one that is 20 feet across.) Usually the most relevant question is "Do we have enough evidence from these pilot runs to make it worthwhile to *try* the modified process on the full scale?" Frequently, small-scale experimentation can bring you fairly close to the best operating conditions. Evolutionary process operation run on the full scale during routine production can bring you even closer. That technique is discussed in Chapter 15.

4.2. RANDOMIZED BLOCK DESIGNS

The experimental arrangement just discussed is sometimes called a randomized *one-way classification.* By general randomization the effect of noise is homogenized between treatment and error comparisons and thus validates the experiment.

However, this one-way design is often not the most sensitive. When you know, or suspect you know, specific sources of undesirable change, you may be able to reduce or eliminate their effects by the use of what is called "blocking." This is a natural extension of the idea of the paired comparisons used in the boys' shoes example in the previous chapter. Randomized block designs use a more limited but equally effective randomization than that needed for the fully randomized design. It is also easier to do and can produce a more sensitive experiment.

Penicillin Yield Example

Table 4.4 shows data from a randomized block experiment in which a process of the manufacture of penicillin was investigated. Yield was the response of primary interest and the experimenters wanted to try four variants of the process, called treatments A, B, C, and D. Unfortunately, the properties of an important raw material (corn steep liquor) varied considerably, and it was believed that this alone might cause considerable differences in yield. It was found, however, that for experimental purposes a blend of the material could be obtained sufficient to make four runs. This supplied the opportunity of running the $k = 4$ treatments within each of $n = 5$ blends (blocks) of the liquor. In a fully randomized one-way treatment classification blend differences could have been randomized away but only at the expense of increasing the experimental noise and making the experiment more difficult to carry out. By randomly assigning the order in which the four treatments were run *within each blend* (block),* validity and simplicity were maintained while blend differences were largely eliminated.

A number of quantities useful for subsequent analysis are recorded in Table 4.4. These are the block (blend) averages, the treatment averages, the grand average, and the deviations of the block and treatment averages from the grand average. The superscripts in parentheses associated with the observations indicate the random order in which the experiments were run *within* each block (blend). To clarify

Table 4.4. Results from Randomized Block Design on Penicillin Manufacture

Block	Treatment A	B	C	D	Block Averages	Block Deviations
Blend 1	$89^{(1)}$	$88^{(3)}$	$97^{(2)}$	$94^{(4)}$	92	+6
Blend 2	$84^{(4)}$	$77^{(2)}$	$92^{(3)}$	$79^{(1)}$	83	−3
Blend 3	$81^{(2)}$	$87^{(1)}$	$87^{(4)}$	$85^{(3)}$	85	−1
Blend 4	$87^{(1)}$	$92^{(3)}$	$89^{(2)}$	$84^{(4)}$	88	+2
Blend 5	$79^{(3)}$	$81^{(4)}$	$80^{(1)}$	$88^{(2)}$	82	−4
Treatment averages	84	85	89	86	Grand average:	
Treatment deviations	−2	−1	+3	0	86	

* It is important to understand that in a randomized block experiment the treatments are randomized *within* the blocks.

Table 4.5. ANOVA Table: Penicillin Example

Source of Variation	Sum of Squares	Degrees of Freedom	Mean Square	F ratio
Between blocks (blends)	$S_B = 264$	$\nu_B = (n-1) = 4$	$m_B = 66.0$	$F_{4,12} = 3.51$
Between treatments	$S_T = 70$	$\nu_T = (k-1) = 3$	$m_T = 23.3$	$F_{3,12} = 1.24$
Residuals	$S_R = 226$	$\nu_R = (n-1)(k-1) = 12$	$m_R = 18.8$	
Deviations from grand average	$S_D = 560$	$nk - 1 = 19$		

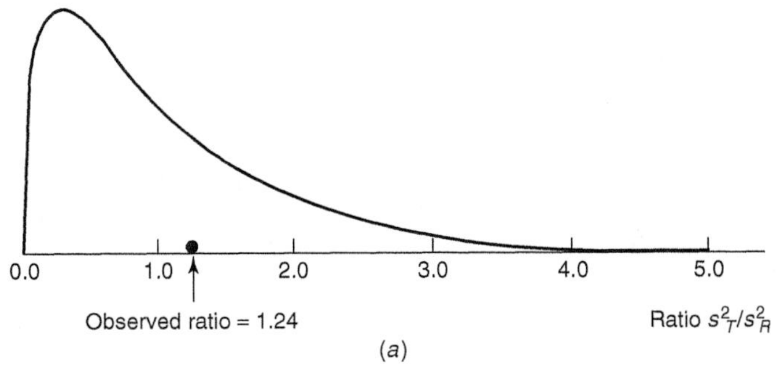

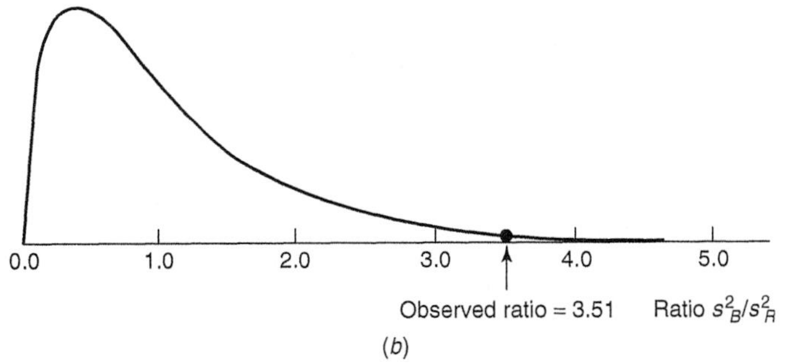

Figure 4.6. Plots of (*a*) $F_{3,12}$ and (*b*) $F_{4,12}$ distributions with observed F ratios.

issues, we have again simplified the data. Using these data, your computer software program should produce an ANOVA table that looks like Table 4.5.

If we suppose for the moment that the NIID assumptions are approximately valid, then the ratio of treatment to residual mean squares $F_{3,12} = 1.24$ yields a significance probability of only about 33%. There is thus no evidence for differences between treatments. However, for blocks, $F_{4,12} = 3.51$ yields a significance probability of about 4%, suggesting that blend differences do occur. To see what these F ratios mean, look at Figures 4.6*a,b*.

Table 4.6. Decomposition of Observations for Randomized Block Experiment

	Observations y_{bt}	Deviations from Grand Average $y_{bt} - \overline{y}$		Block Deviations $\overline{y}_b - \overline{y}$		Treatment Deviations $\overline{y}_t - \overline{y}$		Residuals $y_{bt} - \overline{y}_b - \overline{y}_t + \overline{y}$
	89 88 97 94	3 2 11 8		6 6 6 6		−2 −1 3 0		−1 −3 2 2
	84 77 92 79	−2 −9 6 −7		−3 −3 −3 −3		−2 −1 3 0		3 −5 6 −4
Analysis of observations	81 87 87 85	−5 1 1 −1	=	−1 −1 −1 −1	+	−2 −1 3 0	+	−2 3 −1 0
	87 92 89 84	1 6 3 −2		2 2 2 2		−2 −1 3 0		1 5 −2 −4
	79 81 80 88	−7 −5 −6 2		−4 −4 −4 −4		−2 −1 3 0		−1 0 −5 6
Vectors	**Y**	**D**	=	**B**	+	**T**	+	**R**
Sum of squares		$S_D = 560$	=	$S_B = 264$	+	$S_T = 70$	+	$S_R = 226$
Degrees of freedom		$\nu_D = 19$	=	$\nu_B = 4$	+	$\nu_T = 3$	+	$\nu_R = 12$

To better understand this ANOVA table, consider the decomposition of the data in Table 4.6, which shows the original data **Y**, the deviations **D** from the grand average of 86, the deviations **B** of the block averages from 86, the deviations **T** of the treatment averages from 86, and finally the residuals **R** that remain after subtracting the contribution **B** and **T** from **D**, that is, $\mathbf{R} = \mathbf{D} - \mathbf{B} - \mathbf{T}$. The vectors **B**, **T**, and **R** are mutually orthogonal, and again by an extension of the Pythagorean theorem, their sums of squares are additive, that is, $S_D = S_B + S_T + S_R$. Their degrees of freedom are also additive; $\nu_D = \nu_B + \nu_T + \nu_R$. See Figures 4.7*a,b*.

Increase in Efficiency by Elimination of Block Differences

The ANOVA table shows the advantage of using the randomized block arrangement. Of the total sum of squares not associated with treatments or with the

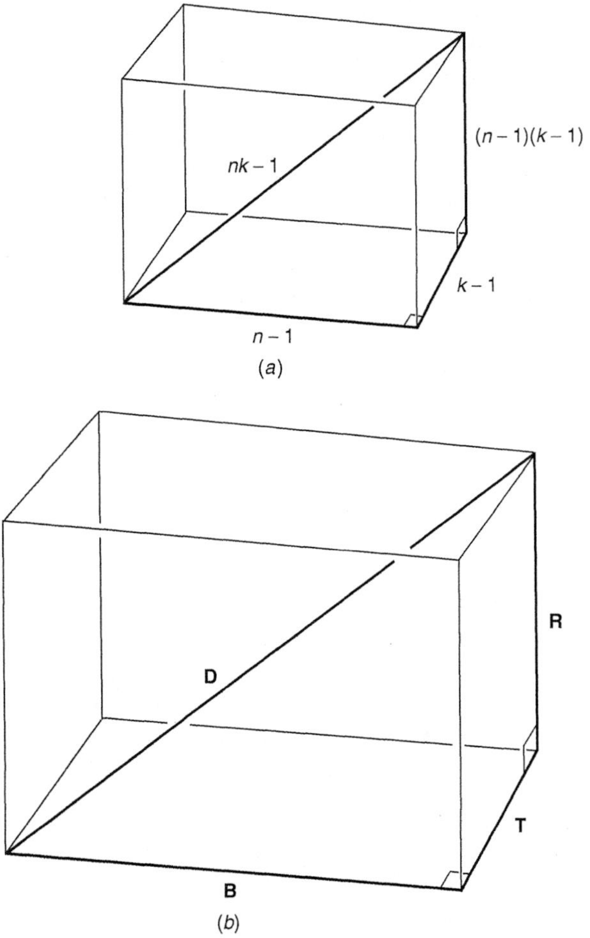

Figure 4.7. Vector decomposition for a randomized block design with $\mathbf{D} = \mathbf{B} + \mathbf{T} + \mathbf{R}$.

mean, almost half is accounted for by block-to-block variation. If the experiment had been arranged on a completely randomized basis with no blocks, the error variance would have been much larger. A random arrangement would have been equally *valid,* where validity implies that data can provide an estimate of the same error that affected the treatment differences. However, with the randomized block design these errors were considerably less. Notice that of the total of $S_D = 560$ a sum of squares $S_B = 264$ (which would otherwise have been ascribed to the error sum of squares) has been removed by blocks. The randomized block design greatly increased the sensitivity of this experiment and made it possible to detect smaller treatment differences had they been present than would otherwise have been possible.

Graphical ANOVA: Randomized Block Experiment

A graphical *analysis of variance* is shown in Figure 4.8 in which the scale factor for the block deviations is $\sqrt{\nu_R/\nu_B} = \sqrt{12/4} = \sqrt{3}$ and that for the treatment deviations is $\sqrt{\nu_R/\nu_T} = \sqrt{12/3} = 2$. The conclusions from this graphical ANOVA are much the same as those from the ANOVA table—that there is no evidence of treatment differences and that blocking has removed a substantial source of variation.

Once again, the graphical ANOVA brings to the attention of the experimenter how big in relation to noise the differences really are. Simple statements of significance levels can be very misleading. In particular, a highly significant result can of course arise from treatment differences that in the given context are too small to have any practical interest.

Exercise 4.4.

		Treatments			
		A	*B*	*C*	*D*
Blends	1	$34^{(1)}$	$22^{(4)}$	$25^{(2)}$	$27^{(3)}$
	2	$22^{(3)}$	$14^{(1)}$	$24^{(2)}$	$24^{(4)}$
	3	$23^{(2)}$	$19^{(3)}$	$14^{(1)}$	$16^{(4)}$
	4	$27^{(2)}$	$30^{(4)}$	$25^{(1)}$	$22^{(3)}$
	5	$28^{(3)}$	$18^{(4)}$	$22^{(2)}$	$20^{(1)}$
	6	$34^{(1)}$	$23^{(2)}$	$22^{(3)}$	$17^{(4)}$

A second series of experiments on penicillin manufacturing employed a randomized block design with four new penicillin treatments and six blends of corn steep liquor. Construct an ANOVA table by data decomposition and using a computer program. Comment.

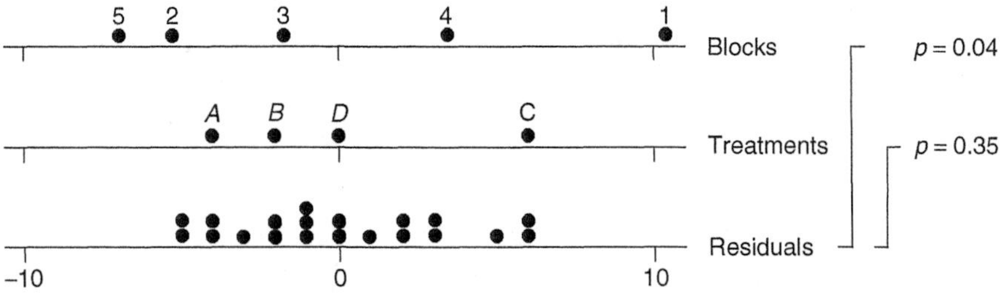

Figure 4.8. Graphical ANOVA for the randomized block experiment.

Implications of the Additive Model

The decomposition of the observations shown in Table 4.6, which leads to the ANOVA table and its graphical counterpart, is a purely algebraic process motivated by a model of the form

$$y_{ti} = \eta + \beta_i + \tau_t + \varepsilon_{ti}$$

Thus the underlying *expected* response model

$$\eta_{ti} = \eta + \beta_i + \tau_t$$

is called *additive* because, for example, if increment τ_3 provided an increase of six units in the response and if the influence of block β_4 increased the response by four units, the increase of both together would be assumed to be $6 + 4 = 10$ units in the response. Although this simple additive model would sometimes provide an adequate approximation, there are circumstances where it would not.

If the block and treatment effects were not additive, an *interaction* would be said to occur between blocks and treatments. Consider, for instance, the comparison of four catalysts A, B, C, and D with five blends of raw material represented by blocks. It could happen that a particular impurity occurring in blend 3 poisoned catalyst B and made it ineffective, even though the impurity did not affect the other catalysts. This would lead to a low response for the observation y_{23} where these two influences came together and would constitute an interaction between blends and catalyst.

Another way in which interactions can occur is when an additive model does apply, but not in the metric (scale, transformation) in which the data are originally measured. Suppose that in the original metric the response relationship was multiplicative, so that

$$\eta_{ti} = \eta\beta_i\tau_t$$

Then, if the response covered a wide range, nonadditivity (interaction) between block effects β_i and treatment effects τ_t would seriously invalidate any linear model you attempted to fit. However, by taking logs and denoting the terms in the transformed model by primes, the model for the observations becomes

$$y'_{ti} = \eta' + \beta'_i + \tau'_t + \varepsilon'_{ti}$$

and assuming the ε'_{ti} were approximately IID, the response $y' = \log y$ could be analyzed using a linear model in which the interactions would disappear.

Interactions may thus be thought of as belonging to two categories: *transformable interactions*, which may be eliminated by analyzing, some transformation such as the log, square root, or reciprocal of the original data, and *nontransformable interactions* such as a blend–catalyst interaction discussed above, which cannot be eliminated in this way.

Diagnostic Checks

All of the residuals and the residuals for each individual block and treatment are shown in Figures 4.9*a,b*. They do not suggest any abnormalities (e.g., differences in treatment variances or the occurrence of outliers or bad values).

Now consider Table 4.7, which displays the best estimates $\hat{y}_{ti}$ for the values in the individual cells of the original randomized block sometimes called

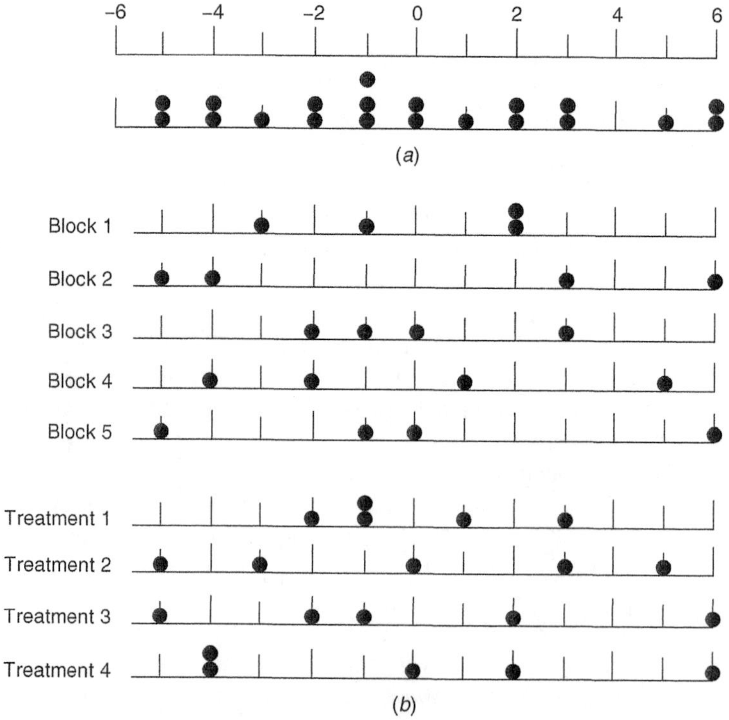

Figure 4.9. (*a*) Dot plots of residuals. (*b*) Residuals identified by block and treatment.

the predicted values. These can be obtained by subtracting the residuals from the original raw data; thus $\hat{y}_{ti} = y_{ti} - r_{ti}$, where the r_{ti} are the elements in **R** in Table 4.6. Figure 4.10 shows the residuals r_{ti} plotted against their predicted values $\hat{y}_{ti}$.

It will be remembered that one discrepancy to look for in such a plot is a funnel shape, suggesting an increase in the variance as the mean increases. This implies the need for data transformation to stabilize the variance. For a two-way analysis, such as that between blocks and treatment effects, a tendency of this plot to show curvature would also have suggested that the data did not support the use of the additive model (and that this might be corrected by data transformation). When the funnel effect and the curvature effect occur together, this produces a plot looking something like a hunting horn. Such a plot would increase suspicion that a data transformation was needed. No tendency of either kind is shown for these data.

Exercise 4.5. Do a graphical ANOVA for the data of Exercise 4.4.

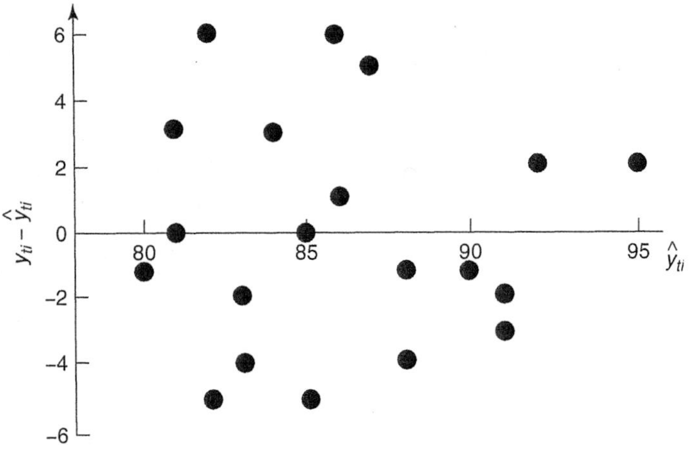

Figure 4.10. Residuals plotted against the predicted values: penicillin experiment.

Table 4.7. Table of Estimated Values $\hat{y}_{ti}$ Randomized Block Example (Penicillin Treatments)

		Treatment			
		A	*B*	*C*	*D*
Block	1	90	91	95	92
	2	81	82	86	83
	3	83	84	88	85
	4	86	87	91	88
	5	80	81	85	82

Exercise 4.6. Do a complete analysis of the predicted values and the residuals for the data given in Exercise 4.4.

Negative Findings

In this penicillin example the four treatments produced no detectable differences in yield. It should not be assumed a finding of this kind tells us nothing. Such a result gives rise to the question "If the treatments are not detectably different, which one is least costly or easiest to run?" If you can find answers to the questions "How much is an increase of one unit of yield worth?" and "How much (more/less) does each modification cost to run?" you can carry out an analysis on cost rather than yield to answer directly the question "Are the *costs* associated with the treatments A, B, C, D detectably different?"

The differences between the blocks (blends of corn steep liquor) could also be informative. In particular, you might speculate about the tantalizingly high average performance of blend 1. Why should that blend be so different in its influence on yield? Perhaps now the experimenters should study the characteristics of the different blends of corn steep liquor.

"As If" with Randomized Blocks

You have perhaps heard it said that experiments should never be run on a process or system that is not in a state of control where "a state of control" would mean that data from the process varied randomly about a fixed mean.* In his earliest thinking about the design of experiments in the 1920s, Fisher had to

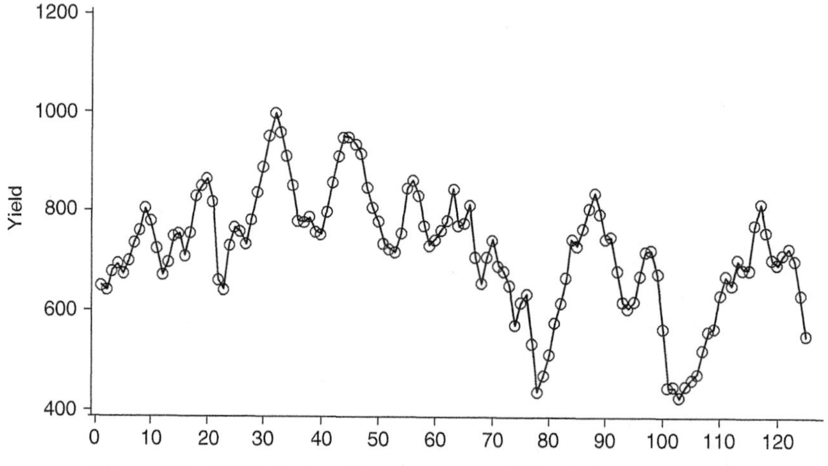

Figure 4.11. Yield of wheat from a sequence of identically treated plots.

*For this to be exactly true would abrogate the second law of thermodynamics and, as the distinguished scientist Sir Arthur Eddington (1935) said, "If your theory is found to be against the second law of thermo-dynamics I can offer you no hope." From an applied point of view, a study by Ryan (1989) found, in a survey of operating quality control systems, that none were in a state of control.

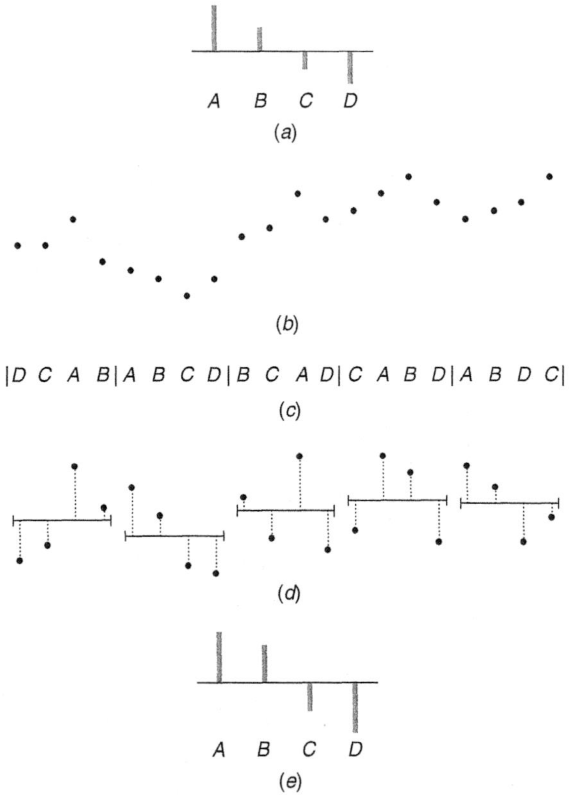

Figure 4.12. Randomized block analysis with nonstationary noise.

discover how to run experiments on processes and systems that were *never* in a state of statistical control. For example, look at Figure 4.11, which is a graph of successive yields of wheat identically treated (from a more extensive series of data due to Wiebe, 1935). You will agree that these data do not look much like the output from a process in a state of control. Fisher's solution to the quandary of how to run such experiments was the invention of randomized blocks.† He showed that it was possible to obtain results that to an adequate approximation could be analyzed "as if" the usual assumptions about IID errors were in fact true. To see how this works, look at Figure 4.12. For illustration suppose you want to compare experimentally four treatments (methods, processes, etc.) A, B, C, D in five replicates. Suppose also that unknown to you the effects, measured as deviations from their mean, are those shown in Figure 4.12a. Together they are the signal. Unfortunately, the system from which this signal is to be retrieved is not in a state of control. That is, the noise (the random variation) might look like that in Figure 4.12b. Ordinarily, the signal would be lost in this noise and

† He later introduced additional block designs such as Latin squares and incomplete blocks employing the same randomized block principle.

not recoverable. But suppose the four treatments are applied randomly in five blocks, as shown in Figure 4.12*c*. Adding the noise to this randomized signal you get Figure 4.12*d*, in which the filled dots are the data you would actually see. In the analysis of such data the variation in the five block averages, indicated by the horizontal lines in Figure 4.12*d*, are eliminated. The best estimate of the A effect would then be obtained by averaging the deviations identified with A, thus averaging the third deviation in block 1, with the first in block 2, the third in block 3, and so on. Repeating these calculations for treatments B, C, and D gives the deviations shown in Figure 4.12*e*, an excellent estimate of the signal. You will see that the process of analysis represented graphically here is precisely equivalent to that employed in the usual ANOVA.

Taking out block differences — a method for removing low frequency noise: One interesting way to think about the problem is to look at it as a communications engineer might. The engineer would most likely have considered the spectrum of the noise. In such a spectrum the time series of deviates are regarded as made up of an aggregate of sine and cosine waves of different amplitudes and frequencies. The variance in each small range of frequencies is called the "power." For the out-of-control series of Figure 4.12*b* most of the power would be at low frequencies. A familiar device applied in this area of expertise is what is called a "bandpass filter." A suitable filter can modify the spectrum by suppressing certain frequencies. In particular, a high-pass filter would allow the passage of high frequencies but reject or attenuate low frequencies. Fisher's blocking procedure is an example of a high-pass filter in which the elimination of the between-blocks component in the ANOVA corresponds to the removal of low-frequency power. The higher frequency randomized signal measuring the differences between the treatments A, B, C, and D can now be separated from the low-frequency noise.

4.3. A PRELIMINARY NOTE ON SPLIT-PLOT EXPERIMENTS AND THEIR RELATIONSHIP TO RANDOMIZED BLOCKS

Later (Chapter 9), after the discussion of factorial designs, a class of designs called *split-plot designs* will be introduced which are of great practical interest in industry. We here briefly look at their relation to randomized blocks.

The randomized block experiment supplies a way of eliminating a known source of variation — differences between blends of corn steep liquor were eliminated in the penicillin example as were differences between boys in the comparison of different types of materials for boys' shoes. The variation between blocks (blends or boys) will be different from and almost certainly larger than the variation within a block.

Now it is easy to imagine situations where additional process factors were deliberately introduced *between* the blocks themselves. For example, if you wanted to compare two types A and B of corn steep liquor, then some of the blends could be of type A and some of type B. Similarly, with boys' shoes you might want to compare the wear for five boys who walked to school with five boys who rode the bus. The blocks (blends, boys) could thus be split to accommodate additional treatments. In such a split-plot experiment you would need to estimate two different error variances, σ_b^2 say, applied to comparisons between blocks, and σ_e^2 (usually considerably smaller) for comparisons within blocks. In this book we will stay with the nomenclature used in agricultural experimentation where these designs were first introduced in which the blocks were called whole plots and the entities within blocks were called subplots. In agricultural field trials you could, for example, compare different depths of plowing on the whole plots (i.e., between blocks) and different varieties of corn on the subplots (i.e., within blocks). The thing to remember is that split-plot designs are like the randomized block design but with factors introduced between the blocks.

4.4. MORE THAN ONE BLOCKING COMPONENT: LATIN SQUARES

Sometimes there is more than one source of disturbance that can be eliminated by blocking. The following experiment was to test the feasibility of reducing air pollution by modifying a gasoline mixture with very small amounts of certain chemicals A, B, C, and D. These four treatments were tested with four different drivers and four different cars. There were thus two block factors—cars and drivers—and the *Latin square* design, shown in Table 4.8, was used to help eliminate from the treatment comparisons possible differences between the drivers, labeled I, II, III, and IV, and between the cars, labeled 1, 2, 3, and 4.

You will see that each treatment A, B, C, or D appears once in every row (driver) and once in every column (car). Adequate randomization can be achieved by randomly allocating the treatments to the symbols A, B, C, and D; the drivers to the symbols I, II, III, and IV; and the cars to the symbols 1, 2, 3, and 4.

You may ask why not standardize the conditions and make the 16 experimental runs with a single car and a single driver for the four different treatments.

Table 4.8. The 4 × 4 Latin Square: Automobile Emissions Data

		Cars 1	Cars 2	Cars 3	Cars 4	Averages Cars	Averages Drivers	Averages Additives
Drivers	I	*A* 19	*B* 24	*D* 23	*C* 26	1: 19	I: 23	*A*: 18
	II	*D* 23	*C* 24	*A* 19	*B* 30	2: 20	II: 24	*B*: 22
	III	*B* 15	*D* 14	*C* 15	*A* 16	3: 19	III: 15	*C*: 21
	IV	*C* 19	*A* 18	*B* 19	*D* 16	4: 22	IV: 18	*D*: 19

Grand average: 20

Table 4.9. Decomposition of the Latin Square: Automobile Emissions Example

Drivers	Observations, Cars 1	2	3	4		Deviations from grand average ($\bar{y} = 20$)		Columns (cars)		Rows (drivers)		Treatments (additives)		Residuals
I	A 19	B 24	C 23	D 26		−1 4 3 6		−1 0 −1 2		3 3 3 3		−2 2 1 −1		−1 −1 2 0
II	D 23	C 24	A 19	B 30	:	3 4 −1 10	+	−1 0 −1 2	+	4 4 4 4	+	−1 1 −2 2	+	1 −1 −2 2
III	B 15	D 14	C 15	A 16		−5 −6 −5 −4		−1 0 −1 2		−5 −5 −5 −5		2 −1 1 −2		−1 0 0 1
IV	C 19	A 18	B 19	D 16		−1 −2 −1 −4		−1 0 −1 2		−2 −2 −2 −2		1 −2 2 −1		1 2 0 −3
Vectors	**Y**				:	**V**	+	**C**	+	**D**	+	**T**	+	**R**
Sum of Squares					:	**312**	+	**24**	+	**216**	+	**40**	+	**32**
Degrees freedom					:	**15**	+	**3**	+	**3**	+	**3**	+	**6**

Such a design could also be statistically valid but the Latin square design has the advantage that, it provides a wider inductive basis for the conclusions drawn—any findings would not just apply to one car and one driver.

Table 4.9 shows the 16 elements of the vector **V** which are the deviations of the observations from the grand average $\overline{y} = 20$. The vector **V** is then partitioned into component vectors **C, D,** and **T**, which are respectively the deviations from the grand average of the averages for cars, drivers, and treatments and the vector of residuals $\mathbf{R} = \mathbf{V} - \mathbf{C} - \mathbf{D} - \mathbf{T}$. The additive ANOVA shown in Table 4.10 once again reflects the fact that the squared length of the vector **V** is equal to the sum of the squared lengths of the component vectors **C, D, T**, and **R**. By reasoning similar to that used for randomized blocks, the associated degrees of freedom are also additive. On NIID assumptions and the null hypothesis that there are no differences between treatments, the ratio of the mean squares for treatments, and residuals is distributed in an $F_{3,6}$ distribution. Inspection of the ANOVA table shows there is no convincing evidence for differences between the treatments but that the Latin square design has been effective in eliminating a large component of variation due to drivers.

The graphical analysis of variance shown in Figure 4.13 further illustrates these findings. Notice that it is assumed in all the above that the effects of

Table 4.10. Analysis of Variance: Latin Square Example

Source of Variation	Sum of Squares	Degrees of Freedom	Mean Square	Ratio of Mean Squares	Significance Probability P
Cars (columns)	$S_C = 24$	3	$m_C = 8.00$	$F_{3,6} = m_C/m_R = 1.5$	0.31
Drivers (rows)	$S_D = 216$	3	$m_D = 72.00$	$F_{3,6} = m_D/m_R = 13.5$	<0.01
Treatments (additives)	$S_T = 40$	3	$m_T = 13.33$	$F_{3,6} = m_T/m_R = 2.5$	0.16
Residuals	$S_R = 32$	6	$m_R = 5.33$		
Total	$S_V = 312$	15			

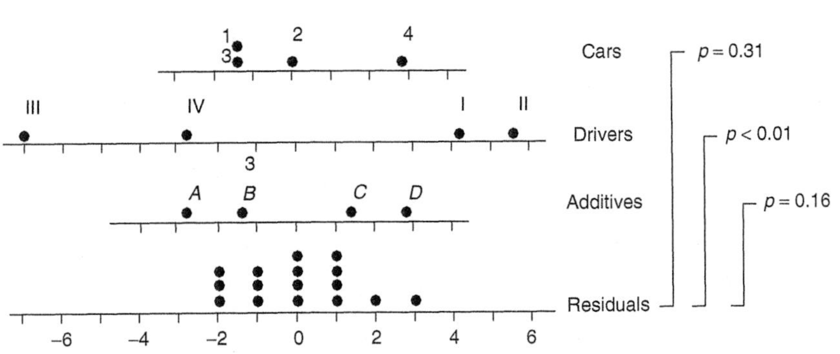

Figure 4.13. Graphical ANOVA for the Latin square example.

treatments, cars, and drivers are all additive so that there are no appreciable interaction effects. The only purpose of cars and drivers—the blocking factors—is to remove identifiable aspects of the *noise.*

For a small Latin square such as this it might be desirable to replicate the design for the purpose of confirmation and to increase the degrees of freedom for the residuals.

Exercise 4.7. Suppose the data in Table 4.8 are averages of two observations each and that the 32 observations displayed below were obtained in random order. Have your computer perform an appropriate ANOVA and make a graphical analysis.

6

		Cars 1	2	3	4
	I	*A* 20.6 21.4	*B* 25.0 27.0	*D* 18.8 19.2	*C* 26.3 25.7
Drivers	II	*D* 20.6 21.4	*C* 25.5 26.5	*A* 22.9 23.1	*B* 25.8 26.2
	III	*B* 17.6 16.4	*D* 14.3 13.7	*C* 14.8 15.2	*A* 13.5 14.5
	IV	*C* 17.3 16.7	*A* 13.8 14.2	*B* 18.2 19.8	*D* 22.3 21.7

The Misuse of Latin Square Designs

The Latin square design has frequently been used inappropriately to study process factors that can interact. In such applications effects of one factor can be inextricably mixed up with interactions of the others. Apparent outliers frequently occur as a result of these interactions. Suppose, for example, that the observation in the second column and third row in the above example was an outlier. This cell is identified with driver III, car 2, and treatment *D*. Such an interaction effect could occur, for example, if driver III was unfamiliar with car 2. But notice that this same effect could just as well be due to an interaction between driver III and treatment *D* or between car 2 and additive *D*. Such ambiguities could sometimes be resolved by adding a few additional runs, for example, by testing driver III with a different car using additive *D*. But when the interactions between factors are a likely possibility, you will need to use the factorial or fractional designs discussed later.

Exercise 4.8. Analyze the following duplicated 3×3 Latin Square design and comment. Can interactions account for these data?

		Columns 1	2	3
Rows	I	A 66 62	B 72 67	C 68 66
	II	B 78 81	C 80 81	A 66 69
	III	C 90 94	B 75 78	A 60 58

Graeco– and Hyper-Graeco–Latin Squares

Other interesting arrangements briefly introduced below that further exploit the idea of blocking are the Graeco–Latin square, balanced incomplete block, and Youden square designs.

A Graeco–Latin square is a $k \times k$ pattern that permits the study of k treatments simultaneously with three different blocking variables each at k levels. For example, the 4×4 Graeco–Latin square shown in Table 4.11 is an extension of the Latin square design used earlier but with one extra blocking variable added. This is labeled α, β, γ, δ and it could be used to eliminate possible differences between, say, four days on which the trials were run. It is constructed from the first two 4×4 Latin squares in Appendix 4B.

Exercise 4.9. Write a 3×3 and a 5×5 Graeco–Latin square.
Answer: See Appendix 4B.

This multiple blocking idea may be further extended using what are called hyper-Graeco–Latin squares.

A Hyper-Graeco–Latin Square Used in a Martindale Wear Tester

The Martindale wear tester is a machine used for testing the wearing quality of types of cloth or other such materials. Four pieces of cloth may be compared

Table 4.11. A 4 x 4 Graeco–Latin Square

		Car 1	2	3	4	
Driver	I	$A\alpha$	$B\beta$	$C\gamma$	$D\delta$	
	II	$B\delta$	$A\gamma$	$D\beta$	$C\alpha$	Additives: A, B, C, D
	III	$C\beta$	$D\alpha$	$A\delta$	$B\gamma$	Days: $\alpha, \beta, \gamma, \delta$
	IV	$D\gamma$	$C\delta$	$B\alpha$	$A\beta$	

simultaneously in one machine cycle. The response is the weight loss in tenths of a milligram suffered by the test piece when it is rubbed against a standard grade of emery paper for 1000 revolutions of the machine. Specimens of the four different *types of cloth* (treatments) A, B, C, D whose wearing qualities are to be compared are mounted in four specimen *holders* 1, 2, 3, 4. Each holder can be in any one of four positions P_1, P_2, P_3, P_4 on the machine. Each emery paper sheet α, β, γ, δ was cut into four quarters and each quarter used to complete a single cycle C_1, C_2, C_3, C_4 of 1000 revolutions. The object of the experiment was twofold: (1) to make a more accurate comparison of the treatments and (2) to discover how much of the total variability was contributed by the various factors—holders, positions, emery papers, and cycles.

The replicated hyper-Graeco–Latin square design employed is shown in Table 4.12. In the first square each of the treatments A, B, C, D occurs once in every cycle C_1, C_2, C_3, C_4 together with each of the four sheets of emery paper α, β, γ, δ and each of the four holders 1, 2, 3, 4 to produce a total of 16 observations. Since there are four versions of each of the five factors—cycles, treatments, holders, positions, and sheets of emery paper—in a single replication, $5 \times 3 = 15$ degrees of freedom are employed in their comparisons, leaving no residual degrees of freedom to provide an estimate of experimental error. For this reason the square was repeated* with four additional sheets of emery paper ε, ξ, θ, κ in four further runs. The ANOVA is given in Table 4.13 and the graphical analysis in Figure 4.14.

The design was effective both in removing sources of extraneous variation and in indicating their relative importance. Because of the elimination of these disturbances, the residual variance was reduced by a factor of about 8, and you could detect much smaller differences in treatments than would otherwise have been possible. Also notice that the graphical analysis points to position P_2 as giving much less wear than the others, a clue toward improvement that might merit further study.

The ratio of mean squares is $F = s_T^2/s_R^2 = 5.39$ with three and nine degrees of freedom. This is significant at about the 2% level. Thus, by using a design which makes it possible to remove the effects of many larger disturbing factors, differences between treatments were made detectable. Also the analysis identified the large contributions to the total variation due to cycles and to emery papers. This suggested improvements which later led to changes in the design of the machine.

4.5. BALANCED INCOMPLETE BLOCK DESIGNS

Suppose that the Martindale wear tester were of a different design which allowed only three, instead of four, samples to be included on each 1000 revolution cycle but that you had four treatments A, B, C, and D you wished to compare. You would then have $t = 4$ treatments but a block *size* of only $k = 3$—too small to

*A better plan might have been to rearrange randomly the design (while retaining its special properties) in the second square, but this was not done.

Table 4.12. Hyper-Graeco–Latin Square Replicated Twice: Martindale Wear Testing Example

Replicate I — Cycles: C_1, C_2, C_3, C_4; Treatments: A, B, C, D; Holders: 1, 2, 3, 4; Emory paper sheets: $\alpha, \beta, \gamma, \delta$

Cycles	Positions P_1	P_2	P_3	P_4
C_1	$\alpha A1$ 320	$\beta B2$ 297	$\gamma C3$ 299	$\delta D4$ 313
C_2	$\beta C4$ 266	$\alpha D3$ 227	$\delta A2$ 260	$\gamma B1$ 240
C_3	$\gamma D2$ 221	$\delta C1$ 240	$\alpha B4$ 267	$\beta A3$ 252
C_4	$\delta B3$ 301	$\gamma A4$ 238	$\beta D1$ 243	$\alpha C2$ 290

Replicate II — Cycles: C_5, C_6, C_7, C_8; Treatments: A, B, C, D; Holders: 1, 2, 3, 4; Emory paper sheets: $\varepsilon, \xi, \theta, \kappa$

Cycles	Positions P_1	P_2	P_3	P_4
C_5	$\varepsilon A1$ 285	$\xi B2$ 280	$\theta C3$ 331	$\kappa D4$ 311
C_6	$\xi C4$ 268	$\varepsilon D3$ 233	$\kappa A2$ 291	$\theta B1$ 280
C_7	$\theta D2$ 265	$\kappa C1$ 273	$\varepsilon B4$ 234	$\xi A3$ 243
C_8	$\kappa B3$ 306	$\theta A4$ 271	$\xi D1$ 270	$\varepsilon C2$ 272

Averages					
Treatments	Holders	Positions	Emery Papers	Cycles	Replicates
A: 270.0	1: 268.9	P_1: 279.0	α: 276.0	C_1: 307.3	I: 276.1
B: 275.6	2: 272.0	P_2: 257.4	β: 264.5	C_2: 248.3	II: 275.8
C: 279.9	3: 274.0	P_3: 274.4	γ: 249.5	C_3: 245.0	
D: 260.4	4: 271.0	P_4: 275.1	δ: 278.5	C_4: 268.0	
			ε: 256.0	C_5: 301.8	
			ξ: 265.3	C_6: 268.0	
			θ: 286.8	C_7: 253.8	
Grand average = 271.5			κ: 295.2	C_8: 279.8	

accommodate all the treatments simultaneously. Table 4.14A shows a *balanced incomplete block* design that you could use. The same design can alternatively be set out as in Table 4.14B. In general, such designs have the property that every

Table 4.13. ANOVA Table for Replicated 4 × 4 Hyper-Graeco–Latin Square: Martindale Wear Testing Example

Source	Degrees of Freedom	Sum of Squares	Mean Squares	Ratio of Mean Squares
$\sum(y-\bar{y})^2$	31	26,463.97		
Replications	1	603.78	$m_D = 603.78$	$m_D/m_R = 5.73$
Cycles	6	14,770.44	$m_C = 2{,}461.74$	$m_C/m_R = 23.35$
Positions	3	2,217.34	$m_P = 739.11$	$m_P/m_R = 7.01$
Emery papers	6	6,108.94	$m_E = 1{,}018.16$	$m_E/m_R = 9.66$
Holders	3	109.09	$m_H = 36.36$	$m_H/m_R = 0.34$
Treatments	3	1,705.34	$m_T = 568.45$	$m_T/m_R = 5.39$
Residuals	9	949.04	$m_R = 105.45$	

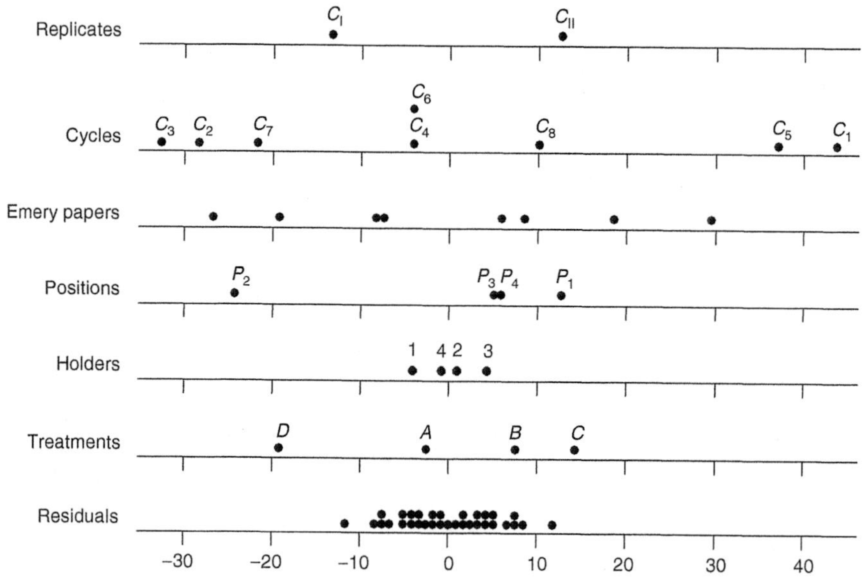

Figure 4.14. Graphical ANOVA for the Martindale wear example.

Table 4.14. A Balanced Incomplete Block Design, $t = 4$ Treatments in $b = 4$ Blocks of Size $k = 3$

A Block (cycle) of 1000 revolutions				
	1	A	B	C
	2	A	B	D
	3	A	C	D
	4	B	C	D

or

B Block (cycle) of 1000 revolutions		A	B	C	D
	1	x	x	x	
	2	x	x		x
	3	x		x	x
	4		x	x	x

Table 4.15. Youden Square, t = 7 Treatments, b = 7 Blocks, Block Size k = 4

		Treatments						
		A	*B*	*C*	*D*	*E*	*F*	*G*
	1		α 627		β 248		γ 563	δ 252
	2	α 344		β 233			δ 442	γ 226
Blocks	3			α 251	γ 211	δ 160		β 297
(cycles)	4	β 337	δ 537			γ 195		α 300
	5		γ 520	δ 278		β 199	α 595	
	6	γ 369			δ 196	α 185	β 606	
	7	δ 396	β 602	γ 240	α 273			

Treatments $t = 7$ Blocks $b = 7$ Block size $k = 4$

Within blocks, every pair of treatments appears twice.

pair of treatments occurs together in a block the same number of times. Thus, in the above design you will see that *A* occurs with *B* twice, with *C* twice, and with *D* twice and the same balance also occurs for *B*, *C*, and *D*. Comparisons between pairs of treatments are made against variability occurring within blocks. A very large number of these designs is available. See, for example, the classic text *Experimental Designs* (Cochran and Cox, 1957).

Youden Squares: A Second Wear Testing Example

A less trivial example of a balanced incomplete block design is shown in Table 4.15. The design is for comparing seven treatments in seven blocks of size 4 (ignore for the moment the Greek letters). The data shown within each of the seven blocks (cycles of 1000 revolutions) represent the weight loss in tenths of a milligram on the Martindale tester. It was necessary to test seven types of cloth *A, B, C, D, E, F*, and *G*, but only four test pieces could be compared simultaneously in a single machine cycle. Thus there is a fixed block size of $k = 4$.

In this particular balanced incomplete block design the number of blocks happens to equal the number of treatments. Because this is so, you have the opportunity to eliminate a second source of block variation. In the experiment set out in Table 4.15 the second source was used to eliminate possible differences in machine positions $\alpha, \beta, \gamma, \delta$. Each treatment was run in each of the four positions, and each of the four positions was represented in each cycle. A doubly balanced incomplete block design of this kind is called a *Youden square* after its inventor, W. J. Youden.

Principles for Conducting Valid and Efficient Experiments

Once more it is emphasized that in running complicated experimental designs of this type you should take special care to:

1. Make use of the specialist's knowledge and experience. Statistical techniques are an adjunct, not a replacement, for special subject matter expertise.
2. Involve the people responsible for operation, testing, and sampling.
3. Be sure that everyone knows what it is they are supposed to do and try to make certain that the experiments are run precisely as required.
4. Use blocking to remove known but uncontrolled sources of variation.
5. Use appropriate randomization so that the effect of noise on the treatment responses and on the residual errors is homogenized.
6. Provide suitable statistical analysis, both computational and graphical, which will make clear what has and has not been established by the experiment and thus help to decide how to proceed.

APPENDIX 4A. THE RATIONALE FOR THE GRAPHICAL ANOVA

In a standard ANOVA table sums of squares of *deviations* having specific numbers of degrees of freedom are compared. Consider, for example, a one-way classification of N observations made up of n data values for each of k treatments so that $nk = N$. Let S_A be the sum of squares of the k deviations of the treatment averages from their grand average. In the ANOVA table the sum of squares for treatments S_T (between treatments) is $n \times S_A$, where $n = N/k$ and has $\nu_T = k - 1$ degrees of freedom. The within-treatments (residual) sum of squares S_R is the sum of squares of the N deviations of the observations from their treatment averages with $\nu_R = k(n-1)$ degrees of freedom. A comparison of the variation between treatments and that within treatments is made by comparing the mean square $m_T = S_T/\nu_T$ with the mean square $m_R = S_R/\nu_R$. On NIID assumptions, if there are no differences between treatments, $E(m_T) = E(m_R)$ and the ratio $(S_T/\nu_T)/(S_R/\nu_R) = m_T/m_R$ is distributed in a F_{ν_T,ν_R} distribution.

Similarly, for other classifications like the randomized block and Latin square designs the mean square m_T of, say, k deviations of averages from the grand average having ν_T degrees of freedom is compared with the mean square m_R of the residual deviations having ν_R degrees of freedom.

Now what is required to make an analysis of the *dots* in the dot plots is to supply *visual comparison* of the k treatment deviations and the n residuals. This is done by comparing the "natural" variances of the dots: $M_A = S_A/k = S_T/N$ for treatments and $M_R = S_R/N$ for residuals. In these expressions the divisors are *not* the number of degrees of freedom but the number of squared deviations and if the null hypothesis is true the natural variance of the treatment dots will be the same as that for the residual dots. Thus the ratio

$$\frac{M_A}{M_R} = \frac{S_A/k}{S_R/N} = \frac{S_T}{S_R} = \frac{\nu_T m_T}{\nu_R m_R}$$

and hence

$$\frac{\nu_R}{\nu_T}\left(\frac{M_A}{M_R}\right) = \frac{m_T}{m_R}$$

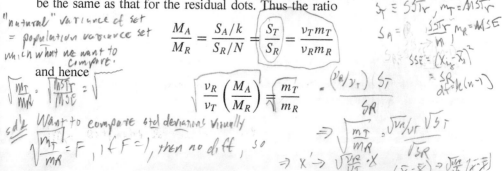

$$\frac{MS_{Tr}}{MSE} = \left(\frac{SS_{Tr}/\nu_T}{SSE/\nu_R}\right) = \frac{SS_{Tr}\,\nu_R}{\nu SE \cdot \nu_T} = SS_{Tr}(\nu_R/\nu_T) \quad \rightarrow \text{scaled deviations}$$

then

$(\text{deviations})^2 \rightarrow$ scaled by $\nu_R/24$

Thus a dot plot made by scaling the treatment deviations by the factor $\sqrt{\nu_R/\nu_T}$ permits visual comparisons with a dot plot of the residuals. And the ratio of the natural variances of these scaled dot distributions reproduces visually the standard F comparison in the ANOVA. It answers the question "Do the treatment deviations when appropriately scaled, look like part of the noise?"

Note: also $\sqrt{k}$ → so need to "enlarge" if treatments.

Unequal Grou

$$S_T = SS_{Tr} = \sum_{i=1}^{k} n_i(\bar{x}_i - \bar{\bar{x}})^2 = n\,\nu A$$
$$\nu_T = k - 1$$

For a one-way … $n_1, n_2, \ldots, n_k$
in the k treatm … usion that the
ith plotted trea $$S_A = \sum_{i=1}^{k}(\bar{x}_i - \bar{\bar{x}})^2$$ e a scale factor
tor $\sqrt{\nu_R n_i/\nu_T \bar{n}}$ … per treatment.
This correctly … $$S_R = \sum_{i=1}^{k}\sum_{j=1}^{n}(x_{ij} - \bar{x}_i)^2 = SSE$$ ments it is the
weighted devia … $\nu_R = k(n-1)$ applied to the
ith squared dev … deviation was
large but was b … $$M_T = MS_{Tr} = \frac{S_T}{\nu_T} = \frac{SS_{Tr}}{\nu_T}$$ less attention
than the same … ervations.

$$M_R = \frac{S_R}{\nu_R} = \frac{SSE}{\nu_R}$$

usually
$\nu_R > \nu_T$
$\nu_R = nk - k$
$\nu_R - \nu_T = nk - k$

APPENDIX 4 … –LATIN SQUARE, AN … NS

Before running … ze the design.
For example, ra … ally randomly
assign the treatments to the letters:

3 × 3:

A	B	C		A	B	C
B	C	A		C	A	B
C	A	B		B	C	A

To form the 3 × 3 Graeco–Latin square, superimpose the two designs using Greek letter equivalents for the second 3 × 3 Latin square; thus

Aα	Bβ	Cγ
Bγ	Cα	Aβ
Cβ	Aγ	Bα

4 × 4:

A	B	C	D		A	B	C	D		A	B	C	D
B	A	D	C		D	C	B	A		C	D	A	B
C	D	A	B		B	A	D	C		D	C	B	A
D	C	B	A		C	D	A	B		B	A	D	C

These three 4 × 4 Latin squares may be superimposed to form a hyper-Graeco–Latin square. Superimposing any pair gives a Graeco–Latin square:

5 × 5 :

A	*B*	*C*	*D*	*E*	*A*	*B*	*C*	*D*	*E*	*A*	*B*	*C*	*D*	*E*	*A*	*B*	*C*	*D*	*E*
B	*C*	*D*	*E*	*A*	*C*	*D*	*E*	*A*	*B*	*D*	*E*	*A*	*B*	*C*	*E*	*A*	*B*	*C*	*D*
C	*D*	*E*	*A*	*B*	*E*	*A*	*B*	*C*	*D*	*B*	*C*	*D*	*E*	*A*	*D*	*E*	*A*	*B*	*C*
D	*E*	*A*	*B*	*C*	*B*	*C*	*D*	*E*	*A*	*E*	*A*	*B*	*C*	*D*	*C*	*D*	*E*	*A*	*B*
E	*A*	*B*	*C*	*D*	*D*	*E*	*A*	*B*	*C*	*C*	*D*	*E*	*A*	*B*	*B*	*C*	*D*	*E*	*A*

These four 5×5 Latin squares may be superimposed to form a hyper-Graeco–Latin square. Also, superimposing any three gives a hyper-Graeco–Latin square design. Similarly, superimposing any pair gives a Graeco–Latin square.

For number of factors $k > 5$ and for a variety of balanced incomplete block designs and the details of their analysis, see Cochran and Cox (1957). Many design of experiments software programs provide large collections of Latin square; Youden square and balanced incomplete block designs.

REFERENCES AND FURTHER READING

Anscombe, F. J. (1973) Graphs in statistical analysis, *The American Statistician*, **27**, 17–21.

Anscombe, F. J., and Tukey, J. W. (1963) The examination and analysis of residuals, *Technometrics* **5**, 141–149.

Cox, D. R. (1958) *Planning for Experiments*, Wiley, New York.

Fisher, R. A. (1935) *The Design of Experiments*, Oliver and Boyd, London.

For the important work of Frank Yates, who made fundamental contributions to the design and analysis of experiments, see:

Pratt, A., and Tort, X. (1990) Case study: Experimental design in a pet food manufacturing company, *Quality Eng.*, **3**(1), 59–73.

Box, G. E. P., and Newbold, P. (1971) Some comments on a paper of Coen, Gomme and Kendall, *J. Roy. Stat. Soc. Series A*, **134**, 229–240.

Ryan, T. P. (1989) *Statistical Methods for Quality Improvement*, Wiley, New York.

Eddington, A. S. (1930) *The Nature of the Physical World*, The University Press, Cambridge, UK.

Wiebe, G. A. (1935) Variation and correlation in grain yield among 1,500 wheat nursery plots, *J. Agri. Res.*, **50**, 331–357.

Fisher, R. A. Personal Communication.

Cochran, W. G., and G. M. Cox (1957). *Experimental Designs*, Wiley, New York.

Shewhart, W. A. (1939) Statistical Method from the Viewpoint of Quality Control, The Graduate School, The Department of Agriculture, Washington.

Deming,W. E. (1975) On Probability as a basis for action, *J. Amer. Stat. Assoc.*, **10**, 146–152.

Yates, F. (1970) *Experimental Design: Selected Papers of Frank Yates, C. B. E., F. R. S.*, Hafner, New York.

Extensive lists of Latin square, Graeco–Latin square, balanced, and partially balanced incomplete block designs can be found in the following classical references. These lists are also available in many design-of-experiments software computer programs.

Bose, R. C., Clatworthy, W. H., and Shrikhande, S. S. (1954) *Tables of the Partially Balanced Designs with Two Associate Classes*, Tech. Bull. 107, North Carolina Agricultural Experiment Station, Raleigh, NC.

Cochran, W. G., and Cox, G. M. (1957) *Experimental Designs*, 2nd Edition Wiley, New York.

NIST/SEMATECH e-Handbook of Statistical Methods, http://www.itl.nist.gov/div898/handbook.

Snedecor, G. W., and Cochran, W. G. (1980) *Statistical Methods*, Iowa State University Press, Ames, IA.

Youden, W. J. (1937) Use of incomplete block replications in estimating tobacco-mosaic virus, *Contributions, Boyce Thompson Inst.*, **9**, 41–48.

Youden, W. J. (1940) Experimental designs to increase the accuracy of greenhouse studies, *Contributions, Boyce Thompson Inst.*, **11**, 219–288.

The influence of interactions in analyzing experimental designs is demonstrated in:

Box, G. E. P. (1990) Do interactions matter? *Quality Eng.*, **2**, 365–369.

Hunter, J. S. (1989) Let's all beware the Latin Square, *Quality Eng.*, **4** 453–466.

Hurley, P. (1994) Interactions: Ignore them at your own risk, *J. Quality Technol.*, **21**, 174–178.

QUESTIONS FOR CHAPTER 4

1. What is a randomized block design?

2. When is it appropriate to use a randomized block design?

3. Can you imagine a situation in which you might want to use a randomized block design but would be unable to do so?

4. What is the usual model for a two-way ANOVA of a randomized block design? What are its possible shortcomings? How can diagnostic checks be made to detect possible inadequacies in the model?

5. With data from a randomized block design, describe the analysis for question 4 using graphical ANOVA?

6. Treating the boys' shoe example as a randomized block design, what would be the ANOVA? Show its essential equivalence to the paired t test. Is every aspect of possible interest obtained from the ANOVA approach?

7. What precautions need to be considered when using a Latin square or Graeco–Latin square design?

8. Yates once said that a randomized block design may be analyzed "as if" standard assumptions were true. Explain.

PROBLEMS FOR CHAPTER 4

1. Paint used for marking lanes on highways must be very durable. In one trial paint from four different suppliers, labeled GS, FD, L, and ZK, were tested on six different highway sites, denoted 1, 2, 3, 4, 5, 6. After a considerable length of time, which included different levels of traffic and weather, the average wear for the samples at the six sites was as follows:

		Paint suppliers			
		GS	FD	L	ZK
	1	69	59	55	70
	2	83	65	65	75
Sites	3	74	64	59	74
	4	61	52	59	62
	5	78	71	67	74
	6	69	64	58	74

The objective was to compare the wear of the paints from the different suppliers.

(a) What kind of an experimental design is this?
(b) Make a graphical analysis and an ANOVA.
(c) Obtain confidence limits for the supplier averages.
(d) Make checks that might indicate departures from assumptions.
(e) Do you think these data contain bad values?
(f) What can you say about the relative resistance to wear of the four paints?
(g) Do you think this experimental arrangement was helpful?

2. Six burn treatments *A*, *B*, *C*, *D*, *E*, *F* were tested on six subjects (volunteers). Each subject has six sites on which a burn could be applied for testing (each arm with two below the elbow and one above). A standard burn was administered at each site and the six treatments were arranged so that each treatment occurred once with every subject once in every position. After treatment each burn was covered by a clean gauze; treatment *C* was a control with clean gauze but without other treatment. The data are the number of hours for a clearly defined degree of partial healing to occur.

		Subjects					
		1	2	3	4	5	6
	I	*A* 32	*B* 40	*C* 72	*D* 43	*E* 35	*F* 50
	II	*B* 29	*A* 37	*F* 59	*E* 53	*D* 32	*C* 53
Positions on arm	III	*C* 40	*D* 56	*A* 53	*B* 48	*F* 37	*E* 43
	IV	*D* 29	*F* 59	*E* 67	*A* 56	*C* 38	*B* 42
	V	*E* 28	*C* 50	*B* 100	*F* 46	*A* 29	*D* 56
	VI	*F* 37	*E* 42	*D* 67	*C* 50	*B* 33	*A* 48

(a) What is this design called? What characteristics does it have?

(b) How can such a design be randomized? Why?

(c) Make an ANOVA and a graphical ANOVA.

(d) State any assumptions you make.

(e) Make an appropriate plot and analysis of the residuals.

3. Three alternative regimes α, β, and γ involving combinations of certain exercises and drugs are being compared for their efficacy in the reduction of overweight in men. Fifteen volunteers were available for the trial. The trials were carried out by first dividing the subjects into "matched" groups; that is, men in any group were chosen to be as alike as possible. The loss of weight after 3 months for the three regimes was as follows:

		Regimes		
		α	β	γ
Groups	1	15	10	8
	2	24	15	17
	3	31	28	34
	4	37	36	34
	5	33	37	39

(a) Make any analysis you feel is appropriate, including a graphical analysis.

(b) Suppose you are told that the average weight in pounds at the beginning of the trial for members in each group is as follows:

Group	1	2	3	4	5
Weight	250	309	327	356	379

How might this affect your analysis and conclusions?

4. Analyze the data shown below obtained at the start of a process. It was known at the time that the process was very unstable. Nevertheless, it was important to compare four variations A, B, C, D of process conditions. The variants A, B, C, D were employed in 32 consecutive runs with results as follows:

Runs	1	2	3	4	5	6	7	8	9	10	11	12	13	14	15	16
Variant	C	B	D	A	B	D	A	C	D	A	B	C	A	D	C	B
Result	56	60	69	61	62	70	65	65	66	63	52	57	58	60	61	66

Runs	17	18	19	20	21	22	23	24	25	26	27	28	29	30	31	32
Variant	A	D	B	C	D	C	A	B	B	D	C	A	C	D	A	B
Result	56	61	53	52	62	57	59	58	60	68	61	65	63	68	61	55

(a) Plot the data. What kind of an experimental design is this?

(b) Make an ANOVA and a graphical ANOVA.

(c) Estimate the mean, with confidence interval, for the four possible process conditions.

(d) Plot the residuals in time order.

(e) Plot the eight averages of the sets of fours runs in time order and comment.

5. It has been said that you should not run experiments unless the system is in a state of statistical control. Do you believe the system described in problem 4 is in a state of control? Do you believe that you are able to make valid comparisons between treatments even though the process is *not* in a state of control? Give an estimate of the reduction in the length of the confidence intervals that were achieved by the design in problem 4 compared with a completely randomized arrangement.

CHAPTER 5

Factorial Designs at Two Levels

5.1. INTRODUCTION

To perform a factorial design, you select a fixed number of "levels" (or "versions") of each of a number of factors (variables) and then run experiments in all possible combinations. This chapter discusses such designs when there are just two levels for each factor.

The factors can be quantitative or qualitative. Thus the two levels of a quantitative variable could be two different temperatures or two different concentrations. For qualitative factors they might be two types of catalysts or the presence and absence of some entity.

Two-level factorial designs are of special importance because:

1. They require relatively few runs per factor studied.
2. The interpretation of the observations produced by the designs can proceed largely by using common sense, elementary arithmetic, and computer graphics.
3. When the factors are quantitative, although unable to fully explore a wide region in the factor space, they often determine a promising direction for further experimentation.
4. Designs can be suitably augmented when a more thorough local exploration is needed—a process we call *sequential assembly*.
5. They form the basis for two-level *fractional* factorial designs discussed in Chapter 6 in which only a carefully chosen part of the full factorial design is performed. As you will see, such fractional designs are particularly valuable for factor-screening purposes—that is, for deciding which of a larger number of factors are the important ones. They are also valuable building blocks in the sequential assembly of experimental designs.

Statistics for Experimenters, Second Edition. By G. E. P. Box, J. S. Hunter, and W. G. Hunter

6. Factorial designs and the corresponding fractional designs fit naturally into a sequential strategy, an essential feature of scientific method discussed throughout the book.

5.2. EXAMPLE 1: THE EFFECTS OF THREE FACTORS (VARIABLES) ON CLARITY OF FILM

The first example concerns an experiment to determine how the cloudiness of a floor wax was affected when certain changes were introduced into the formula for its preparation. The three factors studied were the amount of emulsifier **A**, the amount of emulsifier **B**, and the catalyst concentration **C**. Each of the three factors was studied at two levels so the arrangement is a $2 \times 2 \times 2$, or 2^3, factorial design.* The eight formulations are shown in the table on the left of Figure 5.1, with the factor levels coded by plus and minus signs. For example a run using the *higher* amount of emulsifier A, the *lower* amount of emulsifier B, and the *higher* catalyst concentration would be coded as $+ - +$ (run 6). The design is most easily written down as follows. In the first column single minus and plus signs are alternated. In the second, pairs of minus signs and plus signs, and so on. This produces a two-level factorial design in what is called *standard order.*† As is also shown in the figure, these eight-factor combinations can be conveniently represented geometrically by the vertices of a cube. If you imagine the center of the cube to be the origin of a three-dimensional coordinate system, then the

Formulation	Coded design A	B	C
1	−	−	−
2	+	−	−
3	−	+	−
4	+	+	−
5	−	−	+
6	+	−	+
7	−	+	+
8	+	+	+

(*a*)

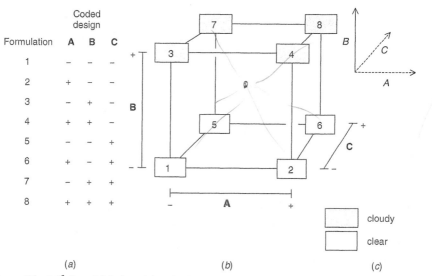

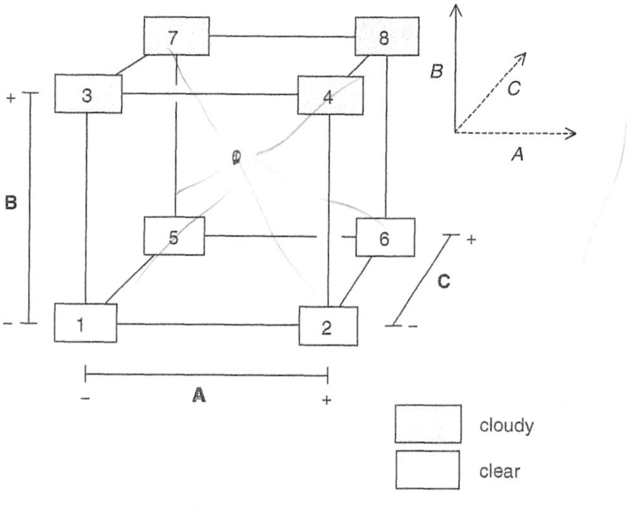

(*b*) (*c*)

Figure 5.1. A 2^3 factorial design: (*a*) coded in standard order; (*b*) used to study the effects of three factors on the clarity of film; (*c*) active and inert factors.

*The same two-level designs are sometimes called orthogonal arrays and the levels indicated by 0 and 1 or by 1 and 2 instead of by minus and plus signs. For our purposes the minus–plus notation is more useful.

†This sequence is also called "Yates order" in honor of Dr. Frank Yates.

eight-factor combinations can be identified by eight points whose coordinates are $(-1, -1, -1), (+1, -1, -1), \ldots, (+1, +1, +1)$ or, suppressing the 1's, by the coordinates $(-, -, -), (+, -, -), \ldots, (+, +, +)$ shown in Figure 5.1*a*.

To assess the results from this experiment, a portion of each formulation was spread thinly on a glass microscope slide and the clarity of the film observed. You will see that a striking visual display of the data can be obtained by placing the slides from the numbered formulations at the eight vertices of the cube, as in Figure 5.1*b*. As was at once evident, it was the presence of emulsifier **B** that produced cloudiness. This conclusion, though simply obtained, represented an important finding and illustrated the value of geometric displays.

One way to describe the results of this experiment would be to say that, in the three-dimensional (3D) design with factors **A**, **B**, and **C**, *activity* in the response "cloudiness" occurred in only a one-dimensional (1D) factor space—that of emulsifier **B**. The factors emulsifier **A** and concentration **C** were without detectable influence. Factors that behave in this way will be called *inert*.* The status of the three factors may be conveniently displayed using the diagram shown on the right of Figure 5.1, in which the one dimension of the active factor **B** is indicated as a solid arrow and the two dimensions of the inert factors **A** and **C** are indicated by dotted arrows.

5.3. EXAMPLE 2: THE EFFECTS OF THREE FACTORS ON THREE PHYSICAL PROPERTIES OF A POLYMER SOLUTION

In the previous example only one response—clarity of film—was of interest, but in the next there were three such responses. Eight liquid polymer formulations were prepared according to a 2^3 factorial design. The three factors studied were **1**, the amount of a reactive monomer; **2**, the type of chain length regulator; and **3**, the amount of chain length regulator, each tested at two levels. As before, eight runs were conveniently coded by plus and minus signs. The three responses observed were whether or not the resulting solution was milky (y_1), whether it was viscous (y_2), and whether it had a yellow color (y_3). The 2^3 factorial design and observed responses are displayed in Table 5.1.

One striking result was the regular alternation between milky and clear solutions. Thus, over the ranges explored the response y_1 was only dependent on factor **1**. Similar visual analysis showed the response y_2 was only dependent on factor **3**, but a change in the response y_3 occurred only if both factors **1** *and* **2** were at their plus levels; that is, a high level of reactive monomer **1** and a high level of chain length regulator **2** were needed. Thus, for this response there was an *interaction* between factors **1** and **2**. Notice that if the factors had been studied in customary "one-factor-at-a-time" fashion this last effect would have escaped detection. The dimensionality of the active factors is diagrammed in as before at the bottom of Figure 5.2.

For this 3D experiment, then, with only eight runs, the identity and dimension of the active subspaces for factors **1**, **2**, and **3** were 1D for milkiness, 1D for

* If the effects of factors could be determined with great accuracy few would be without *any* effect. In this book the word "inert" is applied to factors that are relatively inactive.

Table 5.1. A 2^3 Factorial Design with Three Responses: Polymer Solution Example

Factor Levels	−	+
1 amount of reactive monomer (%)	10	30
2 type of chain length regulator	*A*	*B*
3 amount of chain length regulator (%)	1	3

Formulation	**1**	**2**	**3**	y_1, Milky?	y_2, Viscous?	y_3, Yellow?
1	−	−	−	Yes	Yes	No
2	+	−	−	No	Yes	No
3	−	+	−	Yes	Yes	No
4	+	+	−	No	Yes	Slightly
5	−	−	+	Yes	No	No
6	+	−	+	No	No	No
7	−	+	+	Yes	No	No
8	+	+	+	No	No	Slightly

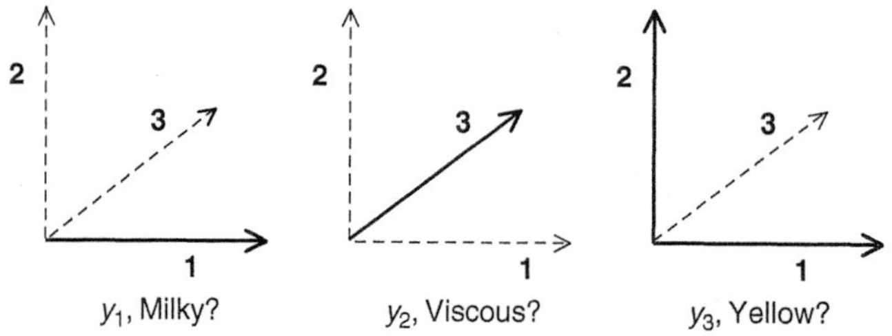

Figure 5.2. Dimensional activity of three factors affecting the three responses milkiness, viscosity, and yellowness.

viscosity, and 2D for yellowness. Now a product was required that was not milky, was of low viscosity, and had no yellowness. The experiment showed that it should be possible to obtain this using the higher level of reactive monomer, a chain length regulator of type **A** and the higher percentage of the chain length regulator as in formulation 6. Later experimentation confirmed this discovery.

Which Factors Do What to Which Responses?

In order to solve problems and make discoveries in science and engineering, it is very important for experimenters to determine *which* factors do *what* to *which* responses. Factorial experiments like the one above, and fractional designs to be discussed later, provide an economical way of finding this out. Notice too that, while such studies can be of great empirical importance, dimensional information of this kind can also suggest physical explanations for the effects, often providing valuable and unexpected directions for further inquiry.

As these two examples illustrate, it is not unusual for a well-designed experiment to almost analyze itself. In both the foregoing studies the responses were qualitative (e.g., cloudy–not cloudy), the effects of the factors could be simply described and experimental error could be largely ignored. Such simplicity is not always to be found. The next example concerns a pilot plant study where the response was quantitative and the effects of the factors not so obvious.

5.4. A 2^3 FACTORIAL DESIGN: PILOT PLANT INVESTIGATION

This experiment employed a 2^3 factorial experimental design with two quantitative factors—temperature **T** and concentration **C**—and a single qualitative factor—type of catalyst **K**. Each data value recorded is for the response yield y averaged over two duplicate runs.* Table 5.2 shows these data with the coded

Table 5.2. A 2^3 Factorial Design: Pilot Plant Investigation

	Temperature, **T** (°C)		Concentration, **C** (%)		Catalyst, **K**		
	−	+	−	+	−	+	
	160	180	20	40	*A*	*B*	

Coded Units of Factors: **T**	**C**	**K**	Average Yield y from Duplicate Runs
−	−	−	60
+	−	−	72
−	+	−	54
+	+	−	68
−	−	+	52
+	−	+	83
−	+	+	45
+	+	+	80

Run Number	Temperature, **T** (°C)	Concentration, **C** (%)	Catalyst, **K** (*A* or *B*)	Yield, **y** (%)
		Operational Levels of Factors		
1	160	20	*A*	60
2	180	20	*A*	72
3	160	40	*A*	54
4	180	40	*A*	68
5	160	20	*B*	52
6	180	20	*B*	83
7	160	40	*B*	45
8	180	40	*B*	80

*The data were from a real example that has, however, been considerably simplified to allow us to concentrate on important issues and avoid distracting fractions and decimals.

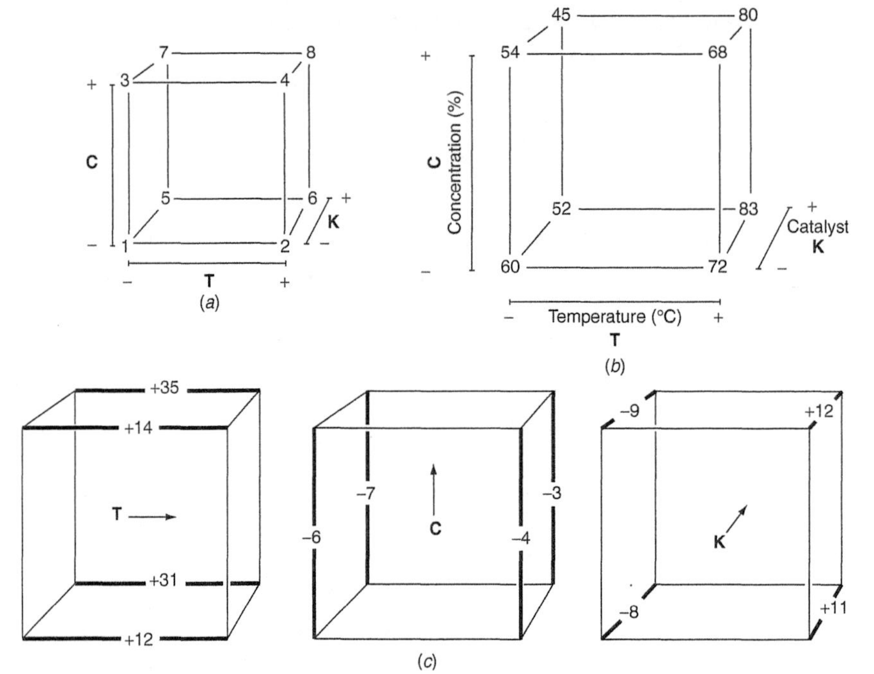

Figure 5.3. Display of the results from a pilot plant investigation employing a 2^3 factorial design to study the effects of **T** (temperature), **C** (concentration), and **K** (catalyst) on yield: (*a*) the 2^3 factorial with runs identified in standard order; (*b*) observed percent yields; (*c*) 12 treatment comparisons.

factor levels and the resulting operational design. (It does not matter which of the two catalysts is associated with the plus and which with the minus sign, so long as the labeling is consistent).

Figure 5.3*a* shows the numbered runs for the various combinations of factors **T**, **C**, and **K** at the corners of a cube. Figure 5.3*b* shows the yields obtained. Such a display is called a *cube plot.* Figure 5.3*c* shows how this eight-run design produces 12 comparisons—along the 12 edges of the cube: four measures of the effect of temperature change, four of the effect of concentration change, and four of the effects of catalyst change. Notice that in this factorial experiment on each edge of the cube only one factor is changed with the other two held constant. (The experimenter who believes that only one factor at a time should be varied is thus amply provided for.)

Most frequently, in factorial experiments factor activity is described in terms of a "main effect–interaction" model. Although this model is extremely valuable, you will see later that factor activity is not always best expressed this way.

5.5. CALCULATION OF MAIN EFFECTS

Averaging Individual Measures of Effects

Consider the results of the first two runs listed in Table 5.2. Aside from experimental error the yields (72 and 60) differ *only* because of temperature; the concentration (20%) and the type of catalyst (A) were identical in this comparison. Thus the difference 72 – 60 = 12 supplies one measure of the temperature effect with the remaining factors held fixed, as is illustrated in Figure 5.3*c*. You will see that there are four such measures of the temperature effect one for each of the four combinations of concentration and catalyst as listed below:

Constant Conditions of **C** and **K** within Which Temperature Comparisons Are Made		
Concentration, **C**	Catalyst, **K**	Effect of Changing Temperature from 160 to 180°C
20	A	$y_2 - y_1 = 72 - 60 = 12$
40	A	$y_4 - y_3 = 68 - 54 = 14$
20	B	$y_6 - y_5 = 83 - 52 = 31$
40	B	$y_8 - y_7 = 80 - 45 = 35$
		Main (average) effect of temperature, $T = 23$

The average of these four measures (+23 for this example) is called the *main effect* of temperature and is denoted by $T = 23$.

The general symmetry of the design (see Fig. 5.3*c*) ensures that there is a similar set of four measures for the effect of concentration **C**. In each of these the levels of the factors **T** and **K** are kept constant:

Constant Conditions of **T** and **K** within Which Concentration Comparisons Are Made		
Temperature, **T**	Catalyst, **K**	Effect of Changing Concentration from 20 to 40%
160	A	$y_3 - y_1 = 54 - 60 = -6$
180	A	$y_4 - y_2 = 68 - 72 = -4$
160	B	$y_7 - y_5 = 45 - 52 = -7$
180	B	$y_8 - y_6 = 80 - 83 = -3$
		Main (average) effect of concentration, $C = -5$

Similarly, there are four measures of the effect of catalyst:

Constant Conditions of **T** and **C** within Which Catalyst Comparisons Are Made		
Temperature, **T**	Concentration, **C**	Effect of Changing from Catalyst *A* to *B*
160	20	$y_5 - y_1 = 52 - 60 = -8$
180	20	$y_6 - y_2 = 83 - 72 = 11$
160	40	$y_7 - y_3 = 45 - 54 = -9$
180	40	$y_8 - y_4 = 80 - 68 = 12$
		Main (average) effect of catalyst, $K = 1.5$

Differences Between Two Averages

The main effects are the difference between two averages:

$$\text{Main effect} = \bar{y}_+ - \bar{y}_-$$

Thus, in Figure 5.4*a*, $\bar{y}_+$ is the average response on one face of the cube corresponding to the plus level of the factor and $\bar{y}_-$ is the average response on the opposite face for its minus level. Thus the main effects of temperature **T**, concentration **C**, and catalyst **K** can equally well be calculated as

$$T = \frac{72 + 68 + 83 + 80}{4} - \frac{60 + 54 + 52 + 45}{4} = 75.75 - 52.75 = 23$$

$$C = \frac{54 + 68 + 45 + 80}{4} - \frac{60 + 72 + 52 + 83}{4} = 61.75 - 66.75 = -5$$

$$K = \frac{52 + 83 + 45 + 80}{4} - \frac{60 + 72 + 54 + 68}{4} = 65.0 - 63.5 = 1.5$$

Notice that *all* the eight observations are being used to estimate *each* of the main effects and that each effect is determined with the precision of a *fourfold replicated difference*. To obtain equal precision, a one-factor-at-a-time series of experiments would have required eight runs *for each factor*.

Advantages of Factorial Experiments over the One-Factor-at-a-Time Method

The one-factor-at-a-time method of experimentation, in which factors are varied one at a time with the remaining factors held constant, was at one time regarded as the correct way to conduct experiments. But this method only provides an estimate of the effect of a single factor at *selected and fixed* conditions of the

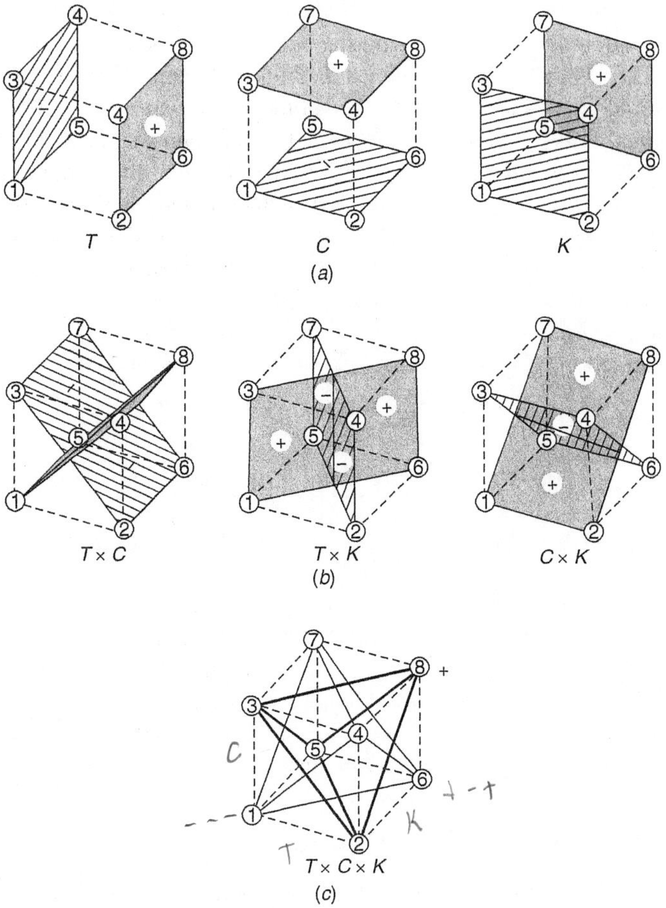

Figure 5.4. Geometric representation of contrasts corresponding to main effects and interactions: (*a*) main effects; (*b*) two-factor interactions; (*c*) three-factor interaction.

other factors. For such an estimate to have more *general relevance* it would be necessary to assume that the effect was the same at all the other settings of the remaining factors — that is, the factors would affect the response additively. But (1) if the factors *do* act additively, the factorial does the job with much more precision, and (2) as is illustrated below, if the factors do *not* act additively, the factorial, unlike the one-factor-at-a-time design, can detect and estimate interactions that *measure* this nonadditivity.

5.6. INTERACTION EFFECTS

Two-Factor Interactions

We have seen that the main effect of temperature $T = 23$ is an average of four individual comparisons. But this is obviously an incomplete description of the influence of temperature on yield. We can see, for example, in Figure 5.3*c* that the temperature effect is much greater on the back face of the cube than on the front face. That is, it is greater with catalyst *B* than with catalyst *A*. Thus the factors temperature and catalyst do not behave additively and are said to "interact"—to have a coupled influence upon the response beyond their main effects. A measure of this interaction is supplied by the difference between the temperature effects at the plus and minus levels of the catalyst factor. One-half of this difference is called the *temperature by catalyst interaction*, denoted by *TK*. Thus $TK = (33 - 13)/2 = 10$. The *TK* interaction may equally well be thought of as one-half the difference in the average *catalyst* effects at the two levels of *temperature* to give, once again, $TK = [11.5 - (-8.5)]/2 = 10$. Using Figure 5.4*b* to check how the interaction estimate *TK* uses the eight observations, you will find that

$$\begin{aligned} TK &= \frac{y_1 + y_3 + y_6 + y_8}{4} - \frac{y_2 + y_4 + y_5 + y_7}{4} \\ &= \frac{60 + 54 + 83 + 80}{4} - \frac{72 + 68 + 52 + 45}{4} \\ &= 69.25 - 59.25 = 10 \end{aligned}$$

Thus, like a main effect, an interaction effect is a difference between two averages, half of the eight results being included in one average and half in the other. Just as main effects may be viewed graphically as a *contrast* between observations on parallel faces of the cube, the *TK* interaction is a contrast between averages on two *diagonal planes*, as shown in Figure 5.4*b*. The *TC* and the *CK* interactions may be viewed in a similar way.

Three-Factor Interaction

Consider the temperature-by-concentration interaction. Two measures of this *TC* interaction are available from the experiment, one for each catalyst. When **K** is at its plus level (catalyst *B*),

$$\begin{aligned} \text{Interaction } TC &= \frac{(y_8 - y_7) - (y_6 - y_5)}{2} = \frac{(80 - 45) - (83 - 52)}{2} \\ &= \frac{35 - 31}{2} = 2 \end{aligned}$$

When **K** is at its minus level (catalyst *A*),

$$\begin{aligned} \text{Interaction } TC &= \frac{(y_4 - y_3) - (y_2 - y_1)}{2} = \frac{(68 - 54) - (72 - 60)}{2} \\ &= \frac{14 - 12}{2} = 1 \end{aligned}$$

The difference between these two measures the consistency of the *temperature-by-concentration interaction* for the two catalysts. Half this difference is defined as the *three-factor interaction* of temperature, concentration, and catalyst, denoted by *TCK*. Thus $TCK = (2 - 1)/2 = 0.5$.

As before, this interaction is symmetric in all the factors. For example, it could equally well have been defined as half the difference between the temperature-by-catalyst interactions at each of the two concentrations. The estimate of this three-factor interaction is again a difference between two averages. If the experimental points contributing to the two averages are isolated, they define the vertices of the two tetrahedra in Figure 5.4*c*, which together comprise the cube.

Again, each of the estimated factorial effects is seen to be a *contrast* between two averages of the form $\overline{y}_+ - \overline{y}_-$. It is helpful to remember that for any main effect or interaction the average $\overline{y}_+$ *always* contains the observation from the run in which all the factors are at their plus levels, for example, for a 2^3 factorial, the $+++$ run.

It would be extremely tedious if effects had to be calculated from first principles in this way. Fortunately, statistical software programs today quickly provide estimates of the effects for any 2^k factorial design and can also compute the standard error of effects.

5.7. GENUINE REPLICATE RUNS

Although there are usually much better ways to employ 16 experimental runs, each of the eight responses displayed in Table 5.2 was the average of the two genuinely replicated runs shown in Table 5.3. By *genuine* run replicates we mean that variation between runs made at the same experimental conditions is a reflection of the *total* run-to-run variability. This point requires careful consideration.

Table 5.3. Estimating the Variance: Pilot Plant Example

Average Response Value (Previously Used in Analysis)	T	C	K	Results from Individual Runs[a]		Difference of Duplicate	Estimated Variance at Each Set of Conditions, $s_i^2 = (\text{Difference})^2/2$
60	−	−	−	$59^{(6)}$	$61^{(13)}$	2	2
72	+	−	−	$74^{(2)}$	$70^{(4)}$	4	8
54	−	+	−	$50^{(1)}$	$58^{(16)}$	8	32
68	+	+	−	$69^{(5)}$	$67^{(10)}$	2	2
52	−	−	+	$50^{(8)}$	$54^{(12)}$	4	8
83	+	−	+	$81^{(9)}$	$85^{(14)}$	4	8
45	−	+	+	$46^{(3)}$	$44^{(11)}$	2	2
80	+	+	+	$79^{(7)}$	$81^{(15)}$	2	2
						Total	64

s^2 = pooled estimate of σ^2 = average of estimated variances
$= \frac{64}{8} = 8$ with $\nu = 8$ degrees of freedom

[a] Superscripts give the order in which the runs were made.

Randomization of run order for the all 16 runs, as indicated in Table 5.3, ensures that the replication is genuine. But replication is not always easy. For this pilot plant experiment a run involved (1) cleaning the reactor, (2) inserting the appropriate catalyst charge, (3) running the apparatus at a given temperature at a given feed concentration for 3 hours to allow the process to settle down at the chosen experimental conditions, and (4) sampling the output every 15 minutes during the final hours of running. A genuine run replicate involved taking all these steps *all over again*. In particular, replication of only the chemical *analysis of a sample from a single run* would provide only an estimate of *analytical variance*. This is usually only a small part of the *experimental run-to-run variance*. Similarly, the replication of samples from the same run could provide only an estimate of sampling plus analytical variance. To obtain an estimate of error that can be used to estimate the standard error of a particular effect, each experimental run must be *genuinely* replicated. The problem of wrongly assessing *experimental* error variance can be troublesome and is discussed more fully in Chapter 12.

Economy in Experimentation

Usually there are better ways to employ 16 independent runs than by fully replicating a 2^3 factorial design as was done here. You will see later how four or five factors (and in some cases more than five) can be studied with a *16-run* two-level design. Also, a means for assessing the significance of effects that does not require replicate runs will later be discussed. For the moment, however, consider the analysis of the replicated data.

Estimate of the Error Variance and Standard Errors of the Effects from Replicated Runs

Look at the pairs of observations recorded at the eight different factor combinations in Table 5.3. From any individual difference d between duplicated runs an estimate of variance with one degree of freedom is $s^2 = d^2/2$. The average of these single-degree-of-freedom estimates yields a pooled estimate $s^2 = 8.0$ with eight degrees of freedom, as shown in Table 5.3. Because each estimated effect $T, C, K, TC, \ldots$ is a difference between two averages of eight observations, the variance of an effect is given as

$$V(\text{effect}) = \left(\frac{1}{8} + \frac{1}{8}\right) s^2 = \frac{8.0}{4} = 2$$

Its square root is the standard error of an effect, SE(effect) = 1.4.

In general, if each factor combination was replicated, a pooled estimate of the experimental run variance from g factor combinations would be

$$s^2 = \frac{\nu_1 s_1^2 + \nu_2 s_2^2 + \cdots + \nu_g s_g^2}{\nu_1 + \nu_2 + \cdots + \nu_g}$$

where $\nu = \nu_1 + \nu_2 + \cdots + \nu_g$ is the number of degrees of freedom of the estimate. In the special case of g duplicate combinations this formula reduces to $s^2 = \sum d^2/2g$ with g degrees of freedom. Thus in this example $s^2 = 128/2(8) = 8.0$ with $\nu = 8$ degrees of freedom.

5.8. INTERPRETATION OF RESULTS

All estimated effects for the pilot plant data along with their standard errors are shown in Table 5.4. It is important to have some method for determining which effects are almost certainly real and which might readily be explained by chance variation. A rough rule is that effects greater than 2 or 3 times their standard error are not easily explained by chance alone. More precisely, on NIID assumptions, each ratio, effect/SE(effect), will be distributed in a t distribution with $\nu = 8$ degrees of freedom. A "significant" value of t at the 5% level is 2.3, that is, $\Pr(|t_8| > 2.3) = 0.05$; thus the 95% confidence interval for an effect in Table 5.4 would be given by the estimated effect $\pm 2.3 \times 1.4$ (i.e., ± 3.2). We prefer to provide the estimated effect and its standard error as is done for example in Table 5.4 and leave the choice of the percentage level of any confidence interval to the judgment of the reader. In Table 5.4 effects almost certainly not due to noise are shown in bold type.

Interpretation of the Table of Estimated Effects

The main effect of a factor should be *individually* interpreted only if there is no evidence that the factor interacts with other factors. When there is evidence of one or more such interactions, the interacting variables must be considered jointly. In Table 5.4 you see that there is a large temperature main effect, 23.0 ± 1.4, but no statement about the effect of temperature alone should be made because temperature interacts with catalyst type (the TK interaction is 10.0 ± 1.4). In the case of concentration, however, the main effect is -5.0 ± 1.4, and there is no

Table 5.4. Calculated Effects and Standard Errors for 2^3 Factorial: Pilot Plant Example

Average	Effect with Standard Error
Main effects	
Temperature, T	**23.0 ± 1.4**
Concentration, C	**-5.0 ± 1.4**
Catalyst, K	1.5 ± 1.4
Two-factor interactions	
$T \times C$	1.5 ± 1.4
$T \times K$	**10.0 ± 1.4**
$C \times K$	0.0 ± 1.4
Three-factor interaction	
$T \times C \times K$	0.5 ± 1.4

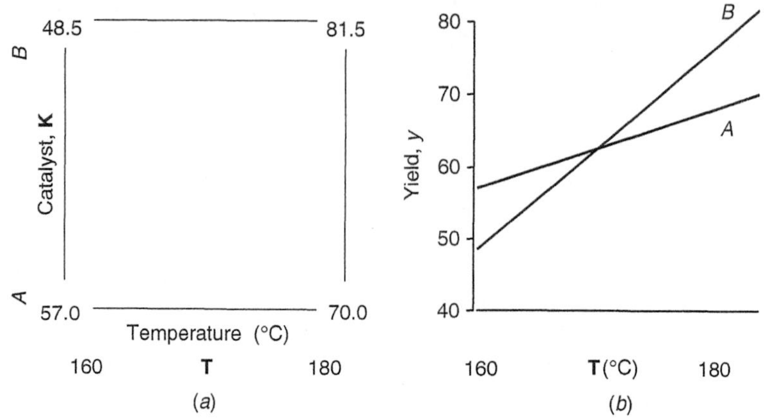

Figure 5.5. Temperature–catalyst (*TK*) interaction displayed by (*a*) a two-way diagram and (*b*) comparing the change in yield produced by temperature for each of the catalysts *A* and *B*.

evidence of any interactions involving concentration. Thus the following tentative conclusions may be drawn:

1. The effect of changing the concentration (**C**) over the ranges studied is to reduce the yield by about five units, and this is approximately so irrespective of the tested levels of the other variables.
2. The effects of temperature (**T**) and catalyst (**K**) cannot be interpreted separately because of the large *TK* interaction. With catalyst *A* the temperature effect is 13 units, but with catalyst *B* it is 33 units. This interaction may be understood by studying the two-way table in Figure 5.5*a*. Equivalently, Figure 5.5*b* shows how the interaction arises from a difference in sensitivity to temperature for the two catalysts.

A result of great practical interest in this experiment was the very different behaviors of the two "catalyst types" in response to temperature. The effect was unexpected, for although obtained from two different suppliers, the catalysts were supposedly identical. Also, the yield from catalyst *B* at 180°C was the highest that had been seen up to that time. This finding led to a careful study of catalysts and catalyst suppliers in a later investigation.

5.9. THE TABLE OF CONTRASTS

To better understand the 2^3 design, look at the table of signs in Table 5.5. The table begins with a column of plus signs followed by three columns labeled **T**, **C**, and **K** that define the *design matrix* of the 2^3 factorial design. This is followed by four additional columns identified as **TC**, **TK**, **CK**, and **TCK**. The yield averages, which are treated as single observations, are shown in the last column of the table.

The first column of plus signs is used to obtain the overall average; we add the observations and divide by 8. The remaining seven columns define a set of

Table 5.5. Table of Contrast Coefficients for 2^3 Factorial Design: Pilot Plant Example

	Mean	T	C	K	TC	TK	CK	TCK	Yield Average
	+	−	−	−	+	+	+	−	60
	+	+	−	−	−	−	+	+	72
	+	−	+	−	−	+	−	+	54
	+	+	+	−	+	−	−	−	68
	+	−	−	+	+	−	−	+	52
	+	+	−	+	−	+	−	−	83
	+	−	+	+	−	−	+	−	45
	+	+	+	+	+	+	+	+	80
Divisor	8	4	4	4	4	4	4	4	

seven contrasts indicating which observations go into $\overline{y}_+$ and which into $\overline{y}_-$ for each effect. Thus, for example, the **T** main effect can be calculated from the column of signs $(- + - + - + -+)$ to give $T = (-60 + 72 - 54 + 68 - 52 + 83 - 45 + 80)/4 = 23.0$ as before. The divisor, 4, transforms the contrast into the difference between the two averages $(\overline{y}_+ - \overline{y}_-)$. The **C** and **K** main effects can be calculated in the same way from the next two columns.

It is a remarkable fact that the columns of signs for the interaction contrasts can be obtained directly by multiplying the signs of their respective factors. Thus, the column of signs for the **TK** interaction is obtained by multiplying together, row by row, the signs for **T** with those for **K**. The estimate of the **TK** interaction effect is therefore $TK = (+60 - 72 + 54 - 68 - 52 + 83 - 45 + 80)/4 = 10.0$. In a similar way, the array of signs for the **TCK** interaction effect requires multiplying together the signs in the columns for **T**, **C**, and **K**. Thus the estimate of the three-factor interaction $TCK = (-60 + 72 + 54 - 68 + 52 - 83 - 45 + 80)/4 = 0.5$. The table of signs for all main-effects and interaction effects may be obtained similarly for any 2^k factorial design. (See Table 5.10b later for the table of contrasts for a 2^4 factorial.) Each contrast column is perfectly balanced with respect to all the others. To see this, choose any one of the contrast columns in Table 5.5 and look at the group of plus signs in that column. Opposite the four plus signs there are two plus signs and two minus signs in *every* one of the other six contrast columns; the same is true for the corresponding group of four minus signs. This remarkable property of perfect balance is called *orthogonality*. Orthogonality* ensures that each estimated effect is unaffected by the magnitudes and signs of the others.

Exercise 5.1. Add any constant to the observations associated with the + levels in any contrast column of Table 5.5 and verify that only that single contrast

* If the eight entries in any column of the table are regarded as defining the signs of the elements ± 1 of a vector, it will be seen that the inner product of any pair of such vectors is zero.

effect will be changed. Similarly, confirm that if a constant is added to all the observations only the average will be influenced.

5.10. MISUSE OF THE ANOVA FOR 2^k FACTORIAL EXPERIMENTS

Many computer programs employ the ANOVA for the analysis of the 2^k designs (and for their fractions, discussed in Chapter 6). However, for the analysis of these particular designs the use of the ANOVA is confusing and makes little sense. Consider the analysis of a two-level design with k factors in n replications and hence $N = n \times 2^k$ observations. In the ANOVA approach the $2^k - 1$ degrees of freedom for *treatments* are partitioned into individual *"sums of squares"* for effects, equal to $N(\text{effect})^2/4$, each of which is divided by its single degree of freedom to provide a mean square. Thus, instead of seeing the effect $(\bar{y}_+ - \bar{y}_-)$ itself, the experimenter is presented with $N(\text{effect})^2/4$, which must then be referred to an F table with one degree of freedom. There is nothing to justify this complexity other than a misplaced belief in the universal value of an ANOVA table. What the experimenter wants to know is the magnitude of the effect $(\bar{y}_+ - \bar{y}_-)$, its standard error, and the ratio $t = \text{effect/SE(effect)}$. Also, since $\sqrt{F_{1,\nu}} = t_\nu$, referring $F_{1,\nu} = [N(\text{effect})^2/4]/s^2$ to an F table is exactly equivalent to referring $(\bar{y}_+ - \bar{y}_-)/2s/\sqrt{N}$ (the effect divided by its standard error) to a t table.

Most computer programs compute p values associated with the null hypothesis at the 5 and 1% level of probability. These p values should not be used mechanically for deciding on a yes-or-no basis what effects are real and to be acted upon and what effects should be ignored. In the proper use of statistical methods *information about the size of an effect and its possible error* must be allowed to interact with the experimenter's subject matter knowledge. (If there were a probability of $p = 0.50$ of finding a crock of gold behind the next tree, wouldn't you go and look?) Graphical methods, soon to be discussed, provide a valuable means of allowing information in the data and in the mind of the experimenter to interact appropriately.

Exercise 5.2. Calculate the main effects and interactions for the yield and taste of popcorn recorded below:

Test Number	Brand of Popcorn, **1**	Ratio of Popcorn/Oil, **2**	Batch Size Cup, **3**	Yield of Popcorn, y_1	Taste Scale 1–10, y_2
1	−(ordinary)	−(low)	$-(\frac{1}{3})$	$6\frac{1}{4}$	6
2	+(gourmet)	−(low)	$-(\frac{1}{3})$	8	7
3	−(ordinary)	+(high)	$-(\frac{1}{3})$	6	10
4	+(gourmet)	+(high)	$-(\frac{1}{3})$	$9\frac{1}{2}$	9
5	−(ordinary)	−(low)	$+(\frac{2}{3})$	8	6
6	+(gourmet)	−(low)	$+(\frac{2}{3})$	15	6
7	−(ordinary)	+(high)	$+(\frac{2}{3})$	9	9
8	+(gourmet)	+(high)	$+(\frac{2}{3})$	17	2

Partial Answer: Yield: $1 = 5.1$, $2 = 1.1$, $3 = 4.8$, $1 \times 2 = 0.7$, $1 \times 3 = 2.4$, $2 \times 3 = 0.4$, $1 \times 2 \times 3 = -0.2$.

What are the important factors for yield and for taste?

What are the dimensions of the active factor subspaces for yield and taste?

Exercise 5.3. For the following data calculate the main effects and interactions and their standard errors. The 24 trials were run in random order.

Test Condition Number	Depth of Planting (in.), **1**	Watering (Times Daily), **2**	Type of Lima Bean, **3**	Yield: Replication 1	Replication 2	Replication 3
1	$-(\frac{1}{2})$	− (once)	− (baby)	6	7	6
2	$+(\frac{3}{2})$	− (once)	− (baby)	4	5	5
3	$-(\frac{1}{2})$	+ (twice)	− (baby)	10	9	8
4	$+(\frac{3}{2})$	+ (twice)	− (baby)	7	7	6
5	$-(\frac{1}{2})$	− (once)	+ (large)	4	5	4
6	$+(\frac{3}{2})$	− (once)	+ (large)	3	3	1
7	$-(\frac{1}{2})$	+ (twice)	+ (large)	8	7	7
8	$+(\frac{3}{2})$	+ (twice)	+ (large)	5	5	4

Answer: $1 = -2.2, 2 = 2.5, 3 = -2.0, 1 \times 2 = -0.3, 1 \times 3 = -0.2, 2 \times 3 = 0.2, 1 \times 2 \times 3 = 0.0$, standard error of effect $= 0.30$.

Exercise 5.4. Repeat Exercise 5.3, assuming that data for replication 3 are unavailable.

Answer: $1 = -2.1$, $2 = 2.6$, $3 = -1.9$, $1 \times 2 = -0.4$, $1 \times 3 = 0.1$, $2 \times 3 = -0.1$, $1 \times 2 \times 3 = -0.1$, standard error of effect $= 0.23$.

Exercise 5.5. Repeat Exercise 5.3 given the observations were randomly run within each replicate. The design now becomes a randomized block with three blocks and eight treatments.

Answer: Estimated effects are identical to those of Example 5.3. Standard error of effect $= 0.25$.

Exercise 5.6. Suppose that in addition to the eight runs in Exercise 5.2 the following popcorn yields were obtained from genuine replicate runs made at the center conditions: **1** = (Brand) a mixture of 50% ordinary and 50% gourmet popcorn, **2** = (popcorn/oil) medium, **3** = (batch size) 1/6 cup and yields $y_1 = 9$, 8, 9.5, and 10 cups. Use these extra runs to calculate standard errors for main effects and interactions. Plot the effects in relation to an appropriate reference distribution and draw conclusions.

Answer: Main effects and interactions remain unchanged. They are given in Exercise 5.2. Estimate of σ^2 is from center point $s^2 = 0.73$ with three degrees of freedom. Standard error of an effect = 0.60. Reference t distribution has three degrees of freedom; scale factor = 0.60.

Exercise 5.7. Show that the sum of squares in the ANOVA table associated with any effect from any two-level factorial design containing (including possible replication) a total of N runs is $N \times (\text{estimated effect})^2/4$.

Exercise 5.8. Given the data in Table 5.3, an alternative way to estimate the variance is to set up a one-way classification ANOVA table. For the 16 observations and 8 treatments of this example, complete the ANOVA table and show that $s^2 = 8$ with $\nu = 8$ degrees of freedom.

5.11. EYEING THE DATA

Testing Worsted Yarn

Table 5.6 shows part of the data from an investigation (by Barella and Sust in Box and Cox 1964) later discussed in greater detail, on the strength of a particular type of yarn under cycles of repeated loading. This is a 2^3 factorial design run with three factors: *length of specimen* (**A**), *amplitude of load cycle* (**B**), and *load* (**C**). In an experimental run a piece of yarn of given length (**A**) was continually stretched a certain amount (**B**) under a given load (**C**) until it broke. The response was $y = \log Y$, where Y is the number of cycles to failure. It is convenient to refer to the quantity y as the *durance* of the fiber.

If you now look at the estimated effects shown in Table 5.7, you will see that all the interactions are negligibly small in relation to the main effects, leading to the conclusion that over the ranges of factors studied changing the length of specimen from 250 to 350 mm increased the durance (by about 8 units), increasing the amplitude from 8 to 10 mm reduced durance (by about −6 units), and increasing the load from 40 to 50 units reduced durance (by about −3.5 units).

Contour Eyeballing

When, as in this yarn example, the main effects are dominant and distinguishable from noise, a helpful device for getting a closer look at what is going on is "contour eyeballing." This uses some elementary ideas from response surface modeling, a subject discussed more fully in Chapter 12. Figure 5.6*a* shows the numbering of the runs (in standard order) and Figure 5.6*b* a cube plot of the yarn data. From the analysis in Table 5.7 you will see that the factors appear to behave approximately independently (with no interactions) in their effects on durance and a reasonable further assumption is that, between the two levels of any given factor, the value of y changes approximately at a constant rate, that is, "linearly."

Table 5.6. A 2^3 Factorial Design to Study Effect of A (Length of Specimen), B (Amplitude of Load Cycle), and C (Load) in Investigation of Strength of Yarn

	Factors	Levels −	Levels +
A	Length, mm	250	350
B	Amplitude, mm	8	10
C	Load, g	40	50

Run Number	Factors **A**	Factors **B**	Factors **C**	Durance **y**
1	−	−	−	28
2	+	−	−	36
3	−	+	−	22
4	+	+	−	31
5	−	−	+	25
6	+	−	+	33
7	−	+	+	19
8	+	+	+	26

Table 5.7. Estimates of Main Effects and Interactions for Yarn Experiment

Mean	27.5
A	**8**
B	**−6**
C	**−3.5**
AB	0
AC	−0.5
BC	−0.5
ABC	−0.5

Now on these assumptions consider the problem of finding *all* the various conditions of the factors **A**, **B**, and **C** that give, say, a *durance of* 25. Look at Figure 5.6*b*. Referring to the edge of the cube joining runs 1 and 3 as the (1, 3) edge, you will see the number 28 at the bottom of this edge and 22 at the top, so you would expect the value 25 to be about midway along this line. This point is marked with a dot in Figure 5.6*b*. Now look at the edge (3, 4). The difference between the values 22 and 31 is 9, so you would expect a value of 25 about one-third of the way along this edge. Again this point is marked with

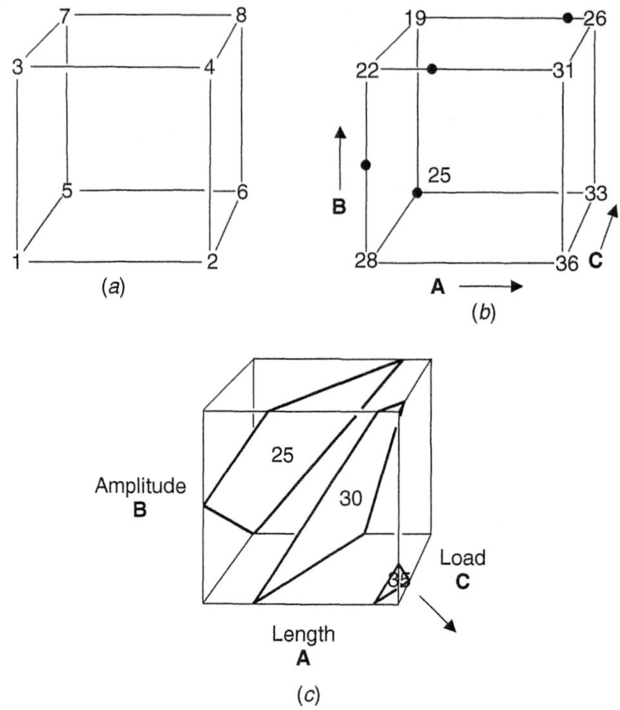

Figure 5.6. (*a*) Cube plot of numbered runs. (*b*) Cube plots of durance results for yarn experiment. (*c*) Approximate contour planes, $y = 25, 30, 35$.

a dot. Similarly, you would expect to find a value of about 25 six-sevenths of the way along the (7, 8) edge and near the bottom of the (5, 7) edge. If you join the dots together, they outline the plane marked 25 in Figure 5.6*c*. The plane thus approximately represents all the various combinations of the factors length, amplitude, and load, where you would expect to get a durance of approximately 25. It is called the 25 *contour* plane. In a similar way, you can obtain the approximate contour planes for y values of 30 and 35. (We discuss model-based methods for estimating contour planes later.)

The contour plot provides a useful summary of what the experiment is telling you about how durance y depends on the three factors length, amplitude, and load. In particular, it illustrates with great clarity how the durance value is increased by increasing length, reducing amplitude, or reducing load.

Exercise 5.9. Make an approximate contour plot using the data in Exercise 5.2.

The Direction of Steepest Ascent

The direction at right angles to the contour planes indicated by an arrow in Figure 5.6 is called the *direction of steepest ascent*. It is followed when, in the experimental design units, changes of factors **A, B**, and **C** are made proportional to the effects they produce. In this case $A = 8$, $B = -6$, and $C = -3.5$, so, for example, for every 0.8 of a unit change in **A** you should simultaneous make a -0.6 unit change in **B** and -0.35 unit change in **C**. This gives the orientation of the arrow in Figure 5.6C.

5.12. DEALING WITH MORE THAN ONE RESPONSE: A PET FOOD EXPERIMENT

When a number of different responses $y_1, y_2, \ldots$ are measured in each experimental run, factorial experiments are of even greater value because they allow the investigator to consider simultaneously how a number of factors affect each of a number of responses. For illustration, we use some of the data from an interesting experiment due to Pratt and Tort (1990).

The manufacturer of pet food had received complaints that packages of food pellets received by the customer contained an unsatisfactorily large amount of powder.

Denote the amount of powder in the product *received by the customer* by y_1. For a batch of routinely manufactured product the value of y_1 was not known but y_2, the amount of powder *produced in the plant*, could be measured rather easily. On the assumption that y_2 was a good substitute indicator for y_1, control of the plant had been attempted by keeping y_2 as low as possible. Unfortunately, control by this means proved ineffective.

Table 5.8 shows data from part of an experiment to study the effect of changing manufacturing conditions. A total of just eight plant runs were made in a 2^3 factorial experiment with the factors *temperature* (**A**), *flow* (**B**), and *zone* (**C**) each varied at two levels.*

In each run, four responses were measured: powder in product y_1, powder in plant y_2, a measure of yield y_3, and energy consumed y_4. Values for the *powder in product* y_1 as it would have been received by the customer were obtained by taking representative samples from the eight experimental batches and subjecting them to the same kind of packaging, shaking, and handling that they would receive before reaching the customer. The values obtained for y_1 and powder in the plant y_2, yield of product produced y_3, and electrical energy consumed y_4 are shown in Table 5.8.†

The experimenters were surprised to find that the eight values of y_1, when plotted against the values of y_2, as in Figure 5.7*a*, showed no positive correlation. Thus the previous policy of attempting to control powder in the product y_1 by observing powder in the plant y_2 was useless.

Therefore, the experimenters decided to concentrate on finding better operating conditions (**A, B, C**) that would reduce the amount of powder in the product y_1

*A fourth factor, not discussed here, was varied but the changes appeared to be without effect.

†The data given by the authors for powder in product, powder in plant, and yield have all been multiplied by 100 to simplify calculations.

Table 5.8. A 2^3 Factorial Design to Study Effect of Temperature, Flow, and Compression Zone on Four Responses

		−	+
A	Conditioning temperature	80% of max	Max
B	Flow	80% of max	Max
C	Compression zone	2	2.5

	Factor Levels (Process Conditions)			Responses			
	A	**B**	**C**	Powder in Product × 100, y_1	Powder in Plant × 100, y_2	Yield ×100, y_3	Energy Consumed, y_4
1	−	−	−	132	166	83	235
2	+	−	−	107	162	85	224
3	−	+	−	117	193	99	255
4	+	+	−	122	185	102	250
5	−	−	+	102	173	59	233
6	+	−	+	92	192	75	223
7	−	+	+	107	196	80	250
8	+	+	+	104	164	73	249

without losing too much yield y_3 or requiring too much electrical energy y_4. The data are plotted on the factorial cubes shown in Figures 5.7*b–e*.

A Rough Estimate of Error

In the original experiment four duplicate runs were made* from which an approximate estimate of the standard deviation of the various responses could be obtained. For example, for powder in the product y_1, the repeated runs gave the results

$$(132, 118) \quad (122, 127) \quad (92, 91) \quad (107, 112)$$

providing an estimate for the standard deviation $\hat{\sigma}_1 = 5.6$ with four degrees of freedom. Using this estimate, the y_1 effects have standard errors $5.6/\sqrt{2} = 4.0$. Similar calculations yield the standard errors for the effects associated with y_3 and y_4. It must be remembered that the standard errors themselves are subject to large errors when based on so few degrees of freedom. For example, by looking at a t table, you will see that such estimates must be multiplied by almost 3 (2.776) to obtain 95% confidence limits for an effect.

*The effects of replication on the estimated effects is ignored here. Later you will see how changes in effect estimates due to partial replication can be easily taken into account. They do not change the conclusions in this example.

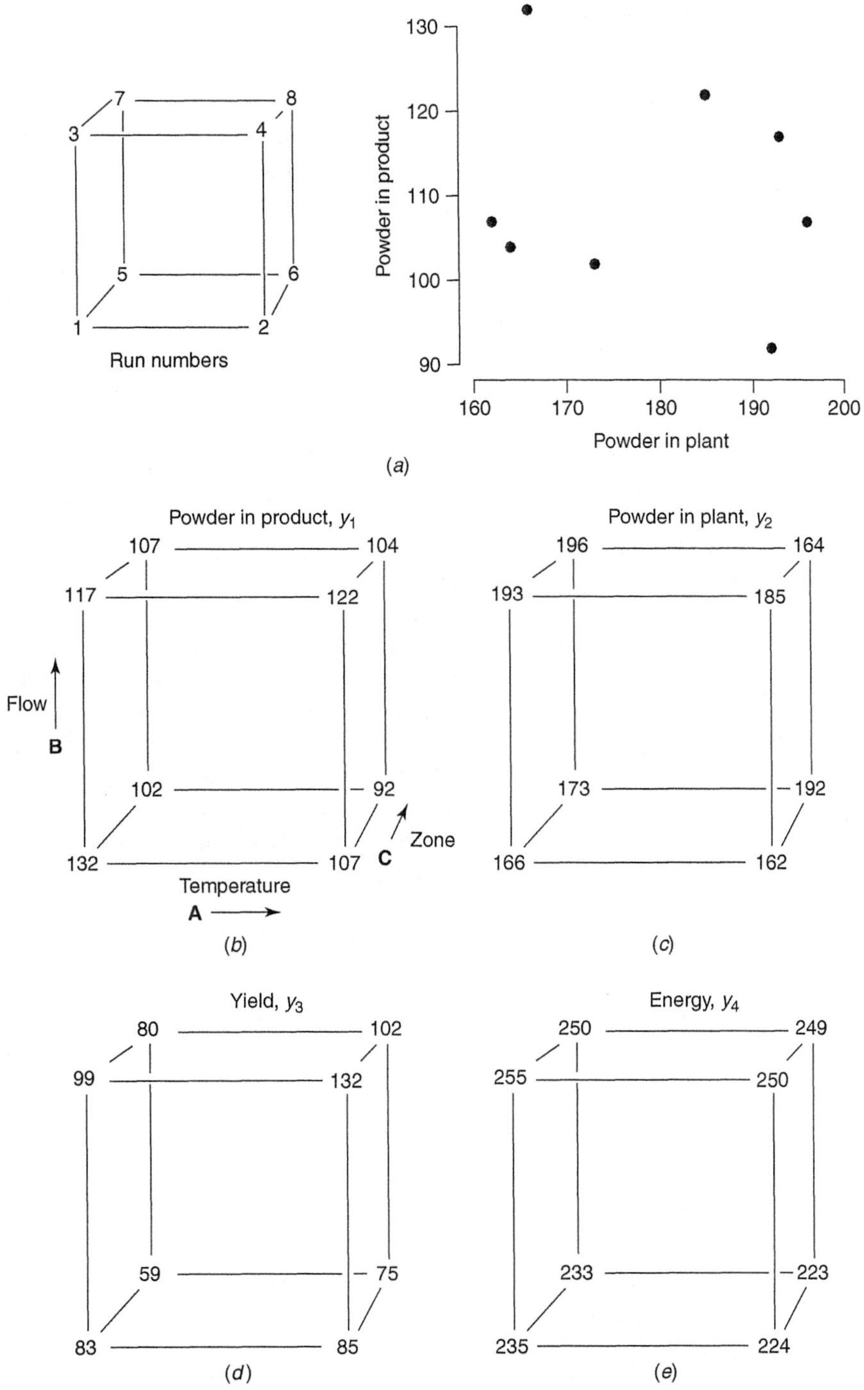

Figure 5.7. (*a*) Plot of y_1 versus y_2. (*b*) Factorial cube plots for y_1, y_2, y_3, and y_4.

Exercise 5.10. What do these data suggest are the active subspaces for y_1, y_2, y_3, and y_4?

Estimated main effects and interactions for each of the responses y_1, y_2, y_3, y_4 are given in Table 5.9. Effects least likely to be due to noise are shown in bold type. Notice that powder in the plant y_2 seems not convincingly related to any of the manufacturing variables. This further confirmed its uselessness for controlling the process.

Ideally the experimenter would have liked to decrease the powder in the product y_1, increase yield y_3, and decrease energy consumption y_4. Often it is possible to see how to satisfy a number of movements by overlaying contour plots for different responses one on the other. The orientations and overlays of the contours of the active factor subspaces for y_1, y_3, and y_4 did not point to any clear solution in this example. It was necessary, therefore, to look at the possibilities for compromise.

The most striking effect on powder in the product y_1 was produced by factor **C** (zone). The change in the level of this factor reduced y_1 by about -18.2 units and had little effect on the energy consumption y_4, but unfortunately it decreased the product yield y_3 by about -20.5. Notice also that while increasing factor **B** (flow) improved the yield by about 13 units, it increased energy consumption by about 22.3 units. The clearest way to perceive the possibilities of compromise was by comparing the approximate contour planes. Consequently, even though there was evidence of nonlinearity (in particular because of a nonnegligible *AB* interactions for y_1 and y_4), results in Figure 5.7 were used to obtain the approximate eyeball contour plots in Figure 5.8. To determine a suitable compromise, the costs of lower yield and higher energy consumption were calculated and alternative possibilities suggested to management, who perhaps had a better idea of the cost of displeasing the customer by having too much powder in the product.

Problems of this kind are common. To illustrate with the present data, here is a typical calculation. Suppose that, per unit package sold, an increase in yield of

Table 5.9. Calculated Effects and *Standard Errors* for Pet Food Experiment

	Powder in Product, y_1	Powder in Plant, y_2	Yield, y_3	Energy, y_4
Average	110.4	178.9	82.0	239.8
Temperature, *A*	-8.2 ± 4.0	-6.3 ± 7.4	3.5 ± 4.9	$\mathbf{-6.8 \pm 1.1}$
Flow, *B*	4.3 ± 4.0	11.3 ± 7.4	$\mathbf{13.0 \pm 4.9}$	$\mathbf{22.3 \pm 1.1}$
Zone, *C*	$\mathbf{-18.2 \pm 4.0}$	4.8 ± 7.4	$\mathbf{-20.5 \pm 4.9}$	-2.3 ± 1.1
AB	$\mathbf{9.3 \pm 4.0}$	-13.7 ± 7.4	-5.5 ± 4.9	$\mathbf{3.8 \pm 1.1}$
AC	1.7 ± 4.0	-0.3 ± 7.4	1.0 ± 4.9	1.3 ± 1.1
BC	4.3 ± 4.0	-13.7 ± 7.4	-3.5 ± 4.9	-0.8 ± 1.1
ABC	-5.7 ± 4.0	-11.7 ± 7.4	-6.0 ± 4.9	0.8 ± 1.1

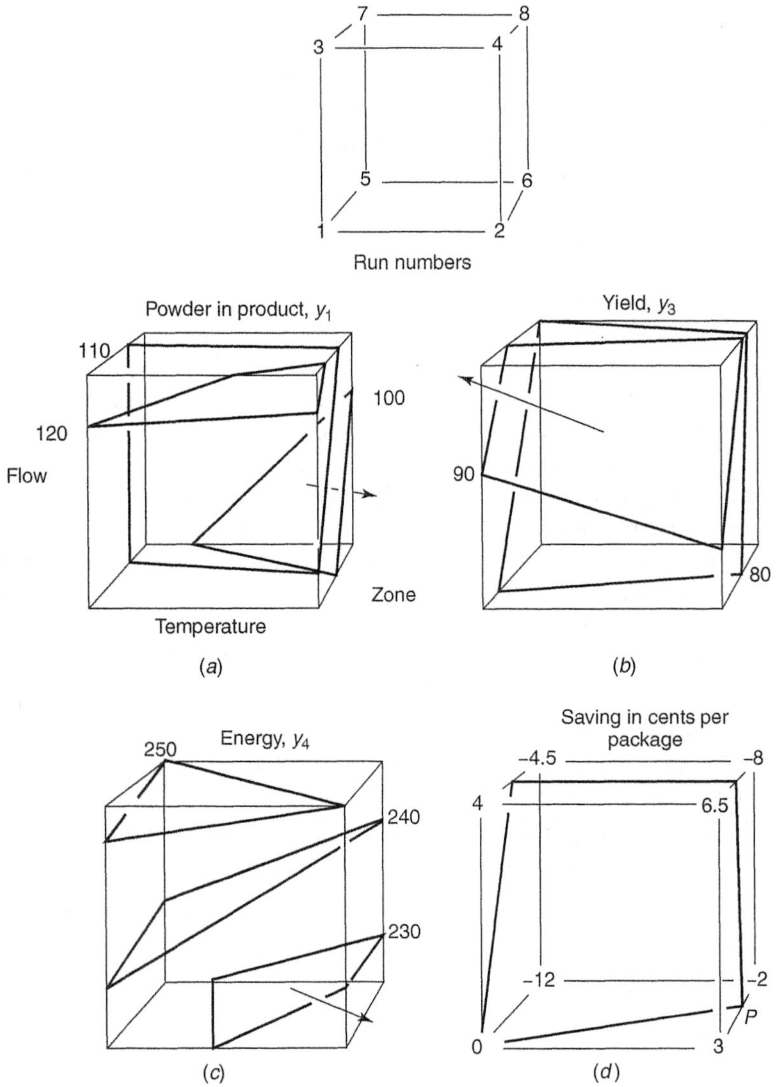

Figure 5.8. "Eyeball contour plots" for (*a*) powder in product, y_1; (*b*) yield, y_3; and (*c*) energy, y_4. (*d*) Savings calculated from (*b*) and (*c*) and the estimated break-even contour.

one unit produces a savings of 0.5 cents and an increase in energy consumption of one unit costs 0.2 cents. Since run number 1 represents standard conditions, this point has been scored zero in Figure 5.8*d*. Changing the conditions from those of run number 1 to those of run 3 gives an observed increase in yield of $99 - 83 = 16$ units, which would produce an estimated savings of 8 cents per package. But such a change produces an increased energy consumption of $255 - 235 = 20$, costing 4 cents per package. The sum $8 - 4 = 4$ is the number shown as "savings in cents per package" for run 3. The other numbers are calculated in the same way. The approximate contours shown in Figure 5.8*d* are for zero savings (break even). By simultaneously studying Figures 5.8*a* and 5.8*d*, management was supplied with information to make an educated choice among the available options. The choice will depend on a number of issues, such as the quality and prices of the competitors' products. In particular, you will notice that at point *P* in Figure 5.8*d* you should be able to achieve about 30 units less powder in the product at no extra cost. Projected process improvements should, of course, be confirmed by further experimentation.

You should not necessarily limit consideration to the region covered by your initial experimental design. Current results frequently suggest where further experimentation might be sufficiently fruitful. For example, it seems clear from the present data that substantial gain might have been possible by increasing temperature above what was previously regarded as the maximum. If there was no objection to running at higher temperatures, then you might decide on a further experimental region like that represented by the bold lines in Figure 5.9.

When several responses must be considered together, a planned design of this kind can be a great incentive to innovation. It will serve to confirm those of our beliefs that are true, discredit those that are mythical, and often point the way on a trail of new discovery.

The present experiment produced surprises and questions for the investigators:

(a) The powder in the plant response y_2 previously used to control the process was not related to the powder in the product y_1 or, indeed, to any of the three factors discussed above. Why was this?

(b) Increasing temperature with low flow rate would have been expected to increase energy consumption but actually reduced it. Why?

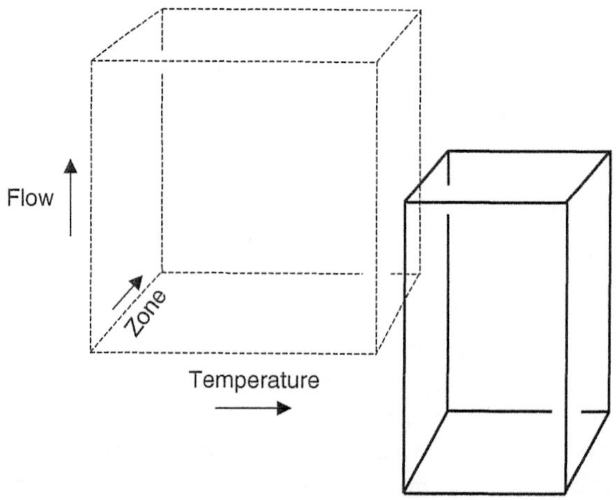

Figure 5.9. Possible additional experiments to be appended to the 2^3 factorial.

5.13. A 2^4 FACTORIAL DESIGN: PROCESS DEVELOPMENT STUDY

Often there are more factors to be investigated than can conveniently be accommodated within the time and budget available, but you will find that usually you can separate genuine effects from noise without replication. Therefore, in the pilot plant study, rather than use 16 runs to produce a replicated 2^3 factorial, it would have been possible to introduce a fourth factor and run an *un*replicated 2^4 design. (Or, as you will see in the next chapter, employ the 16 runs in a half-replicated 2^5 design and study five factors.)

For illustration, a process development experiment is shown in which four factors were studied in a 2^4 factorial design: amount of catalyst charge **1**, temperature **2**, pressure **3**, and concentration of one of the reactants **4**. Table 5.10a shows the design, the random order in which the runs were made, and the response y (percent conversion) at each of the 16 reaction conditions. Table 5.10b displays the contrast coefficients that determine the estimates in Table 5.11 of the four main effects, six two-factor interactions, four three-factor interactions, and one four-factor interaction.

Exercise 5.11. Add nine units to observations 2, 3, 6, 7, 9, 12, 13, 16 in Table 5.10a. Recalculate the effects. Explain what you observe.

Answer: These eight runs are those that correspond to plus signs in the 124 column. Thus the 124 interaction effect is increased by 9 units, but because of orthogonality, none of the other estimates is affected. Notice, however, that interaction 124 is now picked out as deviating from the line.

Table 5.10a. Data Obtained in a Process Development Study Arranged in Standard (Yates) Order

Yates Run Number	1	2	3	4	Conversion (%)	Random Order	Variable	−	+
1	−	−	−	−	70	8	**1**: Catalyst charge (lb)	10	15
2	+	−	−	−	60	2	**2**: Temperature (°C)	220	240
3	−	+	−	−	89	10	**3**: Pressure (psi)	50	80
4	+	+	−	−	81	4	**4**: Concentration (%)	10	12
5	−	−	+	−	69	15			
6	+	−	+	−	62	9			
7	−	+	+	−	88	1			
8	+	+	+	−	81	13			
9	−	−	−	+	60	16			
10	+	−	−	+	49	5			
11	−	+	−	+	88	11			
12	+	+	−	+	82	14			
13	−	−	+	+	60	3			
14	+	−	+	+	52	12			
15	−	+	+	+	86	6			
16	+	+	+	+	79	7			

Table 5.10b. Table of Contrast Coefficients

	I	1	2	3	4	12	13	14	23	24	34	123	124	134	234	1234	conversion (%)
	+	−	−	−	−	+	+	+	+	+	+	−	−	−	−	+	70
	+	+	−	−	−	−	−	−	+	+	+	+	+	+	−	−	60
	+	−	+	−	−	−	+	+	−	−	+	+	+	−	+	−	89
	+	+	+	−	−	+	−	−	−	−	+	−	−	+	+	+	81
	+	−	−	+	−	+	−	+	−	+	−	+	−	+	+	−	69
	+	+	−	+	−	−	+	−	−	+	−	−	+	−	+	+	62
	+	−	+	+	−	−	−	+	+	−	−	−	+	+	−	+	88
	+	+	+	+	−	+	+	−	+	−	−	+	−	−	−	−	81
	+	−	−	−	+	+	+	−	+	−	−	−	+	+	+	−	60
	+	+	−	−	+	−	−	+	+	−	−	+	−	−	+	+	49
	+	−	+	−	+	−	+	−	−	+	−	+	−	+	−	+	88
	+	+	+	−	+	+	−	+	−	+	−	−	+	−	−	−	82
	+	−	−	+	+	+	−	−	−	−	+	+	+	−	−	+	60
	+	+	−	+	+	−	+	+	−	−	+	−	−	+	−	−	52
	+	−	+	+	+	−	−	−	+	+	+	−	−	−	+	−	86
	+	+	+	+	+	+	+	+	+	+	+	+	+	+	+	+	79
Divisor	16	8	8	8	8	8	8	8	8	8	8	8	8	8	8	8	

Table 5.11. Estimated Effects from a 2^4 Factorial Design: Process Development Example

Effects	Estimated Effects ± SE
Average	72.25 ± 0.27
1	**−8.00 ± 0.55**
2	**24.00 ± 0.55**
3	−0.25 ± 0.55
4	**−5.50 ± 0.55**
12	1.00 ± 0.55
13	0.75 ± 0.55
14	0.00 ± 0.55
23	−1.25 ± 0.55
24	**4.50 ± 0.55**
34	−0.25 ± 0.55
123	−0.75
124	0.50
134	−0.25
234	−0.75
1234	−0.25

Assessing the Effects of the Factors Using an Internal Reference Set

A dot plot of the 15 calculated effects is shown in Figure 5.10. No direct estimate of the experimental error variance is available from this 16-run experiment since there are no replicates, but a reasonable assumption might be that the five highest order interactions are largely due to noise and can therefore provide a reference set for the remaining effects.

The estimated interaction effects *123, 124, 134, 234* and *1234* forming this reference set are identified by the open circles in Figure 5.10. Employing these five estimates, a pooled estimate of $V(\overline{y}_+ - \overline{y}_-)$ is given by

$$(\text{SE effect})^2 = \tfrac{1}{5}\{(-0.75)^2 + (0.50)^2 + (-0.25)^2 + (-0.75)^2 + (-0.25)^2\}$$
$$= 0.30$$

provides the square of the standard error for the remaining main effects and two-factor interaction; thus $\text{SE(effect)} = \sqrt{0.30} = 0.55$. On this basis it is highly likely that the main-effect estimates *1*, *2*, and *4* and the interaction *24* represent real effects. Furthermore, on the usual normal theory assumptions the estimated standard error $s = 0.55$ may be used to produce the reference t_ν distribution having five degrees of freedom shown in Figure 5.10.* You would find that if these normal assumptions are exactly true then exactly 95% of the distribution would be contained within the interval $\pm t_{5,0.025}\text{SE (effect)} = \pm 2.57(0.55) = \pm 1.41$. But certainly in this example these assumptions do not need to be made, and you can arrive at approximate and more transparent conclusions simply by regarding the *dot plot* of the estimates of the higher order interactions in Figure 5.10 *as* the basic reference distribution as was done will the residual plot in the graphical ANOVA of Chapter 4. It is a direct plot of what you know without further assumptions.

Interpretation of the Process Development Data

Proceeding on the basis that the main effects of factors **1**, **2**, and **4** and the two-factor interaction **24** are distinguishable from noise you find:

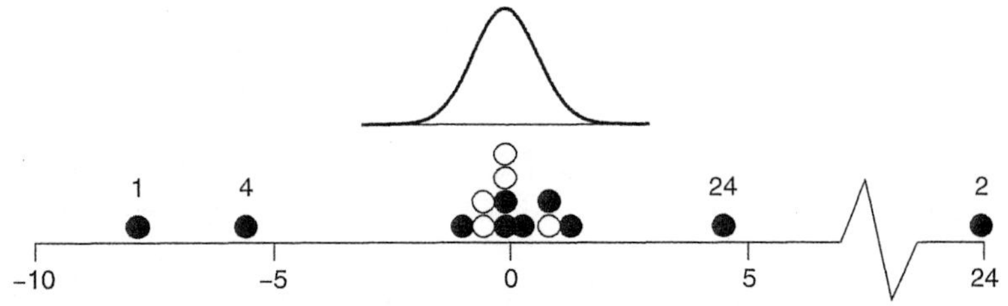

Figure 5.10. Dot plot of the effects from the process development example. High-order interaction effects used to provide a reference distribution for the remaining effects are shown by open circles. A reference $t_\nu = 5$ distribution is also shown.

*The t distribution is drawn with a scale factor equal to the standard error 0.55 of an effect. The ordinates for t distributions are given in Table B2.

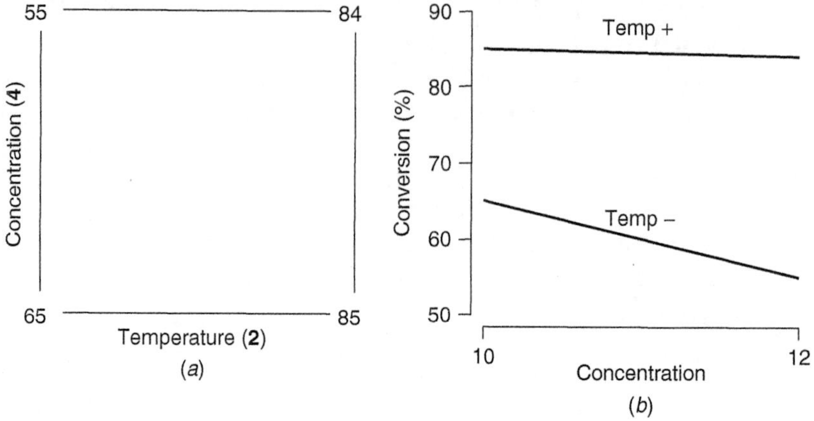

Figure 5.11. Graphical illustrations of temperature–catalyst interaction using (*a*) a two-way table and (*b*) a contrast diagram.

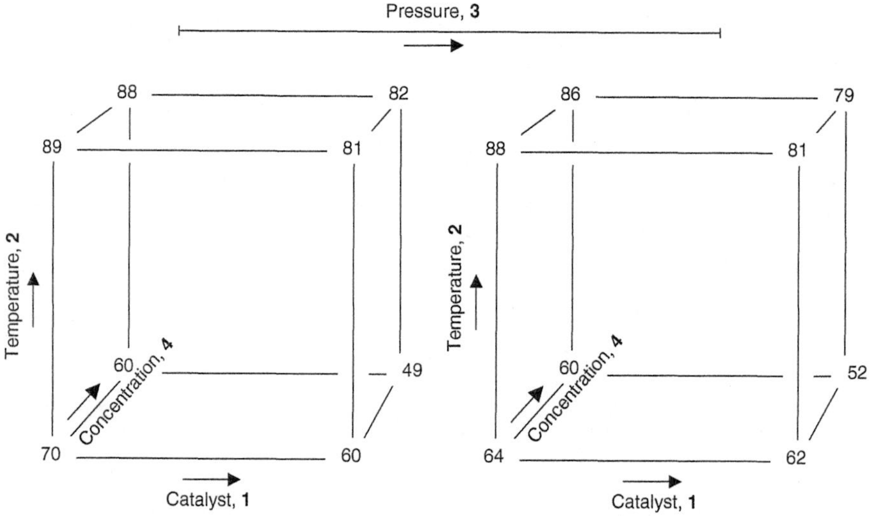

Figure 5.12. Cube plot for factors 1, 2, and 4 at the low and high values of factor 3.

1. An increase in catalyst charge (factor **1**) from 10 to 15 pounds reduces conversion by about 8%, and the effect is consistent over the levels of the other factors tested.
2. Since there is an appreciable interaction between temperature (factor **2**) and concentration (factor **4**), the effects of these factors must be considered jointly. The nature of the interaction is indicated by the two-way table in Figure 5.11*a* obtained by averaging over levels of the other factors or, equivalently, by the contrast diagram in Figure 5.11*b*. These representations make clear that the interaction occurs because an increase in concentration

reduces conversion at the lower temperature but at the higher temperature produces no appreciable effect.

3. Locally, at least for the factors tested, factor **3** (pressure) is essentially inert. This is because neither the main effect 3 nor *any of the interactions 13, 23, 34* involving factor **3** show effects that cannot be readily explained by the noise. Thus you have a 4D experiment revealing a local 3D active subspace in factors **1**, **2**, and **4**. This tells you that locally you should be able to manipulate factor **3** to improve some other response (e.g., cost) without very much affecting conversion.
4. In Figure 5.12 cube plots in factors **1**, **2**, and **4** are made, one for the lower and one for the upper levels of factor **3**. If over this range factor **3** is virtually inert, then the corresponding pairs of data values found on the two cubes should act approximately as replicates. From inspection this appears to be the case.

5.14. ANALYSIS USING NORMAL AND LENTH PLOTS

Two problems arise in the assessment of effects from unreplicated factorials: (a) occasionally meaningful high-order interactions do occur and (b) it is necessary to allow for selection. (The largest member of *any* group is by definition large—but is it *exceptionally* large?) A method due to Cuthbert Daniel* (1959) in which effects were plotted on normal probability paper (discussed in Chapter 2) can provide an effective way of overcoming both difficulties. Today such plots are commonly produced on the computer screen.

Suppose that the data from the 2^4 process development experiment had occurred simply as the result of random variation about a mean and the changes in levels of the factors had no real effect on the response. Then these $m = 15$ effects representing 15 contrasts between pairs of averages would have roughly normal distributions centered at zero and would plot on a normal probability scale as a straight line. To see whether they do, you can order and plot the 15 effects obtained from the process development experiment as shown in Figure 5.13 (If you want to do this manually, the appropriate scaling for the $P(\%)$ axis is from Table E at the end of the book for $m = 15$.) As shown in Figure 5.13, 11 of the estimates fit reasonably well on a straight line called here the "error line" since these estimates are such as might be produced by noise alone. However, the effects *1, 2, 4*, and *24* fall off the line and are not easily explained as chance occurrences. Thus, for this example, the conclusions drawn from the normal plot are precisely those reached earlier.

Some experimenters prefer to use a half-normal plot, that is, to plot on the appropriate scale the *absolute* values of the effects. This diagnostic tool, originally proposed by Daniel (1959), proved useful for detecting defective randomization,

*The reader is urged to study Daniel's book (1976), *Applications of Statistics to Industrial Experimentation*, which takes an appropriately skeptical attitude toward mechanical analysis of data and provides methods for diagnostic checking, including the plotting of factorial effects and *residuals* on normal probability paper.

Order number I	1	2	3	4	5	6	7	8	9	10	11	12	13	14	15
Effect	−8.00	−5.50	−1.25	−0.75	−0.75	−0.25	−0.25	−0.25	−0.25	0.00	0.50	0.75	1.00	4.50	24.00
Identity of effect	*1*	*4*	*23*	*123*	*234*	*33*	*34*	*134*	*1,234*	*14*	*124*	*13*	*12*	*24*	*2*
$P=100(I-1/2)/15$	3.3	10.0	16.7	23.3	30.0	36.7	43.3	50.0	56.7	63.3	70.0	76.7	83.3	90.0	96.7

(*a*)

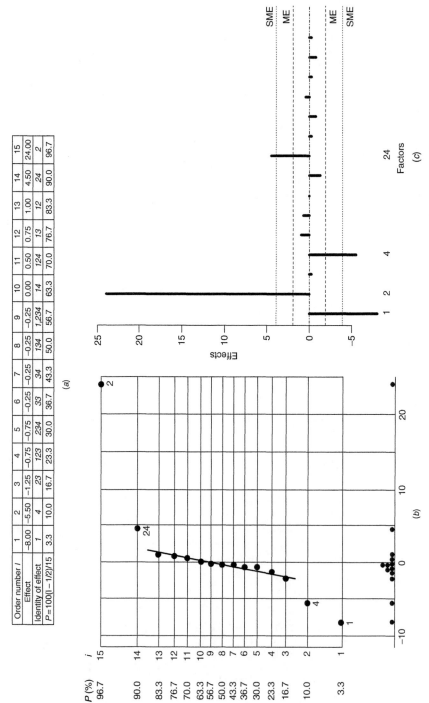

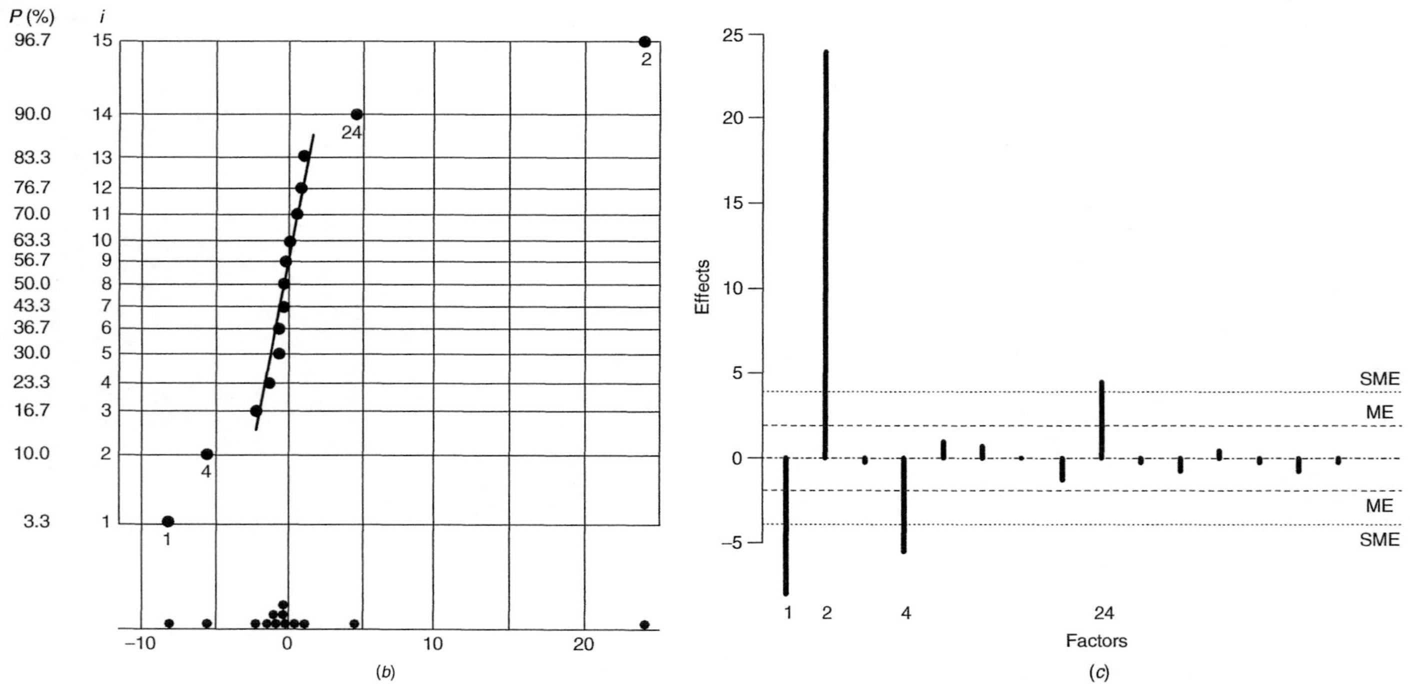

Figure 5.13. Process development example: (*a*) 15 ordered effects; (*b*) a normal plot of the effects; (*c*) Lenth's plot.

bad values, and the dependence of variance on mean. Later Daniel (1976) noted that such plots, since they canceled irregularities in the data, lacked sensitivity to other signals. Many computer programs now produce both plots. A full and valuable discussion on how normal plots may be used to check assumptions is given by Daniel (1976).

Lenth's Analysis as an Alternative to Normal Plots

An alternative procedure to a normal or half-normal plot of the estimated effects is due to Lenth (1989). The median m of the *absolute* values of the k effects is first determined. (The absolute values of the effects also appear on the half normal plot.) An estimate of the standard error of an effect, called a *pseudo* standard error, is then given by $s_0 = 1.5\,m$. Any estimated effect exceeding $2.5s_0$ is now excluded and, if needed, m and s_0 are recomputed. A "margin of error" ME (an approximate 95% confidence interval for an effect) is then given by $\text{ME} = t_{0.975,d} \times s_0$, where $d = k/3$. All estimates greater than the ME may be viewed as "significant," but, of course, with so many estimates being considered simultaneously, some will be falsely identified. Lenth thus next defines a "simultaneous margin of error" $\text{SME} = t_{\gamma,d} \times s_0$, where $\gamma = (1 + 0.95^{1/k})/2$. Estimated effects that exceed the SME are declared active while those that are less than the ME are declared inert.

Using the data from the process development example (see Table 5.11), you will find $s_0 = 1.5(0.50) = 0.75$ with $d = 5$ to give $\text{ME} = 2.57(0.75) = 1.93$ and $\text{SME} = 5.22(0.75) = 3.91$. Thus (see Fig. 5.13*c* as before), the estimated effects *1, 2, 4,* and *24* are identified as active and all other estimates considered as noise. Lenth's table for $t_{0.975,d}$ and $t_{\gamma,d}$ for different numbers of estimated effects is as follows:

Number of Estimates	$t_{0.975,d}$	$t_{\gamma,d}$
7	3.76	9.01
15	2.57	5.22
31	2.22	4.22
63	2.08	3.91
127	2.02	3.84

As with any other technique, it is important not to use this method merely as a brainless machine for determining "significance" or insignificance." The diagram is the most important part of the method, and like a normal plot this diagram should be examined in the light of subject matter knowledge to determine how to proceed.

A Diagnostic Check

Normal plotting of *residuals* provides a diagnostic check for any tentatively entertained model. For example, in the process development example, if it is accepted that all estimated effects with the exception of *1* = −8.0, *2* = 24.0, *4* = −5.5, and *24* = 4.5 could be explained by noise, then the predicted response (percent conversion) at the vertices of the 2^4 design would be given by the fitted model

$$\hat{y} = 72.25 + \left(\frac{-8.0}{2}\right) x_1 + \left(\frac{24.0}{2}\right) x_2 + \left(\frac{-5.5}{2}\right) x_4 + \left(\frac{4.5}{2}\right) x_2 x_4$$

where x_1, x_2, and x_4 take the value −1 or +1 according to the columns of signs in Table 5.10.* Notice that the coefficients that appear in this equation are *half* the calculated effects. This is so because the calculated effects yield changes of *two* units along the x axis from $x = -1$ to $x = +1$.

Exercise 5.12. Confirm that the values of y, $\hat{y}$, and $y - \hat{y}$ are as follows:

y	70	60	89	81	69	62	88	81
$\hat{y}$	69.25	61.25	88.75	80.75	69.25	61.25	88.75	80.75
$y - \hat{y}$	0.75	−1.25	0.25	0.25	−0.25	0.75	−0.75	0.25
y	60	49	88	82	60	52	86	79
$\hat{y}$	59.25	51.25	87.75	79.75	59.25	51.25	87.75	79.75
$y - \hat{y}$	0.75	−2.25	0.25	2.25	0.75	0.75	−1.75	−0.75

Make a plot of these residuals on normal probability paper for $m = 16$. What do you conclude?

Answer: All the points from this residual plot now lie close to a straight line, confirming the conjecture that effects *other than 1, 2, 4,* and *24* are readily explained by the random noise. This residual check is valuable provided that the number of effects eliminated (four in this example) is fairly small compared with n (here $n = 16$).

A First Look at Sequential Assembly

In Chapter 1 the process of investigation was described as iterative, involving deduction and induction. What is gained by running an experiment is not only the immediate possibility of process improvement; the results may also indicate the possibilities for even further advance and thus show where additional runs need to be made. When this is done, the experimental design is built up in pieces—a process we call *sequential assembly*. A simple example occurred in a study by Hill and Wiles (1975).

*The cross-product terms suggest that quadratic terms (involving x_1^2 and x_2^2) might be needed to adequately represent the response at *interpolated* levels of the factors. Such terms cannot be estimated using two-level designs. Additional runs can convert the existing design into, for example, a central composite design allowing the estimation of quadratic effects. These methods are discussed in Chapters 10 and 11.

The objective was to increase disappointingly low yields of a chemical product. Three factorial designs were run in sequence in this investigation, but only the first of these will be described here. In phase I a 2^3 factorial design with factors concentration (**C**), rate of reaction (**R**), and temperature (**T**) was run. Table 5.12 displays the 2^3 design, recorded observations, and estimated effects. Analysis of the effects shows the dominating influence of the main effect of **C**.

In phase II, to check further on the influence of factor **C**, three extra experimental points were run along the concentration axis, as shown in Figure 5.14*a* and Table 5.12. These "axial" experiments were run at the intermediate levels of **R** and **T** but over a wider range of **C**. Figure 5.14 shows the resulting sequentially assembled design and a graph of yield versus concentration for all 11 runs. The conclusions, which were of great practical value, were (1) that over the factor region tested temperature **T** and rate of reaction **R** were essentially inert, that is, without detectable effect, but that (2) reducing the level of factor **C** could produce a much higher yield.

Table 5.12. A 2^3 Factorial with Sequentially Added Runs

Factor Combinations	C	R	T	y_{ic}	Effects		Residuals
Phase I							
1	−1	−1	−1	75.4	Average	74.3	−0.6
2	+1	−1	−1	73.9	*C*	**−3.4**	**+1.3**
3	−1	+1	−1	76.8	*R*	0.6	+0.8
4	+1	+1	−1	72.8	*T*	−0.8	+0.2
5	−1	−1	+1	75.3	*CR*	−0.7	−0.7
6	+1	−1	+1	71.4	*CT*	−0.6	−1.2
7	−1	+1	+1	76.5	*RT*	0.4	+0.5
8	+1	+1	+1	72.3	*CRT*	0.6	−0.3
Phase II							
9	−2	0	0	79			
10	0	0	0	74			
11	2	0	0	69			

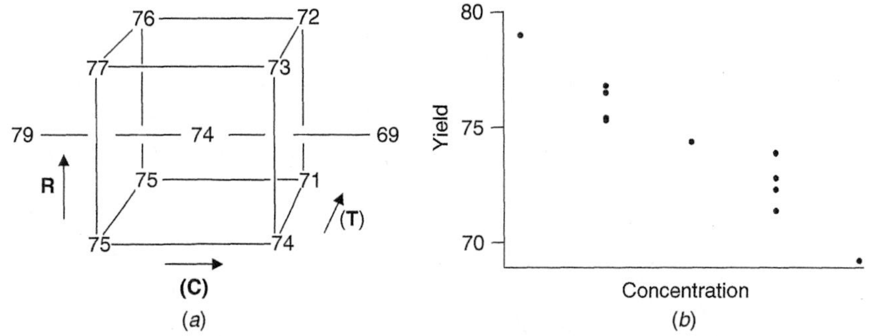

Figure 5.14. (*a*) Sequentially composed design. (*b*) Yield versus levels of concentration **C**.

Importance of Negative Findings

The finding of the large effect of concentration was important, but also of special significance was the discovery that the two factors rate (**R**) and temperature (**T**) had very little influence on yield over the ranges considered and did not appreciably interact with **C**. This 3D experimental space had a 1D active factor space and a 2D locally inert factor space. Knowledge that the system was *robust* to changes in the specific factors **R** and **T** reduced the number of things the plant operators would need to worry about and raised the possibility of using these factors to improve other responses.

Serendipity: Discovering the Unexpected

In a later experimental design in the same investigation four factors were varied. No main effects or interactions were detected but a very large residual noise. The reactant combinations had, of course, been run in random order and when the data were plotted in time order a marked upward shift of about 2% yield was detected part way through the experiment. It was later found that this shift occurred simultaneously with a change in the level of purity of a feed material. When the existence of this assignable cause was later confirmed, the authors declared they believed they had finally found the "ghost" which had for some time been the cause of trouble with routine production. Discovery of the unexpected proved more important than confirming the expected.

5.15. OTHER MODELS FOR FACTORIAL DATA

Always remember that it is the data that are real (they actually happened!). The model ("empirical" or "theoretical") is only a hypothetical conjecture that might or might not summarize and explain important features of the data. But while it is true that *all models are wrong and some are useful*, it is equally true that *no model is universally useful.*

The Heredity Model

The model discussed so far analyzes the factorial design data in terms of main effects and interactions. This model has been widely used with great success since the inception of these designs in the 1920s. It has more recently been elaborated (McCullagh and Nelder, 1989; Chipman, 1996) to include the so-called laws of heredity. These represent beliefs, for example, that a two-factor interaction AB is unlikely without one or both of the corresponding main effects A and B. Although such groupings often do occur, there are important circumstances when they do not.

Consider a two-level design used in an industrial investigation whose object is to increase a response y. The main effects A, B, ... are estimates of the derivatives $\partial y/\partial A$, $\partial y/\partial B$, ... Similarly two-factor interactions are estimates of the mixed second-order derivatives; for example, the interaction AB is an estimate of $\partial^2 y/\partial A\, \partial B$. Suppose we find that the main effects A and B are negligible but that the interaction AB is large: This implies that we have small first-order derivatives but a large second-order derivative. This is exactly what will occur in the neighborhood of a maximum or ridge system. Indeed, the now widely used composite design was first introduced (Box and Wilson, 1951; Box, 1999) to illuminate exactly this circumstance. The location of experiments near a maximum or stationary value would be expected to be rather common because in much experimentation a maximum is precisely what is being sought. In such experimentation, progress is made by exploiting and hence *eliminating* first-order (main) effects.

The Critical Mix Model

You need to remember that sometimes in the analysis of factorials the main effect–interaction model is not relevant at all. When factors are tested together in a factorial design, it sometimes happens that a response occurs *only* when there is a "critical mix" of the factors, qualitative and quantitative.* Thus, for example, a binary mix is required for sexual reproduction, a tertiary critical mix is required to make gun powder, a fourfold critical mix is required for the operation of an internal combustion engine, and various critical mixes are required to produce metallic alloys. While it remains true that mistakes of some kind—copying errors or errors in the conduct of an experiment—are the most likely reason for one or two discrepant runs, nevertheless the possibility of a critical mix must also be considered, as the following example illustrates.

An Experiment with Deep Groove Bearings

An experiment was run by Hellstrand (1989) with the object of reducing the wear rate of deep groove bearings. His 2^3 design employed two levels of osculation **O**,† two levels of heat treatment **H**, and two different cage designs **C**. The response is the failure rate y (the reciprocal of the average time to failure $\times$ 100.) The observations and the calculated main effects and interactions are shown in Table 5.13.

From the normal plot of the estimated *effects* in Figure 5.15a it might be concluded that the data showed nothing of interest. However, look at the normal plot of the *original eight observations* in Figure 5.15b.

This indicates that the data from two of the runs are very different from all the others. An effect occurs only when heat treatment and osculation are both at their plus levels. As illustrated in Figure 5.15c, high osculation and high temperature together produced a critical mix yielding bearings that had an average failure rate *one-fifth* of that for standard bearings! This effect was confirmed in subsequent experimentation, and as Hellstrand reported, this large and totally unexpected effect had profound consequences.

* We encountered a binary critical mix earlier in this chapter in the experiments to determine the effects of three factors on the properties of a polymer solution. (See Figure 5.2.) Yellowness occurred *only* when factors 1 and 2 were *both* at their plus levels.

† Osculation is a measure of the contact between the balls and the casing.

Table 5.13. Bearing Example: 2^3 Design with Data and Calculated Effects

Design			Data		
O	H	C	y	Effects	Estimates
−	−	−	5.9	Mean	4.0
+	−	−	4.0	O	−2.6
−	+	−	3.9	H	−2.0
+	+	−	1.2	C	0.5
−	−	+	5.3	OH	−1.5
+	−	+	4.8	OC	−0.4
−	+	+	6.3	HC	0.5
+	+	+	0.8	OHC	−1.1

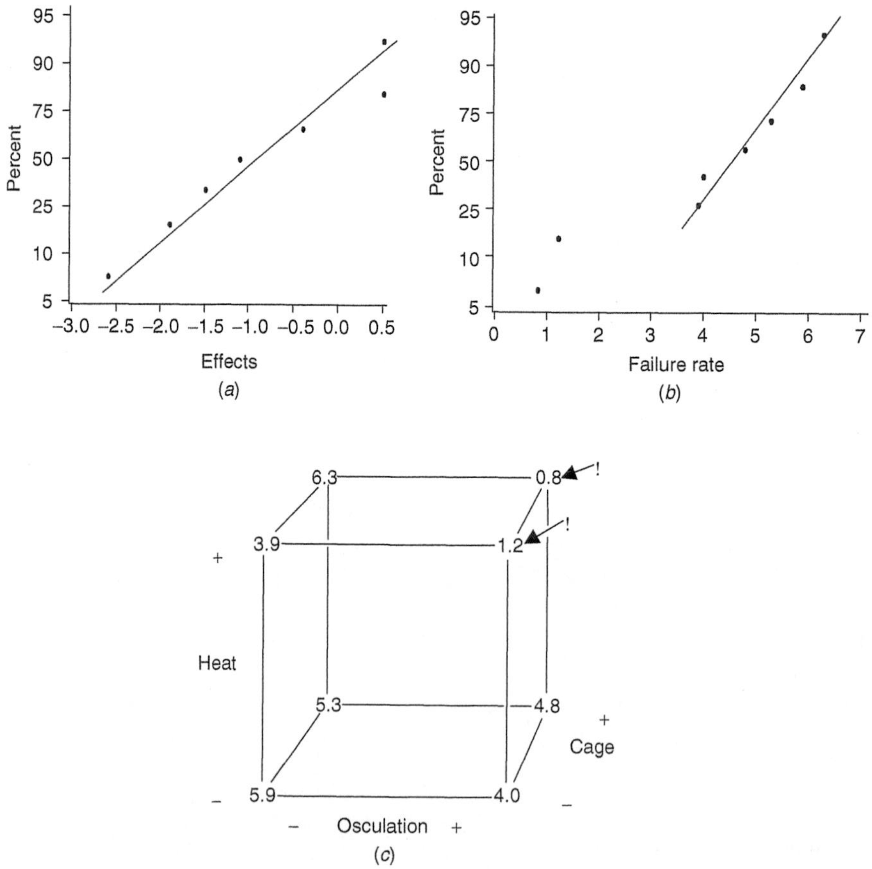

Figure 5.15. (*a*) Normal plot of effects. (*b*) Normal plot of data. (*c*) Cube plot of failure rate.

Remembering that bearings like these had been made for decades, it is surprising that so important a phenomenon had previously escaped discovery. A likely explanation is that until very recently many experimenters had little knowledge of multifactor experimental design. The instruction they received may never have questioned the dogma that factors must be varied one at a time while all others are held constant, a practice that has been discredited now for over 80 years. Clearly, one-factor-at-a-time experimentation could not uncover the phenomenon discovered here. The good news for engineers is that many important possibilities that depend on interaction or a critical mix must be *waiting to be discovered.*

Tue 3.1

5.16. BLOCKING THE 2^k FACTORIAL DESIGNS

In a trial to be conducted using a 2^3 factorial design it might be desirable to use the same batch of raw material to make all eight runs. Suppose, however, that batches of raw material were only large enough to make four runs. Then the concept of blocking referred to in Chapters 3 and 4 might be used. Figure 5.16 shows a 2^3 factorial design run in two blocks of four runs so as to neutralize the effect of possible batch differences. This is done by arranging that the runs numbered 1, 4, 6, and 7 use the first batch of raw material and those numbered 2, 3, 5, and 8 use the second batch.

A 2^3 design is blocked in this way by placing all runs in which the contrast **123** is minus in one block and all the other runs, in which **123** is plus, in the other block. To better understand the idea, suppose all the results in the second block were increased by an amount h, say, then whatever the value of h, this block difference would affect only the *123* interaction; because of orthogonality it would *sum out* in the calculation of effects *1, 2, 3, 12, 13,* and *23*.

Notice how this happens. In Figure 5.16 all three main effects are contrasts between averages on opposite faces of the cube. But there are two filled squares and two open squares on each face so that any *systematic* difference produced by batches is eliminated from the main-effects contrasts. Look again at the figure and consider the diagonal contrasts that correspond to the two-factor interactions. Again two black squares and two white squares appear on each side of the contrast. Thus, any systematic difference between the two blocks of four runs will be eliminated from all main effects and two-factor interactions.

What you have gained is the elimination of any systematic difference between the blocks. You have had to give a little to get this substantial advantage. The three-factor interaction and any batch (block) difference are deliberately *confounded* (i.e., confused). You cannot therefore estimate the three-factor interaction separately from the batch effect. However, often (but not always) this high-order interaction **123** can be assumed unimportant.

Generation of Orthogonal blocks

In the 2^3 factorial example, suppose that you give the block variable the identifying number **4**. Then you could think of your experiment as containing four

Run	1	2	3	12	13	23	123	Block
1	−	−	−	+	+	+	−	I
2	+	−	−	−	−	+	+	II
3	−	+	−	−	+	−	+	II
4	+	+	−	+	−	−	−	I
5	−	−	+	+	−	−	+	II
6	+	−	+	−	+	−	−	I
7	−	+	+	−	−	+	−	I
8	+	+	+	+	+	+	+	II

	1	2	3	Run
Block I	−	−	−	1
	+	+	−	4
	+	−	+	6
	−	+	+	7
Block II	+	−	−	2
	−	+	−	3
	−	−	+	5
	+	+	+	8

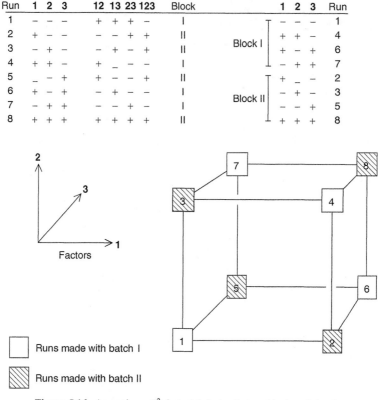

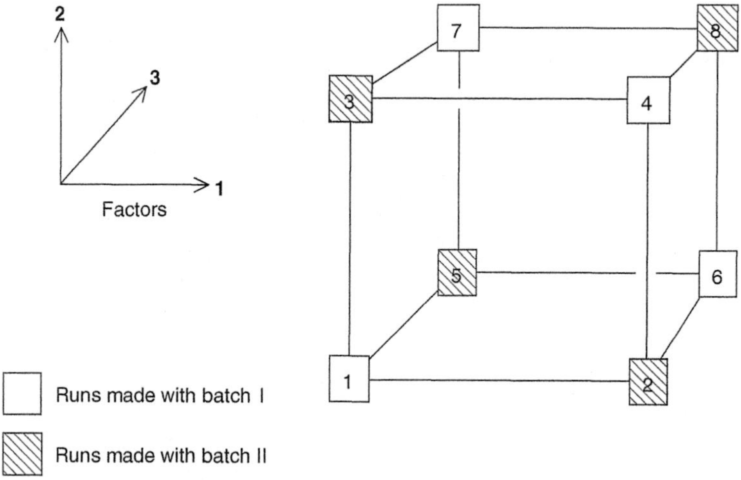

Figure 5.16. Arranging a 2^3 factorial design in two blocks of size 4.

factors, the last of which is assumed to possess the rather special property that it does not interact with the others. If this new factor is introduced by making its levels coincide exactly with the plus and minus signs attributed to the **123** interaction, the blocking may be said to be "generated" by the relationship **4** = **123**. We shall see in Appendices 5A and 5B how this idea may be used to derive more sophisticated blocking arrangements. Similar ideas are used in the next chapter for the construction and blocking of the fractional factorial designs.

Blocks of Size 2

How could the design be arranged in four blocks of two runs so as to do as little damage as possible to the estimates of the main effects?

How Not To Do It

Consider two block factors, which we will call **4** and **5**. At first glance it would seem reasonable to associate block factor **4** with the **123** interaction and to associate block factor **5** with some expendable two-factor interaction, say the **23** interaction. To achieve this, the runs would be placed in different blocks

Table 5.14. (*a*) An *Undesirable* Arrangement: A 2^3 Design in Blocks of Size 2 (*b*) Allocation to Blocks (Variables 4 & 5)

Run Number	Experimental Variable 1	2	3	Block Variable 4 = 123	5 = 23	45 = 1	Experiment Arranged in Four Blocks: Block	1	2	3	Run
					(*a*)						
1	−	−	−	−	+	−	I	+	+	−	4
2	+	−	−	+	+	+		+	−	+	6
3	−	+	−	+	−	−	II	−	+	−	3
4	+	+	−	−	−	+		−	−	+	5
5	−	−	+	+	−	−	III	−	−	−	1
6	+	−	+	−	−	+		−	+	+	7
7	−	+	+	−	+	−	IV	+	−	−	2
8	+	+	+	+	+	+		+	+	+	8

(*b*) Runs

Variable **5**		Variable **4**: −	+
	+	1,7	2,8
	−	4,6	3,5

depending on the signs of the block variables in columns **4** and **5.** Thus runs for which the signs are −− would go in one block, those that have −+ signs in a second block, the +− runs in a third block, and the ++ runs in a fourth. The resulting design is shown in Table 5.14*a*.

A serious weakness in this design can be seen by looking at Table 5.14b. We have confounded block variables **4** and **5** with interactions **123** and **23**. But there are *three* degrees of freedom between the four blocks. With what contrast is the third degree of freedom associated? Inspection of Table 5.14b shows that the third degree of freedom accommodates the **45** interaction. But this is the "diagonal" contrast between runs 1, 7, 3, 5 and runs 2, 8, 4, 6 and also measures the main effect of factor **1**. Thus this arrangement results in confounding of the main effect for factor **1** with block differences!

Generators and Defining Relations

A simple calculus is available to immediately show the consequences of any proposed blocking arrangement. Note that if all the elements of any column in a two-level factorial design are squared (multiplied by themselves pairwise), a column of plus signs is obtained. This is denoted by the symbol **I**. Thus you can write

$$\mathbf{I} = \mathbf{11} = \mathbf{22} = \mathbf{33} = \mathbf{44} = \mathbf{55} \tag{5.1}$$

where, for example, **33** means the product of the elements in column **3** with themselves. Furthermore, the effect of multiplying the elements in any column by the elements in column **I** is to leave those elements unchanged. Thus **I3** = **3**. These are called the *generators* for the blocking arrangements. By multiplying the expression on the left by **4** and that on the right by **5** you get the generators in their standardized form

$$\mathbf{I} = \mathbf{1234}, \qquad \mathbf{I} = \mathbf{235} \tag{5.2}$$

Multiplying these two block generators you obtain a third generator **1223345** = **145** to complete what is called the *defining relation*

$$\mathbf{I} = \mathbf{1234} = \mathbf{235} = \mathbf{145} \tag{5.3}$$

This third generator shows that for this arrangement the estimated interaction *45* and the estimated main effect *1* are confounded.

How To Do It

A better arrangement is obtained by confounding the two block variables **4** and **5** with any two of the two-factor interactions. The third degree of freedom between blocks, the **45** interaction, is then confounded with the third two-factor interaction. Thus setting

$$\mathbf{4} = \mathbf{12}, \qquad \mathbf{5} = \mathbf{13} \tag{5.4}$$

you get the new defining relation

$$\mathbf{I} = \mathbf{124} = \mathbf{135} = \mathbf{2345} \tag{5.5}$$

Thus the estimated block effects *4, 5*, and *45* are associated with the estimated two-factor interactions effects *12, 13*, and *23* and not any main effect. The organization of the experiment in four blocks using this arrangement is shown in Table 5.15, the blocks being typified as before by the pairs of observations for which **4** and **5** take the signatures −−, −+, +−, and ++.

Notice that the two runs comprising each block in the above example are mirror images, complementary in the sense that the plus and minus levels of one run are exactly reversed in the second. Each block is said to consist of a *foldover* pair, for example, in block **I**, the plus and minus signs for the pair of runs + − − and − + +. Any 2^k factorial may always be broken into 2^{k-1} blocks of size 2 by this method. *Such blocking arrangements leave the main effects of the k factors unconfounded with block variables.* All the two-factor interactions, however, are confounded with blocks.

Table 5.15. A 2^3 Design in Blocks of Size 2, a Good Arrangement

Run Number	Experimental Variable			Block Variable		Experiment Arranged in Four Blocks				
	1	**2**	**3**	**4 = 12**	**5 = 13**	Block	**1**	**2**	**3**	Run
1	−	−	−	+	+	I	+	−	−	2
2	+	−	−	−	−		−	+	+	7
3	−	+	−	−	+					
4	+	+	−	+	−	II	−	+	−	3
5	−	−	+	+	−		+	−	+	6
6	+	−	+	−	+					
7	−	+	+	−	−	III	+	+	−	4
8	+	+	+	+	+		−	−	+	5
						IV	−	−	−	1
							+	+	+	8

5.17. LEARNING BY DOING

> One must learn by doing the thing; for though you think you know it, you have no certainty until you try.
>
> SOPHOCLES

You may not be currently involved in a formal project to which you can apply experimental design. In that case we believe you will find it rewarding at this point to plan and perform a home-made factorial experiment and collect and analyze the data. We have regularly assigned such a project to our classes when teaching this material. The students have enjoyed the experience and have learned a great deal from it. We have always left it to the individual to decide what he or she wants to study.

What follows paraphrases a report by a student; Norman Miller, of the University of Wisconsin.

Description of Design

I ride a bicycle a great deal and thought that it might be interesting to see how some of the various parts of the bike affect its performance. In particular, I was interested in the effects of varying the seat height, varying the tire pressure, and using or not using the generator.

In deciding on a design, a 2^3 factorial design immediately came to mind. I expected a large amount of variability, so I decided to replicate the eight data points in order to be able to estimate the variance. Since I wished to do the experiment in a reasonable length of time, I decided to make four runs a day. Although a long run might be desirable in that various types of terrain could be covered, I felt that four long runs a day would be out of the question. I did not have time to make them, I feared that I would tire as the day progressed, making the runs late in the day incompatible with the earlier runs, and I also feared that with the increased exercise my own strength and endurance would increase as the experiment progressed. As a result of the above considerations I decided to do a rather short one and a half block uphill run. My test hill was on Summit Avenue in Madison, starting at Breese Terrace. This hill is of such an inclination that it all must be taken in first gear, thus eliminating gear changes as a possible source of experimental error. I tried to follow a schedule of 9:00 A.M., 11:30 A.M., 2:00 P.M., and 4:30 P.M. As can be seen from the data sheet (Table 5.16), I deviated from the schedule quite a bit; however, I did avoid making runs right after meals. Note that on day 1 I was able to make only three runs because of bad weather. I made up the lost run on day 3. I assigned the variable levels as shown on the data sheet. The seat height of 30 inches is the maximum elevation

Table 5.16. Data from Bicycle Experiment

Setup	Seat Height	Generator	Tire Pressure	Pulse Before	Pulse After	Time (sec)	Day	Hour	Notes
1	−	−	−	76	135	51	3	11:40 A.M.	36°F
2	−	−	−	76	133	54	4	9:40 A.M.	37°F
3	+	−	−	76	132	41	2	12:35 A.M.	23°F, snow on road
4	+	−	−	79	137	43	2	4:05 P.M.	25°F, snow melting
5	−	+	−	77	134	54	3	2:40 P.M.	36°F
6	−	+	−	77	133	60	2	9:50 A.M.	fourth run, 23°F, snow
7	+	+	−	76	130	44	3	9:10 A.M.	26°F, road clear
8	+	+	−	79	139	43	4	3:35 P.M.	40°F
9	−	−	+	76	105	50	1	9:15 A.M.	first run
10	−	−	+	80	144	48	4	6:05 P.M.	40°F
11	+	−	+	77	139	39	3	6:35 P.M.	32°F
12	+	−	+	78	139	39	4	11:40 A.M.	41°F
13	−	+	+	78	137	53	3	4:50 P.M.	35°F, rain
14	−	+	+	79	125	51	1	4:25 P.M.	third run
15	+	+	+	80	122	41	1	11:40 A.M.	second run
16	+	+	+	77	133	44	2	6:35 P.M.	21°F

	Seat height (inches from center of crank)	Generator	Tire pressure (psi)
−	26	off	40
+	30 (maximum height)	on	55

of the seat on my bike. I could ride with it higher, so my bike is in effect too small for me.

I measured two responses. First, I measured the speed in seconds to climb the hill. I naturally was trying to ride as fast as possible. Second, I hoped to get some idea of the amount of energy I was expending during the run by measuring my pulse right before and right after the run. I always rode around two blocks at a slow pace before each run to get warmed up. I then stopped at the bottom of the hill and let my pulse settle to the range of 75 to 80 per minute. I would then make my run, timing it on a watch with a large sweep second hand. At the top of the hill, I took my pulse again. I randomized the experiment by writing the numbers 1 through 16 on pieces of cardboard, putting them in a small box, shaking up the box, and drawing out one number before each experiment to determine the setup of the variables.

Two possible criticisms of the experiment came to mind. First, since I knew the setup of the variables, I might bias the experiment to "come out right." I could not think of a way around this. It would have been quite an imposition to ask a friend to set up the bicycle 16 times! In addition, even if a friend had changed the variables each time, I could have heard the generator operating (or seen that the light was burning) and could have seen or felt the height of the seat. Another rider would probably be just as biased as I would be. In addition, who would ride up a hill 16 times for me? So I just decided to try to be as unbiased as possible. The second criticism involves the rather large temperature range encountered during the experiment. Naturally oil tends to be more viscous at low temperatures. I considered that this problem was probably less acute than it might have been, however, since I used silicone oil to lubricate the bike.

The original data sheet appears in Table 5.16. Notice that there is a problem with the data on pulse. Setup 9, the first run, shows my pulse at the top of the hill to be 105 per minute. Setup 15, the second run, has it at 122 per minute. Setup 14, the third run, has it at 125 per minute, and setup 6, the fourth run, at 133 per minute. What was happening was the following. When I arrived at the top of the hill, my heart was beating quite fast, but it started slowing down quickly as soon as I stopped. With practice, I was simply starting to measure my pulse sooner. Since the pulse data are damaged by this obvious error, I have chosen to ignore them in the following analysis. The readings taken at the bottom of the hill did prove valuable since they served as a gauge of when my body was in a "normal" state.

Analysis of Data

First, I plotted the average times on a cube (Fig. 5.17) and looked at them. Then I calculated:

1. Estimated variance of each of 16 observations = 4.19 square seconds with eight degrees of freedom
2. Main effects and interactions

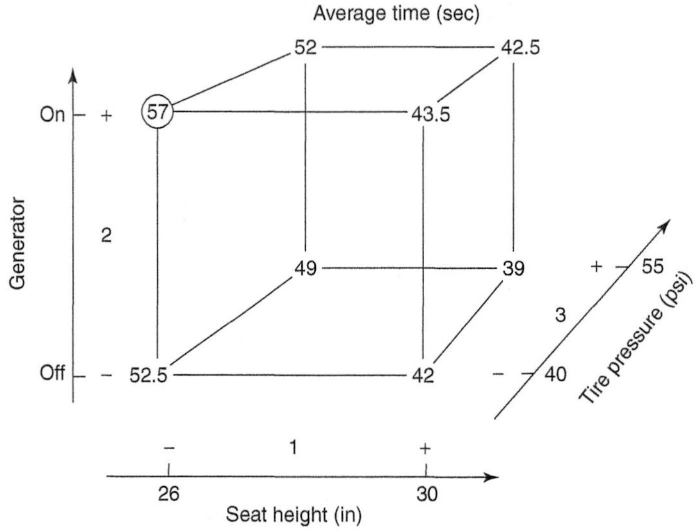

Figure 5.17. Results from bicycle experiment. Response is average time to ride bicycle over set course.

3. Estimated variance of effect = 4.19/4 = 1.05
4. Individual 95% confidence intervals for true values of effects*

Effects		
1 (seat)	=	−10.9 ± 2.4
2 (generator)	=	3.1 ± 2.4
3 (pressure)	=	−3.1 ± 2.4
12 (seat × generator)	=	−0.6 ± 2.4
13 (seat × pressure)	=	1.1 ± 2.4
23 (generator × pressure)	=	0.1 ± 2.4
123 (seat × generator × pressure)	=	0.9 ± 2.4

I believe that the value of 57 circled on the cube is probably too high. Notice that it is the result of setup 6, which had a time of 60 seconds. Its comparison run (setup 5) was only 54 seconds. This is the largest amount of variation in the whole table, by far. I suspect that the correct reading for setup 6 was 55 seconds, that is, I glanced at my watch and thought that it said 60 instead of 55 seconds. Since I am not sure, however, I have not changed it for the analysis. The conclusions would be the same in either case.

Obviously a good stopwatch would have helped the experiment, but I believe that the results are rather good considering the adverse weather and poor timer.

* Notice these provide 95% confidence limits. They are *not* standard errors.

A slightly better design might include a downhill and a level stretch to provide more representative terrain. Probably only two runs could be safely made each day in such a case.

Conclusions

Raising the seat to the high position cuts the time to climb the hill by about 10 seconds. Turning on the generator increases the time by about one-third of that amount. Inflating the tires to 55 psi reduces the time by about the same amount that the generator increases it. The variables appear to be substantially free of interactions.

Since I could ride with the seat at an even higher position, I am planning to modify the bike to achieve this. I also plan to inflate my tires to 60 psi, which is the maximum for which they are designed.

Exercise 5.13. In the bicycle experiment

(a) Identify the two runs that were accidentally switched
(b) Suppose the design had been carried out as intended, if the four days on which the data was obtained are regarded as blocks, would this have been an *orthogonally* blocked experiment?
(c) Make an analysis of variance and show by how much the residual sum of squares would be reduced by blocking.

5.18. SUMMARY

The examples show the value of even very simple factorial experiments. Conclusions from such designs are often best comprehended when the results are displayed geometrically. These designs used in combination with geometric visualization are valuable building blocks in the natural sequential process of learning by experiment. They allow the structure of our thinking to be appropriately modified as investigation proceeds.

APPENDIX 5A. BLOCKING LARGER FACTORIAL DESIGNS

The 2^3 factorial design was used to illustrate the principle of blocking designs by means of confounding. These principles, however, are equally applicable to larger examples. Table 5A.1 provides a list of useful arrangements. To understand how this table is used, consider a 2^6 factorial design containing 64 runs. Suppose that the experimenter wishes to arrange the experiment in eight blocks of eight runs each. The eight blocks, for example, might be associated with eight periods of time or eight batches of raw material.

Then the table suggests you use the arrangement

$$\mathbf{B_1 = 135}, \quad \mathbf{B_2 = 1256}, \quad \mathbf{B_3 = 1234}$$

Remembering the rule **11** = **22** = **33** = **44** = **55** = **66** = **I**, after multiplication of the block variables you obtain the following relationships:

$$\mathbf{B_1 = 135}$$
$$\mathbf{B_2 = 1256}$$
$$\mathbf{B_3 = 1234}$$
$$\mathbf{B_1B_2 = 236}$$
$$\mathbf{B_1B_3 = 245}$$
$$\mathbf{B_2B_3 = 3456}$$
$$\mathbf{B_1B_2B_3 = 146}$$

Thus, only high-order (three-factor and higher order) interactions are confounded with the seven degrees of freedom associated with the eight block comparisons.

Since the generating arrangement is $\mathbf{B_1 = 135}$, $\mathbf{B_2 = 1256}$, $\mathbf{B_3 = 1234}$, to allocate the experimental runs to the eight blocks, you follow the same procedure as before. First, write down a 2^6 factorial in standard order and number the runs. The eight block signatures will be the plus and minus signs corresponding to interactions **135**, **1256**, **1234**. Each signature will identify eight individual runs.

Signs (Signatures) Associated with Block Generators: (135, 1256, 1234)

	(− − −)	(+ − −)	(− + −)	(+ + −)	(− − +)	(+ − +)	(− + +)	(+ + +)
Block number	1	2	3	4	5	6	7	8

The first two blocks are as follows:

Block 1 (**135, 1256, 1234**) = (− − −)							Block 2 (**135, 1256, 1234**) = (+ − −)						
Factors						Run	Factors						Run
1	**2**	**3**	**4**	**5**	**6**	Number	**1**	**2**	**3**	**4**	**5**	**6**	Number
−	+	−	−	−	−	3	+	−	−	−	−	−	2
+	−	+	+	−	−	14	−	+	+	+	−	−	15
−	−	+	−	+	−	21	+	+	+	−	+	−	24
+	+	−	+	+	−	28	−	−	−	+	+	−	25
+	+	+	−	−	+	40	−	−	+	−	−	+	37
−	−	−	+	−	+	41	+	+	−	+	−	+	44
+	−	−	−	+	+	50	−	+	−	−	+	+	51
−	+	+	+	+	+	63	+	−	+	+	+	+	62

Exercise 5.A1. Construct a 2^5 Factorial into Eight Orthogonal Blocks of Four Runs Each

Table 5A.1. Blocking Arrangements for 2^k Factorial Designs

Number of Variables, k	Block Size	Block Generator	Interactions Confounded with Blocks
3	4	$\mathbf{B_1 = 123}$	123
	2	$\mathbf{B_1 = 12, B_2 = 13}$	12, 13, 23
4	8	$\mathbf{B_1 = 1234}$	1234
	4	$\mathbf{B_1 = 124, B_2 = 134}$	124, 134, 23
	2	$\mathbf{B_1 = 12, B_2 = 23, B_3 = 34}$	12, 23, 34, 13, 1234, 24, 14
5	16	$\mathbf{B_1 = 12345}$	12345
	8	$\mathbf{B_1 = 123, B_2 = 345}$	123, 345, 1245
	4	$\mathbf{B_1 = 125, B_2 = 235, B_3 = 345}$	125, 235, 345, 13, 1234, 24, 145
	2	$\mathbf{B_1 = 12, B_2 = 13, B_3 = 34, B_4 = 45}$	12, 13, 34, 45, 23, 1234, 1245, 14, 1345, 35, 24, 2345, 1235, 15, 25, i.e., All 2fi and 4fi *
6	32	$\mathbf{B_1 = 123456}$	123456
	16	$\mathbf{B_1 = 1236, B_2 = 3456}$	1236, 3456, 1245
	8	$\mathbf{B_1 = 135, B_2 = 1256, B_3 = 1234}$	135, 1256, 1234, 236, 245, 3456, 146
	4	$\mathbf{B_1 = 126, B_2 = 136, B_3 = 346, B_4 = 456}$	126, 136, 346, 456, 23, 1234, 1245, 14, 1345, 35, 246, 23456, 12356, 156, 25
	2	$\mathbf{B_1 = 12, B_2 = 23, B_3 = 34, B_4 = 45, B_5 = 56}$	All 2fi, 4fi, and 6fi

* "fi" is an abbreviation for factor interactions.

APPENDIX 5B. PARTIAL CONFOUNDING

When replication is necessary to achieve sufficient accuracy, an opportunity is presented to confound different effects in different replicates. Suppose, for example, that four replicates of a 2^3 factorial system were to be run in 16 blocks of size 2. Then you might run the pattern shown in Table 5B.1 and analyze by averaging unconfounded replicates of the effects

Table 5B.1. Partial Confounding: 2^3 Design in Four Replicates

Replicate	**1**	**2**	**3**	**12**	**13**	**23**	**123**
First	u	u	u	c	c	c	u
Second	c	u	u	u	u	c	c
Third	u	c	u	u	c	u	c
Fourth	u	u	c	c	u	u	c
Number of unconfounded replicates	3	3	3	2	2	2	1
Number of confounded replicates	1	1	1	2	2	2	3

Note: Effects confounded with blocks are indicated by c. Effects not confounded with blocks are indicated by u.

This arrangement would estimate main effects with the greatest precision while providing somewhat less precision in the estimates of two-factor interactions and still less for three-factor interactions. You may find it entertaining to invent other schemes that place different degrees of emphasis on the various effects.

REFERENCES AND FURTHER READING

Barella, and Sust. Unpublished report to the Technical Committee, International Wool Organization. Data given in Box and Cox (1964).

Box, G. E. P. (1999) Statistics as a catalyst to learning by scientific method, part II, A discussion, *J. Qual. Tech.* **31**(1), 16–72.

Box, G. E. P., and Bisgaard, S. (1987) The scientific context of quality improvement, *Quality Progress*, **20** (June), 54–61.

Box, G. E. P. and Cox, D. R. (1964) An analysis of transformations, *J. RSS, Series B* **26**(2), 211–252.

Box, G. E. P. and Wilson K. B. (1951) On the experimental attainment of optimum conditions, *J. Roy. Stat. Soc. Series B*, **13**(1), 1–45.

Chipman, H. (1996) Bayesian variable selection with related predictors, *Can. J. Statist.*, **24**, 17–36.

Daniel, C. (1959) Use of half-normal plot in interpreting factorial two-level experiments, *Technometrics*, **8**, 177.

Daniel, C. (1976) *Applications of Statistics to Industrial Experimentation*, Wiley, New York. (This excellent text thoroughly investigates varied applications of the 2^k factorials, in particular the detection and interpretation of discrepant observations.)

Davies, O. L. (Ed.) (1971) *The Design and Analysis of Industrial Experiments*, Hafner (Macmillan), New York.

Hellstrand, C. (1989) The necessity of modern quality improvement and some experiences with its implementation in the manufacture of roller bearings, *Philos. Trans. Roy. Soc. Ind. Qual. Rel.*, 51–56.

Hill, W. J. and Wiles, R. A., (1975) plant experimentation (PLEX), *J. Qual. Tech.* **7**, 115–122.

Hunter, W. G., and Hoff, M. E. (1967) Planning experiments to increase research efficiency, *Ind. Eng. Chem.*, **59**, 43.

Lenth, R. V. (1989) Quick and easy analysis of unreplicated factorials, *Technometrics*, **31**, 4.

McCullagh, P., and Nelder, J. A. (1989) *Generalized Linear Models*, 2nd ed., Chapman & Hall, London.

Prat, A., and Tort, X. (1990) Case study, design, design of experiments (DOE), fractional factorials, *Quality Eng.*, **3**(1), 59–73.

Yates, F. (1937) *The Design and Analysis of Factorial Experiments*, Bulletin 35, Imperial Bureau of Soil Science, Harpenden Herts, England, Hafner (Macmillan).

QUESTIONS FOR CHAPTER 5

1. What is a factorial design?

2. Why are factorial designs well suited to empirical studies? Suggest an application in your own field.

3. What is a $4^2 \times 3^3 \times 2$ factorial design? How many runs are there in this design? How many variables does it accommodate?

4. What is a two-level factorial design?

5. How many runs are contained in a two-level factorial design for four variables?

6. How many runs does a 2^6 design have? How many variables? How many levels for each variable?

7. How could you write the design matrix for a 2^8 factorial design? How many rows would it have? How many columns? How would you randomize this design?

8. If it could be assumed that each run of a 2^4 design has been independently duplicated, how can the standard error of an effect be calculated? Will the size of the standard error be the same for all the effects? What is meant by "independently duplicated"?

9. If each run of a 2^4 factorial design has been performed only once, describe three methods that might be used to distinguish real effects from noise.

10. How would you set out schemes for running a 2^4 factorial design in (a) two blocks of eight runs? and (b) four blocks of four runs? Mention any assumptions you make.

$9, 10, 11, 12, 13, 14, 15$

PROBLEMS FOR CHAPTER 5

Whether or not specifically asked, the reader should always (1) plot the data in any potentially useful way, (2) state the assumptions made, (3) comment on the appropriateness of these assumptions, and (4) consider alternative analyses.

$?^2$

1. Find the main effects and interaction from the following data:

Nickel (%)	Manganese (%)	Breaking Strength (ft-lb)
N 0 ˜	1 ˜	35
Y 3 +	1 ˜	46
N 0 ˜	2 +	42
Y 3 ˥	2 ˩	40

The purpose of these four trials was to discover the effects of these two alloying elements (nickel and manganese) on the breaking strength of a certain product. Comment on the possible implications of these results.

2. Stating any assumptions you make, plot and analyze the following data:

Temperature	Concentration	Catalyst	Yield
−	−	−	60
+	−	−	77
−	+	−	59
+	+	−	68
−	−	+	57
+	−	+	83
−	+	+	45
+	+	+	85

The ranges for these variables are as follows:

	Temperature (°F)	Concentration (%)	Catalyst
−	160	20	1
+	180	40	2

3. A chemist used a 2^4 factorial design to study the effects of temperature, pH, concentration, and agitation rate on yield (measured in grams). If the standard deviation of an individual observation is 6 grams, what is the variance of the temperature effect?

4. A metallurgical engineer is about to begin a comprehensive study to determine the effects of six variables on the strength of a certain type of alloy.

(a) If a 2^6 factorial design were used, how many runs would be made?

(b) If σ^2 is the experimental error variance of an individual observation, what is the variance of a main effect?

(c) What is the usual formula for a 99% confidence interval for the main effect of a factor?

(d) On the basis of some previous work it is believed that $\sigma = 8000$ pounds. If the experimenter wants 99% confidence intervals for the main effects and interactions whose lengths are equal to 4000 pounds (i.e., the upper limit minus the lower limit is equal to 4000 pounds), how many replications of the 2^6 factorial design will be required?

5. Write a design matrix for a 3^2 factorial design and one for a 2^3 factorial design.

6. A study was conducted to determine the effects of individual bathers on the fecal and total coliform bacterial populations in water. The variables of interest were the time since the subject's last bath, the vigor of the subject's activity in the water, and the subject's sex. The experiments were performed in a 100-gallon polyethylene tub using dechlorinated tap water at 38°C. The bacterial contribution of each bather was determined by subtracting the bacterial concentration measured at 15 and 30 minutes from that measured initially.

A replicated 2^3 factorial design was used for this experiment. The data obtained are presented below. (*Note*: Because the measurement of bacterial populations in water involves a dilution technique, the experimental errors do not have constant variance. Rather, the variation increases with the value of the mean.) Perform analysis using a logarithmic transformation of the data.

(a) Calculate main and interaction effects on fecal and total coliform populations after 15 and 30 minutes.

(b) Construct ANOVA tables.

(c) Do you think the log transformation is appropriate? Can you discover a better transformation?

EXPERIMENTAL VARIABLES

Code	Name	Low Level	High Level
x_1	Time since last bath	1 hour	24 hours
x_2	Vigor of bathing activity	Lethargic	Vigorous
x_3	Sex of bather	Female	Male

Code	Name
y_1	Fecal coliform contribution after 15 minutes (organisms/100 mL)
y_2	Fecal coliform contribution after 30 minutes (organisms/100 mL)
y_3	Total coliform contribution after 15 minutes (organisms/100 mL)
y_4	Total coliform contribution after 30 minutes (organisms/100 mL)

Run Number	x_1	x_2	x_3	Responses (organisms/100 mL) y_1	y_2	y_3	y_4
1	−	−	−	1	1	3	7
2	+	−	−	12	15	57	80
3	−	+	−	16	10	323	360
4	+	+	−	4	6	183	193
5	−	−	+	153	170	426	590
6	+	−	+	129	148	250	243
7	−	+	+	143	170	580	450
8	+	+	+	113	217	650	735
9	−	−	−	2	4	10	27
10	+	−	−	37	39	280	250
11	−	+	−	21	21	33	53
12	+	+	−	2	5	10	87
13	−	−	+	96	67	147	193
14	+	−	+	390	360	1470	1560
15	−	+	+	300	377	665	810
16	+	+	+	280	250	675	795

Source: G. D. Drew, "The Effects of Bathers on the Fecal Coliform, Total Coliform and Total Bacteria Density of Water," Master's Thesis, Department of Civil Engineering, Tufts University, Medford, Mass., October 1971.).

7. Using a 2^3 factorial design, a chemical engineer studied three variables (temperature, pH, and agitation rate) on the yield of a chemical reaction and obtained the following data. Estimate the main effects and interactions. Plot the data. Say what the results might mean, stating the assumptions you make.

x_1	x_2	x_3	y
−1	−1	−1	60
+1	−1	−1	61
−1	+1	−1	54
+1	+1	−1	75
−1	−1	+1	58
+1	−1	+1	61
−1	+1	+1	55
+1	+1	+1	75

where

$$x_1 = \frac{\text{temperature} - 150°\text{C}}{10°\text{C}}$$

$$x_2 = \frac{\text{pH} - 8.0}{0.5}$$

$$x_3 = \frac{\text{agitation rate} - 30 \text{ rpm}}{5 \text{ rpm}}$$

$$y = \text{yield}(\% \text{ theoretical})$$

8. A chemist performed the following experiments, randomizing the order of the runs within each week. Analyze the results.

	Variable			Yield	
Run Number	**Temperature**	**Catalyst**	**pH**	**Week 1**	**Week 2**
1	−	−	−	60.4	62.1
2	+	−	−	75.4	73.1
3	−	+	−	61.2	59.6
4	+	+	−	67.3	66.7
5	−	−	+	66.0	63.3
6	+	−	+	82.9	82.4
7	−	+	+	68.1	71.3
8	+	+	+	75.3	77.1

The minus and plus values of the three variables are as follows:

	Temperature (°C)	**Catalyst (%)**	**pH**
−	130	1	6.8
+	150	2	6.9

The object was to obtain increased yield. Write a report for the chemist, *asking appropriate questions* and giving advice on how the investigation might proceed.

9. Some tests were carried out on a newly designed carburetor. Four variables were studied as follows:

Variable	**−**	**+**
A: tension on spring	Low	High
B: air gap	Narrow	Open
C: size of aperture	Small	Large
D: rate of flow of gas	Slow	Rapid

The immediate object was to find the effects of these changes on the amount of unburned hydrocarbons in the engine exhaust gas. The following results were obtained:

A	*B*	*C*	*D*	Unburned Hydrocarbons
−	+	+	+	8.2
−	−	+	−	1.7
−	−	−	+	6.2
+	−	−	−	3.0
+	−	+	+	6.8
+	+	+	−	5.0
−	+	−	−	3.8
+	+	−	+	9.3

Stating any assumptions, analyze these data. It was hoped to develop a design giving lower levels of unburned hydrocarbons. Write a report, asking appropriate questions and making tentative suggestions for further work.

10. In studying a chemical reaction, a chemist performed a 2^3 factorial design and obtained the following results:

Run Number	Temperature (°C)	Concentration (%)	Stirring Rate (rpm)	Yield (%)
1	50	6	60	54
2	60	6	60	57
3	50	10	60	69
4	60	10	60	70
5	50	6	100	55
6	60	6	100	54
7	50	10	100	80
8	60	10	100	81

The chemist believes that the standard deviation of each observation is approximately unity. Explaining your assumptions, analyze these data.

(a) What are the main conclusions you draw from the data?

(b) If the chemist wants to perform two further runs and his or her object is to obtain high yield values, what settings of temperature, concentration, and stirring rate would you recommend?

11. Write a design matrix for a 4^2 factorial design and draw a geometrical representation of it in the space of the variables. (*Note*: This is *not* a 2^4.)

12. (a) Why do we "block" experimental designs?

(b) Write a 2^3 factorial design in two blocks of four runs each such that no main effect or two-factor interaction is confounded with block differences.

(c) Write a 2^3 factorial design in four blocks of two runs each such that main effects are not confounded with blocks.

13. A group of experimenters is investigating a certain synthetic leather fabric. Set up for them a factorial design suitable for studying four different levels of filler (5, 10, 15, and 20%), two curing agents, and two methods of spreading these agents.

14. Consider the following data:

		Variable			
Run Number	Order in Which Runs Were Performed	Relative Humidity (%)	Ambient Temperature (°F)	Pump	Response: Finish Imperfections
1	4	40	70	On	77
2	7	50	70	On	67
3	1	40	95	On	28
4	2	50	95	On	32
5	3	40	70	Off	75
6	6	50	70	Off	73
7	5	40	95	Off	29
8	8	50	95	Off	28

What inferences can be made about the main effects and interactions? What do you conclude about the production of defects? State assumptions.

15. A 2^3 factorial design has been run in duplicate for the purpose of studying in a research laboratory the effects of three variables (temperature, concentration, and time) on the amount of a certain metal recovered from an ore sample of a given weight. Analyze these data and state the conclusions you reach.

Variable			Response: Weight of Metal Recovered (grams)	
Temperature	Concentration	Time		
−	−	−	80	62
+	−	−	65	63
−	+	−	69	73
+	+	−	74	80
−	−	+	81	79
+	−	+	84	86
−	+	+	91	93
+	+	+	93	93
Code:				
−1600°C	30%	1 hour		
+1900°C	35%	3 hours		

(a) Do you think that these data contain a discrepant value?

(b) Perform an analysis using all the data and plot residuals.

(c) Perform an approximate analysis, omitting the discrepant value.

(d) Report your findings.

16. The following data were obtained from a study dealing with the behavior of a solar energy water heater system. A commercial, selective surface, single-cover collector with a well-insulated storage tank was theoretically modeled. Seventeen computer runs were made, using this theoretical model. The runs were made in accordance with a 2^4 factorial design with an added center point where the four "variables" were four dimensionless groups corresponding to (**1**) the total daily insolation, (**2**) the tank storage capacity, (**3**) the water mass flow rate through the absorber, and (**4**) the "intermittency" of the input solar radiation. For each of these 17 runs two response values were calculated, the collection efficiency (y_1) and the energy delivery efficiency (y_2). Analyze the following results, where L, M, and H designate low, medium, and high values, respectively, for the groups.

Case Number	**1**	**2**	**3**	**4**	y_1	y_2
1	H	H	H	H	41.6	100.0
2	H	H	H	L	39.9	68.6
3	H	H	L	H	51.9	89.8
4	H	H	L	L	43.0	82.2
5	H	L	H	H	39.2	100.0
6	H	L	H	L	37.5	66.0
7	H	L	L	H	50.2	86.3
8	H	L	L	L	41.3	82.0
9	L	H	H	H	41.3	100.0
10	L	H	H	L	39.7	67.7
11	L	H	L	H	52.4	84.1
12	L	H	L	L	44.9	82.1
13	L	L	H	H	38.4	100.0
14	L	L	H	L	35.0	61.7
15	L	L	L	H	51.3	83.7
16	L	L	L	L	43.5	82.0
17	M	M	M	M	45.3	88.9

The motivation for this work was the fact that the theoretical model was too complicated to allow for ready appreciation of the effects of various variables on the responses of interest. This factorial design approach provided the kind of information that was desired.

17. A 2^3 factorial design on radar tracking was replicated 20 times with these results:

1	2	3	$\bar{y}$
−	−	−	7.76 ± 0.53
+	−	−	10.13 ± 0.74
−	+	−	5.86 ± 0.47
+	+	−	8.76 ± 1.24
−	−	+	9.03 ± 1.12
+	−	+	14.59 ± 3.22
−	+	+	9.18 ± 1.80
+	+	+	13.04 ± 2.58

where $\bar{y}$ is the average acquisition time (in seconds) and the number following the plus or minus sign is the calculated standard deviation of $\bar{y}$. The variables were:

1. Modified (−) or standard (+) mode of operation

2. Short (−) or long (+) range

3. Slow (−) or fast (+) plane

Analyze the results. Do you think a transformation of the data should be made?

18. This investigation was concerned with the study of the growth of algal blooms and the effects of phosphorus, nitrogen, and carbon on algae growth. A 2^3 factorial experiment was performed, using laboratory batch cultures of blue-green *Microcystis aeruginosa*. Carbon (as bicarbonate), phosphorus (as phosphate), and nitrogen (as nitrate) were the variable factors. The response was algal population, measured spectrophotometrically.

The levels of the experimental variables were:

Variable		−	+
1:	phosphorus (mg/liter)	0.06	0.30
2:	nitrogen (mg/liter)	2.00	10.00
3:	carbon (mg/liter)	20.0	100.0

The responses (in absorbance units) were:

y_1 Algal population after 4 days incubation
y_2 Algal population after 6 days incubation
y_3 Algal population after 8 days incubation
y_4 Algal population after 10 days incubation

The following data were obtained:

Experimental Design			Response			
1	2	3	y_1	y_2	y_3	y_4
−	−	−	0.312	0.448	0.576	0.326
+	−	−	0.391	0.242	0.309	0.323
−	+	−	0.412	0.434	0.280	0.481
+	+	−	0.376	0.251	0.201	0.312
−	−	+	0.479	0.639	0.656	0.679
+	−	+	0.481	0.583	0.631	0.648
−	+	+	0.465	0.657	0.736	0.680
+	+	+	0.451	0.768	0.814	0.799

Given that $s^2 = 0.003821$ was an independent estimate of variance based on 18 degrees of freedom and appropriate for all four responses, analyze the data from this factorial design.

19. To reduce the amount of a certain pollutant, a waste stream of a small plastic molding factory has to be treated before it is discharged. State laws stipulate that the daily average of this pollutant cannot exceed 10 pounds. The following 11 experiments were performed to determine the best way to treat this waste stream. Analyze these data and list the conclusions you reach.

Order in Which Experiments Were Performed	Chemical Brand[a]	Temperature (°F)	Stirring	Pollutant (lb/day)
5	*A*	72	None	5
6	*B*	72	None	30
1	*A*	100	None	6
9	*B*	100	None	33
11	*A*	72	Fast	4
4	*B*	72	Fast	3
2	*A*	100	Fast	5
7	*B*	100	Fast	4
3	*AB*	86	Intermediate	7
8	*AB*	86	Intermediate	4
10	*AB*	86	Intermediate	3

[a] *AB* denotes a 50–50 mixture of both brands.

20. An experiment is run on an operating chemical process in which four variables are changed in accordance with a randomized factorial plan:

Variable		−	+
1:	concentration of catalyst (%)	5	7
2:	concentration of NaOH (%)	40	45
3:	agitation speed (rpm)	10	20
4:	temperature (°F)	150	180

Order of Running	1	2	3	4	Impurity
2	−	−	−	−	0.38
6	+	−	−	−	0.40
12	−	+	−	−	0.27
4	+	+	−	−	0.30
1	−	−	+	−	0.58
7	+	−	+	−	0.56
14	−	+	+	−	0.30
3	+	+	+	−	0.32
8	−	−	−	+	0.59
10	+	−	−	+	0.62
15	−	+	−	+	0.53
11	+	+	−	+	0.50
16	−	−	+	+	0.79
9	+	−	+	+	0.75
5	−	+	+	+	0.53
13	+	+	+	+	0.54

(a) Make a table of plus and minus signs from which the various effects and interactions could be calculated.

(b) Calculate the effects.

(c) Assuming three-factor and higher order interactions to be zero, compute an estimate of the error variance.

(d) Make normal plots.

(e) Write a report, setting out the conclusions to be drawn from the experiment, and make recommendations. (Assume that the present conditions of operation are as follows: **1** = 5%, **2** = 40%, **3** = 10 rpm, **4** = 180°F.)

(f) Show pictorially what your conclusions mean.

CHAPTER 6

Fractional Factorial Designs

In Chapter 5, as a first introduction to factorial design, there were two experiments on floor waxes. This chapter begins by describing two further experiments from the same study. One objective is again to emphasize the use of the eye and simple sketches for interpreting data.

6.1. EFFECTS OF FIVE FACTORS ON SIX PROPERTIES OF FILMS IN EIGHT RUNS

In this experiment five factors were studied in only eight runs. The factors were the catalyst concentration **A**, the amount of an additive **B**, and the amounts of three emulsifiers **C, D,** and **E**. The polymer solutions were prepared and spread as a film on a microscope slide. The properties of the films were recorded as they dried. The results for six different responses, denoted by y_1, y_2, y_3, y_4, y_5, and y_6, are shown in Table 6.1.

Surprisingly many conclusions can be drawn from these results by visual inspection. On the assumption (reasonable for this application) of dominant main effects, it appears that the important factor affecting haziness y_1 is emulsifier **C** and the important factor affecting adhesion y_2 is the catalyst concentration, factor **A**. Further, the important factors affecting grease on top of film y_3, grease under film y_4, dullness of film when pH is adjusted y_5, and dullness of film with the original pH y_6 are, respectively **D, E, D,** and **D**.

The eight-run design in Table 6.1 was constructed beginning with a standard table of signs for a 2^3 design in the dummy factors **a, b**, and **c**. The column of signs associated with the **bc** interaction was used to accommodate factor **D**. Similarly, the **abc** interaction column was used for factor **E**.

A full factorial for the five factors **A, B, C, D**, and **E** would have needed $2^5 = 32$ runs. In fact, only one-quarter of these were run. Thus the design is a one-fourth fraction of the 2^5 factorial design and is therefore called a quarter fraction,

Statistics for Experimenters, Second Edition. By G. E. P. Box, J. S. Hunter, and W. G. Hunter

Table 6.1. Effects of Five Factors on Six Responses, a 2^{5-3} Design

Run Number	a A	b B	c C	bc D (D = BC)	abc E (E = ABC)	y_1, Hazy?	y_2, Adheres?	y_3, Grease on Top of Film?	y_4, Grease Under Film?	y_5, Dull, Adjusted pH	y_6 Dull, Original pH
1	−	−	−	+	−	No	No	Yes	No	Slightly	Yes
2	+	−	−	+	+	No	Yes	Yes	Yes	Slightly	Yes
3	−	+	−	−	+	No	No	No	Yes	No	No
4	+	+	−	−	−	No	Yes	No	No	No	No
5	−	−	+	−	+	Yes	No	No	Yes	No	Slightly
6	+	−	+	−	−	Yes	Yes	No	No	No	No
7	−	+	+	+	−	Yes	No	Yes	No	Slightly	Yes
8	+	+	+	+	+	Yes	Yes	Yes	Yes	Slightly	Yes

Factors	−	+
A: catalyst (%)	1	1.5
B: additive (%)	1/4	1/2
C: emulsifier *P* (%)	2	3
D: emulsifier *Q* (%)	1	2
E: emulsifier *R* (%)	1	2

or a "quarter replicate," of the full 2^5 factorial and since $1/4 = (\frac{1}{2})^2 = 2^{-2}$, it is referred to as a 2^{5-2} design (a two to the five minus two design). In general, a 2^{k-p} design is a $(\frac{1}{2})^p$ fraction of a 2^k design using 2^{k-p} runs.

6.2. STABILITY OF NEW PRODUCT, FOUR FACTORS IN EIGHT RUNS, A 2^{4-1} DESIGN

This example concerns a chemist in an industrial development laboratory who was trying to formulate a household liquid product using a new process. He was able to demonstrate fairly quickly that the product had a number of attractive properties. Unfortunately, it could not be marketed because it was insufficiently stable.

For months he had been trying many different ways of synthesizing the product in the hope of hitting on conditions that would give stability,* but without success. He had succeeded, however, in identifying four important influences: (**A**) acid concentration, (**B**) catalyst concentration, (**C**) temperature, and (**D**) monomer concentration. He at last decided, somewhat reluctantly, to perform his first statistically planned experiment. His eight-run fractional factorial design is shown in Table 6.2 with the signs of the **abc** interaction used to accommodate factor **D**.

* Stability is one aspect of product robustness discussed in detail in Chapter 13.

In these tests, run in random order, he wanted to achieve a stability value of at least 25. On seeing the data, his first reaction was one of disappointment since none of the individual observations reached the desired stability level.

Following a Trend

From Table 6.2 you will see that the estimated main effects A, B, C, and D were respectively −5.75, −3.75, −1.25, and 0.75. Thus it seemed likely that factors **A** and **B** were dominant in influencing stability—that there was a two-dimensional subspace of maximum activity in acid concentration and catalyst. On this hypothesis the design could now be viewed as a duplicated 2^2 factorial in acid concentration (**A**) and catalyst concentration (**B**), shown in Figure 6.1. This idea was supported by the chemist's judgment that he would not have been surprised to obtain discrepancies in genuine duplications similar in magnitude to those of the pairs of numbers. Sketching the contour lines in by eye, a clear

Table 6.2. Example of a 2^{4-1} Fractional Factorial

	Factors				
Test Number	**a** **A**	**b** **B**	c **C**	**abc** **D**	Stability (R), y
1	−	−	−	−	20
2	+	−	−	+	14
3	−	+	−	+	17
4	+	+	−	−	10
5	−	−	+	+	19
6	+	−	+	−	13
7	−	+	+	−	14
8	+	+	+	+	10

Factors	+	−
A: acid concentration	20	30
B: catalyst concentration	1	2
C: temperature	100	150
D: monomer concentration	25	50

Main Effects	$\bar{y}_+ - \bar{y}_-$
A	11.75 − 17.50 = −5.75
B	12.75 − 16.50 = −3.75
C	14.00 − 12.75 = −1.25
D	15.00 − 14.25 = 0.75

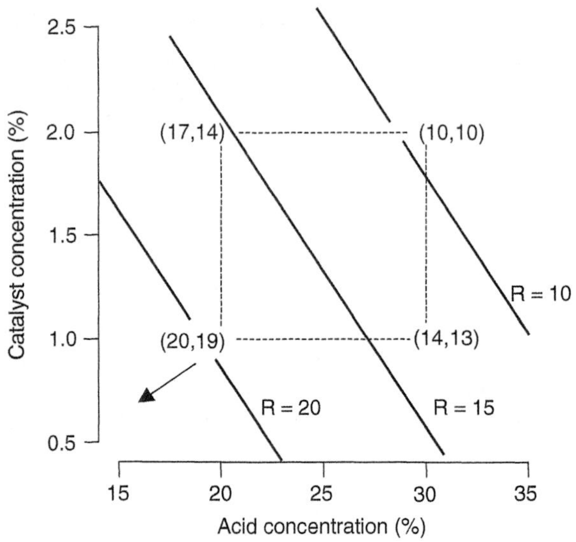

Figure 6.1. Results of a 2^{4-1} fractional factorial design viewed as a duplicated 2^2 factorial design with approximate stability contours: Example 2.

trend was shown.* Thus, although none of the tested conditions produced the desired level of stability, the experiments did show the direction in which such conditions might be found. A few exploratory runs performed in this direction produced for the first time a product with stability even greater than the goal of 25.

Commentary

This example illustrates the following:

1. How a fractional factorial design was used for *screening* purposes to isolate two factors from the original set of four.
2. How a desirable *direction* in which to carry out further experiments was discovered.

All two-level factorial and fractional factorial designs employing eight runs make use of the same fundamental table of signs sometimes called an L_8 orthogonal array. This table is easy to construct. As shown in Table 6.3, one first writes the column of signs for a 2^3 factorial identified by the column headings **a, b**, and **c** along with the four additional columns for the interactions **ab, ac, bc**, and **abc**. You will see later how these same columns of signs may be used to construct eight-run fractional designs for $k = 4, 5, 6, 7$ factors.

* As discussed in Chapter 5, the direction at right angles to the contours is called the *path of steepest ascent*.

Table 6.3. Two-Level Eight-Run Orthogonal Array [a]

Run Number	a A	b B	c C	ab	ac	bc	abc D	FailureRate, y
1	−	−	−	+	+	+	−	16
2	+	−	−	−	−	+	+	7
3	−	+	−	−	+	−	+	14
4	+	+	−	+	−	−	−	5
5	−	−	+	+	−	−	+	11
6	+	−	+	−	+	−	−	7
7	−	+	+	−	−	+	−	13
8	+	+	+	+	+	+	+	4

[a] The factor allocation is for a 2^{4-1} fractional factorial used to investigate the failure rate of a bearing.

6.3. A HALF-FRACTION EXAMPLE: THE MODIFICATION OF A BEARING

In this example a manufacturer of bearings lost an important contract because a competitor had developed a rugged bearing more able to handle ill-balanced loads of laundry in a washing machine. A project team conjectured that a better bearing might result by modifying one or more of the following four factors: **A** a particular characteristic of the manufacturing process for the balls in the bearings, **B** the cage design, **C** the type of grease, and **D** the amount of grease. To check this, a full 2^4 design would have required 16 experimental runs. However a 2^{4-1} half fraction requiring only eight runs was used. The design associating the factors **A, B, C,** and **D** with the orthogonal columns **a, b, c,** and **abc** is shown in Table 6.3. The standard bearing design corresponded to the factor combinations listed for run 1 (Hellstrand (1989)).

The eight runs were made in random order. The response y was the failure rate (the reciprocal of average time to failure). Inspection of the data suggested that factor **D** did not have a large influence. In Figure 6.2*a* cube plots are shown for the runs at the low level of **D** and high level of **D**. Supposing factor **D** to be essentially inert, you can combine the two half fractions to give the full 2^3 factorial design in factors **A, B**, and **C** as in Figure 6.2*b*. You will see that by changing factor **A**—the manufacturing conditions of the balls—a very large reduction in failure rate could be obtained perhaps somewhat enhanced by changing the type of cage **B**. Further tests confirmed these findings and a lost market was regained with a bearing design that exceeded customer expectations.

Exercise 6.1. Redo Figure 6.2 with factor **C** as the splitting factor.

Inert Factor?

In the above example factor **D** was said to be "essentially inert." Reference to "active factors" and "inert factors" occurs frequently in this book. In any

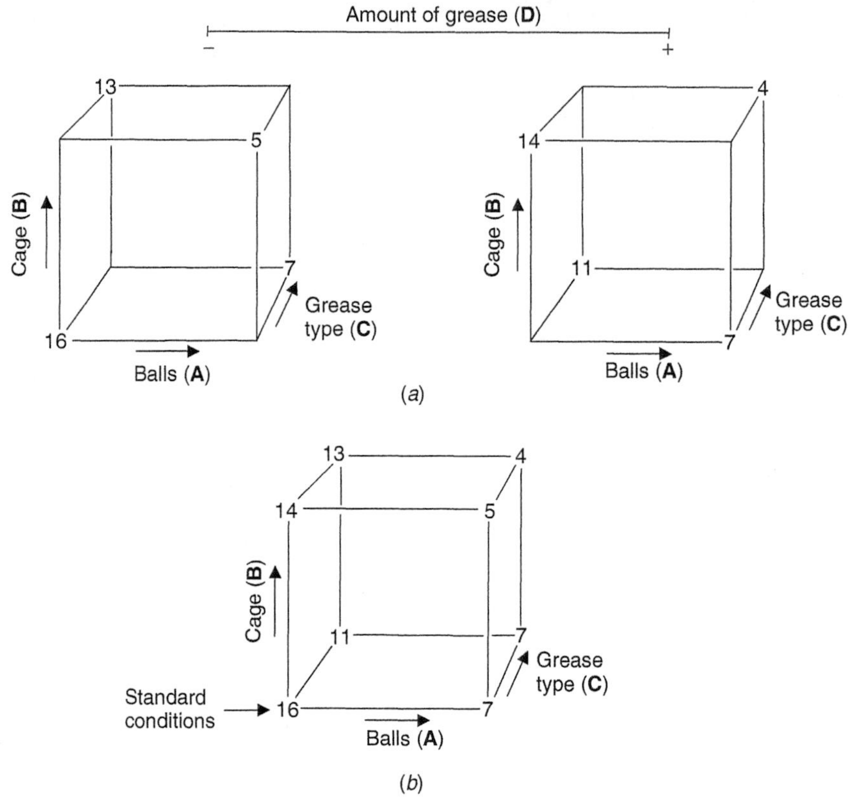

Figure 6.2. (a) Half fraction of the 2^4 factorial with data plotted on the left cube with **D** at the lower (−) level and on the right hand side with **D** at its upper (+) level. (*b*) Combined data plotted on one cube ignoring factor **D**.

investigation it is important to find out which factors have a major influence upon a particular response and which do not. Notice that a major influence might manifest itself via the main effect of the factor or through interactions with other factors. Now, when you change the level of any factor, some effects will almost always occur in some region of the factor space. Thus "inert" implies only that the factor appears *locally* inactive, that is, only in the space of current experimental interest. Notice also that being essentially inert is not necessarily the same as "not statistically significant." You might, for example, have a statistically significant level of change that was of no practical interest.

6.4. THE ANATOMY OF THE HALF FRACTION

With four factors **A, B, C**, and **D**, if you employ a complete 2^4 factorial design, you can estimate 16 independent quantities—the average, four main effects, six two-factor interactions, four three-factor interaction, and one four-factor interaction. Exactly, then, what can you estimate with a 2^{4-1} half fraction? The design in the above example was obtained by using the signs of the **abc** interaction to accommodate factor **D**. Thus, the main effect of factor **D** cannot be distinguished from the **ABC** interaction. In fact, what you really have is an estimate of the sum, **D + ABC**. If the notation l_D is employed to denote the contrast actually computed, that is, $l_D = 1/4(-y_1 + y_2 + y_3 - y_4 + y_5 - y_6 - y_7 + y_8)$, then, using an arrow to mean "is an estimate of," you can write $l_D \to D + ABC$. Using this design, the effects D and ABC are said to be "confounded". They cannot be separately estimated and ABC is called an "alias" of D.

This confounding has consequences for all the estimated effects. By looking at the signs in Table 6.3, you will find that

$$\begin{array}{ll} l_A \to A + (BCD) & l_{AB} \to AB + CD \\ l_B \to B + (ACD) & l_{AC} \to AC + BD \\ l_C \to C + (ABD) & l_{BC} \to BC + AD \end{array}$$
$$l_D \to D + (ABC)$$

Often interactions between three* or more factors would be small enough to be ignored. On this assumption, l_A, l_B, l_C, and l_D would provide estimates of main effects of the four factors **A, B, C**, and **D**.

Here is an explanation of the generation of this design.
The 2^{4-1} fractional factorial was obtained by using the letters of the generating 2^3 factorial so that $\mathbf{A = a, B = b, C = c}$, and $\mathbf{D = abc}$. Thus,

$$\mathbf{D = ABC}$$

is called the *generating relation* for the design. Now if you multiply the signs of the elements in any column by those of the same column you will get a column of plus signs identified by the symbol **I**, called the *identity*. Thus $\mathbf{A \times A = A^2 = I, B \times B = B^2 = I}$, and so forth. Multiplying both sides of the design generator **ABC** by **D** gives

$$\mathbf{D \times D = D^2 = ABCD}$$

that is,

$$\mathbf{I = ABCD}$$

The expression $\mathbf{I = ABCD}$ is a somewhat more convenient form for the generating relation. (As a check, you will find that multiplying together the signs of the columns **A, B, C**, and **D** of the 2^{4-1} design gives a column of plus signs.)
Once you have the generating relation in this convenient form, you can immediately obtain all the alias relationships. For example, multiplying both sides of $\mathbf{I = ABCD}$ by **A** gives $\mathbf{A = A^2BCD = IBCD = BCD}$, the identity **I** acting as a 1. Thus the estimates

*The three factor interactions **BCD, ACD,** ... are estimates of the third-order mixed derivatives $\partial y^3/\partial B\ \partial C\ \partial D,\ \partial y^3/\partial A\ \partial C\ \partial D, \ldots$. It is these third-order effects that are assumed negligible.

of **A** and **BCD** are confounded and **A** and **BCD** are aliases. For the design, then,

$$\mathbf{I} = \mathbf{ABCD}, \quad \mathbf{B} = \mathbf{ACD}, \quad \mathbf{C} = \mathbf{ABD}, \quad \mathbf{D} = \mathbf{ABC}$$

$$\mathbf{AB} = \mathbf{CD}, \quad \mathbf{AC} = \mathbf{BD}, \quad \mathbf{AD} = \mathbf{BC}$$

Design Resolution

The 2^{4-1} fractional design used here is said to be of *resolution* 4. It has main effects aliased with three-factor interactions and two-factor interactions aliased with one another. These alias patterns occur because the generating relation for this design, $\mathbf{I} = \mathbf{ABCD}$, contains four letters. Thus, by using the multiplication rule, we find that any effect identified by a single letter is coupled with a *three-letter effect*; thus $\mathbf{A} = \mathbf{BCD}, \mathbf{B} = \mathbf{ACD}, \mathbf{C} = \mathbf{ABD}$, and $\mathbf{D} = \mathbf{ABC}$. Similarly, every *two-letter effect* is coupled with *another* two-letter effect; thus $\mathbf{AB} = \mathbf{CD}, \mathbf{AC} = \mathbf{BD}$, and $\mathbf{AD} = \mathbf{BC}$. The resolution of a fractional factorial design is denoted by a Roman numeral subscript. The design described here then is a 2_{IV}^{4-1}, or in words a "two to the four minus one, resolution four" design. Notice in this example that the generating relation $\mathbf{I} = \mathbf{ABCD}$ for this design contains four letters. In general, the resolution of *any half-fraction design* equals the number of letters in its generating relation.

If instead you had used a *two-factor* interaction column, say **AB**, to accommodate the signs of factor **D**, you would then have $\mathbf{D} = \mathbf{AB}$ with generating relation $\mathbf{I} = \mathbf{ABD}$. Thus with this design $\mathbf{A} = \mathbf{BD}, \mathbf{B} = \mathbf{AD}, \mathbf{C} = \mathbf{ABCD}, \mathbf{D} = \mathbf{AB}, \mathbf{AC} = \mathbf{BCD}, \mathbf{BC} = \mathbf{ACD}$, and $\mathbf{CD} = \mathbf{ABC}$. Note that with this design three of the main effects are aliased with two-factor interactions. Since the generating relation $\mathbf{I} = \mathbf{ABD}$ contains a three-letter word, this design is of resolution III, a 2_{III}^{4-1} design. To get the most desirable alias patterns, fractional factorial designs of highest resolution would usually be employed; there are, however, important exceptions to the rule.

High-Order Interactions Negligible

Assuming that interactions between three or more factors may be ignored, the analysis of the bearing experiment is shown on the left of the table below. The analysis assumes that **D** is essentially inert whence it follows that all effects containing the letter *D* will be negligible and you obtain the effects shown on the right:

Effects	Estimates	Effects Assuming **D** Is Inert
A	−7.75	A
B	−1.25	B
C	−1.75	C
D	−1.25	D (noise)
$AB + CD$	−1.25	AB
$AC + BD$	1.25	AC
$AD + BC$	0.75	BC

Redundancy

Large factorial designs allow for the estimation of many higher order effects. For example, a 2^5 design with 32 runs provides independent estimates not only of the mean, 5 main effects, and 10 two-factor interactions but also of 10 three-factor interactions, 5 four-factor interactions, and one five-factor interaction. If you make the assumption that you can ignore three-factor and higher order interactions, then you are left with the mean, the 5 main effects, and the 10 two-factor interactions—a 32-run design is being used to estimate just 16 effects. Fractional factorials employ such redundancy by arranging that lower order effects are confounded with those of higher order that are assumed negligible.

Parsimony and Projectivity

A somewhat different justification for the use of fractional factorials is based on the joint use of the principles of *parsimony** and design *projectivity.*

Parsimony

J. M. Juran, a pioneer advocate of the use of statistical methods in quality, refers to the *vital* few factors and the *trivial* many. This is a statement of the Pareto principle, which in this context says that the active factor space is frequently of lower dimension than the design space—that some of the factors are without much effect and may thus be regarded as essentially inert.

Projectivity

Look at the following four-run design in factors **A**, **B**, and **C**:

a	**b**	**ab**
A	B	C
−	−	+
+	−	−
−	+	−
+	+	+

The design is a 2_{III}^{3-1} (resolution III) design sometimes referred to as the L_4 orthogonal array. Since it uses the signs of **ab** to accommodate factor **C**, the defining relation is **C = AB** or **I = ABC**. If you drop out any one factor, you obtain a 2^2 factorial in the remaining two factors. Figure 6.3 shows this geometrically. You can say the 3D design *projects* a 2^2 factorial design in all three subspaces of two dimensions. The 3D design is thus said to be of *projectivity* $P = 2$. In a similar way the 2^{4-1} design of the bearing example is a design of

*The principle, sometimes called "Occam's razor," is very old and seems to have been first stated by a Franciscan monk, William of Occam, sometime in the fourteenth century.

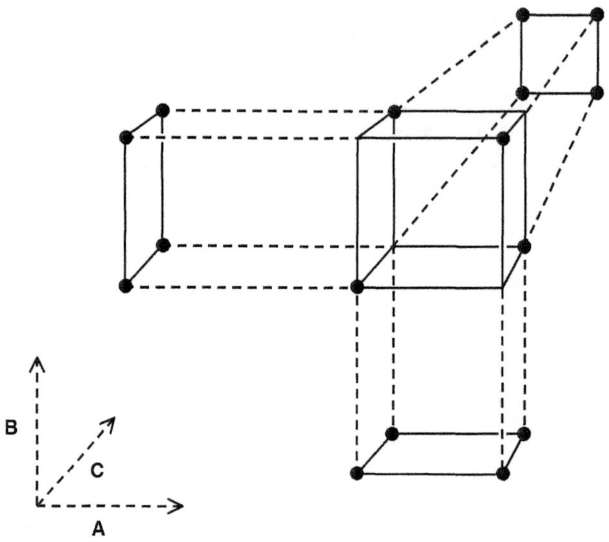

Figure 6.3. A 2_{III}^{3-1} design showing projections into three 2^2 factorials.

projectivity $P = 3$. You saw in Figure 6.2 that by dropping factor **D** a 2^3 factorial was produced in factors **A, B**, and **C**. But if you had dropped *any* one of the four factors **A, B, C, D** you would have a full 2^3 design in the remaining three. In general, for fractional factorial designs the projectivity P is 1 less than the resolution R, that is, every subset of $P = R - 1$ factors produces a complete factorial (possibly replicated) in P factors Thus, if an unknown subset of P or fewer factors are active, you will have a complete factorial design for whichever these active factors happen to be. The fact that a fractional factorial design of resolution R has a projectivity $P = R - 1$ is later used extensively.

6.5. THE 2_{III}^{7-4} DESIGN: A BICYCLE EXAMPLE

As you might expect, in sequential experimentation, unless the total number of runs of a full or replicated factorial is needed to achieve a desired level of precision, it is usually best to start with a fractional factorial design. Such designs can always be augmented later if this should prove necessary. In that case, the first fractional design will indicate how a further fraction should be chosen so as to meld into a larger design in such a way as to answer questions and resolve ambiguities. The previous chapter contained a bicycle example. The following data come from another such experiment conducted by a different (somewhat more ambitious) rider. The investigation began with an eight-run fractional factorial design used to study seven factors.

The data are the times it took to climb a particular hill. The runs were made in random order. The seven factors shown in Table 6.4 indicated by **A, B, C, . . . , G** use the columns of signs obtained from the main effects and interactions of the

Table 6.4. Eight-Run Experimental Design for Studying the Effects of Seven Factors

Run Number	Seat Up/ Down **a** **A**	Dynamo Off/On **b** **B**	handlebars Up/ Down **c** **C**	Gear Low/ Medium **ab** **D**	Raincoat On/Off **ac** **E**	Breakfast Yes/No **bc** **F**	Tires Hard/ Soft **abc** **G**	Climb Hill (sec) y
1	−	−	−	+	+	+	−	69
2	+	−	−	−	−	+	+	52
3	−	+	−	−	+	−	+	60
4	+	+	−	+	−	−	−	83
5	−	−	+	+	−	−	+	71
6	+	−	+	−	+	−	−	50
7	−	+	+	−	−	+	−	59
8	+	+	+	+	+	+	+	88

Table 6.5. Confounding Pattern Between Main Effects and Two-Factor Interactions for the Bicycle Data

	Factors
Seat	$l_A = 3.5 \rightarrow A + BD + CE + FG$
Dynamo	$l_B = \mathbf{12.0} \rightarrow B + AD + CF + EG$
Handlebars	$l_C = 1.0 \rightarrow C + AE + BF + DG$
Gear	$l_D = \mathbf{22.5} \rightarrow D + AB + EF + CG$
Raincoat	$l_E = 0.5 \rightarrow E + AC + DF + BG$
Breakfast	$l_F = 1.0 \rightarrow F + BC + DE + AG$
Tires	$l_G = 2.5 \rightarrow G + CD + BE + AF$
	$l_I = 66.5 \rightarrow$ average

Note: Here and in what follow boldface numbers indicate estimates that are distinguishable from noise.

dummy variables **a**, **b** and **c** shown in Table 6.4. These seven factors identify different dispositions of bike equipment and bike rider. The design, a 2_{III}^{7-4} fractional factorial, is a one-sixteenth "replicate" (i.e., a one-sixteenth fraction) of the 2^7 factorial and is of projectivity $P = 2$. The analysis of the data in Table 6.5 shows the alias structure for the design omitting interactions between three factors or more. Preliminary test runs suggested that the standard deviation for repeated runs up the hill under similar conditions was about 3 seconds. Thus the calculated contrasts $l_A, l_B, \ldots, l_G$ have a standard error of about

$$\sqrt{\frac{3^2}{4} + \frac{3^2}{4}} = 2.1$$

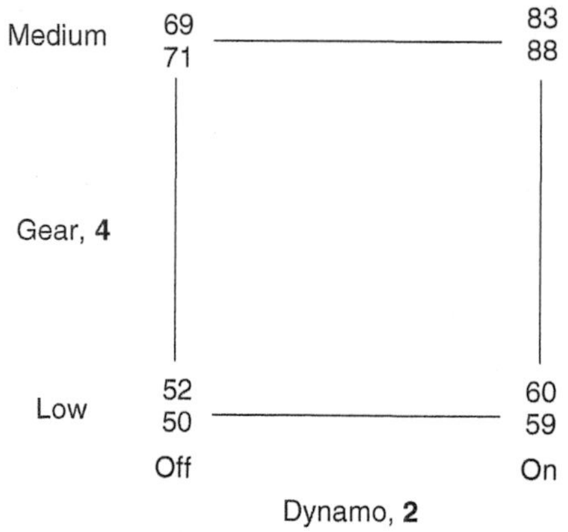

Figure 6.4. Bicycle example: the projection of seven factors into a 2D active subspace.

As indicated in Table 6.5, evidently only two estimates l_B and l_D are distinguishable from noise. The simplest interpretation of the results is that only two of the seven effects, the dynamo **B** and gear **D**, exerted a detectable influence, and they did so by way of their main effects. On this interpretation use of the dynamo adds about 12 seconds to the time, and using the medium gear instead of the low gear adds about 22 seconds. Further, on this interpretation the design projects down to the replicated 2^2 factorial design shown in Figure 6.4. There are of course possible ambiguities in these conclusions, and later in this chapter ways are considered in which they may be resolved.

The Defining Relation

The generators of this 2_{IV}^{7-4} design are $\mathbf{D = AB, E = AC, F = BC, G = ABC}$ or equally the defining relation begins with $\mathbf{I = ABD = ACE = BCF = ABCG}$. To complete the defining relation, we must add all "words" that can be constructed from all multiplications of these four generators. For example, $\mathbf{ABD \times ACE = A^2BCDE = BCDE}$, since $\mathbf{A^2 = I}$. The complete defining relation is thus $\mathbf{I = ABD = ACE = BCF = ABCG = BCDE = ACDF = CDG = ABEF = BEG = AFG = DEF = ADEG = BDFG = CEFG = ABCDEFG}$. Note that the smallest "word" in the defining relation has three letters and thus the 2^{7-4} is a resolution III design. Multiplying through the defining relation by **A** gives $\mathbf{A = BD = CE = ABCF = BCG = ABCDE = CDF = ACDG = BEF = ABEG = FG = ADEF = DEG = ABDFG = ACEFG = BCDEFG}$. Assuming all three-factor and higher order interactions to be nil, we get the *abbreviated* alias pattern $\mathbf{A = BD = CE = FG}$. The abbreviated alias structure for all main effects is given in Table 6.5.

6.6. EIGHT-RUN DESIGNS

Various examples of eight-run fractional designs have been presented, but it is important to understand that all eight-run two-level factorial and fractional factorial designs accommodating up to seven factors can be obtained from the single basic orthogonal array shown in Table 6.6a. This array is obtained by writing all seven columns of signs generated by the main effect and interactions of the three dummy factors **a, b,** and **c**. Table 6.6b shows how this same array may be used to produce the 2^3, the 2_{IV}^{4-1}, or the 2_{III}^{7-4} designs. In particular, associating factors **A, B, . . ., G** with the seven columns **a, b, . . ., abc** results in the "saturated" 2_{III}^{7-4} design. Any other eight-run design may be obtained by dropping appropriate columns from this seven-factor eight-run nodal design. For example, if you drop factors **D, E,** and **F**, you obtain the 2_{IV}^{4-1} nodal design in factors **A, B, C,** and **G**. Similarly, designs for studying only five or six factors in eight runs, the 2_{III}^{5-2} and 2_{III}^{6-3} designs, may be obtained by dropping one or two columns from the 2_{III}^{7-4} nodal design.

The four-factor design (2_{IV}^{4-1}) and the seven-factor design (2_{III}^{7-4}) are called *nodal* designs because for a given number of runs they include the largest number of factors at a given resolution. The alias tables for the four- and seven-factor nodal designs are shown in Table 6.6, c and d. The alias structures for other eight-run designs are derived from these by dropping appropriate letters from the basic nodal alias tables.*

6.7. USING TABLE 6.6: AN ILLUSTRATION

In the early stages of a laboratory experiment five factors listed in Table 6.7 and here labeled **1, 2, . . . , 5** were proposed for further investigation. The best reactor conditions known at that time were thought to be far from optimal, and consequently main effects were believed to be dominant and interactions relatively small. (When you are far down the side of a hill, it is the slopes of the hill that are important.) However, on reflection the experimenters believed that one interaction $\mathbf{1} \times \mathbf{3}$ might be active. The two columns of signs dropped from the 2_{III}^{7-4} nodal design were therefore carefully chosen to ensure that the column of signs of the interaction $\mathbf{1} \times \mathbf{3}$ were not used to accommodate any other effect.

Study of Tables 6.6a and d shows that on the above assumption one way in which this can be done is by dropping columns **E** and **F** from the seven-factor nodal arrangement, and associating the five factors **1, 2, 3, 4, 5** with columns **A, B, C, D, G**. The date obtained from this experiment and the calculated effects of their expected (mean) values are shown in Table 6.7.

*To show the essential unity of the eight-run designs the association of factors **A, B, C, D, E, F, G** with **a, b, c, ab, ac, bc, abc** is maintained here. Earlier the eight runs of the 2_{IV}^{4-1} was shown as a design in factors **A, B, C, D** rather than **A, B, C, G**. This is of course merely a relabeling. Note also that (*a*) switching all the signs in a contrast column, (*b*) reordering the contrast columns, and (*c*) reordering the rows leaves the basic design unchanged. These operations *a, b, c* simply determine (a) which of the two-factor levels is coded by + and which by −, (b) the reallocation of the letters to the fractions, and (c) the change in order in which the runs are listed. So any two designs which are such that one can be obtained from the other by any or all of these operations are regarded as identical.

Table 6.6. Table for Generating Eight-Run Designs

a. Orthogonal Columns

a	**b**	**c**	**ab**	**ac**	**bc**	**abc**
−	−	−	+	+	+	−
+	−	−	−	−	+	+
−	+	−	−	+	−	+
+	+	−	+	−	−	−
−	−	+	+	−	−	+
+	−	+	−	+	−	−
−	+	+	−	−	+	−
+	+	+	+	+	+	+

b. Orthogonal Columns and Factor Identifications for nodal Eight-Run Designs

	a	**b**	**c**	**ab**	**ac**	**bc**	**abc**	Projectivity P
2^3	A	B	C					
2_{IV}^{4-1}	A	B	C				G	3
2_{III}^{7-4}	A	B	C	D	E	F	G	2

Aliases

	c. 2_{IV}^{4-1}		d. 2_{III}^{7-4}
	A B C G		**A B C D E F G**
a	A	**a**	$A + BD + CE + FG$
b	B	**b**	$B + AD + CF + EG$
c	C	**c**	$C + AE + BF + DG$
ab	$AB + CG$	**ab**	$D + AB + EF + CG$
ac	$AC + BG$	**ac**	$E + AC + DF + BG$
bc	$BC + AG$	**bc**	$F + BC + DE + AG$
abc	G	**abc**	$G + CD + BE + AF$

From this analysis it appears that factors **1**, **3**, and **5** all had large *negative* effects. It was concluded that higher yields might be obtained by moving in a direction such that the concentration of γ, the amount of solvent, and the time of reaction were all *reduced.* A series of further experiments in the indicated direction confirmed this possibility and a yield of 84% was eventually obtained.

Exercise 6.2. Write down the alias pattern associated with the design in which column F is used for the fifth factor and column G is left unused.

Table 6.7. A 2_{III}^{5-2} Design Derived from the Nodal 2_{III}^{7-4} and Used to Study the Yield of a Chemical Manufacturing Process

Factors	−1	+1	
1: concentration of γ	94	96	%
2: proportion of γ to α	3.85	4.15	mol/mol
3: amount of solvent	280	310	cm^3
4: proportion of β to α	3.5	5.5	mol/mol
5: reaction time	2	4	hr

	a	**b**	**c**	**ab**	**abc**	
	A	**B**	**C**	**D**	**G**	
Number Run	**1**	**2**	**3**	**4**	**5**	y
1	−	−	−	+	−	77.1
2	+	−	−	−	+	68.9
3	−	+	−	−	+	75.5
4	+	+	−	+	−	72.5
5	−	−	+	+	+	67.9
6	+	−	+	−	−	68.5
7	−	+	+	−	−	71.5
8	+	+	+	+	+	63.7

$\ell_A = \mathbf{-4.6 \to 1}$
$\ell_B = \ 0.2 \to \mathbf{2}$
$\ell_C = \mathbf{-5.6 \to 3}$
$\ell_D = -0.8 \to \mathbf{4}$
$\ell_E = \ 1.0 \to \mathbf{1 \times 3}$
$\ell_F = -0.8 \to$ Noise
$\ell_G = \mathbf{-3.4 \to 5}$

6.8. SIGN SWITCHING, FOLDOVER, AND SEQUENTIAL ASSEMBLY

As we have said earlier, further runs may be needed when fractional designs yield ambiguities. One strategy to employ in this sequential assembly of designs is called "foldover." It is achieved by sign switching.

An Example of a Single-Column Foldover

Look again at the bicycle experiment in Table 6.4. Even though the tentative conclusion from the 8-run design was that factors **B** and **D** (dynamo and gear) explained the data, the cyclist experimenter had previously suspected that changing the gear (factor **D**) from low to medium would interact with, at least, some of the other factors. To check this, he added the second set of 8 runs shown in Table 6.8a in which *only* the column of signs for factor **D** was reversed (each plus sign in this column replaced by a minus sign and each minus by a plus).

Thus in the alias table all factor interactions containing **D** switched signs. (see Table 6.8b.) The 16 runs resulting from the two 2_{III}^{7-4} fractionals may now be considered together by combining the results of Tables 6.5 and 6.8b by addition and subtraction of the estimates obtained from the two 8-run designs. These 16 runs together provide *unaliased* estimates of the main effect of **D** *and all two-factor interactions involving* **D.** In this way we obtain Table 6.9. Equivalently, of course, we can treat the two sets of 8 runs as two blocks in a 16-run design and analyze accordingly. The two sets of 8 runs are here considered separately to illustrate the concept.

The large main effect for factor **D** (gear) is now estimated free of aliases. Also, as required, all two-factor interactions involving **D** are likewise free. For this particular set of data, none of these two-factor interactions proved distinguishable from noise. In this case, then the tentative conclusions from the first 8 runs appear to be borne out.

Table 6.8a. A Second 2_{III}^{7-4} Fractional Factorial Design with Times to Cycle Up a Hill with Defining Relations I = −ABD, I = ACE, I = BCF, I = ABCG

Number Run	Seat Up/ Down **a** **A**	Dynamo Off/On **b** **B**	Handlebars Up/Down **c** **C**	Gear Low/ Medium **-ab** **D**	Raincoat On/Off **ac** **E**	Breakfast Yes/No **bc** **F**	Tires Hard/ Soft **abc** **G**	Climb Hill, y (sec)
9	−	−	−	−	+	+	−	47
10	+	−	−	+	−	+	+	74
11	−	+	−	+	+	−	+	84
12	+	+	−	−	−	−	−	62
13	−	−	+	−	−	−	+	53
14	+	−	+	+	+	−	−	78
15	−	+	+	+	−	+	−	87
16	+	+	+	−	+	+	+	60

Note: **a**, **b**, . . ., **abc** refer to columns; **A**, **B**, . . ., **G** to factors.

Table 6.8b. Calculated Contrasts and Abbreviated Confounding Pattern for the Second Eight-Runs Design in the Bicycle Experiment

Seat	$l'_A = 0.7 \rightarrow A - BD + CE + FG$
Dynamo	$l'_B = \mathbf{10.2} \rightarrow B - AD + CF + EG$
Handlebars	$l'_C = 2.7 \rightarrow C + AE + BF - DG$
Gear	$l'_D = \mathbf{25.2} \rightarrow D - AB - EF - CG$
Raincoat	$l'_E = -1.7 \rightarrow E + AC - DF + BG$
Breakfast	$l'_F = 2.2 \rightarrow F + BC - DE + AG$
Tires	$l'_G = -0.7 \rightarrow G - CD + BE + AF$
	$l_I = 68.125 \rightarrow \text{average}$

Table 6.9. Analysis of Complete Set of 16 runs, Combining the Results of the Two Fractions: Bicycle Example

Seat	$1/2(l_A + l'_A)$	$= 1/2(3.5 + 0.7)$	$= 2.1$	$\rightarrow A + CE + FG$
Dynamo	$1/2(l_B + l'_B)$	$= 1/2(12.0 + 10.2)$	$=$ **11.1**	$\rightarrow B + CF + EG$
Handlebars	$1/2(l_C + l'_C)$	$= 1/2(1.0 + 2.7)$	$= 1.9$	$\rightarrow C + AE + BF$
Gear	$1/2(l_D + l'_D)$	$= 1/2(22.5 + 25.2)$	$=$ **23.9**	$\rightarrow D$
Raincoat	$1/2(l_E + l'_E)$	$= 1/2(0.5 - 1.7)$	$= -0.6$	$\rightarrow E + AC + BG$
Breakfast	$1/2(l_F + l'_F)$	$= 1/2(1.0 + 2.2)$	$= 1.6$	$\rightarrow F + BC + AG$
Tires	$1/2(l_G + l'_G)$	$= 1/2(2.5 - 0.7)$	$= 0.9$	$\rightarrow G + BE + AF$
	$1/2(l_A - l'_A)$	$= 1/2(3.5 - 0.7)$	$= 1.4$	$\rightarrow BD$
	$1/2(l_B - l'_B)$	$= 1/2(12.0 - 10.2)$	$= 0.9$	$\rightarrow AD$
	$1/2(l_C - l'_C)$	$= 1/2(1.0 - 2.7)$	$= -0.9$	$\rightarrow DG$
	$1/2(l_D - l'_D)$	$= 1/2(22.5 - 25.2)$	$= -1.4$	$\rightarrow AB + EF + CG$
	$1/2(l_E - l'_E)$	$= 1/2(0.5 + 1.7)$	$= 1.1$	$\rightarrow DF$
	$1/2(l_F - l'_F)$	$= 1/2(1.0 - 2.2)$	$= -0.6$	$\rightarrow DE$
	$1/2(l_G - l'_G)$	$= 1/2(2.5 + 0.7)$	$= 1.6$	$\rightarrow CD$
	$1/2(l_I + l'_I)$	$= 1/2(66.5 + 68.1)$	$= 67.3$	$\rightarrow$ average
	$1/2(l_I - l'_I)$	$= 1/2(66.5 - 68.1)$	$= -1.6$	$\rightarrow$ block effect

For any given fraction the same procedure could be used to "de-alias" a particular main effect and all its interactions with other effects.

Exercise 6.3. Fold over the last column in the design of Table 6.8a and confirm that, on the usual assumptions concerning three-factor and higher order interactions, the main effect of **G** and its interactions **AG, BG, CG, DG, EG**, and **FG** are free of aliases.

Reprise

The "one-shot" philosophy of experimentation described in much statistical teaching and many textbooks would be appropriate for situations where irrevocable decisions must be made based on data from an individual experiment that cannot be augmented. Particularly in the social sciences, many problems are of this kind and one-shot experimentation is the only option. However, this is much less common in industrial investigations. It is the goal of this book to emphasize the great value of experimental design as a catalyst to the *sequential* process of scientific *learning*. It must be remembered that the framework for an experimental design, fractional or not, and indeed for any investigation is a complex of informed *guesses* that profoundly influence its course. The need for these guesses has nothing to do with the use of statistical experimental design. They must be made whatever the experimental method. These guesses include initially what factors to include, what responses to measure, where to locate the experimental region, by how much to vary the factors, and then once the data are available *how to proceed.* All these guesses are treated as "givens" in a one-shot philosophy. Obviously, different experimenters will make different choices. Thus the aim of any philosophy of continued experimentation cannot be uniqueness, but rather a strategy that is likely to converge to a useful solution. Structured experimentation can produce for the experimenter the clearest possible information on the effects and interactions of the factors for each of the various responses. An opportunity for *second* guessing provides the best chance of understanding what is going on. The subject matter specialist can build on this understanding.* Nothing can eliminate uncertainty, but the skilled interactive use of appropriate tools can enormously reduce its influence. In this context you do not need to answer all questions with one experiment. Smaller experiments that successively reduce the number of possibilities are much more effective. This was illustrated earlier in Chapter 1 in the extreme case of the game "20 questions." In similar fashion, a small number of carefully chosen runs need not necessarily lead to a single interpretation. When they do not themselves supply data that solves the problem, they can serve to eliminate a large number of possibilities and provide a basis for further conjecture.

6.9. AN INVESTIGATION USING MULTIPLE-COLUMN FOLDOVER

The following is a another example of a sequential use of experimental design this time employing multicolumn foldover. A number of similar chemical plants in different locations had been operating successfully for several years. In older plants the time to complete a particular filtration cycle was about 40 minutes. In a newly constructed plant, however, the filtration took almost twice as long, resulting in serious delays. To try to discover the cause of the difficulty, a technical team was called together to consider various possibilities. The many differences between the new and the older production plants were discussed and seven factors were identified that might account for these differences: (i) **A**, the source of the water; (ii) **B**, the origin of the raw material; (iii) **C**, the temperature of filtration; (iv) **D**, the use of a recycle stage; (v) **E**, the rate of addition of caustic soda; (vi) **F**, the type of filter cloth; and (vii) **G**, the length of holdup at a preliminary stage. Little accurate information was available, and considerable disagreement was expressed as to the importance of these various factors.

The Design and the Results

Those responsible for the investigation believed that only one or two of the factors might prove relevant but did not know which ones. To find out, it was decided to use the nodal eight-run 2_{III}^{7-4} design of Table 6.10.

* The Duke of Wellington, who won the Battle of Waterloo against Napoleon Bonaparte, said the business of life is to endeavor to find out what you don't know from what you do—to guess what was at the other side of the hill.

Table 6.10. A Screening Experiment: Filtration Study

Factors	−	+
A: water supply	Town reservoir	Well
B: raw material	On site	Other
C: temperature	Low	High
D: recycle	Yes	No
E: caustic soda	Fast	Slow
F: filter cloth	New	Old
G: holdup time	Low	High

Run Number	**a** A	**b** B	**c** C	**ab** D	**ac** E	**bc** F	**abc** G	Filtration Time y (min)
1	−	−	−	+	+	+	−	68.4
2	+	−	−	−	−	+	+	77.7
3	−	+	−	−	+	−	+	66.4
4	+	+	−	+	−	−	−	81.0
5	−	−	+	+	−	−	+	78.6
6	+	−	+	−	+	−	−	41.2
7	−	+	+	−	−	+	−	68.7
8	+	+	+	+	+	+	+	38.7

Note: **a**, **b**, ..., **abc** are columns; **A**, **B**, ..., **G** are factors.

Table 6.11. Estimated Effects and Abbreviated Confounding Pattern for the Eight-Run 2_{III}^{7-4} Design Filtration Experiment

$$l_A = \mathbf{-10.9} \rightarrow A' + BD + CE' + FG$$
$$l_B = -2.8 \rightarrow B + AD + CF + EG$$
$$l_C = \mathbf{-16.6} \rightarrow C' + AE' + BF + DG$$
$$l_D = 3.2 \rightarrow D + AB + CG + EF$$
$$l_E = \mathbf{-22.8} \rightarrow E' + AC' + BG + DF$$
$$l_F = -3.4 \rightarrow F + AG + BC + DE$$
$$l_G = 0.5 \rightarrow G + AF + BE + CD$$

The runs were made in random order. At this point it was noted that runs 6 and 8 were the runs with the shortest filtration times and these runs were the only runs in which factors **A, C**, and **E** were changed simultaneously. It was likely therefore that if these changes were made the problem would be fixed. However, the experimenters desired to use this opportunity to gain more fundamental understanding of what was going on. Table 6.11 displays the estimated effects with the alias structure obtained from Table 6.6. Three of the estimated effects (l_A, l_C, and l_E) are large but the alias structure in Table 6.11 shows that the results allow four equally good but alternative interpretations. The effects *A, C, E, AC, AE* and *CE* are distinguished by primes in the Table 6.11 and it will be seen that the large effects could be due to

A, C, and *E*, or
A, C, and *AC*, or
A, E, and *AE*, or
C, E, and *CE*.

To distinguish between the four solutions, it was decided to make eight additional runs by "foldover" of all seven columns for the original eight runs. Thus the seven columns of the newly added fraction were mirror images with signs everywhere opposite to those of the original design, they are shown together with the new data in Table 6.12.

The full defining relation for the nodal 2_{III}^{7-4} design is **I** = **ABD** = **ACE** = **BCF** = **ABCG** = **BCDE** = **ACDF** = **CDG** = **ABEF** = **BEG** = **AFG** = **DEF** = **ADEG** = **BDFG** = **CEFG** = **ABCDEFG**. The defining relation for the foldover design, when the signs of all the factors are changed, is identical except that all odd-lettered words now acquire a minus sign. Adding together the 2_{III}^{7-4} with its foldover combines the two defining relations to give **I** = **ABCG** = **ACDF** = **ABEF** = **ADEG** = **BCDE** = **BDFG** = **CEFG**. The smallest "word" in this defining relation has four letters. The combined design is thus a resolution IV design, the 2_{IV}^{7-3}.

Analysis of Sixteen Runs

Combining the data from both 8-run designs gives the estimates in Table 6.13. The large estimated effects are from the factors indicated by **A** and **E** and their

Table 6.12. The Foldover Design and Results of the Second Filtration Experiment

Test Number	−a A	−b B	−c C	−ab D	−ac E	−bc F	−abc G	*y*
9	+	+	+	−	−	−	+	66.7
10	−	+	+	+	+	−	−	65.0
11	+	−	+	+	−	+	−	86.4
12	−	−	+	−	+	+	+	61.9
13	+	+	−	−	+	+	−	47.8
14	−	+	−	+	−	+	+	59.0
15	+	−	−	+	+	−	+	42.6
16	−	−	−	−	−	−	−	67.6

Table 6.13. Estimated Values and Abbreviated Confounding Pattern for the Sixteen-Run Filtration Example
I = ABCG = ACDF = ABEF = ADEG = BCDE = BDFG = CEFG

l_A	=	$\mathbf{-6.7} \rightarrow A$
l_B	=	$-3.9 \rightarrow B$
l_C	=	$-0.4 \rightarrow C$
l_D	=	$2.7 \rightarrow D$
l_{BCD}	=	$\mathbf{-19.2} \rightarrow E$
l_{ACD}	=	$-0.1 \rightarrow F$
l_{ABC}	=	$-4.3 \rightarrow G$
l_{AB}	=	$0.5 \rightarrow AB + CG + EF$
l_{AC}	=	$-3.6 \rightarrow AC + BG + EF$
l_{AD}	=	$1.1 \rightarrow AD + CF + EG$
l_{ABCD}	=	$\mathbf{-16.2} \rightarrow AE + BF + DG$
l_{CD}	=	$4.8 \rightarrow CD + BE + AF$
l_{BC}	=	$-3.4 \rightarrow BC + AG + DE$
l_{BD}	=	$-4.2 \rightarrow BD + CE + FG$
l_{ABD}	=	$3.0 \rightarrow$ noise

interaction is **AE**. It appears therefore that the third interpretation explains the data. On this interpretation factors **B, C, D, F**, and **G** are essentially inert factors and the 16 runs are those of a 2^2 factorial design in **A** and **E** replicated four times, as shown in Figure 6.5. Subsequent testing verified that the problem of excessive filtration time could be eliminated by *simultaneously* adjusting the source of water supply (**A**) and the rate of addition of caustic soda (**E**). The chemical engineers on the team were able to explain the underlying reasons, and this knowledge proved very valuable in further improving results from the older plants as well as the new.

This is a further example of how fractional factorial designs may be used as sequential building blocks The original fractional factorial design allowed for the screening of a large number of factors, but ambiguities remained. The second fraction combined with the first allowed for the separation of main effects and two-factor interactions while simultaneously increasing the precision of the estimates. The discovery of the existence a large *interaction* between factors **A** and **E** proved essential to understanding the problem. A one-factor-at-a-time approach would have failed.

Fix It Or Solve It?

When your television set misbehaves, you may discover that a kick in the right place often fixes the problem, at least temporarily. However, for a long-term solution the reason for the fault must be discovered. In general, problems can be *fixed* or they may be *solved*. Experimental design catalyzes both fixing and solving.

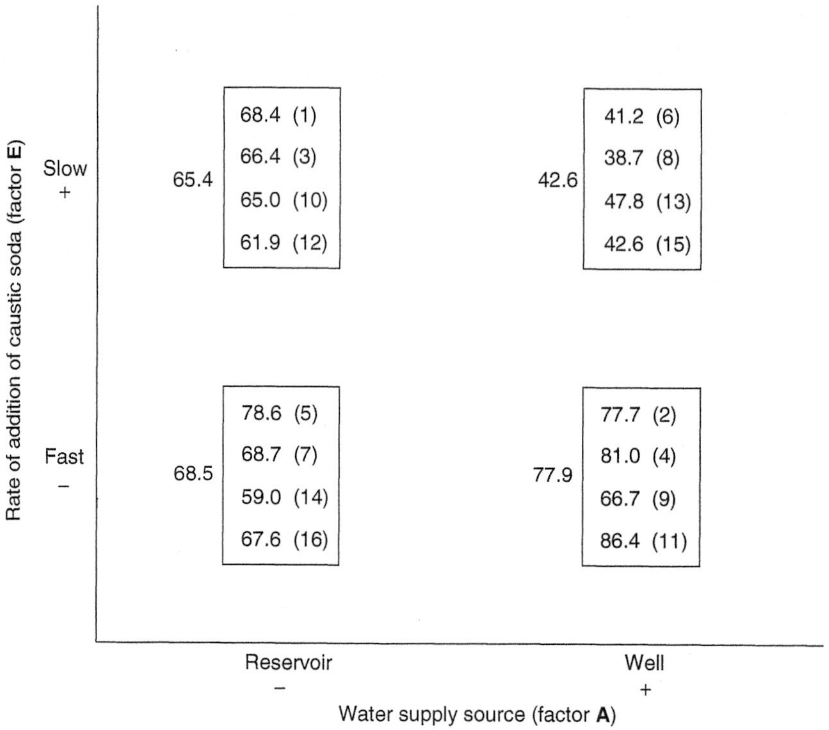

Figure 6.5. Projection into the space of factors **A** and **E**: filtration study.

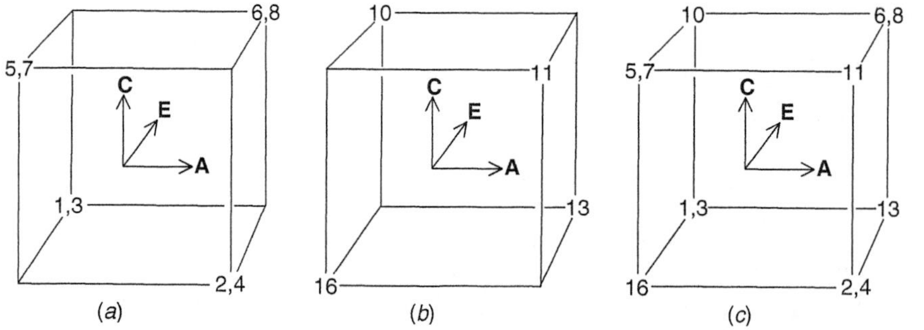

Figure 6.6. Economic alternative to foldover: (*a*) original eight runs; (*b*) four additional runs; (*c*) combined arrays.

An Economical Alternative to Total Foldover

Foldover, like any other technique, should not be used automatically. As always, a little thought may help. At the conclusion of the first eight runs it seemed likely that whatever was going on probably involved factors **A, C**, and **E**. Figure 6.6*a* shows a cube plot of the run numbers for these three factors. You can see that the cause of ambiguity is that the first eight runs produce not a 2^3 design in the three factors but only a duplicated half replicate. Thus an alternative to running the full foldover design would be to make single runs at the four missing points. You could have used just the four additional runs 10, 11, 13, and 16, as in Figure 6.6*b*. Combining these two tetrahedral arrays produces a complete 2^3 factorial in factors **A**, **C**, and **E** with four runs replicated. Staying with the earlier nomenclature you can call this a 1 1/2 replicate of the 2^3 factorial. All main effects and two-factor interactions can now be estimated. An estimate of the experimental error variance may be obtained from the paired data of the first tetrahedron. The estimated main effects and two-factor interactions with standard errors are then as follows:

Using Data from Runs 1–8 and 10, 11, 13, 16
Effects ± SE

$$\begin{aligned} l_A &= -5.0 \pm 2.2 \\ l_C &= 0.7 \pm 2.2 \\ l_E &= -21.7 \pm 2.2 \\ l_{AC} &= -1.1 \pm 2.2 \\ l_{AE} &= -17.3 \pm 2.2 \\ l_{CE} &= -5.8 \pm 2.2 \end{aligned}$$

$$s^2 = 13.30, \quad \nu = 5 \text{ df}$$

This approach would have led more quickly and economically to the same conclusions as before. A more general method is discussed later in which optimally chosen additional runs can be added in sequence until the necessary information is obtained.

Exercise 6.4. (a) Show what happens if you use runs 9, 12, 14, and 15 to augment the first eight runs. (b) Can you see how the problem could be resolved with just three runs? (c) How could you estimate a block effect between the two sets of eight runs, and between the eight-run and four-run designs?

6.10. INCREASING DESIGN RESOLUTION FROM III TO IV BY FOLDOVER

If you look again at Table 6.13 for the filtration experiment you will see that the complete 16-run design generated by augmenting the original 2_{III}^{7-4} design with its mirror image (foldover) is of resolution IV. Thus it is a 2_{IV}^{7-3} design in which main effects are alias free (except for terms of order 3 or higher) and two-factor interactions are aliased with other two-factor interactions. It is true in general that any design of resolution III plus its mirror image becomes a design of resolution IV. Equivalently, an initial design of projectivity 2 becomes one of projectivity 3. More generally, if you replicated with switched signs *any* set of runs, designed or not, in the resulting design the main effects would not be aliased with two-factor interactions.

6.11. SIXTEEN-RUN DESIGNS

All 16-run designs presently discussed are based on the same 15 columns of contrasts set out in Table 6.14a, sometimes called an L_{16} orthogonal array. To obtain the table, we introduce four dummy factors and write down the columns of signs associated with the four main effects **a, b, c**, and **d** their six two-factor interactions **ab, ac, ad, bc, bd, cd**, four three-factor interactions **abc, abd, acd, bcd**, and the four-factor interaction **abcd**.

The nodal 2^4, 2^{5-1}, 2^{8-4}, and 2^{15-11} designs may now be written down as in Table 6.14b using appropriate columns from the 16-run array.

The assignment of the factors **A, B, . . . , P** to the columns **a, b, . . . , abcd** is arranged so that the relationships between the nodal designs can be clearly seen.

Table 6.14a. Table of Signs for the n = 16 Orthogonal Array

a	b	c	d	ab	ac	ad	bc	bd	cd	abc	abd	acd	bcd	abcd
−	−	−	−	+	+	+	+	+	+	−	−	−	−	+
+	−	−	−	−	−	−	+	+	+	+	+	+	−	−
−	+	−	−	−	+	+	−	−	+	+	+	−	+	−
+	+	−	−	+	−	−	−	−	+	−	−	+	+	+
−	−	+	−	+	−	+	−	+	−	+	−	+	+	−
+	−	+	−	−	+	−	−	+	−	−	+	−	+	+
−	+	+	−	−	−	+	+	−	−	−	+	+	−	+
+	+	+	−	+	+	−	+	−	−	+	−	−	−	−
−	−	−	+	+	+	−	+	−	−	−	+	+	+	−
+	−	−	+	−	−	+	+	−	−	+	−	−	+	+
−	+	−	+	−	+	−	−	+	−	+	−	+	−	+
+	+	−	+	+	−	+	−	+	−	−	+	−	−	−
−	−	+	+	+	−	−	−	−	+	+	+	−	−	+
+	−	+	+	−	+	+	−	−	+	−	−	+	−	−
−	+	+	+	−	−	−	+	+	+	−	−	−	+	−
+	+	+	+	+	+	+	+	+	+	+	+	+	+	+

Table 6.14b. Sixteen-Run Nodal Designs

Nodal Designs	a	b	c	d	ab	ac	ad	bc	bd	cd	abc	abd	acd	bcd	abcd
2^4	A	B	C	D	. . .	. . .	. . .	. . .	. . .	. . .	. . .	. . .	. . .	. . .	. . .
2_{V}^{5-1}	A	B	C	D	. . .	. . .	. . .	. . .	. . .	. . .	. . .	. . .	. . .	. . .	P
2_{IV}^{8-4}	A	B	C	D	. . .	. . .	. . .	. . .	. . .	. . .	L	M	N	O	. . .
2_{III}^{15-11}	A	B	C	D	E	F	G	H	J	K	L	M	N	O	P

Table 6.14c. Alias Structure for Sixteen Run Nodal Designs

	2_{V}^{5-1} Five Factors	2_{IV}^{8-4} Eight Factors	2_{III}^{15-11} Fifteen Factors
Column	A B C D P	A B C D L M N O	A B C D E F G H J K L M N O P
a	A	A	$A+BE+CF+DG+HL+JM+KN+OP$
b	B	B	$B+AE+CH+DJ+FL+GM+KO+NP$
c	C	C	$C+AF+BH+DK+EL+GN+JO+MP$
d	D	D	$D+AG+BJ+CK+EM+FN+HO+LP$
ab	AB	$AB+CL+DM+NO$	$E+AB+CL+DM+FH+GJ+KP+NO$
ac	AC	$AC+BL+DN+MO$	$F+AC+BL+DN+EH+GK+JP+MO$
ad	AD	$AD+BM+CN+LO$	$G+AD+BM+CN+EJ+FK+HP+LO$
bc	BC	$AL+BC+DO+MN$	$H+AL+BC+DO+EF+GP+JK+MN$
bd	BD	$AM+BD+CO+LN$	$J+AM+BD+CO+EG+FP+HK+LN$
cd	CD	$AN+BO+CD+LM$	$K+AN+BO+CD+EP+FG+HJ+LM$
abc	DP	L	$L+AH+BF+CE+DP+GO+JN+KM$
abd	CP	M	$M+AJ+BG+CP+DE+FO+HN+KL$
acd	BP	N	$N+AK+BP+CG+DF+EO+HM+JL$
bcd	AP	O	$O+AP+BK+CJ+DH+EN+FM+GL$
abcd	P	$AO+BN+CM+DL$	$P+AO+BN+CM+DL+EK+FJ+GH$

The 2^4, 2_{V}^{5-1}, 2_{IV}^{8-4}, and 2_{III}^{15-11} nodal designs of Table 6.14b obtained by appropriate columns of the 16-run array show the largest number of factors that can be tested at a given resolution or equivalently of a given projectivity: 5 factors at resolution V (the 2_{V}^{5-1}), 8 factors at resolution IV (the 2_{IV}^{8-4}), and 15 factors at resolution III (2_{III}^{15-11}). Each of these designs has projectivity P equal to 1 less than their resolution R, $P = R - 1$. Intermediate numbers of factors can be accommodated by dropping (or adding) other factors from the nodal designs, as illustrated below.

The alias associations of Table 6.14c may be used in like manner. It can provide the appropriate aliases for *any* 16-run two-level orthogonal design by omitting the interaction pairs that contain letters corresponding to the factors that are not used in the derived design. For example, you will find that the alias pattern for the eight-factor 2_{IV}^{8-4} design could be obtained by omitting all aliases containing any of the letters E, F, G, H, J, K, P from the saturated 2_{III}^{15-11} design. For the 2_{V}^{5-1} all aliases are eliminated that include the letters $E, F, G, H, J, K, L, M, N, O$.

Exercise 6.5. Confirm that the last statement is true.

Exercise 6.6. (a) Write down a 16-run design for 10 factors. (b) Give its alias structure. (c) How might you decide which 10 columns to use in an actual design.

6.12. THE 2^{5-1} NODAL HALF REPLICATE OF THE 2^5 FACTORIAL: REACTOR EXAMPLE

Table 6.15 shows the data from a complete 2^5 factorial design in factors **A, B, C, D,** and **E** and the analysis in terms of the estimated effects and interactions. The

Table 6.15. Results from a Thirty-Two-Run 2^5 Factorial Design: Reactor Example

Factor	−	+
A: feed rate (L/min)	10	15
B: catalyst (%)	1	2
C: agitation rate (rpm)	100	120
D: temperature (°C)	140	180
E: concentration	3	6

	A	**B**	**C**	**D**	**E**		Effects		
1 †	−	−	−	−	−	61	Average	=	65.5
2 *	+	−	−	−	−	53	*A*	=	−1.375
3 *†	−	+	−	−	−	63	*B*	=	**19.5**
4	+	+	−	−	−	61	*C*	=	−0.625
5 *	−	−	+	−	−	53	*D*	=	**10.75**
6 †	+	−	+	−	−	56	*E*	=	**−6.25**
7	−	+	+	−	−	54	*AB*	=	1.375
8 *	+	+	+	−	−	61	*AC*	=	0.75
9 *	−	−	−	+	−	69	*AD*	=	0.875
10	+	−	−	+	−	61	*AE*	=	0.125
11	−	+	−	+	−	94	*BC*	=	0.875
12 *†	+	+	−	+	−	93	*BD*	=	**13.25**
13	−	−	+	+	−	66	*BE*	=	2.00
14 *†	+	−	+	+	−	60	*CD*	=	2.125
15 *†	−	+	+	+	−	95	*CE*	=	0.875
16	+	+	+	+	−	98	*DE*	=	**−11.0**
17 *	−	−	−	−	+	56	*ABC*	=	1.50
18 †	+	−	−	−	+	63	*ABD*	=	1.375
19	−	+	−	−	+	70	*ABE*	=	−1.875
20 *	+	+	−	−	+	65	*ACD*	=	−0.75
21	−	−	+	−	+	59	*ACE*	=	−2.50
22 *	+	−	+	−	+	55	*ADE*	=	0.625
23 *†	−	+	+	−	+	67	*BCD*	=	0.125
24 †	+	+	+	−	+	65	*BCE*	=	1.125
25 †	−	−	−	+	+	44	*BDE*	=	−0.250
26 *	+	−	−	+	+	45	*CDE*	=	0.125
27 *	−	+	−	+	+	78	*ABCD*	=	0.0
28 †	+	+	−	+	+	77	*ABCE*	=	1.5
29 *†	−	−	+	+	+	49	*ABDE*	=	0.625
30	+	−	+	+	+	42	*ACDE*	=	1.0
31	−	+	+	+	+	81	*BCDE*	=	−0.625
32 *	+	+	+	+	+	82	*ABCDE*	=	−0.25

Note: Runs from a 2^{5-1} half replicate are denoted by asterisks.* Those runs that were used later to produce a 12-run Plackett–Burman design are identified by daggers.†

normal plot (Fig. 6.7*a*) of the 31 estimated effects indicates that over the ranges studied only the estimates of the main effects *B*, *D*, and *E* and the interactions *BD* and *DE* are distinguishable from noise. These estimates are shown in bold type in Table 6.15.

This full 2^5 factorial required 32 runs. If the experimenter had chosen instead to just make the 16 runs marked with asterisks in Table 6.15, then only the data

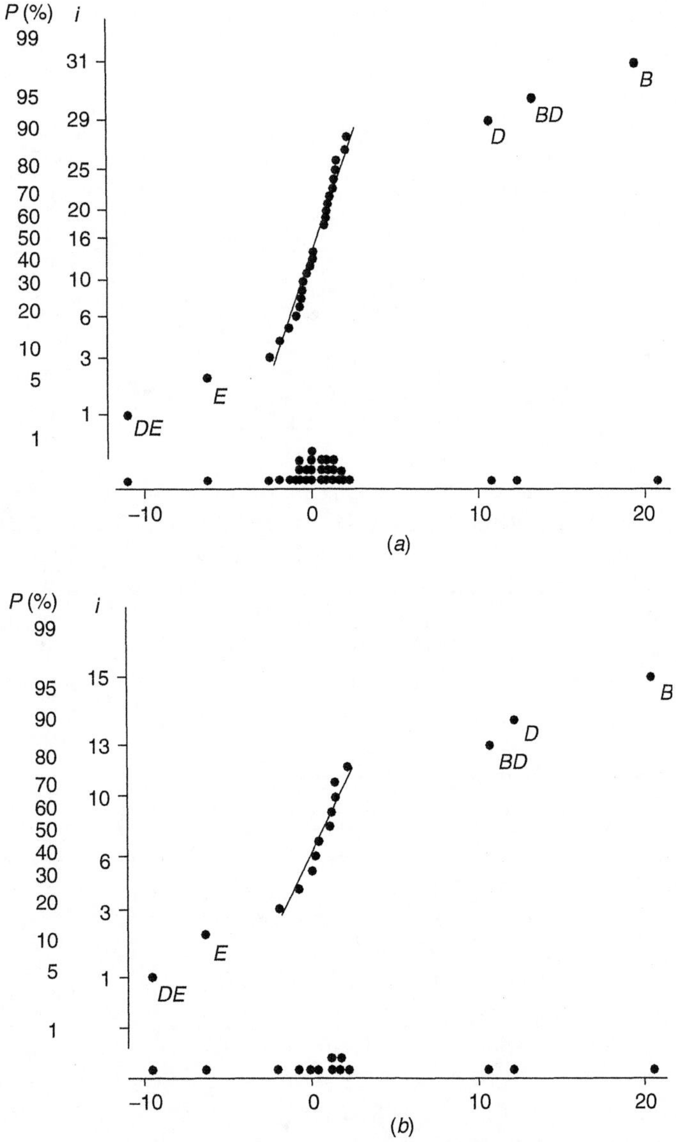

Figure 6.7. Reactor example: (*a*, *b*) normal plots.

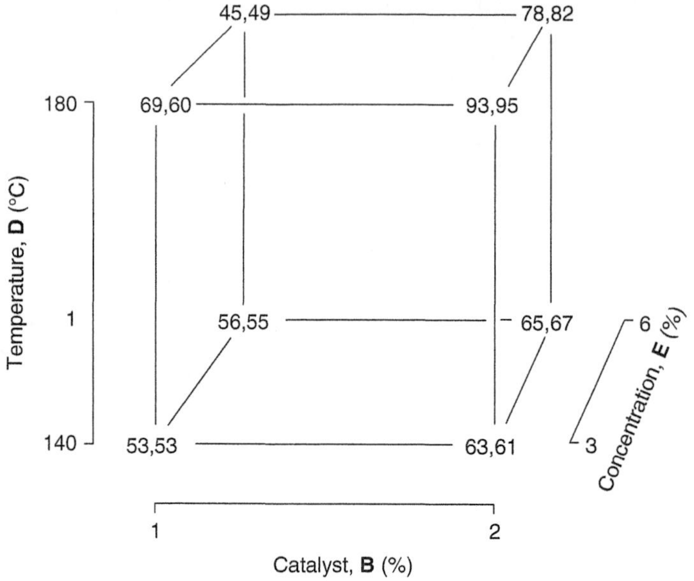

(c)

Figure 6.7. (*continued*) Reactor example: (*c*) cube plot.

of Table 6.16 would have been available. As you can see, these are the 16 runs required for the 2_{V}^{5-1} nodal design given in Table 6.14. (The order of the runs has been changed to clarify the nature of the design.) In Table 6.16 all 5 main effects and 10 two-factor interactions have been estimated from the 16 runs of the reduced design. These estimates are not very different from those obtained from the full 2^5 factorial design. Furthermore, the normal plot of the 15 effects, shown in Figure 6.7b, calls attention to the same outstanding contrasts.

The 2_{V}^{5-1} Design Used as a Factor Screen

The estimates of all 5 main effects and the 10 two-factor interactions obtained from the 2_{V}^{5-1} design are shown in Table 6.16. These estimates are made on the assumption that all interactions between three or more factors can be ignored. This is frequently but not invariably the case. This 16-run 2_{V}^{5-1} design is of resolution IV and hence of projectivity 4. Consequently, if any one of the five factors **A, B, C, D, E** is essentially inert, you will have a complete 2^4 factorial in the remaining factors. (Use Table 6.16 to check it for yourself.) The design may thus be said to provide a factor screen of order [16, 5, 4] since it has 16 runs and 5 factors and is of projectivity 4. If, as would rarely be necessary, you needed to complete the full 2^5 factorial design, the two generated half fractions would be orthogonal blocks, with the second added half the "mirror image" as would be obtained by foldover.

Table 6.16. Results from the Sixteen-Run 2_V^{5-1} Fractional Factorial Design: Reactor Example

Factor	−	+
A: feed rate (L/min)	10	15
B: catalyst (%)	1	2
C: agitation rate (rpm)	100	120
D: temperature (°C)	140	180
E: concentration (%)	3	6

Number Run	2_V^{5-1} **A**	**B**	**C**	**D**	**E**	**AB**	**AC**	**AD**	**AE**	**BC**	**BD**	**BE**	**CD**	**CE**	**DE**	Response (% reacted) observed	Estimated Effects
17	−	−	−	−	+	+	+	+	−	+	+	−	+	−	−	56	$I(+ABCDE)$ = 62.25
2	+	−	−	−	−	−	−	−	−	+	+	+	+	+	+	53	$A(+BCDE)$ = −2.0
3	−	+	−	−	−	−	+	+	+	−	−	−	+	+	+	63	$B(+ACDE)$ = **20.5**
20	+	+	−	−	+	+	−	−	+	−	−	+	+	−	−	65	$C(+ABDE)$ = 0.0
5	−	−	+	−	−	+	−	+	+	−	+	+	−	−	+	53	$D(+ABCE)$ = **12.15**
22	+	−	+	−	+	−	+	−	+	−	+	−	−	+	−	55	$E(+ABCD)$ = **−6.25**
23	−	+	+	−	+	−	−	+	−	+	−	+	−	+	−	67	$AB(+CDE)$ = 1.5
8	+	+	+	−	−	+	+	−	−	+	−	−	−	−	+	61	$AC(+BDE)$ = 0.5
9	−	−	−	+	−	+	+	−	+	+	−	+	−	+	−	69	$AD(+BCE)$ = −0.75
26	+	−	−	+	+	−	−	+	+	+	−	−	−	−	+	45	$AE(+BCD)$ = 1.25
27	−	+	−	+	+	−	+	−	−	−	+	+	−	−	+	78	$BC(+ADE)$ = 1.5
12	+	+	−	+	−	+	−	+	−	−	+	−	−	+	−	93	$BD(+ACE)$ = **10.75**
29	−	−	+	+	+	+	−	−	−	−	−	−	+	+	+	49	$BE(+ACD)$ = 1.25
14	+	−	+	+	−	−	+	+	−	−	−	+	+	−	−	60	$CD(+ABE)$ = 0.25
15	−	+	+	+	−	−	−	−	+	+	+	−	+	−	−	95	$CE(+ABD)$ = 2.25
32	+	+	+	+	+	+	+	+	+	+	+	+	+	+	+	82	$DE(+ABC)$ = **−9.50**

Redundancy of A and C

For this experimental design the estimated effects distinguishable from noise indicated in Table 6.16 by bold numbers, involve only the factors **B, D**, and **E** but not **A** and **C**. It appears, therefore, that although the experimental space has five dimensions in factors **A, B, C, D, E** the locally active subspace has only three dimensions in factors **B, D, E**. As a check of this hypothesis you will see that the numbers at the vertices of the cube plot in Figure 6.7*c* could plausibly be regarded as experimental replicates.

6.13. THE 2_{IV}^{8-4} NODAL SIXTEENTH FRACTION OF A 2^8 FACTORIAL

The 2_{IV}^{8-4} nodal 16-run design obtained from Table 6.14b is especially useful for screening up to eight factors. Because the design is of resolution IV, it has projectivity 3, and hence for any choice of three factors out of eight (there are $C_3^8 = 56$ ways to make this choice) this 16-run design will produce a replicated

2^3 factorial in these three factors. The design can thus be said to provide a [16, 8, 3] factor screen for $N = 16$ runs, $k = 8$ factors at projectivity $P = 3$.

It is sometimes a useful strategy to run this design as a random sequence of eight blocks of size 2, each block a foldover pair of runs. All eight main effects remain clear of two-factor interactions and block effects and differences between blocks then inflate only the variance of the interaction estimates.

Exercise 6.7. Write the 2_{IV}^{8-4} as a design in eight blocks of size 2 such that the eight main effects are unaffected by the block comparisons. State any assumptions you make.

Application of the Sixteen-Run 2_{IV}^{8-4} Design in a Paint Trial

In developing a paint for certain vehicles a customer required that the paint have high glossiness and acceptable abrasion resistance. Glossiness (y_1) was measured on a scale of 1 to 100, abrasion resistance (y_2) on a scale of 1 to 10. It was desired that the abrasion resistance should be at least 5 with glossiness as high as possible. The experimenters believed that only two factors **A** and **B** were important in affecting glossiness and abrasion resistance. However, they were perplexed by the results of a number of ad hoc experiments in which different levels of **A** and **B** produced paints either with high glossiness but low abrasion resistance or with acceptable abrasion resistance but low glossiness. It was suggested that the use of additional ingredients might help. In particular, it might be possible to increase glossiness without influencing abrasion resistance. It was decided to run a 2_{IV}^{8-4} design including six additional factors **C, D, E, F, G,** and **H** suggested by the paint technologists. The design, data, and estimated effects with their aliases are shown in Table 6.17. Corresponding normal plots are given in Figures 6.8*a* and *b*.

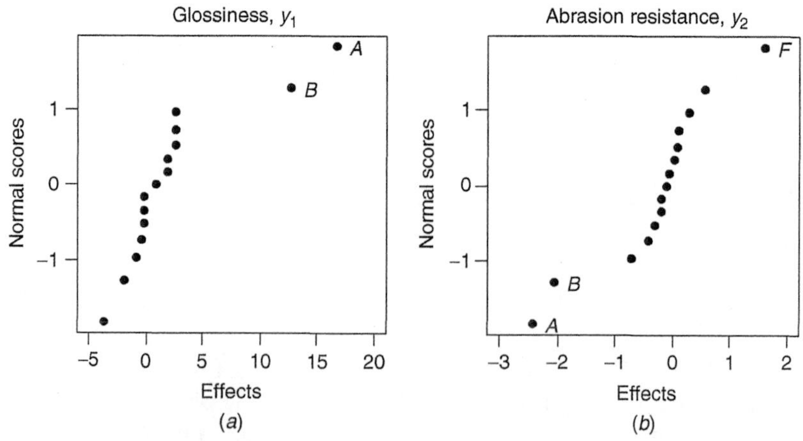

Figure 6.8. Normal plots: glossiness, abrasion effects.

Table 6.17. Paint Trial Example: 2_{IV}^{8-4} Fractional Factorial Data, Estimates, and Aliases

2_{IV}^{8-4} Fractional										Estimated Effects		
a	b	c	d	abc	abd	acd	bcd	Observed			Gloss	Abrasion
A	B	C	D	E	F	G	H	y_1	y_2	Average	y_1	y_2
−	−	−	−	−	−	−	−	53	6.3		64.7	4.8
+	−	−	−	+	+	+	−	60	6.1	$l_A \to A$	**16.6**	**−2.4**
−	+	−	−	+	+	−	+	68	5.5	$l_B \to B$	**12.6**	**−2.1**
+	+	−	−	−	−	+	+	78	2.1	$l_C \to C$	−0.1	−0.2
−	−	+	−	+	−	+	+	48	6.9	$l_D \to D$	2.6	−0.7
+	−	+	−	−	+	−	+	67	5.1	$l_E \to E$	−0.1	0.1
−	+	+	−	−	+	+	−	55	6.4	$l_F \to F$	−0.9	**1.6**
+	+	+	−	+	−	−	−	78	2.5	$l_G \to G$	−3.6	0.6
−	−	−	+	−	+	+	+	49	8.2	$l_H \to H$	1.9	−0.3
+	−	−	+	+	−	−	+	68	3.1	$l_{AB} \to AB + CE + DF + GH$	0.9	0.3
−	+	−	+	+	−	+	−	61	4.3	$l_{AC} \to AC + BE + DG + FH$	2.6	0.0
+	+	−	+	−	+	−	−	81	3.2	$l_{AD} \to AD + BF + CG + EH$	1.9	−0.1
−	−	+	+	+	+	−	−	52	7.1	$l_{AE} \to AE + BC + DH + FG$	−1.9	0.1
+	−	+	+	−	−	+	−	70	3.4	$l_{AF} \to AF + BD + CH + EG$	−0.1	−0.1
−	+	+	+	−	−	−	+	65	3.0	$l_{AG} \to AG + BH + CD + EF$	2.6	−0.4
+	+	+	+	+	+	+	+	82	2.8	$l_{AH} \to AH + BG + CF + DE$	−0.4	0.2

They show that the experimenters were correct in their expectation that **A** and **B** were important ingredients with opposite effects on the two responses. However, an additional factor **F** showed up for abrasion resistance y_2. At the minus level of this factor none of the ingredient **F** was used; at its plus level a (rather arbitrarily chosen) amount of **F** was added. Because **F** produced an active dimension for y_2 that y_1 did not share, there was an opportunity for compromise. This is clarified by the eyeball contours shown in Figure 6.9. In Figure 6.9*b* the shaded regions in the space of the factors **A** and **B**, are those producing acceptable values of abrasion resistance. You will see that the addition of **F**, while producing little difference in glossiness, moved up the acceptable region for adequate abrasion resistance, thus making possible a substantial improvement in glossiness y_1 while maintaining an acceptable level of abrasion resistance y_2.

This experiment again demonstrated the ability of screening designs to seek out active subspaces for *more than one* response and to exploit the way in which these subspaces overlap or fail to overlap. It also showed how data from appropriate designs can clearly point the way to further experiments. In particular, the experiment suggested that by further increasing the amount of ingredient **F** it might be possible to produce even better levels of abrasion resistance. This proved to be the case.

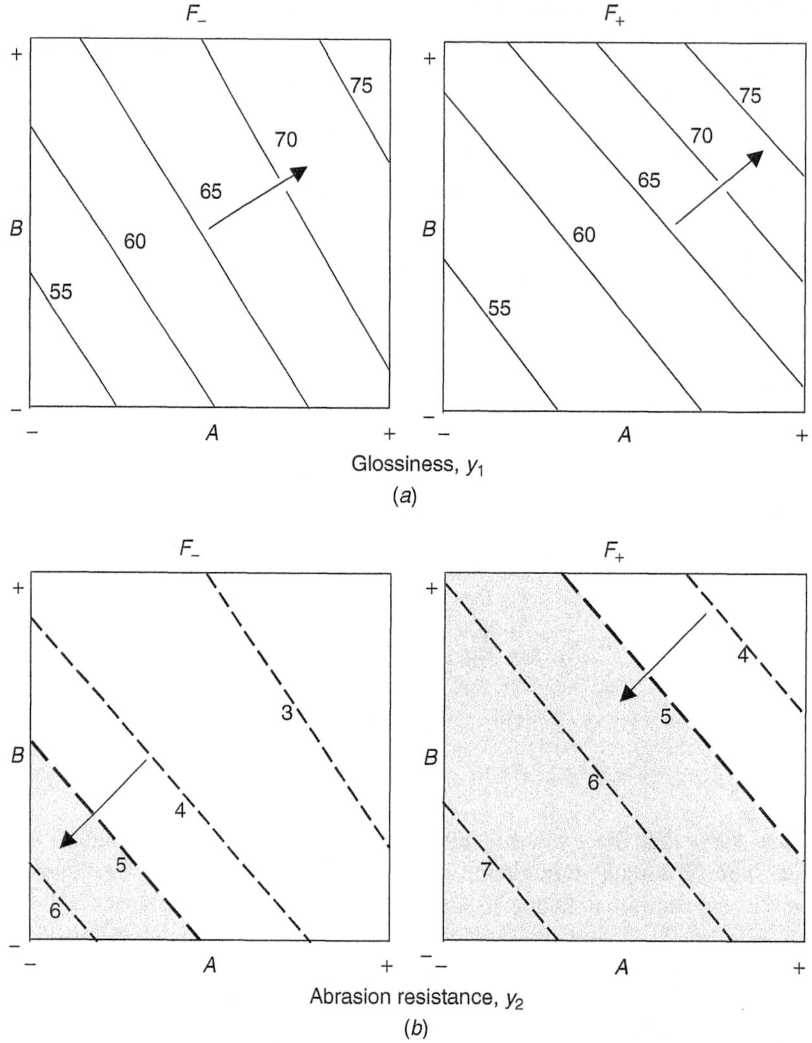

Figure 6.9. Eyeball contours for glossiness and abrasion.

6.14. THE 2^{15-11}_{III} NODAL DESIGN: THE SIXTY-FOURTH FRACTION OF THE 2^{15} FACTORIAL

The 2^{15-11}_{III} design (see Tables 6.14a, 6.14b, and 6.14c) is valuable when we need to screen a large number of factors with the expectation that very few will be of any consequence. The arrangement is a 16-run design of resolution III and hence projectivity 2, thus providing a 2^2 factorial design replicated four times in every one of the 105 choices of 2 factors out of 15. It is a [16, 15, 2] screen with $N = 16$ runs, $k = 15$ factors, and projectivity $P = 2$.

The following is an example due to Quinlan (1985), who rightly emphasized that the running of such a complicated design required very close supervision and management. Postextrusion shrinkage of a speedometer casing had produced undesirable noise. The objective of this experiment was to find a way to reduce this shrinkage. Quinlan used a somewhat different notation, but for consistency we will here translate his design to the pattern we have used in Table 6.14. To reduce decimals, we have multiplied the original shrinkage results by 100. A considerable length of product was made during each run (several hundred meters) and measurements were made at four equally spaced points. Table 6.18 shows the design with the averages and log variances* of the four measurements. Estimated effects are shown at the bottom of the table.

Table 6.18. Shrinkage of Speedometer Cables Under Various Conditions of Manufacture

a	**b**	**c**	**d**	**ab**	**ac**	**ad**	**bc**	**bd**	**cd**	**abc**	**abd**	**acd**	**bcd**	**abcd**	Shrinkage		
A	**B**	**C**	**D**	**E**	**F**	**G**	**H**	**J**	**K**	**L**	**M**	**N**	**O**	**P**	Average	Variance	Factors
−	−	−	−	+	+	+	+	+	+	−	−	−	−	+	48.5	16.3	**A**: liner tension
+	−	−	−	−	−	−	+	+	+	+	+	+	−	−	57.5	4.3	**B**: liner line speed
−	+	−	−	−	+	+	−	−	+	+	+	−	+	−	8.8	2.9	**C**: liner die
+	+	−	−	+	−	−	−	−	+	−	−	+	+	+	17.5	3.0	**D**: liner outside diameter
−	−	+	−	+	−	+	−	+	−	+	−	+	+	−	18.5	20.3	**E**: melt temperature
+	−	+	−	−	+	−	−	+	−	−	+	−	+	+	14.5	5.7	**F**: coating material
−	+	+	−	−	−	+	+	−	−	−	+	+	−	+	22.5	7.0	**G**: liner temperature
+	+	+	−	+	+	−	+	−	−	+	−	−	−	−	17.5	9.0	**H**: braid tension
−	−	−	+	+	+	−	+	−	−	−	+	+	+	−	12.5	19.7	**J**: wire braid type
+	−	−	+	−	−	+	+	−	−	+	−	−	+	+	12.0	58.0	**K**: liner material
−	+	−	+	−	+	−	−	+	−	+	−	+	−	+	45.5	13.7	**L**: cooling method
+	+	−	+	+	−	+	−	+	−	−	+	−	−	−	53.5	0.3	**M**: screen pack
−	−	+	+	+	−	−	−	−	+	+	+	−	−	+	17.0	10.7	**N**: coating die type
+	−	+	+	−	+	+	−	−	+	−	−	+	−	−	27.5	3.7	**O**: wire diameter
−	+	+	+	−	−	−	+	+	+	−	−	−	+	−	34.2	22.9	**P**: line speed
+	+	+	+	+	+	+	+	+	+	+	+	+	+	+	58.2	10.9	

Location effects: estimated effects using run averages:

A	*B*	*C*	*D*	*E*	*F*	*G*	*H*	*J*	*K*	*L*	*M*	*N*	*O*	*P*
6.3	6.2	5.7	6.9	2.6	0	4.2	7.5	**24.4**	9.1	0.5	2.9	678	**−14.2**	0.6

Dispersion effects: estimated effects using run log variances:

A	*B*	*C*	*D*	*E*	*F*	*G*	*H*	*J*	*K*	*L*	*M*	*N*	*O*	*P*
−0.8	−0.8	0.4	0.3	−0.2	0.1	−0.3	1.1	−0.1	−0.2	0.7	−0.9	0.1	0.7	0.6

*The advantage of analyzing log variances are discussed in Bartlett and Kendall (1946). The calculated effects are called *dispersion effects*.

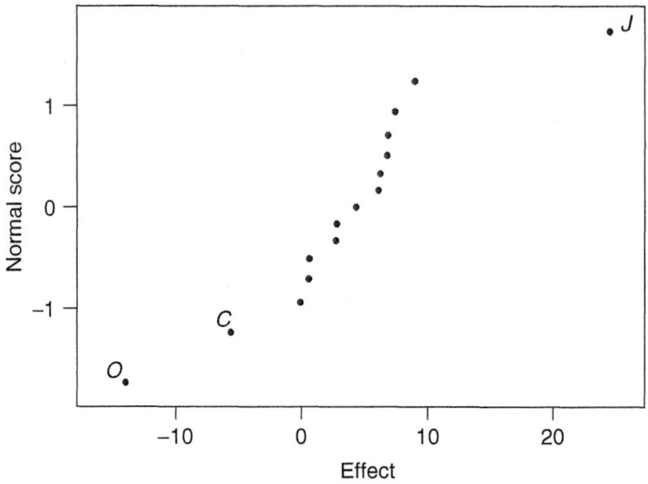

Figure 6.10. Normal Plot: location effects with **C**, **O**, and **J** outstanding.

The estimates of the 15 main effects calculated from the run averages will be referred to as "location effects" to distinguish them from the "dispersion effects" calculated from the logs of the run variances. For the location effects a normal plot in Figure 6.10*a* suggests that two factors, **J** (wire braid type) and **O** (wire diameter), are most likely active in affecting shrinkage. The conjecture is supported by consideration that factor **C**, which also turns up in the normal plot, has an alias of **OJ**, the interaction between **O** and **J**.* For the dispersion effects the logarithms of the within-run variances were also analyzed. But this normal plot does not suggest any outstanding effects. A two-way table for **J** and **O** is shown in Figure 6.11. This suggests that, by using a casing with **J** (wire braid type) at the minus level and **O** (wire diameter) at the plus level, shrinkage could be reduced by about 13 units. Thus it would be appropriate to switch manufacturing to these new conditions. Such discoveries, however, are likely to have other consequences. The minus level for wire diameter **O** was that already in use. Thus attention would be directed to factor **J**, the wire braid. Typically the subject matter specialist might wonder, "What *physical* properties can be responsible for its behavior?" "Can we exploit this knowledge to produce a still better product or perhaps one equally good at less cost?" and so on. Improvement invariably results in further ideas.

Two Components of Variance

There are two quite distinct *components of variance* in this example. The average of the 16 *within-run* variances (the variation obtained from the four

* The four runs 5, 6, 15, 16 with **O** and **J** each at their plus levels give suspiciously variable results and may possibly contain aberrant values. They do not affect the above conclusions but might have been worth further study and experimentation.

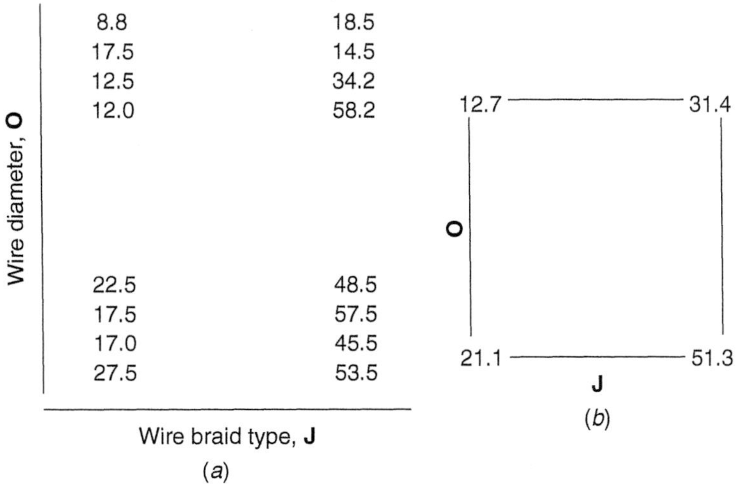

Figure 6.11. Data and averages for the projected 2^2 factorial in **J** and **O**.

measurements on each of the sixteen cables) is 13.0. Thus, the variance of an effect on a per-observation basis equals $\text{Var(effect)} = \left(\frac{1}{32} + \frac{1}{32}\right) 13.0 = 0.8$ and hence $\text{SE(effect)} = 0.9$. The *between-run* $\text{SE(effect)} \cong 4.7$ may be estimated from the gradient of the normal plot. It is almost five times as large. This information could be valuable in thinking further about the process. First, of course, it emphasizes we cannot use the *within*-run variances to test hypotheses concerning *between*-run effects. The data imply that the process can be run very reproducibly *once a particular set of process conditions has been set up* but that the process conditions themselves may be hard to reproduce. The topic of "components of variance" is discussed in detail in Chapter 9.

6.15. CONSTRUCTING OTHER TWO-LEVEL FRACTIONS

The creation of special designs to isolate certain effects from possible aliases can be done by using nodal designs and their associated alias structures. Almost always, your first consideration should be to preserve the highest possible resolution for the design. It may then be possible to isolate the interactions regarded as potentially important by suitably associating the design factors with the letters of the appropriate nodal array. You saw how this could be accomplished in the case of eight-run designs by dropping certain factors from the nodal design, but in some cases, to obtain a suitable arrangement, it is easiest to *add* factors to a nodal design.

For example, suppose you wanted to study nine factors in a 16-run design. You know there is a valuable 2_{IV}^{8-4} nodal array for studying eight factors in 16 runs with the useful property that all eight main effects are free of two-factor

Table 6.19. Alias Structure for the 2_{III}^{9-4} Fractional Factorial Design

2_{III}^{9-4}
A B C D L M N O P
A + OP
B + NP
C + MP
D + LP
L + DP
M + CP
N + BP
O + AP
P + AO + BN + CM + DL
AB + CL + DM + NO
AC + BL + DN + MO
AD + BM + CN + LO
AL + BC + DO + MN
AM + BD + CO + LN
AN + BO + CD + LM

interaction aliases. How can you add a ninth factor to this design doing as little damage as possible? For any two-level design for studying k factors there are $\frac{1}{2}k(k-1)$ two-factor interactions. These all have to appear somewhere in the alias table; you cannot make any of them go away. However, what you can do is rearrange them to best suit your particular situation. Look at Tables 6.14b and 6.14c. The eight-factor 2_{IV}^{8-4} nodal designs in factors **A, B, C, D, L, M, N, O** can be derived by dropping factors **E, F, G, H, J, K, P** from the 2_{III}^{15-11} fifteen-factor nodal design. If you now restore *any* one of these factors, say **P**, as is done in Table 6.19, the estimate of the main effect of the factor **P** will pick up four two-factor interaction aliases, but each of the estimates of the main effects of the original eight factors will pick up only one interaction alias involving the chosen factor **P**. Thus to do least damage to the nodal 2_{IV}^{8-4} design the factor most likely to be inert may be chosen to be the additional factor, here called **P**. (This same basic structure will occur whatever additional factor is chosen.) The eight main effects A, B, C, D, L, M, N, O are now least likely to be affected by two-factor interactions.

Exercise 6.8. Suppose factor **E** is thought least likely to be active. Show the appropriate nine-factor design in 16 runs and its alias pattern.

Exercise 6.9. Add two new columns to the 2_{IV}^{8-4} design and comment.

6.16. ELIMINATION OF BLOCK EFFECTS

Fractional designs may be run in blocks with suitable contrasts used as "block variables." A design in 2^q blocks is defined by q independent contrasts. All effects (including aliases) associated with these chosen contrasts *and all their interactions* are confounded with blocks.

A 2_{V}^{5-1} *Design in Two Blocks of Eight Runs*

Consider the 2_{V}^{5-1} design of Table 6.16. Suppose in planning this experiment the investigator decided that the interaction between factor **A** and factor **C** was most likely to be negligible. The interaction **AC** could then be used for blocking. The eight runs 2, 20, 5, ..., 15, having a minus sign in the **AC** column, would be run in one block and the eight runs 17, 3, 22, ..., 32 in the other. Notice in this design that the effects *AC* and *BDE* are aliased and hence both confounded with blocks.

A 2_{V}^{5-1} Design in Four Blocks of Four Runs

Suppose in the 2_{V}^{5-1} design of Table 6.16 two columns **AC** and **BC** are used to define blocks. The interaction between **AC** and **BC**, $\mathbf{AC} \times \mathbf{BC} = \mathbf{ABC}^2 = \mathbf{AB}$, is then also confounded with blocks. The design would thus be appropriate if we were prepared to confound with blocks all two-factor interactions between the factors **A, B**, and **C** and their aliases. To achieve this arrangement, runs 20, 5, 12, and 29, for which **AC** and **BC** have signs (−−), could be put in the first block; runs 2, 23, 26, and 15, for which the columns **AC** and **BC** have signs (−+), in the second block; and so on. Thus, in terms of a two-way table, the arrangement would be as shown in Table 6.20. Blocks **I, II, III**, and **IV** are run in random order as are the runs within blocks.

Design of experiment software programs commonly provide appropriate blocking arrangements for the 2^k factorials and 2^{k-p} fractionals.

Minimum-Aberration 2^{k-p} Designs

You saw earlier that maximum resolution was an important criteria for two-level fractional factorial designs and that the resolution of a 2^{k-p} design was equal

Table 6.20. A 2_{V}^{5-1} Arranged in Four Blocks of Four Runs Each

		BC −	BC +
AC	−	I (−−)	II (−+)
	+	III (+−)	IV (++)

Table 6.22. Minimal Aberration Two-Level Fractional Factorial Designs for k variables and N Runs (Number in Parentheses Represent Replication)

number of variables k

number runs N	3	4	5	6	7	8	9	10	11	
4	2_{III}^{3-1} $\pm 3 = 12$									
8	2^3	2_{IV}^{4-1} $\pm 4 = 123$	2_{III}^{5-2} $\pm 4 = 12$ $\pm 5 = 13$	2_{III}^{6-3} $\pm 4 = 12$ $\pm 5 = 13$ $\pm 6 = 23$	2_{III}^{7-4} $\pm 4 = 12$ $\pm 5 = 13$ $\pm 6 = 23$ $\pm 7 = 123$					
16	2^3 2 times	2^4	2_{V}^{5-1} $\pm 5 = 1234$	2_{IV}^{6-2} $\pm 5 = 123$ $\pm 6 = 234$	2_{IV}^{7-3} $\pm 5 = 123$ $\pm 6 = 234$ $\pm 7 = 134$	2_{IV}^{8-4} $\pm 5 = 234$ $\pm 6 = 134$ $\pm 7 = 123$ $\pm 8 = 124$	2_{III}^{9-5} $\pm 5 = 123$ $\pm 6 = 234$ $\pm 7 = 134$ $\pm 8 = 124$ $\pm 9 = 1234$	2_{III}^{10-6} $\pm 5 = 123$ $\pm 6 = 234$ $\pm 7 = 134$ $\pm 8 = 124$ $\pm 9 = 1234$ $\pm \overline{10} = 12$	2_{III}^{11-7} $\pm 5 = 123$ $\pm 6 = 234$ $\pm 7 = 134$ $\pm 8 = 124$ $\pm 9 = 1234$ $\pm \overline{10} = 12$ $\pm \overline{11} = 13$	$\left(\frac{1}{128}\right)$
32	2^3 4 times	2^4 2 times	2^5	2_{VI}^{6-1} $\pm 6 = 12345$	2_{IV}^{7-2} $\pm 6 = 1234$ $\pm 7 = 1245$	2_{IV}^{8-3} $\pm 6 = 123$ $\pm 7 = 124$ $\pm 8 = 2345$	2_{IV}^{9-4} $\pm 6 = 2345$ $\pm 7 = 1345$ $\pm 8 = 1245$ $\pm 9 = 1235$	2_{IV}^{10-5} $\pm 6 = 1234$ $\pm 7 = 1235$ $\pm 8 = 1245$ $\pm 9 = 1345$ $\pm 10 = 2345$	2_{IV}^{11-6} $\pm 6 = 123$ $\pm 7 = 234$ $\pm 8 = 345$ $\pm 9 = 134$ $\pm \overline{10} = 145$ $\pm \overline{11} = 245$	$\left(\frac{1}{64}\right)$
64	2^3 8 times	2^4 4 times	2^5 2 times	2^6	2_{VII}^{7-1} $\pm 7 = 123456$	2_{V}^{8-2} $\pm 7 = 1234$ $\pm 8 = 1256$	2_{IV}^{9-3} $\pm 7 = 1234$ $\pm 8 = 1356$ $\pm 9 = 3456$	2_{IV}^{10-4} $\pm 7 = 2346$ $\pm 8 = 1346$ $\pm 9 = 1245$ $\pm \overline{10} = 1235$	2_{IV}^{11-5} $\pm 7 = 345$ $\pm 8 = 1234$ $\pm 9 = 126$ $\pm \overline{10} = 2456$ $\pm \overline{11} = 1456$	$\left(\frac{1}{32}\right)$
128	2^3 16 times	2^4 8 times	2^5 4 times	2^6 2 times	2^7	2_{VIII}^{8-1} $\pm 8 = 1234567$	2_{VI}^{9-2} $\pm 8 = 13467$ $\pm 9 = 23567$	2_{V}^{10-3} $\pm 8 = 1237$ $\pm 9 = 2345$ $\pm \overline{10} = 1346$	2_{V}^{11-4} $\pm 8 = 1237$ $\pm 9 = 2345$ $\pm \overline{10} = 1346$ $\pm \overline{11} = 1234567$	
	(16)	(8)	(4)	(2)	(1)	$\left(\frac{1}{2}\right)$	$\left(\frac{1}{4}\right)$	$\left(\frac{1}{8}\right)$	$\left(\frac{1}{16}\right)$	

Table 6.21. Three Choices for a 2_{IV}^{7-2} Fractional Factorial in Seven Factors 1, 2, 3, 4, 5, 6, 7 in 32 Runs

	Design (a)	Design (b)	Design (c)
Generators	6 = 123, 7 = 234	6 = 123, 7 = 145	6 = 1234, 7 = 1235
Defining relation	I = 1236 = 2347	I = 1236 = 1457	I = 12346 = 12357
	= 1467	= 234567	= 4567
Strings of aliased	12 + 36	12 + 36	45 + 67
two-factor	13 + 26	13 + 26	46 + 57
interactions	14 + 67	14 + 57	47 + 56
(assuming 3-factor	17 + 46	15 + 47	
and higher order	24 + 37	16 + 23	
interactions	24 + 34	17 + 45	
negligible)	16 + 23 + 47		

to the shortest word in its defining relation. However, it was pointed out by Fries and Hunter (1980) that there can be more than one choice for a design of highest resolution and one choice can be better than another. Table 6.21 shows three choices considered by these authors for a quarter replicate of a 2^7 factorial. All the designs are of resolution IV. Thus, on the customary assumption that all three-factor and higher order interactions are negligible, the table supplies a summary of the confounding that exists among the two-factor interactions. Unconfounded estimates are obtained for all two-factor interactions *not* shown in the table. Design (c) is the minimum-aberration design that minimizes the number of words in the defining relation having minimum length. Only three pairs of two-factor interactions are confounded in this design. In general, minimum aberration ensures that main effects will be confounded with the minimum number of words of length $R - 1$; the smallest number of two-factor interactions will be confounded with interactions of order $R - 2$. A useful listing of minimum-aberration designs for two-level fractional factorial designs up to 128 runs is given in Table 6.22.

REFERENCES AND FURTHER READING

Bartlett, M. S., and Kendall D. G. (1946) The statistical analysis of variance heterogeneity and the logarithmic transformation, *J. Roy. Statist. Soc. Ser. B*, **8** 128–150.

Bisgaard, S. (1993) Iterative analysis of data from two-level factorials, *Quality Eng.*, **6**, 149–157.

Box, G. E. P., and Bisgaard, S. (1987) The scientific context of quality improvement, *Quality Progress*, **20** (June), 54–61.

Box, G. E. P., and Hunter, J. S. (1961) The 2^{k-p} fractional factorial designs, Part 1, *Technometrics* **3**, 311–351; Part 2, **4**, 449–458.

Box, G. E. P., and Tyssedal, J. (1996) Projective properties of certain orthogonal arrays. *Biometrika*, **83**, 950–955.

Box, G. E. P., and Tyssedal, J. (2001) Sixteen run designs of high projectivity for factor screening. *Comm. Stat. Part B*, **30**, 217–228.

Box, G. E. P., and Wilson, K. B. (1951) On the experimental attainment of optimum conditions. *J. Roy. Stat. Soc. Series B*, **13**, 1–45.

Daniel, C. (1962) Sequences of fractional replicates in the 2^{k-p} series, *J. Am. Statist. Assoc.*, **58**, 403–408.

Daniel, C. (1976) *Applications of Statistics to Industrial Experimentation*, Wiley, New York.

Finney, D. J. (1945) The fractional replication of factorial arrangements, *Ann. Eugen.*, **12**, 291–301.

Fries, A., and Hunter, W. G. (1980) Minimum aberration 2^{k-p} designs, *Technometrics*, **22**, 601–608.

Haaland, P. D., and O'Connell, M. R. (1995) Inference for effect-saturated fractional factorials, *Technometrics*, **37**, 82–93.

Hellstrand, C. (1989) The necessity of modern quality improvement and some experiences with its implementation in the manufacture of roller bearings, *Philos. Trans. Roy. Soc. London, Series A, Math and Physical Science*, **327**, 529–537.

Juran, J. M., and Godfrey A. B. (1998) *Juran's Quality Handbook*, 5th ed., McGraw-Hill, New York.

Lenth, R. S. (1989) Quick and easy analysis of unreplicated factorials, *Technometrics*, **31**, 469–473.

Menon, A., Chakrabarti, S., and Nerella, N. (2002) Sequential experimentation in product development, *Pharmaceut. Devel. Technol.*, **7**(1), 33–41.

Quinlan, J. (1985) Product improvement by application of Taguchi methods, in *Third Supplier Symposium on Taguchi Methods*, American Supplier Institute, Dearborn, MI.

Rao, C. R. (1947) Factorial experiments derivable from combinatorial arrangements of arrays, *J. Roy. Statist. Soc. Ser. B*, **9** 128–139.

QUESTIONS FOR CHAPTER 6

1. Why were no computations necessary for the data on cast films in Table 6.1?

2. What was the experimental problem facing the chemist in the example concerning the development of a new product? (Data in Table 6.2.) What were the key factors in finding a solution to the problem?

3. In the filtration problem (see Table 6.10) why would a one-variable-at-a-time approach have failed?

4. What are different ways in which fractional factorial designs have been used in the examples in this chapter?

5. In which of the examples can you identify the iterative cycle described in Chapter 1?

6. What is a fractional factorial design?

7. What is a half fraction and how do you construct such a design?

8. What is a saturated design and how can you construct such a design?

9. What is meant by the sequential use of fractional factorials. By sequential assembly?

10. A 2^{8-3} design has how many runs? How many factors? How many levels for each factor? Answer the same questions for the 2^{k-p} design.

11. Other things being equal, why would a resolution IV design be preferred to a resolution III design?

12. Is it possible to construct a 2^{8-4} design of resolution III? Resolution IV? Resolution V? How?

13. What are three justifications for fractional factorials designs? Discuss.

14. How can fractional factorial designs be blocked? How should they be randomized?

15. Why is it necessary to know the alias structure associated with a fractional factorial design?

16. How would you design a 2^{5-1} design in eight blocks of size 2 so that main effects are clear of blocks? Comment on the two runs that comprise each block.

17. What is achieved by using a design of "minimum oberration."

PROBLEMS CHAPTER 6

1. In this experiment the immediate objective was to find the effects of changes on the amount of unburned carbons in engine exhaust gas. The following results were obtained:

A	*B*	*C*	*D*	**Unburned Carbon**
−	+	+	+	8.2
−	−	+	−	1.7
−	−	−	+	6.2
+	−	−	−	3.0
+	−	+	+	6.8
+	+	+	−	5.0
−	+	−	−	3.8
+	+	−	+	9.3

Stating any assumptions, analyze these data. What do you conclude? It was hoped to develop an engine giving lower levels of unburned carbon. Write a report asking appropriate questions and making tentative suggestions for further work.

2. Make an analysis for the following 2^{5-1} fractional factorial design, stating your assumptions:

A	B	C	D	E	Observed, y
−	−	−	−	+	14.8
+	−	−	−	−	14.5
−	+	−	−	−	18.1
+	+	−	−	+	19.4
−	−	+	−	−	18.4
+	−	+	−	+	15.7
−	+	+	−	+	27.3
+	+	+	−	−	28.2
−	−	−	+	−	16.0
+	−	−	+	+	15.1
−	+	−	+	+	18.9
+	+	−	+	−	22.0
−	−	+	+	+	19.8
+	−	+	+	−	18.9
−	+	+	+	−	29.9
+	+	+	+	+	27.4

3. Consider a 2^{8-4} fractional factorial design:
   (a) How many factors does this design have?
   (b) How many runs are involved in this design?
   (c) How many levels for each factor?
   (d) How many independent generators are there for this design?
   (e) How many words in the defining relation (counting **I**).
4. Repeated problem 3 for a 2^{9-4} design.
5. Repeated problem 3 for a 2^{6-1} design.
6. Write generators for a 2_{IV}^{8-4} design and for a 2_{IV}^{6-2} design. Does a 2_{V}^{6-2} design exist?
7. **(a)** Construct a 2^{6-2} fractional factorial with as high a resolution as possible.
   (b) What are the generators of your design?
   (c) What is the defining relation of your design?
   (d) What is confounded with the main effect 3 of your design?
   (e) What is confounded with the 1×2 interaction?
8. Derive the generators for a 2^{11-4} design and show how, using these generators, you could set out the design.
9. Analyze the following experimental results from a welding study. Do you notice any relationships among y_1, y_2, and y_3? How might such relationships be put to use?

Order of Runs	A B C D E	y_1	y_2	y_3
12	− − − − +	23.43	141.61	3318
1	+ − − − −	25.70	161.13	4141
2	− + − − −	27.75	136.54	3790
6	+ + − − +	31.60	128.52	4061
15	− − + − −	23.57	145.55	3431
8	+ − + − +	27.68	123.75	3425
7	− + + − +	28.76	121.93	3507
4	+ + + − −	31.82	120.37	3765
11	− − − + −	27.09	95.24	2580
14	+ − − + +	31.28	78.31	2450
3	− + − + +	31.20	74.34	2319
16	+ + − + −	33.42	91.76	3067
13	− − + + +	29.51	65.23	1925
10	+ − + + −	31.35	78.65	2466
5	− + + + −	31.16	79.76	2485
9	+ + + + +	33.65	72.80	2450

The factors and their settings were as follows:

Factors	−	+
A: open circuit voltage	31	34
B: slope	11	6
C: electrode melt-off rate (ipm)*	162	137
D: electrode diameter (in)	0.045	0.035
E: electrode extension (in)	0.375	0.625

[*Source*: Stegner, D. A. J., Wu, S. M., and Braton, N. R., *Welding J. Res Suppl.*, **1** (1967).]

*inches per minute

10. **(a)** Write a design of resolution III in seven factors and eight runs

(b) From (a) obtain a design in eight factors in 16 runs of resolution IV arranged in two blocks of 8.

(c) How could this design be used to eliminate time trends?

(d) What are the generators of the 16-run design?

(e) If no blocking is used, how can this design be employed to study three principal factors, the main effects of eight minor factors not expected to interact with the major factors or with each other?

11. **(a)** Write an eight-run two-level design of resolution III in seven factors and give its generators.

(b) How may members of this family of designs be combined to isolate the main effects and interactions of a single factor?

(c) How may this design be used to generate a 2^{8-4} design of resolution IV in two blocks of eight runs with no main effects confounded with blocks.

12. **(a)** Construct a 2^{7-1} design.

(b) Show how the design may be blocked into eight blocks of eight runs each so that no main effect or two-factor interaction is confounded with any block effect.

(c) Show how the blocks can themselves accommodate a 2^{4-1} fractional factorial design.

13. Suppose an engineer wants to study the effects of seven factors and is willing to make 32 runs. He elects to perform a 2^{7-2} fractional factorial design with the generators **I = 1237** and **I = 126.** The main effect of **1** is confounded with what other effects? Suggest an alternative plan that produces less confounding of low-order terms.

14. A consulting firm engaged in road-building work is asked by one of its clients to carry out an experimental study to determine the effects of six factors on the physical properties of a certain kind of asphalt. Call these factors **A, B, C, D, E,** and **F.**

(a) If a full factorial design is used, how many runs are needed?

(b) Construct a two-level resolution IV design requiring 16 runs.

(c) What is the defining relation for this 16-run design?

(d) What effects are confound with the main effect A, with the BD interaction?

15. Suppose that after you have set up a 2^{14-10} resolution III design for an engineer she says it is essential to split the design into two blocks of equal size and she fears a real difference between the blocks. Make recommendations.

16. An experimenter performs a 2^{5-2} fractional factorial design with generators **I = 1234** and **I = 135.** After analyzing the results from this design he decides to perform a second 2^{5-2} design exactly the same as the first but with signs changed in column **3** of the design matrix.

(a) How many runs does the first design contain?

(b) Give a set of generators for the second design

(c) What is the resolution of the second design?

(d) What is the defining relation of the combined design?

(e) What is the resolution of the combined design?

(f) Give a good reason for the unique choice of the second design.

17. A mechanical engineer used the following experimental design:

1	**2**	**3**	**4**	**5**
−	−	−	+	−
+	−	−	−	−
−	+	−	−	+
+	+	−	+	+
−	−	+	−	+
+	−	+	+	+
−	+	+	+	−
+	+	+	−	−

(a) Write a set of generators for this design

(b) Is this set unique? If not, write a different set of generators for this design.

18. (a) Write a 2^{6-1} fractional factorial of the highest possible resolution for factors **1, 2, 3, 4, 5**, and **6**. Start with a full factorial in the first five factors.

(b) Write the generator, the defining relation, and the resolution of this design.

(c) What effects do the following contrasts estimate? l_2, l_{23}, l_{234}, l_{2345}.

19. Design an eight-run fractional factorial design for an experimenter with the following five factors: temperature, concentration, pH, agitation rate, and catalyst type (A or B). She tells you she is particularly concerned about the two-factor interactions between temperature and concentration and between catalyst type and temperature. She would like a design, if it is possible to construct one, with main effects unconfounded with one another. Can you help her?

CHAPTER 7

Additional Fractionals and Analysis

7.1. PLACKETT AND BURMAN DESIGNS

In the early days of the bombing of London in World War II it was said that the main purpose of keeping the antiaircraft guns firing (they usually made more noise than the bombs) was to keep everyone in good spirits. In those days the chance of actually hitting an enemy aircraft was small. The experimental arrangements discussed in this chapter were originally invented to help in the development of a proximity fuse. With such a fuse, the trajectory of anti-aircraft shell needed only to pass close to an enemy plane to destroy it. To develop this weapon in time for it to be useful, it was necessary to screen a large number of factors as quickly as possible. Fractional factorials could be used as screening designs when the number of runs n was a power of 2, that is, 4, 8, 16, 32, 64, ..., but you will notice that the gaps between these numbers become wider and wider as n increases. Experimental arrangements were needed to fill the gaps. Two scientists, Robin Plackett and Peter Burman, derived a new class of two-level orthogonal designs in time for them to be used. These designs were available for any n that was a multiple of 4, in particular, arrangements for $n = 12$, 20, 24, 28, 36, ... runs were obtained.* An arrangement of this kind is also called an orthogonal array and is denoted by L_n.

A Twelve-Run Plackett and Burman Design

For illustration the 12-run Plackett and Burman (PB) design may be written down once you know its first row of signs:

$$+ - + - - - + + + - +$$

* It was later realized that these arrangements were Hadamard matrices. Tables of such matrices can now be found using computer search programs.

Table 7.1. The Twelve-Run PB_{12} Design

	Factors										
Run Number	A	B	C	D	E	F	G	H	J	K	L
1	+	−	+	−	−	−	+	+	+	−	+
2	+	+	−	+	−	−	−	+	+	+	−
3	−	+	+	−	+	−	−	−	+	+	+
4	+	−	+	+	−	+	−	−	−	+	+
5	+	+	−	+	+	−	+	−	−	−	+
6	+	+	+	−	+	+	−	+	−	−	−
7	−	+	+	+	−	+	+	−	+	−	−
8	−	−	+	+	+	−	+	+	−	+	−
9	−	−	−	+	+	+	−	+	+	−	+
10	+	−	−	−	+	+	+	−	+	+	−
11	−	+	−	−	−	+	+	+	−	+	+
12	−	−	−	−	−	−	−	−	−	−	−

The next 10 rows are obtained by successively shifting all the signs one step to the right (or left) and moving the sign that falls off the end to the beginning of the next row. The twelfth row consists solely of minus signs. Such a PB_{12} design (also called the L_{12} orthogonal array) is shown in Table 7.1. You will find that, just as with the fractional factorial designs, all the columns of this design are mutually orthogonal. There are six plus signs and six minus signs in every contrast column and opposite any six plus signs or six minus signs you will find three plus and three minus signs in every other contrast column.

Thus, using Table 7.1 you can calculate the main effect of factor **A** from

$$A = [(y_1 + y_2 + y_4 + y_5 + y_6 + y_{10}) - (y_3 + y_7 + y_8 + y_9 + y_{11} + y_{12})]/6 = \bar{y}_+ - \bar{y}_-$$

It turns out, however, that unless you can assume *only* main effects are likely to occur (which would only rarely be a safe supposition) the PB designs will give misleading results if they are analyzed in a manner appropriate for the 2^{k-p} fractional factorials. In particular if they are analyzed using Daniel's normal plot or Lenth's procedure. So two questions arise: (a) Are PB designs potentially useful? (b) If so, can a means be found for analyzing them?

Usefulness of Plackett and Burman Designs

Plackett and Burman designs are useful for screening because of their remarkable projective properties. For illustration, compare the screening capabilities of a 16-run fractional factorial with those of a 12-run PB design. Figure 7.1 shows a 16-run 2_{IV}^{8-4} fractional factorial. Since this design is of resolution IV, it has projectivity 3, and as illustrated in Figure 7.1*a*, it will project a duplicated 2^3 factorial for any one of the 56 choices of 3 factors out of 8. Thus it is a 16-run design capable of screening 8 factors with projectivity 3. It is a [16, 8, 3] screen.

2_{IV}^{8-4}

A	B	C	D	E	F	G	H
−	−	−	−	−	−	−	−
+	−	−	−	+	+	+	−
−	+	−	−	+	+	−	+
+	+	−	−	−	−	+	+
−	−	+	−	+	−	+	+
+	−	+	−	−	+	−	+
−	+	+	−	−	+	+	−
+	+	+	−	+	−	−	−
−	−	−	+	−	+	+	+
+	−	−	+	+	−	−	+
−	+	−	+	+	−	+	−
+	+	−	+	−	+	−	−
−	−	+	+	+	+	−	−
+	−	+	+	−	−	+	−
−	+	+	+	−	−	−	+
+	+	+	+	+	+	+	+

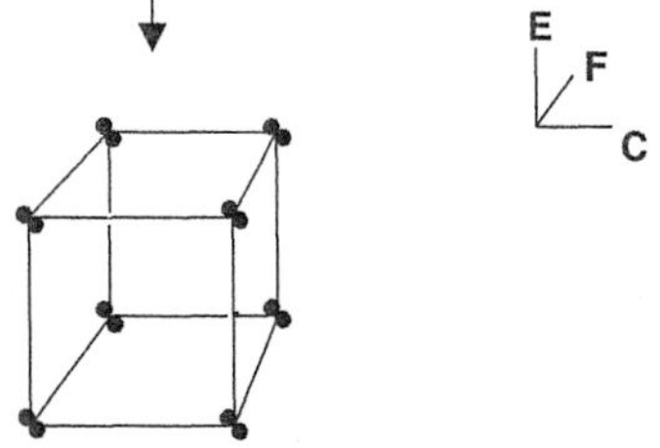

[16, 8, 3] screen, 56 choices

(a)

PB_{12}

A	B	C	D	E	F	G	H	J	K	L
+	−	+	−	−	−	+	+	+	−	+
+	+	−	+	−	−	−	+	+	+	−
−	+	+	−	+	−	−	−	+	+	+
+	−	+	+	−	+	−	−	−	+	+
+	+	−	+	+	−	+	−	−	−	+
+	+	+	−	+	+	−	+	−	−	−
−	+	+	+	−	+	+	−	+	−	−
−	−	+	+	+	−	+	+	−	+	−
−	−	−	+	+	+	−	+	+	−	+
+	−	−	−	+	+	+	−	+	+	−
−	+	−	−	−	+	+	+	−	+	+
−	−	−	−	−	−	−	−	−	−	−

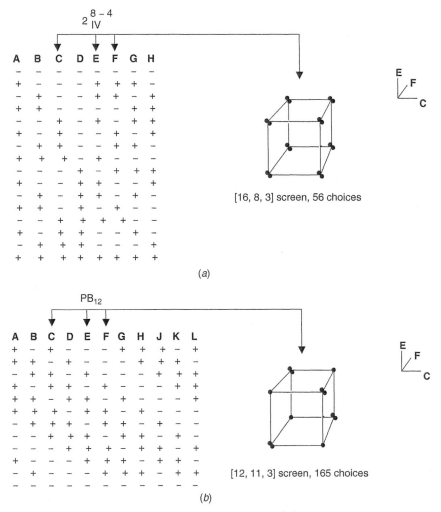

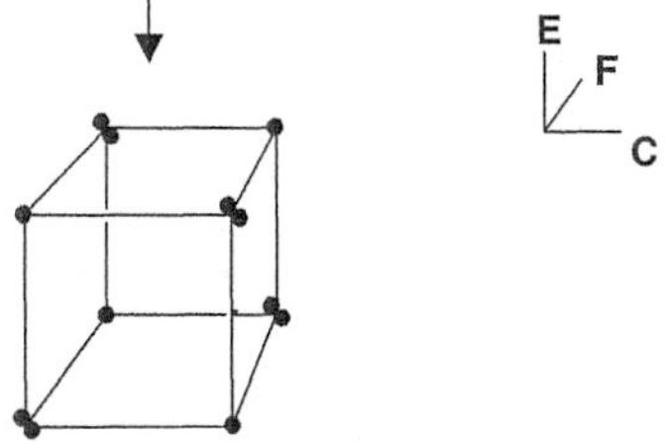

[12, 11, 3] screen, 165 choices

(b)

Figure 7.1. Projectivity 3 attributes of (*a*) 2^{8-4} and (*b*) PB_{12}.

Now look at the PB_{12} design in Figure 7.1*b*, which has four fewer runs. For this design every one of the 165 choices of 3 factors out of 11 produces a $1\frac{1}{2}$ replicate design, a full 2^3 factorial, plus a tetrahedral half-replicate 2^{3-1} fractional as illustrated in Figure 7.1*b* (Lin and Draper, 1992, Box and Bisgaard, 1993, Box and Tyssedal, 1996). The first rows of the PB_{12}, PB_{20} and PB_{24} designs are shown in Table 7.2. Each of these three designs may be obtained by cyclic substitution* in the manner described for the PB_{12} design, and they have similar useful projective properties. The PB designs are thus of considerable general interest for factor screening.

*It is not true that all order PB designs can be generated by cyclic substitution.

Table 7.2. Generator Rows for Constructing PB designs

PB_{12}	$n = 12$	+ + − + + + − − − + −
PB_{20}	$n = 20$	+ + − − + + + + − + − + − − − − + + −
PB_{24}	$n = 24$	+ + + + + − + − + + − − + + − − + − + − − − −

When n is a power of two, PB designs are identical to the corresponding fraction & factorials.

Exercise 7.1. Using cyclic substitution, obtain the PB design for $n = 20$ and satisfy yourself that all the columns are mutually orthogonal.

Exercise 7.2. The $n = 16$ Plackett and Burman design can be obtained by cyclic substitution using the initial row

$$+ + + + - + - + + - - + - - -$$

Use this to generate an L_{16} orthogonal array. Show that it is identical to the 2_{III}^{15-11} design previously discussed. In establishing the identity, remember that the basic design is not influenced by rearranging the order of either rows or columns or by switching the signs in one or more of the columns. These operations correspond to merely changing the order of the runs, renaming the factors, or deciding which level of a factor is to be denoted by a minus or plus sign.

The Projective Properties of the PB_{20} Design

Many PB designs have remarkable screening properties; in particular, since the designs for 12, 20, and 24 runs are of projectivity $P = 3$, they are respectively [12,11,3], [20,19,3], and [24,23,3] screens. For illustration, Table 7.3 displays a PB_{20} design where, for the purposes of exposition, the columns **A** and **B** have been rearranged to reproduce a 2^2 factorial replicated five times. If any third column except column **T** is added, you will generate a duplicated 2^3 design with a single added 2_{III}^{3-1} half replicate (a 2 1/2 replicated design). If **T** is chosen for the third column, the projected design is a less desirable arrangement which consists of single 2^3 factorial plus a 2_{III}^{3-1} tetrahedral design replicated three times. Thus, of the 969 ways of choosing 3 factors out of 19, only 57 produce this latter type and the remaining 912 are all 2 1/2 replicates.

Analysis of PB Designs

Normal plots and Lenth's procedure work well as diagnostic tools for the 2^k and 2^{k-p} fractionals because in the alias patterns of these designs every main effect and two-factor interaction appear only *once*. This property further allows these designs to be categorized in terms of their resolution R. However, these methods of analysis are inappropriate for PB designs because these arrangements have much more complicated alias structures. For instance, suppose you generate a design for the five factors **A, B, C, D,** and **E** from the first five columns of the

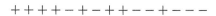

Table 7.3. The PB_{20} Design, [20,19,3] Screen Rearranged to Illustrate its Projectivity Attributes

Run Number	A	B	C	D	E	F	G	H	J	K	L	M	N	O	P	Q	R	S	T
1	+	+	+	−	−	−	+	+	−	−	+	−	−	+	+	−	+	−	−
2	+	+	−	+	−	−	−	+	+	−	−	+	−	+	−	+	−	+	−
3	+	+	−	−	+	−	+	−	−	+	−	+	+	−	+	−	−	+	−
4	+	+	−	−	−	+	−	−	+	+	+	−	+	−	−	+	+	−	−
5	+	+	+	+	+	+	−	−	−	−	−	−	−	−	−	−	−	−	+
6	+	−	+	−	−	+	−	+	+	+	−	−	+	+	+	−	−	+	+
7	+	−	−	+	+	−	+	−	+	+	+	−	−	+	+	+	−	−	+
8	+	−	+	+	−	−	+	+	−	+	−	+	+	−	−	+	+	−	+
9	+	−	−	−	+	+	+	+	+	−	+	+	−	−	−	−	+	+	+
10	+	−	+	+	+	+	−	−	−	−	+	+	+	+	+	+	+	+	−
11	−	+	−	+	−	+	+	−	+	−	−	+	+	+	+	−	+	−	+
12	−	+	−	+	+	−	−	+	−	+	+	−	+	+	−	−	+	+	+
13	−	+	+	−	−	+	+	−	−	+	+	+	−	+	−	+	−	+	+
14	−	+	+	−	+	−	−	+	+	−	+	+	+	−	+	+	−	−	+
15	−	+	+	+	+	+	+	+	+	+	−	−	−	−	+	+	+	+	−
16	−	−	−	+	−	+	−	+	−	+	+	+	−	−	+	−	−	−	−
17	−	−	−	−	+	+	+	+	−	−	−	−	+	+	−	+	−	−	−
18	−	−	+	−	+	−	−	−	+	+	−	+	−	+	−	−	+	−	−
19	−	−	+	+	−	−	+	−	+	−	+	−	+	−	−	−	−	+	−
20	−	−	−	−	−	−	−	−	−	−	−	−	−	−	+	+	+	+	+

PB_{12} design in Table 7.1. Then the alias structure for the 11 orthogonal linear contrasts of the PB_{12} design are shown in Table 7.4.

For instance, each of the main effects of each of the five factors is aliased to a string containing *all* the two-factor interactions not identified with that factor. Thus a single two-factor interaction between say **A** & **B**, will bias the remaining main effects and will appear in all the interaction strings. Thus, unless it is assumed that no interactions occur, normal plots and Lenth's procedure are of limited use for the analysis of PB designs.

Bayesian Analysis

Now, although the *methods* of analysis so far used for fraction factorials are inappropriate for the analysis of PB design, the *model* on which these methods are based *is* appropriate. As you saw—it uses the principle of parsimony—in the context of factor screening there are likely to be few factors that are active and a number that are essentially inert. This provides an appropriate structure for a *Bayesian analysis* (Box and Meyer 1986, 1993) that takes direct account of

Table 7.4. Alias Structure, Twelve-Run PB Design Employing Five Factors

Contrasts		Alias Structure
l_A	$\rightarrow$	$\mathbf{A} + \frac{1}{3}(-BC + BD + BE - CD - CE - DE)$
l_B	$\rightarrow$	$\mathbf{B} + \frac{1}{3}(-AC + AD + AE - CD + CE - DE)$
l_C	$\rightarrow$	$\mathbf{C} + \frac{1}{3}(-AB - AD - AE - BD + BE - DE)$
l_D	$\rightarrow$	$\mathbf{D} + \frac{1}{3}(+AB - AC - AE - BC - BE - CE)$
l_E	$\rightarrow$	$\mathbf{E} + \frac{1}{3}(+AB - AC - AD + BC - BD - CD)$
l_F	$\rightarrow$	$\frac{1}{3}(-AB + AC - AD + AE + BC - BD - BE + CD - CE - DE)$
l_G	$\rightarrow$	$\frac{1}{3}(-AB - AC - AD + AE - BC + BD - BE + CD - CE + DE)$
l_H	$\rightarrow$	$\frac{1}{3}(+AB + AC - AD - AE - BC - BD - BE - CD + CE + DE)$
l_J	$\rightarrow$	$\frac{1}{3}(-AB - AC - AD - AE + BC + BD - BE - CD - CE - DE)$
l_K	$\rightarrow$	$\frac{1}{3}(-AB - AC + AD - AE - BC - BD - BE + CD + CE - DE)$
l_L	$\rightarrow$	$\frac{1}{3}(-AB + AC + AD - AE - BC - BD + BE - CD - CE + DE)$

parsimony. In this analysis the data obtained from an experiment are analyzed in the light of what was known before the experiment was carried out. This may be formalized* as follows:

Effects calculated for inactive factors (i.e., noise) may be represented approximately as items from a normal distribution with mean zero and standard deviation σ.

For a proportion π of active factors the resulting effects are represented as items from a normal distribution with mean zero and a larger standard deviation $\gamma\sigma$.

A survey of a number of authors' published analyses of two-level factorial and fractional factorial designs showed that γ, the ratio of the standard deviation of the active to the inactive (noise) effects, was on average between 2 and 3. Also the percentage π of active factors was about 0.25. Therefore these values were used to represent prior information in the analysis and recent investigation has confirmed that when active factors were present the results were not very sensitive to moderate changes in γ and π [see Barrios, (2004b)]. Mathematical details of the Bayesian analysis based on these precepts are given in Appendix 7A. All you really need to know, however, is how to use one of the available computer programs that can be used to make these calculations. [For the calculations in this chapter we used the BsMD package for R language based on Daniel Meyer's code mdopt. See Barrios (2004a) and Meyer (1996).]

A PB_{12} Design Used to Study Five Factors

For illustration and comparison look again at the data in Table 6.15 for a complete 2^5 factorial design in five factors **A, B, C, D**, and **E**. The analysis in Chapter 6

* As usual in this analysis it is also supposed that before the experiment is carried out the experimenter has no information about the overall mean or the standard deviation (see e.g., Box and Tiao, 1973, 1992)

for this full 32-run design led to the conclusion that the effects *B, D, BD, E,* and *DE* were almost certainly real. Later it was shown that precisely the same conclusions could be reached with half the effort using the 16-run 2_{V}^{5-1}—the half-replicate design set out in Table 6.16.

A natural question is, "Could a twelve-run PB design have been used to obtain a similar result?" To shed light on this, the first five columns of the PB_{12} design in Table 7.1 were used to accommodate the five factors **A, B, C, D**, and **E** as shown in Table 7.5. The appropriate "observations" for the 12 runs were taken from the 32-run design. They are indicated by daggers in Table 6.15. These data are reproduced in Table 7.5.

The calculated contrasts $l_A, l_B, \ldots, l_L$ are shown in the last row of the table and a normal plot for these values $(\bar{y}_+ - \bar{y}_-)$ is shown in Figure 7.2. Notice that, although the contrast identified with the main-effect factor **B** falls off the line, the plot fails to show any effects due to factors **D** and **E**. Now look at the alias structure for the five-factor PB_{12} design given in Table 7.4. Using the results from the 32-run design, you will see that the large interactions $BD = 13.25$ and $DE = -11.0$ will bias the estimates of D and E.

The Bayes analysis of the PB_{12} data, making use of the concept of factor sparsity (parsimony), proceeds as follows. Regarding as active any factor that produces a main effect *and/or is involved in any interaction*, the important questions are as follows: (a) Which are the active factors? (b) What do they do? The questions are decided separately. To find the active factors, the analysis assesses all of the alternative possibilities in the following manner (Box and Meyer 1993). One possibility is that just a single factor is responsible for all that is going on—there is a 1D active subspace. To explore this possibility, the procedure

Table 7.5. PB_{12} Twelve Runs Abstracted from the 2^5 Design of Table 6.15 to Form Five Columns of a PB_{12} Design Reactor Example

Run Numbers		Factors					Columns from PB_{12} Not Used						
PB_{12}	2^5	**A**	**B**	**C**	**D**	**E**	**F**	**G**	**H**	**J**	**K**	**L**	Observed, y
1	6	+	−	+	−	−	−	+	+	+	−	+	56
2	12	+	+	−	+	−	−	−	+	+	+	−	93
3	23	−	+	+	−	+	−	−	−	+	+	+	67
4	14	+	−	+	+	−	+	−	−	−	+	+	60
5	28	+	+	−	+	+	−	+	−	−	−	+	77
6	24	+	+	+	−	+	+	−	+	−	−	−	65
7	15	−	+	+	+	−	+	+	−	+	−	−	95
8	29	−	−	+	+	+	−	+	+	−	+	−	49
9	25	−	−	−	+	+	+	−	+	+	−	+	44
10	18	+	−	−	−	+	+	+	−	+	+	−	63
11	3	−	+	−	−	−	+	+	+	−	+	+	63
12	1	−	−	−	−	−	−	−	−	−	−	−	61
Contrast		l_A	l_B	l_C	l_D	l_E	l_F	l_G	l_H	l_J	l_K	l_L	
effects		5.8	21.2	−1.5	7.2	−10.5	−2.2	2.2	−8.8	7.2	−0.5	−9.8	

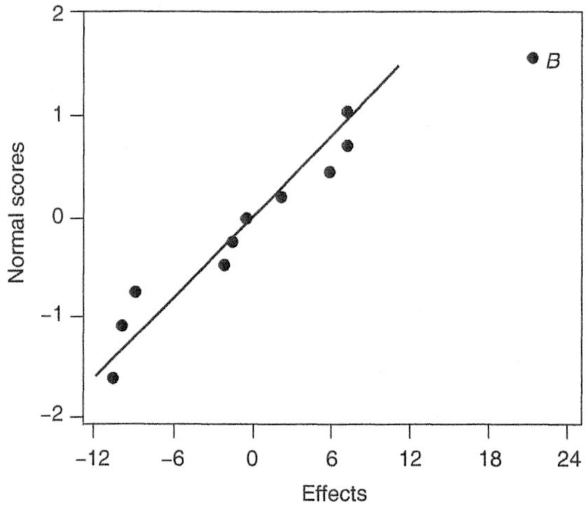

Figure 7.2. Normal plot with only **B** outstanding.

considers individually the five separate main effects *A, B, C, D, E*. Another possibility is that the activity occurs in a 2D space. Suppose for example, you were currently looking at the *subspace* of factors **A** and **B**, denoted by [**AB**]. You would need to consider whether the changes found could be explained by the action of the main effects *A* and *B* and the two-factor interaction *AB*. Similarly, for the space [**AC**] you would consider the joint effects of *A*, *C*, and *AC*. In all there would be 10 (2D) subspaces [**AB**], [**AC**], ..., [**DE**] to be explored. For a 3D subspace, say [**ABC**], you would need to consider jointly the effects *A, B, C, AB, AC, AD* and *ABC*, and so on.

In this Bayesian procedure *all* the models for all factors under consideration are individually evaluated, including the possibility that no factor is active. By this means an appropriate marginal *posterior probability* is calculated that each of the factors is active.* For this example the marginal probabilities obtained in this way for the activity of each of the five factors are plotted in Figure 7.3. This points to the conclusion that the active subspace determined from this 12-run design is that of [**BDE**], which agrees with the results already obtained from the 32- and 16-run designs.

Adding More Columns

To make the problem of identifying the active factors more difficult, you can include the remaining columns **F, G, H, J, K**, and **L** in the design (more hay added to the haystack). As shown in Figure 7.4, the probabilities obtained from what is now an 11-column analysis indicates the same subspace as before.

*By "marginal" is meant that the probabilities involved for each factor have been appropriately combined. These probabilities are "posterior" probabilities because they represent the situation after the data have been taken into account.

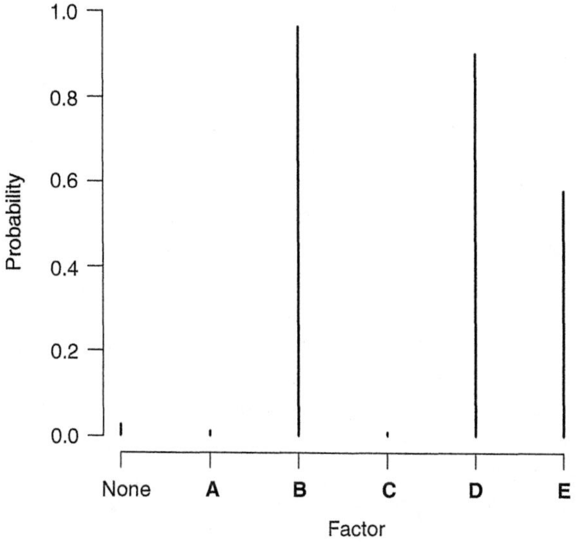

Figure 7.3. Posterior probabilities for factor activity: PB_{12} design.

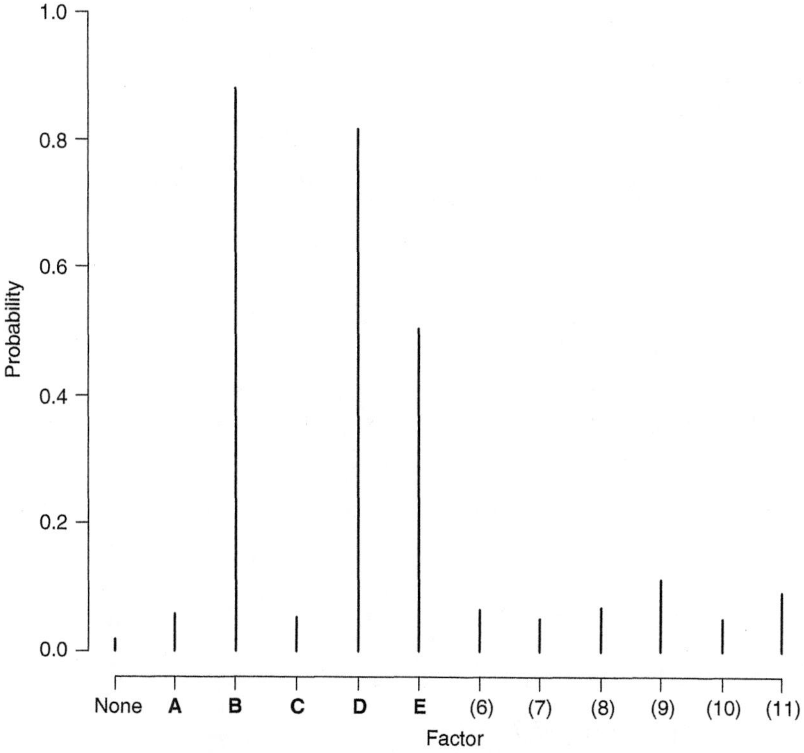

Figure 7.4. Posterior probabilities for 11 factors: PB_{12} design.

Further Analysis

The analysis so far has indicated *which* factors are important; to find out *what* they do, a factorial analysis of the identified active factors **B, D**, and **E** may now be done in the usual way. In this particular case three factors have been identified as active and the design happens to be of projectivity 3 and a 2^3 design plus a 2^{3-1} half replicate is available in the factors. Notice, however, that this circumstance is not required by the analysis. More generally, once the active factors are known, a more detailed analysis in terms of main effects and interactions can be carried out using least squares (regression) discussed in Chapter 10.

In this example in the subspace [**BDE**] four of the runs are duplicated. It turns out that least squares estimates of the effects (see Chapter 10) are such that the averages of the duplicated points are treated as individual observations. For this example, the projected design, data, and estimated effects are shown in Table 7.6. An estimate of the standard error of each effect obtained using the "replicated" runs is ± 1.55. It will be seen that an analysis of the 12-run design supports the earlier conclusion that the effects B, D, E, BD, and BE are those which stand out from the noise.

Thus the designs using 32, 16, and 12 runs have all identified the same effects, *B, D, E, BD*, and *DE,* set out below. From Table 7.7 you will see that the estimated magnitudes of the effects are in reasonably good agreement. (For further comparison, a set of estimates marked "sequential" have been added to the table. These come from runs chosen in a run-by-run manner to be described later.)

Exercise 7.3. Choose a different set of five columns from Table 7.1 with which to associate the design factors **A, B, C, D, E**. Select the appropriate responses from the full 2^5 factorial of Table 6.15. Analyze your data as above. What do you conclude?

Table 7.6. A 2^3 Design in the Active Subspace of Factors B, D, and E

Run Number	B	D	E	BD	BE	DE	BDE	Observed, y	Average
1, 12	−	−	−	+	+	+	−	56, 61	58.5
11	+	−	−	−	−	+	+	63	63
4	−	+	−	−	+	−	+	60	60
2, 7	+	+	−	+	−	−	−	93, 95	94
10	−	−	+	+	−	−	+	63	63
3, 6	+	−	+	−	+	−	−	67, 65	66
8, 9	−	+	+	−	−	+	−	49, 44	46.5
5	+	+	+	+	+	+	+	77	77
Effects:	**18.00**,	**6.75**,	**−5.75**,	**14.25**,	−1.25,	**−9.50**,	−0.50	±2.4	

Bold numerals indicate effects of active factors.

Table 7.7. Comparison of Estimated Effects

Designs	B	D	E	BD	DE
2^5	19.5	10.8	−6.3	13.3	−11.0
2^{5-1}	20.5	12.2	−6.3	10.8	−9.5
PB	18.0	6.8	−5.8	14.3	−9.5
Sequential	20.0	13.8	−3.4	13.0	−12.4

Augmenting Plackett and Burman Designs to Form a 2^k or 2^{k-p} Design

The PB_{12} designs can be augmented to clarify specific uncertainties. For example, suppose up to 11 factors have been tested in the 12-run PB design of Table 7.1 and the Bayes analysis points to, say, **A, B, C**, and possibly **K** as active factors. Then, as shown in Table 7.8, a complete 2^4 factorial in **A, B, C**, and **K** with one factor combination duplicated, in this case [− − −−], can be obtained by adding five runs to the PB_{12} design. This is a further example of sequential assembly in which the data from the initial 12-run design suggest which runs to add.

Adding a Few Extra Runs to Resolve Ambiguities

The idea that every experimental design should by itself lead to an unambiguous conclusion is of course false. If you want to reduce experimentation to a

Table 7.8. The PB_{12} Design Augmented by Five Runs to Form a Complete 2^4 Factorial

	Factors Suggested as Possibly Active					2^4 Design				
Run Number	**A**	**B**	**C**	**K**	Five Extra Runs	**A**	**B**	**C**	**K**	Runs from PB_{12}
1	+	−	+	−		−	−	−	−	9, 12
2	+	+	−	+	•	+	−	−	−	
3	−	+	+	+	•	−	+	−	−	
4	+	−	+	+		+	+	−	−	5
5	+	+	−	−	•	−	−	+	−	
6	+	+	+	−		+	−	+	−	1
7	−	+	+	−		−	+	+	−	7
8	−	−	+	+		+	+	+	−	6
9	−	−	−	−	•	−	−	−	+	
10	+	−	−	+		+	−	−	+	10
11	−	+	−	+		−	+	−	+	11
12	−	−	−	−		+	+	−	+	2
						−	−	+	+	8
						+	−	+	+	4
						−	+	+	+	3
					•	+	+	+	+	

minimum, it is frequently best to begin with a small fractional design. Sometimes it is true that such a design may by itself solve (or at least fix) a problem. But more likely it will greatly reduce the number of models that need to be considered. (See the discussion of 20 questions in Chapter 1). It can also help identify additional runs that may be needed.

Fatigue Life of Repaired Castings

In the following experiment (Hunter, Hodi, and Eager, 1992) seven factors were investigated in a 12-run PB design.* The PB_{12} design, the data, and the 11 calculated column contrast effects are shown in Table 7.9. The normal plot of these effects made by the original authors (see Fig. 7.5*a*) seems, perhaps rather dubiously, to support their conclusion that factor **F** and possibly factor **D** have

Table 7.9. A PB_{12} Design: Fatigue Life Data

Factor	Level (+)	Level (−)
A: initial structure	As received	β Treated
B: bead size	Small	Large
C: pressure treatment	None	HIP
D: heat treatment	Anneal	Solution treat/age
E: cooling rate	Slow	Rapid
F: polish	Chemical	Mechanical
G: final treatment	None	Peen

Run	Factors							Unused columns				Log Life
Number	**A**	**B**	**C**	**D**	**E**	**F**	**G**	8	9	10	11	×100
1	+	+	−	+	+	+	−	−	−	+	−	6.06
2	+	−	+	+	+	−	−	−	+	−	+	4.73
3	−	+	+	+	−	−	−	+	−	+	+	4.63
4	+	+	+	−	−	−	+	−	+	+	−	5.90
5	+	+	−	−	−	+	−	+	+	−	+	7.00
6	+	−	−	−	+	−	+	+	−	+	+	5.75
7	−	−	−	+	−	+	+	−	+	+	+	5.68
8	−	−	+	−	+	+	−	+	+	+	−	6.61
9	−	+	−	+	+	−	+	+	+	−	−	5.82
10	+	−	+	+	−	+	+	+	−	−	−	5.92
11	−	+	+	−	+	+	+	−	−	−	+	5.86
12	−	−	−	−	−	−	−	−	−	−	−	4.81
Effects	0.33	0.30	−0.25	−0.52	0.15	0.92	0.18	0.45	0.45	0.08	−0.25	

*This PB_{12} design is in the form given by the authors. It is a rearrangement of that given in Table 7.1.

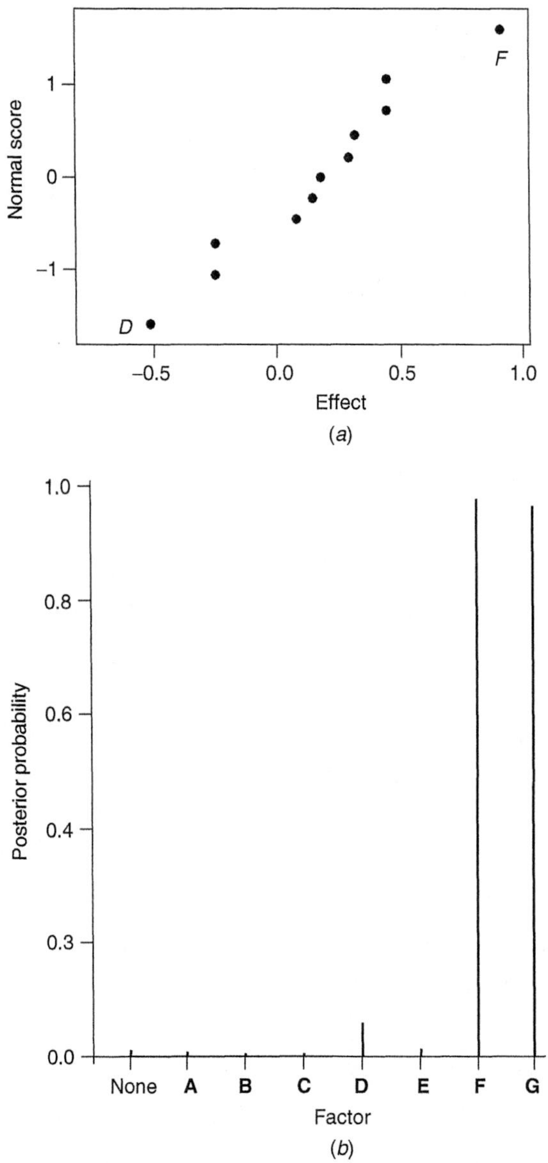

Figure 7.5. (*a*) Normal plot **F** and **D** possibly real. (*b*) Bayes plot identifying factors **F** and **G**.

significant main effects. However, by the Bayes analysis (see Fig. 7.5*b*), factors **F** and **G,** with probabilities of 0.979 and 0.964, respectively, are identified as the active factors and *not* **F** and **D**.

The reason is not hard to find. There is a very large interaction between **F** and **G** that confuses standard analysis. To show this, the data in Table 7.10 are arrayed as a 2^2 factorial design. The graphical analysis in Figure 7.6 shows (a) a two-way

Table 7.10. A 2^2 Factorial Analysis of the Active Factors F and G

Run Numbers	F	G	FG	Observations	Average
2, 3, 12	−	−	+	4.73, 4.63, 4.81	4.72
1 5, 8	+	−	−	6.06, 7.00, 6.61	6.56
4, 6, 9	−	+	−	5.90, 5.75, 5.82	5.82
7, 10, 11	+	+	+	5.68, 5.92, 5.86	5.82
Effects	0.92	0.18	−0.92		
	F	*G*	*FG*		

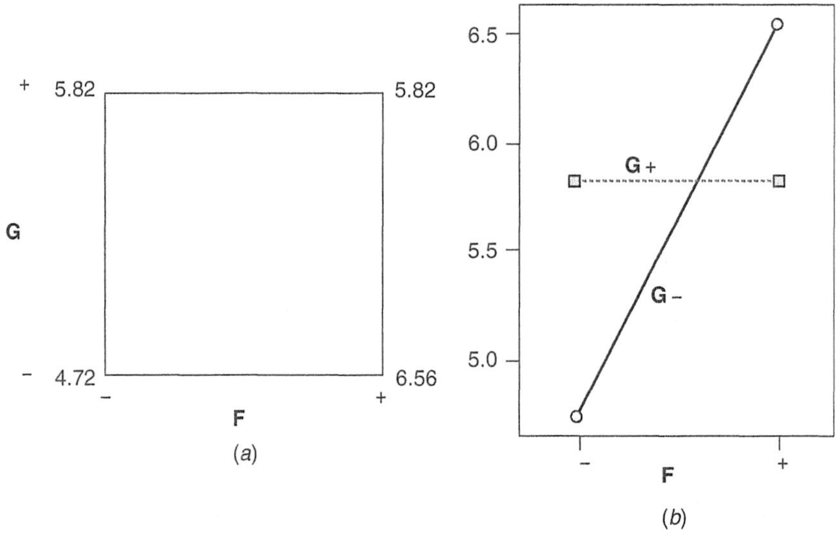

Figure 7.6. (*a*) A 2^2 factorial in **F** and **G**. (*b*) Contrast plot for averages.

plot and (b) a contrast plot for the averages. This example dramatically illustrates how interactions can defeat any single factor analysis.*

7.2. CHOOSING FOLLOW-UP RUNS

Suppose for illustration you had six alternative models 1, 2, ..., 6 and that before any experiments were performed the models were equally probable, as in Figure 7.7*a*. Suppose now an experiment was run which produced the probability distribution in Figure 7.7*b*. Obviously, the experiment would have produced a great change in information about the models.

* See also Wang and Wu (1995) and Hamada and Wu (2000).

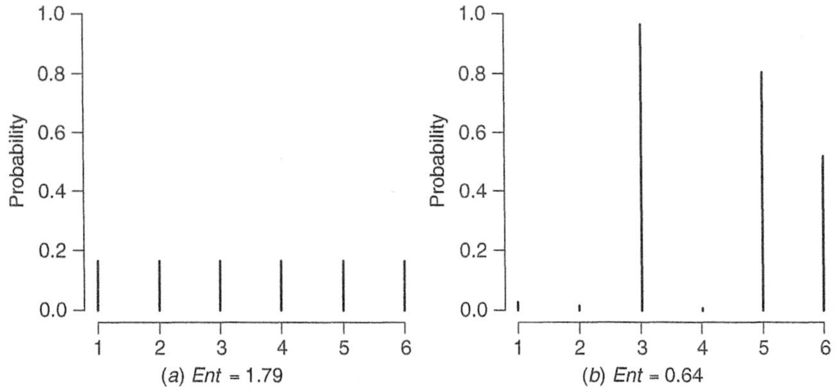

Figure 7.7. Entropy: (*a*) before and (*b*) after the experiment.

Entropy

As a measure of information Shannon (1948) introduced the concept of *entropy*. This measures the state of disorganization of a system. In the above example this would be a measure of the degree of evenness of the probabilities, with maximum entropy occurring when all the probabilities P_i were equal. As a measure of entropy Shannon employed

$$Ent = -\sum_{i=1}^{n} P_i \ln P_i.$$

For the equal probabilities shown in Figure 7.7*a* you will find that the entropy $Ent = 1.79$. After the experiment the probabilities in Figure 7.7*b* produce the value $Ent = 0.64$. The gain in information supplied by the experiment is represented by the *reduction* in entropy.

Adding Runs to Reduce Entropy

A method for obtaining one or more informative follow-up runs uses the idea of Box and Hill (1967) further developed by Meyer, Steinberg and Box (1996) choosing those runs that produce the greatest expected reduction in entropy (for details see Appendix 7A). As an illustration an example discussed in the earlier edition of this book is reanalyzed. The experiment concerned various factors affecting shrinkage in a plastics injection molding process. The 2_{IV}^{8-4} design in eight factors **A, B, . . ., H** and the resulting data are given in Table 7.11a. The calculated contrasts and aliases are in Table 7.11b. A normal plot of the effects is shown in Figure 7.8. The associated normal plot showed that the estimated contrasts l_C, l_E, and l_{AE}, indicated by bold type in Table 7.11b, were distinguishable from noise. However, the contrast l_{AE} estimated a string of four aliased two-factor interactions $AE + BF + CH + DG$. In the previous analysis an attempt was made to identify four additional runs that would unravel this alias string. This analysis took account of interactions up to second order only. The same

Table 7.11a. A 2_{IV}^{8-4} Design: Molding Example (I = ABDH, I = ACEH, I = BCFH, I = ABCG)

Run Number	Mold Temperature, A	Moisture Content, B	Hold Press, C	Cavity Thickness, D	Booster Pressure, E	Cycle Time, F	Gate Size, G	Screw Speed, H	Shrinkage, y
1	−	−	−	+	+	+	−	+	14.0
2	+	−	−	−	−	+	+	+	16.8
3	−	+	−	−	+	−	+	+	15.0
4	+	+	−	+	−	−	−	+	15.4
5	−	−	+	+	−	−	+	+	27.6
6	+	−	+	−	+	−	−	+	24.0
7	−	+	+	−	−	+	−	+	27.4
8	+	+	+	+	+	+	+	+	22.6
9	+	+	+	−	−	−	+	−	22.2
10	−	+	+	+	+	−	−	−	17.1
11	+	−	+	+	−	+	−	−	21.5
12	−	−	+	−	+	+	+	−	17.5
13	+	+	−	−	+	+	−	−	15.9
14	−	+	−	+	−	+	+	−	21.9
15	+	−	−	+	+	−	+	−	16.7
16	−	−	−	−	−	−	−	−	20.3

Note: The design is *not* displayed in Yates order.

Table 7.11b. Contrasts, Estimated Effects, and Alias Strings

l_A	=	−0.7	→	A
l_B	=	−0.1	→	B
l_C	=	**5.5**	→	$\boldsymbol{C}$
l_D	=	−0.3	→	D
l_E	=	**−3.8**	→	$\boldsymbol{E}$
l_F	=	−0.1	→	F
l_G	=	0.6	→	G
l_H	=	1.2	→	H
l_{AB}	=	−0.6	→	$AB + CG + DH + EF$
l_{AC}	=	0.9	→	$AC + BG + DF + EH$
l_{AD}	=	−0.4	→	$AD + BH + CF + EG$
l_{AE}	=	**4.6**	→	$\boldsymbol{AE + BF + CH + DG}$
l_{AF}	=	−0.3	→	$AF + BE + CD + GH$
l_{AG}	=	−0.2	→	$AG + BC + FH + DE$
l_{AH}	=	−0.6	→	$AH + BD + CE + FG$

experiment may now be used to choose the four best discriminating runs from entropy reduction.

The Bayes analysis of this 2_{IV}^{8-4} design displayed in Figure 7.9*a*, shows that the posterior probabilities for **B, D, F**, and **G** are negligible while those for factors

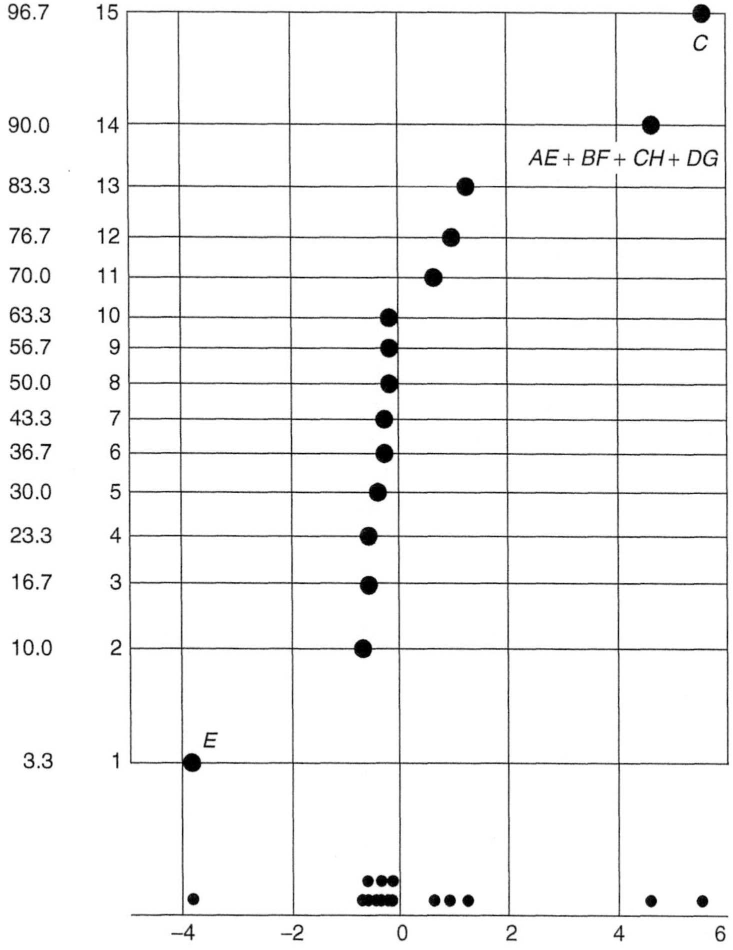

Figure 7.8. Normal plot of effect shrinkage data.

A, C, E, and **H** are not only large but equal. The reason they are equal can be seen from Figure 7.9*b*, where each projection of the four active factors into a three dimensional space produces the same data pairs at the vertices of a cube. Since **B, D, F**, and **G** are essentially inert, the design is a duplicated 2^{4-1} arrangement with defining contrast **I = ACEH** shown in Table 7.12a.

The five models of highest probability, M_1, M_2, M_3, M_4, and M_5, are shown in Table 7.12b. A possible block effect between this initial design and the follow-up runs was included in the model. Table 7.12c shows the predicted values $\hat{y}$ obtained for each of the five models and the 16 possible follow-up runs numbered 1 to 16. All 16 runs were included to allow the possibility that duplicates of one or more of the first 8 runs might be required. It would be expected that of these possible 16 runs the best to discriminate among the models would be those in

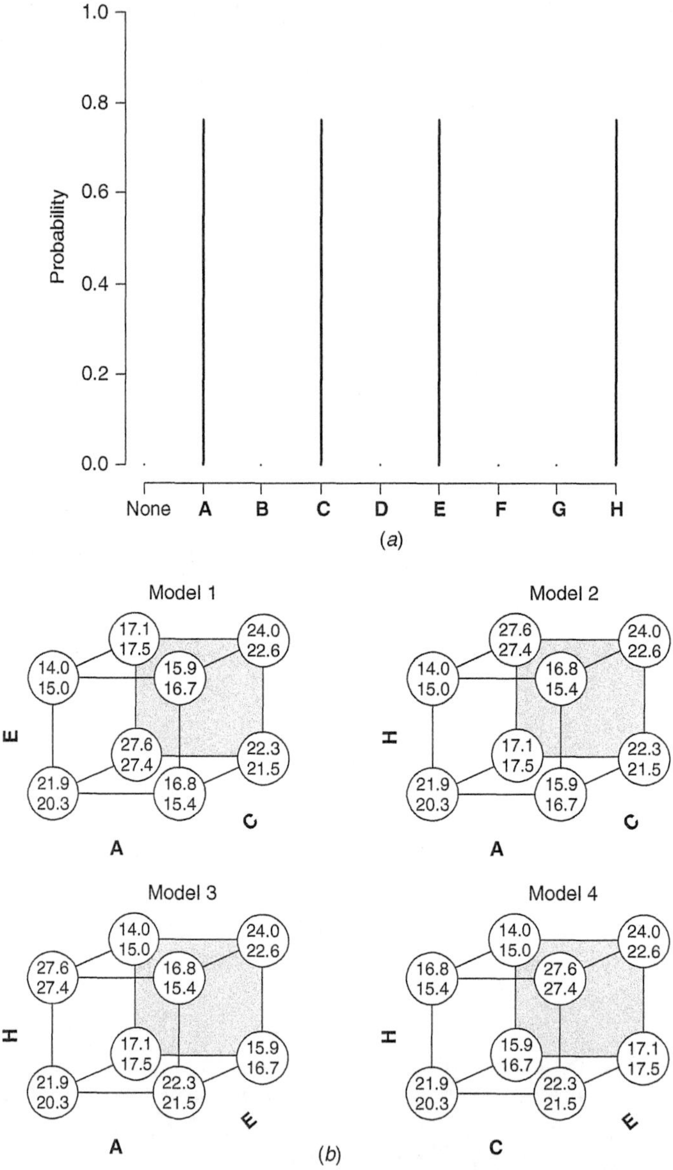

Figure 7.9. (*a*) Equal probabilities **A, C, E, H**. (*b*) Four 3D projections.

a row where the five model predictions $\hat{y}$ *differed* the most in some way. The entropy concept leads to a discrimination criterion of this kind, called here the model discrimination (MD*) score.

*A theoretical explanation will be found in Appendix 7A. For the practitioner suitable computer programs are of course used to perform the necessary calculations.

Table 7.12. MD Scores for Different Models using $\pi = 0.25$ and $r = 2.0$

(a)

Point	Run Numbers	A	C	E	H	Observed, y	
1	14, 16	−	−	−	−	21.9	20.3
2	1, 3	−	−	+	+	14.0	15.0
3	5, 7	−	+	−	+	27.6	27.4
4	10, 12	−	+	+	−	17.1	17.5
5	2, 4	+	−	−	+	16.8	15.4
6	13, 15	+	−	+	−	15.9	16.7
7	9, 11	+	+	−	−	22.3	21.5
8	6, 8	+	+	+	+	24.0	22.6

(b)

Model	Factors	Probability
M_1:	**A, C, E**	0.2356
M_2:	**A, C, H**	0.2356
M_3:	**A, E, H**	0.2356
M_4:	**C. E. H**	0.2356
M_5:	**A, C, E, H**	0.0566

(c)

Run Number	A	C	E	H	Predicted Values, $\hat{y}$ M_1	M_2	M_3	M_4	M_5	Standard Deviation
1	−	−	−	−	21.08	21.08	21.08	21.08	21.09	0.0
2	−	−	+	+	14.58	14.59	14.58	14.58	14.54	0.0
3	−	+	−	+	27.38	27.38	27.38	27.38	27.44	0.0
4	−	+	+	−	17.34	17.34	17.34	17.34	17.32	0.0
5	+	−	−	+	16.16	16.16	16.16	16.16	16.13	0.0
6	+	−	+	−	16.35	16.35	16.35	16.35	16.13	0.0
7	+	+	−	−	21.87	21.87	21.87	21.87	21.88	0.0
8	+	+	+	+	23.25	23.25	23.25	23.25	23.27	0.0
9	−	−	−	+	21.08	14.58	27.38	16.16	19.75	5.0
10	−	−	+	−	14.58	21.08	17.34	16.35	19.75	2.6
11	−	+	−	−	27.38	17.34	21.08	21.87	19.75	3.7
12	−	+	+	+	17.34	27.38	14.58	23.25	19.75	5.0
13	+	−	−	−	16.16	16.35	21.87	21.08	19.75	2.7
14	+	−	+	+	16.35	16.16	23.25	14.58	19.75	3.5
15	+	+	−	+	21.87	23.25	16.16	27.38	19.75	4.2
16	+	+	+	−	23.25	21.87	16.35	17.34	19.75	2.9

(d)

Possible Additional Runs				MD Score
9	9	12	15	85.7
9	12	14	15	84.4
9	11	12	15	83.6
9	11	12	12	82.2
9	9	12	12	79.6
9	11	12	14	77.1
9	9	11	12	77.1
9	12	15	16	77.1
9	12	13	15	76.7
9	12	12	14	76.6

The computer program calculated the MD criterion for each of the 3876 possible sets of 4 follow-up runs chosen from the 16 candidates. The eight choices giving the highest MD scores are shown in Table 7.12d. For example the 4 runs numbered 9, 9, 12, 15 give the largest MD score and would normally be chosen. The next highest MD score is obtained from the 4 runs 9, 12, 14, 15, and so on. A very rough idea of the nature of the analysis can be obtained by looking at the last column of Table 7.12c where, instead of the MD criterion, the standard deviation of each set of predictions for the models M_1, M_2, M_3, M_4, and M_5 is given. You will see that runs 9, 12, 15, 11, 14 produce the largest standard deviations and these runs are favored by the MD criterion.

Adding Runs One at a Time

When the result of each run is available before the next is planned, it can be convenient and economical to make additional runs one at a time.* For illustration,

*The strategy of making additional runs one at a time will not always be a good one since there are situations where, for example, making a block of experiments is no more expensive than making one individual run.

Table 7.13a. A 2^{5-2} factorial design with additional runs added one at a time.

Run Number in 2^5 design	A	B	C	D	E	Observed, y
	Factors					
25	−	−	−	+	+	44
2	+	−	−	−	−	53
19	−	+	−	−	+	70
12	+	+	−	+	−	93
13	−	−	+	+	−	66
22	+	−	+	−	+	55
7	−	+	+	−	−	54
32	+	+	+	+	+	82

AB = D, AC = E

Sequence of Chosen Follow-Up Runs

Run Number	A	B	C	D	E	y
12	+	+	−	+	−	93
10	+	−	−	+	−	61
11	−	+	−	+	−	94
15	−	+	+	+	−	95

suppose you wished to study five factors **A, B, C, D, E** and initially you had only the data from an 8-run 2^{5-2} design of Table 7.13a abstracted from the full 2^5 factorial given earlier in Table 6.15. From the aliasing generators **AB** = **D**, **AC** = **E** in Table 7.13a, it will be seen that this particular design was chosen so as *not* to favor the unraveling of the effects **B, D, E, BD** and **DE** inherent in the data source. The initial analysis of the 8 runs gave the posterior probability distribution shown in Figure 7.10*a*, so at this stage the largest probability was for activity in none of the factors: not surprisingly there was no evidence favoring any particular choice of model. Using estimates from this initial design, the computed MD scores for each of the 32 candidate runs of the full 2^5 factorial are shown in the column of Table 7.13b marked stage 1. The largest MD score of 11.12 was for run 12, indicating the next run should be made at factor levels [+ + − + −]. The result $y_{12} = 93$ was obtained from the complete 32-run design. The probability distribution calculated after 9 runs was now that shown in Figure 7.10*b*. The possibility of no active factors had now become small. However, there was essentially no indication about *which* if any of the effects were active. By refitting the 9 runs now available, the MD scores were again calculated for each of the 32 runs, as shown

Table 7.13b. MDs for Possible Follow-Up Runs with MDs at Various Stages

	Factors						MD at Stages			
Run Number	A	B	C	D	E	Observed, y	1	2	3	4
1	−	−	−	−	−	61	2.74	15.55	13.58	3.80
2	+	−	−	−	−	53	1.93	6.61	6.92	2.72
3	−	+	−	−	−	63	3.29	25.90	14.81	26.79
4	+	+	−	−	−	61	6.20	48.30	22.83	26.95
5	−	−	+	−	−	53	1.89	11.29	13.37	12.99
6	+	−	+	−	−	56	1.95	9.45	9.28	12.52
7	−	+	+	−	−	54	2.31	9.73	12.72	7.66
8	+	+	+	−	−	61	4.50	33.94	21.11	8.81
9	−	−	−	+	−	69	4.81	37.01	11.93	2.57
10	+	−	−	+	−	61	7.18	**55.39**	2.49	1.76
11	−	+	−	+	−	94	7.67	51.58	**29.76**	6.41
12	+	+	−	+	−	93	**11.12**	18.78	24.22	6.28
13	−	−	+	+	−	66	0.42	1.80	2.10	1.53
14	+	−	+	+	−	60	3.67	25.00	7.44	2.13
15	−	+	+	+	−	95	3.62	27.11	28.18	**50.35**
16	+	+	+	+	−	98	6.51	30.66	23.82	49.79
17	−	−	−	−	+	56	3.48	22.26	14.56	13.22
18	+	−	−	−	+	63	2.45	12.65	9.29	12.80
19	−	+	−	−	+	70	0.64	1.45	2.05	1.36
20	+	+	−	−	+	65	3.60	24.42	14.78	2.21
21	−	−	+	−	+	59	2.46	13.91	11.39	2.92
22	+	−	+	−	+	55	1.33	4.74	4.92	1.84
23	−	+	+	−	+	67	2.49	17.18	10.99	10.78
24	+	+	+	−	+	65	3.51	24.53	15.92	10.72
25	−	−	−	+	+	44	6.03	17.66	20.35	8.30
26	+	−	−	+	+	45	6.52	47.62	16.98	9.75
27	−	+	−	+	+	78	5.63	43.18	18.40	18.34
28	+	+	−	+	+	77	7.20	37.59	18.41	18.11
29	−	−	+	+	+	49	3.80	26.26	17.50	21.98
30	+	−	+	+	+	42	3.14	28.75	17.46	22.53
31	−	+	+	+	+	81	3.30	27.59	17.92	3.07
32	+	+	+	+	+	82	4.09	5.80	7.73	1.91

in the column marked stage 2 in Table 7.13b. It will be seen that now the largest MD score is 55.39 for run 10 at levels [+ − − + −] for which $y_{10} = 61$. The new data were now added to the experimental design, new predictions obtained, new MD scores computed, and the probabilities recomputed. The successive probability distributions for stages 3 and 4 are shown in Figures 7.10*d* and 7.10*e*. The process shows the gradual convergence to the previous conclusion that factors **B, D**, and **E** are the active factors. If now a least squares analysis is made for

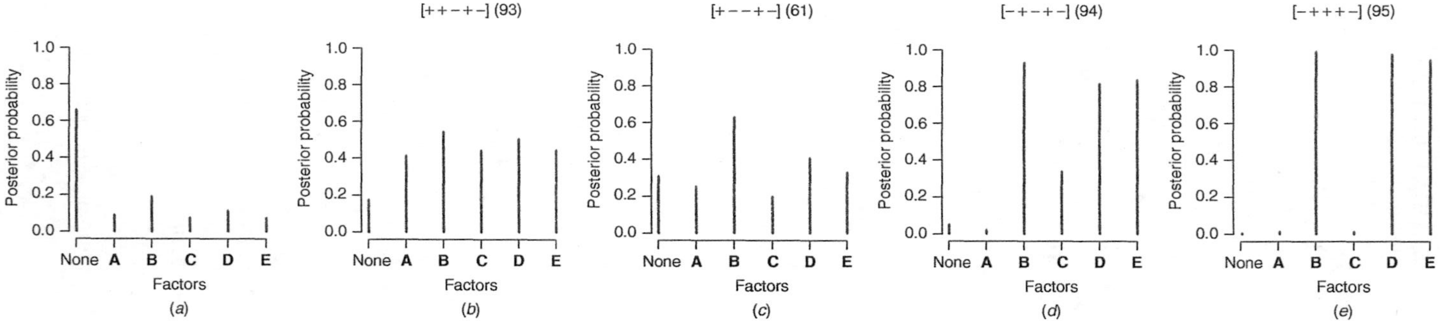

Figure 7.10. Probabilities associated with added runs: (*a*) 8, (*b*) 9, (*c*) 10, (*d*), 11, and (*e*) 12 runs.

main effects and two-factor interactions from the augmented design for all 12 runs the estimated effects are

$$B = 20.0, \qquad D = 13.8, \qquad E = -3.4, \qquad BD = 13.0, \qquad DE = -12.4$$

For comparison the estimates are also listed in Table 7.7
These effects are labeled as "sequential" estimates in Table 7.7 and are in good agreement with the previous estimates obtained respectively for the complete 32-run factorial, the 16-run half fraction, and the 12-run PB designs.

It may at first seem strange that the first new run selected is a repeat of a run used in the original 2^{5-2} design. Consideration shows, however, that this is not particularly surprising. The individual probabilities in Figure 7.10*a* suggest that on the data available from the initial design the possibility of no active effects is large, while there is almost nothing to choose between the small probabilities associated with **A, B, C, D, E.** It is reasonable therefore to first check whether the possibility of no active effects is viable. You see that after this repetition the probability of activity of some kind among the factors is large compared with no activity.

7.3. JUSTIFICATIONS FOR THE USE OF FRACTIONALS

Fractional designs and other orthogonal arrays often seem to work better than might be expected. Different rationales have been proposed to justify their use, some more persuasive than others.

Main Effects Only

In the analysis of designed experiments the (usually dubious) assumption of "main effects only" can sometimes approximate reality. Main effects (first-order effects) are likely to dominate, for example, in the early stages of an investigation if the experimental region is remote from a stationary region and in particular remote from a region containing a maximum*.

You Know Important Interactions?

Another rationale is that the subject matter expert can guess in advance which few of the large number of interactions is likely to exist. On such assumptions it is frequently possible to choose a fractional factorial design so as to allow the chosen interactions to be estimated free from aliases. We have given an example of this in Chapter 6, Section 10, where one interaction was thought to be likely in the design of an eight-run experiment. In general, however, this concept of knowing in advance of experimentation which few of the large number of interactions are likely to be real [for k factors there are $k(k-1)/2$ two-factor interactions] raises

* When fitted to happenstance data, regression equations frequently make the assumption that only linear (first-order main effects) terms occur.

some difficult questions. In particular, if the subject matter expert can decide in advance which are the likely *second-order* effects, this suggests she should have no difficulty in deciding what are the likely *first-order* effects.

It is when things do not go the way you expect that you often learn the most. An investigator is most intrigued by counterintuitive discoveries and will try to figure out the mechanism underlying these unexpected effects. If successful, such new knowledge may be useful for solving not only the immediate problem but others as well. Thus the most important effects may be those that have *not* been allowed for.

Projective Properties

For any two-level design of projectivity P every choice of P factors produces a complete 2^P factorial design, possibly replicated. If we now apply Juran's concept (also that of William of Occam: *Entia non sunt multiplicanda practer necessitatem*) of parsimony, if the number of active factors is P or less, then a full factorial will be available in these factors whichever they happen to be.

Sequential Assembly

When ambiguities and uncertainties of interpretation remain after the completion of a fractional design, you can employ additional runs to resolve the questions raised while simultaneously increasing the precision of your estimates.

Symmetric Filling of the Experimental Space

It has been argued that fractionation should be regarded merely as a means of sampling the experimental space symmetrically and economically with a limited number of experimental points. Associated with this idea is the "pick-the-winner" method of analysis whereby the factor combination providing the best response is chosen. Such procedures are sometimes helpful, but since they make no use of the structure of the design, they are inefficient.

We Know Higher Order Interactions Are Negligible?

It is common to assume that interactions between three or more factors will be negligible. While this is frequently true, we should always be on the lookout for exceptions*. The tentative assumptions that high order interactions are negligible is also bedeviled by the tacit idea that we should be concerned only about effects we can estimate and not be concerned with those of the same order that we cannot. For illustration consider the 2^3 factorial design in factors x_1, x_2, and x_3. The quantities we can estimate with this design are shown in boldface and those of the same order that we cannot estimate in ordinary type:

* In the first edition of the book a different set of four follow-up runs was chosen for the 2^{8-4}_{IV} experiment on injection molding in Section 7.2. One reason for this was the assumption that three- and higher-order interactions were negligible.

First-order effects: **1, 2, 3**
Second-order effects: **12, 13, 23** but not 11, 22, 33
Third-order effects: **123** but not 111, 222, 333, 112, 113, 223, 331, 332

If we are interested in a two-factor interaction such as **12**, which tells us how the effect of factor **1** changes as factor **2** is changed, why should we not be equally interested in quadratic terms that tell how the effect of factor **1** changes as factor **1** is changed. In terms of the calculus, if one is concerned about the derivative $\partial^2 y/\partial x_1 \partial x_2$, should we not be equally concerned with $\partial^2 y/\partial x_1^2$?

That terms of the same order are equally important is further justified by the fact that a linear transformation of the original factors produces only linear transformations in effects of the same order. For example, suppose a 2^2 factorial design is used to test the effect of changing the concentrations of two catalysts x_1 and x_2. One experimenter might think separately of x_1 and x_2 and plan to identify the main effects and two-factor interaction in the model

$$\hat{y} = b_0 + b_1 x_1 + b_2 x_2 + b_{12} x_1 x_2$$

But to another experimenter it might seem natural to consider as factors of the total catalyst concentration $X_1 = x_1 + x_2$ and the difference $X_2 = x_1 - x_2$ and similarly seek to estimate the main effects and interaction. The same model in the X's becomes

$$\hat{y} = B_0 + B_1 X_1 + B_2 X_2 + B_{12} X_1 X_2$$

or in terms of x_1 and x_2

$$\hat{y} = b_0 + (b_1 + b_2) x_1 + (b_1 - b_1) x_2 + b_{12}(x_1^2 - x_2^2)$$

Thus, the interaction of the first experimenter becomes the difference of the quadratic effects for the second experimenter.

As is discussed more fully in Chapters 11 and 12, the two-level factorial designs are ideal instruments for uncovering first-order and interaction effects and with the addition of a center point can provide information on overall curvature. Consequently, they can point out when it is likely necessary to employ a full second-order model, say

$$\hat{y} = b_0 + b_1 x_1 + b_2 x_2 + b_{11} x_1^2 + b_{22} x_2^2 + b_{12} x_1 x_2$$

Sequential assembly in particular, as with composite designs, can make possible the estimation of missing quadratic terms when this appears to be necessary.

APPENDIX 7A. TECHNICAL DETAILS

In this Appendix we give some technical details of the Bayesian model–discrimination method and the design criterion MD. For additional information on this approach to analyzing the data, see Box and Meyer (1993) and Meyer, Steinberg and Box (1996).

Bayes Model Discrimination

Let $\mathbf{Y}$ denote the $n \times 1$ vector of responses from a some experimental arrangement with k factors. The model for $\mathbf{Y}$ depends on which factors are active, and our analysis considers all the possible sets of active factors. Let M_i label the model in which a particular combination of f_i factors, $0 \leq f_i \leq k$, is active. Conditional on M_i being the true model, we assume the usual normal linear model $\mathbf{Y} \sim N(\mathbf{X}_i \beta_i, \sigma^2 \mathbf{I})$.

Here $\mathbf{X}_i$ is the regression matrix for M_i and includes main effects for each active factor and interaction terms up to any desired order. Let t_i denote the number of effects (excluding the constant term) in M_i. We will let M_0 label the model with no active factors.

We assign noninformative prior distributions to the constant term β_0 and the error standard deviation σ, which are common to all the models. Thus $p(\beta_0, \sigma) \propto 1/\sigma$. The remaining coefficients in β_i are assigned independent prior normal distributions with mean 0 and standard deviation $\gamma\sigma$.

We complete the model statement by assigning prior probabilities to each of the possible models. Following the notion of factor sparsity, we assume that there is a probability π, $0 < \pi < 1$, that any factor is active and that our prior assessment of any one factor being active is independent of our beliefs about the other factors. Then the prior probability of model M_i is $P(M_i) = \pi^{f_i}(1-\pi)^{k-f_i}$.

Having observed the data vector $\mathbf{Y}$, we can update the parameter distributions for each model and the probability of each model being valid. In particular, the posterior probability that M_i is the correct model will be

$$P(M_i|\mathbf{Y}) \propto \pi^{f_i}(1-\pi)^{k-f_i}\gamma^{-t_i}|\Gamma_i + \mathbf{X}_i'\mathbf{X}_i|^{-1/2} S_i^{-(n-1)/2} \tag{A.1}$$

where

$$\Gamma_i = \frac{1}{\gamma^2}\begin{pmatrix} 0 & \mathbf{0} \\ \mathbf{0} & \mathbf{I}_{t_i} \end{pmatrix} \tag{A.2}$$

$$\hat{\beta}_i = (\Gamma_i + \mathbf{X}_i'\mathbf{X}_i)^{-1}\mathbf{X}_i'\mathbf{Y} \tag{A.3}$$

and

$$S_i = (\mathbf{Y} - \mathbf{X}_i\hat{\beta}_i)'(\mathbf{Y} - \mathbf{X}_i\hat{\beta}_i) + \hat{\beta}_i'\Gamma_i\hat{\beta}_i \tag{A.4}$$

The constant of proportionality in (A.1) is that value that forces the probabilities to sum to unity. The probabilities $P(M_i|\mathbf{Y})$ can be summed over all models that include factor j, say, to compute the posterior probability P_j that factor j is active,

$$P_j = \sum_{M_i:\,\text{factor}\,j\,\text{active}} P(M_i|\mathbf{Y}) \tag{A.5}$$

The probabilities $\{P_j\}$ provide a convenient summary of the importance, or activity, of each of the experimental factors.

Often available data do not unambiguously identify a particular model. Building on the Bayes approach outlined above Meyer, Steinberg, and Box (1996) developed a procedure for identifying a set of one or more follow-up runs best calculated to clarify factor activity.

Choosing Follow up Runs

The MD design criterion proposed by Box and Hill (1967) has the following form. Let $\mathbf{Y}^*$ denote the data vector that will be obtained from the additional runs and let $p(\mathbf{Y}^*|M_i, \mathbf{Y})$ denote the predictive density of $\mathbf{Y}^*$ given the initial data $\mathbf{Y}$ and the model M_i. Then

$$\text{MD} = \sum_{0 \le i \ne j \le m} P(M_i|\mathbf{Y})P(M_j|\mathbf{Y}) \int_{-\infty}^{\infty} p(\mathbf{Y}^*|M_i, \mathbf{Y}) \ln\left(\frac{p(\mathbf{Y}^*|M_i, \mathbf{Y})}{p(\mathbf{Y}^*|M_j, \mathbf{Y})}\right) d\mathbf{Y}^* \tag{A.6}$$

(Box and Hill used the letter D for the criterion. We use the notation MD for model discrimination.)

The MD criterion has a natural interpretation in terms of Kullback–Leibler information. Let p_i and p_j denote the predictive densities of $\mathbf{Y}^*$ under models M_i and M_j, respectively. The Kullback–Leibler information is $I(p_i, p_j) = \int p_i \ln(p_i/p_j)$ and measures the mean information for discriminating in favor of M_i against M_j when M_i is true. Then

$$\text{MD} = \sum_{0 \le i \ne j \le m} P(M_i|\mathbf{Y})P(M_j|\mathbf{Y})I(p_i, p_j) \tag{A.7}$$

Thus MD is the Kullback–Leibler information averaged over pairs of candidate models with respect to the posterior probabilities of the models.

We now simplify MD, proceeding conditionally on σ^2 and then integrating out σ^2 at the last step. We first derive the predictive distribution of $\mathbf{Y}^*$ for each model. Let $\mathbf{X}_i$ and $\mathbf{X}_i^*$ denote the regression matrices for the initial runs and the additional runs, respectively, when M_i is the model. Then

$$\mathbf{Y}^*|M_i, Y, \sigma^2 \sim N(\hat{\mathbf{Y}}_i^*, \sigma^2\mathbf{V}_i^*)$$

where $\hat{\mathbf{Y}}_i^* = \mathbf{X}_i^*\hat{\beta}_i$, $\hat{\beta}_i$ was defined by (A.3), and

$$\mathbf{V}_i^* = \mathbf{I} + \mathbf{X}_i^*(\Gamma_i + \mathbf{X}_i'\mathbf{X}_i)^{-1}\mathbf{X}_i^{*'} \tag{A.8}$$

The log ratio of the predictive densities for two models, M_i and M_j, is then

$$\ln\left(\frac{p(\mathbf{Y}^*|M_i, \mathbf{Y}, \sigma^2)}{p(\mathbf{Y}^*|M_j, \mathbf{Y}, \sigma^2)}\right) = \frac{1}{2}\ln\left(\frac{|\mathbf{V}_j^*|}{|\mathbf{V}_i^*|}\right) - \frac{1}{2\sigma^2}((\mathbf{Y}^* - \hat{\mathbf{Y}}_j^*)'\mathbf{V}_j^{*^{-1}}(\mathbf{Y}^* - \hat{\mathbf{Y}}_j^*)$$
$$- (\mathbf{Y}^* - \hat{\mathbf{Y}}_i^*)'\mathbf{V}_i^{*^{-1}}(\mathbf{Y}^* - \hat{\mathbf{Y}}_i^*))$$

Integrating with respect to $p(\mathbf{Y}^*|M_i, \mathbf{Y}, \sigma^2)$ yields

$$\frac{1}{2}\ln\left(\frac{|\mathbf{V}_j^*|}{|\mathbf{V}_i^*|}\right) - \frac{1}{2\sigma^2}(n^*\sigma^2 - \sigma^2\mathrm{tr}\{\mathbf{V}_j^{*^{-1}}\mathbf{V}_i^*\} - (\hat{\mathbf{Y}}_i^* - \hat{\mathbf{Y}}_j^*)'\mathbf{V}_j^{*^{-1}}(\hat{\mathbf{Y}}_i^* - \hat{\mathbf{Y}}_j^*))$$

Finally, integrating with respect to $p(\sigma^2|\mathbf{Y}, M_i)$, we have the MD criterion defined by

$$\mathrm{MD} = \frac{1}{2}\sum_{0\le i\ne j\le m} P(M_i|\mathbf{Y})P(M_j|\mathbf{Y})(-n^* + \mathrm{tr}\{\mathbf{V}_j^{*^{-1}}\mathbf{V}_i^*\} + (n-1)$$
$$\times ((\hat{\mathbf{Y}}_i^* - \hat{\mathbf{Y}}_j^*)'\hat{\mathbf{V}}_j^{*^{-1}}(\hat{\mathbf{Y}}_i^* - \hat{\mathbf{Y}}_j^*))/S_i) \qquad \text{(A.9)}$$

where S_i was given by (A.4). (The term involving determinants sums to 0 across $0 \le i \ne j \le m$ and so was removed.)

Note that, for the models and prior distributions we use, the quadratic form in (A.9) is 0 for starting designs. Starting designs are thus evaluated by the trace term in (A.9). For mathematical convenience, we place a very small value (0.00001) in the (1, 1) position of the matrix Γ_i. This is equivalent to approximating the locally uniform prior for the intercept β_0 with a normal distribution, with mean 0 and large variance.

To compute MD-optimal augmentation designs, when the required number of follow up runs is not small we use an exchange algorithm like those proposed by Mitchell and Miller (1970) and Wynn (1972). First, all candidate design points must be listed. Typically this is the full factorial list of all factor-level combinations. An initial augmentation design is chosen at random, with replacement, from among the candidate points. The initial design is then improved by adding that point, from among all the candidates, which most improves the MD criterion, followed by removing that point from all those currently in the design, which results in the smallest reduction in the criterion. This add/remove procedure is continued until it converges, with the same point being added and then removed. The algorithm can get trapped in local optima, so we typically make from 20 to 50 or more random starts.

APPENDIX 7B. AN APPROXIMATE PARTIAL ANALYSIS FOR PB DESIGNS

As always, the Bayes approach focuses attention on essentials. You will find (see Appendix 7A) that the posterior probability for any given model is a product of two distinct expressions. The first is a penalty function that allows comparison of models containing different numbers of factors (subspaces of different dimensions). The second is a function mainly of the sums of squares associated with the models under consideration. For comparison of different models containing the same number of factors (e.g., all three-factor models), Kulahci and Box (2003) consider, in order of magnitude, the ratios

$$\varphi = \frac{\sum_{i=1}^{n} (\hat{y}_i - \bar{y})^2 / k}{\sum_{i=1}^{n} (y_i - \hat{y}_i)^2 / (n - k - 1)}$$

where the notation φ is used for this ratio rather than F to emphasize that, because of the selection process, the standard F significance levels do not apply. The partial analysis is then equivalent to a regression analysis except in the nature of the models considered.

For illustration consider again the data set out in Table 7.5 for a PB_{12} design abstracted from the 2^5 design of Table 6.15. The values of φ are arranged in order of magnitude in Table 7B.1. For the two-factor models φ *is* much larger for the model space [**BD**]. Also notice that the next most extreme two-factor model is for [**BE**]. It is not surprising, therefore, that for the three-factor models the factor space [**BDE**] involving all main effects and interactions between three factors is strongly favored, agreeing with the fuller Bayes analysis.

Table 7B.2 shows the same analysis for the Hunter, Hodi, and Eager (1982) fatigue data on repaired castings. Of the two-factor models that of the subspace [**FG**] clearly stands out from the others. Also, for the three-factor models the first five contain the space **FG** with each one of the other five factors. Again the analysis points to the earlier conclusion that **FG** is the active subspace.

Table 7B.1. Alternative Models for Reactor Example

Two-Factor Models			Three-Factor Models			
Model		φ	Model			φ
B	*D*	16.8	*B*	*D*	*E*	51.5
B	*E*	4.9	*A*	*B*	*D*	8.2
A	*B*	3.3	*B*	*C*	*D*	7.3
B	*C*	2.8	*A*	*C*	*E*	5.0
D	*E*	2.2	*A*	*D*	*E*	2.1
A	*C*	1.3	*A*	*B*	*E*	1.9
A	*E*	0.9	*B*	*C*	*E*	1.8
A	*D*	0.6	*A*	*B*	*C*	1.7
C	E	0.4	*C*	*D*	*E*	0.7
C	*D*	0.2	*A*	*C*	*D*	0.7

Table 7B.2. Alternative Models for Fatigue Data

Two-Factor Models			Three-Factor Models			
Model		φ	Model			φ
F	*G*	27.1	*D*	*F*	*G*	26
D	*F*	3.9	*E*	*F*	*G*	15.8
A	*E*	2.7	*C*	*F*	*G*	6.5
A	*F*	2.7	*B*	*F*	*G*	6.5
B	*F*	2.6	*A*	*F*	*G*	6.0
C	*F*	2.5	*A*	*C*	*E*	4.0
E	*F*	2.4	*C*	*D*	*F*	3.5
C	*D*	1.2	*A*	*E*	*G*	3.0
D	*G*	1.1	*A*	*D*	*D*	2.6
A	*B*	1.0	*B*	*D*	*D*	2.6

APPENDIX 7C. HALL'S ORTHOGONAL DESIGNS

Hall (1961) produced a number of two-level orthogonal arrays yielding saturated designs for $k = 15$ and $k = 19$ factors in 16 and 20 runs that were different from fractional factorials and the designs of Plackett and Burman. The five 16×16 arrays are of special interest and will be called $\mathbf{H}_1$, $\mathbf{H}_2$, $\mathbf{H}_3$, $\mathbf{H}_4$, and $\mathbf{H}_5$*, where $\mathbf{H}_1$ is the standard saturated 16-run 2_{III}^{15-11} fractional factorial discussed earlier (see Table 6.15) and $\mathbf{H}_2$, $\mathbf{H}_3$, $\mathbf{H}_4$, and $\mathbf{H}_5$ are distinct new 16-run designs. These designs, after convenient rearrangements, Box and Tyssedal (1996) are shown in Tables 7C.1 to 7C.4.

Look at Table 7C.1. For this $\mathbf{H}_2$ design you will see that the first eight columns are those of the familiar 2_{IV}^{8-4} fractional factorial which is of projectivity 3 and a [16,8,3] screen. If the remaining seven columns $J_2, K_2, \ldots, R_2$ are added, you obtain the $\mathbf{H}_2$ saturated orthogonal design that is fundamentally different† from the saturated 2_{III}^{15-11} fractional factorial design considered earlier. By adding only the four columns J, K, L, and M, you obtain a remarkable 12-factor design of projectivity 3 and thus a [16,12,3] screen—it can screen with projectivity 3 four more factors than the 2_{IV}^{8-4} design. The same is true of $\mathbf{H}_3$ and $\mathbf{H}_4$ in Tables 7C.2, and 7C.3, which however are distinct designs. Design $\mathbf{H}_5$ in Table 7C.4 can be obtained by using only the first six columns of the 2_{IV}^{8-4} and adding the further nine columns shown. This design is also remarkable because adding only the eight columns $G_5, H_5, J_5, \ldots, Q_5$ provides a [16,14,3] screen. It can screen 14 factors at projectivity 3. A fuller discussion will be found in Box and Tyssedal (2001). See also Sloane (1999) for an extensive list of Hadamard matrices.

* This notation differs somewhat from that used by Hall.
† Different in the sense that it cannot be obtained by sign switching and renaming rows and columns.

Table 7C.1. Design H_2

	a	b	c	d	abc	abd	acd	bcd							
Run Number	A	B	C	D	E	F	G	H	J_2	K_2	L_2	M_2	P_2	Q_2	R_2
1	−	−	−	−	−	−	−	−	+	−	−	−	+	+	+
2	+	−	−	−	+	+	+	−	−	+	−	−	−	−	+
3	−	+	−	−	+	+	−	+	−	−	+	−	−	+	−
4	+	+	−	−	−	−	+	+	−	−	−	+	+	−	−
5	−	−	+	−	+	−	+	+	+	+	+	−	+	−	−
6	+	−	+	−	−	+	−	+	+	+	−	+	−	+	−
7	−	+	+	−	−	+	+	−	+	−	+	+	−	−	+
8	+	+	+	−	+	−	−	−	−	+	+	+	+	+	+
9	−	−	−	+	−	+	+	+	−	+	+	+	+	+	+
10	+	−	−	+	+	−	−	+	+	−	+	+	−	−	+
11	−	+	−	+	+	−	+	−	+	+	−	+	−	+	−
12	+	+	−	+	−	+	−	−	+	+	+	−	+	−	−
13	−	−	+	+	+	+	−	−	−	−	−	+	+	−	−
14	+	−	+	+	−	−	+	−	−	−	+	−	−	+	−
15	−	+	+	+	−	−	−	+	−	+	−	−	−	−	+
16	+	+	+	+	+	+	+	+	+	−	−	−	+	+	+

Table 7C.2. Design H_3

	a	b	c	d	abc	abd	acd	bcd							
Run Number	A	B	C	D	E	F	G	H	J_3	K_3	L_3	M_3	P_3	Q_3	R_3
1	−	−	−	−	−	−	−	−	+	−	−	−	+	+	+
2	+	−	−	−	+	+	+	−	−	+	−	−	−	+	−
3	−	+	−	−	+	+	−	+	−	−	+	−	+	−	−
4	+	+	−	−	−	−	+	+	−	−	−	+	−	−	+
5	−	−	+	−	+	−	+	+	+	+	+	−	−	−	+
6	+	−	+	−	−	+	−	+	+	−	+	+	−	+	−
7	−	+	+	−	−	+	+	−	+	+	−	+	+	−	−
8	+	+	+	−	+	−	−	−	−	+	+	+	+	+	+
9	−	−	−	+	−	+	+	+	−	+	+	+	+	+	+
10	+	−	−	+	+	−	−	+	+	+	−	+	+	−	−
11	−	+	−	+	+	−	+	−	+	−	+	+	−	+	−
12	+	+	−	+	−	+	−	−	+	+	+	−	−	−	+
13	−	−	+	+	+	+	−	−	−	−	−	+	−	−	+
14	+	−	+	+	−	−	+	−	−	−	+	−	+	−	−
15	−	+	+	+	−	−	−	+	−	+	−	−	−	+	−
16	+	+	+	+	+	+	+	+	+	−	−	−	+	+	+

Table 7C.3. Design H_4

Run Number	a A	b B	c C	d D	abc E	abd F	acd G	bcd H	J_4	K_4	L_4	M_4	P_4	Q_4	R_4
1	−	−	−	−	−	−	−	−	+	+	+	+	+	+	+
2	+	−	−	−	+	+	+	−	+	−	−	+	−	−	+
3	−	+	−	−	+	+	−	+	−	−	+	+	+	−	−
4	+	+	−	−	−	−	+	+	+	−	+	−	−	+	−
5	−	−	+	−	+	−	+	+	−	−	−	−	+	+	+
6	+	−	+	−	−	+	−	+	−	+	−	+	−	+	−
7	−	+	+	−	−	+	+	−	−	+	+	−	−	−	+
8	+	+	+	−	+	−	−	−	+	+	−	−	+	−	−
9	−	−	−	+	−	+	+	+	+	+	−	−	+	−	−
10	+	−	−	+	+	−	−	+	−	+	+	−	−	−	+
11	−	+	−	+	+	−	+	−	−	+	−	+	−	+	−
12	+	+	−	+	−	+	−	−	−	−	−	−	+	+	+
13	−	−	+	+	+	+	−	−	+	−	+	−	−	+	−
14	+	−	+	+	−	−	+	−	−	−	+	+	+	−	−
15	−	+	+	+	−	−	−	+	+	−	−	+	−	−	+
16	+	+	+	+	+	+	+	+	+	+	+	+	+	+	+

Table 7C.4. Design H_5

Run Number	a A	b B	c C	d D	abc E	abd F	G_5	H_5	J_5	K_5	L_5	M_5	P_5	Q_5	R_5
1	−	−	−	−	−	−	+	+	+	+	+	+	−	−	+
2	+	−	−	−	+	+	+	−	+	−	+	−	+	−	−
3	−	+	−	−	+	+	−	+	+	−	−	+	−	+	−
4	+	+	−	−	−	−	−	−	−	−	+	+	+	+	+
5	−	−	+	−	+	−	+	+	−	−	−	−	+	+	+
6	+	−	+	−	−	+	−	+	−	+	−	+	+	−	−
7	−	+	+	−	−	+	+	−	−	+	+	−	−	+	−
8	+	+	+	−	+	−	−	−	+	+	−	−	−	−	+
9	−	−	−	+	−	+	−	−	+	+	−	−	+	+	+
10	+	−	−	+	+	−	+	−	−	+	−	+	−	+	−
11	−	+	−	+	+	−	−	+	−	+	+	−	+	−	−
12	+	+	−	+	−	+	+	+	−	−	−	−	−	−	+
13	−	−	+	+	+	+	−	−	−	−	+	+	−	−	+
14	+	−	+	+	−	−	−	+	+	−	+	−	−	+	−
15	−	+	+	+	−	−	+	−	+	−	−	+	+	−	−
16	+	+	+	+	+	+	+	+	+	+	+	+	+	+	+

REFERENCES AND FURTHER READING

Akaike, H. (1973) Information theory and an extension of the maximum likelihood principle, *Proc. 2nd Int. Symp., Inform. Theory*, 267–281.

Akaike, H. (1985) Prediction and entropy, in *A Celebration of Statistics* (Eds. A. C. Atkinson and S. E. Feinberg), Springer-Verlag, New York.

Barrios, E (2004a) Bayesian Screening and Model Selection. BsMD: R-package URL: http://cran.r-project.org/src/contrib/Descriptions/BsMD.html

Barrios, E (2004b) Topics in Engineering Statistics Unpublished Ph.D thesis. Department of Statistics, University of Wisconsin-Madison.

Bose, R. C., and Bush K. A. (1952) Orthogonal arrays of strength two and three, *Ann. Math. Statist.*, **4**, 508–524.

Box, G. E. P. (1952) Multi-factor designs of first order, *Biometrika*, **39**, 49–57.

Box, G. E. P. and Bisgaard, S. (1993) What can you find out from 12 Experimental runs, *Quality Engineering* **5**, 663–668.

Box, G. E. P., and Hill, W. J. (1967) Discrimination among mechanistic models, *Technometrics*, **9**, 57–71.

Box, G. E. P., and Meyer, R. D. (1986) An analysis for unreplicated factorial designs, *Technometrics*, **28**, 11–18.

Box, G. E. P., and Meyer, R. D. (1993) Finding the active factors in fractionated screening experiments, *J. Quality Technol.*, **25**, 94–105.

Box, G. E. P., and Tiao, G. C. (1992) *Bayesian Inference in Statistical Analysis*, Wiley Classics, New York.

Box, G. E. P., and Tyssedal, J. (1996) Projective properties of certain orthogonal arrays, *Biometrika*, **83**(4), 950–955.

Box, G. E. P., and Tyssedal, J. (2001) Sixteen run designs of high projectivity for factor screening, *Communi. Statist.: Simulation and Computation*, **30**, 217–228.

Chipman, H. (1998) Fast model search for designed experiments with complex aliasing, in *Quality Improvement through Statistical Methods* (Eds. B. Abraham and N. U. Nair), Birkhauser, Boston, MA, pp. 207–220.

Cornell, J. A. (2002) *Experiments with Mixtures: Designs, Models and the Analysis of Mixture Data*, 3rd. ed., Wiley, New York.

Fisher, R. A. (1945) A system of confounding for factors with more than two alternatives, giving completely orthogonal cubes and higher powers, *Ann. Eugen.*, **12**, 283–290.

Hahn, G. J. (1984) Experimental design in a complex world, *Technometrics*, **26**, 19–31.

Hall, M. J. (1961) Hadamard matrices of order 16, *Jet Propulsion Lab. Summary*, **1**, 21–26.

Hamada, M., and Wu, C. F. J. (1992) Analysis of designed experiments with complex aliasing, *J. Quality Technol.*, **24**, 130–137.

Hill, W. J., and Wiles, R. A. (1975) Plant experimentation, *J. Quality Technol.*, **7**, 115.

Hunter, G. B., Hodi, F. S., and Eager, T. W. (1982) High-cycle fatigue of weld repair cast Ti-6Al-4V, *Metall. Trans.*, **13A**, 1589–1594.

Hunter, J. S., and Naylor, T. H. (1971) Experimental designs for computer simulations, *Manag. Sci.*, **16**, 422.

John, P. W. M. (1996) Nesting Plackett-Burman designs, *Statist. Prob. Lett.*, **27**, 221–223.

Kulahci, M., and Box, G. E. P (2003) Catalysis of discovery and development in engineering and industry. *Quality Eng.*, **15**(3), 509–513.

Kullback, S., and Leibler, R. A. (1951) On information and sufficiency, *Ann. Math. Statist.*, **22**, 79–86.

Lin, D. K. J., and Draper, N. R. (1992) Projection properties of Plackett and Burman Designs, *Technometrics*, **34**, 423–428.

Lindley, D. V. (1956) On a measure of the information provided by an experiment, *Ann. Math. Statist.*, **7**, 986–1005.

McCullagh, P., and Nelder, J. A. (1989) *Generalized Linear Models*, 2nd ed., Chapman & Hall, London.

Meyer, R. D. and Box, G. E. P. (1992) *Technical Report No. 80*, Center for Quality and Productivity, Univ. of Wisconsin, Madison.

Meyer, R. D. (1996) mdopt: FORTRAN programs to generate MD-optimal screening and follow-up designs, and analysis of data. *Statlib.* URL: http://lib.stat.cmu.edu

Meyer, R. D., Steinberg, D. M., and Box, G. E. P. (1996) Follow-up designs to resolve confounding in multi-factor experiments, *Technometrics*, **38**, 303–313.

Mitchell, T. J., and Miller, F. L., Jr. (1970) *Use of Design Repair to Construct Designs for Special Linear Models*, Mathematical Division Annual Progress Report (ORNL-4661), Oak Ridge National Laboratory, Oak Ridge, TN.

Plackett, R. L., and Burman, J. P. (1946) The design of optimal multi-factor experiments, *Biometrika*, **34**, 255–272.

Rao, C. R. (1947) Factorial experiments derivable from combinatorial arrangements of arrays, *J. Roy. Statist. Soc.*, **9**, 128–139.

Rao, C. R. (1973) Some combinatorial problems of arrays and applications to design of experiments, in *A Survey of Combinatorial Theory*, (Ed. J. N. Srivastava *et al.*), North Holland, Amsterdam, pp. 349–359.

Seiden, E. (1954) On the problem of construction of orthogonal arrays, *Ann. Math. Statist.*, **25**, 151–156.

Shannon, C. E. (1948) A mathematical theory of communication *Bell Syst. Tech. J.*, **27** 379–423, 623–656.

Sloane, N. J. A. (1999) A library of Hadamard Matrices URL: http://www.research.att.com/~njas/hadamard

Snee, R. D. (1985) Computer aided design of experiments—some practical experiences, *J. Quality Technol.*, **17**, 222–236.

Stewart, W. E., Henson, T. L., and Box, G. E. P. (1996) Model discrimination and criticism with single response data, *AICHE J.*, **42**, (11), 3055.

Taguchi, G. (1987) *System of Experimental Design*, Unipub/Kraus International Publications, White Plains, NY.

Wang, J. C. and Wu, C. F. J. (1995) A hidden projection quality of Plackett and Burman and related designs, *Statistica Sinica*, **5**, 235–250.

Wu, C. F. J., and Hamada, M. (2000) *Experiments, Planning, Analysis and Parameter Design Optimization*. Wiley, New York.

Wynn, H. P. (1972), "Results in the Theory and Construction of *D*-Optimum Experimental Designs," *Journal of the Royal Statistical Society*, Ser. B, **34**, 133–147 (with discussion 170–185).

QUESTIONS FOR CHAPTER 7

1. Here is the first row of a PB design: − + − − − + + + − + +. Write down the 12 run design?
2. What is meant by "projectivity," "factor sparsity," and ""'parsimony"? What are the characteristics of a design of projectivity 3 design? Projectivity 4?
3. In terms of n, the number of runs, what distinguishes the PB designs in the family of two-level designs? Which of the designs with $n = 8$, 12, 16, 20 have projectivity 3?
4. When might normal plotting be appropriate in the analysis of PB designs?
5. Why might blocking be important in the sequential assembly of designs?
6. State the essential *ideas* of the Bayesian analysis of the PB designs.
7. What is entropy? How is it defined? How is it used in the selection of follow-up runs?
8. Suppose that after running a design it was found that more than one model could explain the data. How can the principle of entropy be used to add additional runs to resolve the ambiguity?
9. How many distinct two-level 16-run statistical orthogonal designs are available? What is meant by "distinct"?
10. When are Hall's designs useful?

PROBLEMS FOR CHAPTER 7

1. Consider the following two-level experimental design for seven factors in 12 runs:

P	Q	R	S	T	U	V	Observed
+	−	−	−	+	+	+	37
+	+	−	+	+	−	+	23
+	−	+	+	−	+	−	40
−	+	−	−	−	+	+	37
−	+	+	+	−	+	+	5
−	−	−	−	−	−	−	39
+	+	−	+	−	−	−	7
+	−	+	−	−	−	+	44
−	−	+	+	+	−	+	51
+	+	+	−	+	+	−	35
−	−	−	+	+	+	−	56
−	+	+	−	+	−	−	33

Illustrate that (a) the design is orthogonal, (b) the design is of projectivity 3, and (c) the seven columns are those of a PB design.

2. Calculate the main effects for the factors **P, Q, R, S, T, U,** and **V**. Make a normal plot and comment.

3. Make a more sophisticated analysis of the data in problem 1 using an available computer program.

4. Suppose on using the design in Problem 1 you find suspect factors **P, Q, R,** and **U** are active. How could you augment the design to make it a 2^4 factorial?

5. What is meant by cyclic substitution? Apply cyclic substitution to the following signs 11 times:

$$- + + + - + + - + - -$$

If a row of minus signs is added, is this a PB design? How can you tell? Can all PB designs be obtained by cyclic substitution?

6. When is a PB design the same as a 2^{k-p} fractional factorial? What is meant by "the same as" in the preceding statement? What is different about the PB design and a 2^{k-p} fractional factorial in (a) their alias structure and (b) their analysis?

7. Randomize the design in problem 1.

8. Explain the meaning of "factor sparsity" and "design projectivity." How can these properties be combined to form a factor screen? What is meant by an [8,3,16] and a [11, 3, 12] factor screen? Give examples of each.

9. An experimenter is concerned with the possible effects of 11 trace elements on a particular system. He thinks it likely that only a small number, say none, one, two, or three elements, might be found active but he does not know which ones. Explain in detail how he might use a design that provided an appropriate factor screen.

10. Write down a 12-run PB design for three factors **A, B, C**. Obtain the alias structure of the design assuming the three-factor interaction is negligible.

11. Suppose you run a fractional two-level design in six factors **A, B, C, D, E, F** and find five alternative models M_1, M_2, M_3, M_4, M_5 each of which plausibly explains the data. Outline a way to choose one extra run drawn from the 2^6 possibilities ($\pm1, \pm1, \pm1, \pm1, \pm1, \pm1$) which best discriminates between the models. Without going into mathematical detail, explain how it works.

CHAPTER 8

Factorial Designs and Data Transformation

In Chapters 5 and 6 factorial designs have been considered only in the special but important case where each factor is tested at two levels. In this chapter factorial designs are considered more generally and methods of data transformation are discussed which in certain cases can greatly increase their efficiency.

In general, to perform a factorial design, the investigator selects a fixed number of "levels" (or "versions") of each of a number of factors (variables) and then runs experiments with all possible combinations. If, for example, there were three levels for the first factor, four for the second, and two for the third, the complete arrangement would be called a $3 \times 4 \times 2$ factorial design and would require 24 runs, a design involving two factors each at three levels would be a 3×3, or 3^2, factorial design and would require 9 runs, and so on. An experiment in which a 3×4 factorial design was employed to test survival times will be used as an illustration.

Exercise 8.1. How many factors and how many runs are there in a $2 \times 4 \times 3 \times 2$ factorial design? *Answer*: four factors, 48 runs.

Exercise 8.2. If four factors are to be studied, each of which is to be tested at three levels in a factorial design, how would you designate the design and how many runs would it require? *Answer*: 3^4 design, 81 runs.

8.1. A TWO-WAY (FACTORIAL) DESIGN

Look at the data of Table 8.1. The measured responses are the survival times of groups of four animals randomly allocated to each of the 12 combinations

Statistics for Experimenters, Second Edition. By G. E. P. Box, J. S. Hunter, and W. G. Hunter

Table 8.1. Survival Times (unit, 10 hours) of Animals in a 3 x 4 Factorial Experiment: Toxic Agents Example

	Treatment			
Poison	*A*	*B*	*C*	*D*
I	0.31	0.82	0.43	0.45
	0.45	1.10	0.45	0.71
	0.46	0.88	0.63	0.66
	0.43	0.72	0.76	0.62
II	0.36	0.92	0.44	0.56
	0.29	0.61	0.35	1.02
	0.40	0.49	0.31	0.71
	0.23	1.24	0.40	0.38
III	0.22	0.30	0.23	0.30
	0.21	0.37	0.25	0.36
	0.18	0.38	0.24	0.31
	0.23	0.29	0.22	0.33

of three poisons and four treatments. The experiment was part of an investigation to combat the effects of certain toxic agents. However, the same approach could be appropriate for other life-testing studies as, for example, in the study of the useful life of machine tools under various manufacturing conditions. In this particular example you need to consider the effects of both poisons and treatments, remembering that their influences may not be additive; that is, the difference in survival times between specific treatments may be different for different poisons—poisons and treatments might interact.

The analysis of the data is best considered in two stages. If poison and treatment labels are temporarily ignored, the design becomes a one-way arrangement, such as the one in Chapter 4 with 12 groups of four animals. The total sum of squares of deviations from the grand average (i.e., "corrected" for the mean) may then be analyzed as follows:

Source of Variation	Sum of Squares × 1000	Degrees of Freedom	Mean Square × 1000
Between groups	2205.5	11	200.5
Within groups (error)	800.7	36	22.2
Total (corrected)	3006.2	47	

Alternately, the error mean square $s^2 = 22.2$ can be obtained by estimating the variance in each of the 12 cells separately and then pooling these estimates.

Now consider a model for the data taking account of the two factors. Let η_{ti} be the mean survival time in the cell which receives the tth treatment and the ith poison. Also let y_{tij} be the jth observation in that cell. There are four treatments (columns) and three poisons (rows) and 4 observations per cell for a total of 48 observations. If poisons and treatments act additively, then $\eta_{ti} = \eta + \tau_t + \pi_i$, where τ_t is the mean increment in survival time associated with the tth treatment and π_i is the corresponding increment associated with the ith poison. If, however, there is *interaction*, an additional increment $\omega_{ti} = \eta_{ti} - \eta - \tau_t - \pi_i$ is needed to make the equation balance. The model and the related data breakdown are therefore as follows:

$$\eta_{ti} = \eta + \tau_t + \pi_i + \omega_{ti}$$

$$\overline{y}_{ti} = \overline{y} + (\overline{y}_t - \overline{y}) + (\overline{y}_i - \overline{y}) + (\overline{y}_{ti} - \overline{y}_t - \overline{y}_i + \overline{y})$$

In this decomposition τ_t and π_i are called the *main effects* of treatments and poisons and the ω_{ti} the *interaction effects*.

The between-group sum of squares with 11 degrees of freedom can now be analyzed as follows:

Source of Variation	Sum of Squares ×1000	Degrees of Freedom	Mean Square ×1000
Poisons	1033.0	2	516.5
Treatments	922.4	3	307.5
Interaction	250.1	6	41.7
Between groups	2205.5	11	

In this table the interaction sum of squares is obtained by subtracting the poisons and treatment sums of squares from the between-groups sum of squares. Combining the two analyses the complete ANOVA table shown in Table 8.2 is obtained. The "total corrected" sum of squares — the sum of squared deviations from the grand average — has $48 - 1 = 47$ degrees of freedom.

Table 8.2. Analysis of Variance: Toxic Agents Example

Source of Variation	Sum of Squares × 1000	Degrees of Freedom	Mean Square × 1000	F
Poisons (P)	1033.0	2	516.5	23.3
Treatments (T)	922.4	3	307.5	13.9
$P \times T$ interaction	250.1	6	41.7	1.9
Within groups (error)	800.7	36	22.2	
Total (corrected)	3006.2	47		

Exercise 8.3. Make a decomposition of the observations for the poisons–treatments example. There should be six tables of elements associated with the original observations—the grand average, the poison main effects, the treatment main effects, the interaction, and the residuals. Confirm to your own satisfaction that (a) the various components are orthogonal, (b) the sums of squares of the individual elements in the tables correctly give the sums of squares of the ANOVA table, and (c) the allocated numbers of degrees of freedom are logical.

8.2. SIMPLIFICATION AND INCREASED SENSITIVITY FROM TRANSFORMATION

On the assumption that the model was adequate and in that the errors ε_{tij} were NIID with, in particular, a *constant* variance, then using the F table, the effects of both poisons and treatments in Table 8.2 would be judged highly significant. In addition the F value for the interaction would be suspiciously large. However, if you look at Table 8.3b, you will see that the assumption of the constancy of variance appears very dubious.

Now there are two kinds of variance inhomogeneity: (a) *inherent* inhomogeneity—an example would be the smaller variance in a manufacturing operation achieved by an experienced rather than by an inexperienced operator—and (b) *transformable* inhomogeneity, which comes about because the untransformed observations give rise to a needlessly complicated model which induces variance inhomogeneity and unnecessary interaction. A vital clue that transformation is needed to correct transformable inhomogeneity is when the standard deviation increases with the mean. In Figure 8.1 the plot of the residuals $y_{tij} - \bar{y}_{ti}$ versus the cell averages $\bar{y}_{ti}$ has a funnel shape, strongly suggesting that, contrary to the NIID assumption, the standard deviation increases as the mean is increased.

Variance Stabilizing Transformations

When the standard deviation σ_y is a function of η, you can frequently find a data transformation $Y = f(y)$ that has more constant variance. It turns out, for

Table 8.3. Cell Averages $\bar{y}$ and Variances s^2 for the 3 x 4 Factorial

(a) Cell Averages: $\bar{y} \times 100$

		Treatments			
		A	*B*	*C*	*D*
	I	41	88	57	61
Poisons	II	32	82	38	67
	III	21	34	24	33

(b) Cell Variances: $s^2 \times 10{,}000$

		Treatments			
		A	*B*	*C*	*D*
	I	48	259	246	127
Poisons	II	57	1131	32	734
	III	5	22	2	7

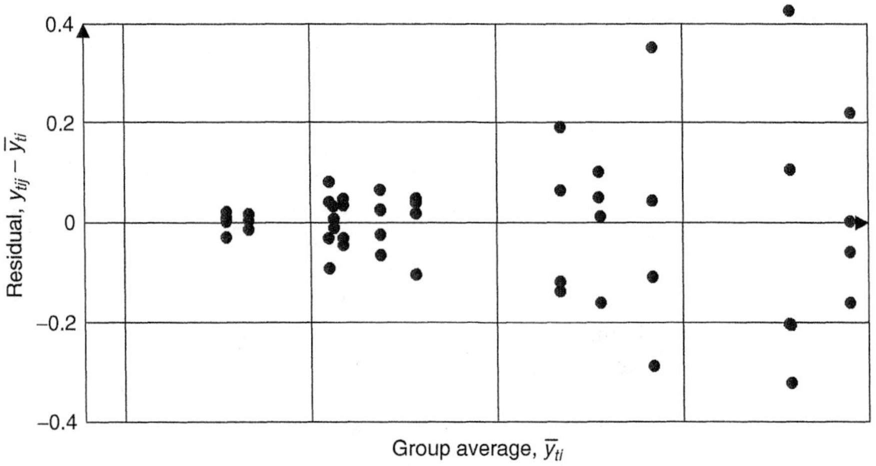

Figure 8.1. Survival data, plot of within-group residuals versus group averages.

example, that if the standard deviation was proportional to the mean, a data transformation $Y = \log y$ would stabilize the variance. In general (see Appendix 8A for details), if σ is proportional to some power of η, so that $\sigma \propto \eta^{\alpha}$, then the standard deviation of $Y = y^{\lambda}$ where $\lambda = 1 - \alpha$ will be made approximately constant and independent of the mean. When $\alpha = 1$, by taking the mathematical limit appropriately, the value $\lambda = 1 - \alpha = 0$ produces the log transformation. Equivalently, you may work with the form $Y = (y^{\lambda} - 1)/\lambda$; later you will see that this later form has advantages. Some values of α with corresponding transformations are summarized in Table 8.4.

For example, the square root transformation is appropriate for Poisson data because, for that distribution, the standard deviation is the square root of the mean. For an observed binomial proportion $p = y/n$ the mean of p is denoted by π; the standard deviation is $\sqrt{\pi(1-\pi)}$. Consequently, the standard deviation is related to the mean but not by a power law. It was shown by Fisher that the

Table 8.4. Variance Stabilizing Transformations When $\sigma_y \propto \eta^{\alpha}$

Dependence of σ_y on η	α	$\lambda = 1 - \alpha$	Variance Stabilizing Transformation	Example
$\sigma \propto \eta^2$	2	-1	Reciprocal	
$\sigma \propto \eta^{3/2}$	$1\frac{1}{2}$	$-\frac{1}{2}$	Reciprocal square root	
$\sigma \propto \eta$	1	0	Log	Sample variance
$\sigma \propto \eta^{1/2}$	$\frac{1}{2}$	$\frac{1}{2}$	Square root	Poisson frequency
$\sigma \propto$ const	0	1	No transformation	

appropriate variance stabilizing transformation for the binomial is $\sin Y = \sqrt{y/n}$. Thus $Y = \sin^{-1}\sqrt{y/n}$ has an approximately constant variance independent of π.

Empirical Determination of α

For replicated data such as appear in Table 8.1, an estimate of α can be found empirically. Suppose that for the jth set of experimental conditions $\sigma_j \propto \eta_j^\alpha$; then $\log \sigma_j = \text{const} + \alpha \log \eta_j$ and $\log \sigma_j$ would yield a straight line plot against $\log \eta_j$ with slope α. In practice, you do not know σ_j and η_j, but estimates s_j and $\bar{y}_j$ may be substituted. For the toxic agents data a plot of $\log s_j$ versus $\log \bar{y}_j$ for the 12 cells is shown in Figure 8.2*a*. The slope of the line drawn through the points is close to 2, indicating (see Table 8.4) the need for a reciprocal transformation $Y = 1/y$. Figure 8.2*b* shows the corresponding plot for $Y = 1/y$, demonstrating that in this "metric" the standard deviation is essentially independent of the mean. Since y is the time to failure, $Y = y^{-1}$ will be called the failure rate.

Advantages of the Transformation

The mean squares and degrees of freedom for the ANOVA for the untransformed and reciprocally transformed poisons data are shown in Table 8.5. That the data have been used to choose the transformation is approximately allowed for (see Box and Cox, 1964) by reducing the number of degrees of freedom in the within-group mean square by unity (from 36 to 35). Notice that transformed and untransformed data Y and y are in different units; consequently, to show the effect of transformation, it is the F ratios that must be compared.

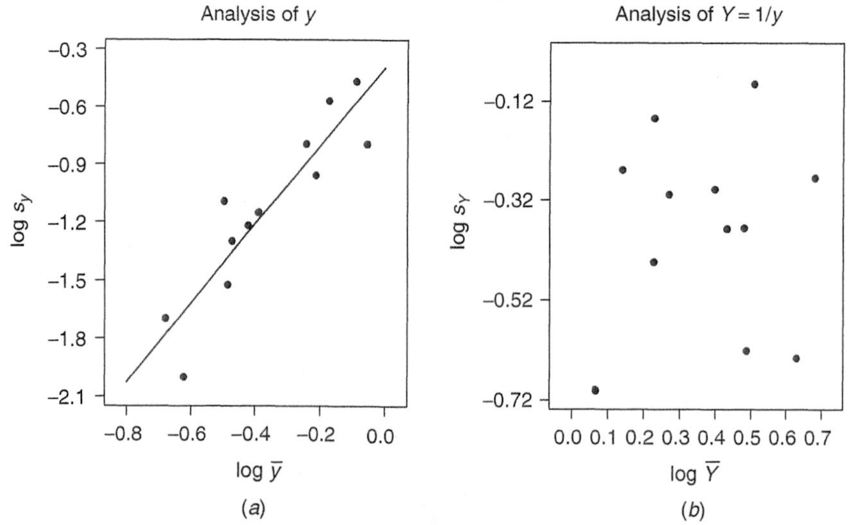

Figure 8.2. Survival data, plot of the log of the cell standard deviation versus the log of the cell averages: (*a*) before transformation; (*b*) after the transformation $Y = 1/y$.

Table 8.5. Analyses of Variance for Untransformed and Transformed Data: Toxic Agents Example

	(a) Untransformed			(b) Transformed, $Y = y^{-1}$		
Effects	Degrees of Freedom	Mean Square ×1000	F	Degrees of Freedom	Mean Square ×1000	F
Poisons (P)	2	516.5	23.3	2	1743.9	70.6
Treatments (T)	3	307.5	13.9	3	680.5	27.6
$P \times T$	6	41.7	1.9	6	26.2	1.1
Interaction within groups (error)	36	22.2		35	24.7	

The analysis on the reciprocal scale is much more sensitive and easier to interpret. Specifically:

1. You see from Table 8.5 that the F values for both poisons and treatments are greatly increased. In fact, the increases in efficiency achieved by making the transformation is equivalent to increasing the size of the experiment by a factor of almost 3.
2. The poison–treatment interaction mean square that previously gave some suggestion of significance is now very close to the error mean square. Thus, in the transformed scale there is no evidence of interaction and the effects of poison and treatments measured as *rates of failure* ($Y = 1/y$) are roughly additive. Thus you obtain a very simple interpretation of the data in which the effects of poisons and treatments can be entirely described in terms of their additive main effects.
3. For the untransformed data, Bartlett's test for variance homogeneity, as defined in Appendix 8B, gave a χ^2 value of 44.0 based on 11 degrees of freedom.* After transforming the data the value of χ^2 is 9.7 with a nominal significance probability of about 0.60 as compared with the value before transformation of less than 0.0001. Thus all the inhomogeneity of variances seems to have been removed by the transformation and there is now no evidence of residual *inherent* inhomogeneity.

The greater sensitivity of the analysis in the transformed metric is further illustrated in Figure 8.3, which shows reference t distributions for the comparisons of poisons I, II, and III before and after transformation. To facilitate comparison, the averages are drawn so as to cover the same ranges on both scales.

*It is well known that Bartlett's statistic is itself extremely sensitive to nonnormality. However, it is unlikely that such an extreme significance level could be explained in this way. As indicated in Appendix 8B, when the NIID assumptions are true, for k groups of observations the mean value of M is $k - 1$. In this example $k - 1 = 11$, which is close to the value 9.7.

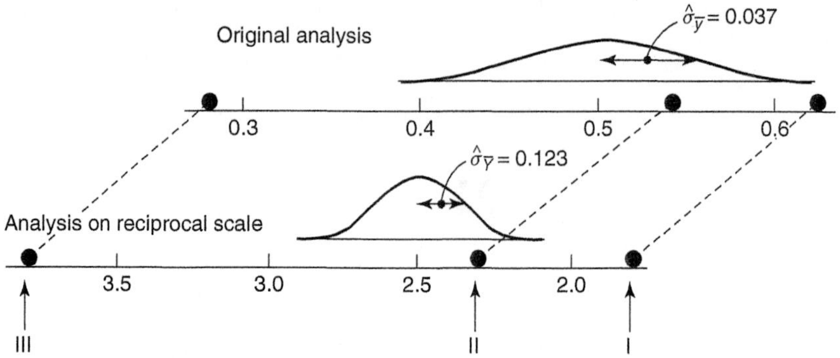

Figure 8.3. Poison averages with reference distributions for analyses using the original and reciprocal scales; toxic agents examples.

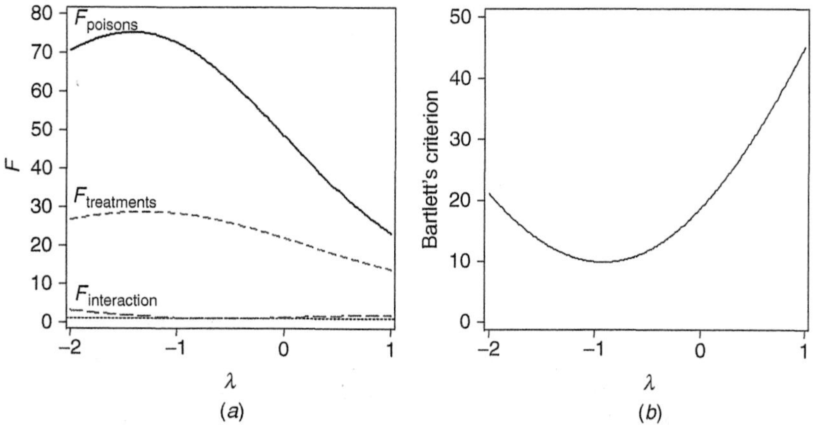

Figure 8.4. Lambda plots for (*a*) F values and (*b*) Bartlett's criterion: toxic agent example.

Lambda Plots

Another way of finding an appropriate transformation is to construct "lambda plots" like those in Figure 8.4. To do this, the computer is set the task of analyzing the data for a series of values of the transformation parameter λ and then plotting a relevant statistic, for example the F value, against λ. Figure 8.4*a* shows λ plots of F ratios for poisons, treatments, and their interactions and Figure 8.4*b* a corresponding plot for Bartlett's M' statistic. You will see in Figure 8.4*a* that for $\lambda = -1$ the F ratios for poisons and treatments are close to their maximum values while the F ratio for interaction becomes negligible. Also Bartlett's criterion is close to its minimum value.

Graphical ANOVA

The graphical ANOVA shown in Figures 8.5*a* and *b* display the data in both their original y and transformed $1/y$ forms. The comparison again illustrates

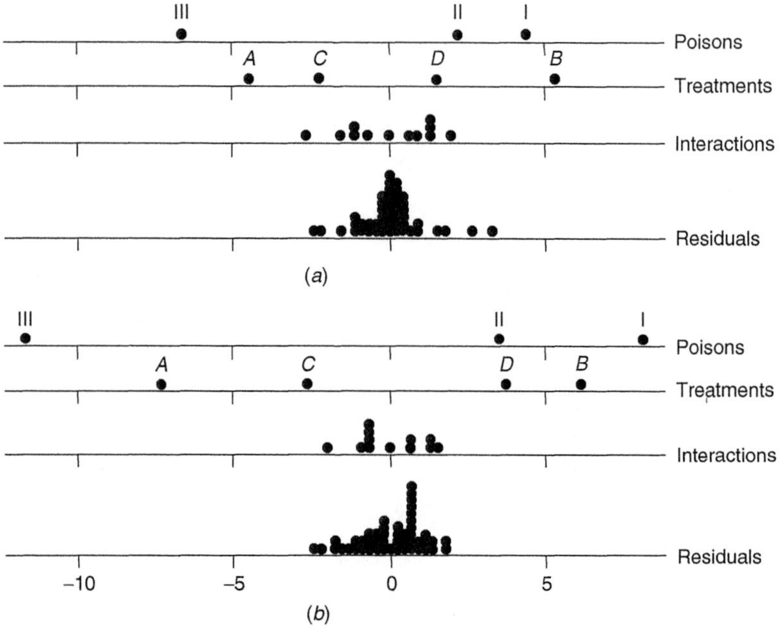

Figure 8.5. Survival data, graphical ANOVA for y and $1/y$: (a) untransformed; (b) transformed.

the greatly increased sensitivity of the analysis after transformation and the disappearance of any suggestion of interaction. To make comparisons easier, the standard deviations of the two dot diagrams of residuals have been scaled to be equal.

Rationale for Using a Reciprocal Transformation for Survival Data

Figure 8.6 shows why a reciprocal transformation can make sense for life-testing data (whether it is the "life" of an animal, a machine tool, or a computer chip). The diagram illustrates a hypothetical situation where the amount of wear w of a specimen (animal, chip, or tool) is a linear function of time t. It is imagined that the critical level of wear at which failure occurs has a value close to w_0. Now suppose that you have three specimens A, B, and C treated in different ways and that the amount of wear at any fixed time t has a roughly normal distribution with the same standard deviation. Although you do not know the actual rates of wear, if your hypothesis is roughly correct, the rate of wear r (say) will be proportional to the reciprocal of the time taken to reach the critical level w_0. So $r = w_0/t_0$ and, by taking the reciprocal of time to failure t_0, you are in effect analyzing the *rate* of wear. In this case the distributions for A', B', C' of times to failure will be skewed and the spreads of the distributions will increase as the time to failure increases, so that treatments yielding longer life will be associated with greater variability, as is the case for this example. In fact, the standard deviation of time to failure t will be proportional to t_0^2.*

*If log w had been a linear function of time with a constant standard deviation, then the appropriate transformation to produce a constant standard deviation in t_0 would be log t_0. In general, if w^α

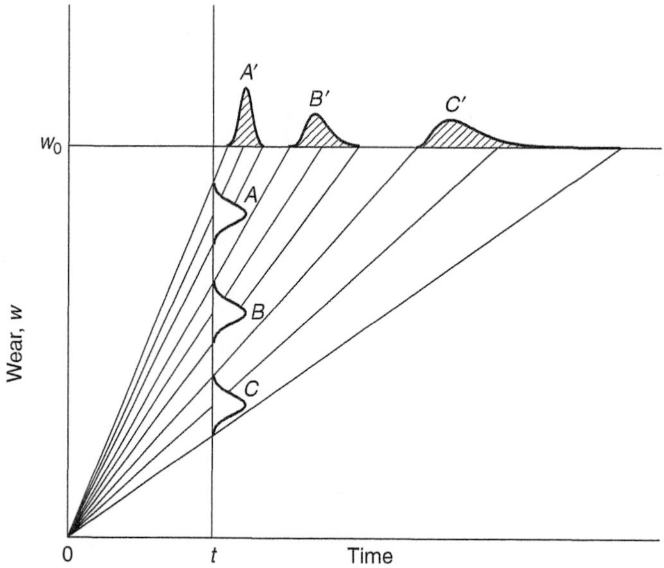

Figure 8.6. Hypothetical example of wear versus time relationship for three specimens A, B, and C with rates of wear w/t having equal standard deviations.

When Does Transformation Make a Difference?

Look again at the data in Table 8.1. Denote the smallest observation by y_S and the largest by y_L. For these data the ratio of the largest to the smallest observation $y_L/y_S = 1.24/0.18$ is about 7. The large value for this ratio is the reason why transformation has such a profound effect in this example. For smaller ratios, $y_L/y_S < 2$ say, transformation of the data is not likely to affect the conclusions very much. The reason is that for smaller values of this ratio the use of power transformations approximately corresponds to a linear recoding which makes little difference to F and t ratios or to the conclusions drawn. (For example, a graph of $Y = \log y$ against y over the range $y_L/y_S = 2$ is not all that different from a straight line.)

Transformation of Data from a 2^4 Factorial Design

The cube plots Figure 8.7a show data from a 2^4 factorial due to Daniel (1976, p. 72).

There were four factors, three quantitative. These were **A**, the load of a small stone drill; **B**, the flow rate through it; **C**, its rotational speed; and one qualitative **D**, the type of "mud" used in drilling. The response y was the advance of the drill. The plot shows that simple response comparisons made along the edges of the cube have a clear tendency to increase as their averages increase. As we have seen, such nonadditivity can often be removed by a power transformation

were a linear function of time with constant standard deviation, the stabilizing transformation would be $t^{-\alpha}$.

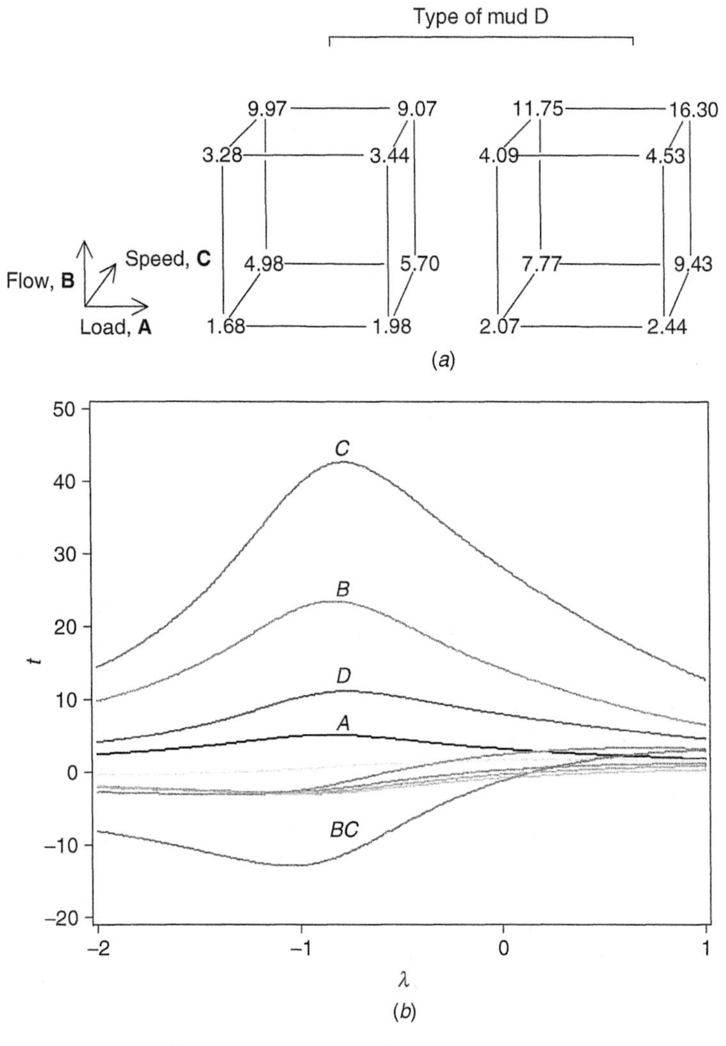

Figure 8.7. (*a*) Original 2^4 data. (*b*) The λ plots of t values for estimated effects with power transformations of the data.

$Y = y^\lambda$ of the data. Also, since the largest and smallest observations have a ratio $y_L/y_S = 16.30/1.68$ of almost 10, this is an example where transformation could have a profound effect on the analysis.

In a very interesting and thorough treatment of these data, Daniel shows how such a transformation may be chosen by using normal probability plotting techniques. We here consider the appropriate choice of a power transformation by plotting the t values against λ. To do this, we tentatively assume that the mean squares associated with the five higher order interactions *ABC, ABD, ACD, BCD,*

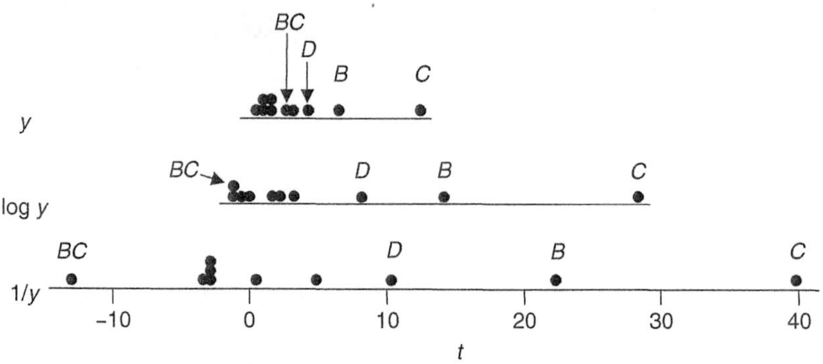

Figure 8.8. Stone drill data: dot plots for the t values of estimated effects using the original data y, $\log\ y$, and $1/y$.

and *ABCD* may be used as an estimate of the error variance having five degrees of freedom. In making a λ plot for t values it is best to work with the power transformation in the form $Y = (y^\lambda - 1)/\lambda$ because with this alternative form, although the same in absolute value, t does not change sign for negative values of λ.* Figure 8.7*b* displays such a lambda plot showing these t values for a series of values for λ for each of the main effects and two-factor interactions. It will be seen that a value close to $\lambda = -1$ (a reciprocal transformation) produces the largest values for t. Today's computers make these calculations easy.

Dot plots of the calculated t values for the effects using the data y, $\log\ y$, and $1/y$ are displayed in Figure 8.8. Using the reciprocal transformation, the estimated effects C, B, and D and the interaction *BC* are more clearly distinguishable from the noise. Finding the proper data transformation is much like viewing an object using a microscope. The lens must be set to get the clearest image. A rationale for using the reciprocal transformation in this example might be that the data y measured the *actual* advance of the drill but it was the *rate* of advance for which the data were approximately IID and the model was linear in the effects.

Beware Transforming Just to Simplify the Model

Transformation of data often leads to model simplification. However, transforming *just to simplify the model* can be dangerous. For example, suppose you *really had* IID observations $y = \eta + \varepsilon$ and a theoretical model like $\eta = 1/(\alpha + \beta x)$. You might be tempted to use a reciprocal transformation of the observations y because $y^{-1} = a + bx$ is an easy model to fit by ordinary least squares. (Least squares and regression are discussed in Chapter 9.) But if the original data $y = \eta + \varepsilon$ had constant variance σ_y^2 and if you made the transformation $Y = y^{-1}$, the variance of Y would be approximately proportional to σ_y^2/η^4. Thus

*Whatever form of the transformation is used, y^λ or $(y^\lambda - 1)/\lambda$, you will obtain the same values of F and of Bartlett's M' statistic. This is because these statistics depend upon the squares of the observations and thus do not change sign for negative values of lambda.

if your data y covered a range where the largest data value y_L was 10 times the smallest y_S, transformation would change the variance from one end of the scale of measurement to the other by a factor of $(y_L/y_S)^4 = 10^4$! By analyzing the reciprocal of the data by ordinary least squares, your observation would be discounted more and more as you moved across the scale—it would be as if you had 10,000 observations at one end and only one observation at the other. Clearly, simplification for its own sake can be a dangerous undertaking.

APPENDIX 8A. RATIONALE FOR DATA TRANSFORMATION

Suppose the standard deviation σ_y of y is proportional to some *power* of the mean η of y, that is, $\sigma_y \propto \eta^\alpha$, and that you make a power transformation of the data, $Y = y^\lambda$. Then to a first approximation

$$\sigma_Y = \theta\sigma_y \propto \theta\eta^\alpha$$

and θ, the gradient of the graph of Y plotted against y, depends on the mean η of y and so might more appropriately be denoted by θ_η.

Now, if $Y = y^\lambda$, the gradient θ_η is proportional to $\eta^{\lambda-1}$. Thus

$$\sigma_Y \propto \eta^{\lambda-1}\eta^\alpha = \eta^{\lambda+\alpha-1}$$

Thus σ_Y will be (approximately) constant if $\lambda + \alpha - 1$ is zero, that is if $\lambda = 1 - \alpha$.

APPENDIX 8B. BARTLETT'S χ^2_ν FOR TESTING INHOMOGENEITY OF VARIANCE

For k groups of n observations Bartlett's statistic may be written as

$$M' = \frac{k(n-1)\ln(\bar{s}^2/\dot{s}^2)}{1 + (k+1)/3k(n-1)}$$

Where $\bar{s}^2$ is the arithmetic mean of the estimated variances and $\dot{s}^2$ is their geometric mean so that $\ln \dot{s}^2 = \sum_{i=1}^{k} \ln s_i^2/k$. Given NIID$(0, \sigma^2)$ assumptions, M' is, to a close approximation, distributed as χ^2 with $\nu = k - 1$ degrees of freedom and the mean value of M' is thus $k - 1$. As mentioned earlier, the *distribution* of M' is greatly affected by nonnormality. However, independent of distribution considerations, M' is a convenient *comparative measure* of lack of equality of k variances.

REFERENCES AND FURTHER READING

Bartlett, M. S. (1937) Properties of sufficiency and statistical tests, *Proc. Roy. Statist. Soc.*, **A160**, 268–282.

Bartlett, M. S., and Kendall, D. G. (1946) The statistical analysis of variance-heterogeneity and the log arithmetic transformation, *J. Roy. Statist. Soc. Suppl*, **8**, 128.

Box, G. E. P., and Fung, C. (1995) The importance of data transformation in designed experiments for life testing, *Quality Eng.*, **7**, 625–638.

Box, G. E. P. (1988) Signal to noise ratios, performance criteria and transformations, *Technometrics*, **30**, 1–17.

Box, G. E. P., and Andersen, S. L. (1954) Permutation theory in the derivation of robust criteria and the study of departures from assumptions, *J. Roy. Statist. Soc. Ser. B*, **17**, 1–34.

Box, G. E. P., and Cox, D. R. (1964) An analysis of transformations, *J. Roy. Statist. Soc. Ser. B*, **26**, i211–252.

Daniel, C. (1976) *Applications of Statistics to Industrial Experimentation*, Wiley, New York.

Fung, C. (1986) Statistical Topics in Off-Line Quality Control, PhD thesis, University of Wisconsin, Madison.

Hartley, H. O. (1940) Testing the homogeneity of a set of variances, *Biometrika*, **31**, 249–255.

Sprott, D. A. (2000) *Statistical Inference in Science*, Springer-Verlag, New York.

QUESTIONS FOR CHAPTER 8

1. By looking at the magnitudes of the data values, when might you expect data transformation to be beneficial?
2. Why should a data transformation be considered when (a) the data cover a very wide range and (b) the standard deviation increases with the mean?
3. If you were able to obtain averages and standard deviations for data taken at different factor combinations in a factorial design, how might this help you determine whether and what transformation was needed?
4. What is a λ plot? How might it help you choose a data transformation?
5. What are the differences between inherent and transformable interaction?
6. How may the need for interaction terms and data transformation be linked?
7. In comparing the analyses of data from a factorial design before and after a data transformation, what do you look for?
8. If $y = 1/(a + bx_1 + cx_2)$, what might be the advantage or disadvantage of using a reciprocal transformation of the data?
9. When appropriate, why is a data transformation desirable?
10. What is Bartlett's test statistic? When is it helpful?
11. Suppose you have a reason for believing that the nonlinear model $\eta = kx_1^{\alpha_1}x_2^{\alpha_2}x_3^{\alpha_3}$ might be appropriate. (a) How could the model be transformed into a linear model? (b) How might analysis after such a transformation be misleading? (c) How could you find out?

12. Why is a reciprocal transformation sometimes useful in the analysis of survival data (e.g., on the useful life of a machine tool)? When might analysis with a log transformation be preferable for such data? Not preferable?

13. What data transformations are made by doing the following?

(a) Using the Richter scale to measure the intensity of earthquakes

(b) Transforming the wavelength of a sine wave to a frequency

(c) Transforming a variance to a standard deviation

(d) Determining the weight of a solid sphere from a measure of its diameter

PROBLEMS FOR CHAPTER 8

1. (a) Perform an ANOVA on the following data and make a graphical ANOVA. Comment on any anomalies you find.

		Machines		
		A	*B*	*C*
	1	382	427	399
	2	151	167	176
Materials	3	182	209	320
	4	85	123	127
	5	40	52	75

(b) Do you think a transformation might be useful for these data? Why? Do a λ plot and decide.

(c) Make a graphical ANOVA for your chosen transformation.

2. (a) A 3×3 factorial design was used to explore the effects of three sources of clay I, II, III and three ceramic molds *A*, *B*, *C*. Five replicates of each clay–mold combination were prepared and all 45 randomly placed within a furnace before firing. The response is a measure of surface roughness:

Position I			Position II			Position III		
A	*B*	*C*	*A*	*B*	*C*	*A*	*B*	*C*
0.53	0.58	0.43	1.18	1.31	0.56	0.81	0.95	0.61
0.68	0.51	0.42	1.52	0.94	0.65	1.14	1.55	0.69
0.69	0.63	0.39	1.26	0.79	0.66	1.08	1.14	0.62
0.66	0.44	0.44	1.07	1.69	0.55	1.03	0.72	0.65
0.64	0.54	0.42	1.26	1.89	0.61	1.01	1.09	0.65

Might a transformation be useful for these data? Why?

(b) Make calculations to show if an appropriate transformation might change the interpretation of these data.

3. Below are data on the treatment of burns. In this experiment six treatments $A, B, \ldots, F$ were employed on six willing subjects 1, 2, . . ., 6 each inflicted with a standard burn at six different arm positions I, II, . . ., VI. The response is a measure of the degree of healing after forty eight hours, 100 being a perfect score.

Positions on arms	Subjects 1	2	3	4	5	6
I	*A* 32	*B* 40	*C* 72	*D* 43	*E* 35	*F* 50
II	*B* 29	*A* 37	*F* 59	*E* 52	*D* 32	*C* 53
III	*C* 42	*D* 56	*A* 83	*B* 48	*F* 37	*E* 43
IV	*D* 29	*F* 59	*E* 47	*A* 56	*C* 38	*B* 42
V	*E* 28	*C* 50	*B* 100	*F* 46	*A* 29	*D* 56
VI	*F* 37	*E* 42	*D* 67	*C* 50	*B* 33	*A* 48

From an initial inspection of the data and the residuals do you think a data transformation might be helpful? Make a λ plot appropriate for a power transformation of these data. Do you find any outliers in these data? What do you conclude?

4. The following are production data for four different machines 1, 2, 3, 4 cutting four different types of laminated plastics A, B, C, D:

Plastic	Machines *A*	*B*	*C*	*D*
1	1271	4003	1022	2643
2	1440	1651	372	5108
3	612	1664	829	1944
4	605	2001	258	2607

Make an ANOVA and a graphical ANOVA.

Make a λ plot for

$$F = \frac{\text{machine mean square}}{\text{interaction mean square}} \quad \text{and} \quad F = \frac{\text{type of plastic mean square}}{\text{interaction mean square}}$$

What conclusions do you draw? Do you find any evidence of bad values in these data? Explain.

5. What is meant by "transformable" and "nontransformable" interactions? Give examples that illustrate the difference.

6. The following 2^{5-1} fractional factorial design was part of a study to find high-permeability conditions for fly ash. Analyze these data. Is data transformation helpful?

Type of Ash	Fly Ash, %	Wet/Dry Cycle	Freeze/Thaw Cycle	Bentonite, %	Permeability Response
−1	−1	−1	−1	1	1020
1	−1	−1	−1	−1	155
−1	1	−1	−1	−1	1550
1	1	−1	−1	1	75
−1	−1	1	−1	−1	1430
1	−1	1	−1	1	550
−1	1	1	−1	1	340
1	1	1	−1	−1	25
−1	−1	−1	1	−1	1390
1	−1	−1	1	1	410
−1	1	−1	1	1	570
1	1	−1	1	−1	5
−1	−1	1	1	1	2950
1	−1	1	1	−1	160
−1	1	1	1	−1	730
1	1	1	1	1	20

CHAPTER 9

Multiple Sources of Variation

9.1. SPLIT-PLOT DESIGNS, VARIANCE COMPONENTS, AND ERROR TRANSMISSION

Data from industry and other sources are frequently affected by more than one source of variation. This must be properly taken into account in planning investigations and analyzing results. In this chapter you will see how to identify and exploit multiple components of variation in three different but related ways:

1. To produce what are called *split-plot designs*. These designs have important applications in industrial experimentation; they enable comparisons to be made with smaller error and are easier to carry out, compared with arrangements previously described. In particular, they are employed in Chapter 13 for designing robust products and processes.
2. Estimation of individual *components of variance* to determine, for example, how much of the observed variation is due to testing, how much to sampling the product, and how much to the process itself.
3. Calculating how variability from a number of different sources is *transmitted* to an outcome. (Such calculations are used later in the discussion of the design of products to minimize transmitted component variation.)

9.2. SPLIT-PLOT DESIGNS

These designs, like so many statistical devices, were originally developed for agricultural field trials by R. A. Fisher and F. Yates. As was suggested by D. R. Cox, because of their usefulness outside the agricultural context, they might be better called split-*unit* designs. However, even in this wider context, the term "split plot" has become commonplace and is used here.

Statistics for Experimenters, Second Edition. By G. E. P. Box, J. S. Hunter, and W. G. Hunter

Cuthbert Daniel, an engineer and statistician of great experience and originality once remarked (perhaps with slight exaggeration) that "All industrial experiments are split-plot experiments." Table 9.1 shows the results from a split-plot experiment designed to study the corrosion resistance of steel bars treated with four different coatings C_1, C_2, C_3, and C_4 at three duplicated furnace temperatures 360, 370, and 380°C. Table 9.1a shows the results as the experiment was run with positions of the four differently coated bars randomly arranged in the furnace within each heat. Table 9.1b shows the same data rearranged in a two-way table with six rows and four columns. The positions of the coated steel bars in the furnace were randomized within each heat. However, because the furnace temperature was difficult to change, the heats were run in the systematic order shown (although it would have been preferable to randomize the order had this been possible).

What was of primary interest were the comparison of coatings C and how they interacted with temperature T. The split-plot design supplied this information efficiently and economically. The arrangement appears similar to the two-way factorial design of Chapter 8 in which the survival times of a number of animals were analyzed. In that experiment, however, the experimental units were animals randomly assigned to *all* of the factor combinations of poisons and treatments. There was therefore only one source of error influencing the differences in survival times of animals treated in the same way. By contrast, in this corrosion

Table 9.1. Split-Plot Design to Study Corrosion Resistance of Steel Bars Treated with Four Different Coatings Randomly Positioned in a Furnace and Baked at Three Different Replicated Temperatures

(a) Data Arranged as Performed

Heats (Whole Plots), °C	Coating–Position (Subplots)			
360	C_2	C_3	C_1	C_4
	73	83	67	89
370	C_1	C_3	C_4	C_2
	65	87	86	91
380	C_3	C_1	C_2	C_4
	147	155	127	212
380	C_4	C_3	C_2	C_1
	153	90	100	108
370	C_4	C_1	C_3	C_2
	150	140	121	142
360	C_1	C_4	C_2	C_3
	33	54	8	46

(b) Data Rearranged in Rows and Columns

	C_1	C_2	C_3	C_4	Averages	
360	67	73	83	89	78.00	
	33	8	46	54	35.25	56.63
370	65	91	87	86	82.25	
	140	142	121	150	138.25	110.25
380	155	127	147	212	160.25	
	108	100	90	153	112.75	135.50
Average	94.67	90.17	95.67	124.00	—	101.125

resistance experiment, to have run each combination of temperature and coating in random order could have required the readjustment of furnace temperature up to 24 times and would have resulted in estimates with much larger variance. Also it would have been prohibitively difficult to do and expensive. The split plot is like a randomized block design (with whole plots as blocks) in which the opportunity is taken to introduce additional factors *between blocks*.

In the split-plot design in Table 9.1 the "whole plots" were the six different furnace heats. In each of these heats there were four subplots consisting of four steel bars with coatings C_1, C_2, C_3, C_4 randomly positioned within the furnace and heated. It would have been quite misleading to treat this experiment as if there were only one error source and only one variance. There were in fact two different experimental units: the six different furnace heats, generically called *whole plots*, and the four positions, generically called *subplots*, in which the differently coated bars could be placed in the furnace. With these are associated two different variances: σ_W^2 for whole plots and σ_S^2 for subplots. Achieving and maintaining a given temperature in this furnace were very imprecise and the whole-plot variance σ_W^2, measuring variation from one heat to another, was therefore expected to be large. However, the subplot variance σ_S^2, measuring variation from position to position within a given heat, was likely to be relatively small. A valuable characteristic of the split-plot design is that all subplot effects *and* subplot–main plot interaction effects are estimated with the same subplot error. This design was ideal for this experiment, the main purpose of which was to compare the coatings and determine possible interactions between coatings and heats. Thus the effects of principal interest could be measured with the small subplot error.

Two considerations important in choosing any statistical arrangement are convenience and efficiency. In industrial experimentation a split-plot design is often convenient and is sometimes the only practical possibility. This is so whenever there are certain factors that are difficult to change and others that are easy. In this example changing the furnace temperature was difficult; rearranging the positions of the coated bars in the furnace was easy. Notice that the cost of generating information for these arrangements is *not* proportional to the total number of observations made as it was for the animal experiment of Chapter 8. There complete randomization was convenient and inexpensive, but this would certainly not have been the case for this corrosion coating experiment. Not only would a fully randomized experiment have been much more costly, it also would result in the effects of interest having a much larger variance.

The numerical calculations for the ANOVA of a split-plot design [sums of squares (SS), degrees of freedom, and mean squares (MS)] are the same as for any other balanced design (you can temporarily ignore the split plotting), and they can be performed by any of the excellent computer programs now available. But having made these calculations experimenters sometimes have difficulty in identifying appropriate error terms. Such difficulties are best overcome by rearranging the items of the ANOVA table into two parallel columns: one for whole plots and one for subplots. This has been done in Table 9.2.

Table 9.2. Parallel Column Layout of ANOVA Showing How Various Combinations Are Associated with Their Appropriate Error

	Whole Plots–Between Heats (Averaged over Coatings)		Subplots–Within Heats (Interactions with Coatings)	
	Source	df	Source	df
Average	I	1	$I \times C$	3
Temperature	T	2	$T \times C$	6
Error	E_W	3	E_S	9

In the whole-plot analysis, the contribution from the grand average (sometimes called "the correction for the mean" or the "correction factor") is denoted by the identity I and has one degree of freedom. The contributions from differences associated with the three temperatures denoted by T have two degrees of freedom. The whole-plot error E_W with three degrees of freedom measures the differences between the replicated heats at the three different temperatures.

In the subplot analysis all the effects can be thought of as "interactions" of the subplot factor (C) with the corresponding whole-plot effects. Thus, $I \times C$, measuring the differences in the corrosion resistances of the four different coatings (the main effect of C), is positioned opposite I because it shows how different are the *means* for the different coatings. It can be thought of as the interaction $IC = C$. Similarly, TC, which measures the interaction between T and C, is shown opposite T. It shows the extent to which coating effects are different at different temperatures. Also, the subplot error E_S can be thought of as the interaction of E_W with C. It measures to what extent the coatings give dissimilar results within each of the replicated temperatures. This analysis is shown in Table 9.3.

Because there are four coatings in each heat, the whole-plot error mean square 4813 is an estimate of $4\sigma_W^2 + \sigma_S^2$. The subplot error mean square 125 is an estimate of σ_S^2. Estimates of the individual whole-plot and subplot error variance

Table 9.3. ANOVA for Corrosion Resistance Data Showing Parallel Column Layout Identifying Appropriate Errors for Whole-Plot and Subplot Effects

Heats (Whole Plots)					Coatings (Subplots)				
Source	df	SS	MS		Source	df	SS	MS	
Average, I	1	245430	245430		C	3	4289	1430	$F_{3,9} = 11.5^{**}$
Temperature	2	26519	13260	$F_{2,3} = 2.8$	$T \times C$	6	3270	545	$F_{6,9} = 4.4^{*}$
Error E_w	3	14440	4813		Error E_s	9	1121	125	

Note: The convention is used that a single asterisk indicates statistical significance at the 5% level and two asterisk statistical significance at the 1% level.

components are therefore

$$\hat{\sigma}_W^2 = \frac{(4813 - 125)}{4} = 1172, \qquad \hat{\sigma}_S^2 = 125$$

and

$$\hat{\sigma}_W = 34.2, \qquad \hat{\sigma}_S = 11.1$$

Thus, in this experiment $\hat{\sigma}_W$, the estimated standard deviation for furnace heats, is three times as large as $\hat{\sigma}_S$, the standard deviation for coatings.

The whole-plot analysis for differences in temperature T produces a value of $F_{2,3} = 2.8$ with a significance probability of only about 25%. Thus, little can be said about the overall effect of temperature. This comparison is, of course, extremely insensitive because of the large whole-plot error and also the very small number of degrees of freedom involved. However, what were most important were the effects of coatings and the possible interactions of coatings with temperature. For this purpose, even though it required much less effort than a fully randomized design, the split-plot design produced very sensitive comparisons revealing statistically significant differences for C and for $T \times C$.

There are eight observations in each temperature (whole-plot) average and six in each coating (subplot). The associated standard errors (SE) are therefore

$$SE\,(\text{temperature}) = 34.2/\sqrt{8} = 12.1, \qquad SE\,(\text{coatings}) = 11.1/\sqrt{6} = 4.53$$

Since there are only two observations in each temperature-by-coating combination, these (interaction) cell averages have standard error

$$SE\,(\text{temperature} \times \text{coating}) = 11.1/\sqrt{2} = 7.85$$

Graphical Analysis

Table 9.4 shows the original 6×4 table of observations in Table 9.1(b) for the three duplicate temperatures T and the four treatments C broken into component tables for the average, temperature, and whole-plots residuals. Below are shown the component subplot tables for the coatings, temperature-by-coating interactions, and subplot residuals. As before, the various sums of squares in the ANOVA table are equal to the corresponding sums of squares of the elements in these component tables. The graphical ANOVA in Figure 9.1 complements the analysis in Table 9.4 and provides a striking visual display of what the experiment has to tell you. For example, it shows the much larger spread for the whole-plot error in relation to the subplot error. It also brings to your attention that the differences between coatings is almost entirely due to the superior performance of C_4 and that a large interaction with temperature is probably associated with T_3C_4.

Table 9.4. Corrosion Resistance Data with Component Tables

Whole Plots

$$
\underset{\text{Corrosion resistance}}{\begin{pmatrix} 67 & 73 & 83 & 89 \\ 33 & 8 & 46 & 54 \\ 65 & 91 & 87 & 86 \\ 140 & 142 & 121 & 150 \\ 155 & 127 & 147 & 212 \\ 108 & 100 & 90 & 153 \end{pmatrix}}
=
\underset{\text{average}}{\begin{pmatrix} 101.1 & 101.1 & 101.1 & 101.1 \\ 101.1 & 101.1 & 101.1 & 101.1 \\ 101.1 & 101.1 & 101.1 & 101.1 \\ 101.1 & 101.1 & 101.1 & 101.1 \\ 101.1 & 101.1 & 101.1 & 101.1 \\ 101.1 & 101.1 & 101.1 & 101.1 \end{pmatrix}}
+
\underset{\text{temperature}}{\begin{pmatrix} -44.5 & -44.5 & -44.5 & -44.5 \\ -44.5 & -44.5 & -44.5 & -44.5 \\ 9.1 & 9.1 & 9.1 & 9.1 \\ 9.1 & 9.1 & 9.1 & 9.1 \\ 35.4 & 35.4 & 35.4 & 35.4 \\ 35.4 & 35.4 & 35.4 & 35.4 \end{pmatrix}}
+
\underset{\text{whole-plot residuals}}{\begin{pmatrix} 21.4 & 21.4 & 21.4 & 21.4 \\ -21.4 & -21.4 & -21.4 & -21.4 \\ -28.0 & -28.0 & -28.0 & -28.0 \\ 28.0 & 28.0 & 28.0 & 28.0 \\ 23.8 & 23.8 & 23.8 & 23.8 \\ -23.8 & -23.8 & -23.8 & -23.8 \end{pmatrix}}
$$

Subplots

$$
+
\underset{\text{Coatings}}{\begin{pmatrix} -6.5 & -11.0 & -5.5 & 22.9 \\ -6.5 & -11.0 & -5.5 & 22.9 \\ -6.5 & -11.0 & -5.5 & 22.9 \\ -6.5 & -11.0 & -5.5 & 22.9 \\ -6.5 & -11.0 & -5.5 & 22.9 \\ -6.5 & -11.0 & -5.5 & 22.9 \end{pmatrix}}
+
\underset{\text{temperatures} \times \text{coatings}}{\begin{pmatrix} -0.2 & -5.2 & 13.3 & -8.0 \\ -0.2 & -5.2 & 13.3 & -8.0 \\ -1.3 & 17.2 & -0.8 & 15.1 \\ -1.3 & 17.2 & -0.8 & 15.1 \\ 1.5 & -12.0 & -12.5 & 23.1 \\ 1.5 & -12.0 & -12.5 & 23.1 \end{pmatrix}}
+
\underset{\text{subplot residuals}}{\begin{pmatrix} -4.4 & 11.1 & -2.9 & -3.9 \\ 4.4 & -11.1 & 2.9 & 3.9 \\ -9.5 & 2.5 & 11.0 & -4.0 \\ 9.5 & -2.5 & -11.0 & 4.0 \\ -0.3 & -10.3 & 4.8 & 5.8 \\ 0.3 & 10.3 & -4.8 & -5.8 \end{pmatrix}}
$$

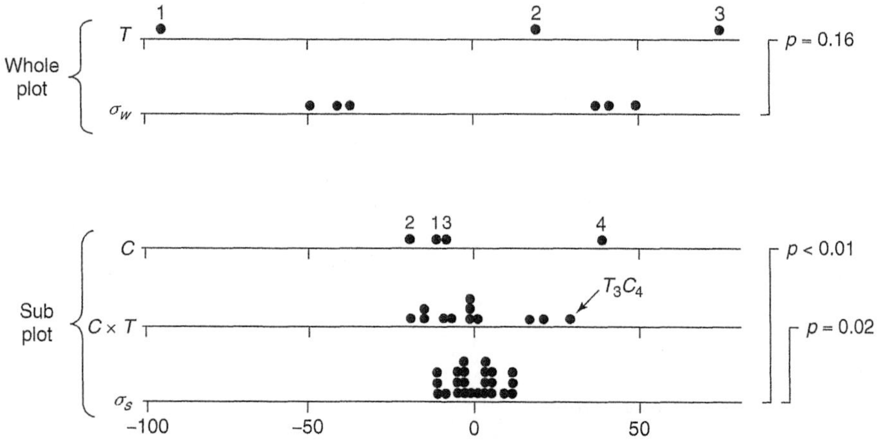

Figure 9.1. Graphical ANOVA, corrosion data.

Further Analysis for the Coatings

Two factors (not previously mentioned) which characterized the four coatings were the base treatment **B** and finish **F**, both at two levels. The four coatings tested in this experiment thus formed a 2^2 factorial design as shown below:

F_2	C_3	C_4
F_1	C_1	C_2
	B_1	B_2

The subplot analysis can thus be further broken down to produce Table 9.5, from which it will be seen that B, F, BF, and BTF provide statistically significant effects.

Table 9.5. Further Analysis of Subplots

Source	SS		df	SS	MS	F
		B	1	852	852	$F_{1,9} = 6.9^*$
C	4289	F	1	1820	1820	$F_{1,9} = 14.7^{**}$
		BF	1	1617	1617	$F_{1,9} = 13.0^{**}$
		TB	2	601	300	$F_{2,9} = 2.4$
$T \times C$	3270	TF	2	788	394	$F_{2,9} = 3.2$
		TBF	2	1881	940	$F_{1,9} = 15.2^{**}$
Error	E_s		9	1121	125	

Finish	B_1	B_2
F_2	96	124
F_1	95	90

Base

(*a*)

Temperature 360°C

	B_1	B_2
F_2	65	72
F_1	50	41

370°C

	B_1	B_2
F_2	104	118
F_1	103	117

380°C

	B_1	B_2
F_2	119	183
F_1	132	114

(*b*)

Figure 9.2. Tables of average corrosion resistance: (*a*) overall averages for base and finish; (*b*) base and finish averages for different temperatures.

Exercise 9.1. Extend the graphical analysis to take account of the **B** and **F** factors.

The meaning of these main effects and interactions is best understood by inspecting the appropriate two-way tables of averages. Figure 9.2*a* makes clear the meaning of the B, F, and BF effects. The base and finish factors act synergistically to produce a coating having a large additional anticorrosion effect when level 2 of finish and level 2 of base are used *together*.

Figure 9.2*b*, showing the two-way B-by-F tables for each temperature, exemplifies the TBF interaction and illustrates how the synergistic BF interaction is greatly enhanced by the use of the high temperature.

Actions and Conclusions from the Corrosion Experiment

As always, the actions called for as a result of the experiment were of two kinds:

1. *"Cashing in" on new knowledge:* Coating C_4 (combination B_2F_2) with a temperature of 380°C should be adopted for current manufacture.
2. *Using the new knowledge to look for further possibilities of improvement*:
   (i) Coating C_4 should be tested at higher temperatures.
   (ii) Since getting the best out of coating C_4 depended critically on temperature, which was very poorly controlled, considerable effort might be justified in getting better control of the furnace.
   (iii) By consulting expert opinion, possible reasons (physical, chemical, metallurgical) for the discovered effects might be elicited. Specialists could gain an appreciation of the size and nature of these effects by looking at Figures 9.1 and 9.2. Promising conjectures could be investigated by further experimentation.*

* Subject matter specialists, engineers, and scientists can almost always rationalize unexpected effects. Their speculations may be right or wrong. But in any case, they should be further discussed and will often be worth exploring experimentally.

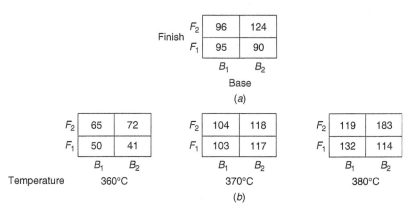

Comparison of Machine Heads

As a second example the data shown in Table 9.6*a* are from an experiment designed to seek out the sources of unacceptably large variability which, it was suspected, might be due to small differences in five supposedly identical *heads* on a machine tool. To test the idea, the engineer arranged that material from each of the five heads was sampled on two days in each of six successive 8-hour *periods* corresponding to the three shifts. The order of sampling within each period was random and is indicated by the numerical superscripts in the table of data.

An Unintended Split-Plot Design

A standard ANOVA for these data is shown in Table 9.6*b*. It will be seen that the F ratio $18.4/13.9 = 1.3$ suggests that there are no differences in head means that could not be readily attributed to chance. The investigator thus failed to find what had been expected. A partial graphical analysis of the data is shown in Figure 9.3, and on further consideration it was realized that the design was a split-plot arrangement with "periods" constituting the main plots. The analysis strongly suggested that real differences in means occurred between the six 8-hour periods during which the experiment was conducted. These periods corresponded almost exactly to six successive shifts labeled A, B, C for day 1 and A', B', C' for day 2. Big discrepancies are seen at the high values obtained in both night shifts C and C'. A further analysis in Table 9.6*b* of *periods* into effects associated with *days* (D), *shifts* (S), and $D \times S$ confirms that the differences seem to be associated with *shifts*. This clue was followed up and led to the discovery that there were substantial differences in operation during the night shift—an unexpected finding because it was believed that great care had been taken to ensure the operation did not change from shift to shift.

Exercise 9.2. Complete the data breakdown, the ANOVA table, and graphical analysis for the data shown in Table 9.6*a*. Show that this is a split-plot design and rearrange the ANOVA table. Check the analysis and identify the variances associated with the "whole plots" and "subplots." Do you see anything unusual in your analysis?

Summary

1. By the deliberate use of a split-plot design, it is possible to achieve greater precision in studying what are often the effects of major interest—namely the subplot factors and their interactions with whole-plot factors. This increase in precision can be obtained even though whole-plot factors are highly variable.
2. A split-plot design can produce considerable economy of effort. Thus, in the first example a fully randomized design would require 24 heats tested with one coating in each heat and would have been almost impossible to perform. Also, because of the large value of the whole-plot variance $\sigma_W{}^2$, it would have been far less efficient than the spilt-plot design actually used.

Table 9.6. (*a*) **Data from a Quality Improvement Experiment: Machine Heads** (*b*) **Standard ANOVA** (*c*) **Additive Basic for the Analysis**

Periods	Heads 1	2	3	4	5	
A	20(2)	14(1)	17(4)	12(5)	22(3)	17.0
B	16(2)	19(3)	16(5)	17(1)	21(4)	17.8
C	25(5)	32(2)	31(3)	24(4)	24(1)	27.2
A′	18(5)	22(4)	9(1)	15(3)	17(2)	16.2
B′	21(1)	21(2)	16(4)	20(5)	17(3)	19.2
C′	19(4)	19(3)	22(5)	29(1)	29(2)	25.4
	19.8	22.8	18.5	19.5	21.7	20.5

(*a*)

Source	df	SS	MS	
Heads(*H*)	4	73.5	18.4	1.3
Periods(*P*)	5	543.1	108.6	
Residual	20	278.9	13.9	
Days	1	1.2	1.2	
Shifts	2	528.5	264.2	39.4
D × S	2	13.4	6.7	
		543.1		

(*b*)

Data						Mean						Heads(*H*)						Periods(*P*)						Residual(*H* × *P*)				
20	14	17	12	22		20.5	20.5	20.5	20.5	20.5		−0.6	2.4	−2.0	−1.0	1.2		−3.5	−3.5	−3.5	−3.5	−3.5		3.6	−5.4	2.0	−4.0	3.8
16	19	16	17	21		20.5	20.5	20.5	20.5	20.5		−0.6	2.4	−2.0	−1.0	1.2		−2.7	−2.7	−2.7	−2.7	−2.7		−1.2	−1.2	0.2	0.2	2.0
25	32	31	24	24	=	20.5	20.5	20.5	20.5	20.5	+	−0.6	2.4	−2.0	−1.0	1.2	+	6.7	6.7	6.7	6.7	6.7	+	−1.6	2.4	5.8	−2.2	4.4
18	22	9	15	17		20.5	20.5	20.5	20.5	20.5		−0.6	2.4	−2.0	−1.0	1.2		−4.3	−4.3	−4.3	−4.3	−4.3		2.4	3.4	−5.2	−0.2	−0.4
21	22	16	20	17		20.5	20.5	20.5	20.5	20.5		−0.6	2.4	−2.0	−1.0	1.2		−1.3	−1.3	−1.3	−1.3	−1.3		2.4	0.4	−1.2	1.8	3.4
19	28	22	29	29		20.5	20.5	20.5	20.5	20.5		−0.6	2.4	−2.0	−1.0	1.2		4.9	4.9	4.9	4.9	4.9		−5.8	0.2	−1.4	4.6	2.4

	Data		Mean		Heads(*H*)		Periods(*P*)		Residual(*H* × *P*)
S/S	13,462.0	=	12,566.5	+	73.5	+	543.1	+	278.9
D/F	30	=	1	+	4	+	5	+	20

(*c*)

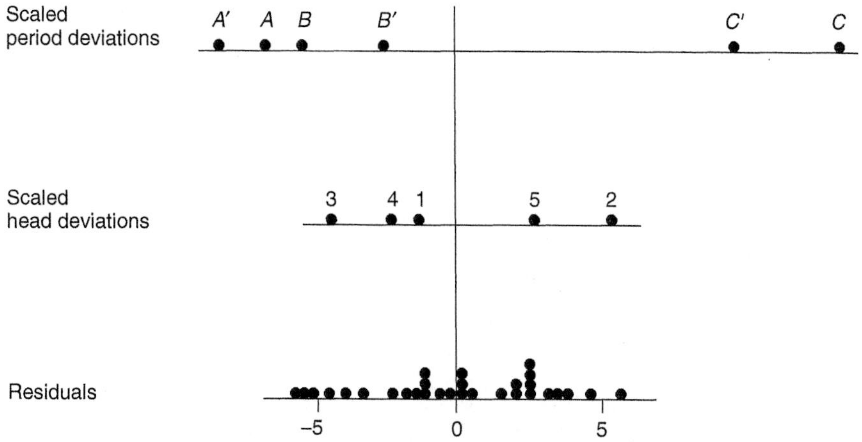

Figure 9.3. Machine heads experiment: Partial graphical analysis.

3. It is important to be on the outlook for a split-plot structure (*often unintended*, as was the case in the machine tool experiment) and to take it into account in the subsequent data analysis. You can be seriously misled by failure to take into account "split plotting" (see Fisher, 1925; Daniel, 1976).
4. Experimenters sometimes have trouble in identifying which effects should be tested against which error source with designs of this kind. The arrangement of the ANOVA table into parallel columns can alleviate this problem. Whole-plot effects are shown in one column and their interactions with subplot effects in the other.
5. The graphical ANOVA provides a visual appreciation of what the experiment shows you.
6. Many of the "inner and outer array" experiments popularized by Taguchi (1987) for developing robust processes and products are examples of split-plot designs (Box and Jones, 1992b) but usually have not been recognized or analyzed as such. The great value, correct use, and analysis of split-plot arrangements for designing robust industrial products were demonstrated in a paper by Michaels in 1964.

9.3. Estimating Variance Components

Consider a batch process used for the manufacture of a pigment paste where the quality of the immediate interest is the *moisture content*. Suppose a batch of the product has been made, a chemical sample taken to the laboratory, and a portion of the sample has been tested to yield an estimate of its moisture content. The variance of y will reflect variation due to *testing*, variation due to *sampling* the product, and the inherent variation of the *process* itself. Variance component analysis separates this total variation into parts assignable to the various causes. This is important because of the following:

1. Such analysis shows where efforts to reduce variation should be directed. It tells, for example, whether you can reduce variation most economically by improving the testing method or modifying the sampling procedure or by working on the process itself.
2. It clarifies the question of what is the appropriate estimate of error variance in comparative experiments.
3. It shows how to design a minimum-cost sampling and testing procedure for any particular product.

First consider a model that can represent the creation of the observations. If η is the long-term process mean for the moisture content, then, as illustrated in Figure 9.4, the overall error $e = y - \eta$ will contain three separate components: $e = e_T + e_S + e_P$, where e_T is the analytical testing error, e_S is the error in taking the chemical sample, and e_P is the error due to inherent process variation.

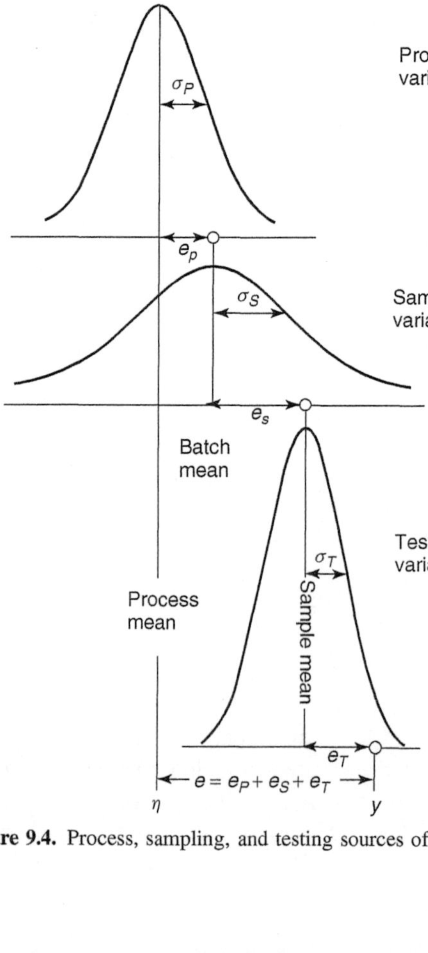

Figure 9.4. Process, sampling, and testing sources of variation.

Moisture Content Data

Table 9.7 shows three sets of data all from the same process: (a) from 10 tests made on a single sample, (b) from 10 samples each tested once, and (c) from a single test made on a single sample from each of 10 different process batches. Dot plots of the data are shown in Figure 9.5, from which it is obvious that the variances of the three sets of data are very different.

The calculated averages and variances V_T, V_S, and V_P for each of these three sets of data are shown in Table 9.7. From these you can calculate estimates of the three variance components σ_T^2, σ_S^2 and σ_P^2, which represent the individual variances contributed by the process alone, by sampling alone, and by testing alone, respectively. Testing error, sampling error, and process error will vary

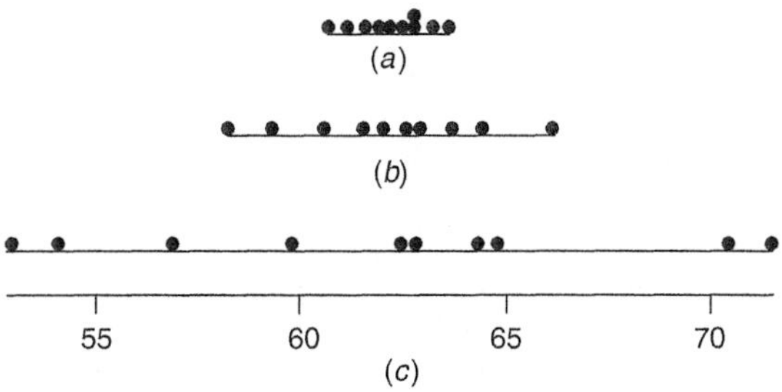

Figure 9.5. Dot diagrams of process, sampling, and testing sources of variation.

Table 9.7. Data Plotted in Figure 9.5

	(a) Ten Tests on One Sample	(b) Ten Samples Each Tested Once	(c) Ten Batches Each Sampled and Tested Once
	59.6	57.0	62.2
	59.2	63.2	56.8
	61.3	61.4	64.5
	60.4	64.9	70.3
	59.7	57.9	54.1
	60.8	62.5	53.3
	58.8	59.1	64.2
	59.9	61.3	59.7
	60.4	60.5	62.4
	60.1	60.2	71.5
Averages	60.0	60.8	61.9
Variances	$V_T = 0.55 \rightarrow \sigma_T^2$	$V_S = 5.81 \rightarrow \sigma_S^2 + \sigma_T^2$	$V_P = 37.82 \rightarrow \sigma_P^2 + \sigma_S^2 + \sigma_T^2$
Variance components	$\hat{\sigma}_T^2 = 0.55$	$\hat{\sigma}_S^2 = 5.81 - 0.55 = 5.26$	$\hat{\sigma}_P^2 = 37.82 - 5.81 = 32.01$
Standard deviations	$\hat{\sigma}_T = 0.74$	$\hat{\sigma}_S = 2.29$	$\hat{\sigma}_P = 5.66$

independently, and hence, in Table 9.7, V_T is an estimate of the testing variance alone, $V_T \to \sigma_T^2$; V_S is an estimate of the sampling plus testing variance, $V_S \to \sigma_S^2 + \sigma_T^2$; and V_P is an estimate of the process plus sampling plus testing variance, $V_P \to \sigma_P^2 + \sigma_S^2 + \sigma_T^2$. Separate variance component estimates shown in Table 9.7 may then be obtained by subtraction.

Examination of records of plant data should always raise the question, *where* does the variation arise? For the data of Table 9.7, since $\hat{\sigma}_P^2 = 32.01$, a very large part of the overall variance was coming from the process itself. The contribution $\hat{\sigma}_S^2 = 5.26$ arising from the method of sampling product was also considerable. However, the component due to testing $\hat{\sigma}_T^2 = 0.55$ was very small. Thus in this example efforts to reduce the overall variance should be directed at reducing process and sampling variability. Reduction of testing variance will make very little difference to the overall variation of data.

Getting the Correct Estimate of Error

An understanding of variance components is important in experimental design. Consider the simple problem of comparing two or more averages. The following data were supposed to show that a process modification denoted by B gave a higher process mean than the standard process denoted by A:

A	B
58.3	63.2
57.1	64.1
59.7	62.4
59.0	62.7
58.6	63.6
$\bar{y}_A = 58.54$	$\bar{y}_B = 63.20$

A test to compare the means of these two sets of observations produced a value of Student's $t_{\nu=8} = 8.8$, which, if there were no real difference, would occur less than once in a thousand times by chance. However, this did not demonstrate that the mean of *process B* was *higher* than that of *process A*. It turned out that the data were obtained by making five tests on a single sample from a single batch from process A followed by five tests from a single sample from a single batch from process B. Any *difference* between the two process *averages* was subject to testing plus sampling plus process errors. But the variation estimated *within* the two sets of five observations measured only testing error. About all that you could conclude from these data is that the difference between the two averages is almost certainly not due to *testing* error.

Two Testing Methods. If you wished to compare two *testing* methods A and B, then you would need to compare determinations using test method A with determinations using test method B on the *same sample* with all tests made in random order.

Two Sampling Methods. To compare two different *sampling* methods, you would need to compare determinations using sampling method A with those made on a set of different *samples* using method B with samples A and B chosen randomly from the *same batch.*

Two Processing Methods. To compare a standard *process* A with a modified *process* B, you would need to compare data from randomly chosen *batches* taken from the standard process A with data from randomly chosen *batches* made by the modified process B.

Variance Component Estimation Using a Nested (Hierarchical) Design

An alternative experimental arrangement often employed to estimate components of variance is called a nested or hierarchical design. Suppose you chose B batches of product with each batch sampled S times and each sample tested T times. You would have a hierarchical design. Figure 9.6 illustrates a hierarchical design with $B = 5$, $S = 3$, and $T = 2$.

Table 9.8 shows 60 determinations obtained from an investigation of pigment paste using a hierarchical design. There were $B = 15$ batches, each sampled $S = 2$ times, and each sample tested $T = 2$ times. The associated ANOVA table is also shown.

The multipliers used in the expected mean square column in the ANOVA in Table 9.8 are chosen so that, if there were no variation due to batches or to samples, each mean square would provide a separate estimate of σ_T^2. Thus, for example, the expected mean square for batches is $4\sigma_P^2 + 2\sigma_S^2 + \sigma_T^2$ and the multiplier for σ_P^2 is 4 because each batch average is based on four tests. Also σ_S^2 is multiplied by 2 because there are two tests for each sample. The mean square for tests as it stands provides an estimate of the variance component due to testing.

In this case $\hat{\sigma}_T^2 = 0.9$ and the remaining components of variance obtained by subtraction are $\hat{\sigma}_S^2 = (58.1 - 0.9)/2 = 28.6$ and $\hat{\sigma}_P^2 = (86.9 - 58.1)/4 = 7.2$. The corresponding estimated standard deviations are respectively $\hat{\sigma}_T = 0.95$, $\hat{\sigma}_S = 5.35$, and $\hat{\sigma}_P = 2.68$.

For these data the analysis showed that the largest individual source of variation arose from sampling the product. Two possible causes were as follows:

1. The standard sampling procedure was satisfactory *but was not being carried out.*
2. The standard sampling procedure was not satisfactory.

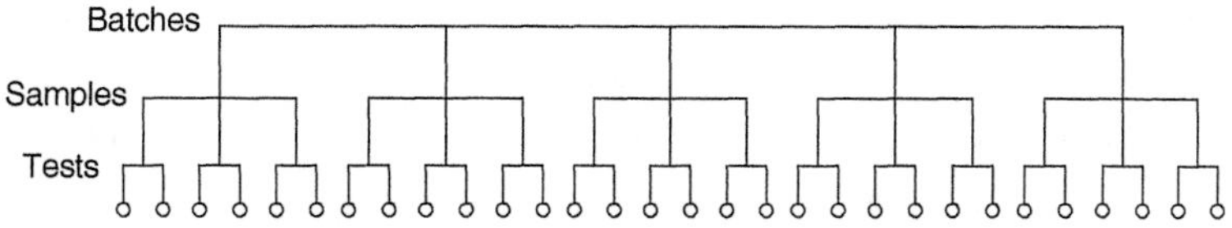

Figure 9.6. A $5 \times 3 \times 2$ hierarchical design.

Table 9.8. Data and Analysis of a 15 x 2 x 2 Hierarchical Design for Components of Variance

Batch	1		2		3		4		5		6		7	
Sample	1	2	3	4	5	6	7	8	9	10	11	12	13	14
Tests	40	30	26	25	29	14	30	24	19	17	33	26	23	32
	39	30	28	26	28	15	31	24	20	17	32	24	24	33

Batch	8		9		10		11		12		13		14		15	
Sample	15	16	17	18	19	20	21	22	23	24	25	26	27	28	29	30
Tests	34	29	27	31	13	27	25	25	29	31	19	29	23	25	39	26
	34	29	27	31	16	24	23	27	29	32	20	30	23	25	37	28

Source of Variation	Sum of Squares	Degrees of Freedom	Mean Square	Expected Value of Mean Square
Average	42987.3	1		
Batches	1216.2	14	86.9	$4\sigma_p^2 + 2\sigma_s^2 + \sigma_T^2$
Samples	871.5	15	58.1	$2\sigma_s^2 + \sigma_T^2$
Tests	27.0	30	0.9	σ_T^2
Total	45,102.0	60		

A batch of pigment paste consisted of approximately 80 drums of material. The supposedly standard procedure for taking a sample consisted of (1) randomly selecting five drums of the material from the batch, (2) carefully taking material from each selected drum using a special, vertically inserted sampling tube to obtain a representative sample from all layers within the drum, (3) thoroughly mixing the five tube-fulls of material, and finally (4) taking a sample of the aggregate mixture for testing.

By walking down to the plant, it was discovered that this procedure was not being followed. In fact, the operators responsible for conducting the sampling were unaware that there *was* any such procedure; they believed that all that was required was to take a sample from one drum and put it into the sampling bottle. The special sampling tube was nowhere to be found. Over a period of time, poor communication, laxness in supervision, and changes in personnel had brought about this situation. When the test was conducted properly using the recommended sampling procedure, a dramatic reduction in variance was found. The action taken as a result of the study was

1. to explain to the operators what was expected of them,
2. to introduce adequate supervision in the taking of samples, and
3. to initiate studies to discover whether further improvement in sampling was possible.

Choice of Numbers of Samples and Tests

In a follow-up investigation a second hierarchical design was run with the correct sampling procedure except that each sample was a specimen from *one* drum rather than five. We denote the corresponding single-drum sample variance component drum by σ_U^2. The experiment yielded these estimates:

$$\hat{\sigma}_B^2 = 4.3, \qquad \hat{\sigma}_U^2 = 16.0, \qquad \hat{\sigma}_T^2 = 1.3$$

Exercise 9.3. Construct a hierarchical design capable of estimating the three components of variance $\hat{\sigma}_B^2$, $\hat{\sigma}_U^2$, and $\hat{\sigma}_T^2$.

Deciding on a Routine Sampling Scheme

How can the results of this special experiment be used to set up a good routine sampling procedure? Suppose the procedure consists of taking one sample from each of U randomly selected drums and performing T tests on each of these samples. Then a total of $A = UT$ analyses will be made. The average of the tests performed on the U samples would then have variance

$$V = \frac{\sigma_U^2}{U} + \frac{\sigma_T^2}{A}$$

A number of alternative schemes are listed in Table 9.9, where the fourth column refers to cost and is discussed later. The third column shows the value of the variance V for various choices of U and A. Schemes in which A, the total number of analyses, is less than the number of samples taken can be physically realized if it is possible to use "bulked" sampling. For example, the scheme $U = 10$ and $A = 1$ can be accomplished by taking 10 samples, mixing them together, and performing one test on the aggregate mixture.

You can see that in this example, because of the large relative size of the sampling error to the analytical error, little is gained by replicate testing. However, a large reduction in variance can be obtained by replicate sampling.

Minimum-Cost Sampling Schemes

The fourth column in Table 9.9 shows the cost of each alternative procedure on the assumption that it costs \$1 to take a sample and \$20 to make an analytical test.

Let C_U be the cost of taking one sample and C_T the cost of making one test. The cost of the sampling scheme is then

$$C = UC_u + AC_T$$

Minimization of C for fixed variance V leads to the conclusion that any desired value of variance can be achieved for minimum cost by choosing the ratio U/A

Table 9.9. Alternative Sampling and Testing Schemes: Pigment Paste Example

Number of Samples, U	Number of Test Observations, $A = TU$	Variance of Batch Mean, $V = 16.0/U + 1.3/A$	Cost of Procedure ($)
1	1	17.3	21
	10	16.1	201
	100	16.0	2001
10	1	2.9	30
	10	1.7	210
	100	1.6	2010
100	1	1.5	120
	10	0.3	300
	100	0.2	2100

Table 9.10. Variances and Costs Associated with Alternative Sampling Schemes

Scheme Number	Number of Samples, U	Total Number of Tests, A	Variance, V	Cost ($/batch)
1	1	1	17.3	21
2	5	1	4.5	25
3	8	1	3.3	28
4	16	1	2.3	36
5	32	2	1.1	72

of samples to tests so that

$$\frac{U}{A} = \sqrt{\frac{\sigma_U^2}{\sigma_A^2}\frac{C_T}{C_A}}$$

If, for instance, we substitute the estimates $\hat{\sigma}_U^2 = 16.0$ and $\hat{\sigma}_T^2 = 1.3$ and $C_T/C_U = 20$, we obtain $U/A = 15.7$. In this case there is no point in making more than a single analytical test unless more than 16 samples are to be taken. Results of this kind of study are usually best presented to management in terms of a series of alternative possible schemes with associated variances and costs. Five schemes for the present case are shown in Table 9.10. The table makes clear that the previously recommended standard procedure (scheme 2) with $U = 5$ and $A = 1$ would be good if properly executed. It also appears, however, that by bulking more samples further reduction in variance is possible at little extra cost by using scheme 3 with $U = 8$ and $A = 1$.

9.4. Transmission of Error

Section 9.3 was about how the overall variation in a response may be *analyzed* by estimating individual *variance components* contributing to the overall variation. The present section is about how the effects of *known* component error variances may be *synthesized* to determine their combined effect on the final response.

Transmitted Error in the Determination of the Volume of a Bubble

Suppose that the volume Y of a spherical bubble is calculated by measuring its diameter x from a photograph and applying the formula

$$Y = \frac{\pi}{6}x^3 = 0.524x^3$$

Suppose that the diameter of a particular bubble is measured as $x = 1.80\,\text{mm}$ and the error of measurement is known to have a standard deviation $\sigma_x = 0.02\,\text{mm}$. Approximately how big will be the consequent standard deviation σ_Y in the estimate $Y = 3.05\,\text{mm}^3$ of the volume? Figure 9.7 shows a plot of Y against x

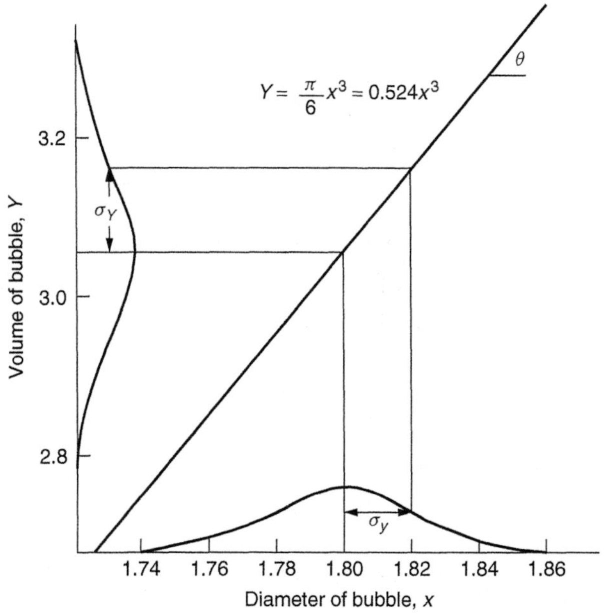

Figure 9.7. Plot of the nonlinear function: volume Y versus diameter x.

over the range $x = 1.74$ to $x = 1.86$, that is, over the range $x \pm 3\sigma_x$ which is chosen so as to include most of the error distribution. Now x^3 is certainly a nonlinear function of x, yet *over this narrow range* the plot of Y versus x is almost linear. Also, since Y increases by about 5.1 units per unit change in x and the graph passes through the point $x = 1.80$, $Y = 3.05$, their approximate linear relation is

$$Y - 3.05 = 5.1(x - 1.80), \qquad 1.74 < x < 1.86$$

But for any linear relationship

$$Y - Y_0 = \theta(x - x_0)$$

in which Y_0, x_0, and θ are constants, we have for the variance of Y

$$V(Y) = \theta^2 V(x)$$

Taking the square root of both sides of this equation gives

$$\sigma_Y = \theta\sigma_x$$

Thus, as illustrated in Figure 9.7, the gradient or slope θ (dY/dx) is the magnifying factor that converts σ_x to σ_Y. Since for this example the gradient θ is about 5.1, $\sigma_Y \cong 5.1 \times 0.02 \cong 0.102$. The diameter measurement of 1.80 ± 0.02 therefore produces a volume measurement of about 3.05 ± 0.10. Notice that the *percentage* error in the measurement of volume is about three times that for the diameter.

Exercise 9.4. Given $V(x) = 3$, find the $V(Y)$ for these two cases: $Y = 10x$, $Y = 4x + 7.0$. *Answer*: (a) 300, (b) 48.

Checking Linearity and Estimating the Gradient

A check on linearity and an estimate of the gradient are conveniently obtained numerically from a table of the following kind:

x	Y	Differences of Y
$x_0 - 3\sigma_x = 1.74$	2.76	
		0.30
$x_0 = 1.80$	3.06	
		0.31
$x_0 + 3\sigma_x = 1.86$	3.37	

In this table the values of Y are recorded at x_0, $x_0 - 3\sigma_x$, and $x_0 + 3\sigma_x$. The close agreement between the differences $3.06 - 2.76 = 0.30$ and $3.37 - 3.06 = 0.31$ indicates (without help of the graph) that a straight line provides a good approximation in this particular example. Also, a good average value for the gradient θ is provided by the expression

$$\frac{\text{Overall change in } Y}{\text{Overall change in } x} = \frac{3.37 - 2.76}{1.86 - 1.74} = \frac{0.61}{0.12} = 5.1$$

which is also the value read from the graph.

The gradient can of course be obtained by differentiating* the function $Y = 0.542x^3$ with respect to x, but notice that the numerical procedure offers two advantages:

It can be applied even when the function $Y = f(x)$ is not known explicitly; for example, Y might be the output from a complex simulation or numerical integration and x one of the variables in the system.

It provides a convenient check on linearity.

Error in a Nonlinear Function of Several Variables

Figure 9.8 shows how a nonlinear function of two variables $Y = f(x_1, x_2)$ [as, e.g., $Y = 0.366(x_1)^{0.5}(x_2)^{-0.8}$] can be represented graphically by a curved surface. Over a sufficiently small region in the neighborhood of some point 0 of interest (x_{10}, x_{20}), it may be possible to approximate this surface by a plane

$$Y - Y_0 = \theta_1(x_1 - x_{10}) + \theta_2(x_2 - x_{20})$$

where θ_1 and θ_2 are gradients in the direction of axes x_1 and x_2. The equation $Y - Y_0 = \theta_1(x_1 - x_{10}) + \theta_2(x_2 - x_{20})$ is referred to as a "local linearization" of the function $Y = f(x_1, x_2)$. As before, the gradients $\theta_1(\partial Y/\partial x_1)$ and $\theta_2(\partial Y/\partial x_2)$ may be obtained from calculus, but for our purpose it is better to get them numerically. In Figure 9.8 consider the points labeled 0, 1, 2, 3, 4. If d_1 and d_2 are the changes in x_1 *and* x_2 measured from the center 0, the gradient in the direction of axis x_1 is approximated by

$$\theta_1 = \frac{Y_1 - Y_3}{2d_1}$$

*The value given by direct differentiation is 5.089.

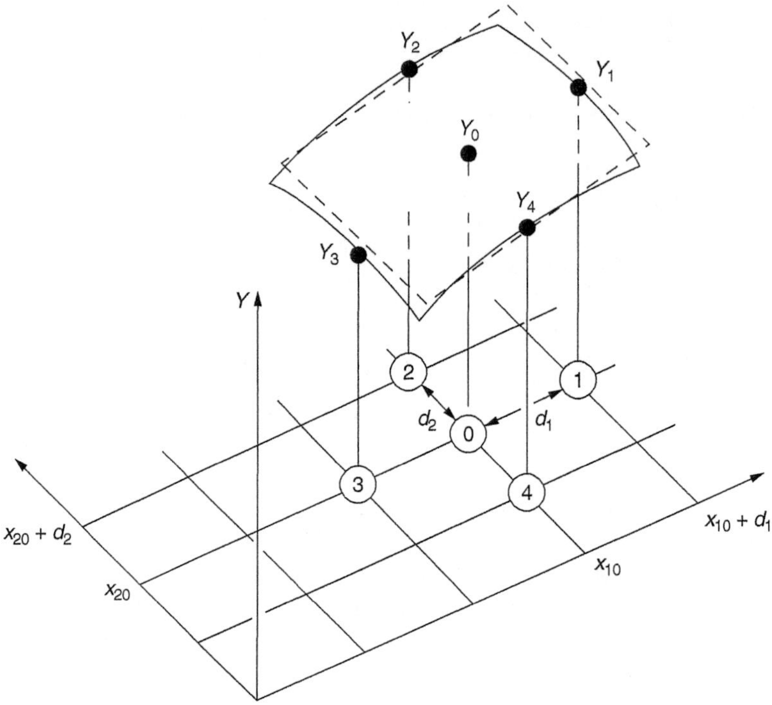

Figure 9.8. Local linearization of the function $Y = f(x_1, x_2)$.

and the gradient in the direction of the axis x_2 is approximated by

$$\theta_2 = \frac{Y_2 - Y_4}{2d_2}$$

Also, linearity can be checked in the x_1 direction by comparing $Y_1 - Y_0$ with $Y_0 - Y_3$ and in the x_2 direction by comparing $Y_2 - Y_0$ with $Y_0 - Y_4$.

Very conservatively, the approximating plane will cover the region of interest if d_1 and d_2 are set equal to $3\sigma_1$ and $3\sigma_2$, respectively, where σ_1^2 and σ_2^2 are the error variances of x_1 and x_2. Then if the linear approximation is adequate and if x_1 and x_2 varying independently, the variance of Y is approximated by

$$V(Y) \approx \theta_1^2\sigma_1^2 + \theta_2^2\sigma_2^2$$

The method is readily generalized to more than two variables, as is illustrated in the following example.*

*In most applications component errors vary independently. However, if the errors were correlated, the error transmission formula would need to take into account these correlations. Thus for two components $V(Y) \approx \theta_1^2\sigma_1^2 + \theta_2^2\sigma_2^2 + 2\theta_1\theta_2\sigma_1\sigma_2\rho_{12}$, where ρ_{12} is the correlation coefficient between x_1 and x_2.

A Nonlinear Example with Four Variables: Molecular Weight of Sulfur

The following problem appeared in the St. Catherine's College Cambridge Scholarship Papers:

> A sample of pure prismatic sulphur melted initially at 119.25°C, but in the course of a few moments the melting point fell to 114.50°C. When the sulphur had completely melted at this temperature, the liquid sulphur was plunged into ice water: 3.6 percent of the resultant solid sulphur was then found to be insoluble in carbon disulfide. Deduce the molecular formula of the type of sulfur insoluble in carbon disulfide. The latent heat of fusion of sulphur is 9 cals. per gram.*

Denote the initial melting point of the pure sulfur in degrees Kelvin by x_1, the value in degrees Kelvin to which the melting point fell by x_2, the percentage in the result of the allotropic form (the part insoluble in carbon disulfide) by x_3, and the latent heat of sulfur (in calories per gram) by x_4.

The molecular weight Y of sulfur in the allotropic form is then given by the nonlinear expression

$$Y = f(x_1, x_2, x_3, x_4) = \frac{0.02x_3(100 - x_3)x_1^2}{100x_4(x_1 - x_2)}$$

Substituting the values $x_{10} = 392.25$, $x_{20} = 387.5$, $x_{30} = 3.6$, and $x_{40} = 9$, we obtain

$$Y_0 = \frac{0.02 \times (392.25)^2 \times 3.6 \times 96.4}{9 \times 4.74 \times 100} = 249.8$$

from which the answer required for this question can be deduced.

This equation is typical of nonlinear functions that occur in practical work, and we use it to illustrate the transmission of error formula. Suppose the standard deviations of x_1, x_2, x_3, and x_4 to be $\sigma_1 = 0.1$, $\sigma_2 = 0.1$, $\sigma_3 = 0.2$, and $\sigma_4 = 0.5$. Calculation of the required gradients is shown in Table 9.11.

There is evidence of some nonlinearity. As we have said, however, checking linearity over so wide a range as $\pm 3\sigma$ is a conservative procedure, and linearization should yield a reasonably good approximation for $V(Y)$ even in this example. If nonlinearity were very serious, the computer could, by sampling from distributions of x_1, x_2, x_3, and x_4, numerically estimate $V(Y)$.

Assuming, as might be expected, that x_1, x_2, x_3, and x_4 vary independently, we have

$$\begin{aligned} V(Y) &\approx (-51.5)^2(0.01) + (52.7)^2(0.01) + (66.8)^2(0.04) + (-28.6)^2(0.25) \\ &= 26.52 + 27.77 + 178.48 + 204.50 = 437.27 \end{aligned}$$

*Quoted by permission of the Cambridge University Press.

Table 9.11. Numerical Computation of Gradients and Check for Linearity for Four Variables: Molecular Weight Example

		Value of Y	Difference	Numerical Value for Gradient
$x_{10} + 3\sigma_1$	392.55	235.3		
			−14.5	
x_{10}	392.25	249.8		$\theta_1 \cong -51.5$
			−16.4	
$x_{10} - 3\sigma_1$	391.95	266.2		
$x_{20} + 3\sigma_2$	387.8	266.6		
			16.8	
x_{20}	387.5	249.8		$\theta_2 \cong 52.7$
			14.8	
$x_{20} - 3\sigma_2$	387.2	235.0		
$x_{30} + 3\sigma_3$	4.2	289.6		
			39.8	
x_{30}	3.6	249.8		$\theta_3 \cong 66.8$
			40.3	
$x_{30} - 3\sigma_3$	3.0	209.5		
$x_{40} + 3\sigma_4$	10.5	214.1		
			−35.7	
x_{40}	9.0	249.8		$\theta_4 \cong -28.6$
			−50.0	
$x_{40} - 3\sigma_4$	7.5	299.8		

whence

$$\sigma(Y) = 20.9$$

Thus the estimated molecular weight, $Y = 249.8$, would be subject to a standard deviation of about 21.

Where does most of the variation come from? It is seen from the size of the individual contributions in the equation for $V(Y)$ that the greater part of the variation arises because of uncertainty in x_3 and x_4. If substantial improvement in the estimation of Y were required, it would be essential to find ways of obtaining more precise measurements of these quantities. No great improvement in $V(Y)$ can be expected by determining temperatures x_1 and x_2 more precisely since their contribution to $V(Y)$ is only $26.52 + 27.77 = 54.29$—only about one-fifth of the total. Notice that the main contributors to $V(Y)$ are not necessarily variables having large variances. Error transmission depends critically on the values of the gradients.

REFERENCES AND FURTHER READING

Airy, G. B. (1861) *On the Algebraical and Numerical Theory of Errors of Observations and the Combination of Observations*, MacMillan, London.

Daniel, C. (1976) *Applications of Statistics to Industrial Experimentation*, Wiley, New York.

Daniels H. E. (1938) Some topics of statistical interest in wool research, *J. Roy. Statist. Soc. Ser. B*, **5** 89–112.

Searle, S. R. (1977) *Variance Component Estimation, a Thumbnail Review*, Biometrics Unit Mineo Series, BU-612-M, Cornell University, Ithaca, NY.

Tippett, L. H. C. (1935) Some applications of statistical methods to the study of variation of quality in the production of cotton yarn, *Suppl. J. Roy. Statist. Soc.*, **11**, 27–55.

For more information on variance components and their uses see:

Box, G. E. P., and Tiao, G. C. (1973) *Bayesian Inference in Statistical Analysis*, Addison-Wesley, Reading, MA.

Box G. E. P., and Jones, S. (1992) Split-plot designs for robust product experimentation, *J. App. Stat.*, **19**, 3–26.

Daniel C. (1976). *Applications of Statistics to Industrial Experiments*, Wiley, New York.

Davies, O. L. (1960) *Design and Analysis of Industrial Experiments*, Hafner (Macmillan) New York.

Fisher R. A. (1925). *Statistical Methods for Research Workers*, Oliver & Boyd, Edinburgh.

Michaels S. E. (1964) The usefulness of experimental designs. *Appl. Stat.*, **13**, 221–235.

Satterwaith, F. E. (1946) An approximate distribution of estimates of variance components, *Biometrics Bull.*, **2**, 110–114.

Searle, S. R., Casella, G., and McCulloch C. E. (1992) *Variance Components*, Wiley, New York.

Snee, R. D. (1983) Graphical analysis of process variation studies, *J. Quality Technol.*, **15**, 76–88.

Taguchi G. (1987). *System of Experimental Design*, Kraus International Publications, White Plains, NY.

QUESTIONS FOR CHAPTER 9

1. What is a split-plot design?

2. Do you know or can you imagine circumstances in which split plotting might be overlooked?

3. When is a split-plot design likely to be most useful?

4. What are variance components?

5. Can you find or imagine an example from your own field of a system with four different components of variance?

6. How might a variance component analysis be helpful in an investigation directed toward reducing variation as economically as possible?

7. What is meant by transmission of error?

8. Consider a 20-cm by 10-cm rectangle. If both dimensions have a standard deviation of 0.1 cm, what is the standard deviation of the area of the rectangle?

9. Consider the function $Y = f(x_1, x_2) = [x_1]/[x_2]$ with standard deviations σ_1 and σ_2, respectively. Approximately, what is $V(Y)$ at the point x_{10} and x_{20}?

PROBLEMS FOR CHAPTER 9

1. **(a)** Analyze the results from the hierarchical design:

Batch Number	Sample Number	Tests		
1	1	41	40	38
	2	32	31	33
	3	35	36	35
2	4	26	27	30
	5	25	24	23
	6	35	33	32
3	7	30	29	29
	8	14	16	19
	9	24	23	25
4	10	30	31	31
	11	25	24	24
	12	24	26	25
5	13	19	18	21
	14	16	17	17
	15	27	26	29

(b) Criticize the experiment.

2. With appropriate assumptions the number of moles of oxygen in a fixed volume of air is given by the formula $Y = 0.336x_1/x_2$, where Y = moles of oxygen, x_1 = atmospheric pressure (centimeters of mercury), and x_2 = atmospheric temperature (degrees Kelvin). Suppose that average laboratory conditions are $x_{10} = 76.0$ centimeters of mercury and $x_{20} = 295$ K.

(a) Obtain a linear approximation for Y at average laboratory conditions.

(b) Assuming $\sigma_1 = 0.2$ cm and $\sigma_2 = 0.5$ K, check the linearity assumption over the ranges $x_1 \pm 3\sigma_1$ and $x_2 \pm 3\sigma_2$.

(c) Obtain an approximate value for the standard deviation of Y at average laboratory conditions.

3. The following data were collected in a study of error components:

Five Tests in One Sample	Five Samples Each Tested Once	Five Batches Each Sampled and Tested Once
9.7	7.9	4.1
10.4	14.9	20.3
11.3	11.4	14.5
9.2	13.2	6.8
9.6	7.0	12.2

Make dot plots and estimate components of variance for testing, sampling, and process variation.

4. Suppose you had three machines A, B, and C and you were asked to compare them using seven data values which had been obtained for each of the machines. What questions would you ask about the data? What answers would you need to justify the required comparisons?

5. Eight subjects 1, 2, 3, 4, 5, 6, 7, and 8 are tested with six different hearing aids A, B, C, D, E, and F. Two subjects were tested in each category of hearing loss denoted by "severe," "marked," "some," and "light." The data are the number of mistakes made in a hearing test using the various hearing aids. Analyze these data. What do you conclude? What type of experimental design is this? How many component variances are there?

		Hearing Aids					
	Subjects	A	B	C	D	E	F
Severe	1	7	16	13	19	14	11
	2	4	15	14	26	12	10
Marked	3	0	9	10	15	2	3
	4	3	6	12	13	3	0
Some	5	4	3	9	8	2	1
	6	7	2	1	4	6	8
Slight	7	2	0	4	3	0	1
	8	3	1	0	5	2	1

6. The following formula was used in the estimation of a physical constant:

$$y = \frac{(x_1 - x_3)(5x_1 - x_2)x_2}{x_4 - x_2}$$

Assuming that $x_{10} = 10, x_{20} = 40, x_{30} = 7$, and $x_{40} = 70$, calculate y. Estimate the approximate standard deviation of y given that the standard deviations of the x's are $\sigma_1 = 5, \sigma_2 = 7, \sigma_3 = 4, \sigma_4 = 10$. Obtain the necessary derivative (a) algebraically and (b) numerically. Using the numerical method, check the linearity assumptions.

7. To estimate the height of a vertical tower, it is found that a telescope placed 100 ± 1.0 meters from the foot of the tower and focused on the top of the tower makes an angle of $56° \pm 0.5°$ with the horizontal. If the numbers following the $\pm$ signs are standard deviations, estimate the height of the tower and its standard deviation.

CHAPTER 10

Least Squares and Why We Need Designed Experiments

Objectives

These days the computer can make least squares* calculations almost instantaneously. This chapter then will not concentrate on computational details, but rather will show how these calculations may be used correctly and appropriately. The explanation that follows contains simple examples to illustrate essential points in both the theory and applications of least squares. It also shows what assumptions are being made, which are innocuous, and which are not.

We have another intention. Some least squares (regression) computer programs fail to include all of the essential items and sometimes produce other output that is redundant and confusing. Also, some make automatic and invisible decisions about matters that you yourself should make. Discussions in the chapter should enable you to choose a program wisely and to inform software suppliers when their products are inadequate. Most important, you must have sufficient knowledge so that you, and not the computer, are in charge of the calculations. The study of least squares and an awareness of what can go wrong will also enable you to see why you need to design experiments. Discussing first some simple examples, you will see how to apply least squares and the design of experiments to a wide class of problems.

* For reasons long since irrelevant, least squares analysis is often called *regression analysis* and least squares computer programs are called regression programs.

Models Linear in the Parameters

A very important class of models* you will want to be able to fit to data are of the form

$$y = \beta_0 x_0 + \beta_1 x_1 + \beta_2 x_2 + \cdots + \beta_k x_k + e \qquad (10.1)$$

In this equation the β's are unknown constants to be estimated, and the x's have known values. One common example is where $x_1, x_2, \ldots$ are the levels of k factors, say temperature x_1, line speed x_2, concentration x_3, and so on, and y is a measured response such as yield. Suppose data on n runs are available (not necessarily in any particular design) in each of which the k factors have been set at specific levels. The method of least squares is used to obtain estimates of the coefficients $\beta_0, \beta_1, \ldots, \beta_k$ and hence to estimate the model of Equation 10.1. The computer can handle the analysis for any value of k, but you can fully understand least squares analysis by studying the case where $k = 2$. Here is the first example.

10.1. ESTIMATION WITH LEAST SQUARES

The Development of an Impurity

Table 10.1 shows a small† illustrative set of data from an experiment to determine how the initial rate of formation of an undesirable impurity y depended on two factors: (1) the concentration x_0 of monomer and (2) the concentration of dimer x_1. The mean rate of formation y was zero when both components x_0 and x_1 were zero, and over the relevant ranges of x_0 and x_1 the relationship was expected to be approximated by

$$y = \beta_0 x_0 + \beta_1 x_1 + e \qquad (10.2)$$

Now consider the equation for the sum of squares of the discrepancies, between the data values of the values calculated from the model

$$S(\boldsymbol{\beta}) = \sum (y - \beta_0 x_0 - \beta_1 x_1)^2 \qquad (10.3)$$

*If in Equation 10.1 the quantity x_0 is set equal to 1, then β_0 will be a constant. If you have n observations the averaging of the corresponding n equations each having the form of 10.1 produces

$$\bar{y} = \beta_0 + \beta_1 \bar{x}_1 + \beta_2 \bar{x}_2 + \cdots + \beta_k \bar{x}_k + \bar{e}$$

By subtracting, we get

$$y - \bar{y} = \beta_0 + \beta_1 (x_1 - \bar{x}_1) + \beta_2 (x_2 - \bar{x}_2) + \cdots + \beta_k (x_k - \bar{x}_k) + e - \bar{e}$$

where the y's and the x's are now related as deviations from their respected averages. In this form the equation is often referred to as a *regression equation.* However for a general understanding of linear least squares it is best to stay for the time being with the fundamental model of Equation 10.1.

†For clarity the examples in this chapter use much smaller sets of data than would usually be encountered.

Table 10.1. Initial Rate of Impurity Investigation

Observed Run Number	Order in Which Experiments Were Performed	Concentration of Monomer, x_0	Concentration of Dimmer, x_1	Initial Rate of Formation of Impurity y
1	3	0.34	0.73	5.75
2	6	0.34	0.73	4.79
3	1	0.58	0.69	5.44
4	4	1.26	0.97	9.09
5	2	1.26	0.97	8.59
6	5	1.82	0.46	5.09

Quantities needed in subsequent calculations: $\sum x_0^2 = 7.0552$, $\sum x_1^2 = 3.6353$, $\sum x_0x_1 = 4.1782$, $\sum x_0y = 38.2794$, $\sum x_1y = 30.9388$, $\sum y^2 = 267.9245$

You will see that for any particular set of trial values of the parameters β_0 and β_1 you could calculate $S(\boldsymbol{\beta})$. For example, for the data of Table 10.1, if you set $\beta_0 = 1$ and $\beta_1 = 7$, you would get

$$S(1, 7) = \sum (y - 1x_0 - 7x_1)^2 = 1.9022$$

Thus in principle you could obtain the minimum value of S by repeated calculation for a grid of trial values. If you did this, you would eventually be able to construct Figure 10.1, a 3D plot of the sum of squares surface of $S(\boldsymbol{\beta})$ versus β_0 and β_1. The coordinates of the minimum value of this surface are the desired least squares estimates

$$b_0 = 1.21 \qquad \text{and} \qquad b_1 = 7.12 \tag{10.4}$$

Thus the *fitted* least squares equation

$$\hat{y} = b_0x_0 + b_1x_1$$

is

$$\hat{y} = 1.21x_0 + 7.12x_1$$

with $S_{\min} = 1.33$. This fitted equation is represented by the solid diagram of Figure 10.2. Contours of the fitted equation in relation to the data are shown in Figure 10.3. Contours of the plane, shown in Figure 10.3, are obtained by substitution in the prediction equation. For example, setting $\hat{y} = 6$ produces the equation of the contour line $1.21x_0 + 7.12x_1 = 6$ in the x_0, x_1 coordinate system. All values of x_0 and x_1 that satisfy this equation provide the contour line for $\hat{y} = 6$.

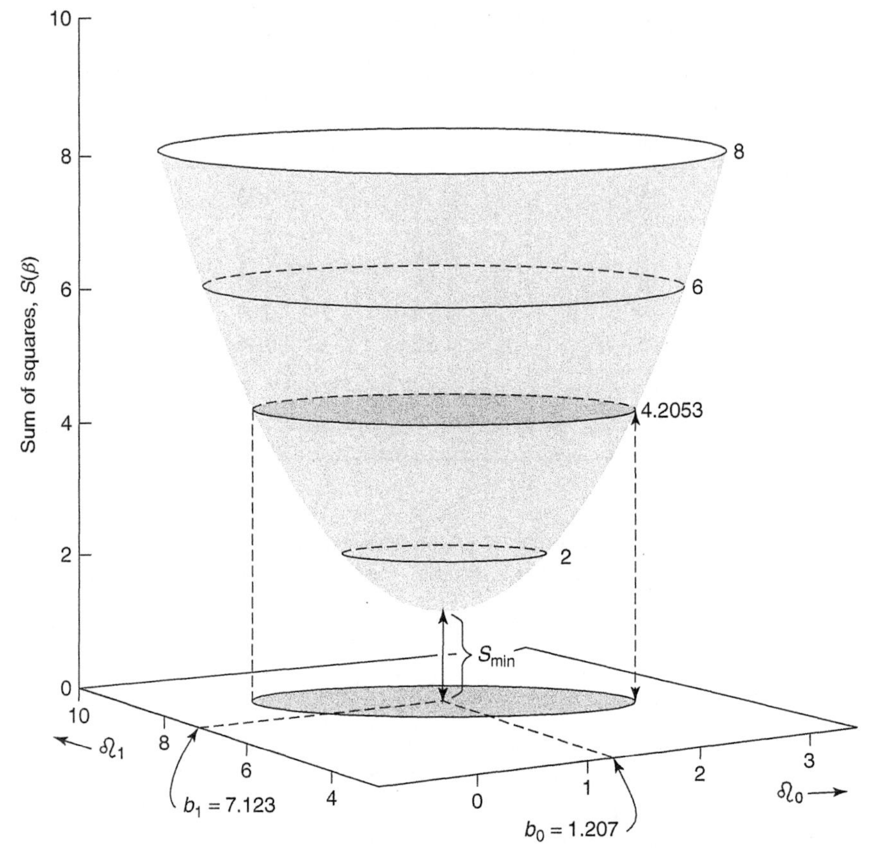

Figure 10.1. Sum of squares surface: $S(\beta) = \sum(y - \beta_0 x_0 - \beta_1 x_1)^2$.

The Normal Equations

You would not of course need to carry out this tedious calculation because the minimum can be found by setting the first derivatives of $S(\boldsymbol{\beta})$ with respect to β_0 and β_1 equal to zero. This gives the two simultaneous equations

$$\frac{\partial S(\beta)}{\partial \beta_0} = -2\left(\sum y - \beta_0 x_0 - \beta_1 x_1\right) x_0 = 0$$

$$\frac{\partial S(\beta)}{\partial \beta_1} = -2\left(\sum y - \beta_0 x_0 - \beta_1 x_1\right) x_1 = 0$$

After simplification these become what are called the *normal* equations

$$\begin{aligned} b_0 \sum x_0^2 + b_1 \sum x_0 x_1 &= \sum y x_0 \\ b_0 \sum x_0 x_1 + b_1 \sum x_1^2 &= \sum y x_1 \end{aligned} \tag{10.5}$$

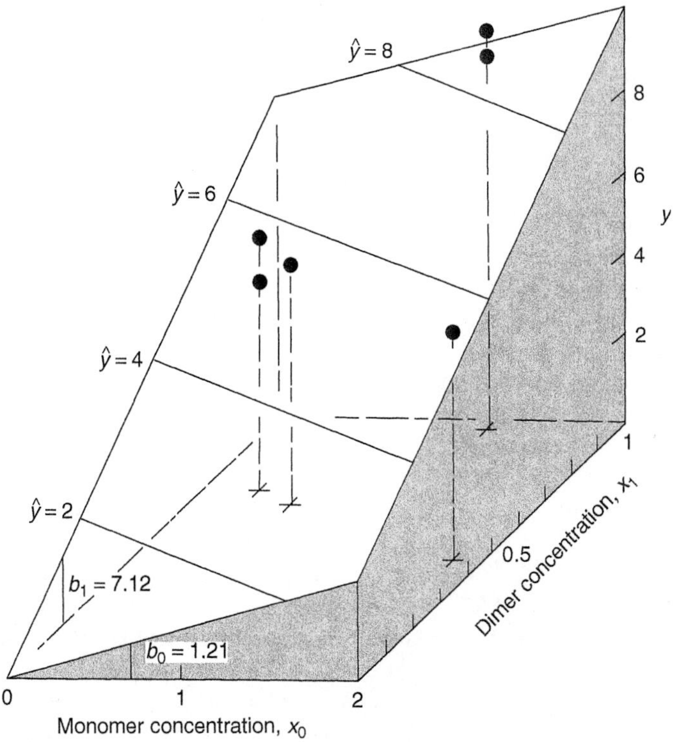

Figure 10.2. Data shown by dots with least squares fitted plane, $\hat{y} = 1.21x_0 + 7.12x_1$, passing through the origin, representing the formation of the impurity y for values of x_0 and x_1.

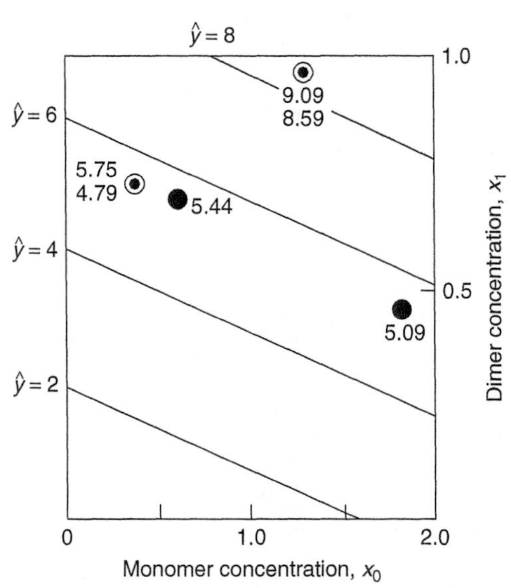

Figure 10.3. Contour representation: impurity example.

Substituting the sums of squares and cross products shown in Table10.1, you obtain

$$7.055b_0 + 4.178b_1 = 38.279$$

$$4.178b_0 + 3.635b_2 = 30.939$$

Solving this pair of linear simultaneous equations for b_0 and b_1 gives the least squares estimates $b_0 = 1.21$ and $b_1 = 7.12$ as before. It is best to carry a large number of digits in solving the normal equations, which are sometimes "ill-conditioned," meaning that the solutions are greatly affected by rounding errors. Ill conditioning and the possibility of adequately checking the form of the model depend very much on where the observations are located (on the experimental design). In this example the experimenter might feel uneasy at the absence of an observation close to the origin.

Residuals

There is of course no guarantee that the tentative model you have chosen to fit to the impurity data can provide an adequate approximation to the true response. It must be checked. One way to do this is by careful examination of *residuals*, that is, the differences $y - \hat{y}$ between the value of y actually obtained and the value $\hat{y}$ predicted by the fitted model. The idea is much as when you buy yourself a suit. The salesperson will ask you to try on a suit she thinks is about the right size. Her choice of size depends of course on her experience and prior knowledge. She will then look at the "residuals," the difference between your dimensions and those of the suit. Are the sleeves too short, the trousers (or skirt) too long? By considering such questions, she will know whether her first choice was right and, if not, how she should modify the suit. In this analogy, you are the data and the suit is the model. The model can be modified, but the data are fixed (unless you go on a diet.) To illustrate, the impurity data residuals are shown in the last column of Table 10.2.

Table 10.2. Calculation of Estimates $\hat{y}$ and Residuals $y - \hat{y}$ for the Impurity Data

Run Number	x_0	x_1	$1.207x_0$	$7.123x_1$	Observed, y	Predicted, $\hat{y}$	Residuals, $y - \hat{y}$
1	0.34	0.73	0.410	5.200	5.75	5.61	0.14
2	0.34	0.73	0.410	5.200	4.79	5.61	−0.82
3	0.58	0.69	0.700	4.915	5.44	5.62	−0.18
4	1.26	0.97	1.521	6.909	9.09	8.43	0.66
5	1.26	0.97	1.521	6.909	8.59	8.43	0.16
6	1.82	0.46	2.197	3.277	5.09	5.47	−0.38
		Sums of squares		$S_T = 267.92$		$S_M = 266.59$	$S_R = 1.33$

With so few data only gross inadequacies of the fitted model would be detectable and none such appear, but if you had more data, you could run checks on residuals similar to those discussed in Chapter 4. Most important is whether the fitted model is useful for the purpose you have in mind. As has been mentioned, if you wanted to use it for points close to the origin then you should first make one or two further experiments in the appropriate region.

Analysis of Variance and a Check of Model Fit

At the bottom of Table 10.2 are shown the sums of squares for the observed values y, the estimated values $\hat{y}$, and the residuals $y - \hat{y}$. The sum of squares of the residuals $\sum(y - \hat{y})^2$ obtained from the least squares fit and the sum of squares of the fitted values $\sum \hat{y}^2$ have the following important additive property:

$$\sum y^2 = \sum \hat{y}^2 + \sum (y - \hat{y})^2 \tag{10.6}$$

which may be written $S_T = S_M + S_R$. For these particular data Equation 10.6 becomes $267.92 = 266.59 + 1.33$, and the components are shown in the form of an ANOVA in Table 10.3.* When some of the experimental runs have been genuinely replicated† as are runs 1 and 2 and runs 4 and 5, we can make a further analysis, checking the adequacy of the model. To do this, you extract from the residual sum of squares that portion associated with replicated runs called the "error" sum of squares S_E. Thus $S_E = (5.75 - 4.79)^2/2 + (9.09 - 8.59)^2/2 = 0.59$ and has two degrees of freedom. The remaining part of the residual sum of squares, called the "lack-of-fit" sum of squares S_L, is therefore

$$S_L = S_R - S_E = 1.33 - 0.59 = 0.74$$

and measures any contribution from possible inadequacy of the model. A check of lack of fit can be made in an extended ANOVA as shown in Table 10.4.

Estimating Variances, Standard Errors, and Confidence Intervals

In this example the mean squares for error and for lack of fit are not very different, and we conclude that the data support the adequacy of the model. (Although with

Table 10.3. The ANOVA Table for Least Squares

Source	Sum of Squares	Degrees of Freedom
Model	$S_M = 266.59$	2
Residual	$S_R = 1.33$	4
Total	$S_T = 267.92$	6

*Remember that the distribution of derived criteria such as F and t referred to later are based on the assumption that errors are distributed normally and independently with constant variance.

†Recall what we mean by "genuine replicates"; these are repetitions which are subject to all the sources of error that affect unreplicated observations.

Table 10.4. The Lack-of-Fit Test

Source	Sum of Squares	Degrees of Freedom	Mean Square
Model	$S_M = 266.59$	2	
Residual	$S_R = 1.33 \begin{cases} \text{lack of fit } S_L = 0.74 \\ \text{pure error } S_E = 0.59 \end{cases}$	$4 \begin{cases} 2 \\ 2 \end{cases}$	$m_R = 0.33 \begin{cases} m_L = 0.37 \\ m_E = 0.30 \end{cases}$

so few data such a test would, of course, be very insensitive.) On the assumption then that the model provides an adequate fit, the residual mean square may be used as an estimate of the error variance σ^2. If there is lack of fit, this estimate will give an inflated estimate of σ^2.

For this two-parameter model, with $m_R = s^2 = 0.33$ supplying an estimate of σ^2, the variances of b_0 and b_1 may be written as

$$\hat{V}(b_0) = \frac{1}{(1-\rho^2)} \frac{s^2}{\sum x_0^2} = \frac{1}{0.3193} \frac{0.3309}{7.055} = 0.1469$$
$$\hat{V}(b_1) = \frac{1}{(1-\rho^2)} \frac{s^2}{\sum x_1^2} = \frac{1}{0.3193} \frac{0.3309}{3.635} = 0.2851 \qquad (10.7)$$

where ρ, which measures the correlation* between b_0 and b_1, is

$$\rho = \frac{-\sum x_0 x_1}{\sqrt{\sum x_0^2 \sum x_1^2}} = -0.8250^*$$

In this example, the standard errors of b_0 and b_1 are thus

$$\text{SE}(b_0) = 0.383, \qquad \text{SE}(b_1) = 0.534$$

The best fitting equation is often conveniently written with standard errors attached as follows:

$$\hat{y} = \underset{\pm 0.38}{1.21 x_0} + \underset{\pm 0.53}{7.12 x_1}$$

Individual 90% confidence limits for the regression coefficients β_0 and β_1 are therefore

$$\text{For } \beta_0: \quad 1.21 \pm (2.132 \times 0.38) = 1.21 \pm 0.83$$
$$\text{For } \beta_1: \quad 7.12 \pm (2.132 \times 0.53) = 7.12 \pm 1.13$$

*Notice that this is the *negative* of the correlation between x_0 and x_1. As is discussed in greater detail later, if $\sum x_0 x_1 = 0$, then $\rho = 0$ and the x's are said to be orthogonal. In that case, as you will see, considerable simplification occurs.

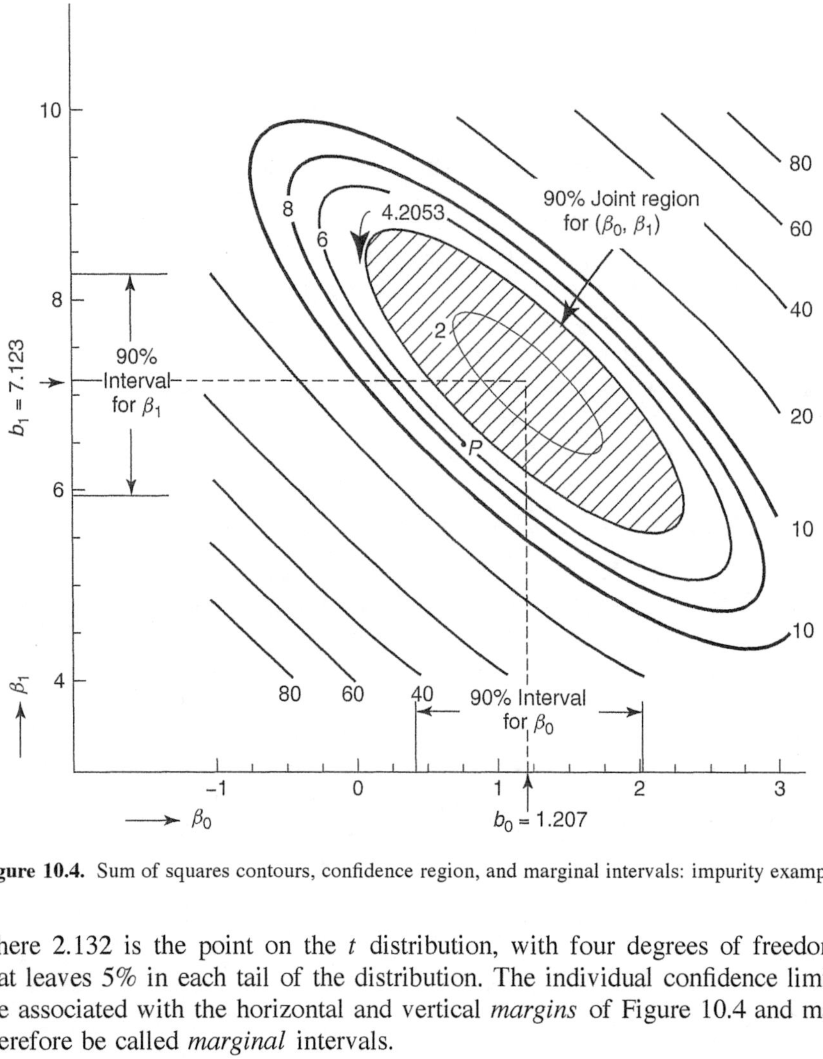

Figure 10.4. Sum of squares contours, confidence region, and marginal intervals: impurity example.

where 2.132 is the point on the t distribution, with four degrees of freedom, that leaves 5% in each tail of the distribution. The individual confidence limits are associated with the horizontal and vertical *margins* of Figure 10.4 and may therefore be called *marginal* intervals.

A Joint Confidence Region for β_0 and β_1

Confidence regions for different probabilities are given by areas (volumes) contained within specific sum of squares contours. On the assumption that the model form is adequate, a $1 - \alpha$ joint confidence region for p fitted parameters is bounded by a sum of squares contour S such that

$$S = S_R \left[1 + \frac{p}{n-p} F_\alpha(p, n-p) \right] \tag{10.8}$$

where S_R is the sum of squares of the residuals and $F_\alpha(p, n-p)$ is the α significance point of the F distribution with p and $n - p$ degrees of freedom.

In the present case, for a 90% confidence region we would have $S_R = 1.3308$, $p = 2$, $n - p = 6 - 2 = 4$, and $F_{0.10}(2, 4) = 4.32$. Thus

$$S = 1.3308[1 + (2/4) \times 4.32] = 4.2053$$

The boundary of this confidence region is thus the sum of squares contour for β_0 and β_1 given by $S(\beta) = \sum(y - \beta_0 x_0 + \beta_1 x_1)^2 = 4.2053$. Expanding this quadratic expression and substituting the sums of squares and cross products shown in Table 10.1 supplies the required 90% confidence region—the elliptical contour enclosing the shaded region in Figure 10.4. Its oblique orientation implies that the estimates are (negatively) correlated. The formal interpretation of this confidence region is that, if you assume the exactness of the model and imagine the experiment repeated many times and regions calculated in the above manner each time, 90% of these regions will include the true parameter point (β_0, β_1) and 10% will exclude it.

Individual (Marginal) Limits and Joint Confidence Regions

You may ask what the relation is between the shaded joint confidence *region* in Figure 10.4 and the individual marginal confidence *limits* also shown in the figure. Consider the point P with coordinates $\beta_0 = 0.75$ and $\beta_1 = 6.50$. The value $\beta_0 = 0.75$ lies well within the marginal 90% confidence limits for β_0, and the value $\beta_1 = 6.50$ lies well within the marginal confidence limits for β_1, but the point P does *not* lie within the *joint* confidence region. This means, for example, that although the value $\beta_0 = 0.75$ is not discredited by the data, if *any* value of the other parameter β_1 is allowed, the hypothesized *pair* of values $(\beta_0, \beta_1) = (0.75, 6.50)$ *is* discredited by the data.

Fitting a Straight Line

In the first two-parameter example discussed above x_0 and x_1 were the concentrations of two chemicals. If you make x_0 a column of 1's, the theoretical model $y = \beta_0 x_0 + \beta_1 x_1 + e$ becomes

$$y = \beta_0 \times 1 + \beta_1 x_1 + e$$

or simply

$$y = \beta_0 + \beta_1 x_1 + e \tag{10.9}$$

and the fitted least squares model is

$$\hat{y} = b_0 + b_1 x_1$$

This represents a straight line with b_0 its intercept on the y axis (when $x_I = 0$) and b_1 its slope (see Figure 10.5). The "variable" $x_0 = 1$ is introduced to indicate the presence of the constant β_0 and is called an "indicator" variable.

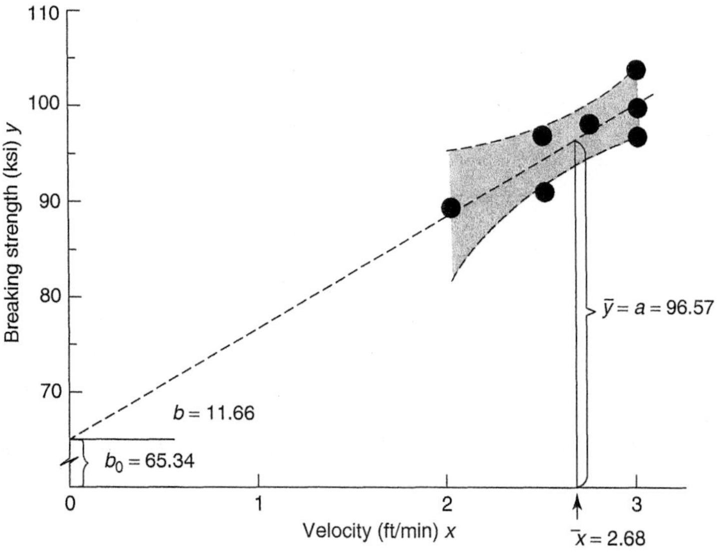

Figure 10.5. Least squares fitted straight line: welding data example.

Table 10.5. Welding Experiment Data with Indicator Variable $x_0 = 1$

Run Number	Indicator Variable, $x_0(=1)$	Velocity, $x_1(=x)$	Observed Strength, y	Estimated from Model, $\hat{y}$	Residual, $y - \hat{y}$
1	1	2.00	89	88.66	0.34
2	1	2.50	97	94.49	2.51
3	1	2.50	91	94.49	−3.49
4	1	2.75	98	97.40	0.60
5	1	3.00	100	100.32	−0.32
6	1	3.00	104	100.32	3.68
7	1	3.00	97	100.32	−3.32

A Welding Experiment

The data listed in Table 10.5 and plotted in Figure 10.5 are from an inertia welding experiment. One part of a rotating workpiece was brought to a standstill by forcing it into contact with another part. The heat generated by friction at the interface produced a hot-pressure weld. Over the experimental region studied the breaking strength of the weld was expected to have a nearly linear relationship to the velocity x of the rotating workpiece.

Your computer will tell you that the fitted least squares straight line is

$$\hat{y} = b_0 + bx = 65.34 + 11.66x \tag{10.10}$$

(Here computed values have been rounded to two places.) Also from Figure 10.5 you see that $b_0 = \overline{y} - b\overline{x}$ so an equivalent form of the fitted model is

$$\hat{y} = \overline{y} - b\overline{x} + bx \qquad \text{or} \qquad \hat{y} = \overline{y} + b(x - \overline{x})$$

that is

$$\hat{y} = 96.57 + 11.66(x - 2.68)$$

where $96.57 = \overline{y}$ is the "intercept" at $x = \overline{x}$ and $b = 11.66$ is the slope of the line. (Every straight line fitted by least squares passes through the point $\overline{y}$, $\overline{x}$). Referring the fitted line to the origin $\overline{x}$ is useful because it serves to reminds us that the fitted line is only expected to apply *over the range of the x actually used*, which is centered at $\overline{x}$ and is shaded in Figures 10.5 and 10.6.

The value $b_0 = 65.34$, the intercept at $x_1 = 0$, can only be regarded as a construction point for the purpose of drawing the line. In general, then, the model for the straight line is most conveniently written with the average $\overline{x}$ clearly in mind as

$$y = \alpha + \beta(x - \overline{x}) + e$$

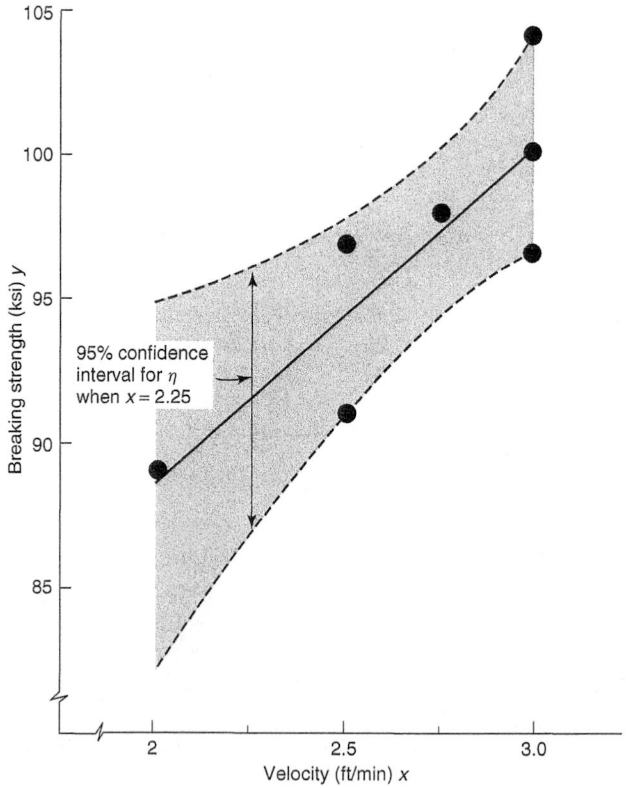

Figure 10.6. Confidence limits for $\hat{y} = b_0 + bx = 65.34 + 11.66x$.

and the fitted equation as

$$\hat{y} = a + b(x - \overline{x}) \tag{10.11}$$

where $a = \overline{y}$ is the value of the response when x is at its average value.

If it could be assumed that the model provided an adequate approximation over the range of the data, an estimate of σ^2 would be

$$s^2 = \frac{S_R}{n-2} = \frac{43.6170}{5} = 8.7234$$

On the same assumption, the standard errors for a, b_0, and b would be

$$\text{SE}(a) = \frac{s}{\sqrt{n}} = 1.12$$

$$\text{SE}(b_0) = s\left[\frac{1}{n} + \frac{\overline{x}^2}{\sum(x-\overline{x})^2}\right]^{1/2} = 8.71 \tag{10.12}$$

$$\text{SE}(b) = \frac{s}{\sqrt{\sum(x-\overline{x})^2}} = 3.22$$

Exercise 10.1. Construct an ANOVA table and test the goodness of fit of the straight line using the data in Table 10.5. (*Hint:* Refer to Tables 10.4 and 10.5.) The ratio of the lack of fit and pure error mean squares is $0.48/14.22 = 0.03$. Lack of fit would be indicated by a significantly *large* value of this ratio.

The observed value is actually unusually *small*. Can you think of reasons that might explain this fact?

Variance of an Estimated Value and Confidence Intervals

Now look at Figure 10.6. If the model is assumed adequate, we can calculate $\hat{y}$, the estimated breaking strength of the weld, for any given velocity x in the relevant range. Also using the formula $\hat{y} = \overline{y} + b(x_0 - \overline{x})$, since $\overline{y}$ and b are uncorrelated, the variance of the prediction $V(\hat{y}_X)$ at some specified value X of x is

$$V(\hat{y}_X) = V(\overline{y}) + (X - \overline{x})^2 V(b)$$

Thus

$$\hat{V}(\hat{y}_X) = \left[\frac{1}{n} + \frac{(X-\overline{x})^2}{\sum(x-\overline{x})^2}\right] s^2 \tag{10.13}$$

For the welding example

$$\hat{V}(\hat{y}_X) = \left[\frac{1}{7} + \frac{(X - 2.679)^2}{0.8393}\right] 8.7234$$

A $1-\alpha$ confidence interval for the response at some particular X is now given by

$$\hat{y}_X \pm t_{\alpha/2}\sqrt{\hat{V}(\hat{y}_X)} \tag{10.14}$$

where t has five degrees of freedom. Confidence bands calculated by this formula are shown by dashed lines in Figure 10.6. Such limits are, of course, meaningful only over the range (shaded) of the data itself.

The Geometry of Least Squares and Its Extension to Any Number of Variables

A deeper understanding of least squares can be gained from its geometric representation. In general $\mathbf{y}$, $\hat{\mathbf{y}}$, $\mathbf{y}-\hat{\mathbf{y}}$, $\mathbf{x}_0$, and $\mathbf{x}_1$ can represent *vectors* in the n-dimensional space of the n observations. However, basic ideas can be understood by studying the case where $n=3$, that is, in the 3D space you can actually visualize. (If you need to brush up on vectors and coordinate geometry, there is a brief review in Appendix 10A.)

For two parameters and three observations the fitted model can be written in vector form as

$$\hat{\mathbf{y}} = b_0\mathbf{x}_0 + b_1\mathbf{x}_1$$

$$\begin{bmatrix}\hat{y}_1\\ \hat{y}_2\\ \hat{y}_3\end{bmatrix} = b_0\begin{bmatrix}x_{01}\\ x_{02}\\ x_{03}\end{bmatrix} + b_1\begin{bmatrix}x_{11}\\ x_{12}\\ x_{13}\end{bmatrix}$$

In Figure 10.7, $\mathbf{x}_0$ and $\mathbf{x}_1$ are vectors of specific lengths and directions which define the coordinates of a plane. The above equation says that $\hat{\mathbf{y}}$ must lie in this plane. Notice that the two vectors define a plane whose coordinates are *not* in general orthogonal (at right angles) to one another. Thus what delineates the plane in Figure 10.7 are *not* the squares of ordinary (orthogonal) graph paper but parallelograms.

Since the squared length of a vector is the sum of squares of its elements, the principle of least squares says the best value for $\hat{\mathbf{y}}$ is the point in the plane which is closest to the point $\mathbf{y}$ representing the actual data. Thus $\hat{\mathbf{y}}$ is obtained by dropping a perpendicular from $\mathbf{y}$ onto the plane formed by $\mathbf{x}_0$ and $\mathbf{x}_1$. Thus looking at Figure 10.7 you will see that the least squares estimates b_0 and b_1 in this example are the coordinates 2.2 and 0.8 of $\hat{\mathbf{y}}$ measured on the plane in units of $\mathbf{x}_0$ and $\mathbf{x}_1$ and the fitted least squares equation is $\hat{y} = 2.2x_0 + 0.8x_1$. Also since the vectors $\mathbf{y}$, $\hat{\mathbf{y}}$, and $\mathbf{y}-\hat{\mathbf{y}}$ form a right triangle,

$$\sum y^2 = \sum \hat{y}^2 + \sum (y-\hat{y})^2$$

which is the basic equation in the ANOVA.

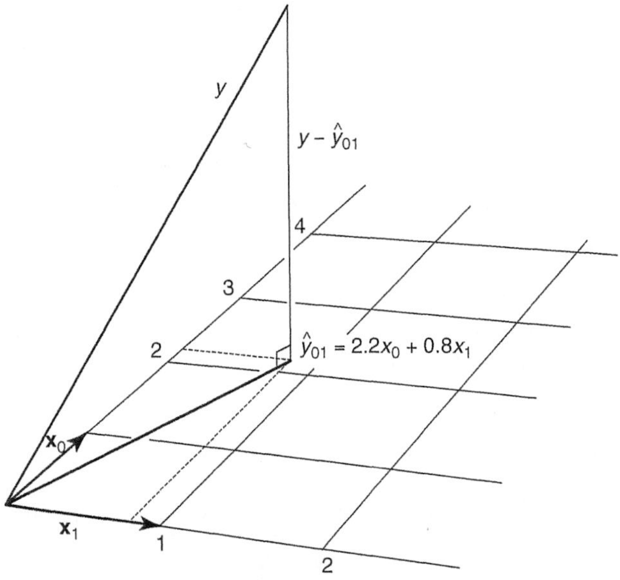

Figure 10.7. Geometry of two-parameter least squares.

The Normal Equations

The reason for the name normal equations now becomes clear. A vector normal to a plane is normal to every line in the plane. [When two n-dimensional vectors, say $\mathbf{p}$ and $\mathbf{q}$, are at right angles (normal, perpendicular) to one another, the "inner product" of their elements must equal zero, that is, $\sum_{i=1}^{n} p_i q_i = 0$.] Thus the inner products of the vector $\mathbf{y} - \hat{\mathbf{y}}$ with both $\mathbf{x}_0$ and $\mathbf{x}_1$ must be zero.* The normal equations thus express this fact that $\mathbf{y} - \hat{\mathbf{y}}$ is normal to $\mathbf{x}_0$ and $\mathbf{x}_1$ and $\sum (y - \hat{y})x_0 = 0$ and $\sum (y - \hat{y})x_1 = 0$.
That is,

$$\sum (y - b_0 x_0 - b_1 x_1) x_0 = 0, \qquad \sum (y - b_0 x_0 - b_1 x_1) x_1 = 0$$

Thus, as before in Equation 10.5,

$$\sum y x_0 = b_0 \sum x_0^2 + b_1 \sum x_0 x_1 \qquad \sum y x_1 = b_0 \sum x_0 x_1 + b_1 \sum x_1^2$$

A convenient notation is to write $[ij]$ for $\sum x_i x_j$ and $[yi]$ for $\sum y x_i$ so that the normal equations become

$$[y0] = b_0[00] + b_1[01] \qquad [y1] = b_0[01] + b_1[11]$$

*In general matrix notation $(\mathbf{y} - \hat{\mathbf{y}})^{\mathrm{T}} \mathbf{x}_0 = 0$ and $(\mathbf{y} - \hat{\mathbf{y}})^{\mathrm{T}} \mathbf{x}_1 = 0$.

These equations may be generalized to encompass any k-parameter linear model. Thus for $k = 3$ the fitted model is $\hat{y} = b_0x_0 + b_1x_1 + b_2x_2$ and the normal equations are

$$[00]b_0 + [01]b_1 + b_2[02] = [y0]$$
$$[01]b_0 + [11]b_1 + b_2[12] = [y1]$$
$$[02]b_0 + [12]b_1 + b_2[22] = [y2]$$

Your computer will fit least squares models so you do not need to know all the details of the calculations. For those who are interested, a matrix summary of the mathematics is given in Appendix 10B.

10.2. THE VERSATILITY OF LEAST SQUARES

Least Squares for Estimating Any Model Linear in Its Parameters

Least squares estimates may be obtained for any model *linear in the unknown parameters* $\beta_0, \beta_1, \ldots$. Such a model can in general be written

$$y = \beta_0 + \beta_1x_1 + \beta_2x_2 + \cdots + \beta_kx_k + \varepsilon$$

As mentioned earlier, this model is written in the equivalent and sometimes more convenient form in which the variables are the deviations from their average values; thus

$$y = \alpha + \beta_1(x_1 - \overline{x}_1) + \beta_2(x_2 - \overline{x}_2) + \cdots + \beta_k(x_k - \overline{x}_k) + \varepsilon$$

and $\alpha = \overline{y}$. To fit these models by ordinary least squares, it only needs to be assumed that the x's are *quantities known* for each experimental run and are not functions of the *unknown* β *parameters.* In particular, the derivative of y with respect to any of the unknown β's, $dy/d\beta$, must be a function with all β's absent. Some examples of models linear in their parameters which can be fitted by least squares are as follows:

1. Polynomial models of the form

$$y = \beta_0 + \beta_1x + \beta_2x^2 + \beta_3x^3 + \cdots + \beta_kx^k + \varepsilon$$

This is of the same form as the general model with $x_0 = 1, x_1 = x, x_2 = x^2$, and so on.

2. Sinusoidal models such as

$$y = \beta_0 + \beta_1 \sin\theta + \beta_2 \cos\theta + \varepsilon$$

where $x_0 = 1$, $x_1 = \sin\theta$, $x_2 = \cos\theta$.

3. Models such as

$$y = \beta_0 + \beta_1 \log \xi_1 + \beta_2 \frac{e^{\xi_2}}{\xi_3} + \varepsilon$$

where $x_0 = 1$, $x_1 = \log \xi_1$, $x_2 = e^{\xi_2}/\xi_3$, and ξ_1, ξ_2, and ξ_3 are supposed known for each experimental run.

An example of a *nonlinear* model (i.e., nonlinear in its parameters) is

$$y = \beta_0(1 - e^{-\beta_1 \xi}) + \varepsilon$$

where ξ is known for each experimental run. This exponential model is *not* a linear model since it cannot be written in the form of Equation 10.1 and its derivatives with respect to the unknown β's contain the unknown parameters, for example $dy/d\beta_0 = 1 - e^{-\beta_1 \xi}$. However, the idea of least squares still applies and the model may be fitted by numerical search to find the minimum residual sum of squares.

An illustration (a biochemical oxygen demand example) and a short discussion of the fitting of a nonlinear model are given later The subject is, however, a specialized one and interested readers are referred, for example, to Bates and Watts (1988).

More on Indicator Variables

In the model

$$y = \beta_0 x_0 + \beta_1 x_1 + \beta_2 x_2 + \cdots + \beta_k x_k + e$$

as you have seen, the variables (factors) $x_0, x_1, \ldots, x_k$ are simply quantities that are known *for each observation.* They need not be measured variables. Indicator variables taking the values 0 or 1 can, for example, indicate the presence or absence of some factor.

The Welding Experiment with Two Suppliers

For illustration consider again the welding experiment for the data shown in Table 10.5 but suppose you have obtained workpieces from two different suppliers and you wished to find out whether the material from each supplier gave equally strong welds. For example, suppose that runs 1, 3, and 6 were made with product from supplier I and runs 2, 4, 5, and 7 from supplier II. Then the data could be characterized as in Table 10.6a. Now consider the three-parameter model

$$\hat{y} = b_{01} x_{01} + b_{02} x_{02} + bx$$

If you write out the rows of Table 10.6a line by line, you will see that the model says that for runs made with supplier I material the estimated intercept constant is b_{01} but for runs made with supplier II material it is b_{02}. Thus $b_{02} - b_{01}$ represents

Table 10.6. Welding Data with Two Indicator Variables x_{01} and x_{02}

	(a)				(b)		
Run Number	Indicator Supplier I, x_{01}	Indicator Supplier II, x_{02}	Velocity, x	Observed Strength, y	x_{01}	x_{02}	x
1	1	0	2.00	89	1	0	2.00
2	0	1	2.50	97	1	1	2.50
3	1	0	2.50	91	1	0	2.50
4	0	1	2.75	98	1	1	2.75
5	0	1	3.00	100	1	1	3.00
6	1	0	3.00	104	1	0	3.00
7	0	1	3.00	97	1	1	3.00

the *difference* of weld strengths associated with the two suppliers. The data can be analyzed as an ordinary regression to give least squares estimates and their *standard errors* as follows:

$$b_{01} = 64.9 \pm 10.2, \qquad b_{02} = 64.5 \pm 11.4, \qquad b = 11.9 \pm 4.0$$

The variance of the observations is estimated by $s^2 = \sum(y - \hat{y})^2/4 = 43.41/4 = 10.85$

An Alternative Form for the Model

There is usually more than one choice for expressing the indicator (dummy) variables. Thus you might use the alternative arrangement shown in Table 10.6b with x_{01} a series of 1's and x_{02} having a 1 only for material from supplier II. In this model b_{02} now represents the added contribution of supplier II, that is, the *difference* in weld strength for the two supplier materials I and II. The analyses are equivalent but the second allows for the estimation of the difference and its standard error directly. The estimates and standard errors are $b_{01} = 64.9 \pm 10.2$, $b_{02} = -0.4 \pm 2.8$, $b = 11.9 \pm 4.0$. In this form of the model the estimated difference b_{02} for suppliers is -0.4 ± 2.8, and you can see at once there is no reason to prefer one supplier over the other.

Exercise 10.2. Redo the analysis for the welding example supposing that a third supplier contributed work pieces for runs 6 and 7.

Exercise 10.3. Suppose that two operators A and B work at the job of packing boxes. You have records of the number of boxes packed on the days when A is working, indicated by $x_0 = 1$, when B is working, indicated by $x_1 = 1$, and when A and B are working together, indicated by $x_2 = 1$:

x_0	x_1	x_2	Number Packed
1	0	0	32
1	0	0	35
1	1	1	67
1	1	1	69
1	1	1	72
0	1	0	29
0	1	0	25
0	1	0	31

Estimate the number of boxes packed for *A*, for *B*, and for *A* and *B* together. Do you think that there is any evidence that when *A* and *B* work together they individually pack more boxes or fewer?

Does It Fit? Does It matter? Growth Rate Example

The data in Table 10.7 is for the growth rate of rats fed various doses of a dietary supplement. From similar investigations it was believed that over the range tested the response would be approximately linear. Once the data were in hand, a plot such as that in Figure 10.8 would have shown that a straight line was clearly an inadequate model. Nevertheless, suppose that on routinely employing a standard software program a straight line was fitted to the data. This would give

$$\hat{y} = 86.44 - 0.20x$$

Inadequacy of Straight Line Model

A method for checking model adequacy using the ANOVA was given earlier in Tables 10.3 and 10.4. If you apply this method to the growth rate data,

Table 10.7. Growth Rate Data

Observation Number	Amount of Supplement, x (g)	Growth Rate, y (coded units)
1	10	73
2	10	78
3	15	85
4	20	90
5	20	91
6	25	87
7	25	86
8	25	91
9	30	75
10	35	65

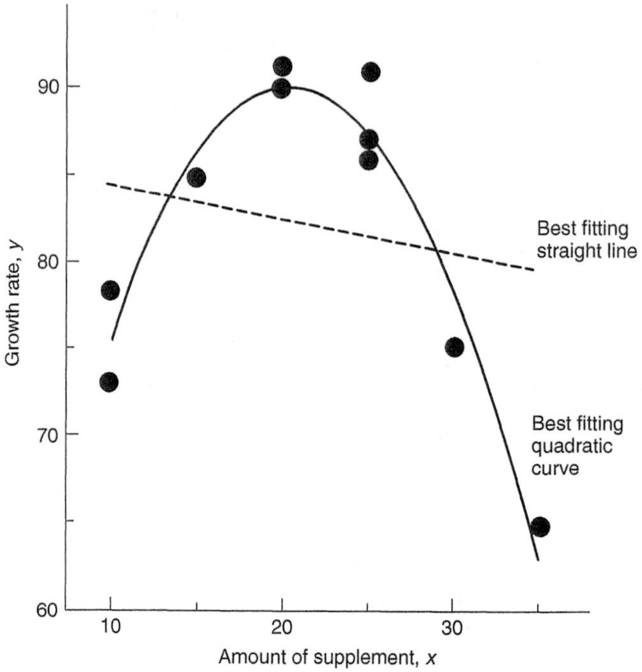

Figure 10.8. Plot of data, best fitting straight line, and best fitting quadratic curve: growth rate example.

Table 10.8. Analysis of Variance for Growth Rate Data: Straight Line Model

Source	Sum of Squares	Degrees of Freedom	Mean Square
Straight line model	$S_M = 67,428.6 \left\{ \begin{array}{l} \text{mean}(b_0) = 67,404.1 \\ \text{extra for } b = \quad 24.5 \end{array} \right\}$	$2 \left\{ \begin{array}{l} 1 \\ 1 \end{array} \right.$	$\left\{ \begin{array}{l} 67,404.1 \\ \quad 24.5 \end{array} \right.$
Residual	$S_R = \quad 686.4 \left\{ \begin{array}{l} \text{lack of fit } S_L = 659.4 \\ \text{pure error } S_E = \quad 27.0 \end{array} \right\}$	$8 \left\{ \begin{array}{l} 4 \\ 4 \end{array} \right.$	$\left\{ \begin{array}{l} 164.85 \\ \quad 6.75 \end{array} \right\}$ Ratio = 24.4
Total	$S_T = 68,115.0$		

you obtain the ANOVA for the fitted straight line model $\hat{y} = b_0 + bx$ shown in Table 10.8. The model sum of squares $S_M = \sum \hat{y}^2$ has been split into two additive parts: a sum of squares $N\bar{y}^2 = 67,404.1$ associated with the simple model $\hat{y} = b_0$ representing a horizontal line and an *extra* sum of squares for $b = S_M - N\bar{y}^2 = 67,428.6 - 67,404.1 = 24.5$ due to the linear (slope) term in the model. Noting that runs 1 and 2, 4 and 5, and 6, 7, 8 are replicates, a sum of squares S_E 27.0 having 4 degrees of freedom may be calculated.

The value 24.42 for the mean square ratio, lack of fit/pure error, is large and significant at the 0.005 level. From the data plot in Figure 10.8 it is obvious that the inadequacy arises from the curvilinear nature of the data. Increasing amounts of the additive up to about 20 grams produces increasing growth rate, but more additive reduces growth rate.

A better approximation might be expected from a quadratic (second-order) model

$$y = \beta_0 + \beta_1 x + \beta_2 x^2 + \varepsilon$$

where β_2 is a measure of curvature. This model fitted to the data gives the equation

$$\hat{y} = 35.66 + 5.26x - 0.128x^2$$

The ANOVA (Table 10.9) for this quadratic model shows no evidence of lack of fit. The model sum of squares S_M is now split into *three* parts, the third being the extra sum of squares obtained by including the quadratic term $\beta_2 x^2$ in the fitted model. Its large mean square (641.2) relative to the error mean square (6.75) reflects the important contribution that this term makes in providing an adequate explanation of the data. The residuals plotted in Figure 10.9 now appear

Table 10.9. Analysis of Variance for Growth Rate Data, Quadratic (Second-Order) Model

Source	Sum of Squares			Degrees of Freedom		Mean Square	
Quadratic model	$S_M = 68{,}071.8$	mean(b_0)	= 67,404.1	3	1	67,404.1	
		extra for linear (b_1)	= 24.5		1	24.5	
		extra for quadratic (b_2)	= 641.2		1	641.2	
Residual	$S_R = 45.2$	lack of fit S_L	= 18.2	7	3	6.07	Ratio = 0.90
		pure error S_E	= 27.0		4	6.75	
Total	$S_T = 68{,}115.0$						

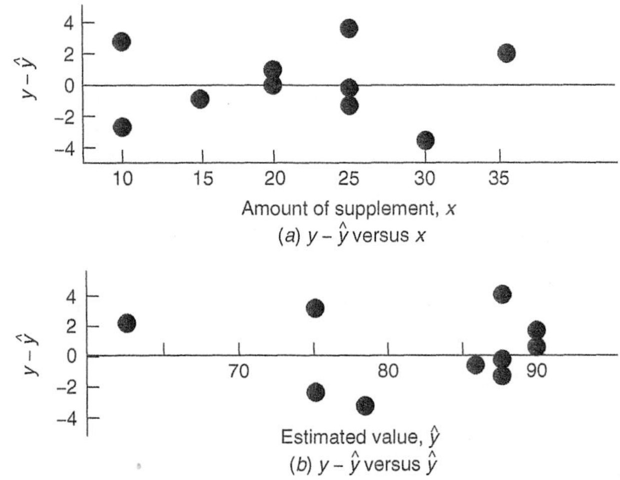

Figure 10.9. Plots of residuals from quadratic model: (*a*) $y - \hat{y}$ versus x; (*b*) $y - \hat{y}$ versus $\hat{y}$.

to display no systematic tendencies in relation to either x or $\hat{y}$. Taken together, the plotted data, the fitted second-order curve in Figure 10.8, the residuals plot, and the ANOVA table suggest that the quadratic equation supplies an adequate representation over the region studied.

A Closer Look at Lack of Fit

Since all models are wrong but some models are useful, consider what action you might take after you have made a test of goodness of fit. On the one hand, the fact that you can detect a significant lack of fit from a particular model does not necessarily mean that the model is useless. On the other hand, the fact that you cannot detect any lack of fit does not mean that the fitted model is correct. The result you get depends largely on the size of the experimental error in relation to the bias, estimated by $\eta - \hat{y}$, the difference between the actual response of the fitted value $\hat{y}$.

Look at Figure 10.10:

Figure 10.10*a* shows the actual but unknown relationship $\eta = f(x)$.

Figure 10.10*b* shows the generation of the data $y = \eta + e$; thus the true value plus error produces the observed value y of the response. Notice that the error e is negative in the case illustrated.

Figure 10.10*c* shows the estimated response $\hat{y}$ obtained by fitting a straight line by least squares to the data with the residual $r = y - \hat{y}$.

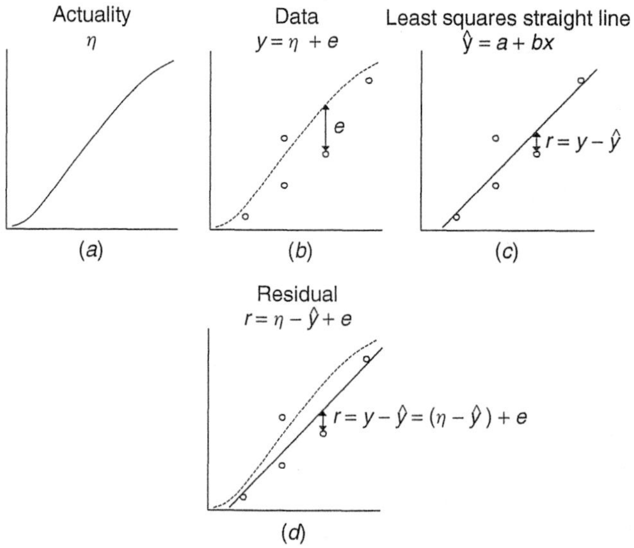

Figure 10.10. Plots of (*a*) the actual response $\eta = f(x)$, (*b*) the observed responses y, (*c*) the fitted straight line, and (*d*) the structure of the residual $r = y - \hat{y}$.

Figure 10.10*d* compares the actual and the estimated response and shows the structure of the residual.

Thus the residual r is given as

$$r = y - \hat{y} = \eta + e - \hat{y} = (\eta - \hat{y}) + e$$

the difference between the true value and the model's least squares estimate plus random error. Whether you can *detect* lack of fit depends on how large is the error e relative to the difference $(\eta - \hat{y})$ due to lack of fit. Whether you should be greatly *concerned* about lack of fit depends on the purpose to which the approximating function is to be put. If you had a calibration problem in which the experimental error e was small, it would be important that the fitted function be close to reality. But a straight line approximation such as that illustrated in Figure 10.10*c* would be good enough for some purposes whether you could detect lack of fit or not.

Orthogonality of Regressors: Experimental Design

The factorial, fractional factorial, and PB designs discussed in Chapters 5, 6, and 7 were orthogonal designs. The very profound effect of orthogonality can be understood by studying once again just two regressors x_0 and x_1 in the fitted two-parameter model $\hat{y} = b_0x_0 + b_1x_1$. When the regressors are orthogonal, $\sum x_0x_1 = 0$. Earlier in this chapter you saw that estimated variances for the coefficients b_0 and b_1 in a two-parameter model were

$$\hat{V}(b_0) = \frac{1}{(1-\rho^2)}\frac{s^2}{\sum x_0^2} \qquad \text{and} \qquad V(b_1) = \frac{1}{(1-\rho^2)}\frac{s^2}{\sum x_1^2}$$

with $\rho = -\sum x_0x_1/\sqrt{\sum x_0^2 \sum x_1^2}$ and with s^2 an estimate of the error variance σ^2. It will be seen that the greater is the correlation ρ between the x's and hence between b_0 and b_1 the larger will be the variances of the estimates. For example, if $\rho = 0.9$, then the variance of the estimates b_0 and b_1 would be five times as great as they would be if ρ were zero. Thus, other things being equal, when the sum of the cross products $\sum x_0x_1$ is zero so that vectors x_0 and x_1 are orthogonal, $\rho = 0$ and the estimates of b_0 and b_1 will have smallest possible variances and will be uncorrelated. Thus one of the reasons why orthogonal designs such as the factorials, fractional factorials, and PB designs are so valuable is that, given $\sum x_i^2$ and $\sum x_j^2$ are fixed, the variances of the estimates are minimized. Further, the normal equations

$$\sum yx_0 = b_0\sum x_0^2 + b_1\sum x_0x_1$$
$$\sum yx_1 = b_0\sum x_0x_1 + b_1\sum x_1^2$$

become

$$\sum yx_0 = b_0 \sum x_0^2 + 0$$
$$\sum yx_1 = 0 + b_1 \sum x_1^2$$

so that $b_0 = \sum yx_0 / \sum x_0^2$ and $b_1 = \sum yx_1 / \sum x_1^2$ and each estimate can be calculated as if the other were not present.

The basis for the ANOVA is the identity

$$\sum y^2 = \sum \hat{y}^2 + \sum (y - \hat{y})^2$$

and if the x's are orthogonal, the regression sum of squares is

$$\begin{aligned}\sum \hat{y}^2 &= \sum (b_0 x_0 + b_1 x_1)^2 \\ &= b_0^2 \sum x_0^2 + b_1^2 \sum x_1^2 \\ &= b_0 \sum yx_0 + b_1 \sum yx_1\end{aligned}$$

The regression sum of squares can then be split into two independent components each showing the separate contributions of x_0 and x_1.

If x_0 is set equal to 1 for all observations, then

$$\sum \hat{y}^2 = n\bar{y}^2 + b_1 \sum x_1 y$$

When in addition to x_0 there are k regressors all mutually orthogonal, the sum of squares of predictions $\sum \hat{y}^2$ can be partitioned into the $k + 1$ independent components

$$\sum \hat{y}^2 = n\bar{y}^2 + b_1 \sum x_1 y + b_2 \sum x_2 y + \cdots + b_k \sum x_k y$$

the basis for the ANOVA for orthogonal designs. You see that, in general, when the x's are orthogonal, variances are minimized and calculations and understanding are greatly simplified. Thus orthogonality is a property that is still important even in this age of computers.

Extra Sum of Squares Principle and Orthogonalization

Table 10.10a shows the percentages x_1, x_2, x_3 of three ingredients and the hardness y of 20 samples of a mineral ore. The least squares linear regression equation and associated analysis are shown in Table 10.10b. From the ANOVA

Table 10.10. Analysis for Twenty Hardness Samples

(a)

	x_1	x_2	x_3	y
1	10.2	3.5	16.0	27.6
2	9.2	3.3	15.2	14.8
3	8.3	2.5	15.5	17.1
4	7.7	2.2	15.2	13.9
5	7.5	2.4	15.5	20.0
6	8.0	1.3	14.6	26.6
7	6.9	1.9	16.0	20.3
8	9.9	3.6	15.9	19.7
9	8.6	2.3	15.6	25.7
10	9.7	3.0	16.0	23.2
11	11.1	5.0	15.2	25.8
12	11.0	4.3	16.8	33.6
13	10.1	4.2	15.5	24.2
14	10.1	3.2	16.5	28.6
15	9.5	3.3	16.0	29.3
16	8.5	3.0	16.8	26.0
17	8.6	2.6	14.9	18.0
18	10.3	3.2	16.4	39.2
19	9.7	3.4	16.0	25.9
20	6.9	1.3	14.6	25.3

(b)

$\hat{y} = 24.2$	$+\ 5.9(x_1 - 9.1)$	$-\ 6.4(x_2 - 3.0)$	$+\ 3.7(x_3 - 15.7)$
SE	5.9 ± 1.9	-6.4 ± 2.5	3.7 ± 1.8
t	3.1	−2.5	2.0

(c) ANOVA

Source	SS	df	MS	F	Probability
Regression (x_1, x_2, x_3)	393.9	3	131.3	6.34	0.005
Residual	331.6	16	20.7		
Total SS $\sum(y - \bar{y})^2$	725.6	19			

table 10.10c it is evident that this relationship allows you to make an estimate $\hat{y}$ of hardness for any combination of the three ingredients within the ranges tested.

Suppose now that the chemical analyses that were needed to obtain x_1, x_2, x_3 were successively more difficult and time consuming and the experimenter wanted to know whether x_2 and x_3 were adding information about the hardness response not already supplied by x_1. An equation containing only x_1 fitted to the data gave

$$\hat{y} = 24.2 + 2.50(x_1 - 9.1)$$

with a sum of squares for regression using x_1 alone of 191.8 and residual sum of squares of 533.7. For x_1 and x_2 together the fitted equation was

$$\hat{y} = 24.2 + 6.6(x_1 - 9.1) - 6.1(x_2 - 3.0)$$

with a regression sum of squares of 311.4, that is, 191.8 for x_1 and an additional 119.6 for x_2. The residual sum of squares is now 414.1. Table 10.10b gives the equation with all three regressors x_1, x_2, and x_3. The regressor x_3 adds 82.5 to the regression sum of squares and reduces the residual sum of squares to 331.6 as before.*

*Computer programs can readily supply for each added regressor the additional orthogonal (independent) sum of squares that accrues to the regression sum of squares.

Table 10.11. Analysis of Variance for Regression Components

			SS	df		MS		F		Probability
Regression	393.9	Due to x_1	191.8	3	1	131.3	191.8	6.34	9.26	0.008
		Extra due to x_2	119.6		1		119.6		5.77	0.029
		Extra due to x_3	82.5		1		82.5		3.98	0.063
Residual			331.6		16		20.7			
$\sum(y-\bar{y})^2$			725.6		19					

Table 10.12. Average Variance of $\hat{y}$ for Various Models

$\hat{y}_1 = b_0 + b_1x_1$		$\hat{y}_2 = b_0 + b_1x_1 + b_2x_2$		$\hat{y}_3 = b_0 + b_1x_1 + b_2x_2 + b_3x_3$	
SS x_1	191.8	SS $x_{2.1}$	119.6	SS $x_{3.21}$	82.5
$\sum(y-\hat{y}_1)^2$	533.7	$\sum(y-\hat{y}_2)^2$	414.1	$\sum(y-\hat{y}_3)^2$	331.6
df	18	df	17	df	16
MS	29.7	MS	24.4	MS	20.7
p/n	2/20	p/n	3/20	p/n	4/20
$\overline{V}(\hat{y}_1) = p\hat{\sigma}^2/n$	2.97	$\overline{V}(\hat{y}_2) = p\hat{\sigma}^2/n$	3.66	$\overline{V}(\hat{y}_3) = p\hat{\sigma}^2/n$	4.15
$\overline{\sigma}_{\hat{y}_1}$	1.72	$\overline{\sigma}_{\hat{y}_2}$	1.91	$\overline{\sigma}_{\hat{y}_3}$	2.04

Table 10.11 provides a convenient arrangement for considering the question about the need for x_2 and x_3. The regression sums of squares for x_1 alone, extra due to x_2, and extra due to x_3 are all distributed independently. The ANOVA table using the residual mean square 20.7 with 16 degrees of freedom and the 5% level of F as a convenient reference point tells you therefore that x_2 provides significant additional information to that supplied by x_1 but x_3 does not.

Statistical significance does not imply necessarily that you should include x_2 in a predictive model. The objective is to make an estimate $\hat{y}$ of the hardness of a specimen of mineral ore with a prediction equation over the ranges of values of x_1, x_2, x_3, and y covered by the experiment. To answer the question whether you need x_2 and x_3 once you have x_1, a criterion to consider, as discussed earlier, is the *average value* of the variance of the $n = 20$ predictions $\hat{y}$, that is, $p\sigma^2/n$, where p is the number of regressors. Table 10.12 shows the values of $\overline{V}\hat{y} = p\hat{\sigma}^2/n$ for the various fitted models. On this argument, since the average variance increases with x_2 and x_3, nothing is gained by using more than the single regressor x_1.

Exercise 10.4. Repeat the ANOVA in Table 10.11 using the regressor sequence x_1 followed by x_3 followed by x_2.

Orthogonalization

The independence of the extra sums of squares is the result of what may be called orthogonalization. For example, to determine that part of the vector $\mathbf{x}_2$

Table 10.13. Orthogonalized x's

	x_1	x_2	x_3	x_1	$x_{2.1}$	$x_{3.21}$	y
1	1.11	0.53	0.29	1.11	−0.22	0.04	27.6
2	0.11	0.33	−0.51	0.11	0.25	−0.56	14.8
3	−0.79	−0.48	−0.21	−0.79	0.05	−0.02	17.1
4	−1.39	−0.78	−0.51	−1.39	0.16	−0.19	13.9
5	−1.59	−0.58	−0.21	−1.59	0.49	0.13	20.0
6	−1.09	−1.68	−1.11	−1.09	−0.94	−0.76	26.6
7	−2.19	−1.08	0.29	−2.19	0.39	0.78	20.3
8	0.81	0.63	0.19	0.81	0.08	−0.01	19.7
9	−0.49	−0.68	−0.11	−0.49	−0.35	0.04	25.7
10	0.61	0.03	0.29	0.61	−0.38	0.18	23.2
11	2.01	2.03	−0.51	2.01	0.68	−1.05	25.8
12	1.91	1.33	1.09	1.91	0.04	0.63	33.6
13	1.01	1.23	−0.21	1.01	0.55	−0.50	24.2
14	1.01	0.23	0.79	1.01	−0.45	0.59	28.6
15	0.41	0.33	0.29	0.41	0.05	0.19	29.3
16	−0.59	0.03	1.09	−0.59	0.42	1.19	26.0
17	−0.49	−0.38	−0.81	−0.49	−0.05	−0.69	18.0
18	1.21	0.23	0.69	1.21	−0.59	0.45	39.2
19	0.61	0.43	0.29	0.61	0.02	0.14	25.9
20	−2.19	1.68	−1.11	−2.19	−0.21	−0.56	25.3

that is orthogonal to $\mathbf{x}_1$, you regress $\mathbf{x}_2$ on $\mathbf{x}_1$ to get $b = \sum(x_1x_2/\sum x_1^2)$. Then $\mathbf{x}_2$ can be represented by two orthogonal components $\mathbf{x}_{2.1}$ (read as x_2 given x_1) and $b\mathbf{x}_1$, where the residual component $\mathbf{x}_{2.1} = \mathbf{x}_2 - b\mathbf{x}_1$ is orthogonal to $\mathbf{x}_1$. In a similar way you can regress $\mathbf{x}_3$ on $\mathbf{x}_1$ and $\mathbf{x}_2$ to get the residual component $\mathbf{x}_{3.21} = \mathbf{x}_3 - c_1\mathbf{x}_1 - c_2\mathbf{x}_{2.1}$. The vector $\mathbf{x}_{3.21}$ (read as x_3 given x_2 and x_1) then represents the part of $\mathbf{x}_3$ which is orthogonal to (independent of) both $\mathbf{x}_1$ and $\mathbf{x}_2$. Table 10.13 displays the orthogonal vectors $\mathbf{x}_1$, $\mathbf{x}_{2.1}$, and $\mathbf{x}_{3.21}$. In terms of the orthogonalized regressors x_1, $x_{2.1}$, and $x_{3.21}$ the least squares equation is

$$\hat{y} = 24.2 + 2.5x_1 - 6.1x_{2.1} + 3.7x_{3.21}$$

This equation produces the same estimates $\hat{y}$ as that given earlier in Table 10.10b, but since x_1, $x_{2.1}$, and $x_{3.21}$ are now orthogonal, the coefficients are different. Notice that when regressors are orthogonal their coefficients may be calculated individually (as if the other x's were not present) Thus, for example, $b_{2.1} = \sum x_{2.1}y/\sum x_{2.1}^2 = -19.67/3.23 = -6.1$.

Allowing for Block Effects and Concomitant Factors

The extra sum of squares principle can be useful in other contexts.

The importance of *blocking* was emphasized earlier in this book. However, there the experimental designs used blocking factors orthogonal to the other factors. This produced considerable simplification since block factor sums of squares were then distributed independently of the other factor sums of squares. To illustrate how allowance can be made for blocks in the more general case, suppose in the example of the hardness of 20 samples of ore that the analyses had been carried out in three different laboratories. The first eight determinations were made in laboratory 1, the next five in laboratory 2, and the last seven in laboratory 3. Suppose now you ask the same question as before after allowing for possible block effects (differences between laboratories): Are x_2 and x_3 supplying additional information not provided by x_1?

You can represent the blocks by dummy variables as shown in Table 10.14. The model for the predicted values is $\hat{y} = c_1 z_1 + c_2 z_2 + c_3 z_3 + b_1 x_1 + b_2 x_2 + b_3 x_3$. Reading across rows in Table 10.14 you will see that the coefficient c_1 is the estimated mean and c_2 and c_3 are the amounts by which laboratories 2 and 3 differ from laboratory 1. Thus fitting the model first for z_1 and then for z_2, z_3 you will obtain first the sum of squares due to the mean and then the extra sums of squares due to blocks. If you now fit the model with z_1, z_2, z_3 and x_1, you can get the extra sum of squares for x_1 allowing for blocking. After fitting with z_1, z_2, z_3, x_1, x_2 you can get the extra sum of squares

Table 10.14. Hardness Data from Three Different Laboratories

Sample Number	z_1	z_2	z_3	x_1	x_2	x_3	y	$y - \bar{y}$
1	1	0	0	10.2	3.5	16.0	27.6	7.60
2	1	0	0	9.2	3.3	15.2	14.8	−5.20
3	1	0	0	8.3	2.5	15.5	17.1	−2.90
4	1	0	0	7.7	2.2	15.2	13.9	−6.10
5	1	0	0	7.5	2.4	15.5	20.0	0.00
6	1	0	0	8.0	1.3	14.6	26.6	6.60
7	1	0	0	6.9	1.9	16.0	20.3	0.30
8	1	0	0	9.9	3.6	15.9	19.7	−0.30
9	1	1	0	8.6	2.3	15.6	25.7	−0.80
10	1	1	0	9.7	3.0	16.0	23.2	−3.30
11	1	1	0	11.1	5.0	15.2	25.8	−0.70
12	1	1	0	11.0	4.3	16.8	33.6	7.10
13	1	1	0	10.1	4.2	15.5	24.2	−2.30
14	1	0	1	10.1	3.2	16.5	28.6	1.13
15	1	0	1	9.5	3.3	16.0	29.3	1.83
16	1	0	1	8.5	3.0	16.8	26.0	−1.47
17	1	0	1	8.6	2.6	14.9	18.0	−9.47
18	1	0	1	10.3	3.2	16.4	39.2	11.73
19	1	0	1	9.7	3.4	16.0	25.9	−1.57
20	1	0	1	6.9	1.3	14.6	25.3	−2.17

for x_2. Finally by fitting the entire model with z_1, z_2, z_3, x_1, x_2 and x_3 you can get the contribution of x_3 clear of block effects and also the residual sum of squares.
Similar to that in Table 10.11 the ANOVA is now as follows:

Analysis of Variance for Regression Components Adjusted for Blocks

		SS	df		MS		F	Probability
Blocks	Extra due to z_2, z_3	242.4	2		121.2		6.74	0.009
Regression	Extra due to x_1	86.8	1	77.1	86.8	4.28	4.82	0.045
	Extra due to x_2	87.7	1		87.7		4.87	0.044
	Extra due to x_3	56.8	1		56.8		3.16	0.097
Residual		251.7	14		17.98			
$\sum(y-\bar{y})^2$		725.6	19					

Corresponding to Table 10.12, we get Table 10.15.

To answer the question whether you need x_2 and x_3 to estimate y, once you have x_1, after *allowing for blocks*, you use results in Table 10.15. Making the same calculations as before, you see that for this data nothing is gained by using the additional variable x_2 and x_3.

The Geometry of the Extra Sum of Squares Principle

Suppose there were just $n = 3$ observations and you contemplated two linear models, one with only x_0 and the other involving both x_0 and x_1. In Figure 10.11 the observations are presented by the vector $\mathbf{y}$ and the variables x_0 and x_1 by the vectors $\mathbf{x}_0$ and $\mathbf{x}_1$. For the fitted least squares model

$$\hat{\mathbf{y}} = \hat{\beta}_0\mathbf{x}_0$$

Table 10.15. Average Variance of $\hat{y}$ for Various Models

$\hat{y}_1 = b_0 + b_1x_1$ Adjusted for Blocks		$\hat{y}_2 = b_0 + b_1x_1 + b_2x_2$ Adjusted for Blocks		$\hat{y}_3 = b_0 + b_1x_1 + b_2x_2 + b_3x_3$ Adjusted for Blocks	
SS x_1	86.78	SS $x_{2.1}$	87.74	SS $x_{3.21}$	56.85
$\sum(y-\hat{y}_1)^2$	396.33	$\sum(y-\hat{y}_2)^2$	308.59	$\sum(y-\hat{y}_3)^2$	251.74
df	16	*df*	15	*df*	14
MS	24.7	MS	20.6	MS	18.0
p/n	4/20	p/n	5/20	p/n	6/20
$\overline{V}(\hat{y}_1) = p\hat{\sigma}^2/n$	4.95	$\overline{V}(\hat{y}_2) = p\hat{\sigma}^2/n$	5.14	$\overline{V}(\hat{y}_3) = p\hat{\sigma}^2/n$	5.39
$\overline{\sigma}_{\hat{y}_1}$	2.23	$\overline{\sigma}_{\hat{y}_2}$	2.27	$\overline{\sigma}_{\hat{y}_3}$	2.32

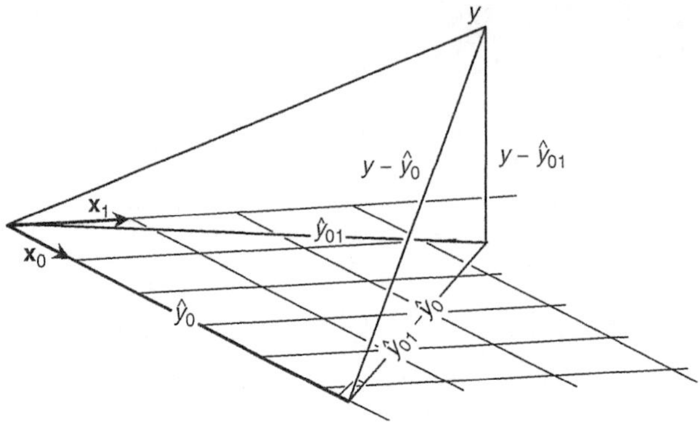

Figure 10.11. Geometry: extra sum of squares principle.

containing $\mathbf{x}_0$ only, a perpendicular $\mathbf{y} - \hat{\mathbf{y}}_0$ is dropped onto the *line* of the $\mathbf{x}_0$ axis. The intersection is 5.4 units along the x_0 axis. Thus

$$\hat{\mathbf{y}}_0 = 5.4\mathbf{x}_0$$

is the model of best fit. Now denote the squared length of the vector $\hat{\mathbf{y}}_0$ by S_0.

For the fitted least squares model containing both x_0 and x_1 a perpendicular $\mathbf{y} - \hat{\mathbf{y}}_{01}$ is dropped onto the plane generated by the vectors $\mathbf{x}_0$ and $\mathbf{x}_1$. This intersects the plane at the point (1.2, 3.4) that is 1.2 units along the x_0 axis and 3.4 units along the x_1 axis. Thus

$$\hat{\mathbf{y}}_{01} = 1.2\mathbf{x}_0 + 3.4\mathbf{x}_1$$

is the model of best fit. Now denote the squared length of the vector $\hat{\mathbf{y}}_{01}$ by S_{01}. The extra sum of squares is the squared length of the vector $\hat{\mathbf{y}}_{01} - \hat{\mathbf{y}}_0$ and by the theorem of Pythagoras is $S_{01} - S_0$. Since $\hat{\mathbf{y}}_{01} - \hat{\mathbf{y}}_0$ is orthogonal to $\hat{\mathbf{y}}_0$. The ANOVA can then be written in terms of the two additive components S_0 and S_{01} and on NIID assumptions they are independently distributed.

Other Applications

The same device may be employed to eliminate effects of any other regressors (concomitant variables). For example, suppose you wished to eliminate possible effects due to weather conditions. The analysis could be used with z_1, the barometric pressure; z_2, the temperature; and so on. Similarly, if you believed there might be a linear trend to your experiments, you could use the sequence $1, 2, 3, \ldots$ as the concomitant variable.

Do Not Introduce Concomitant Variables Needlessly

Remember when the IID assumptions are true, the average variance of the predictions $\hat{y}$ is $p\sigma^2/n$; *therefore each time you add a new regressor the value of* p *increases and so does the average variance of the* $\hat{y}$'s. *There is no such thing as a free lunch.*

Orthogonal Polynomials

Sometimes you need to fit a polynomial $y = b_0x^0 + b_1x + b_{11}x^2 + b_{111}x^3 + \ldots$ (where $x^0 = 1$) in which y has been observed at a number of levels of x and you have to answer the question "What is the appropriate degree of the fitted polynomial?" Least squares can of course be applied in the usual way, but the question can be answered more directly. Instead of employing the regressors x^0, x^1, x^2, x^3, orthogonal functions x^0, $x^{1.0}$, $x^{2.10}$, $x^{3.210}$ are used where the x's are successively orthogonalized in the manner described in the previous section. When the levels of x are equidistant, the problem can be further simplified by working with orthogonal components already available, for example, in the Fisher and Yates tables (Table XXIII). The idea may be illustrated by an example.

Suppose the expansion of a workpiece has been determined at three equally spaced levels of temperature. A sample of 12 items all believed to be of the same length have been randomly allocated in groups of four and their lengths measured at 100°C, 150°C, and 200°C with the results shown in Table 10.16 (the measurements are in millimeters). A linear effect of temperature is anticipated, but is there any evidence of curvature? If we code the three levels of temperature as $x = 1, 2, 3$, the necessary data and calculations are shown in Table 10.16a; the ANOVA is shown in Table 10.16b. Over the range of temperature studied, no significant curvature effect is found.

Table 10.16a. Example of the Application of Orthogonal Polynomials: Equally Spaced Data

	100°C (1)	150°C (2)	200°C (3)	
	6.213	7.541	7.794	
	6.724	7.372	7.013	
	7.086	7.075	7.189	
	6.649	6.923	7.889	
Average, $\bar{y}$	6.668	7.228	7.889	
Variance, s^2	0.1284	0.0784	0.1894	Pooled $s^2 = 0.1321$, $\nu = 9$

Model to be fitted: $\hat{y} = b_0x_1 + b_Lx_{2.1} + b_Qx_{3.21}$

	Original Vectors			Orthogonal Vectors			
	$\mathbf{x}^0$	$\mathbf{x}$	$\mathbf{x}^2$	Mean	Linear	Quadratic	
Temperature (°C)	[$\mathbf{x}_1$	$\mathbf{x}_2$	$\mathbf{x}_3$]	$\mathbf{x}_1$	$\mathbf{x}_{2.1}$	$\mathbf{x}_{3.21}$	$\bar{y}$
100	1	1	1	1	−1	1	6.692
150	1	2	4	1	0	−2	7.295
200	1	3	9	1	1	1	7.436
				$\sum x\bar{y} = 21.367$	0.803	− 0.317	
				$b = \sum x\bar{y}/\sum x^2 =$ 7.122	0.402	− 0.053	

Table 10.16b. ANOVA for Three Levels of Temperature

$SS = b\sum x\bar{y}$	$b\sum xy$	df	MS	F
$b_0 = 152.1829$	608.7316	1		
$b_L = 0.3224$	1.2896	1	1.2896	9.8
$b_Q = 0.0167$	0.0670	1	0.0670	0.5
Error SS =	1.1892	9	0.1321	
Total SS =	611.2774	12		

Using Least Squares to Analyze Designed Experiments

The estimates of the main effects and interactions discussed earlier for factorials and fractional factorials are all least squares estimates. For example, suppose you were studying two factors A and B each at two levels in a 2^2 factorial design. If you fitted the model $\hat{y} = b_0x_0 + b_1x_1 + b_2x_2$ where x_0 is the indicator variable for the constant term b_0 and x_1 and x_2 are the usual coded factors, you would have

x_0	x_1	x_2
1	−1	−1
1	1	−1
1	−1	1
1	1	1

Since the regressors are mutually orthogonal the sum of all the cross-product terms between the x's are zero, and the normal equations reduce to

$$\sum x_0^2 b_0 + 0 + 0 = \sum yx_0$$

$$0 + \sum x_1^2 b_1 + 0 = \sum yx_1$$

$$0 + 0 + \sum x_2^2 b_2 = \sum yx_2$$

Thus the *coefficients* of the fitted model $\hat{y} = b_0x_0 + b_1x_1 + b_2x_2$ are

$$b_0 = \frac{\sum yx_0}{\sum x_0^2} = \frac{\sum y}{4} = \bar{y}$$

$$b_1 = \frac{\sum yx_1}{\sum x_1^2} = \frac{\sum yx_1}{4} = \frac{1}{2}A$$

$$b_2 = \frac{\sum yx_2}{\sum x_2^2} = \frac{\sum yx_2}{4} = \frac{1}{2}B$$

where A and B are the estimated *effects* $(\bar{y}_+ - \bar{y}_-)$ as they were calculated earlier in Chapter 5. The multiple of 1/2 comes about because, for example, the *effect* A represents the change in $\hat{y}$ when x_1 changes from -1 to $+1$, that is, over two units. The *coefficient* b_1, however, measures the change in $\hat{y}$ as x_1 changes by one unit. A similar argument applies to interactions. In general, the least squares *coefficients* for the two-level orthogonal designs are always one-half the corresponding *effects*. Thus $b = (\bar{y}_+ - \bar{y}_-)/2$, where $\bar{y}_+ - \bar{y}_-$ is the contrast associated with the main effect or interaction. Correspondingly, the standard error of the coefficient b_1 calculated by the regression program would be half that of the A effect.

You can, of course, use a "dummy" model that is not intended to approximate any physical reality but that simply includes terms you want to estimate. For example, if you were analyzing a 2^3 factorial, you could obtain the estimates of all the main effects and two factor interactions by fitting the model

$$\hat{y} = b_0x_0 + b_1x_1 + b_2x_2 + b_3x_3 + b_{12}x_1x_2 + b_{13}x_1x_3 + b_{23}x_2x_3$$

As an approximating polynomial this fitted model is incomplete since quadratic terms $b_{11}x_1^2$, $b_{22}x_2^2$, $b_{33}x_3^2$ of the same order as the interaction terms are absent, and since you only have two levels for each factor, you cannot estimate them. Nevertheless, the 2^3 factorial is a good way-station that allows you to see whether you should perhaps simplify the model or elaborate it.

Analysis of Automobile Emissions Data

The following example uses least squares to analyze a 3^2 factorial used to study the emissions from a new automobile engine design. The three coded levels (-1, 0, $+1$) for x_1 and x_2 represent three equispaced levels of added ethanol and air–fuel ratio, respectively. The response y was carbon monoxide concentration in micrograms/per cubic meter. The data are given in Table 10.17. A plot of the data displayed considerable nonplanarity and the model

$$y = \beta_0 + \beta_1x_1 + \beta_2x_2 + \beta_{11}x_1^2 + \beta_{22}x_2^2 + \beta_{12}x_1x_2 + e$$

in factors x_1 and x_2 was employed. The fitted model and corresponding ANOVA are as follows:

$$\hat{y} = 78.6 + 4.4x_1 - 6.9x_2 - 4.6x_1^2 - 4.1x_2^2 - 9.1x_1x_2$$

	SS			df		MS		
Regression	1604.7	First-order terms	795.9	5	2	320.9	398.0	$F_{5,12} = 50.1$
		Second-order terms	808.8		3		269.6	
Residual	76.5	Lack of fit	31.7	12	3	6.4	10.6	$F_{3,9} = 2.1$
		Error	44.8		9		5.0	
Total	1681.2			17				

Table 10.17. A 3^2 Factorial Design Repeated, All Eighteen Runs in Random Order

Observations			
x_1	x_2	y	
−	−	61.9	65.6
0	−	80.9	78.0
+	−	89.7	93.8
−	0	72.1	67.3
0	0	80.1	81.4
+	0	77.8	74.8
−	+	66.4	68.2
0	+	68.9	66.0
+	+	60.2	57.9

There was no detectable lack of fit and over the studied ranges of x_1 and x_2 and the fitted model was effective in predicting the CO concentration in terms of added ethanol and air–fuel ratio. The implications of this model are studied further in Chapter 11.

Advantages of Taking the Least Squares Route for the Analysis of Experimental Data

1. The experimental arrangements you may use may not be orthogonal designs.
2. Sometimes it is difficult to run the required conditions exactly. For instance, the design may require a run at temperature 60°C and time 2 hours, but because of experimental difficulties, the nearest you were able to get is temperature 63°C and time 2 1/2 hours. Using least squares, you can estimate the parameters from the actual levels you ran rather than those you intended to run.
3. Even simple two-level factorial designs are often botched: You discover, for example, that run 7 that should have been made at the (+ + − + −) levels was actually made at the (+ − − + +) level. It will often still be possible to estimate the effects you need, although the botched experimental design will give estimates with somewhat larger variances (though often not much larger). In any case, least squares gives you the appropriate estimates and tells you what their variances are.
4. You can use least squares for estimating the parameters in more sophisticated models, where the designs do not necessarily follow a simple pattern.
5. When you add additional runs, you can readily combine the new data with that from the initiating design.

6. Some runs may be missing or there may be certain runs that are almost certainly bad values. Using least squares you can readily check (run an analysis with and without the suspect values) the influence of bad values. Also you can estimate the missing values. For illustration, see Appendix 10C.

In every case you will get a correct analysis by employing valid runs as they were performed and using a least squares computer program.

10.3. THE ORIGINS OF EXPERIMENTAL DESIGN

This book is primarily about experimental design—choosing the values of the levels of the factors (the x's) and conducting the experiment so as to produce required information with the least amount of effort. It is interesting to consider how experimental design came to be invented.

At the beginning of the last century the agriculturalists at Rothamsted Agricultural Experiment Station had records extending over decades for crop yields from extensive plots of land each of which had been treated every year with the same particular fertilizer. They also had records of rainfall, temperature, and so on, for the same periods. In 1918 the director of the station brought in R. A. Fisher (originally as a temporary employee) to see whether he could extract additional information from these records by statistical methods. Thus Fisher had to deal with happenstance data and he attempted their analysis by fitting various models by least squares (regression analysis). In spite of many original developments* which he introduced along the way, he became aware that some critical questions could not be answered from such data, however ingenious the analysis. With the encouragement of W. S. Gosset he invented the design of experiments specifically to overcome the difficulties encountered when attempting to draw conclusions from such happenstance data.

Analyzing Happenstance Data by Least Squares

All this work predated experimental design, so let us look at some of the difficulties Fisher found (and so will you) in analyzing unplanned "happenstance" data. In industry happenstance data often take the form of hourly or daily readings on various process variables, as illustrated in Table 10.18. The computer is a wonderful device for compiling such data arrays—for storing one's "autumn leaves"—although it may not tell you what to do with them. Table 10.18 shows the first six rows of such a data array. Here there are four separately recorded factors and only one response. In practice, there could be a much larger number of factors and responses and observations.

*Developments such as the distribution of the multiple correlation coefficient leading directly to the distribution of t and F, the examination of residuals, the use of distributed lag models, and the analysis of variance. (See Fisher, R.A., Studies in crop variation, 1921, 1923.)

Table 10.18. Some Typical Happenstance Data

(Input) Factor Levels				(Output) Response
Temperature, T (°C)	Pressure, P (psi)	Concentration, C (%)	Air Flow, F (10^5 m^3/ min)	Plant Yield, y
159	173	82.1	10.3	86.2
161	180	89.5	10.4	84.7
160	178	87.9	10.3	85.9
162	181	83.4	10.3	85.3
166	179	85.4	10.2	84.5
164	181	85.3	10.3	85.4
.	.	.	.	.
.	.	.	.	.
.	.	.	.	.

In such regression applications the input factor levels (the regressors) are often called the "independent variables" and the output responses the "dependent variables." The word *independent* has other important connotations and it is best not to use this confusing nomenclature. As with designed experiments it is usually most convenient to code the data; for example, you may prefer to analyze $x_1 = (T - \overline{T})/s_T$, where s_T is some suitable scale factor. In particular, this will lead to less trouble with rounding errors and simplify calculations. Often, without much thought, happenstance data are fitted using the linear regression model

$$\hat{y} = \overline{y} + b_1(x_1 - \overline{x}_1) + b_2(x_2 - \overline{x}_2) + \cdots + b_k(x_k - \overline{x}_k)$$

In favor of such a model it is sometimes argued that, over the comparatively small changes of the factors $x_1, x_2, \ldots$ that typically occur, the model should be approximately linear in the x's.

If happenstance data are *really* all you can get, such analyses may be better than nothing. But they can be downright misleading, as is reflected by the acronym PARC (practical accumulated records computations) and its inverse. We consider some of the reasons for this below.

1. *Inconsistent Data.* It is quite rare that a long record of data is consistent and comparable. For example, for data from an industrial process, standards are often modified over time, instruments change, calibrations drift, operators come and go, changes occur in raw materials, and processes age and in some cases are affected by weather. We can try to take some of these effects into account by using dummy variables, blocking, and so on, but much that is relevant is usually unknown and not recorded.

2. *Range of Factors Limited by Control.* Figure 10.12 shows a plot of yield against temperature as it might appear if temperature were freely varied. Suppose,

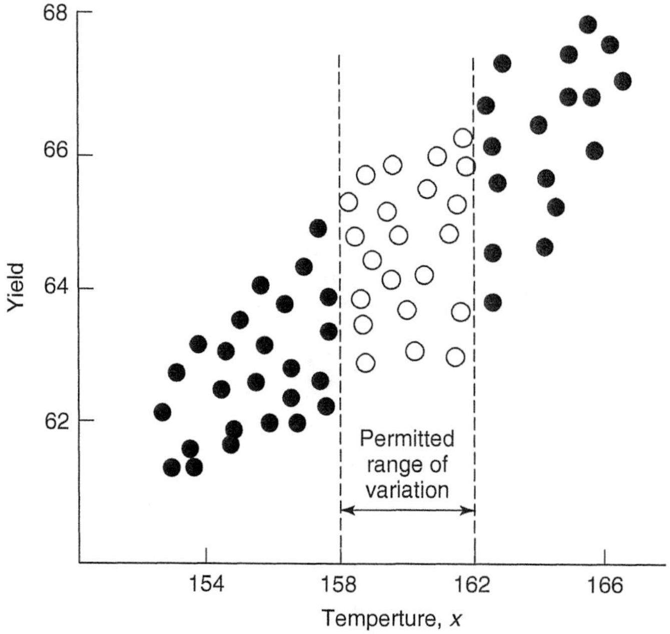

Figure 10.12. Plot of yield versus temperature where permissible range of variation in temperature is restricted.

however, that during normal operation temperature was controlled to say 160 ± 2°C. Over this more limited range (open dots) it could well be that no relationship was apparent, either visually or statistically. The variance of the slope of a fitted line to such restricted data can be many times what it would have been for unrestricted data. If you fit a regression equation to the available restricted data and tell the process engineer that temperature has no (statistically) significant effect on yield, you may well be laughed to scorn and told, "But of course it has an effect, that's why we control temperature so carefully." (Here, of course, the engineer misunderstands the meaning of "statistically significant".)

3. *Semiconfounding of Effects.* Processes often operate so that a particular change in one input variable is usually (sometimes invariably) accompanied by a change in another. Such dependence may be produced by, for example, an automatic controller or operating policy. Suppose that a process operator's instructions are that whenever the temperature increases by 1° he must increase the flow rate by one-tenth of a unit. This will produce a highly unsatisfactory design in temperature and flow rate, such as that in Figure 10.13*a*. At the five different conditions shown there is a marked increase in yield as temperature and flow are increased. But assuming some kind of causal relationship, is the increase produced by higher temperature, by faster flow rate, or by both? Common sense rightly says that the question may be impossible to answer. So does statistics.

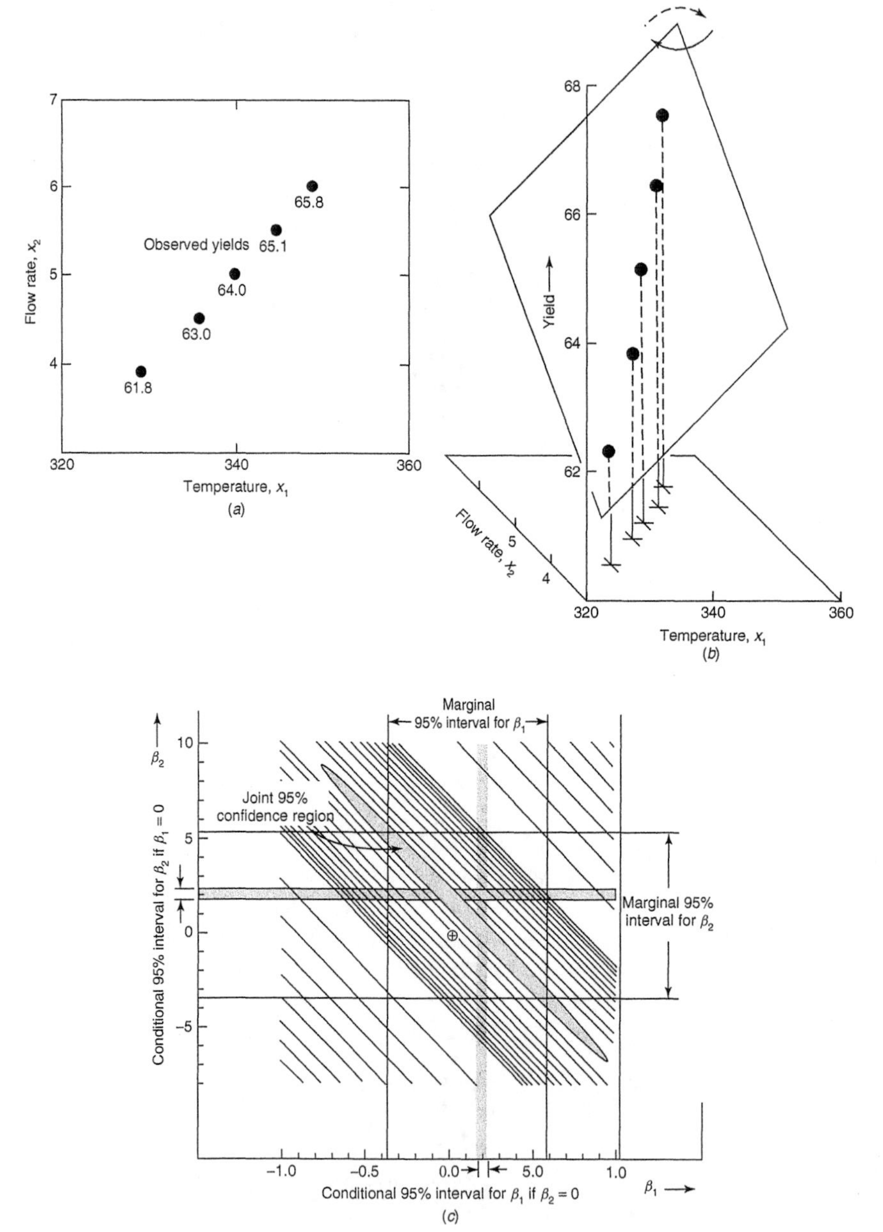

Figure 10.13. Influence of semiconfounding.

The best fitting plane for these data is

$$\hat{y} = 63.94 + \underset{(\pm 0.67)}{0.10}\ (x_1 - \overline{x}_1) + \underset{(\pm 4.48)}{1.01}\ (x_2 - \overline{x}_2)$$

The quantities in parentheses below the coefficients are the standard errors. Looking at these you might at first conclude that there was no relationship between y and x_1 and x_2. The reason this happens can be understood by looking at the fitted plane shown in Figure 10.13*b*. Now look at the 95% confidence limits and regions for the estimated coefficients in Figure 10.13*c*. Notice first that the point ($\beta_1 = 0$, $\beta_2 = 0$) shown by a circle with cross falls *outside* the joint region represented by the diagonally placed ellipse. The hypothesis that *neither* factor is related to yield is obviously not acceptable. However, as the standard errors for the coefficients in the fitted model suggest, both *marginal* confidence intervals cover zero. That is, the data can be readily explained by either a plane with zero slope in the x_1 direction *or* a different plane with zero slope in the x_2 direction.

This state of affairs arises because the factor levels x_1 and x_2 are almost linearly related. The fitted plane in Figure 10.13*b* sits unstably on the obliquely situated points provided by the "design" and has therefore a great tendency to wobble. A whole range of planes in which increase in the slope b_1 is compensated for by a corresponding decrease in slope b_2 would fit the data almost equally well. Thus highly correlated estimates for b_1 and b_2 are produced with correspondingly large standard errors. These standard errors would be infinite if x_1 and x_2 were exactly linearly dependent, that is, if the design points lay exactly along a straight line. In that case the estimates b_1 and b_2 would be perfectly confounded and you could estimate only some linear combination $k_1\beta_1 + k_2\beta_2$ of the two coefficients.

Think how much better off you would be if you had a "four-legged table" on which to balance the plane as provided by a 2^2 factorial design. For such an arrangement small errors in the length of the "legs" would have minimum effect on the stability of the tabletop. The unstable tendency of the present design is reflected in the oblique contours of the sum of squares surface shown in Figure 10.13*c*. Notice that, although neither b_1 nor b_2 is significantly different from zero when *both* temperature and flow rate are included in the regression equation, when *only one* is included, it appears to have a highly significant effect. The reason for this corresponds to fitting a straight line representing y versus x_1 alone or alternatively a line representing y versus x_2 alone. Fitting $y = \beta_0 + \beta_1 x_1 + \varepsilon$ to the data gives 0.21 ± 0.02 for the 95% confidence limits for β_1 and fitting $y = \beta_0 + \beta_2 x_2 + \varepsilon$ gives 1.94 ± 0.22 for the β_2 interval. In Figure 10.13*c* these are called *conditional* intervals because omission of one of the factors is equivalent to the assumption that its effect is *exactly zero. On this assumption* the effect of the other factor is then precisely estimated.

In summary, it will be impossible to identify the separate contributions of x_1 and x_2 if the data points in the (x_1, x_2) plane form a highly correlated pattern, yielding a poor experimental design.

4. *Nonsense Correlation—Beware the Lurking Variable.* Inevitably you cannot observe all the variable factors that affect a process. You tacitly admit this fact as soon as you write a model containing the error term e. The error term is a catchall for all the other "lurking" variables that affect the process but are

not observed or even known to exist. A serious difficulty with happenstance data is that, when you establish a *significant* correlation between y and factor x_1, this does not provide evidence that these two variables are necessarily *causally* related. (The storks and population example of Figure 1.4 illustrates the difficulty.) In particular, such a correlation can occur if a change in a "lurking" factor x_2 produces a change in *both* y and x_1.

Impurity, Frothing, and Yield

For illustration consider the hypothetical reactor data in Figure 10.14*a* relating yield y and pressure x_1. The 95% confidence interval for the slope of the regression line in Figure 10.14*a* is -1.05 ± 0.17, so there is little question as to its statistical significance. However, the cause (unknown to the investigator) might be as follows:

1. There exists an unknown impurity x_2 (a lurking variable) which varies from one batch of raw material to another.
2. High levels of the impurity cause low yield.
3. High levels of impurity *also cause frothing*.
4. The standard operating procedure is to reduce frothing by increasing the reactor pressure.
5. Over the relevant range, pressure does influence frothing but has no influence on yield.
6. Only yield and pressure are recorded.

First look at Figure 10.14*a*, where the recorded yield and pressure data are plotted and a straight line is fitted to the data. Next look a Figure 10.14*b*, where the true relationships are shown schematically. The actual but unknown casual relationship between yield y and the level of the unsuspected impurity x_2 (the lurking variable) is shown in Figure 10.14*c*. But increases in x_2 also cause increased frothing and management requires the process operator to increase pressure to reduce frothing. Figure 10.14*c*(i) represents the real functional dependence of yield on both pressure x_1, pressure, and x_2 impurity where the plane has *no slope in the* x_1, *direction (pressure)*—changing pressure with a fixed level of impurity does not change yield. Management operating policy thus produces the nonsense correlation between y and x_1 shown in Figure 10.14*a*.

Correlations of this kind are sometimes called *nonsense correlations* because they do not imply direct causation. Just as shooting storks will not reduce the birth rate, decreasing pressure in the present example will not increase yield. The happenstance data that provided the relationship shown in Figure 10.14*a* are seen as gossip begging for an explanation.

5. *Two Ways to Use Regression Equations.* It is important to understand what can be and cannot be done with regression relationships.

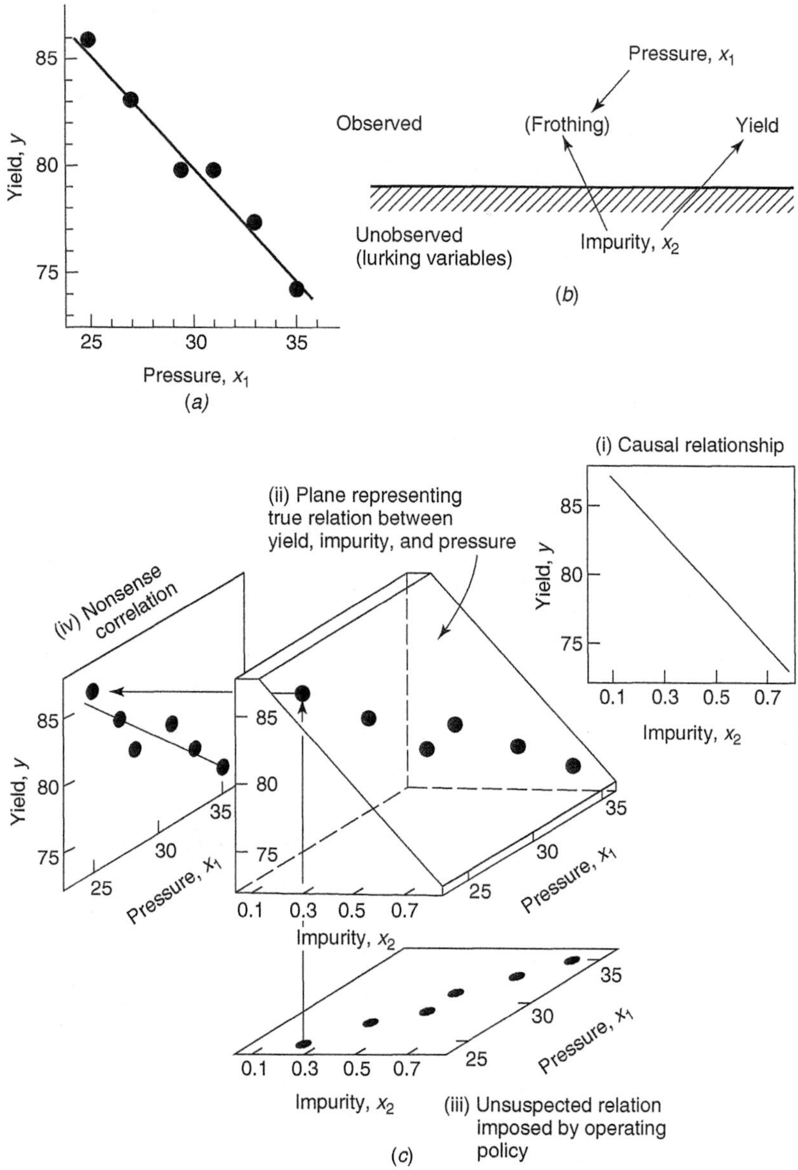

Figure 10.14. Generation of a nonsense correlation: (*a*) fitted line; (*b*) causal links; (*c*) actual yield dependence.

Use for Prediction

Even though there is no casual relationship in the above example between yield and pressure x_1, *if the system continues to be operated in the same fashion as when the data were recorded, then* the regression equation can still be used to *predict* the yield y from the pressure x_1. It would be true that, for example, on days when high pressure was recorded the yield would tend to be low. However, y could not be changed by changing x_1. To manipulate a factor x so as to *influence* a response y, one needs a *causal* relationship. The monthly sales x of fuel oil in Madison, Wisconsin, is a good predictor of the monthly temperature y during the winter months, but experience has shown that it is not possible to induce warm weather by canceling the order for oil.

With the introduction of various types of "best subset" regression programs for the analysis of happenstance data, it remains important to keep a clear head about the difference between correlation and causation. Suppose there are q regressors in all and $p < q$ in the subset used for prediction. Then, as has already been discussed, the average variance of the predicted value $\hat{y}$ taken over the n points of the design will always equal $p\sigma^2/n$. So far as prediction using happenstance data is concerned, the questions are "What should be the size p of the subset?" and "Which regressors should be included to provide a prediction that is most closely *connected* with the outcome y?"

But for a well-designed experiment the question is "Do changes in x *cause* changes in y?" You hope to answer the critical question "What does what to what?" The ideas of prediction and of seeking cause relationships must be clearly separated. For example, the most important factor for prediction (e.g., pressure in the last example) may not be important, or even be present, in a causal model.

Need for Intervention

To safely infer causality, the experimenter cannot rely on natural happenings to choose the design. She must intervene, choose the factors, design the experiment for herself, and, in particular, introduce randomization to break the links with possible lurking factors.

> ***To find out what happens when you change something, it is necessary to change it.***

Autocorrelated Errors

Estimation with the method of least squares and the calculation of associated confidence limits is appropriate if the assumed NIID model for the errors e is adequate. The most important element of the NIID assumption is not the N (normality) part but the II part. That is, the errors are *identically* distributed (in particular have constant variance*) and are *independently* distributed. For sets of data that occur as a sequence in time it should be *expected* that the errors will be dependent, not independent. Most often successive errors are positively

*An important way in which they may not be identically distributed is that they have different variances. If these are *known* (or can be estimated), Gauss provided the necessary extensions to the theory with what is now called "weighted" least squares discussed in Appendix 10.D.

autocorrelated — when the error e_t is high, the error e_{t+1} also tends to be high; although very occasionally, as was the case with the industrial data discussed in Chapter 3, successive errors are negatively correlated. In a designed experiment randomization breaks these links. But without it, when such dependencies occur, the use of ordinary least squares is inappropriate and can be disastrous. In particular, the estimates of the standard errors of the regression coefficients can be wrong by an order of magnitude. For example (Coen, Gomme and Kendall, 1969), in an study of a possible relationship between current stock prices and lagged data on car sales x_1 and a consumer price index x_2, ordinary least squares produced an apparent relationship in which the regression coefficients were more than eight times their standard errors and hence tremendously significant. On the basis of this regression equation it was argued that current car sales and values of the consumer index could foretell stock prices. However, from a subsequent study of the data in which the model allowed for dependence between the errors (see Box and Newbold, 1971) it was clear that no such relationship existed.

Notice also that the *detectability* of departures from assumptions and *sensitivity* to these assumptions are unrelated concepts. Thus, autocorrelation between errors of a magnitude not easily detectable can seriously invalidate least squares calculations. Conversely, large and detectable departures from distributional normality may have little effect on tests to compare means. (See, e.g., Table 3A.2).

Dynamic Relationships

Another problem that arises in the analysis of serially collected data concerns the proper treatment of dynamic relationships. Consider a simple example. The viscosity of a product is measured every 15 minutes at the outlet of a continuous reactor. There is just one input causative factor, temperature, which has a value x_t at time t and a value x_{t+1} at time $t+1$ one interval later, and so on. If we denote the corresponding values of viscosity by $y_t, y_{t+1}, \ldots,$ the usual regression model

$$y_t = \alpha + \beta x_t + e$$

says that the viscosity y_t at time t is related only to the temperature at that same time τ. Now in this situation the viscosity at time t is in fact likely to be dependent, not just on the current temperature, but also on the recent past history of temperature. Indeed, for the output from a stirred tank reactor you can justify the model

$$y_t = \alpha + \beta(x_{t-1} + \delta x_{t-2} + \delta^2 x_{t-3} + \delta^3 x_{t-4} + \cdots) + e$$

where δ is between zero and unity. This is a dynamic model in which the output is related via a down-weighted "exponential memory" to recent past inputs. If such a model is appropriate, standard least squares (regression) can produce totally misleading results. See, for example, Box, Jenkins, and Reinsel (1994), who discuss such memory functions and how to deal with them.

Feedback

Quite frequently data arise from processes where feedback control is in operation. For example, suppose that, to maintain the level of viscosity y_t at some desired level L, the temperature x_t is adjusted every 15 minutes with the following "control equation":

$$x_t = c + k(y_t - L)$$

where c and k are constants. Turning this equation around you will see that such control *guarantees* a linear relationship between y_t and x_t of the form

$$y_t = L + \frac{1}{k}(x_t - c)$$

Thus, under feedback control this exact relationship would occur in the data *whether or not* a change in temperature really affected viscosity. The regression equation would reflect information about the (known) control equation and nothing about the relation between viscosity y_t and temperature x_t.

Experimental Design Procedures for Avoiding Some of the Problems That Occur in the Analysis of Happenstance Data

What can be done to avoid the problems associated with happenstance data? Fisher's ideas for experimental design solved many of these problems:

Problems in the Analysis of Happenstance Data	Experimental Design Procedures for Avoiding Such Problems
Inconsistent data	Blocking and randomization
Range limited by control	Experimenter makes own choice of ranges for the factors
Semiconfounding of effect	Use of designs such as factorials that provide uncorrelated estimates
Nonsense correlations due to lurking variables	Randomization
Serially correlated errors	Randomization
Dynamic relations	Where only steady-state characteristics are of interest, it may be possible to allow sufficient time between successive experimental runs for the process to settle down. If not, specific dynamics must be incorporated into the model (Box, Jenkins, and Reinsel, 1994).
Feedback	In some cases it may be possible to temporarily disconnect the feedback system. Otherwise the introduction of a "dither signal" can make estimation possible (Box and MacGregor, 1974).

Time Series Analysis

Many historical data records are necessarily happenstance in nature, for example, various economic time series such as unemployment records. Here the use of experimental design (particularly randomization) is impossible, but frequently the investigator can cope with serial correlation by including an appropriate time series model for the errors which allows for estimation of the autocorrelations. Some aspects of this topic are discussed in Chapter 14, but the general analysis of dynamic and feedback relationships is beyond the scope of this book (see, e.g., Box, Jenkins, and Reinsel, 1994).

10.4. NONLINEAR MODELS

Biochemical Oxygen Demand Example

All the models so far discussed in this chapter have been *linear in the parameters* and appropriate analysis was possible by standard least squares (regression) methods. The following example shows how the method of least squares can also be used to fit models that are nonlinear in the parameters.

Biochemical oxygen demand (BOD) is used as a measure of the pollution produced by domestic and industrial wastes. In this test a small portion of the waste is mixed with pure water in a sealed bottle and incubated usually for 5 days at a fixed temperature. The reduction of the dissolved oxygen in the water allows the calculation of the BOD occurring during the incubation period. The data given in Table 10.19 were obtained from six separate bottles tested over a range of incubation days. Physical considerations suggest that the exponential model

$$\eta = \beta_1(1 - e^{-\beta_2 x})$$

should describe the phenomenon. The parameter β_2 would then be the overall rate constant and β_1 would be the ultimate BOD.

Sum of Squares Surface

As before, the problem can be understood by calculating the sums of squares for a grid of values of β_1 and β_2 and plotting the contours in the (β_1, β_2) plane of

Table 10.19. Biochemical Oxygen Demand Data

Observation Number, u	BOD (mg/L) y	incubation, x (days)
1	109	1
2	149	2
3	149	3
4	191	5
5	213	7
6	224	10

the sum of squares surface

$$S = \sum_{u=1}^{n} [y_u - \beta_1(1 - e^{-\beta_2 x_u})]^2$$

as in Figure 10.15. The minimum of the sum of squares surface yields the least squares estimates $b_1 = 213.8$ and $b_2 = 0.5473$. Substituting these values in $\hat{y} = b_1(1 - e^{-b_2 x})$ allows the calculation of the fitted least squares line shown in Figure 10.16. The residuals listed in Table 10.20 provide no reason to suppose the model is suspect. The sum of squares of these residuals is $S_R = 1168.01$. In

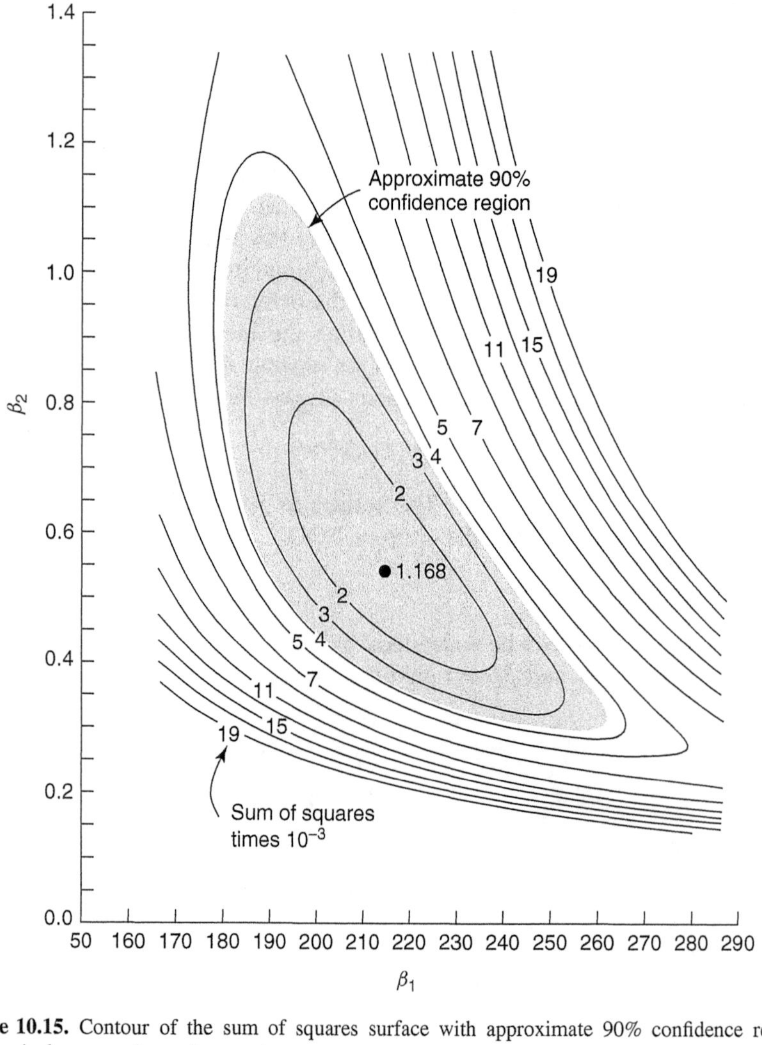

Figure 10.15. Contour of the sum of squares surface with approximate 90% confidence region: biochemical oxygen demand example.

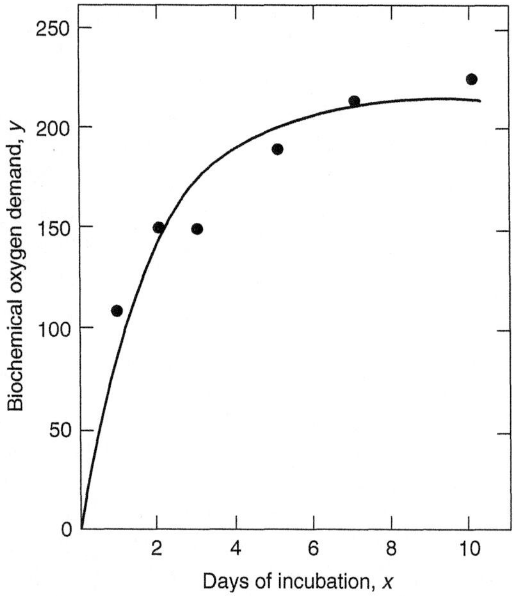

Figure 10.16. Plot of data and fitted least squares line: biochemical oxygen demand example.

Table 10.20. Calculation of the Predicted Values $\hat{y}_u$ and the Residuals $y_u - \hat{y}_u$: BOD Data

u	x_u	y_u	$\hat{y}_u = 213.8(1 - e^{-0.5473x_u})$	$y_u - \hat{y}_u$
1	1	109	90.38	18.62
2	2	149	142.53	6.47
3	3	149	172.62	−23.62
4	5	191	200.01	−9.01
5	7	213	209.11	3.89
6	10	224	212.81	11.19

general, a $1 - \alpha$ joint confidence region for β_1 and β_2 may be obtained *approximately* using Equation 10.8, which is appropriate for linear models:

$$S = S_R \left[1 + \frac{p}{n - p} F_{\alpha,(p,n-p)}\right]$$

where the contour S encloses the region. In this formula, p is the number of parameters and $F_{\alpha,(p,n-p)}$ is the upper $100\alpha\%$ point of the F distribution with p and $n - p$ degrees of freedom. For the BOD data the shaded area in Figure 10.15 is an approximate 90% region for β_1 and β_2. In this case $S_R = 1168.01$, $p = 2$,

$n = 6$, and $F_{0.10,(2,4)} = 4.32$, so the boundary of the region shown in Figure 10.15 lies on the contour for which $S = 3690.91$.

Obviously, drawing contour diagrams such as in Figure 10.15 is impractical for models with a number of parameters greater than 2. However, computer programs exist specifically to carry out iterative minimization calculations for nonlinear models in many parameters. Using these programs, you supply the data, the model, and initial guesses for the parameter values. The computer then sets out to search for the sum of squares minimum. Some programs also obtain approximate standard errors and confidence regions which should, however, be treated with some reserve.

Simplifying Calculations

A method that can be used in other contexts of breaking these calculations into "bite-size" pieces is to take advantage of the fact that in this example, when β_2 is fixed, you have a simple linear model. Thus calculating the sum of squares graph for a series of fixed values of β_2 is equivalent to calculating the surface in Figure 10.15 in horizontal sections. It is often possible to simplify by splitting nonlinear models into linear and nonlinear parts and to proceed accordingly as was done in the "λ plots" of Chapter 8.

APPENDIX 10A. VECTOR REPRESENTATION OF STATISTICAL CONCEPTS

In this section a vector representation least squares was used. Here is a brief review of vectors and the corresponding coordinate geometry and their use in explaining statistical tests.

Three-Dimensional Vectors and Their Geometric Representation

1. *A set of* $n = 3$ numbers $[y_1, y_2, y_3]$ arranged in a row or column that might, for example, be three data values can be regarded as the elements of a *vector* in an $n = 3$ dimensional space. A vector may be represented by a single letter in boldface type; thus $\mathbf{y} = [y_1, y_2, y_3]$.
2. A vector has *magnitude* and *direction* and can be represented geometrically by a line in space pointing from some origin to the point with coordinates (y_1, y_2, y_3). Thus, if the coordinates are labeled as in Figure 10A.1 and you set $y_1 = 3$, $y_2 = 5$, and $y_3 = 4$, the line $\mathbf{y}$ represents the vector $\mathbf{y} = [3, 5, 4]$. However, a vector holds no particular position in space. Any other line having the same length and pointing in the same direction represents the same vector. A vector multiplied by a constant has its elements multiplied by the same constant; thus $-2\mathbf{y} = [-6, -10, -8]$.
3. The *length* of a vector is the square root of the sum of squares of its elements. For example, the length of the data vector $\mathbf{y} = [3, 5, 4]$ in Figure 10A.1 is $\sqrt{9 + 25 + 16} = \sqrt{50} = 7.07$.

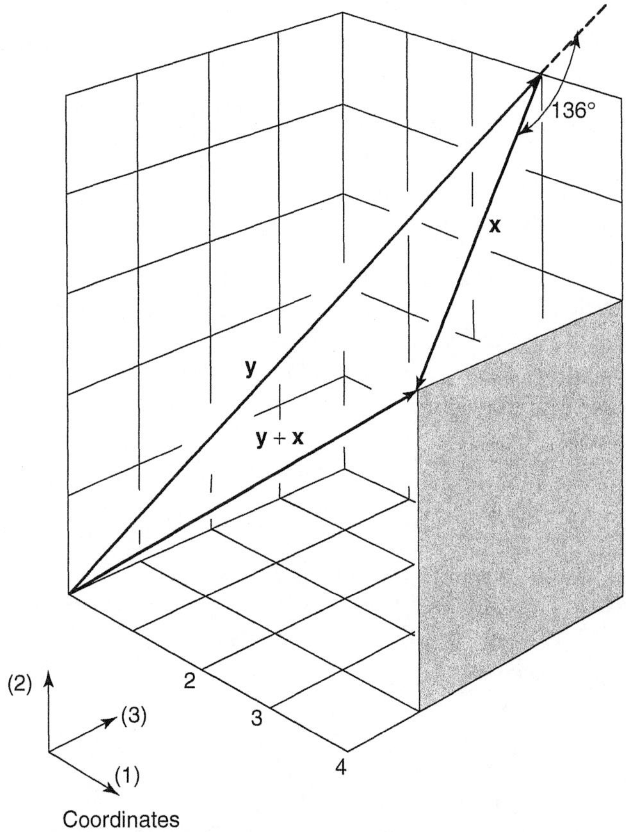

Figure 10A.1. Geometric representation of y, x, and $y + x$.

4. Two vectors having the same number of dimensions are *summed* by summing their corresponding elements. For example,

$$[\mathbf{y}] + [\mathbf{x}] = [\mathbf{y} + \mathbf{x}]$$

$$\begin{bmatrix} 3 \\ 5 \\ 4 \end{bmatrix} + \begin{bmatrix} 1 \\ -2 \\ -3 \end{bmatrix} = \begin{bmatrix} 4 \\ 3 \\ 1 \end{bmatrix}$$

The geometrical representation of this addition is shown in Figure 10A.1. The origin of the vector **x** is placed at the end of vector **y**; the vector from the origin of **y** to the end of **x**, signed as shown and the line completing the triangle, now represents **y** + **x**.

5. The *inner product* (sometimes called the "dot" product) of two vectors **y** and **x** is the sum of the products $\sum yx$ of their corresponding elements. Thus the inner product of the vectors $\mathbf{y} = [3, 5, 4\}$ and $\mathbf{x} = [1, -2, -3]$ is $\{3 \times 1\} + \{5 \times (-2)\} + \{4 \times (-3)\} = -19$. The *normalized* inner product $\sum yx/(\sum y^2 \sum x^2)^{0.5} = -19/(50 \times 14)^{0.5} = -0.72$ is the *cosine* of the

angle between the two vectors. Thus in Figure 10A.1 the angle between **y** and **x** is 136° whose cosine is −0.72. Notice in particular that when $\sum yx = 0$ the cosine of the angle between the two vectors is zero, and consequently they are at right angles. The vectors are then said to be "orthogonal" or "normal."

Extensions to *n* Dimensions

Although it is not easy to imagine a space of more than three dimensions, all the properties listed above can be shown to have meaning for vectors in n-dimensional space. As was first demonstrated by R. A. Fisher, vector representation can provide striking insights into the nature of statistical procedures such as the t test and the ANOVA. Again, however, demonstration must be limited to three dimensions.

The *t* Test

Suppose that a random sample of three manufactured items is measured for conformance with a certain standard and the data [3, 5, 4] are recorded as *deviations* in micrometers from specifications. Can you say these are chance deviations? To test the hypothesis that their mean η equals zero, we might calculate the t statistic

$$t_0 = \frac{\bar{y} - \eta_0}{s/\sqrt{n}} = \frac{4 - 0}{1/\sqrt{3}} = 4\sqrt{3} = 6.9$$

where $s^2 = \sum(y - \bar{y})^2/(n - 1) = 1.0$. The t table with $\nu = n - 1 = 2$ degrees of freedom gives $\Pr(t > 6.9) \cong 0.01$, which discredits the null hypothesis that $\eta = 0$. If the mean could equally well have deviated in either direction, the appropriate two-sided significance level would be $\Pr(|t| > 6.9) \cong 0.02$.

Alternatively, the hypothesis could be tested using the ANOVA table shown in Table 10A.1. The significance level $\Pr(F > 48.0) \cong 0.02$ is exactly the same as that for the two-sided t. For when the F ratio has a single degree of freedom in the numerator, $t^2 = F$.

Table 10A.1. Analysis of Variance for Conformance Data

	Sum of Squares	Degrees of Freedom	Mean Square	
Average	$S_{\bar{y}} = n\bar{y}^2 = 48$	1	48	$F = 48.0$
Residuals	$S_{y-\bar{y}} = \sum(y - \bar{y})^2 = 2$	2	1	
Total	$S_y = \sum y^2 = 50$	3		

Specifically

$$t^2 = \frac{n\overline{y}^2}{s^2} = \frac{n\overline{y}^2}{\sum(y-\overline{y})^2/(n-1)} = F$$

Thus

$$\Pr(F > 48.0) = \Pr(t^2 > 6.9^2) = \Pr(|t| > 6.9)$$

Orthogonal Vectors and Addition of Sums of Squares

Table 10A.2 shows the relevant analysis of the observation vector into an average vector and a residual vector. Figure 10A.2*a* shows the same analysis geometrically. Since $\sum \overline{y}(y-\overline{y}) = 0$, the vectors $\overline{\mathbf{y}}$ and $\mathbf{y}-\overline{\mathbf{y}}$ are orthogonal, and as is indicated in the diagram, the three vectors $\mathbf{y}$, $\overline{\mathbf{y}}$, and $\mathbf{y}-\overline{\mathbf{y}}$ form a right triangle. Now the *length* of the vector $\overline{\mathbf{y}}$, indicated by $|\overline{\mathbf{y}}|$, is

$$|\overline{\mathbf{y}}| = \left(\overline{y}_1^2 + \overline{y}_2^2 + \overline{y}_3^2\right)^{1/2} = \sqrt{3}\overline{y} = \sqrt{n}\overline{y}$$

and the length of the vector $\mathbf{y}-\overline{\mathbf{y}}$ is

$$|\mathbf{y}-\overline{\mathbf{y}}| = [(y_1-\overline{y})^2 + (y_2-\overline{y})^2(y_3-\overline{y})^2]^{1/2} = \left[\sum(y-\overline{y})^2\right]^{1/2}$$

The algebraic identity

$$\sum y^2 = n\overline{y}^2 + \sum(y-\overline{y})^2 \qquad S_y = S_{\overline{y}} + S_{y-\overline{y}}$$

may be written as

$$|\mathbf{y}|^2 = |\overline{\mathbf{y}}|^2 + |\mathbf{y}-\overline{\mathbf{y}}|^2$$

From Figure 10A.2*a* it is seen that this is merely a statement of Pythagoras' theorem.

Table 10A.2. Data for Illustration

	Observation Vector	=	sample average vector	+	residual vector
	$\begin{bmatrix}3\\5\\4\end{bmatrix}$	=	$\begin{bmatrix}4\\4\\4\end{bmatrix}$	+	$\begin{bmatrix}-1\\+1\\0\end{bmatrix}$
Sum of squares	50	=	48	+	2
Degrees of freedom	3	=	1	+	2

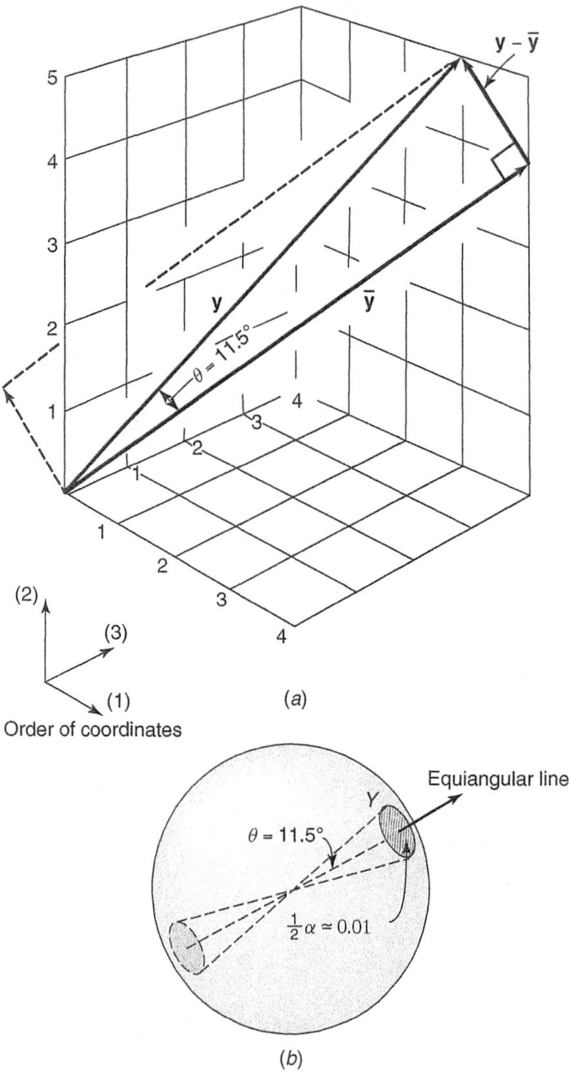

Figure 10A.2. (*a*) Geometric decomposition of [3, 5, 4]. (*b*) Relation to *t* and the ANOVA.

Geometry of the *t* Test

As an early example of the use that Fisher made of vectors and geometry, below is his derivation of the distribution of the *t* statistic. The expression for t_0 can be written as

$$t_0 = \sqrt{n-1}\left\{\frac{n\bar{y}^2}{\sum(y-\bar{y})^2}\right\}^{1/2} = \sqrt{2}\left\{\frac{3\bar{y}^2}{\sum(y-\bar{y})^2}\right\}^{1/2} = \frac{\sqrt{2}|\bar{\mathbf{y}}|}{|\mathbf{y}-\bar{\mathbf{y}}|}$$

Now (see Fig. 10A.2*a*), the length of the vector $\overline{\mathbf{y}}$ divided by the length of the vector $\mathbf{y} - \overline{\mathbf{y}}$ is the cotangent of the angle θ (*theta*) that $\mathbf{y}$ makes with $\overline{\mathbf{y}}$. Thus

$$t_0 = \sqrt{2}\cot\theta \qquad \text{or in general} \qquad t_0 = \sqrt{n-1}\cot\theta$$

Now cot θ gets larger as θ gets smaller, and we see that the quantity t_0, which we routinely refer to the t table, is a measure of how *small* the angle is between $\mathbf{y}$ and $\overline{\mathbf{y}}$. In general, the smaller the angle between $\mathbf{y}$ and its best representation $\hat{\mathbf{y}}$ in the model's plane or hyperplane, the more we would disbelieve the null hypothesis according to which $\mathbf{y}$ and $\hat{\mathbf{y}}$ could lie in any direction with respect to each other.

The data vector $\mathbf{y} = [3, 5, 4]$ strongly suggests the existence of a consistent nonzero component because this vector makes an angle of only 11.5° with the sample average vector [4, 4, 4] on the equiangular line, that is, $[\cot^{-1}(6.9/\sqrt{n-1}) = \cot^{-1}(6.9/\sqrt{2}) = 11.5°]$. In this case the angle is small, t_0 is large, and so the significance level is small.

On the other hand, an inconsistent vector like [−6, 14, 4] would have the same average and hence the same vector $\overline{\mathbf{y}} = [4, 4, 4]$ but the angle between these two vectors is 63.9°, yielding a much smaller value for t_0 and a much larger probability.

Exercise 10A.1. Obtain t_0, θ, and the significance level for the data sets [3.9, 4.1, 4.0] and [−6, 14, 4].

Answer: (69.3, 1.2°, < 0.0005), (0.69, 63.9°, 0.28).

Significance Tests

From the geometry of t it is seen that a test of the hypothesis that the mean $\eta = 0$ against $\eta > 0$ amounts to asking what the probability is that a vector $\mathbf{y}$ would, by chance, make so small an angle as θ with the equiangular line. Now (see Fig. 10A.2*b*), if $\eta = 0$ and the elements of the vector $\mathbf{y}$ are a set of independent normally distributed errors, the vector $\mathbf{y}$ has the property that its tip would be equally likely to lie anywhere on a sphere drawn with its center at the origin and radius $|\mathbf{y}|$.

Now consider the shaded cap shown in the figure on the surface of the sphere and included in the cone obtained by rotating $\mathbf{y}$ about the equilateral line, the vector representing $\overline{y}$. The required probability is evidently equal to the surface area of this cap, expressed as a fraction of the total surface area of the sphere. For a two-sided test we must include the surface area of a second complementary cap on the other side of the sphere, shown in the figure. These are the probabilities in the t table.

Exercise 10A.2. Using the data in Table 10A.2 for illustration, show geometrically the generation of a 95% confidence interval.

Analysis of Variance

You have seen that the sums of squares of the various elements in Table 10A.1 indicating the squared lengths of the corresponding vectors are additive because the components $\overline{\mathbf{y}}$ and $\mathbf{y} - \overline{\mathbf{y}}$ are orthogonal and consequently Pythagoras' theory applies. The degrees of freedom indicate the number of dimensions in which the vectors are free to move. The vector $\mathbf{y}$ is unconstrained and has three degrees of freedom; the vector $\overline{\mathbf{y}}$, which has elements $[\overline{y}, \overline{y}, \overline{y}]$, is constrained to lie on the equiangular line and has only one degree of freedom; and the vector $\mathbf{y} - \overline{\mathbf{y}}$ whose elements must sum to zero has $n - 1 = 2$ degrees of freedom. The ANOVA of Table 10A.2 conveniently summarizes these facts.

In general, each statistical model discussed in this book determines a certain line, plane, or space on which, *if there were no error*, the data *would have* to lie. For the example above, for instance, the model is $y = \eta + e$. Thus, were there no errors ε the data would have to lie on the equiangular line at some point $[\eta, \eta, \eta]$. The t and F criteria measure the angle that the actual data vector, which is subject to error, makes with the appropriate line, plane, or space dictated by the model. The corresponding tables indicate probabilities that angles as small or smaller could occur by chance. These probabilities are dependent on the dimensions of the model and of the data through the degrees of freedom in the table and, of course, on the crucial assumption that the errors are normal and independent.

The geometry and resulting distribution theory for the general case are essentially an elaboration of that given above.

APPENDIX 10B. MATRIX VERSION OF LEAST SQUARES

Matrices provide a convenient shorthand method for writing the important equations used in least squares calculations. In matrix notation the model at the n distinct sets of conditions can be written as

$$\eta = \mathbf{X}\beta$$

where η is the $n \times 1$ vector of expected values for the response, $\mathbf{X}$ is the $n \times p$ matrix of independent variables (regressors), and $\boldsymbol{\beta}$ is the $p \times 1$ vector of parameters.

Since $\mathbf{x}_1, \mathbf{x}_2, \ldots, \mathbf{x}_p$ are the columns of $\mathbf{X}$, the normal equations

$$\mathbf{x}_1^{\mathrm{T}}(\mathbf{y} - \hat{\mathbf{y}}) = \mathbf{0},\ \mathbf{x}_2^{\mathrm{T}}(\mathbf{y} - \hat{\mathbf{y}}) = \mathbf{0}, \ldots,\ \mathbf{x}_p^{\mathrm{T}}(\mathbf{y} - \hat{\mathbf{y}}) = \mathbf{0}$$

can be written succinctly as

$$\mathbf{X}^{\mathrm{T}}(\mathbf{y} - \hat{\mathbf{y}}) = \mathbf{0}$$

where T stands for transpose (exchanging the rows and columns of the matrix to allow for matrix multiplication). Upon substitution of $\hat{\mathbf{y}} = \mathbf{Xb}$

$$\mathbf{X}^{\mathrm{T}}(\mathbf{y} - \mathbf{Xb}) = \mathbf{0}$$

or

$$\mathbf{X}^{\mathrm{T}}\mathbf{Xb} = \mathbf{X}^{\mathrm{T}}\mathbf{y}$$

If we suppose that the columns of $\mathbf{X}$ are linearly independent, $\mathbf{X}^{\mathrm{T}}\mathbf{X}$ has an inverse and

$$\mathbf{b} = [\mathbf{X}^{\mathrm{T}}\mathbf{X}]^{-1}\mathbf{X}^{\mathrm{T}}\mathbf{y}$$

The variance–covariance matrix for the estimates is

$$V(\mathbf{b}) = [\mathbf{X}^{\mathrm{T}}\mathbf{X}]^{-1}\sigma^2$$

If σ^2 is not known, assuming the model form is appropriate, we can substitute the estimate s^2, where $s^2 = (\mathbf{y} - \hat{\mathbf{y}})^{\mathrm{T}}(\mathbf{y} - \hat{\mathbf{y}})/(n-p) = S_R/(n-p)$.

Example

To illustrate this use of matrices, consider the problem of fitting the quadratic model $\eta = \beta_0 + \beta_1 x + \beta_2 x^2$ to the growth data in Table 10.7. The matrices needed are as follows:

$$\mathbf{X} = \begin{bmatrix} 1 & 10 & 100 \\ 1 & 10 & 100 \\ 1 & 15 & 225 \\ 1 & 20 & 400 \\ 1 & 20 & 400 \\ 1 & 25 & 625 \\ 1 & 25 & 625 \\ 1 & 25 & 625 \\ 1 & 30 & 900 \\ 1 & 35 & 1225 \end{bmatrix}, \quad \mathbf{y} = \begin{bmatrix} 73 \\ 78 \\ 85 \\ 90 \\ 91 \\ 87 \\ 86 \\ 91 \\ 75 \\ 65 \end{bmatrix}, \quad \mathbf{b} = \begin{bmatrix} b_0 \\ b_1 \\ b_2 \end{bmatrix}$$

$$\mathbf{X}^{\mathrm{T}}\mathbf{X} = \begin{bmatrix} n & \sum x & \sum x^2 \\ \sum x & \sum x^2 & \sum x^3 \\ \sum x^2 & \sum x^3 & \sum x^4 \end{bmatrix} = \begin{bmatrix} 10 & 215 & 5{,}225 \\ 215 & 5{,}225 & 138{,}125 \\ 5{,}225 & 138{,}125 & 3{,}873{,}125 \end{bmatrix}$$

$$\mathbf{X}^{\mathrm{T}}\mathbf{y} = \begin{bmatrix} \sum y \\ \sum xy \\ \sum x^2 y \end{bmatrix} = \begin{bmatrix} 821 \\ 17{,}530 \\ 418{,}750 \end{bmatrix}$$

Thus $\mathbf{b} = [\mathbf{X}^T\mathbf{X}]^{-1}\mathbf{X}^T\mathbf{y} =$*

$$\begin{bmatrix} b_0 \\ b_1 \\ b_2 \end{bmatrix} = \begin{bmatrix} 4.888449 & -0.468434 & 0.010111 \\ -0.468434 & 0.04823 & -0.001088 \\ 0.010111 & -0.001088 & 0.00002542 \end{bmatrix} \begin{bmatrix} 821 \\ 17530 \\ 418750 \end{bmatrix} = \begin{bmatrix} 35.66 \\ 5.26 \\ -0.128 \end{bmatrix}$$

that is, $b_0 = 35.66$, $b_1 = 5.26$, and $b_2 = -0.128$, and the fitted model is

$$\hat{y} = 35.66 + 5.26\, x - 0.128\, x^2$$

APPENDIX 10C. ANALYSIS OF FACTORIALS, BOTCHED AND OTHERWISE

As already noted, the calculation of main effects and interactions for factorial and fractional factorial designs in the manner discussed in Chapters 5 and 6 is equivalent to the use of least squares. When the runs are not perfectly carried out, as in the example shown in Table 10C.1, analysis can still be performed using least squares.

Suppose that the first-order model

$$\eta = \beta_0 + \beta_1 x_1 + \beta_2 x_2 + \beta_3 x_3 \tag{10C.1}$$

is to be fitted. The intended design is shown in Table 10C.1a, but the design actually carried out appears in Table 10C.1b.

The normal equations for the botched design are

$$\begin{aligned} 8.00b_0 + 0.40b_1 + 1.60b_2 + 0.00b_3 &= 445.50 \\ 0.40b_0 + 7.36b_1 - 0.16b_2 - 0.40b_3 &= 30.06 \\ 1.60b_0 - 0.16b_1 + 8.82b_2 + 0.40b_3 &= 123.68 \\ 0.00b_0 - 0.40b_1 + 0.40b_2 + 8.00b_3 &= 445.50 \end{aligned} \tag{10C.2}$$

yielding

$$b_0 = 54.831, \quad b_1 = 1.390, \quad b_2 = 3.934, \quad b_3 = -3.685 \tag{10C.3}$$

* Specifically, the normal equations are

$$\begin{aligned} 10b_0 + 215b_1 + 5{,}225b_2 &= 821 \\ 215b_0 + 5{,}225b_1 + 138{,}125b_2 &= 17{,}530 \\ 5{,}225b_0 + 138{,}125b_1 + 3{,}873{,}125b_2 &= 418{,}750 \end{aligned}$$

Table 10C.1. Two Versions of a Factorial Design

	Natural Units			Coded Units			
Number Run	Temperature (°C)	pH	Water (gm)	x_1	x_2	x_3	Resiliency, y
		(a) As Planned					
1	100	6	30	−1	−1	−1	
2	150	6	30	+1	−1	−1	
3	100	8	30	−1	+1	−1	
4	150	8	30	+1	+1	−1	
5	100	6	60	−1	−1	+1	
6	150	6	60	+1	−1	+1	
7	100	8	60	−1	+1	+1	
8	150	8	60	+1	+1	+1	
		(b) As Executed					
1	100	6.3	30	−1	−0.7	−1	55.7
2	150	6.5	30	+1	−0.5	−1	56.7
3	100	8.3	30	−1	+1.3	−1	61.4
4	150	7.9	30	+1	+0.9	−1	64.2
5	110	6.1	60	−0.6	−0.9	+1	45.9
6	150	5.9	60	+1	−1.1	+1	48.7
7	100	8.0	60	−1	+1.0	+1	53.9
8	150	8.6	60	+1	+1.6	+1	59.0

If the factorial design had been actually carried out as planned, the entries in Equation 10C.2 would have been

$$\sum_{u=1}^{n} x_{iu} = 0, \qquad i = 1, 2, 3$$

$$\sum_{u=1}^{n} x_{iu} x_{ju} = 0, \qquad i \neq j$$

$$\sum_{u=1}^{n} x_{iu}^2 = 8$$

Thus but the diagonal terms in the normal equations would have vanished, and the design would have been *orthogonal*. Suppose now it was assumed that the design in Table 10C.1a had been executed perfectly and the same data obtained. In that case you would have had equations and estimates as follows:

$$b_0 = \frac{445.5}{8} = 55.688; \qquad b_1 = \frac{11.7}{8} = 1.463$$

$$b_2 = \frac{31.5}{8} = 3.938; \qquad b_3 = \frac{-30.5}{8} = -3.813$$

These estimates differ somewhat from those obtained previously, although perhaps not so much as you might have anticipated. (If you have adequate data you can also deal with missing values and explore the effects of omitting suspect observations in this way.)

APPENDIX 10D. UNWEIGHTED AND WEIGHTED LEAST SQUARES

The ordinary method of least squares produces parameter estimates having smallest variance* if the supposedly independent observations have equal variance. In some instances, however, the variances may not be equal but may differ in some known manner. Suppose that the observation $y_1, y_2, \ldots, y_n$ have variances $\sigma^2/w_1, \sigma^2/w_2, \ldots, \sigma^2/w_n$, where σ^2 is unknown but the constants $w_1, w_2, \ldots, w_n$ that determine the *relative* precision of the observations are known. These constants $w_1, w_2, \ldots, w_n$ are then called the *weights* of the observations. They are proportional to the reciprocals of the variances of the observations.† If the observations are distributed independently, Gauss showed that the procedure yielding estimates with smallest variance is not ordinary least squares but *weighted least squares.*

Assume $\eta_u = \eta(\boldsymbol{\beta}, \mathbf{x})$ is the mean (expected) value of the response at the uth set of experimental conditions. The weighted least squares estimates are obtained by minimizing the weighted sum of squares of discrepancies between the observations observed y_u and the predicted values $\hat{y}_u$,

$$S_w = \sum w_u (y_u - \hat{y}_u)^2$$

instead of the corresponding unweighted sum of squares $S = \sum (y_u - \hat{y}_u)^2$. If the model is nonlinear in the parameters $\boldsymbol{\beta}$, then S_w can be minimized just as in the unweighted case by iterative search methods.

In the particular case of a model linear in the parameters the minimum can be found algebraically by substituting the weighted sums of squares and products in the normal equations. For example, for the first model

$$\hat{y}_u = b_1 x_1 + b_2 x_2$$

the normal equations are

$$b_1 \sum w_u x_{1u}^2 + b_2 \sum w_u x_{1u} x_{2u} = \sum w_u y_u x_{1u}$$

$$b_1 \sum w_u x_{1u} x_{2u} + b_2 \sum w_u x_{2u}^2 = \sum w_u y_u x_{2u}$$

* Strictly, smallest variance among all unbiased estimates that are linear functions of the data.

† You will see in Chapter 11 how they are measures of the amount of relative "information" supplied by each observation.

Also the variances and covariances of the estimates are obtained by substituting weighted for unweighted sums of squares and products in the formulas for ordinary least squares,

$$V(b_1) = \frac{\sigma^2}{\sum w_u x_{1u}^2}$$

For example

$$V(b_1) = \frac{\sum w_u x_{2u}^2}{\sum w_u x_{1u}^2 \sum w_u x_{2u}^2 - \left(\sum w_u x_{1u} x_{2u}\right)^2}$$

Finally, with p fitted parameters the estimate of σ^2 is provided by substituting the minimum weighted sum of squares of the weighted residuals in the formula for s^2. Thus

$$s^2 = \frac{S_w \min}{n-p} = \frac{\sum w_u (y_u - \hat{y}_u)^2}{n-p}$$

Some examples follow.

Example 1: Replicated Data

Suppose you had some data for which the variance was constant but in which certain observations were genuine replicates of each other, as in the following example.

	Experimental conditions		
	1	**2**	**3**
Observations	63, 67, 62	71, 75	87, 89, 83, 85
Averages	64	73	86
Number of observations	3	2	4

The three experimental conditions could be three levels of temperature to which it was believed that the response was linearly related or three different treatments. In either case, for any given values of the model parameters, consider the mean value for *each* replicated set of experimental conditions. Let it be η_1 for the first set, η_2 for the second, and η_3 for the third. Then, (correctly) using ordinary least squares for the individual observations, we estimate the model parameters, η_1, η_2, η_3 by minimizing

$$S = (63 - \hat{y}_1)^2 + (67 - \hat{y}_1)^2 + (62 - \hat{y}_1)^2 + (71 - \hat{y}_2)^2 + (75 - \hat{y}_2)^2 + (87 - \hat{y}_3)^2 + (89 - \hat{y}_3)^2 + (83 - \hat{y}_3)^2 + (85 - \hat{y}_3)^2$$

Alternately you can regard the data as consisting of only the three averages 64, 73, 86, and since these averages have variances $\sigma^2/3$, $\sigma^2/2$, $\sigma^2/4$, appropriate weights are $w_1 = 3$, $w_2 = 2$, and $w_3 = 4$. Fitting these by weighted least squares, you minimize

$$S_w = 3(64 - \hat{y}_1)^2 + 2(73 - \hat{y}_2)^2 + 4(86 - \hat{y}_3)^2$$

For any fitted model this will give exactly the same parameter values as minimizing S with the individual data points.

Consider for example the first three terms in S; these have an average 64 so that

$$(63 - \eta_1)^2 + (67 - \eta_1)^2 + (62 - \eta_1)^2 = [(63 - 64)^2 + (67 - 64)^2 + (62 - 64)^2] + 3(64 - \eta_1)^2$$

The right-hand side of the expression consists of the within-group sum of squares in brackets *not containing* η_1 (hence not containing the parameters of the model) and a second part that is the first term in S_w. If the pooled within-group sum of squares is denoted by S_W, the total sum of squares is therefore

$$S = S_W + S_w$$

Since S_W contains only the deviations from group averages that are independent of the model used, minimization of S implies minimization of S_w and vice versa.

This example shows that weighted least squares is like ordinary least squares, in which observations were replicated in proportion to their weights $w_1, w_2, \ldots, w_n$. Suppose, say, that a straight line is fitted to data $(x_1, y_1), (x_2, y_2), \ldots, (x_n, y_n)$ and the weights are $w_1, w_2, \ldots, w_n$. This is equivalent to fitting ordinary least squares to the averages from runs replicated $w_1, w_2, \ldots, w_n$ times.

If the weights do not greatly differ, the weighted estimates will not be greatly different from unweighted estimates. The Example 3 illustrates what can be a contrary situation.

Example 2: Weighted Least Squares by the Use of Pseudovariates

Consider any model that is linear in the parameters

$$Y_u = \sum_{i=1}^{p} \beta_i x_{iu} + \varepsilon_u$$

and suppose the weight of the uth observation is w_u, that is, $V(y_u) = \sigma^2/w_u$. It is sometimes convenient to perform the weighted analysis by carrying out ordinary least squares on a constructed set of pseudovariates:

$$Y_u = \sqrt{w_u} y_u, \qquad X_{iu} = \sqrt{w_u} x_{iu}$$

Thus the model becomes

$$Y_u = \sum_{i=1}^{p} \beta_i X_{iu} + e_u$$

in which the errors e_u have a constant variance since $e_u = \sqrt{w_u}\varepsilon_u$ and $V(e_u) = \sigma^2$. An ordinary unweighted least squares analysis using Y_u and X_u will now yield the correct *weighted* least squares estimates and their standard errors.

Exercise 10D.1. Confirm the above statement by substituting $Y_u = \sqrt{w_u} y_u$ and $X_{iu} = \sqrt{w_u} x_{iu}$ in the ordinary least squares formulas and noting that they correctly give the weighted least squares formulas.

Example 3: How Weighting Changes the Estimate of Slope of a Straight Line Passing Through the Origin

The effect of moderate changes in weights is usually small. Thus, however, is not true for major changes in weighting as the following examples show.

Consider a model expressing the fact that the mean response is ***proportional*** to x:

$$y = \beta x + e$$

with the errors e independently distributed. Then, if the variance of e is a constant, we know that the least squares estimate of β is $b = \sum xy / \sum x^2$. It quite often happens, however, that σ_e^2 increases as the mean value η increases; or, for this particular model, as x increases. As you see profound differences then occur in the least squares estimates

Relationship Between σ and η	**Weighting**	**Least Squares Estimate, $b = \sum wxy / \sum wx^2$**
I: σ_e^2 constant	$w = \text{const}$	$b_{\text{I}} = \sum xy / \sum x^2$
II: $\sigma_e^2 \propto \eta \propto x$	$w \propto x^{-1}$	$b_{\text{II}} = \bar{y}/\bar{x}$
III: $\sigma_e^2 \propto \eta^2 \propto x^2$	$w \propto x^{-2}$	$b_{\text{III}} = \sum (y/x)/n$

The estimate b_{I} is the usual least squares estimate, appropriate if the variance is independent of the level of the response. If, however, the variance is proportional to η (and hence to x), substitution in the appropriately weighted least squares model gives b_{II} is $\bar{y}/\bar{x}$. Thus the best fitting line is obtained by joining the origin to the point ($x = \bar{x}$ and $y = \bar{y}$). Finally, if the standard deviation is proportional to η, so that the variance is proportional to η^2 and hence to x^2, then $b_{\text{III}} = \sum (y/x)/n$. The best fitting straight line has a slope that is the average of all the slopes of the lines joining the individual points to the origin.

Exercise 10D.2. Justify the statements made above.

REFERENCES AND FURTHER READING

Aitken, A. C. (1951) *Determinants and Matrices*, University Mathematical Texts, Interscience, New York.

Box, G. E. P., and Hunter, W. G. (1965) The experimental study of experimental mechanisms, *Technometrics*, **7**, 23–42.

Box, G. E. P., and MacGregor, J. F. (1974) The analysis of closed loop dynamic stochastic systems, *Technometrics*, **16**, 391–401.

Box, G. E. P., and Wetz, J. (1973) *Criteria for Judging Adequacy of Estimation by an Approximating Response Function*, Tech. Report 9, Statistics Department, University of Wisconsin, Madison.

Daniel, C., and Wood, F. S. (1980) *Fitting Equations to Data*, 2nd ed., Wiley, New York.

Fisher, R. A. (1921). Studies in crop variation. I. An examination of the yield of dressed grain from Broadbalk. *J. Agric. Sci.* **11**, 107–135.

Fisher, R. A., and Mackenzie (1923). Studies in crop variation. II. The manurial response of different potato varieties. *J. Agric. Sci.* **13**, 311–320.

Fisher, R. A. (1924). The influence of rainfall on the yield of wheat at Rothamsted. *Phil. Trans., B*, **213**, 89–142.

Fisher, R. A., and Yates, F. (1963) *Statistical Tables for Biological, Agricultural and Medical Research*, 6th ed., Hafner, New York.

Hunter, W. G., and Mezaki, R. (1964) A model building technique for chemical engineering kinetics, *Am. Inst. Chem. Eng. J.*, **10**, 315–322.

Marske, D. M., and Polkowski, L. B. (1972) Evaluation of methods for estimating biochemical oxygen demand parameters, *J. Water Pollut. Control Fed.*, **44**.

A thorough exposition of least squares accompanied by an extensive bibliography is provided by:

Draper, N. R., and Smith, H. (1998) *Applied Regression Analysis*, 3rd. ed., Wiley, New York.

Plackett, R. L. (1960) *Regression Analysis*, Clarendon, Oxford.

If errors are correlated, statistical methods based on the assumption of independence of errors can give highly misleading results. A dramatic illustration is provided by the following:

Box, G. E. P., and Newbold P. (1971) Some comments on a paper of Coen, Gomme and Kendall, *J. Roy. Statist. Soc. Ser. A*, **134**, 229–240.

Coen, P. G., Gomme, E. D., and Kendall, M. G. (1969) Lagged relationships in economic forecasting, *J. Roy. Statist. Soc. Ser. A*, **132** 133–141.

Estimation when the distribution is not normal is elucidated in:

McCullagh, P., and Nelder, J. A. (1983) *Generalized Linear Models*, Chapman & Hall, London.

Box, G. E. P., Jenkins, G. M., and Reinsel, G. C. (1994) *Time Series Analysis, Forecasting and Control*, 3rd ed. Prentice Hall, Englewood Cliffs, NJ.

Box, G. E. P., and MacGregor, J. F. (1974). The analysis of closed loop dynamic stochastic system, *Technometrics*, **16**, 391–398.

Bates, D. M., and Watts, D. G. (1988) *Nonlinear Regression Analysis and Its Applications*, Wiley, New York.

QUESTIONS FOR CHAPTER 10

1. What is an example of a model that is linear in the parameters and one that is not?
2. What is meant by an ANOVA obtained from a least squares fit?
3. What is meant by a test of goodness of fit?
4. What is meant by a confidence region?
5. What is meant by a confidence band about a fitted line?
6. What is meant by an indicator variable? Give an example in which there are two indicator variables.
7. What are some of the advantages of least squares analyses of experimental designs?
8. What is meant by nonsense correlation?
9. Might you be able to employ a nonsense correlation to make a prediction? Explain.
10. What is meant by orthogonality?
11. What are some of the advantages of orthogonal designs?
12. Give a graphical representation of the fitting the model $y = \beta x + e$ to the data y_1, y_2, y_3 and x_1, x_2, x_3.
13. What is meant by the extra sum of squares principle? How may it be used?

PROBLEMS FOR CHAPTER 10

Whether or not specifically asked, the reader should always (1) plot the data in any potentially useful way, (2) state the assumptions made, (3) comment on the appropriateness of these assumptions, and (4) consider alternative analyses.

1. In a physics laboratory experiment on thermal conductivity a student collected the following data:

Time, x (sec)	$y = \log I$
300	0.79
420	0.77
540	0.76
660	0.73

(a) Plot these data, fit a straight line by eye, and determine its slope and intercept.

(b) By a least squares fit the model $y = \beta_0 + \beta x + \varepsilon$ to these data. Plot the least squares line $\hat{y} = b_0 + bx$ with the data.

(c) Compare the answers to parts (*a*) and (*b*).

2. **(a)** An electrical engineer obtained the following data from six randomized experiments:

Dial Setting, x	Measured Voltage, y
1	31.2
2	32.4
3	33.4
4	34.0
5	34.6
6	35.0

Suggest a simple empirical formula to relate y to x, assuming that the standard deviation of an individual y value is $\sigma = 0.5$.

(b) Would your answer change if $\sigma = 0.05$? Why?

3. A metallurgist has collected the following data from work on a new alloy:

x	Temperature (°F)	Amount of Iron Oxidized (mg)
1	1100	6
2	1200	6
3	1300	10
4	1400	30

The first column gives

$$x = \frac{\text{temperature} - 1000}{100}$$

If the prediction equation $\hat{y} = bx^2$ is used (where $\hat{y}$ is the predicted value of the amount of iron oxidized), what is the least squares estimate for the amount of iron oxidized at 1500°F? Comment on the assumptions necessary to validate this estimate.

4. **(a)** Using the method of least squares, fit a straight line to the following data. What are the least squares estimates of the slope and intercept of the line?

x	10	20	30	40	50	60
y	2.7	3.6	5.2	6.1	6.0	4.9

(b) Calculate 99% confidence intervals for the slope and intercept.

(c) Comment on the data and analysis and carry out any further analysis you think is appropriate.

5. Fit the model $y = \beta_1 x + \beta_2 x^2 + \varepsilon$ to these data, which were collected in random order:

x	1	1	2	2	3	3	4	4
y	15	21	36	32	38	49	33	30

(a) Plot the data.

(b) Obtain least squares estimates b_1 and b_2 for the parameters β_1 and β_2.

(c) Draw the curve $\hat{y} = b_1 x + b_2 x^2$.

(d) Do you think the model is adequate? Explain your answer.

6. Using the method of least squares, an experimenter fitted the model

$$\eta = \beta_0 + \beta_1 x \qquad \text{(I)}$$

to the data below. It was known that σ was about 0.2. A friend suggested it would be better to fit the following model instead:

$$\eta = \alpha + \beta(x - \overline{x}) \qquad \text{(II)}$$

where $\overline{x}$ is the average value of the x's.

x	10	12	14	16	18	20	22	24	26	28	30	32
y	80.0	83.5	84.5	84.8	84.2	83.3	82.8	82.8	83.3	84.2	85.3	86.0

(a) Is $\hat{\alpha} = \hat{\beta}_0$? Explain your answer. (The caret indicates the least squares estimate of the parameter.)

(b) Is $\hat{\beta} = \hat{\beta}_1$? Explain your answer.

(c) Are $\hat{y}_{\text{I}}$ and $\hat{y}_{\text{II}}$, the predicted values of the responses for the two models, identical at $x = 40$? Explain your answer.

(d) Considering the two models above, which would you recommend the experimenter use: I or II or both or neither? Explain your answer.

7. The residual mean square is often used to estimate σ^2. There are circumstances, however, in which it can produce (*a*) a serious underestimate of σ^2 or (*b*) a serious overestimate of σ^2. Can you give examples of such circumstances?

8. The following data, which are given in coded units, were obtained from a tool life-testing investigation in which x_1 = measure of the feed, x_2 = measure of the speed, and y = measure of the tool life:

x_1	x_2	y
−1	−1	61
1	−1	42
−1	1	58
1	1	38
0	0	50
0	0	51

(a) Obtain the least squares estimates of the parameters β_0, β_1, and β_2 in the model

$$\eta = \beta_0 + \beta_1 x_1 + \beta_2 x_2$$

where $\eta =$ the mean value of y.

(b) Which variable influences tool life more over the region of experimentation, feed, or speed?

(c) Obtain a 95% confidence interval for β_2, assuming that the standard deviation of the observations $\sigma = 2$.

(d) Carry out any further analysis that you think is appropriate.

(e) Is there a simpler model that will adequately fit the data?

9. Using the data below, determine the least squares estimates of the coefficients in the model

$$\eta = \beta_0 + \beta_1 x_1 + \beta_2 x_2 + \beta_3 x_3$$

where η is the mean value of the drying rate.

Temperature (°C)	Feed rate (kg/min)	Humidity (%)	Drying Rate (coded units)
300	15	40	3.2
400	15	40	5.7
300	20	40	3.8
400	20	40	5.9
300	15	50	3.0
400	15	50	5.4
300	20	50	3.3
400	20	50	5.5

Coding for the variables is as follows:

x_i	$i = 1$ Temperature (°C)	$i = 2$ Feed rate (kg/min)	$i = 3$ Humidity (%)
−1	300	15	40
+1	400	20	50

The coding equation for temperature is

$$x_1 = \frac{\text{temperature in units of °C} - 350}{50}$$

(a) What are the coding equations for feed rate and humidity?

(b) The design is a 2^3 factorial. What are the relations between b_0, b_1, b_2, b_3 and the factorial effects calculated in the usual way?

10. Suppose that the first seven runs specified in problem 9 had been performed with the results given but in the eighth run the temperature had been set at 350°C, the feed rate at 20 kg/min, and the humidity at 45% and the observed drying rate had been 3.9.

(a) Analyze these data.

(b) Compare the answers to those obtained for problem 9.

11. An experimenter intended to perform a full 2^3 factorial design, but two values were lost as follows:

x_1	x_2	x_3	y
−1	−1	−1	3
+1	−1	−1	5
−1	+1	−1	4
+1	+1	−1	6
−1	−1	+1	3
+1	−1	+1	Data missing
−1	+1	+1	Data missing
+1	+1	+1	6

He now wants to estimate the parameters in the model

$$\eta = \beta_0 + \beta_1 x_1 + \beta_2 x_2 + \beta_3 x_3$$

With the data above, is this possible? Explain your answer. If your answer is yes, set up the equation(s) he would have to solve. If your answer is no, what is the minimum amount of additional information that would be required?

12. Mr. A and Ms. B, both newly employed, pack crates of material in the basement. Let x_1 and x_2 be indicator variables showing which packer is on duty and y be the number of crates packed. On 9 successive working days these data were collected:

Day, t	**Mr. A, x_{1t}**	**Ms. B, x_{2t}**	y_t
1	1	0	48
2	1	1	51
3	1	0	39
4	0	1	24
5	0	1	24
6	1	1	27
7	0	1	12
8	1	0	27
9	1	1	24

Denote by β_1 and β_2 the number of crates packed per day by A and B, respectively. First entertain the model

$$\eta = \beta_1 x_1 + \beta_2 x_2$$

Carefully stating what assumptions you make, do the following:

(a) Obtain least squares estimates of β_1 and β_2.

(b) Compute the residuals $y_t - \hat{y}_t$.

(c) Calculate $\sum_{t=1}^{9} x_{1t}(y_t - \hat{y}_t)$ and $\sum_{t=1}^{9} x_{2t}(y_t - \hat{y}_t)$.

(d) Show an analysis of variance appropriate for contemplation of the hypothesis that A and B pack at a rate of 30 crates per day, the rate laid down by the packers' union.

(e) Split the residual sum of squares into a lack-of-fit term $S_L = \sum_{u=1}^{3} 3(\hat{y}_u - \overline{y}_u)^2$ and an "error" term $\sum_{u=1}^{3}\sum_{i=1}^{3}(y_{ui} - \overline{y}_u)^2$.
[*Note*: The above design consists of three distinct sets of conditions, $(x_1, x_2) = (0, 1), (1, 0), (1, 1)$, which are designated by the subscripts $u = 1, 2, 3$, respectively. The subscripts $i = 1, 2, 3$ are used to designate the three individual observations included in each set.]

(f) Make a table of the following quantities, each having nine values: x_1, x_2, y_{ui}, $\hat{y}_u$, $\overline{y}_u$, $\hat{y}_u - \overline{y}_u$, and $y_{ui} - \overline{y}_u$. Make relevant plots. Look for a time trend in the residuals.

(g) At some point it was suggested that the presence of B might stimulate A to do more (or less) work by an amount β_{12} while the presence of A might stimulate B to do more (or less) work by an amount β_{21}. Furthermore, there might be a time trend. A new model was therefore tentatively entertained:

$$\eta = (\beta_1 + \beta_{12} x_2) x_1 + (\beta_2 + \beta_{21} x_1) x_2 + \beta_4 t$$

that is,

$$\eta = \beta_1 x_1 + \beta_2 x_2 + (\beta_{12} + \beta_{21}) x_1 x_2 + \beta_4 t$$

The β_{12} and β_{21} terms cannot be separately estimated, but the combined effect $\beta_{12} + \beta_{21} = \beta_3$ can be estimated from this model:

$$\eta = \beta_1 x_1 + \beta_2 x_2 + \beta_3 x_3 + \beta_4 t$$

where $x_1 x_2 = x_3$. Fit this model, examine the residuals, and, if the model appears adequate, report least squares estimates of the parameter together with their standard errors.

(h) What do you think is going on? Consider other ways in which the original model could have been wrong.

13. **(a)** Fit a first-order model to the following data on the strength of welds:

Ambient Temperature, x_1 (°C)	Wind Velocity, x_2 (km/hr)	Ultimate Tensile Stress, y (coded units)
5	2	83.4
20	2	93.5
5	30	79.0
20	30	87.1

(b) Plot contours of the fitted surface in the region of experimentation.

(c) Stating assumptions, say how ultimate tensile stress might be increased. The standard deviation of an individual y reading is approximately one unit.

14. Using least squares, fit a straight line to the following data:

x	12	13	14	15	16
y	404	412	413	415	422

Carefully stating *all* assumptions you make, obtain a 95% confidence interval for the mean value of the response at $x = 10$.

15. The following results were obtained by a chemist (where all the data are given in coded units):

Temperature, x_1	pH, x_2	Yield of Chemical Reaction, y
−1	−1	6
+1	−1	14
−1	+1	13
+1	+1	7
−1	−1	4
+1	−1	14
−1	+1	10
+1	+1	8

(a) Fit the model $\eta = \beta_0 + \beta_1 x_1 + \beta_2 x_2$ by the method of least squares (η is the mean value of the yield).

(b) Obtain an estimate of the experimental error variance of an individual yield reading, assuming this model is adequate.

(c) Try to fit the model $\eta = \beta_0 + \beta_1 x_1 + \beta_2 x_2 + \beta_{11} x_1^2 + \beta_{22} x_2^2 + \beta_{12} x_1 x_2$. What difficulties arise? Explain these. What parameters and linear combinations of parameters can be estimated with this design?

(d) Consider what experimental design is being used here and make an analysis in terms of factorial effects. Relate the two analyses.

16. An engineer wants to fit the model $\eta = \beta x$ to the following data by the method of least squares. The tests were made in the random order indicated below. Do this work for him. First make a plot of the data.

Test Order, u	Temperature, x_u (°C)	Surface Finish, y_μ
1	51	61
2	38	52
3	40	58
4	67	57
5	21	41
6	13	21

(a) Obtain the least squares estimate for the parameter β.

(b) Find the variance of this estimate (assuming that the variances of the individual y readings are constant and equal to 16).

(c) Obtain a 90% confidence interval for β.

(d) Make any further analysis you think appropriate and comment on your results. State any assumptions you make.

17. Which of the following models from physics, as written, are linear in the parameters?

(a) $s = \frac{1}{2}at^2$ (a is the parameter)

(b) $d = (1/Y)(FL^3/12\pi r^4)$ ($1/Y$ is the parameter)

(c) $d = CL^n$ (C and n are parameters)

(d) $\ln d = \ln C + n \ln L$ ($\ln C$ and n are parameters)

(e) $\log I = \log I_0 - (KAL/2.303MLS)$ ($\log I_0$ and K are parameters)

(f) $\Delta I_p = g_m \Delta E_g$ (g_m is the parameter)

(g) $N = N_0 e^{-\lambda t}$ (λ is the parameter)

(h) $\log T = -kn$ (k is the parameter)

18. Which of the following models are linear in the parameters? Explain your answer.

(a) $\eta = \beta_1 x_1 + \beta_2 \log x_2$
(b) $\eta = \beta_1 \beta_2 x_1 + \beta_2 \log x_2$
(c) $\eta = \beta_1 \sin x_2 + \beta_2 \cos x_2$
(d) $\eta = \beta_1 \log \beta_1 x_1 + \beta_2 x_1 x_2$
(e) $\eta = \beta_1 (\log x_1) e^{-x_2} + \beta_2 e^{x_3}$

19. State whether each of the following models is linear in β_1, linear in β_2, and so on. Explain your answers.
(a) $\eta = \beta_1 e^{x_1} + \beta_2 e^{x_2}$
(b) $\eta = \beta_1 \beta_2 x_1 + \beta_3 x_2$
(c) $\eta = \beta_1 + \beta_2 x + \beta_3 x^2$
(d) $\eta = \beta_1 (1 - e^{-\beta_2 x})$
(e) $\eta = \beta_1 x_1^{\beta_2} x_2^{\beta_3}$

20. Suppose that a chemical engineer uses the method of least squares with the data given below to estimate the parameters β_0, β_1, and β_2 in the model

$$\eta = \beta_0 + \beta_1 x_1 + \beta_2 x_2$$

where η is the mean response (peak gas temperature °R). She obtains the following fitted equation:

$$\hat{y} = 1425.8 + 123.3 x_1 + 96.7 x_2$$

where $\hat{y}$ is the predicted value of the response and x_1 and x_2 are given by the equations

$$x_1 = \frac{\text{compression ratio} - 12.0}{2}$$

and

$$x_2 = \frac{\text{cooling water temperature} - 550°\mathbf{R}}{10°\mathbf{R}}$$

Test	**Compression Ratio**	**Cooling Water Temperature (°R)**	**Peak gas Temperature, y (°R)**	**Residual, $y - \hat{y}$**
1	10	540	1220	14
2	14	540	1500	48
3	10	560	1430	41
4	14	560	1650	4
5	8	550	1210	31
6	16	550	1700	29
7	12	530	1200	−32
8	12	570	1600	−29
9	12	550	1440	14
10	12	550	1450	24
11	12	550	1350	−76
12	12	550	1360	−66

The tests were run in random order. Check the engineer's calculations. Do you think the model form is adequate?

21. **(a)** Obtain least squares estimates of the parameters in the model

$$y = \beta_1(1 - e^{-\beta_2 x}) + \varepsilon$$

given the data

x (day)	1	2	3	4	5	6
y (average BOD, mg/L)	4.3	8.2	9.5	10.35	12.1	13.1

(b) Construct an approximate 95% confidence region for the parameters. [*Source*: L. B. Polkowski and D. M. Marske (1972), *J. Water Pollut. Control Fed.*, **44**, 1987.]

22. The following data were obtained from a study on a chemical reaction system:

Trial Number	**Temperature (°F)**	**Concentration (%)**	**pH**	**Yield**
1	150	40	6	73, 70
2	160	40	6	75, 74
3	150	50	6	78, 80
4	160	50	6	82, 82
5	150	40	7	75
6	160	40	7	76, 79
7	150	50	7	87, 85, 82
8	160	50	7	89, 88

The 16 runs were performed in random order, trial 7 being run three times, trial 5 once, and all the rest twice. (The intention was to run each test twice, but a mistake was made in setting the concentration level on one of the tests.)

(a) Analyze these data. One thing you might wish to consider is fitting by weighted least squares the following model to them: $\eta = \beta_0 + \beta_1 x_1 + \beta_2 x_2 + \beta_3 x_3$, where x_1, x_2, and x_3 (preferably, though not necessarily, in coded units) refer to the variables temperature, concentration, and pH, respectively, and η is the mean value of the yield. Alternatively, a good indication of how elaborate a model is needed will be obtained by averaging the available data and making a preliminary factorial analysis on the results.

(b) Compare these two approaches.

23. If suitable assumptions are made, the following model can be derived from chemical kinetic theory for a consecutive reaction with η the yield of intermediate product, x the reaction time, and β_1 and β_2 the rate constants to be estimated from the experimental data:

$$\eta = \frac{\beta_1}{\beta_1 - \beta_2}(e^{-\beta_2 x} - e^{-\beta_1 x})$$

(a) A chemist wants to obtain the most precise estimates possible of β_1 and β_2 from the following data and asks you what he should do. What would you ask him? What might you tell him?

Reaction Time, x (min)	**Concentration of Product, y (g-mol/L)**
10	0.20
20	0.52
30	0.69
40	0.64
50	0.57
60	0.48

The reaction is this type: reactant → product → by-product; β_1 is the rate constant of the first reaction and β_2 of the second reaction.

(b) Obtain the least squares estimates for β_1 and β_2.

24. As part of a pollution study, a sanitary engineer has collected some BOD data on a particular sample of water taken from a certain stream in Wisconsin. The BOD equation is

$$\eta = L_a(1 - e^{-kt})$$

where η = mean value of BOD, t = time, and L_a and k are two parameters to be estimated. Her data consist of one BOD reading for each of several times t. She wants to estimate the parameters in the equation.

(a) Under what circumstances would it be appropriate to use the method of least squares?

(b) Why might it be inappropriate to use the ordinary method of least squares in this situation?

(c) Assuming that appropriate circumstances exist, explain how the method of least squares could be used to solve this problem. Your explanation should be understandable to an engineer who has had no experience with statistical methods.

25. **(a)** Fit a suitable model to the following data:

Coded Values of Three Adjustable Variables on Prototype Engine			Measured Noise (coded units)
x_1	x_2	x_2	
−1	−1	−1	10
+1	−1	−1	11
−1	+1	−1	4
+1	+1	−1	4
−1	−1	+1	9
+1	−1	+1	10
−1	+1	+1	5
+1	+1	+1	3
0	0	0	6
0	0	0	7
0	0	0	8
0	0	0	7

(b) Draw any conclusions that you consider appropriate. State assumptions

CHAPTER 11

Modeling, Geometry, and Experimental Design

Chapter 12 will discuss the application of response surface methods (RSM). The present chapter is a preliminary discussion of the basic tools of RSM. Some readers with previous experience may prefer to read Chapter 12 first and refer back to whatever they may need in this chapter.

Graphical Representation

Our minds are such that graphical representation can greatly help understanding and stimulate discovery. The solid line in Figure 11.1*a* is a graph displaying the relationship between the mean value η of a response such as yield and a single quantitative factor temperature. The influence of a second factor, concentration, may be shown by displaying additional graphs such as those indicated by the dotted lines in the same figure. Such a relationship can also be represented by a curved surface such as that in Figure 11.1*b* for which the graphs in Figure 11.1*a* are sections at various levels of concentration. Alternatively, the same relationship may be represented by Figure 11.1*c*, which shows yield *contours*, that is, sections of Figure 11.1*b* for various levels of the response η. The relationship between the response η and *three* quantitative factors such as temperature, pressure, and concentration may be represented by a series of contour *surfaces* such as those shown in Figure 11.2*a*. Notice that, for example, a section of this display taken at a specific pressure should produce the 2D contour diagram in temperature and concentration for that pressure.

Exercise 11.1. (a) Draw graphs similar to those in Figure 11.1 for yield versus concentration. (b) How might you represent geometrically the relationship between a response and four factors?

Quantitative and Qualitative Factors

Among the design factors in this book some have been quantitative and some qualitative. Thus, for the 2^3 factorial design for the pilot plant experiment in

Statistics for Experimenters, Second Edition. By G. E. P. Box, J. S. Hunter, and W. G. Hunter

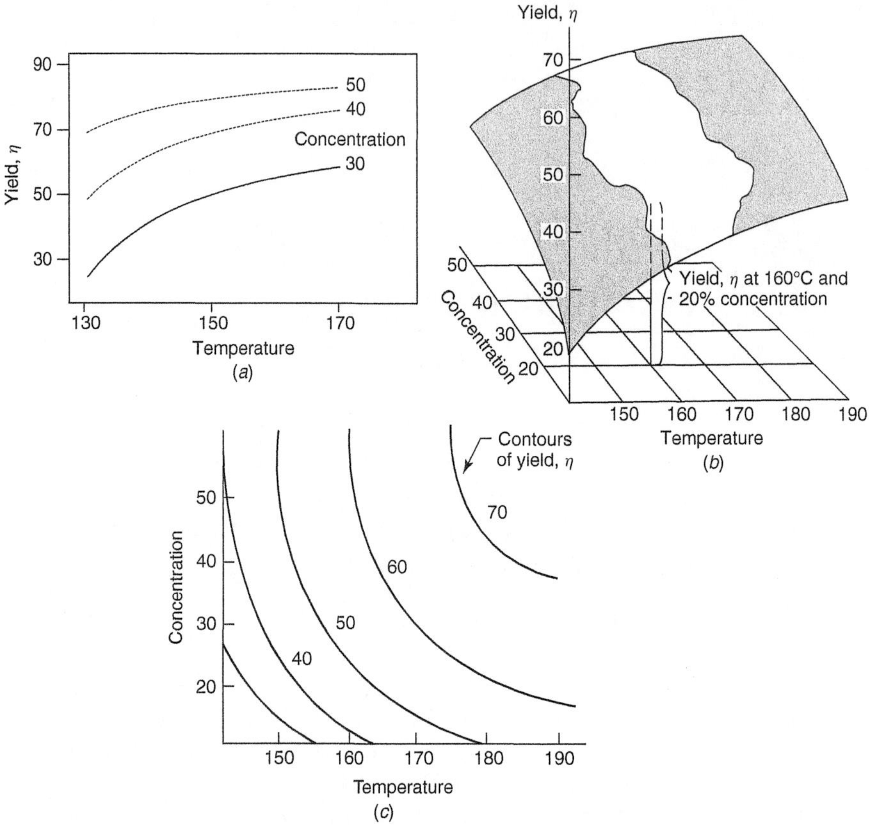

Figure 11.1. Three graphical representations of the relationship between mean yield and temperature and concentration.

Chapter 5, two of the factors, temperature (T) and concentration (C), were quantitative and one, type of catalyst (K), was qualitative. Quantitative factors have different *levels* and qualitative factors different *versions*. (In this book "levels" is commonly used to identify both.) One reason for testing two or more levels or versions together in an experimental design is that the effects can reinforce each other. That they do so implies the characteristic of *smoothness* for the quantitative factors and *similarity* for the qualitative factors.

Smoothness

Over the ranges in which one or more of the quantitative factors are changed, it is implicitly assumed that the response changes in a regular manner that can be represented by a smooth response curve or a smooth response surface. It is not difficult to imagine circumstances where there this might not be true. Materials can go through a change of state—water can freeze and it can boil; crystals may or may not appear in an alloy. In most cases, however, the experimenter will

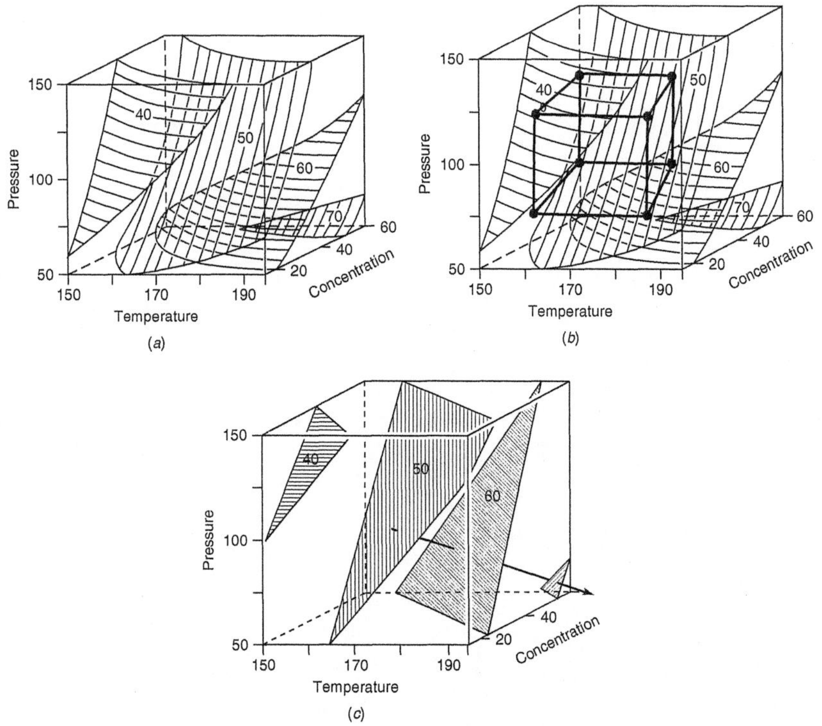

Figure 11.2. Contour representation of the relationship between yield η and three quantitative factors; (*a*) the true relationship; (*b*) a 2^3 factorial pattern of experimental points; (*c*) of a first-order approximating model and path of steepest ascent.

know enough about the physical state in the factor–response relationship to tell when smoothness of the response is a reasonable assumption.

Similarity

For qualitative factors the levels (versions) that are to be compared should be of the same essential nature. In an experiment concerning dogs' preferences for particular dog foods, although one type of dog food may be better than another, all are assumed to be edible by dogs.* Sometimes an experiment demonstrates that versions thought to be similar are not. For illustration Figure 11.3 shows a 2^3 experiment in which the response to changes in temperature and concentration are totally different for two catalysts A and B. In general, this situation would be exemplified by large interactions between the relevant qualitative and quantitative factors and would imply that, in further experimentation, the levels of the qualitative factor, for example the type of catalyst, may need to be considered separately. Notice also in Figure 11.3 that, although, locally, catalyst B gave lower yields than A, you could not safely conclude from these data that

* It is just conceivable that this would need to be checked by experimentation.

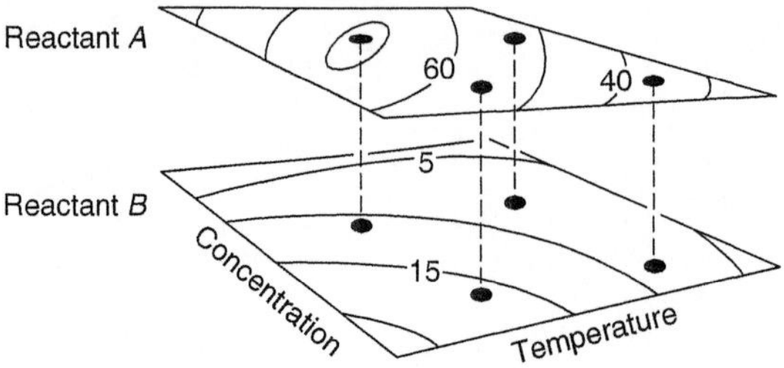

Figure 11.3. A 2^3 design. Response surfaces smooth over temperature and concentration but dissimilar for different catalysts.

no further trials should be made with catalyst B. It could still be true that at levels of temperature and concentration other than those tested in the experiment catalyst B could be better than catalyst A.

Models and Approximations

The most that can be expected from any model is that it can supply a useful approximation to reality:

All models are wrong; some models are useful.

For example, suppose the *true but unknown* relationship between a response η and the factors temperature T, pressure P, and concentration C was represented by the contour diagram of Figure 11.2*a*. A pattern of experimental points that might be used to probe the space of the factors to obtain some idea of the nature of the unknown relationship is shown in Figure 11.2*b*.

As in earlier chapters experimental designs such as this 2^3 factorial are usually expressed generically in coded form. In this case the eight-run design may be written in standardized (coded) form obtained from the eight experimental combinations

$$x_1 = \pm 1, \qquad x_2 = \pm 1, \qquad x_3 = \pm 1$$

Thus

x_1	x_2	x_3
−1	−1	−1
1	−1	−1
−1	1	−1
1	1	−1
−1	−1	1
1	−1	1
−1	1	1
1	1	1

where

$$x_1 = \frac{T - 172}{20}, \qquad x_2 = \frac{P - 100}{25}, \qquad x_3 = \frac{C - 36}{15} \tag{11.1a}$$

and $T = 172$, $P = 100$, and $C = 36$ defines the center of the experimental region where $x_1 = x_2 = x_3 = 0$. The numbers 20, 25, and 15 in the denominators of these expressions are suitable scale factors chosen by the experimenter.

Suppose that after running these eight experiments in some random sequence and recording the observed yields $y_1, y_2, \ldots, y_8$ the fitted linear relationship was

$$\hat{y} = 53 + 6x_1 - 5x_2 + 7x_3 \tag{11.1b}$$

Figure 11.2*c* represents this relationship. It has equally spaced parallel planar contours whose closeness indicates the rate of change of $\hat{y}$. This *first-order approximation* to the response η could be adequate if your main objective was to gain a general idea of what was happening to η within the current experimental region and discover what direction you might take to further improve things. For example, in Figure 11.2*c* the direction to possibly higher yields is indicated by the arrow at right angles to the planar contours. But there are aspects of response relationship that the first-order approximation did not capture—for instance that the true contour surfaces were not planar but somewhat curved and that they were not exactly parallel or equally spaced. The approximating equation is merely a local "graduating function." It can be of great value for representing a response system that (at least initially) must be studied empirically. Furthermore, such an empirical approximation can be helpful to the subject matter specialist attempting to gain a better understanding of the system and occasionally lead to a mechanistic rather than an empirical model. Remember, however, that no model, mechanistic or empirical, can ever *exactly* represent any real phenomenon. Even the theoretical models of physics, chemistry, or of any other science are really approximations good enough until a better one comes along (see Kuhn, 1996).

11.1. SOME EMPIRICAL MODELS

Linear Relationships

The fitted approximate model displayed in Figure 11.2*c* is

$$\hat{y} = 53 + 6x_1 - 5x_2 + 7x_3 \tag{11.2}$$

A general expression for such a linear or *first-order* model is

$$\hat{y} = b_0 + b_1x_1 + b_2x_2 + b_3x_3 \tag{11.3}$$

where in this example $b_0 = 53$ is the response at the local origin (supposed to be at the center of the design). Also the *first-order* coefficients $b_1 = 6$, $b_2 = -5$, and $b_3 = 7$ are the gradients or rates of change of $\hat{y}$ as x_1, x_2, and x_3 are increased—they are estimates of the first derivatives $\partial\eta/\partial x_1$, $\partial\eta/\partial x_2$, and $\partial\eta/\partial x_3$ at the center of the system.

Trends and Steepest Ascent

In addition to showing what is happening in the immediate region of interest, the changes implied by the first-order model can indicate in what direction you might find a higher response. (We temporarily suppose that higher is better.) It might then be helpful to do a few exploratory runs following such an indicated trend. A vector radiating from the center of the design that is normal (perpendicular) to the planar contours is indicated by the arrow in Figure 11.2*c*. This is called the direction of *steepest ascent* and the relative values of the first-order coefficients (6, −5, 7)* determine the direction in which, given the present coding of the factors, a maximum increase can be expected. This concept applies to any first-order equation in k factors,

$$\hat{y} = b_0 + b_1x_1 + b_2x_2 + \cdots + b_kx_k \tag{11.4}$$

Runs on the Path of Steepest Ascent

For the fitted model

$$\hat{y} = 53 + 6x_1 - 5x_2 + 7x_3$$

with the scaling shown in Equation 11.1a,

$$x_1 = \frac{T - 172}{20}, \qquad x_2 = \frac{P - 100}{25}, \qquad x_3 = \frac{C - 36}{15}$$

the steepest ascent path requires *relative* changes calculated as follows:

Change in x's	Unit Conversion	Planned Change
$x_1 + 6$	Temperature: $6 \times 20 = 120$	+20°C
$x_2 - 5$	Pressure: $-5 \times 25 = -125$	−20.8 psi
$x_3 + 7$	Concentration: $+7 \times 15 = 105$	+17.5%

To follow the steepest ascent path, you must therefore make the *proportional* changes indicated in the table. For example if you decided as a first step to increase the temperature T by +20°C the corresponding changes for pressure P would be −20.8 psi and for concentration C+17.5%. These steps are made from an origin located at the center of the experimental design at $T = 172$, $P = 100$,

* These are proportional to the direction cosines of this vector.

and $C = 36$. Thus equal steps along the steepest ascent path would be as follows:

	Step 0	Step 1	Step 2
T	172	192	212
P	100	79.2	58.4
C	36	53.5	71.0

The results would tell you whether to stop or to explore further.* You will be helped in deciding how far to go by taking into account any other subsidiary responses that might change either for the better or worse along the path.

One-Factor Curved Relationships

Frequently a graph is needed to represent the curved relationship between a mean response η as yield and a single factor such as temperature coded as x_1. Figure 11.4 displays such a true relationship between a response η and factor x_1. Also shown by shading under the curve is on approximating quadratic relationship between $\hat{y}$ and x_1. The quadratic approximation is called a *second-order* model, obtained by adding a term x_1^2 to the first-order model, so that

$$\hat{y} = b_0 + b_1 x_1 + b_{11} x_1^2 \tag{11.5}$$

(Note how the quadratic coefficient b_{11} has a double subscript.) The maximum value of the approximating quadratic curve is at the point $x_1 = -b_1/2b_{11}$ with $\hat{y}_{\max} = b_0 - b_1^2/4b_{11}$. It is important to remember that such an approximation might be good locally e.g., in the unshaded region but become inadequate over

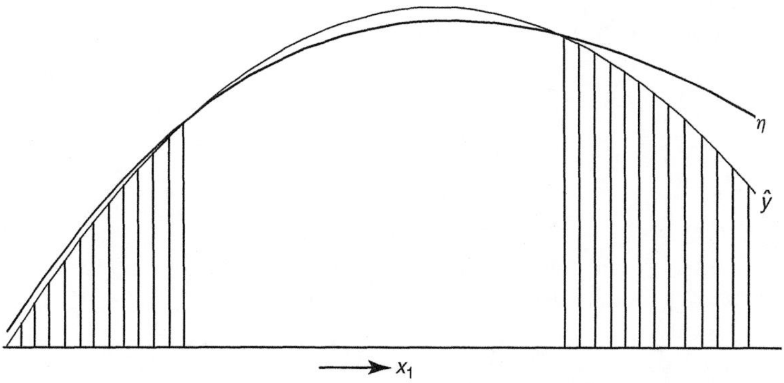

Figure 11.4. True curved relationship and quadratic approximation.

* The direction of steepest ascent is scale dependent. It is the direction of greatest increase assuming the factors are scaled as in Equation 11.1a. Of course, the experimenter picks the scales and thus, in the experimenter's chosen metrics, the path is steepest.

a larger range. In particular, the fitted quadratic approximation must fall away symmetrically on either side of its maximum.

Two-Factor Curved Relationships: Maxima, Stationary Regions, and Related Phenomena

Sometimes the objective of experimentation is to find the levels of several factors that identify a region of greatest response. To better understand the problem, consider a situation where there is a single response and just two factors.

Suppose the mean value η of a response such as yield has a curved relationship with two factors x_1 and x_2, say reaction time and reaction temperature. In the region of the maximum this true relationship might be represented by a smooth hill, shown with its contour representation in Figure 11.5*a*. The gradients with respect to x_1 and x_2 will be zero at the maximum point. More generally, the local region about such a stationary point may be called a *stationary* region. Figure 11.5 illustrates a number of naturally occurring surfaces associated with stationary regions. For example, for the surface shown in Figure 11.5*b* there is no single maximum point but rather a *line* of alternative maxima all giving just over 80% yield. With this type of surface an increase in x_1 (e.g., reaction time) could be compensated by reducing the level of x_2 (e.g., temperature) to give the same yield. This kind of surface is called a *stationary ridge*. If you now imagine this ridge tilted, you obtain a sloping ridge like that in Figure 11.5*c*. A series of increasing maxima occur along the dotted line shown in the contour plot. Thus, on a surface of this kind higher values of the response would be obtained by simultaneously reducing x_1 while increasing x_2. This kind of surface is called a *rising* (sloping) ridge. The lines in Figure 11.5*c* indicated by *AEB* and *CED* are included to demonstrate a common error associated with the change-one-factor-at-a-time philosophy. For example, changing x_1 at a fixed value of x_2 along the line *AB* would lead to the conclusion that a conditional maximum occurred close to the point *E*. If x_1 was now fixed at this maximizing value and x_2 varied along the line *CD*, you would again find a maximum close to *E*, which could wrongly lead to the conclusion that *E* is an overall maximum.

Another kind of curved surface which is less common but can occur is the minimax (saddle or col), shown in Figure 11.5*d* with its contour plot.

Second-Order Models: Two Factors, $k = 2$

It is frequently possible to represent systems like those in Figure 11.5 by empirical approximations. Thus the approximating systems depicted in Figures 11.6 *a, b, c, d* were all obtained by appropriately choosing the coefficients in the second-degree polynomial equation

$$\hat{y} = b_0 + b_1x_1 + b_2x_2 + b_{11}x_1^2 + b_{22}x_2^2 + b_{12}x_1x_2$$

as follows:*

$$\text{(a)} \quad \hat{y} = 83.6 + 9.4x_1 + 7.1x_2 - 7.4x_1^2 - 3.7x_2^2 - 5.8x_1x_2 \qquad (11.6\text{a})$$

*One interpretation of Equation 11.6 is that it is a second-order Taylor expansion of the unknown theoretical function truncated after second-order terms.

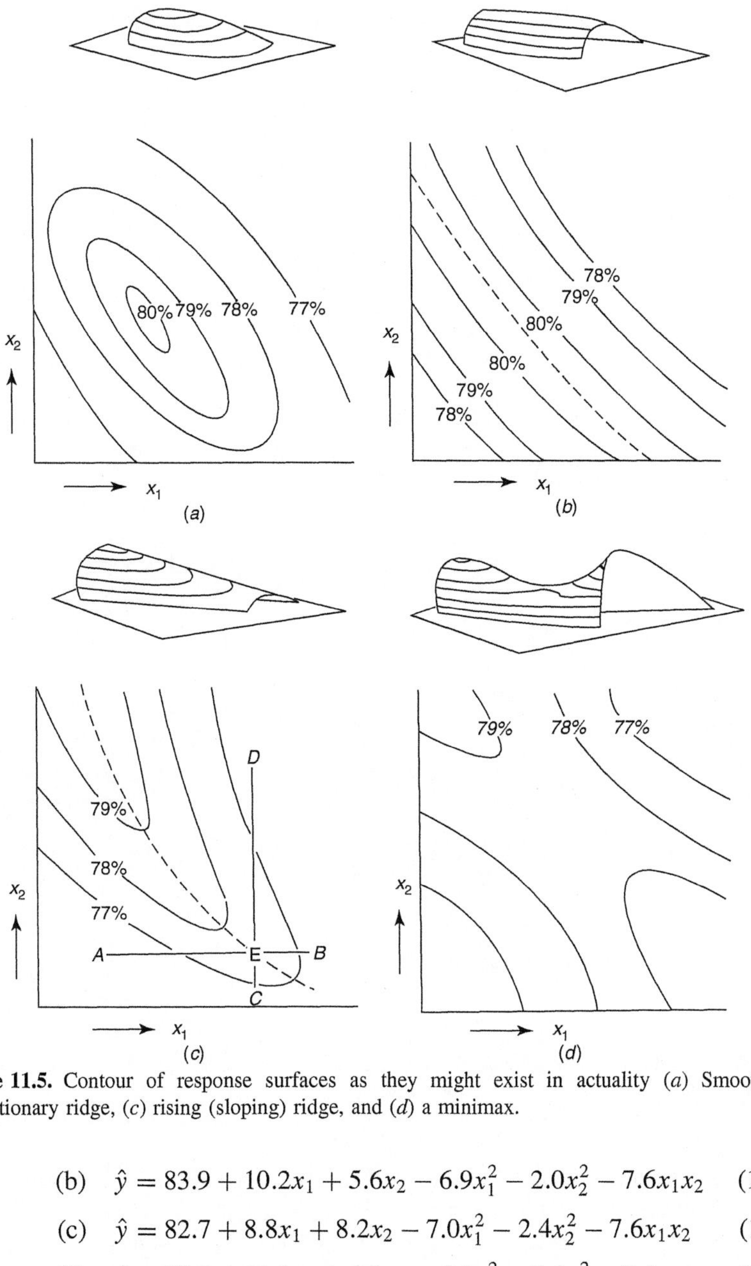

Figure 11.5. Contour of response surfaces as they might exist in actuality (*a*) Smooth hill, (*b*) stationary ridge, (*c*) rising (sloping) ridge, and (*d*) a minimax.

$$\text{(b)} \quad \hat{y} = 83.9 + 10.2x_1 + 5.6x_2 - 6.9x_1^2 - 2.0x_2^2 - 7.6x_1x_2 \quad \text{(11.6b)}$$

$$\text{(c)} \quad \hat{y} = 82.7 + 8.8x_1 + 8.2x_2 - 7.0x_1^2 - 2.4x_2^2 - 7.6x_1x_2 \quad \text{(11.6c)}$$

$$\text{(d)} \quad \hat{y} = 83.6 + 11.1x_1 + 4.1x_2 - 6.5x_1^2 - 0.4x_2^2 - 9.4x_1x_2 \quad \text{(11.6d)}$$

In their raw form these second-order approximations are not easy to interpret. But for two factors x_1, x_2 you can easily see the nature of the fitted response surfaces by computing their contour plots, and when more than one response is studied, it can be very informative to use the computer or an overhead projector to overlay

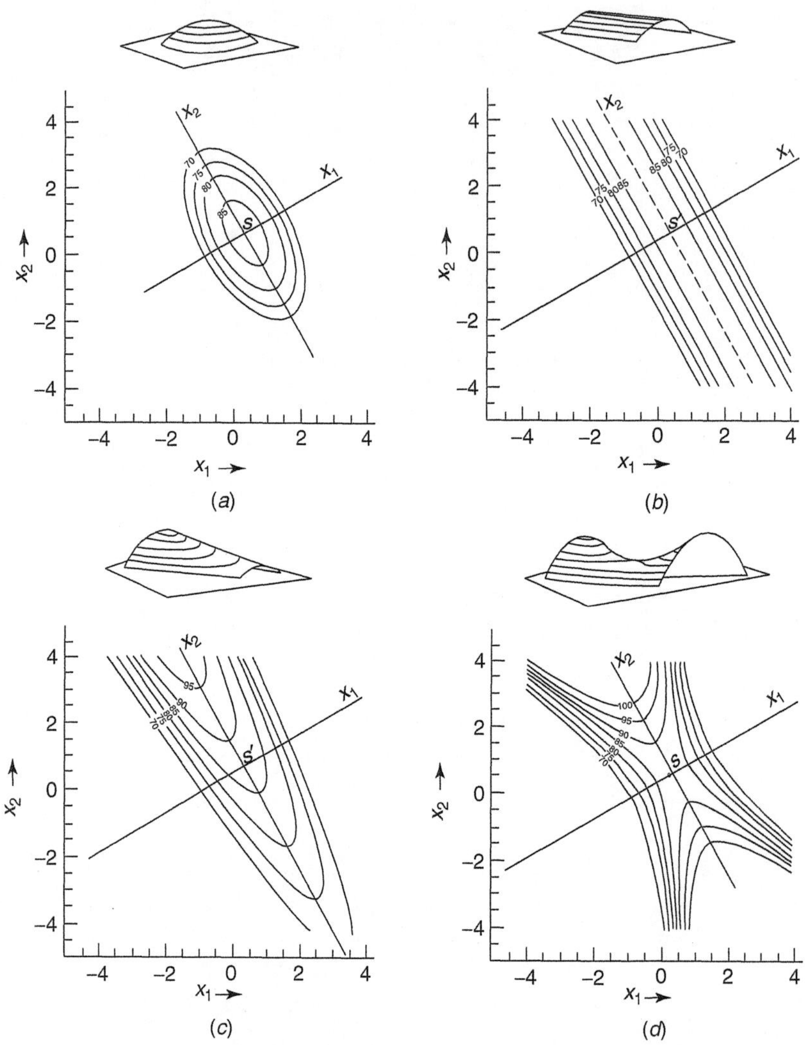

Figure 11.6. Contour representations using approximating second-order models: (*a*) maximum; (*b*) a stationary ridge; (*c*) a rising ridge; (*d*) a minimax.

contour plots and so locate regions that may produce the best compromise for a number of responses. For illustration Figure 11.7 shows contour overlays for three responses color index w, viscosity z, and yield y. The problem was to set the process factors reaction time and temperature to obtain a product with a color index w no less than 8, with viscosity z between 70 and 80, and with maximum yield. The overlays suggest that such conditions can be found close to the point P. That is, using a temperature of 147°C and a reaction time of about 82 minutes will give a product with a yield of about 88%. In a similar way your computer may be able to construct pictures of 3D contour systems and overlay these.

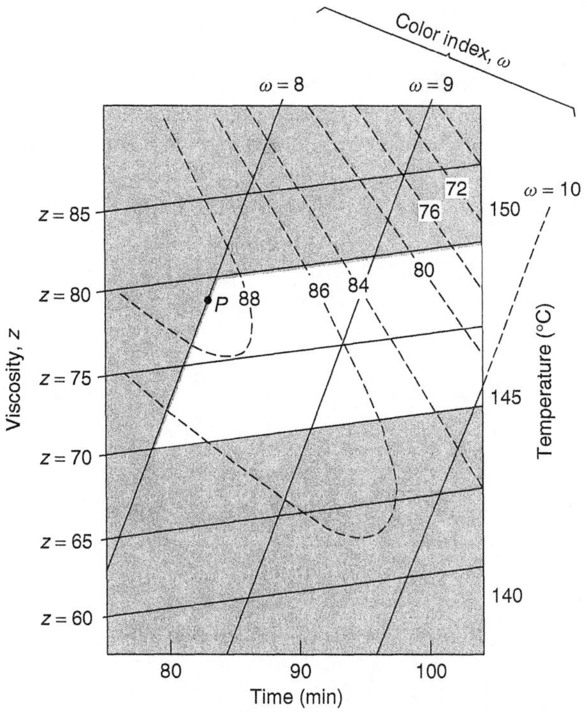

Figure 11.7. Contour overlay for three responses.

For problems involving more than two or three factors it is often useful to make contour plots of *sections* of the higher dimensional surfaces. For example, with four factors x_1, x_2, x_3, and x_4 the analysis might suggest the primary importance of, say, the factors x_2 and x_4. Contour plots in the space defined by x_2 and x_4 for a series of fixed values of x_1 and x_3 could then be very enlightening. Later in this chapter you will see how canonical analysis of second-degree equations can provide deeper understanding in such analyses.

11.2. SOME EXPERIMENTAL DESIGNS AND THE DESIGN INFORMATION FUNCTION

Consider now some experimental designs appropriate for fitting the approximating models of first and second order. The problem is to find a pattern of experimental conditions $(x_1, x_2, \ldots, x_k)$ that will tell you most about the response in the current region of interest.

First-Order Experimental Designs

When locally the response function is approximately planar, first-order orthogonal designs can generate a great deal of information with only a small expenditure of experimental effort. In particular, such designs are provided by the orthogonal two-level factorials, fractional factorials, and Plackett–Burman designs of Chapter 5, 6, and 7. To better understand their capability of *information generation*, look at Figure 11.8, which illustrates the information-gathering ability of a 2^2 factorial design in relation to the first-order model with IID errors

$$y = \beta_0 + \beta_1 x_1 + \beta_2 x_2 + e \tag{11.7}$$

Suppose you had obtained observations y_1, y_2, y_3, y_4 at the four design points in Figures 11.8*a* and *b*. Then the least squares estimates of the coefficients in Equation 11.7 would be

$$b_0 = \tfrac{1}{4}(y_1 + y_2 + y_3 + y_4), \qquad b_1 = \tfrac{1}{4}(-y_1 + y_2 - y_3 + y_4),$$
$$b_2 = \tfrac{1}{4}(-y_1 - y_2 + y_3 + y_4)$$

and the least squares estimate $\hat{y}$ of the response at some point (x_1, x_2) would be

$$\hat{y} = b_0 + b_1 x_1 + b_2 x_2$$

Also, since the estimated coefficients are distributed independently, the variance of $\hat{y}$ would be

$$V(\hat{y}) = V(b_0) + x_1^2 V(b_1) + x_2^2 V(b_2)$$

where $V(b_0) = V(b_1) = V(b_2) = \frac{1}{4}\sigma^2$. Thus the standardized variance $V(\hat{y})/\sigma^2$ for $\hat{y}$ is

$$\frac{V(\hat{y})}{\sigma^2} = \tfrac{1}{4}(1 + x_1^2 + x_2^2) = \tfrac{1}{4}(1 + r^2)$$

where $r^2 = x_1^2 + x_2^2$ is the distance of the point (x_1, x_2) from the center of the design.

Design Information Function

The standardized variance $V(\hat{y})/\sigma^2$ of the estimated response will therefore be constant on circles and so will its reciprocal, $I(\hat{y}) = \sigma^2/V(\hat{y})$. The quantity $I(\hat{y})$ is called the *design information function*, and it measures the amount of information generated by a design at any point (x_1, x_2). When you say the information generated at point A is twice that at point B, it is as though you had twice as many replicate observations at A as at B. For the 2^2 factorial in Figure 11.8*b* the information function is

$$I(\hat{y}) = \frac{4}{1 + r^2}$$

The *information surface* generated by this equation is shown in Figure 11.8*c*. The information falls off rapidly as you move away from the neighborhood

$$\begin{array}{cc} x_1 & x_2 \end{array}$$
$$\begin{bmatrix} -1 & -1 \\ 1 & -1 \\ -1 & 1 \\ 1 & 1 \end{bmatrix}$$

(*a*)

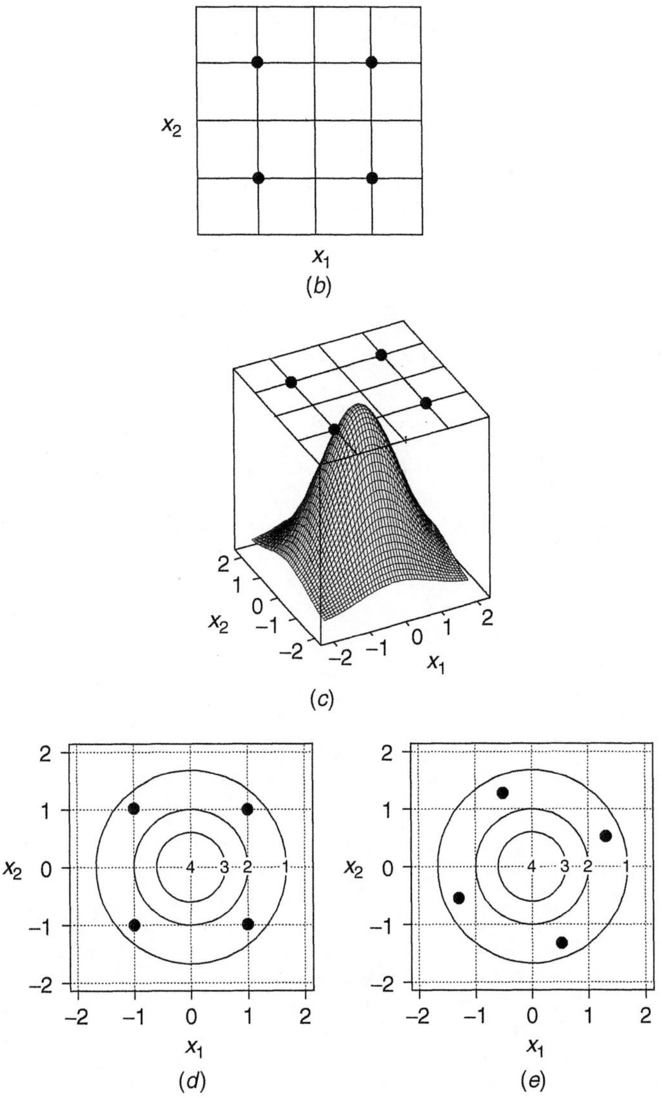

Figure 11.8. The 2^2 factorial design, information surface, information contours, and rotatability.

of the design. For this design the information contours are circles. These are shown in Figures 11.8*d* and *e* for $I(\hat{y}) = 4, 3, 2, 1$. Figure 11.8*e* shows how the information-gathering capacity of the 2^2 design remains the same whatever the design's orientation. The design is therefore called a *rotatable* design. This quality of rotatability extends to any appropriately scaled first-order orthogonal design in any number of factors. For example, in k dimensions the information will be constant for all points at the same distance r from the center of the design where $r = (x_1^2 + x_2^2 + \cdots + x_k^2)^{1/2}$. In expressing this idea, generally we say that the k-factor first-order orthogonal design has "spherical" information contours.*

Second-Order Experimental Designs

The two-level factorial designs are valuable for fitting first-order approximating models but they cannot take account of curvature in a response. To do this, you need to be able to fit a second-order model. For example, for $k = 2$ the fitted second-order model is

$$\hat{y} = b_0 + b_1x_1 + b_2x_2 + b_{11}x_1^2 + b_{22}x_2^2 + b_{12}x_1x_2$$

and for $k = 3$ the fitted second-order approximating model would be

$$\hat{y} = b_0 + b_1x_1 + b_2x_2 + b_3x_3 + b_{11}x_1^2 + b_{22}x_2^2 + b_{33}x_3^2 + b_{12}x_1x_2 + b_{13}x_1x_3 + b_{23}x_2x_3$$

We will illustrate for $k = 3$, but the concepts are readily extended to higher values of k.

The Central Composite Design

Initially you will not know how elaborate a model is necessary to provide an adequate approximation. Suppose you were studying three factors and, as illustrated in Figure 11.9*a*, you ran a 2^3 factorial design, preferably with a center point (whose purpose is discussed later). With this design you could obtain estimates of the constant term b_0, the first-order coefficients b_1, b_2, b_3, and the cross-product (interaction) coefficients b_{12}, b_{13}, b_{23} but not the quadratic coefficients b_{11}, b_{22}, b_{33}. If the estimated cross-product terms were small, it would be likely

*It might at first seem reasonable to seek designs that, if this were possible, generated square information contours for $k = 2$ and for $k = 3$ cubic information contours, and so on. On reflection, we see that this is not reasonable. For example, with three factors—temperature, pressure, and concentration—cubic contours would imply that we should treat the point $(+1, +1, +1)$ representing the extreme conditions for these factors as equivalent to a point $(+1, 0, 0)$ with only one factor at its extreme condition. The point $(+1, +1, +1)$ is a distance $\sqrt{3} = 1.73$ from the center of the experimental region while the point $(+1, 0, 0)$ is at a distance equal to 1. The discrepancy becomes greater and greater as k is increased. The property of generating information that is constant on spheres is of course not invariant to scale changes, However, at any given stage of experimentation the coding that the experimenter chooses (the ranges of the x's) is expressing his or her current understanding of the best scaling.

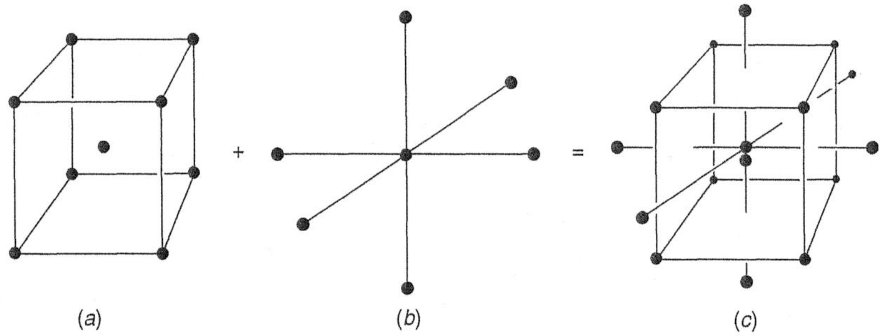

Figure 11.9. (*a*) A 2^3 design with center point; (*b*) star; (*c*) central composite design.

that the remaining second-order quadratic terms b_{11}, b_{22}, b_{33} would also be small, suggesting you could approximate the response surface locally by fitting the first-order model $\hat{y} = b_0 + b_1x_1 + b_2x_2 + b_3x_3$. If, however, second-order interaction terms were not negligible, you could expect that the quadratic terms b_{11}, b_{22}, b_{33} would also be needed.

To estimate quadratic terms, you can enhance an initial 2^3 design by adding the six axial points of the "star" design in Figure 11.9*b* along with one or more center points. Suppose that the 2^3 design is coded in the usual way with the vertices of the cube at (±1, ±1, ±1) and denote the distance of the axial points from the center by α. The axial points then have coordinates ($\pm\alpha$, 0, 0), (0, $\pm\alpha$, 0), and (0, 0, $\pm\alpha$) and the center point (0, 0, 0). The resulting design, as in Figure 11.9*c*, is called a *central composite design*. The choice of α depends on k, the number of factors studied. If there are three factors, any value between 1.5 and 2.0 gives a reasonably satisfactory design. For example, setting $\alpha = \sqrt{3}$ places all the points on a sphere. This choice of α is discussed more carefully later.

Illustration: A Central Composite Design Used in an Autoclave Process

The following example concerns a process where the effect on percent yield y was studied as a function of temperature T, concentration C, and reaction time t and from previous experience the error standard deviation σ_y was thought to be about one unit. The coding of the factors is shown in Table 11.1a. An initial 2^3 design (in this early example without a center point) yielded the results shown in Table 11.1b. The analysis in Table 11.1c shows that the interaction terms b_{12}, b_{13}, b_{23} were large in relation to the first-order terms b_1, b_2, b_3. A star design was therefore added with a center point and axial points set at $\alpha = \pm2$ to produce the 15-run composite design Figure 11.1d. An initial analysis of these 15 runs led to the conclusion that the response surface was approximated by a stationary ridge discussed more fully later. If this was so confirmatory runs (16, 17) and (18, 19) chosen widely at different conditions should produce yields in the region of a yield of 60 — the maximum so far obtained. This proved to be the case. A least squares analysis was now applied to the complete 19-run design the regressors for which are shown in Table 11.1e. This produced the fitted second-order model given in Table 11.1f.

Table 11.1. Data for Autoclave Process Example

(a) $x_1 = (T - 167)/5$, $x_2 = (C - 27.5)/2.5$, $x_3 = (t - 6.5)/1.5$

(b)

x_1	x_2	x_3	y
−1	−1	−1	45.9
1	−1	−1	60.6
−1	1	−1	57.5
1	1	−1	58.6
−1	−1	1	53.3
1	−1	1	58.0
−1	1	1	58.8
1	1	1	52.4

(c) $b_1 = 1.8$, $b_2 = 1.2$, $b_3 = 0.0$

$b_{12} = -3.1$, $b_{13} = -2.2$, $b_{23} = -1.2$

(d)

x_1	x_2	x_3	y
−1	−1	−1	45.9
1	−1	−1	60.6
−1	1	−1	57.5
1	1	−1	58.6
−1	−1	1	53.3
1	−1	1	58.0
−1	1	1	58.8
1	1	1	52.4
−2	0	0	46.9
2	0	0	55.4
0	−2	0	55.0
0	2	0	57.5
0	0	−2	50.3
0	0	2	58.9
0	0	0	56.9

(e)

x_0	x_1	x_2	x_3	x_1^2	x_2^2	x_3^3	x_1x_2	x_1x_3	x_2x_3	y
1	−1	−1	−1	1	1	1	1	1	1	45.9
1	1	−1	−1	1	1	1	−1	−1	1	60.6
1	−1	1	−1	1	1	1	−1	1	−1	57.5
1	1	1	−1	1	1	1	1	−1	−1	58.6
1	−1	−1	1	1	1	1	1	−1	−1	53.3
1	1	−1	1	1	1	1	−1	1	−1	58.0
1	−1	1	1	1	1	1	−1	−1	1	58.8
1	1	1	1	1	1	1	1	1	1	52.4
1	−2	0	0	4	0	0	0	0	0	46.9
1	2	0	0	4	0	0	0	0	0	55.4
1	0	−2	0	0	4	0	0	0	0	55.0
1	0	2	0	0	4	0	0	0	0	57.5
1	0	0	−2	0	0	4	0	0	0	50.3
1	0	0	2	0	0	4	0	0	0	58.9
1	0	0	0	0	0	0	0	0	0	56.9
1	2	−3	0	4	9	0	−6	0	0	61.1
1	2	−3	0	4	9	0	−6	0	0	62.9
1	−1.4	2.6	0.7	1.96	6.76	0.49	−3.6	−1	1.82	60.0
1	−1.4	2.6	0.7	1.96	6.76	0.49	−3.6	−1	1.82	60.6

(f) $\hat{y} = 58.78 + 1.90x_1 + 0.97x_2 + 1.06x_3 - 1.88x_1^2 - 0.69x_2^2 - 0.95x_3^2 - 2.71x_1x_2 - 2.17x_1x_3 - 1.24x_2x_3$

11.3. IS THE SURFACE SUFFICIENTLY WELL ESTIMATED?

An ANOVA table for the fitted model in Table 11.1f is as follows:

Source	SS	df	MS	
Second-order model	371.4	9	41.3	F ratio = 12.5
Residual	29.5	9	3.3	
Total	400.9	18		

The Interpretability Multiplier

Reference to an F table with nine and nine degrees of freedom shows that using the residual mean square to estimate the error variance a value of $F = 3.2$ would make the model component significant at the 5% level. But such an F value would merely discredit the null hypothesis that the x's had *no* influence on y. There is of course a great deal of difference between detecting that *something* is going on and adequately estimating what it is. To decide when a fitted function has been estimated with sufficient accuracy to make further analysis worthwhile, (Box and Wetz, 1973, Box and Draper, 1987) it is reasonable to require that the estimated change in the level of the response produced by the factors should be two or three times the average standard error of estimation of $\hat{y}$ itself. This is called the *interpretability multiplier*. An approximate argument then leads to the requirement that the value of F should be at least 4 times, and perhaps 10 times, the tabulated significance value. In the present example the F value actually obtained, 12.5, is about 4 times the value of F required for 5% significance, 3.2, and just satisfies the criterion. It appears then that the full equation is worth further analysis. To get some idea of the fitted approximation, three 2D contour plots are shown in Figure 11.10 at $x_2 = 2, 0, -2$, that is, at concentrations 32.5, 27.5, and 22.5%. These appear to show the existence of a ridge of alternate conditions giving a response close to 60. A fuller understanding of this response system can be gained by canonical analysis, to be discussed later in this chapter.

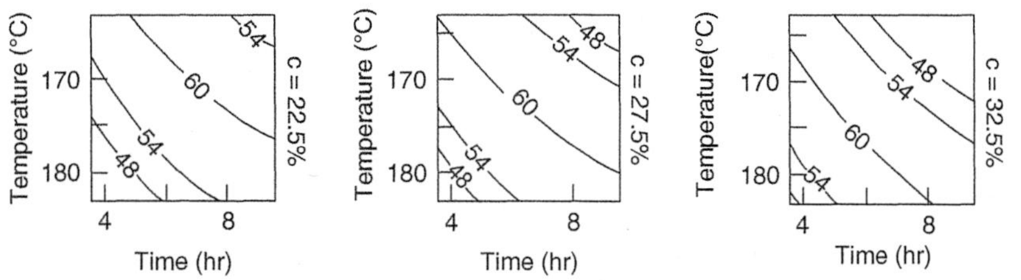

Figure 11.10. Yield contour plots at three concentrations, autoclave process example.

11.4. SEQUENTIAL DESIGN STRATEGY

Sequential assembly of a design means any procedure where the choice of a further design increment depends upon previous data. The sequential assembly of a central composite design in two stages was illustrated in Figure 11.9. As illustrated in Figure 11.11*a*, an even more tentative approach would be to build the design in three stages. You might begin with a first-order design such as the 2^{3-1} arrangement with a center point requiring only five runs.

A "Try-It-and-See" Check

If relatively large estimates of the first-order coefficients were found, you might adapt a try-it-and-see policy and make one or two exploratory runs along a path of steepest ascent calculated from these first-order terms. If these runs gave substantially better results, you would consider moving to a different experimental region. For an ongoing investigation a try-and-see policy is the preferred approach. But it is possible to check for the linearity before you venture forth.

A Linear Check, the "Droop" and "Bump" Functions

The quantity $q = \overline{y}_P - \overline{y}_O$ compares the average response $\overline{y}_O$ at the center of the design with the average of the peripheral fractional factorial runs $\overline{y}_P$. If q was large, this would suggest that the fitted surface was likely curved. If q was close to zero, this would suggest that you had a locally linear surface or, less likely, a minimax, in which the positive and negative quadratic coefficients canceled. If you have an estimate s of the standard deviation of the data, then q may be

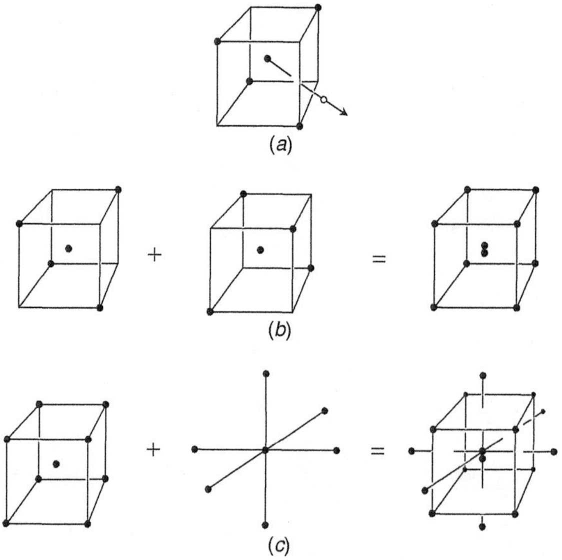

Figure 11.11. A 2^3 system with center points + star = central composite.

compared with its standard error. Suppose more generally there are n_p peripheral runs and n_o center runs; then

$$\mathrm{SE}(q) = s\sqrt{\frac{1}{n_o} + \frac{1}{n_p}}$$

For obvious reasons in the region of a minimum Cuthbert Daniel called q a measure of "droop." In the region of a maximum it measures the size of the "bump."

Adding a Second Half Fraction

To gain further information, the other half fraction could be run as a second block completing the full 2^3 factorial design as in Figure 11.11*b*. This would clear up aliases with two-factor interactions and would provide separate estimates of the first-order and the interactions terms.

Adding Axial Points

Following the previous discussion, if two-factor interactions are relatively large, the design could be further augmented by adding the star design shown in Figure 11.11*c*, producing the complete composite design.

Choosing α for a Central Composite Design

In the example the value of $\alpha = 2$ was arbitrarily chosen. Analysis shows that the choice of the distance α of the axial points from the design center involves two considerations:

(a) The nature of the desired information function $I(\hat{y})$. A circular (spherical) information function may be obtained by appropriate choice of α and the design will then be rotatable.
(b) Whether or not you want the cube part of the design and the axial part to form orthogonal blocks.

Orthogonal blocking ensures that in a sequential assembly an inadvertent shift in the level of response between running the first part of the design and the second has no effect on the estimated coefficients. Suppose n_c is the number of cubic (2^k factorial) points and n_{c0} the number of center points added to this first part of the design. And suppose n_a is the number of axial points and n_{a0} the number of center points added to this second part of the design. Then you can choose α as follows:

For rotatability $\alpha^2 = \sqrt{n_c}$.
For orthogonal blocking $\alpha^2 = k(1 + n_{a0}/n_a)/(1 + n_{c0}/n_c)$.

Table 11.2. Useful Second-Order Composite Designs

	$k = 2$	$k = 3$	$k = 4$	$k = 5$*
Cube points, n_c	4	8	16	16
Added center points, n_{c0}	1	1	1	1
Axial points, n_a	4	6	8	10
Added center points, n_{a0}	1	1	1	1
Total	10	16	26	28
α for rotatability	1.4	1.7	2.0	2.0
α for orthogonal blocks	1.4	1.8	2.1	2.1

* For $k = 5$ the 2_{V}^{5-1} fraction is used in place of the 2^5.

Table 11.2 provides some useful designs for up to $k = 5$ factors. Somewhat surprisingly, the values of α for rotatability and those for orthogonal blocking are very close. Thus for these designs orthogonal blocking also ensures approximate rotatability.

Parallel Chord Checks for the Central Composite Designs

Just as there is a linearity checking function a "droop" or "hump" function for first-order designs containing one or more center points, there are quadratic checking functions for the central composite designs. For $k = 3$ Figure 11.12*a* shows the points of a central composite design with $\hat{\alpha} = 2$. In Figure 11.12(*b*) these are projected onto the horizontal axis of x_1. Write $\bar{y}_+$ for the average of the four points on the right-hand face of the cube and $\bar{y}_-$ for the average on the left-hand face. To test the quadratic approximation in the x_1 direction, you can calculate the checking function

$$q = \frac{1}{3}\left(\frac{y_{\alpha+} - y_{\alpha-}}{4} - \frac{\bar{y}_+ - \bar{y}_-}{2}\right)$$

where $\bar{y}_{\alpha+}$ and $\bar{y}_{\alpha-}$ are the average responses at the axial points. Similar checking functions can be calculated for the x_2 and x_3 directions.

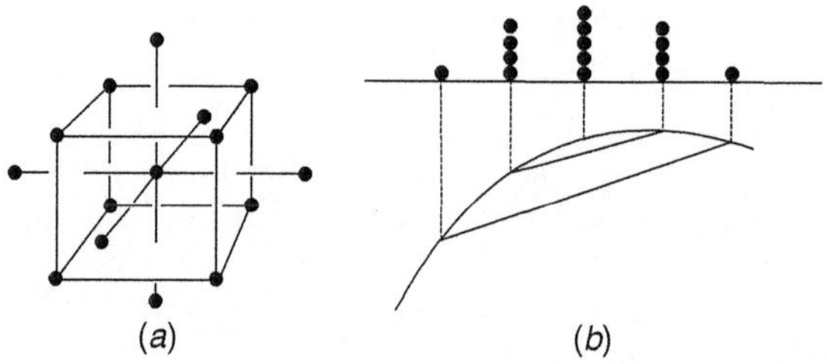

Figure 11.12. Parallel chords enclosing a maximum point, second-order model.

Now it is a known characteristic of any quadratic curve that the slopes of two chords joining points equidistant from an arbitrary point must be parallel, as in Figure 11.12*b*. Thus perfect "quadraticity" would require that the expected value of q be zero. It is easy to show that the standard deviation for each q is $\sigma/6$. In this example $k = 3$ and $\alpha = 2$. Thus, if an estimate $\hat{\sigma}$ of σ is available, q_1, q_2, and q_3 can provide checks in all three directions. A similar check is available more generally for central composite designs for k factors in which the cubic part has n_c values. (This could be a full factorial or a suitable fractional replicate.) Suppose also that the axial points are at any distance α from the center of the design and might be replicated r times (so if they are unreplicated $r = 1$). Then a general measure of nonquadraticity is

$$q = \frac{1}{\alpha^2 - 1}\left(\frac{\overline{y}_{\alpha+} - \overline{y}_{\alpha-}}{2\alpha} - \frac{\overline{y}_{+} - \overline{y}_{-}}{2}\right)$$

with standard deviation

$$\sigma_q = \frac{1}{\alpha^2 - 1}\left(\frac{1}{n_c} + \frac{1}{2r\alpha^2}\right)^{1/2} \sigma$$

where $\overline{y}_{\alpha+}$ and $\overline{y}_{\alpha-}$ refer to averages at the axial points.

An Equivalent Approach to the Quadratic Check

Alternatively, to detect the possible lack of fit of the second-order approximation, it is natural to consider the magnitude of the third-order terms. For example, consider the 15 runs of the central composite design in Table 11.3. You can obtain estimates of combinations of third-order terms by adding columns x_1^3, x_2^3, x_3^2 to the **X** matrix. Alternatively, you can add the columns, identified in Table 11.3 as $\dot{x}_1$, $\dot{x}_2$, $\dot{x}_3$, that are orthogonal to the first 10 columns of **X**. (See Section 10.3 on orthogonalization.) These orthogonal components have an interesting structure. You will see that they exactly correspond to the lack of fit measures q_1, q_2, q_3.

In relation to the full third-order model their alias structure is

$$q_1 = \beta_{111} + \beta_{112} + \beta_{113}$$

$$q_2 = \beta_{221} + \beta_{222} + \beta_{223}$$

$$q_3 = \beta_{331} + \beta_{332} + \beta_{333}$$

Thus, for example, q_1 measures to what extent the estimate of the quadratic term β_{11}, is affected by the levels of the three factors.

Noncentral Composite Designs

You saw above that an initial two-level factorial design that produced large two-factor interaction terms suggested the need for a second-order model. If at this stage you found that first-order terms were small, your experiments might already be spanning a stationary region. In that case adding axial points to form a central

Table 11.3. A $k = 3$ Central Composite Regression Matrix X for Fitting a Second-Order Model with Additional Cubic Effects Identified by the Orthogonalized Vectors $\dot{x}_1, \dot{x}_2, \dot{x}_3$

Run Number	x_0	x_1	x_2	x_3	x_1^2	x_2^2	x_3^2	x_1x_2	x_1x_3	x_2x_3	$\dot{x}_1$	$\dot{x}_2$	$\dot{x}_3$
1	1	−1	−1	−1	1	1	1	1	1	1	1.5	1.5	1.5
2	1	1	−1	−1	1	1	1	−1	−1	1	−1.5	1.5	1.5
3	1	−1	1	−1	1	1	1	−1	1	−1	1.5	−1.5	1.5
4	1	1	1	−1	1	1	1	1	−1	−1	−1.5	−1.5	1.5
5	1	−1	−1	1	1	1	1	1	−1	−1	1.5	1.5	−1.5
6	1	1	−1	1	1	1	1	−1	1	−1	−1.5	1.5	−1.5
7	1	−1	1	1	1	1	1	−1	−1	1	1.5	−1.5	−1.5
8	1	1	1	1	1	1	1	1	1	1	−1.5	−1.5	−1.5
9	1	−2	0	0	4	0	0	0	0	0	−3	0	0
10	1	2	0	0	4	0	0	0	0	0	3	0	0
11	1	0	−2	0	0	4	0	0	0	0	0	−3	0
12	1	0	2	0	0	4	0	0	0	0	0	3	0
13	1	0	0	−2	0	0	4	0	0	0	0	0	−3
14	1	0	0	2	0	0	4	0	0	0	0	0	3
15	1	0	0	0	0	0	0	0	0	0	0	0	0

composite design would make good sense and could help elucidate things further. However, suppose you found that the two-factor interactions and *also* the first-order terms were large. This would suggest that there remained the possibility of exploitable first-order gradients and that a stationary region might lie just outside the present design region. You could explore this possibility by using a *noncentral* composite design.

For illustration, consider the following constructed example. Suppose you had data from the first four factorial runs in Table 11.4 along with some preliminary estimate of the standard deviation and your analysis showed two large negative linear coefficients $b_1 = -2.3 \pm 0.4$ and $b_2 = -3.8 \pm 0.4$ together with a nonnegligible interaction term $b_{12} = 1.3 \pm 0.4$. This would suggest that further exploration of a region just beyond the $(-1, -1)$ vertex could be profitable. If now the run at $(-1, -1)$ was duplicated and further runs were made at the points $(-3, -1)$ and $(-1, -3)$, you would have the *noncen*tral composite design (see Box and Wilson, 1951) illustrated in Figure 11.13*a*. Such a design makes possible the fitting of the second-order model with data added in the relevant region. For the data of Table 11.4, the fitted equation adjusted for blocks, is $\hat{y} = 63.75 - 3.75x_1 - 2.25x_2 - 1.50x_1^2 - 1.50x_2^2 + 1.25x_1x_2$.

The Information Function for a Noncentral Composite Design

The noncentral composite design shown in Table 11.4 is of considerable interest. From the design information function it will be seen how, by adding a few further

Table 11.4. The Sequential Assembly of a Noncentral Composite Design

Run Number	x_1	x_2	y
	Block 1		
1	−1	−1	66
2	+1	−1	56
3	−1	+1	59
4	+1	+1	54
	Block 2		
5	−1	−1	68
6	−3	−1	66
7	−1	−3	63

runs in a direction indicated by the parent design (here a 2^2 factorial), (Fung, 1986) the information-gathering power of that design can be extended in the direction desired. Figures 11.13*a* and *b* show two possible arrangements of this kind. For $k = 3$ factors Figures 11.14*a* and *b* show the extension of the parent design from a point and from an edge.

There are a number of "optimality" measures for designs. These usually refer to a single aspect of design. For example, there is a D optimality criterion that relates to the size of the determinant of the variance–covariance matrix of the estimated coefficients. When D is maximized, the design is said to be "D optimal" and the volume of the confidence region is minimized. The

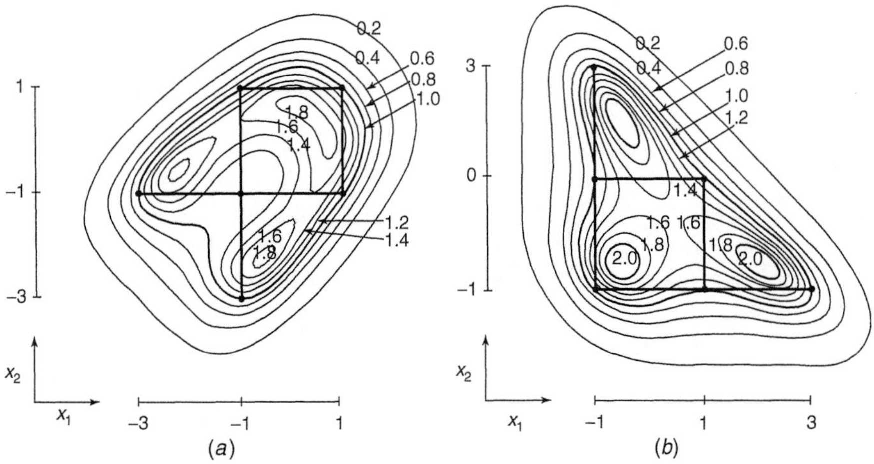

Figure 11.13. Information contours of two noncentral composite designs, $k = 2$.

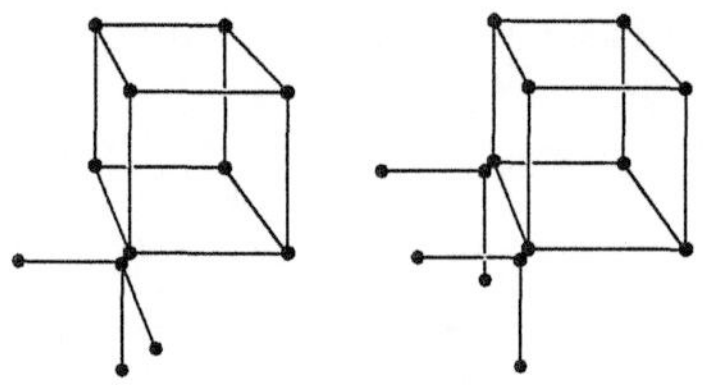

Figure 11.14. Two noncentral composite designs, $k = 3$.

attempt to compress the information about a design into a single number seems mistaken.

A Noncentral Composite Design Example: Producing a Chemical Used in Car Tires

Car tires contain a number of chemicals that improve the performance of the tire. For one such chemical it was believed that less expensive operating conditions might be found. Experimentation was costly and it was essential that the number of experimental trials be kept to a minimum. Three factors were studied: temperature (x_1), concentration (x_2), and molar ratio (x_3); with the coding shown in Table 11.5. The response y to be *minimized* was the calculated cost per pound (measured from a convenient reference value). The currently known best conditions were Temp 142°C, Conc 22% and Molar Ratio 6.5. Initially, it was uncertain whether these "best" conditions were sufficiently far from a minimum to require a preliminary application of steepest descent (*descent* since the response was cost). It was decided therefore to run the 2^3 factorial design shown in Table 11.5a centered close to the best known conditions. Analysis of this 2^3 factorial produced the following estimated coefficients: $b_0 = 12.5$, $b_1 = 0.8$, $b_2 = 2.8$, $b_3 = 2.5$, $b_{12} = -3.0$, $b_{13} = -1.3$, and $b_{23} = 2.8$. From the relative sizes of the linear effects and interactions it did not seem likely that steepest descent would be effective, but as a check, runs 9 and 10 were made along a steepest descent path determined from the estimated first-order effects giving costs of $y_1 = 13$ and $y_{10} = 27$.

Not surprisingly no reduction in cost was obtained. However, since the first-order estimated coefficients had all been positive, it seemed quite likely that a local minimum might exist beyond the vertex $(-1, -1, -1)$. The response beyond this vertex was investigated by making three additional runs 11, 12, and 13 shown in Table 11.5c. Runs 1 through 8 together with runs 11, 12, and 13 produced a noncentral composite design. Using these 11 runs, the estimated second-order equation for approximating cost was

$$\hat{y} = 6.8 + 0.8x_1 + 2.7x_2 + 2.5x_3 + 3.0x_1^2 + 1.6x_2^2 + 1.1x_3^2 - 3.0x_1x_2 - 1.3x_1x_3 + 2.7x_2x_3$$

Table 11.5. Sequential Construction of a Noncentral Composite Design

Temperature, °C		*Concentration, %*		*Molar Ratio, A/B*	
140	145	20	24	6	7
x_1		x_2		x_3	
−1	+1	−1	+1	−1	+1

Run Number	x_1	x_2	x_3	y
		(a) 2^3 Factorial		
1	−1	−1	−1	5
2	+1	−1	−1	15
3	−1	+1	−1	11
4	+1	+1	−1	9
5	−1	−1	+1	7
6	+1	−1	+1	12
7	−1	+1	+1	24
8	+1	+1	+1	17
		(b) Steepest Descent		
9	−0.5	−2.0	−1.8	[13]
10	−0.8	−3.0	−2.7	[27]
		(c) Added Points		
11	−3	−1	−1	19
12	−1	−3	−1	12
13	−1	−1	−3	12
		(d) Confirmatory Runs		
14	−1.13	−2.51	1.00	[3]
15	0.07	1.68	−3.00	[5]

The coordinates of the stationary point obtained by equating the first-order derivatives to zero are

$$x_{1S} = -0.25, \qquad x_{2S} = -0.55, \qquad x_{3S} = -1.93$$

The estimated cost at this point was $\hat{y}_s = 5.0$. You will see later that further consideration of this example using canonical analysis provided a clearer understanding of the system leading to further process improvements.

11.5. CANONICAL ANALYSIS

The calculations required for canonical analysis of a fitter second-order model can best be understood by first considering an example with just two factors coded as

x_1 and x_2. In this simple case canonical analysis cannot tell you anything that is not obvious from contour plots. However, if you understand canonical analysis for $k = 2$, you gain an understanding for larger values of k.

Look again at the four examples for which the equations appear in Equation 11.6a, b, c, and d for which the contours are plotted in Figure 11.6*a*, *b*, *c*, and *d*. As a first illustration consider the canonical analysis of the fitted second-order equation 11.6a, that is,

$$\hat{y} = 83.6 + 9.4x_1 + 7.1x_2 - 7.4x_1^2 - 3.7x_2^2 - 5.8x_1x_2$$

For clarity the various operations are summarized and shown graphically in Figure 11.15. In what follows the letter O is used to denote the point (0, 0) at the center of the experimental design) and the letter S denotes the location of the stationary value (e.g., the maximum).

The fitted second-order model will be of the form

$$\hat{y} = b_0 + b_1x_1 + b_2x_2 + b_{11}x_1^2 + b_{22}x_2^2 + b_{12}x_1x_2$$

Location of the Stationary Point

Differentiation of the second-order model with respect to x_1 and x_2 gives the $k = 2$ simultaneous equations

$$2b_{11}x_1 + b_{12}x_2 = -b_1 \qquad -14.8x_1 - 5.8x_2 = -9.4$$

$$b_{12}x_1 + 2b_{22}x_2 = -b_2 \qquad -5.8x_1 - 7.4x_2 = -7.1$$

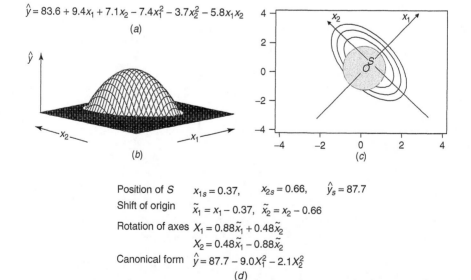

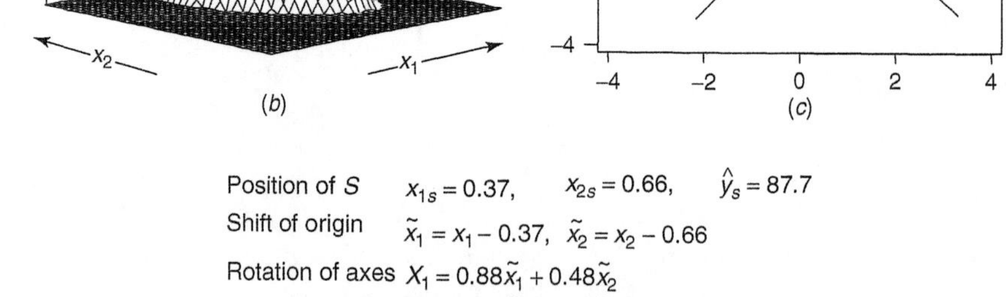

Position of S	$x_{1S} = 0.37$,	$x_{2S} = 0.66$,	$\hat{y}_S = 87.7$
Shift of origin	$\tilde{x}_1 = x_1 - 0.37$,	$\tilde{x}_2 = x_2 - 0.66$	
Rotation of axes	$X_1 = 0.88\tilde{x}_1 + 0.48\tilde{x}_2$		
	$X_2 = 0.48\tilde{x}_1 - 0.88\tilde{x}_2$		
Canonical form	$\hat{y} = 87.7 - 9.0X_1^2 - 2.1X_2^2$		

(*d*)

Figure 11.15. Attributes of the canonical analysis: (*a*) fitted second-order model; (*b*) geometric representation; (*c*) fitted contours and canonical representation; the shaded circle indicates the location of the experimental design (*d*) shows calculations for the canonical representation.

The solution of these equations gives the location of the stationary value S,

$$x_{1S} = 0.37, \qquad x_{2S} = 0.67$$

The shift of design coordinates x_1, x_2 from O to the design coordinates $\tilde{x}_1$, $\tilde{x}_2$ for S is such that

$$\tilde{x}_1 = x_1 - 0.37, \qquad \tilde{x}_2 = x_2 - 0.67$$

The estimated response at the point S is 87.7. This can be found by direct substitution in the equation, or equivalently from

$$\hat{y}_S = b_0 + \tfrac{1}{2}(b_1 x_{1S} + b_2 x_{2S}) = 87.7$$

Obtaining the Canonical Form

The canonical form is obtained by analysis of the following matrix,* whose elements are obtained from the second-order coefficients in the fitted equation

$$\begin{bmatrix} b_{11} & \frac{1}{2}b_{12} \\ \frac{1}{2}b_{12} & b_{22} \end{bmatrix} = \begin{bmatrix} -7.4 & -2.9 \\ -2.9 & -3.7 \end{bmatrix}$$

The quantities now required are obtained by getting your computer software to produce what are called the latent roots and latent vectors for this matrix.† The $k = 2$ latent roots and latent vectors for this example

Latent roots	−9.0	−2.1
Latent vectors	$\begin{bmatrix} 0.88 \\ 0.48 \end{bmatrix}$	$\begin{bmatrix} 0.48 \\ -0.88 \end{bmatrix}$

The latent roots are the coefficients in the canonical form of the fitted equation

$$\hat{y} = 87.7 - 9.0X_1^2 - 2.1X_2^2$$

where the X's are obtained by rotation about S (see Figure 11.5c). The transformation from the $\tilde{x}$'s to the X's producing this rotation has the elements of the eigenvectors for its coefficients. That is,

$$X_1 = 0.88\tilde{x}_1 + 0.48\tilde{x}_2, \qquad X_2 = +0.48\tilde{x}_1 - 0.88\tilde{x}_2$$

You will see that what is being done in this case is to switch to a new coordinate system that is centered at S and has for its axes the principal axes of the elliptical

* Here a matrix is simply treated as a table of numbers. You do not need to know matrix theory.

† Latent roots are also called eigenvalues or characteristic roots. Latent vectors are also called eigenvectors or characteristic vectors.

contours. The system is in this canonical form, is much easier to interpret. In general, the canonical form is

$$\hat{y} = B_0 + B_{11}X_1^2 + B_{22}X_2^2$$

In these equations, $B_0 = 87.7$ is the estimated response at S and the coefficients $B_{11} = -9.0$ and $B_{22} = -2.1$ measure the quadratic curvatures along the new X_1 and X_2 axes. Since in this example they are both negative, the point S (at which $\hat{y}_S = 87.7$) must be a maximum. The relative magnitudes of the coefficients, $B_{11} = -9.0$ and $B_{22} = -2.1$, also tell you that movement away from S along the X_1 axis will cause $\hat{y}$ to decrease much more rapidly than will similar movement along the X_2 axis. Thus, the long dimension of the elliptical response contours falls along the axis that has the smaller coefficient.

A Minimax

For the purpose of exposition it is easiest to consider next the analysis for Equation 11.6d:

$$\hat{y} = 84.3 + 11.1x_1 + 4.1x_2 - 6.5x_1^2 - 0.4x_2^2 - 9.4x_1x_2$$

The same calculations as before are illustrated in Figure 11.16. The canonical form for this equation is

$$\hat{y} = 87.7 - 9.0X_1^2 + 2.1X_2^2$$

Since the coefficient of X_1^2, $B_{11} = -9.0$, is negative while the coefficient of X_2^2, $B_{22} = 2.1$, is positive, this tells you at once that the resultant surface contains a minimax. Thus the center of the system S is a maximum along the X_1 axis but a minimum along the X_2 axis. The relative sizes of the coefficients -9.0 and $+2.1$ indicate the surface is attenuated in the X_2 direction. Minimax surfaces are rather rare but, as you will see in Chapter 12, they can occur and may suggest the existence of more than one maximum.

A Stationary Ridge

In Figure 11.15 there is an analysis of an equation containing a maximum. The resulting surface has a local origin with the canonical form

$$\hat{y} = 87.7 - 9.0X_1^2 - 2.1X_2^2$$

The system is attenuated along the X_2 axis and, specifically, intercepts of X_2 and X_1 onto any given elliptical contour are in the ratio $\sqrt{9.0/2.1} = 2.1$. But now suppose the coefficient B_{22} had been not -2.1 but -0.2 so that the canonical equation was

$$\hat{y} = 87.7 - 9.0X_1^2 - 0.2X_2^2$$

$$\hat{y} = 84.3 + 11.1x_1 + 4.1x_2 - 6.5x_1^2 - 0.4x_2^2 - 9.4x_1x_2$$

(a)

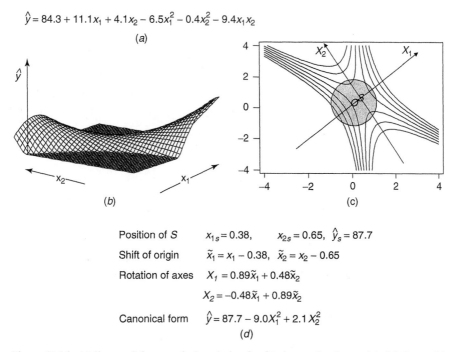

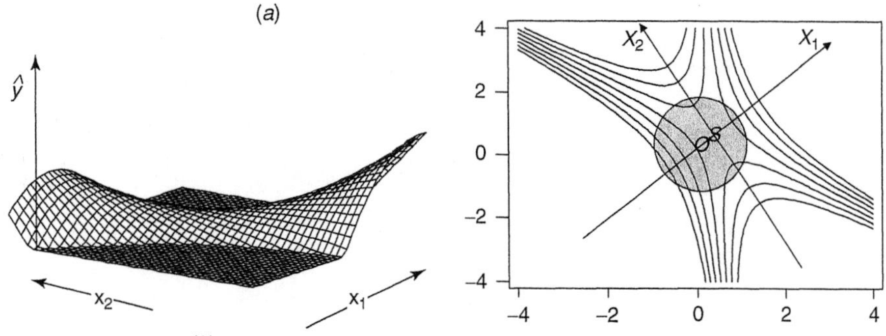

(b) (c)

Position of S	$x_{1s} = 0.38, \quad x_{2s} = 0.65, \quad \hat{y}_s = 87.7$
Shift of origin	$\tilde{x}_1 = x_1 - 0.38, \quad \tilde{x}_2 = x_2 - 0.65$
Rotation of axes	$X_1 = 0.89\tilde{x}_1 + 0.48\tilde{x}_2$
	$X_2 = -0.48\tilde{x}_1 + 0.89\tilde{x}_2$
Canonical form	$\hat{y} = 87.7 - 9.0X_1^2 + 2.1X_2^2$

(d)

Figure 11.16. Attributes of the canonical analysis of a fitted second-order model: (*a*) the model; (*b*) geometric representation; (*c*) the canonical fit; (*d*) attributes of the canonical fit.

The above ratio would now be $\sqrt{9.0/0.2} = 6.7$. The contours would then be much more attenuated along the X_2 axis and in the neighborhood of O would approximate those of a stationary ridge.

An alternative construction of the stationary ridge might begin in Figure 11.6*d* with the canonical equation $\hat{y} = 87.7 - 9.0X_1^2 + 2.1X_2^2$, which implies the existence of a minimax. Now suppose the coefficient B_{22} was not 2.1 but 0.2. Then the hyperbolic contours around the minimax would be greatly attenuated about the X_2 axis, and again in the neighborhood of O the contours would approach those of a stationary ridge. If B_{22} were exactly zero, then the contours would be parallel straight lines as in Figure 11.6*b*. Thus the signature of an approximately stationary ridge in a canonical analysis is a *local* center S and a canonical coefficient that can be positive or negative and is *small in absolute magnitude* in relation to the other coefficient.

The reason why it is important to discover such ridges when they exist is that they show that there are a series of choices along the axis of the ridge producing process conditions giving close to maximal response. This can be very important when certain of these choices are less expensive or more convenient. Also, when contour maps representing different responses can be superimposed, a ridge in one response can offer great freedom of choice in the factor settings for the other responses.

Exercise 11.2. Carry out the canonical analysis for Equation 11.6b in the same detail as was done for Equation 11.6a and 11.6d.

A Sloping Ridge

For illustration consider again the automobile emissions data given in Table 10.17, representing the dependence of CO emissions y on ethanol concentration x_1 and air–fuel ratio x_2. In the earlier analysis the second-order model was fitted by least squares to give

$$\hat{y} = 78.6 + 4.4x_1 - 6.9x_2 - 4.6x_1^2 - 4.1x_2^2 - 9.1x_1x_2$$

Figure 11.17 shows its contours. The surface is a sloping ridge predicting lower CO concentration as ethanol is reduced and air–fuel ratio is simultaneously increased. Such sloping ridges are important for they can tell you when, even though no improvement is possible by varying factors separately, a higher (or lower) response can be obtained by appropriately changing them *together*.

Canonical Analysis for the CO Emissions Data

The calculations for the canonical form go precisely as before:

Fitted model	$\hat{y} = 78.6 + 4.4x_1 - 6.9x_2 - 4.6x_1^2 - 4.1x_2^2 - 9.1x_1x_2$
Position of S	$\tilde{x}_1 = -14.64, \quad \tilde{x}_2 = 15.29$ *
Shift of origin	$\tilde{x}_1 = x_1 + 14.64, \quad \tilde{x}_2 = x_2 - 15.29$
Rotation of axes	$X_1 = 0.72\tilde{x}_1 + 0.69\tilde{x}_2, \quad X_2 = -0.69\tilde{x}_1 + 0.72\tilde{x}_2$
Canonical form	$\hat{y} = -5.98 - 8.89X_1^2 + 0.18X_2^2$

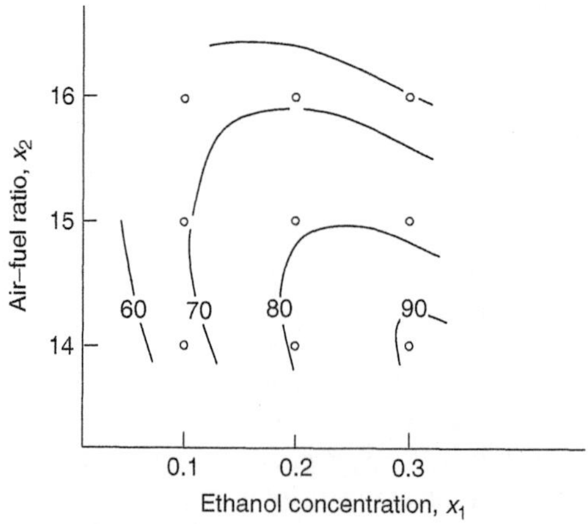

Figure 11.17. A 3^2 experimental design and contours of response η (CO concentration).

Because the coefficient $B_{22} = 0.18$ is small in magnitude compared with $B_{11} = -8.89$, you know that you have some kind of ridge system. But notice that the coordinates of S show that the system center is remote from the region of the experimentation centered at O. Rewriting the coordinates of O in terms of X_1 and X_2, you get $X_{10} = -0.01$ and $X_{20} = -21.10$. Thus O is close to the X_2 axis ($X_{10} = -0.01$) but remote ($X_{20} = -21.10$) from S. To obtain an expression that is relevant in the experimental region around O, the canonical form is rewritten in terms of $X'_1 = X_1 + 0.01$ and $X'_2 = X_2 + 21.10$, that is, $X_2 = X'_2 - 21.10$. The canonical form now becomes

$$\hat{y} = -5.98 - 8.89(X'_1 - 0.01)^2 + 0.18(X'_2 - 21.10)^2$$

which simplifies to

$$\hat{y} = 74.2 - 8.9X_1^2 - 7.6X'_2 + 0.2X_2'^2$$

Ignoring the small squared term, $0.2X_2'^2$ gives

$$\hat{y} = 74.2 - 8.9X_1^2 - 7.6X'_2$$

The *linear* coefficient -7.6 of X'_2 shows the rate of diminution of the CO response along the X_2 axis—it measures the slope of the ridge.

Geometry of the Sloping Ridge

In the fitted canonical equation $\hat{y} = B_0 + B_{11}X_1^2 + B_{22}X_2^2$ assume B_{11} is large relative to B_{22}. If S is local, then you have approximately a stationary ridge like that in Figure 11.6*b* which can occur either with B_{22} small and negative or with B_{22} small and positive. If S is remote, however, you will almost certainly have a sloping ridge. As you see from Figure 11.18, the contours of a sloping ridge can be produced either by an elliptical system (B_{22} small and negative) with a remote center at S_1 or by a hyperbolic system (B_{22} small and positive) with a remote center at S_2. There is a theoretically limiting case when $B_{22} = 0$ and the contours of the sloping ridge become exact parabolas. In practice, experimental error produces a small negative or positive value for B_{22}.

Second-Order Models, $k = 3$ Factors

Consider the second-order model for $k = 3$ factors:

$$\hat{y} = b_0 + b_1x_1 + b_2x_2 + b_3x_3 + b_{11}x_1^2 + b_{22}x_2^2 + b_{33}x_3^2 + b_{12}x_1x_2 + b_{13}x_1x_3 + b_{23}x_2x_3$$

*When S is remote from O, you will find calculations of the position of S extremely sensitive to rounding errors. However after transforming back to the origin O the conclusions are affected very little.

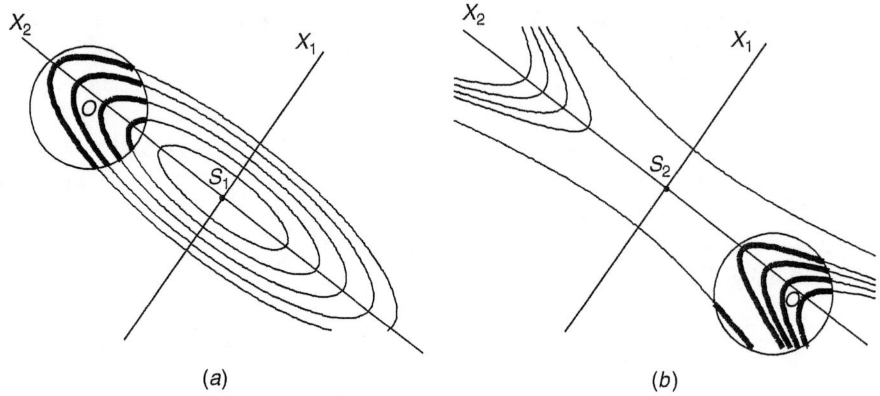

Figure 11.18. Alternative representations of a sloping ridge.

Such a fitted equation can represent many different and interesting surfaces, as illustrated in Figure 11.19. A maximum in three factors will be represented by a single fixed point surrounded by football-shaped contour surfaces such as that in Figure 11.19*a*. These are arranged rather like the skins of an onion. If X_1, X_2, and X_3 are the coordinates of the canonical variables in Figure 11.19*a*, you will see that the response would fall away from the central maximum point most rapidly along the X_2 axis, more slowly along the X_1 axis, and slowest in the direction of X_3.

The diagrams illustrate other possibilities. If in Figure 11.19*a* you imagine the ellipsoidal contours greatly attenuated along the X_3 axis, the contour surfaces will be approximated by tubes, as shown in Figure 11.19*b* and in the example in Figure 11.15. There then will be not a point maximum but a line of alternative maxima in the direction of the X_3 axis. If you consider the contour surfaces in Figure 11.19*a* strongly attenuated in both the X_1 and the X_2 directions, they will approximate a family of planar contours such as those in Figure 11.19*c*. The contour planes shown in this figure are like a ham sandwich where the maximal plane in the center is the layer of ham. In Figure 11.19*d* the level of the maximum in X_1 and X_2 will increase as the level of X_3 is increased. This is a generalization of the sloping ridge illustrated earlier in Figure 11.6*c*. A planar sloping ridge is shown in Figure 11.19*e*. Minimax saddles are shown in Figures 11.19*f* and 11.19*g*.

Canonical analysis is valuable for the interpretation of all such systems. In particular, some kind of sloping ridge system is signaled whenever the center of the system S is remote from the region of experimentation O and one or more of the coefficients in the canonical equation are close to zero. For example, suppose there were k factors and two of the coefficients, say B_{11} and B_{22}, were small and that O was remote from S. Rewriting the canonical equation in terms of $X_1' = X_1 - X_{10}$ and $X_2' = X_2 - X_{20}$ will then produce an equation with large linear terms in X_1 and X_2 denoting a sloping planar ridge within the design region. (See, e.g., Fig. 11.19*e* for $k = 3$.)

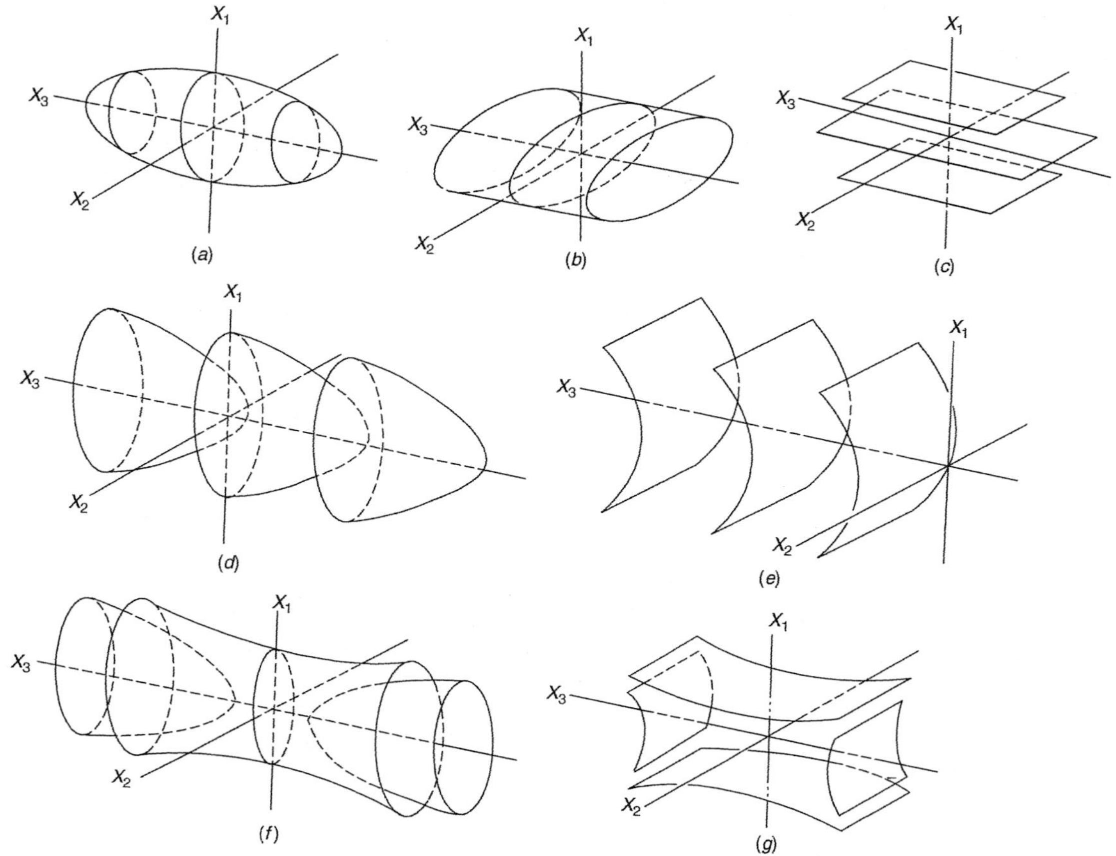

Figure 11.19. Some 3D contour systems generated by the second-order model.

Exercise 11.3. Write down the form of the canonical equations for systems in Figures 11.19*a, b, c, d, e, f, g*.

Illustration with the Autoclave Data

For illustration consider again the autoclave example for which the least squares analysis is given in Table 11.1 The canonical analysis of the fitted equation for the autoclave data goes as before but for $k = 3$ factors. The calculations are set out below step by step. As before, the letter O is used to denote the center of the experimental design (the point 0, 0, 0 in this example) and the letter S denotes the location of the stationary value (e.g., the maximum). The fitted second-order model

$$\hat{y} = b_0 + b_1x_1 + b_2x_2 + b_3x_3 + b_{11}x_1^2 + b_{22}x_2^2 + b_{33}x_3^2 + b_{12}x_1x_2 + b_{13}x_1x_3 + b_{23}x_2x_3$$

for this example is

$$\hat{y} = 58.78 + 1.90x_1 + 0.97x_2 + 1.06x_3 - 1.88x_1^2 - 0.69x_2^2 - 0.95x_3^2 - 2.71x_1x_2 - 2.17x_1x_3 - 1.24x_2x_3$$

Location of the Stationary Point

Differentiation of the second-order model with respect to x_1, x_2, and x_3 gives the $k = 3$ simultaneous equations

$$2b_{11}x_1 + b_{12}x_2 + b_{13}x_3 = -b_1, \qquad -3.75x_1 - 2.71x_2 - 2.17x_3 = -1.90$$

$$b_{12}x_1 + 2b_{22}x_2 + b_{33}x_3 = -b_2, \qquad -2.71x_1 - 1.37x_2 - 1.24x_3 = -0.97$$

$$b_{13}x_1 + b_{23}x_2 + 2b_{33}x_3 = -b_3, \qquad -2.17x_1 - 1.24x_2 - 1.91x_3 = -1.06$$

The solution of these equations gives the location of the stationary value S,

$$x_{1S} = -0.02, \qquad x_{2S} = 0.55, \qquad x_{3S} = 0.23$$

The shift of the design coordinates x_1, x_2, x_3 from O to $\tilde{x}_1$, $\tilde{x}_2$, $\tilde{x}_3$ for S is such that

$$\tilde{x}_1 = x_1 + 0.02, \qquad \tilde{x}_2 = x_2 - 0.55, \qquad \tilde{x}_3 = x_3 - 0.23$$

The estimated response at the point S is 59.15. This can be found by direct substitution or equivalently by using the equation

$$\hat{y}_S = b_0 + \tfrac{1}{2}(b_1x_{1S} + b_2x_{2S} + b_3x_{3S}) = 59.15$$

Obtaining the Canonical Form

As before, the canonical form is obtained by analysis of a matrix whose elements are obtained from the second-order coefficients in the fitted equation:

$$\begin{bmatrix} b_{11} & \frac{1}{2}b_{12} & \frac{1}{2}b_{13} \\ \frac{1}{2}b_{12} & b_{22} & \frac{1}{2}b_{23} \\ \frac{1}{2}b_{13} & \frac{1}{2}b_{23} & b_{33} \end{bmatrix} = \begin{bmatrix} -1.88 & -1.35 & -1.09 \\ -1.35 & -0.69 & -0.62 \\ -1.09 & -0.62 & -0.95 \end{bmatrix}$$

The $k = 3$ latent roots and latent vectors (eigenvalues and eigenvectors) are as follows:

$$\begin{array}{lccc} \text{Latent roots} & -3.40 & -0.33 & 0.21 \\ \text{Latent vectors} & \begin{bmatrix} 0.75 \\ 0.48 \\ 0.46 \end{bmatrix} & \begin{bmatrix} 0.31 \\ 0.36 \\ -0.88 \end{bmatrix} & \begin{bmatrix} 0.58 \\ -0.80 \\ -0.12 \end{bmatrix} \end{array}$$

The latent roots are the coefficients in the canonical form

$$\hat{y} = 59.15 - 3.40X_1^2 - 0.33X_2^2 + 0.21X_3^2$$

The latent vectors give the transformation from the $\tilde{x}$'s to the X's as follows:

$$X_1 = 0.75\tilde{x}_1 + 0.48\tilde{x}_2 + 0.46\tilde{x}_3$$

$$X_2 = 0.31\tilde{x}_1 + 0.36\tilde{x}_2 - 0.88\tilde{x}_3$$

$$X_3 = 0.58\tilde{x}_1 - 0.80\tilde{x}_2 - 0.12\tilde{x}_3$$

The Origin O in Terms of X_1, X_2, and X_3

The design origin O in terms of the origin of the canonical coordinates X_1, X_2, X_3 is found by substituting $x_1 = 0$, $x_2 = 0$, and $x_3 = 0$ into these equations, remembering that $\tilde{x}_1 = x_1 + 0.02$, $\tilde{x}_2 = x_2 - 0.55$, and $\tilde{x}_3 = x_3 - 0.23$. Thus

$$X_{10} = -0.35, \qquad X_{20} = 0.01, \qquad X_{30} = 0.48$$

In this example you have two of the canonical coefficients, $B_{22} = -0.33$ and $B_{33} = 0.21$, that are small compared with $B_{11} = -3.40$. Thus, approximately,

$$\hat{y} = 60 - 3.4X_1^2$$

with

$$X_1 = 0.75(x_1 + 0.02) + 0.48(x_2 - 0.55) + 0.46(x_3 - 0.23)$$

or

$$X_1 = -0.38 + 0.75x_1 + 0.48x_2 + 0.46x_3$$

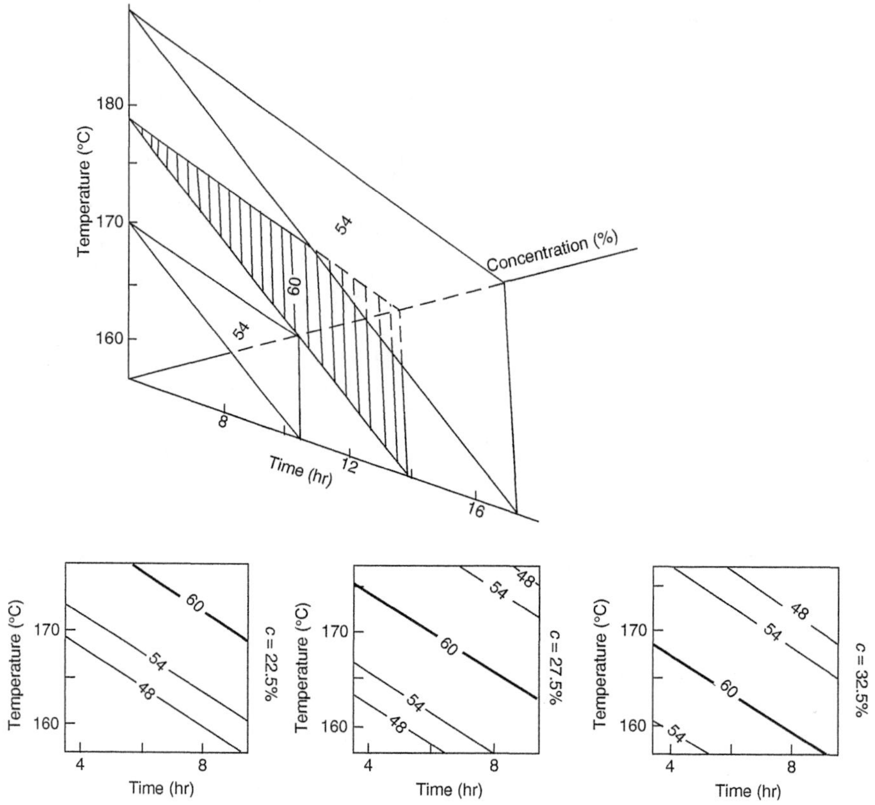

Figure 11.20. Contours of an empirical yield surface with sections at three levels of concentration.

a stationary ridge system in two dimensions such as that shown in Figure 11.20. The implications of this kind of sandwichlike system is explored in detail in Chapter 12.

What Do Maxima Look Like?

In the 1950s, when the basic ideas of response surface methods were being worked out, it was commonly supposed that the usual type of surface containing a maximum would look like the smooth hill in Figure 11.6*a*. Canonical analysis became necessary because in many examples it was found that this was frequently not the case. It became clear that ridge systems were common and moreover that these could be exploited commercially. To a good approximation, for the autoclave example locally there is an entire plane of process conditions that produces the maximum yield of about 60%. That plane is obtained by setting X_1 equal to zero. Thus you can get the maximum yield for any combination of conditions for which

$$0.75x_1 + 0.48x_2 + 0.41x_3 = 0.38$$

or in terms of the factors T, C, and t where

$$x_1 = \frac{T - 167}{5}, \qquad x_2 = \frac{C - 27.5}{2.5}, \qquad x_3 = \frac{t - 6.5}{1.5}$$

where you will remember that T, C, and t are the reaction temperature, concentration, and time. Not all these alternative conditions had equal cost or produced product whose other properties were equally desirable. Thus an opportunity was provided to bring these other properties to better values.

The Car Tire Example Revisited

A second example is provided by the data shown earlier in Table 11.5, where a sequentially assembled non–central composite design was used to study the relation between the cost y of producing a chemical used in car tires and the levels of the factors temperature, concentration, and molar ratio coded as x_1, x_2, x_3. The estimated second-order equation fitted to the data from 11 runs was

$$\hat{y} = 6.8 + 0.8x_1 + 2.7x_2 + 2.5x_3 + 3.0x_1^2 + 1.6x_2^2 + 1.1x_2^3$$
$$- 3.0x_1x_2 - 1.3x_1x_3 + 2.7x_2x_3$$

Proceeding as before, a canonical analysis of the fitted equation produced the following results:

Position of stationary value S	$x_{1S} = -0.25,\ x_{2S} = -0.55,\ x_{3S} = -1.93$
Shift from origin O to origin S	$\tilde{x}_1 = x_1 + 0.25, \tilde{x}_2 = x_2 - 0.55, \tilde{x}_3 = x_3 + 1.93$
Rotation of axes about S	$X_1 = -0.20\tilde{x}_1 - 0.71\tilde{x}_2 + 0.68\tilde{x}_3$ $X_2 = 0.65\tilde{x}_1 + 0.42\tilde{x}_2 + 0.64\tilde{x}_3$ $X_3 = 0.74\tilde{x}_1 + 0.57\tilde{x}_2 - 0.37\tilde{x}_3$
Canonical form	$\hat{y} = 5.01 - 0.12X_1^2 + 1.40X_2^2 + 4.47X_3^2$

The small relative value of the X_1^2 coefficient (−0.12) in the canonical equation implies that the fitted contour surfaces are approximated by tubes centered about the X_1 axis. The arrangement of the points in space as they appear looking down the X_1 axis—"down the tube"—is shown in Figure 11.21. Contours of cost of 10 and 20 as they should appear on the plane if $X_1 = 0$ are also shown in that figure. You will see that when you look at the design from this aspect the numbers agree reasonably well with the contours. The equation was now refitted

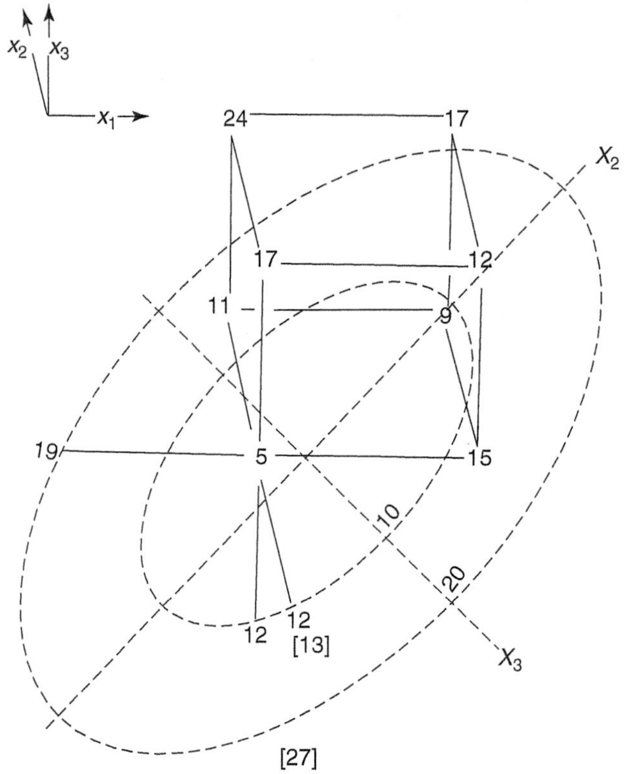

Figure 11.21. Projection of canonical surface centered on the X_1 axis.

after including the steepest ascent runs 9 and 10 and two further runs labelled 14 and 15 in Table 11.5.

For the complete set of fifteen runs the canonical form of the second degree equation was

$$\hat{y} = 3.76 + 0.72X_1^2 + 1.98X_2^2 + 5.47X_3^2$$

The coefficient of X_1^2 has now changed from small negative value to a small positive value. This did not change very much the appearance of the contours near the center of the system that is now estimated as

$$x_{1S} = -0.54 \quad x_{2S} = -0.42 \quad x_{3S} = -0.97$$

The predicted cost at this point was $\hat{y}_S = 0.86$

Further experimentation at these conditions produced costs between 2 and 3. Thus the investigation succeeded in reducing cost by about ten units. It also showed that moderated changes along the X_1 axis would not alter the cost very much.

11.6. BOX–BEHNKEN DESIGNS

Sometimes it is necessary or desirable that factors be run at *only* three levels. Box–Behnken designs are incomplete three-level factorials where the experimental points are specifically chosen to allow the efficient estimation of the coefficients of a second-order model. Table 11.6 shows Box–Behnken designs for $k = 3, 4, 5$ factors. To help appreciate the symmetries of such designs, Figures 11.22*a*, *b* show two different representations of the same 13-run three-level Box–Behnken design for $k = 3$ factors. The notation ± 1 is used to indicate the *four* runs of a 2^2 factorial., the notation 0 means a parallel column of 0's. For each k-factor design the last row of 0's identifies one or more center points.

Table 11.6. Box–Behnken Designs for $k = 3, 4, 5$ Factors

Three factors:

$$\begin{bmatrix} \pm 1 & \pm 1 & 0 \\ \pm 1 & 0 & \pm 1 \\ 0 & \pm 1 & \pm 1 \\ 0 & 0 & 0 \end{bmatrix}$$

Four factors:

$$\begin{bmatrix} \pm 1 & \pm 1 & 0 & 0 \\ 0 & 0 & \pm 1 & \pm 1 \\ 0 & 0 & 0 & 0 \\ \cdots & \cdots & \cdots & \cdots \\ \pm 1 & 0 & 0 & \pm 1 \\ 0 & \pm 1 & \pm 1 & 0 \\ 0 & 0 & 0 & 0 \\ \cdots & \cdots & \cdots & \cdots \\ \pm 1 & 0 & \pm 1 & 0 \\ 0 & \pm 1 & 0 & \pm 1 \\ 0 & 0 & 0 & 0 \end{bmatrix}$$

Five factors:

$$\begin{bmatrix} \pm 1 & \pm 1 & 0 & 0 & 0 \\ 0 & 0 & \pm 1 & \pm 1 & 0 \\ 0 & \pm 1 & 0 & 0 & \pm 1 \\ \pm 1 & 0 & \pm 1 & 0 & 0 \\ 0 & 0 & 0 & \pm 1 & \pm 1 \\ \cdots & \cdots & \cdots & \cdots & \cdots \\ 0 & \pm 1 & \pm 1 & 0 & 0 \\ \pm 1 & 0 & 0 & \pm 1 & 0 \\ 0 & 0 & \pm 1 & 0 & \pm 1 \\ \pm 1 & 0 & 0 & 0 & \pm 1 \\ 0 & \pm 1 & 0 & \pm 1 & 0 \\ 0 & 0 & 0 & 0 & 0 \end{bmatrix}$$

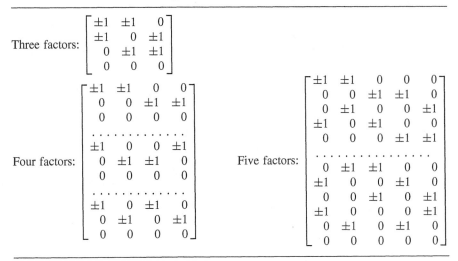

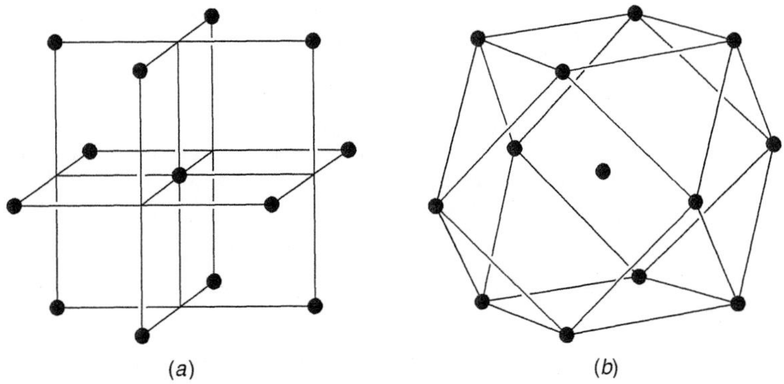

Figure 11.22. Two representations of the Box–Behnken for three factors.

An Example with $k = 4$ Factors

In the data of Table 11.7a four factors were tested in three orthogonal blocks of nine (Box and Behnken, 1960). The second-order model fitted to these data with two additional indicator variables x_{10} and x_{20} for contributions associated with blocks 1 and 2 was

$$\begin{aligned}\hat{y} = 90.8 &+ 2.2x_{10} - 2.7x_{20} + 1.9x_1 - 2.0x_2 + 1.1x_3 \\ &- 3.7x_4 - 1.4x_1^2 - 4.3x_2^2 - 2.2x_3^2 - 2.6x_4^2 - 1.7x_1x_2 \\ &- 3.8x_1x_3 + 1.0x_1x_4 - 1.7x_2x_3 - 2.6x_3x_4 - 4.3x_3x_4\end{aligned}$$

The coefficients +2.2 and −2.7 are the estimated deviations from the mean of blocks 1 and 2.* The predicted values $\hat{y}$ and residuals $y - \hat{y}$ are given in Table 11.7b. The ANOVA is given in Table 11.7d. Of the residual sum of squares $\sum(y - \hat{y})^2 = 126.71$ a large portion, 105.53, is accounted for by the blocks. On the assumption that the model is adequate, the error mean square 2.12 then provides an estimate of σ^2.

In the normal plot of the residuals shown in Figure 11.23*a* runs 10 and 13 appear to be highly discrepant. Thus, as was previously discussed, if you were the data analyst, your reaction to these possibly aberrant observations should be to discover how they occurred. To show the influence of such discrepant values on the least squares analysis, a reanalysis omitting the suspect runs 10 and 13 produced the fitted equation where now 2.1 and −3.8 were revised estimates of block effects

$$\begin{aligned}\hat{y} = 91.21 &+ 2.1x_{10} - 3.8x_{20} + 2.6x_1 - 2.0x_2 + 1.1x_3 - 3.0x_4 \\ &- 2.2x_1^2 - 3.9x_2^2 - 1.8x_3^2 - 3.4x_4^2 - 1.7x_1x_2 - 3.8x_1x_3 - 1.5x_1x_4 \\ &- 1.7x_2x_3 - 2.67x_2x_4 - 4.3x_3x_4\end{aligned}$$

The new predicted values and residuals are shown in Table 11.7c and the associated ANOVA in Table 11.7e. The coefficients in the new fitted equation are not greatly different, but the residual mean square is much reduced from 2.12 to 0.15 and the normal plot of the residuals (Fig. 11.23*b*) shows much more regularity.

Other Three-Level Factorials and Fractions

In the automobile emissions experiment set out in Table 10.17 a 3^2 factorial design was used. The question of interest was how the carbon monoxide from an automobile engine depended on two factors: ethanol concentration x_1 and air–fuel ratio x_2. The two columns forming the design matrix of the 3^2 factorial are shown

*The estimate +2.2 is the deviation of the first block average from 90.8, the −2.7 the deviation from 90.8 of the second block. Since the block deviations must sum to zero, the deviation for the third block is +0.5.

Table 11.7. Analysis of a Three Level Design (a) Box–Behnken design in $k = 4$ factors randomly run within three blocks. (b) Observations y, predicted values $\hat{y}$ and residuals $y - \hat{y}$. (c) Predicted values $\hat{y}$ and residuals $y - \hat{y}$ with runs 10 and 13 omitted. (d) Analysis of variance for the second-order model, 27 runs. (e) Analysis of variance for the second-order model, omitting runs 10 and 13.

Run	(a)							(b)			(c)	
Number	x_0	x_{10}	x_{20}	x_1	x_2	x_3	x_4	y	$\hat{y}$	$y - \hat{y}$	$\hat{y}$	$y - \hat{y}$
1	1	1	0	−1	−1	0	0	84.7	85.53	−0.83	84.86	−0.16
2	1	1	0	1	−1	0	0	93.3	92.75	0.55	93.33	−0.03
3	1	1	0	−1	1	0	0	84.2	84.97	−0.77	84.29	−0.09
4	1	1	0	1	−1	0	0	86.1	85.48	0.62	86.06	0.04
5	1	1	0	0	0	−1	−1	85.7	86.40	−0.07	85.72	−0.02
6	1	1	0	0	0	1	−1	96.4	97.17	−0.77	96.49	−0.09
7	1	1	0	0	0	−1	1	88.1	87.55	0.55	88.13	−0.03
8	1	1	0	0	0	1	1	81.8	81.32	0.48	81.89	−0.09
9	1	1	0	0	0	0	0	93.8	92.93	0.87	93.33	0.47
10	1	0	1	−1	0	0	−1	89.4	86.79	2.61		
11	1	0	1	1	0	0	−1	88.7	88.76	−0.06	88.86	−0.16
12	1	0	1	−1	0	0	1	77.8	77.54	0.26	77.64	0.16
13	1	0	1	1	0	0	1	80.9	83.31	−2.41		
14	1	0	1	0	−1	−1	0	80.9	80.68	0.22	80.77	0.13
15	1	0	1	0	1	−1	0	79.8	80.11	−0.31	80.21	−0.41
16	1	0	1	0	−1	1	0	86.8	86.29	0.51	86.39	0.41
17	1	0	1	0	1	1	0	79.0	79.03	−0.03	79.12	−0.12
18	1	0	1	0	0	0	0	87.3	88.10	−0.80	87.30	0.00
19	1	0	0	0	−1	0	−1	86.1	86.87	−0.77	86.20	−0.10
20	1	0	0	0	1	0	−1	87.9	88.21	−0.31	87.53	0.37
21	1	0	0	0	−1	0	1	85.1	84.77	0.33	85.35	−0.25
22	1	0	0	0	1	0	−1	76.4	75.61	0.79	76.18	0.22
23	1	0	0	−1	0	−1	0	79.7	80.22	−0.52	79.55	0.15
24	1	0	0	1	0	−1	0	92.5	91.74	0.76	92.32	0.18
25	1	0	0	−1	0	1	0	89.4	90.14	−0.74	89.46	−0.06
26	1	0	0	1	0	1	0	86.9	86.36	0.54	86.93	−0.03
27	1	0	0	0	0	0	0	90.7	90.77	−0.07	91.17	−0.47

			SS		df		MS
(d)	Due to first and second order terms		563.28		14		40.23
	Residual	Blocks	126.71	105.53	12	2	52.77
		Error		21.18		10	2.12
	Total		689.99		26		

Table 11.7. (*continued*)

			SS		df		MS
(e)	Due to first and second order terms		539.23		14		38.51
	Residual	Blocks	123.41	121.91	10	2	52.77
		Error		1.50		8	0.19
	Total		652.64		24		

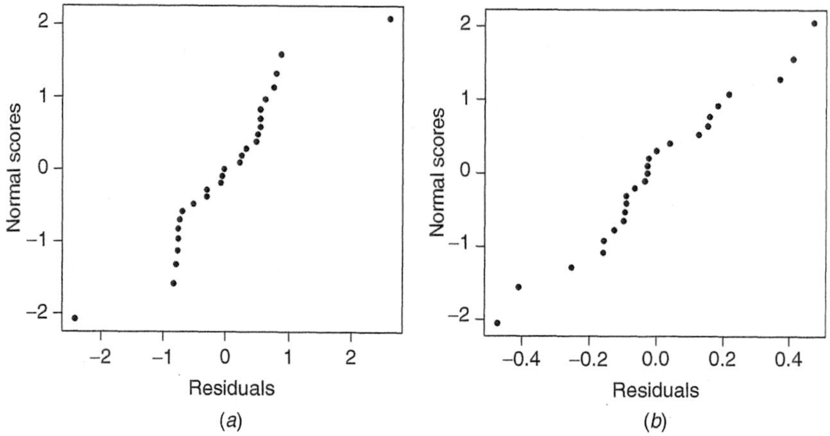

Figure 11.23. Normal plots of residuals, $n = 27$, Box–Behnken design.

here in Table 11.8a along with the observed average yields. A regression matrix for fitting a second-order model is given in Table 11.8b. The *orthogonalized* regression matrix is given in Table 11.8c. As you saw earlier, this produced the fitted relationship

$$\hat{y} = 78.0 + 4.4x_1 - 6.9x_2 - 4.6x_1^2 - 4.1x_2^2 - 9.1x_1x_2$$

which on analysis turned out to be that of a rising ridge whose implications were discussed.

Exercise 11.4. The same fitted model will result using the regression matrix from either Table 11.8b or Table 11.8c. Why would one be preferred over the other? How does one obtain the orthogonal vectors $\dot{x}_1^2$ and $\dot{x}_2^2$

The 3^{4-2} Fractional Factorial

Using the properties of the nine-run 3×3 hyper-Graeco–Latin square it is possible to produce a 3^{4-2} fractional factorial with a different orthogonal breakdown. Table 11.9a displays a 3×3 Graeco–Latin square as it might appear in an

Table 11.8. Regression Matrices, 3×3 Factorial, Co emissions Data

(a)			(b)						(c)					
3^2 Factorial		Average Yield,	Regression Matrix, Second-Order Model						Orthogonal Regression Matrix, Second-Order Model					
x_1	x_2	$\bar{y}$	x_0	x_1	x_2	x_1^2	x_2^2	x_1x_2	x_0	x_1	x_2	$\dot{x}_1^2$	$\dot{x}_2^2$	x_1x_2
−	−	63.75	1	−1	−1	1	1	1	1	−1	−1	−1	−1	1
0	−	79.45	1	0	−1	0	1	0	1	0	−1	2	−1	0
+	−	91.75	1	1	−1	1	1	−1	1	1	−1	−1	−1	−1
−	0	69.70	1	−1	0	1	0	0	1	−1	0	−1	2	0
0	0	80.75	1	0	0	0	0	0	1	0	0	2	2	0
+	0	73.60	1	1	0	1	0	0	1	1	0	−1	2	0
−	+	67.30	1	−1	1	1	1	−1	1	−1	1	−1	−1	−1
0	+	67.45	1	0	1	0	1	0	1	0	1	2	−1	0
+	+	59.05	1	1	1	1	1	1	1	1	1	−1	−1	1

Table 11.9. Design Matrices for the 3 × 3 Graeco-Latin Square and 3^{4-2} Fractional

(a) The 3 × 3 Graeco–Latin Square

	Columns (c) I	II	III
Rows (r) 1	$A\alpha$	$B\beta$	$C\gamma$
2	$C\beta$	$A\gamma$	$B\alpha$
3	$B\gamma$	$C\alpha$	$A\beta$

Design Matrix (b)				The 3^{4-2} Fractional Factorial (c)				Orthogonal Quadratic Vectors (d)				
R x_1	**G** x_2	**r** x_3	**c** x_4	x_1	x_2	x_3	x_4	$\dot{x}_1^2$	$\dot{x}_1^2$	$\dot{x}_3^2$	$\dot{x}_4^2$	x_0
A	α	1	I	−1	−1	−1	−1	−1	−1	−1	−1	1
B	α	2	III	0	−1	0	+1	2	−1	2	−1	1
C	α	3	II	+1	−1	+1	0	−1	−1	−1	2	1
A	β	3	III	−1	0	+1	+1	−1	2	−1	−1	1
B	β	1	II	0	0	−1	0	2	2	−1	2	1
C	β	2	II	+1	0	0	−1	−1	2	2	−1	1
A	γ	2	II	−1	+1	0	0	−1	−1	2	2	1
B	γ	3	I	0	+1	+1	−1	2	−1	−1	−1	1
C	γ	1	III	+1	+1	−1	+1	−1	−1	−1	−1	1

agricultural field trial with the Roman and Greek letters identifying two factors each at three versions, with the design blocked by rows and columns. Table 11.9b is an alternative display of the same nine-run arrangement in which the four factors are now identified as $x_1 = \mathbf{R}$ (Roman), $x_2 = \mathbf{G}$ (Greek), $x_3 = \mathbf{r}$ (rows) and $x_4 = \mathbf{c}$ (columns). Table 11.9c is again the same arrangement but with the three levels of each factor identified by $[-1, 0, +1]$ producing four orthogonal vectors that together produce the 3^{4-2} fractional factorial design with four factors each at three levels in nine runs. Table 11.9d displays the four orthogonalized quadratic terms $\dot{x}_1^2$, $\dot{x}_2^2$, $\dot{x}_3^2$, and $\dot{x}_4^2$ useful for estimating quadratic terms. The vectors $\dot{\mathbf{x}}_1^2$, $\dot{\mathbf{x}}_2^2$, $\dot{\mathbf{x}}_3^2$, and $\dot{\mathbf{x}}_4^2$ are not only mutually orthogonal but also orthogonal to the linear term vectors $\mathbf{x}_1$, $\mathbf{x}_2$, $\mathbf{x}_3$, and $\mathbf{x}_4$. These eight column vectors along with a column vector $\mathbf{x}_0$ together comprise the L9 orthogonal array. These orthogonal vectors have remarkable properties, but they must be used with some care.

To explain, consider the data from the automobile emission example, repeated again in Table 11.10a along with the fitted second-order model. The contours of the fitted response surface appear in Figure 11.17. The design employs the nine runs of the 3^2 factorial. Suppose you now add two *extraneous* factors x_3 and x_4 to produce the 3^{4-2} design shown in Table 11.10b. Clearly the factors x_3 and x_4, added to the design *after* the data had been collected, have no influence whatever. Suppose, however, that all four factors were included in an analysis of the data. What would you have learned? A quick analysis of the 3^{4-2} is to plot the average response at each level for each factor as displayed in Figure 11.24. Each average is based on six observations and has an estimated standard error $\hat{\sigma}_{\bar{y}} = 0.9$. On viewing these figures it might be concluded that over the range chosen factor x_4 has a profound curvature effect and x_3 a linear effect. But, of course, for these data you know that both x_3 and x_4 are inert! For the 3^{4-2} design fitted model for linear and quadratic terms is

$$\hat{y} = 87.3 + 4.4x_1 - 6.8x_2 + 4.2x_3 + 0.1x_4 - 4.6x_1^2 - 4.1x_2^2 - 2.3x_3^2 - 10.7x_4^2$$

The coefficients of x_3 and x_4^2 are both large and "statistically significant." This suggests that both x_3 and x_4 have important roles in influencing the response. But, again, you know that with these data both factors are without influence.

Table 11.10. A 3^2 Factorial with Data and a 3^{4-2} Fractional Factorial

(a) 3^2 Factorial Design					(b) 3^{4-2} Fractional Factorial			
x_1	x_2	Observed, y		Average	x_1	x_2	x_3	x_4
−	−	61.9	65.6	63.75	−	−	−	−
0	−	80.9	78.0	79.45	0	−	0	+
+	−	89.7	93.8	91.75	+	−	+	0
−	0	72.1	67.3	68.70	−	0	+	+
0	0	80.1	81.4	80.75	0	0	−	0
+	0	77.8	74.8	73.60	+	0	0	−
−	+	66.4	68.2	67.30	−	+	0	0
0	+	68.9	66.0	67.45	0	+	+	−
+	+	60.2	57.9	59.05	+	+	−	+

$\hat{y} = 78.6 + 4.4x_1 - 6.9x_2 - 4.6x_1^2 - 4.1x_2^2 - 9.1x_1x_2$

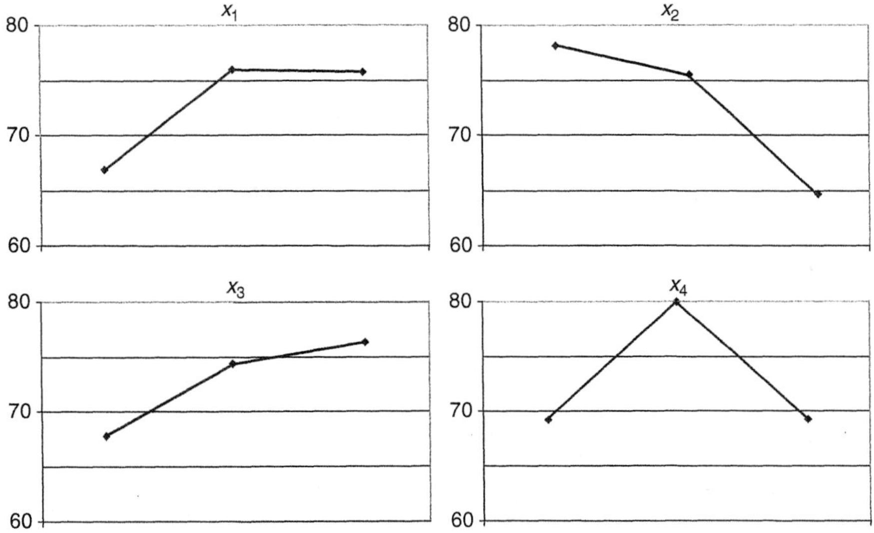

Average response

–	0	+
66.9	75.9	75.7
78.3	75.6	64.6
67/9	74.4	76.3
69.2	79.9	69.4

Figure 11.24. Plots of averages: four factors each at three levels.

The fault lies of course in the choice of the experimental design. The Graeco–Latin square (L9 orthogonal array) used as a 3^{3-1} or 3^{4-2} fractional factorial requires the assumption that *all* two-factor interactions are zero. (Recall, in this example the original 3^2 factorial displayed a large x_1x_2 interaction with $b_{12} = -9.1$.) For these nine-run designs, if two-factor interactions exist, they bias *both* the estimates of the first order and the quadratic estimates. An experimenter who needs to estimate second-order effects and is restricted to only three levels for each factor should use a design in which interactions as well as quadratic effects may be separately estimated. The Box-Behnken arrangements were designed for this purpose.

The 3^3 Factorial and Its Fractions

There are similar difficulties associated with the fractional factorials that can be constructed from the 27 runs generated by the 3^3 factorial, that is, three factors each at three levels. Table 11.11a displays 13 mutually orthogonal "linear" columns $x_1, x_2, \ldots, x_{13}$ that together form the 27-run 3^{13-10} fractional factorial. Table 11.11b displays the associated 13 mutually orthogonal quadratic columns $\dot{\mathbf{x}}_1^2, \dot{\mathbf{x}}_2^2, \ldots, \dot{\mathbf{x}}_{13}^2$ that are themselves orthogonal to the linear columns and to each

Table 11.11. The 3^{13-3} Factorial with Quadratic Columns $\dot{x}_1, \ldots, \dot{x}_{13}$

	(a)													(b)												
Run	x_1	x_2	x_3	x_4	x5	x6	x7	x8	x9	x10	x11	x12	x13	$\dot{x}_1^2$	$\dot{x}_2^2$	$\dot{x}_3^2$	$\dot{x}_4^2$	$\dot{x}_5^2$	$\dot{x}_6^2$	$\dot{x}_7^2$	$\dot{x}_8^2$	$\dot{x}_9^2$	$\dot{x}_{10}^2$	$\dot{x}_{11}^2$	$\dot{x}_{12}^2$	$\dot{x}_{13}^2$
1	−1	−1	−1	−1	−1	−1	−1	−1	−1	−1	−1	−1	−1	1	1	1	1	1	1	1	1	1	1	1	1	1
2	−1	−1	−1	−1	0	0	0	0	0	0	0	0	0	1	1	1	1	−2	−2	−2	−2	−2	−2	−2	−2	−2
3	−1	−1	−1	−1	1	1	1	1	1	1	1	1	1	1	1	1	1	1	1	1	1	1	1	1	1	1
4	−1	0	0	0	−1	−1	−1	0	0	0	1	1	1	1	−2	−2	−2	1	1	1	−2	−2	−2	1	1	1
5	−1	0	0	0	0	0	0	1	1	1	−1	−1	−1	1	−2	−2	−2	−2	−2	−2	1	1	1	1	1	1
6	−1	0	0	0	1	1	1	−1	−1	−1	0	0	0	1	−2	−2	−2	1	1	1	1	1	1	−2	−2	−2
7	−1	1	1	1	−1	−1	−1	1	1	1	0	0	0	1	1	1	1	1	1	1	1	1	1	−2	−2	−2
8	−1	1	1	1	0	0	0	−1	−1	−1	1	1	1	1	1	1	1	−2	−2	−2	1	1	1	1	1	1
9	−1	1	1	1	1	1	1	0	0	0	−1	−1	−1	1	1	1	1	1	1	1	−2	−2	−2	1	1	1
10	0	−1	0	1	−1	0	1	−1	0	1	−1	0	1	−2	1	−2	1	1	−2	1	1	−2	1	1	−2	1
11	0	−1	0	1	0	1	−1	0	1	−1	0	1	−1	−2	1	−2	1	−2	1	1	−2	1	1	−2	1	1
12	0	−1	0	1	1	−1	0	1	−1	0	1	−1	0	−2	1	−2	1	1	1	−2	1	1	−2	1	1	−2
13	0	0	1	−1	−1	0	1	0	1	−1	1	−1	0	−2	−2	1	1	1	−2	1	−2	1	1	1	1	−2
14	0	0	1	−1	0	1	−1	1	−1	0	−1	0	1	−2	−2	1	1	−2	1	1	1	1	−2	1	−2	1
15	0	0	1	−1	1	−1	0	−1	0	1	0	1	−1	−2	−2	1	1	1	1	−2	1	−2	1	−2	1	1
16	0	1	−1	0	−1	0	1	1	−1	0	0	1	−1	−2	1	1	−2	1	−2	1	1	1	−2	−2	1	1
17	0	1	−1	0	0	1	−1	−1	0	1	1	−1	0	−2	1	1	−2	−2	1	1	1	−2	1	1	1	−2
18	0	1	−1	0	1	−1	0	0	1	−1	−1	0	1	−2	1	1	−2	1	1	−2	−2	1	1	1	−2	1
19	1	−1	1	0	−1	1	0	−1	1	0	−1	1	0	1	1	1	−2	1	1	−2	1	1	−2	1	1	−2
20	1	−1	1	0	0	−1	1	0	−1	1	0	−1	1	1	1	1	−2	−2	1	1	−2	1	1	−2	1	1
21	1	−1	1	0	1	0	−1	1	0	−1	1	0	−1	1	1	1	−2	1	−2	1	1	−2	1	1	−2	1
22	1	0	−1	1	−1	1	0	0	−1	1	1	0	−1	1	−2	1	1	1	1	−2	−2	1	1	1	−2	1
23	1	0	−1	1	0	−1	1	1	0	−1	−1	1	0	1	−2	1	1	−2	1	1	1	−2	1	1	1	−2
24	1	0	−1	1	1	0	−1	−1	1	0	0	−1	1	1	−2	1	1	1	−2	1	1	1	−2	−2	1	1
25	1	1	0	−1	−1	1	0	1	0	−1	0	−1	1	1	1	−2	1	1	1	−2	1	−2	1	−2	1	1
26	1	1	0	−1	0	−1	1	−1	1	0	1	0	−1	1	1	−2	1	−2	1	1	1	1	−2	1	−2	1
27	1	1	0	−1	1	0	−1	0	−1	1	−1	1	0	1	1	−2	1	1	−2	1	−2	1	1	1	1	−2

Note: All columns are mutually orthogonal.

other. These 26 vectors together with a column vector of 1's form the L27 orthogonal array. Of the 13 factors all pairs project into a 3^2 factorial replicated three times and thus provide second-order designs, that is, designs capable of providing estimates of all linear, quadratic, and two-factor interaction terms. There are 286 sets of three factors that project into complete 3^3 factorials and thus also second-order designs. But in this context, the design must be used with some caution because there are 52 projections that do not provide second-order designs, specifically any set of three factors falling within the following subsets of four: [1, 2, 3, 4], [1, 5, 6, 7], [1, 8, 9, 10], [1, 11, 12, 13], [2, 5, 8, 11], [2, 6, 9, 12], [2, 7, 10, 13], [3, 5, 9, 13], [3, 6, 10, 11], [3, 7, 8, 12], [4, 5, 10, 12], [4, 6, 8, 13], [4, 7, 9, 11].

REFERENCES AND FURTHER READING

Response surface methodology was first developed and described in:

Box, G. E. P. (1954) The exploration and exploitation of response surfaces: Some general considerations and examples, *Biometrics*, **10**, 16–60.

Box, G. E. P., and Coutie, G. A. (1956) Application of digital computers in the exploration of functional relationships, *Proc. Inst. Electrical Eng.*, **103**(1) 100–107.

Box, G. E. P., and Hunter, J. S. (1957) Multi-factor designs for exploring response surfaces, *A. Math. Statist.*, **28**, 195–241.

Box, G. E. P., and Wilson, K. B. (1951) On the experimental attainment of optimum conditions, *J. Roy. Statist. Soc. Ser. B*, **13**, 1.

Box, G. E. P., and Youle, P. V. (1955) The exploration and exploitation of response surfaces: An example of the link between the fitted surface and the basic mechanism of the system, *Biometrics*, **11**, 287–323.

Fung, C. A. (1986) *Statistical Topics in Off-line Quality Control*, Unpublished Ph.D Thesis, University of Wisconsin-Madison.

Textbooks devoted to the subject of response surface models, design, and analyses are:

Box, G. E. P., and Draper, N. R. (1987) *Empirical Model Building and Response Surfaces*, Wiley, New York.

Box, G. E. P., and Draper, N. R. (2007) *Response Surfaces, Mixtures, and Ridge Analyses*, 2nd ed., Wiley, Hoboken, N.J.

Myers, R. H., and Montgomery, D. C. (2002) *Response Surface Methodology*, 2nd ed., Wiley, New York.

The technical literature concerning response surface methodology is vast. The following useful article contains some 270 references:

Myers, R. H., Montgomery, D. C., Vining, G. G., Borror, C. M., and Kowalksi, S. M. (2004) Response surface methodology: A retrospective and literature survey, *J. Quality Technol.*, **36**, 53–78.

Papers of particular interest are:

Alexsson, A.-K., and Carlson, R. (1995) Links between the theoretical rate law and the response surface. An example with a solvolysis reaction, *Acta Chem. Scand.*, **49**, 663–667.

Barella, A., and Sust A. Unpublished report to the Technical Committee, International Wool Organization. See G. E. P. Box and D. R. Cox, (1964) An analysis of transformations, *J. Roy. Statist. Soc. Ser. B*, **26**, 211–252.

Box, G. E. P. (1982) Choice of response surface design and alphabetic optimality, *Utilitas Math.*, **21B**, 11–55.

Box, G. E. P. (1992) Sequential experimentation and sequential assembly of designs, *Quality Eng.*, **5**, 321–330.

Box, G. E. P., and Behnken, D. W. (1960) Some three level designs for the study of quantitative variables, *Technometrics*, **2**, 455–476.

Box, G. E. P., and Hunter, W. G. (1965) The experimental study of physical mechanisms, *Technometrics*, **7**, 23–42.

Box, G. E. P., and Liu, P. Y. T (1999) Statistics as a catalyst to learning by scientific method, Part 1, An example, *J. Qual. Technol.*, **31**, 1–15.

Box, G. E. P., and Lucas, H. L. (1959) Design of experiments in non-linear situations, *Biometrika*, **46**, 77–90.

Box, G. E. P., and Wetz, J. (1973) *Criteria for Judging Adequacy of Estimation by Approximating Response Function*, Technical Report 9, Department of Statistics, University of Wisconsin, Madison.

Carlson, R. (2002) Canonical analysis of response surface models as tools for exploring organic reactions and processes, in *Proceedings of 1st International Symposiu: Optimising Organic Reaction Processes*, Oslo, Norway.

Hahn, G. J. (1984) Experimental design in a complex world, *Technometrics*, **26**, 19–31.

Hunter, W. G., and Mezaki, R. (1964) A model building technique for chemical engineering kinetics, *Am. Inst. Chem. Eng. J.*, **10**, 315–322.

Kim, K.-J., and Lin D. K. J. (2000) Dual response surface optimization: A fuzzy modeling approach, *J. Qual. Technol.*, **28**, 1–10.

Kuhn, T. S. (1996) *The Structure of Scientific Revolutions*, 3rd Edition, Univ. of Chicago Press, Chicago.

Rogers, K. (1998) Personal communication.

Snee, R. D. (1985) Computer aided design of experiments — some practical experiences, *J. Quality*, **17**, 222–236.

QUESTIONS FOR CHAPTER 11

1. What are response contours? Illustrate for a k-factor problem with $k = 2, 3$.

2. Suppose $k = 2$. What is a first-order model?

3. Draw contours for a two-factor problem in which the response might be approximated by a first-order model. What is the path of steepest ascent?

4. What is a second-order model?

5. For a $k = 2$ factor problem how would you draw a set of contours in which a second-order model might locally approximate (a) a maximum, (b) a minimum, (c) a minimax, (d) a stationary ridge, and (e) a sloping ridge?

6. A response that is a smooth function over the levels of two factors may or may not be similar when viewed over different levels of a third factor. Can you explain this statement? Provide examples.

7. A response found to be smooth over different levels of a factor may or may not be smooth over the levels of another factor. Can you explain why and provide examples?

8. Assume $k = 2$. Can you explain diagrammatically the canonical analysis of a second-order model which represents a maximum?

9. For a rising ridge why would the use of a one-factor-at-a-time experimentation be a mistaken strategy?

10. How would you sketch a central composite design for $k = 3$ factors?

11. Imagine a 2^3 design in eight runs represented by the vertices of a cube in 3-space. When would a statistical analysis suggest that there might be a stationary region just outside the cube?

12. How would you illustrate how a non–central composite design might be used to shed further light on the problem in question 11 if the stationary region is believed to be (a) near an edge and (b) near a vertex of the cube?

13. What is a Box–Behnken design? Make a sketch of the design for $k =$ 3 factors.

14. What is meant by the information function of a design? What property has the function for a rotatable design?

PROBLEMS FOR CHAPTER 11

1. Write down the design matrix for a 29-run central composite design having five factors with three center points.

2. Suppose you had fitted a second-order model by least squares with a three-factor central composite design with axial and center points duplicated and obtained the following partial ANOVA table:

Source	**SS**		**df**	
Due to model	746		9	
Lack of fit		170		5
Residual	154		12	
Replicate error		47		7

(a) Complete the ANOVA table.

(b) Is there evidence of lack of fit of the fitted model?

(c) Do you think the model is well enough estimated to justify further analysis?

3. A 2^3 factorial design with center point was used in an experiment to reduce cost. The following partial ANOVA table resulted:

Source	SS	df
First-order terms	85.8	3
Higher order terms	12.7	3

Do you think a first-order model would be adequate to approximate this system? Given that the estimated model was $\hat{y} = 45.2 + 13.9x_1 - 6.3x_2 + 21.3x_3$ calculate four points on the path of steepest descent for which the x_1 coordinates are 1, 2, 3, 4.

4. Describe two checks of adequacy of fit for a first-order model.

5. Suppose you have a design with $k = 1$ factor and $n = 3$ points located at $x = -1, 0, +1$. Given a straight line model, sketch the information function.

6. Suppose a second-order model in three factors had been fitted. Using canonical analysis, how could you detect and analyze a sloping ridge system? Describe and illustrate two different kinds of limiting models for a sloping ridge and their corresponding canonical form.

7. What is meant by the "sequential assembly of an experimental design"? Illustrate the sequential assembly of a three-factor central composite design beginning with a half fraction of a 2^3 system. Suppose you suspected that shifts in response might occur between the addition of parts of the design. How would this affect your design and analysis?

8. An experimenter with limited resources included 11 factors in a 12-run Plackett and Burman design with a center point. Analysis pointed to three factors of greatest importance. The experimenter then added a center point and six axial points for these three particular factors and fitted a second-order model. What do you think of this approach? Name some good and some bad aspects.

9. The following data are due to Baker, who used a 27-run Box–Behnken design in three blocks of nine runs each to study the influence of four factors on nylon flake blender efficiency. Analyze the experiment. Do you find any evidence of bad values?

Experimental Design for Flake Blender Efficiency

Run Order	Particle Size, X_1 (in.)	Screw Lead Length, X_2 (in.)	Screw Rotation, X_3 (rpm)	Blending Time, X_4 min	Measured Blending Efficiency, y (%)
			Block 1		
5	1/4	4	200	60	84.5
6	1	4	200	60	62.9
3	1/4	8	200	60	90.7
7	1	8	200	60	63.2
4	5/8	6	175	45	70.9

Experimental Design for Flake Blender Efficiency

Run Order	Particle Size, X_1 (in.)	Screw Lead Length, X_2 (in.)	Screw Rotation, X_3 (rpm)	Blending Time, X_4 min	Measured Blending Efficiency, y (%)
9	5/8	6	225	45	69.2
1	5/8	6	175	75	80.1
2	5/b	6	225	75	79.8
8	5/8	6	200	60	75.1
			Block 2		
17	1/4	6	200	45	81.8
13	1	6	200	45	61.8
11	1/4	6	200	75	92.2
14	1	6	200	75	70.7
10	5/8	4	175	60	72.4
12	5/8	8	175	60	76.4
18	5/8	4	225	60	71.9
15	5/8	8	225	60	74.9
16	5/8	6	200	60	74.5
			Block 3		
25	1/4	6	175	60	88.0
27	1	6	175	60	63.0
22	1/4	6	225	60	86.7
19	1	6	225	60	65.0
26	5/8	4	200	45	69.4
21	5/8	8	200	45	71.2
24	5/8	4	200	75	77.3
23	5/8	8	200	75	81.1
20	5/8	6	200	60	73.0

CHAPTER 12

Some Applications of Response Surface Methods

In this chapter the following aspects of RSM (response surface methodology) are illustrated:

1. Using RSM to improve a product design
2. Simplification of a complicated response function by data transformation
3. Using RSM to determine and exploit active and inert factor spaces for multiple-response data
4. Using RSM to exploit inert canonical spaces
5. Using RSM to move from empiricism to mechanism

12.1. ITERATIVE EXPERIMENTATION TO IMPROVE A PRODUCT DESIGN

Iterative Experimentation

In the analysis of data it is frequently emphasized that, though observations are subject to error, probability statements can be made and conclusions drawn. Statistics tells us how. Frequently overlooked, however, is how heavily conditional* on the experimental setup are most such statements and conclusions. This aspect of investigation has been emphasized throughout this book, and in RSM it is especially important.

*One reason this is overlooked is that in one-shot experiments the experimental design is treated as a "given," the only one that could have been run.

For example, consider an identical problem faced by a number of different experimenters:

Different factors could have been chosen for the study.
Different ranges for the factors could have been selected.
Different choices could have been made for qualitative and blocking factors.
Different transformations for the factors might have been employed.
Different responses and their metrics might have been chosen.
Different models could have been considered.

These arbitrary choices and educated guesses will affect conclusions drawn from a single experiment far more than observational error or statistical niceties. However, in an iterative *sequence* of experiments, although different experimenters may begin from different starting points and take different routes, they may nevertheless arrive at similar or equally satisfactory, solutions. Like mathematical iteration, scientific iteration tends to be self-correcting.

IMPROVING THE DESIGN OF A PAPER HELICOPTER

Traditional training in statistics tends to be imbedded in the one-shot approach. To appreciate the process of the sequential unfolding of a problem, it is very important for *you* yourself to *take part* in the process of using statistical design for scientific discovery in a real investigation. You grow when your imagination is challenged and you experience the excitement of conceiving and testing your own ideas. The following is a suggestion of how you may get some practice at this.

Since readers will be from many different areas of endeavor, the learning exercise had to be for some simple product available to everyone. Also it needed to be capable of many different kinds of modification. The following is one history (Box and Liu, 1999) of the improvement of the design of a paper helicopter.* *Please do not regard what follows as supplying just one more dead data set to be further analyzed and reanalyzed.* It is a sample of what we hope *you* will experience when *you* perform your own experiments using experimental design and discovering your own iterative path to a solution. The art of investigation cannot be found just by playing with someone else's data. Also remember that the "best" helicopters arrived at in this particular investigation were certainly *not* optimal. You should be able to find a better helicopter design when you try out some of your own ideas. There is usually more than one answer to a problem and you may discover a good product design and perhaps a much better one very different from those found here.

* You may prefer to use some other device, but if you do, it is important to choose something that can be changed in a variety of ways, limited only by your imagination, not preordained. Devices with only a fixed number of built-in factors and levels that can be changed, though perhaps useful for teaching one-shot experimentation, are useless for the purpose of illustrating iterative investigation.

There is one important respect in which these helicopter experiments do not provide a picture of a real investigation. In this example, progress is made almost entirely empirically. If you were really in the business of making helicopters, aerodynamicists and other engineers would be on the team supplying subject matter knowledge. Their help in designing the experiments and in interpreting the results would undoubtedly produce a better helicopter quicker. Also, a number of responses in addition to the mean and variance of the flight times would be measured, recorded, and jointly considered.

An Initial Screening Experiment

The prototype helicopter design displayed in Figure 12.1 was made available by a colleague. Starting with this prototype, the objective was to find an improved design giving longer flight times. The designs were limited to those that could be constructed from readily available office supplies and, in particular, from standard 11 × 8.5-inch paper. The test flights described were carried out in a room with a ceiling 8.5 feet (102 inches, or 2.6 meters) from the floor. The average flight time for the prototype was about 200 centiseconds.

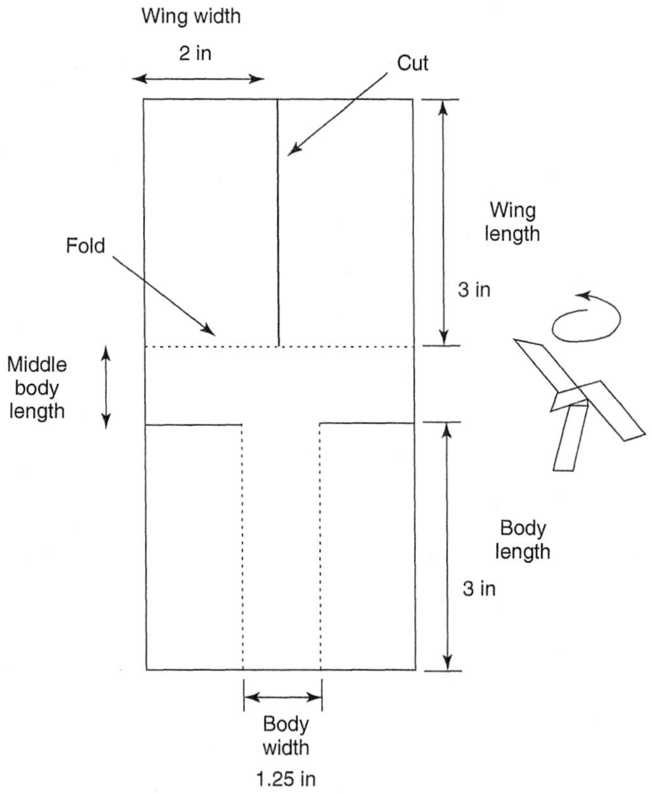

Figure 12.1. Initial prototype helicopter design.

Table 12.1. Factor Levels (+1, −1) for Eight Factors

Factors	−1	+1
P: paper type, x_1	Regular	Bond
l: wing length, x_2	3.00 in.	4.75 in., $x_2 = (l - 3.875)/0.875$
L: body length, x_3	3.00 in.	4.75 in., $x_3 = (L - 3.875)/0.875$
W: body width, x_4	1.25 in.	2.00 in., $x_4 = (W - 1.625)/0.375$
F: fold, x_5	No	Yes
T: taped body, x_6	No	Yes
C: paper clip, x_7	No	Yes
M: taped wing, x_8	No	Yes

Table 12.2. Design 1: Layout and Data for 2_{IV}^{8-4} Screening Design

Run Number	x_1 P	x_2 l	x_3 L	x_4 W	x_5 F	x_6 T	x_7 C	x_8 M	$\overline{y}$	s	100 log(s)
1	−1	−1	−1	−1	−1	−1	−1	−1	236	2.1	32
2	+1	−1	−1	−1	−1	+1	+1	+1	185	4.7	67
3	−1	+1	−1	−1	+1	−1	+1	+1	259	2.7	43
4	+1	+1	−1	−1	+1	+1	−1	−1	318	5.3	72
5	−1	−1	+1	−1	+1	+1	+1	−1	180	7.7	89
6	+1	−1	+1	−1	+1	−1	−1	+1	195	7.7	89
7	−1	+1	+1	−1	−1	+1	−1	+1	246	9.0	96
8	+1	+1	+1	−1	−1	−1	+1	−1	229	3.2	51
9	−1	−1	−1	+1	+1	+1	−1	+1	196	11.5	106
10	+1	−1	−1	+1	+1	−1	+1	−1	203	10.1	100
11	−1	+1	−1	+1	−1	+1	+1	−1	230	2.9	16
12	+1	+1	−1	+1	−1	−1	−1	+1	261	15.3	118
13	−1	−1	+1	+1	−1	−1	+1	+1	168	11.3	105
14	+1	−1	+1	+1	−1	+1	−1	−1	197	11.7	107
15	−1	+1	+1	+1	+1	−1	−1	−1	220	16.0	120
16	+1	+1	+1	+1	+1	+1	+1	+1	241	6.8	83

To get some idea as to which factors might be important for increasing flight times, the experimenters, after careful discussion, decided to first use a 2_{IV}^{8-4} fractional factorial arrangement to test eight chosen factors at levels listed in Table 12.1. It will be remembered (see Chapter 6) that this experimental arrangement provides a [16,8,3] screen. Specifications for the 16 constructed helicopters are shown in Table12.2 together with their average flight times $\overline{y}$, estimated standard deviations s, and values of 100 $\log_{10}(s)$. In a flight test the wings were held against the ceiling and the helicopter dropped. The time taken to reach the floor was recorded in units of one-hundredth of a second. The helicopter designs were dropped in random order four times giving the average flight times

$\overline{y}$ and the standard deviations s shown in the table. It is well known (Bartlett and Kendall, 1946) that there is considerable advantage in analyzing the logarithm of the standard deviation rather than s itself. To avoid decimals, 100 log(s) was used here. (See also the discussion of transformations in Chapter 7.) The effects calculated from the average flight time $\overline{y}$ are called *location effects* and those calculated using 100 log(s) called *dispersion effects*. The effects shown in Table 12.3 are regression coefficients (one-half the usual effect $\overline{y}_+ - \overline{y}_-$) with the constant the overall average.

The normal plot for the location coefficients shown in Figure 12.2*a* suggested that the three tested dimensions of the helicopter—wing length l, body length L, and body width W—or equivalently the factors of the 2^{8-4} arrangement x_2, x_3, and x_4—had substantial effects on average flight time. Of the remaining five "qualitative" factors only C (paper clip, x_7) produced an effect distinguishable from noise, but since adding the paper clip reduced flight time, all subsequent tests were run without it. The normal plot for the dispersion coefficients in Figure 12.2*b* suggested that the main effects of factors L, W, and C and a single aliased string of two-factor interactions might be distinguishable from noise. No attempt was made at this stage however, to unravel the dispersion interaction string. The signs of the location effects were such that the changes in the helicopter dimensional factors l, L, and W, which gave increases in flight time, were also associated with reductions in dispersion. It was decided therefore to explore further the possibility of increasing the flight time by changing the dimensional factors l, L, and W along the path of steepest ascent for location.

Table 12.3. Design 1: Estimates (Coefficients) for the 2^{8-4}_{IV} Design

Coefficients	Location	Dispersion
Constant	222.8	80.9
$x_1 = P$	5.9	5.0
$x_2 = l$	**27.8**	−6.0
$x_3 = L$	**−13.3**	**11.6**
$x_4 = W$	**−8.3**	**13.5**
$x_5 = F$	3.8	6.9
$x_6 = T$	1.4	−1.4
$x_7 = C$	**−10.9**	**−11.6**
$x_8 = M$	−3.9	7.5
$Pl + LC + WM + FT$	5.9	1.1
$PL + lC + WT + FM$	0.1	**−15.0**
$PW + lM + LT + FC$	5.1	2.6
$PF + lT + LM + WC$	6.9	−6.8
$PT + lF + LW + CM$	5.3	−2.3
$PC + lL + WF + TM$	−3.3	1.0
$PM + lW + LF + TC$	−4.3	−4.1

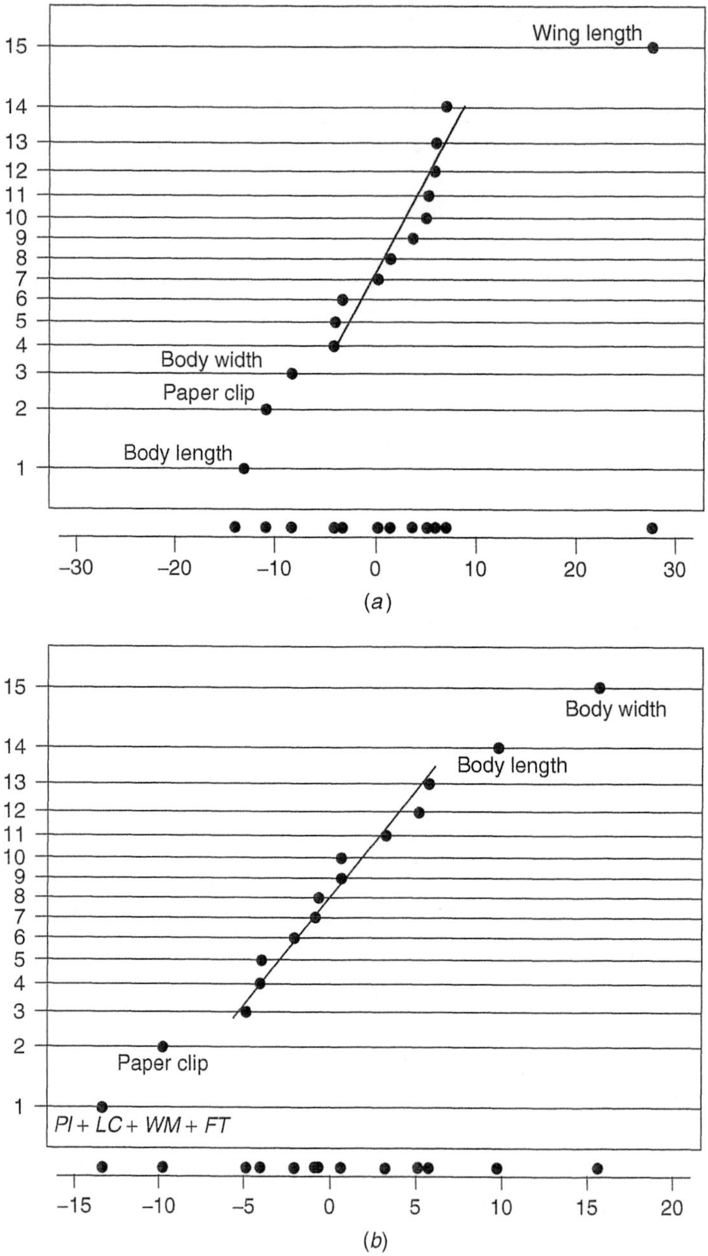

Figure 12.2. Normal plots of coefficients: (*a*) location effects for $\overline{y}$; (*b*) dispersion effects for 100 log *s*.

The linear model for estimating the mean flight times was then

$$\hat{y} = 223 + 28x_2 - 13x_3 - 8x_4 \tag{12.1}$$

where the coefficients are rounded values from Table 12.3 and x_2, x_3, and x_4 are the factors in the 2^{8-4} factorial for helicopter dimensions l, L, and W. The contour diagram in Figure 12.3 is a convenient way of conveying visually what is implied by Equation 12.1. The path of steepest ascent is shown at right angles to the contour planes.

Steepest Ascent

A series of helicopters were now constructed along the steepest ascent path. To calculate points on the path, the factors were changed simultaneously in proportion to the coefficients of the fitted equation. Thus, the *relative* changes in x_2, x_3, and x_4 were such that for every increase in 28 units in x_2, x_3 was reduced by 13 units and x_4 by 8 units. The units were the scale factors $l_{\text{scale}} = 0.875$, $L_{\text{scale}} = 0.875$, and $W_{\text{scale}} = 0.375$ (see Table 12.1), which were the changes in l, L, and W corresponding to a change of one unit in x_2, x_3, and x_4, respectively. A helicopter with a 4-inch wing length l was first tested on the steepest ascent path and then additional helicopters were built along this path with wing length l increased by 3/4-inch increments and the other dimensions adjusted accordingly. The resulting specifications are shown in Figure 12.4 listed as points 2, 3, 4, and 5. Ten repeat flights were made for each of the five new helicopter designs with design 3 giving the longest average flight time of 347 centiseconds—an impressive improvement with flight times increased by more than 50%. (In the practical development of a manufactured product, if a new product design had been discovered that was this much ahead of current competition, a management

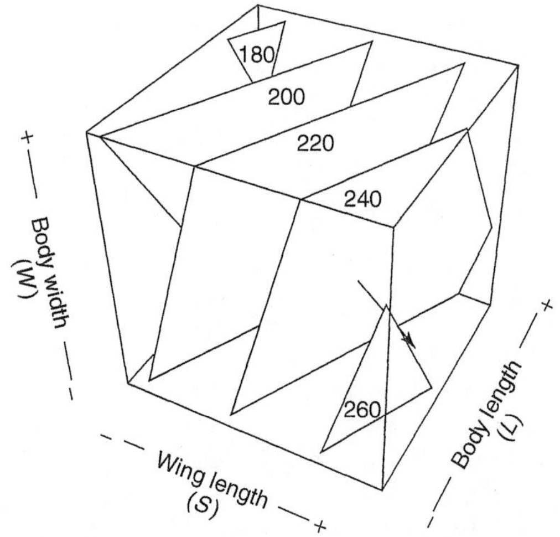

Figure 12.3. Contours of average fight times.

Helicopter	1	2	3	4	5
Wing length, l	4.00	4.75	5.50	6.25	7.00
Body length, L	3.82	3.46	3.10	2.75	2.39
Body width, W	1.61	1.52	1.42	1.33	1.24

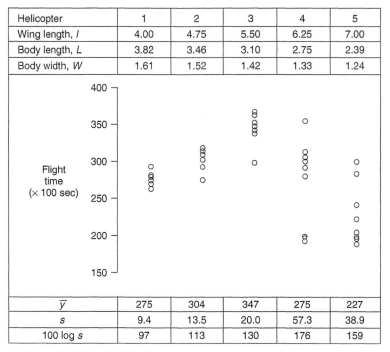

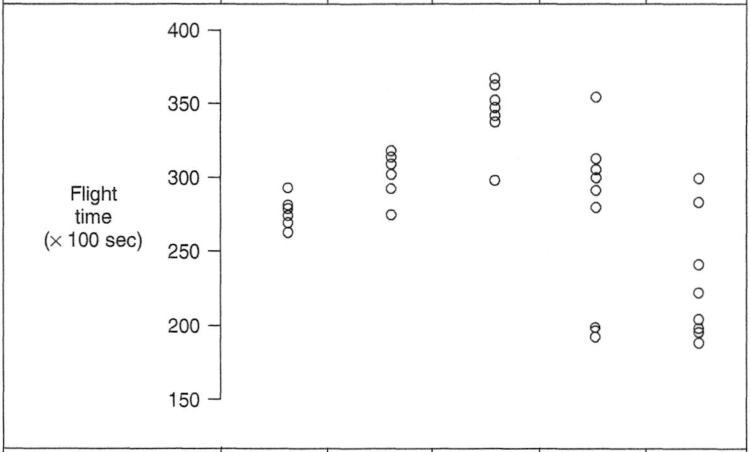

	1	2	3	4	5
$\overline{y}$	275	304	347	275	227
s	9.4	13.5	20.0	57.3	38.9
100 log s	97	113	130	176	159

Figure 12.4. Data for fight helicopters built along the path of steepest ascent.

decision might be made to "cash in" on what had already been discovered and the investigation temporarily suspended. However, as soon as the competition began to catch up again, further investigation would be needed.)

In the case of this experiment the investigation was continued in an effort to discover an even better design. Since none of the qualitative factors so far tried seemed to produce any positive effects, it was decided for the present to fix these features and explore four helicopter dimensions—wing length l, wing width w, body length L, and body width W. Also about this time a discussion with an engineer led to the suggestion that a better characterization of the wing dimensions might be in terms of *wing area* $A = lw$ and the length-to-width ratio $R = l/w$ of the wing. This reparameterization was adopted in subsequent experiments.

A Sequentially Assembled Composite Design

It was uncertain whether further advance might be possible using first-order steepest ascent, and if not, then a full second-order approximating equation might be needed. Subsequently, therefore, a 2^4 factorial in A, R, W and L was run with two added center points with the expectation that, if necessary, additional runs could be added to the design to allow the fitting of a second-order model.

Table 12.4 lists the levels of the four factors A, R, W, and L used in a 2^4 factorial experiment with two added center points. The averages $\overline{y}$ and the

Table 12.4. Helicopter Data for a 2^4 Factorial Design with Two Center Points

Factor	Symbol	−1	0	1
Wing area (lw)	A	11.80	12.40	13.00
Wing length ratio (l/w)	R	2.25	2.52	2.78
Body width (in.)	W	1.00	1.25	1.50
Body length (in.)	L	1.50	2.00	2.50

Run	$2^4 + 2$ Center Points						Estimated Coefficients		
Number	A	R	W	L	$\bar{y}$	100 log(s)		$\bar{y}$	100 log(s)
1	−1	−1	−1	−1	367	72	Constant	367.1	79.6
2	+1	−1	−1	−1	369	72	A	−0.5	1.9
3	−1	+1	−1	−1	374	74	R	**5.4**	4.5
4	+1	+1	−1	−1	370	79	W	0.5	−2.9
5	−1	−1	+1	−1	372	72	L	**−6.6**	4.1
6	+1	−1	+1	−1	355	81			
7	−1	+1	+1	−1	397	72	AR	**−2.9**	2.1
8	+1	+1	+1	−1	377	99	AW	**−3.7**	−0.7
9	−1	−1	−1	+1	350	90	AL	**4.4**	−3.2
10	+1	−1	−1	+1	373	86	RW	**4.6**	−0.1
11	−1	+1	−1	+1	358	92	RL	−1.5	1.1
12	+1	+1	−1	+1	363	112	WL	−2.1	−6.2
13	−1	−1	+1	+1	344	76			
14	+1	−1	+1	+1	355	69	ARW	0.1	−1.5
15	−1	+1	+1	+1	370	91	ARL	−1.8	−0.7
16	+1	+1	+1	+1	362	71	AWL	0.6	−4.6
17	0	0	0	0	377	51	RWL	0.2	0.2
18	0	0	0	0	375	74	$ARWL$	−0.2	−3.1

dispersions 100 log(s) obtained from 10 replicated flights are shown in Table 12.4 together with the estimated coefficients. Nothing of special interest was revealed by the dispersion analysis so further analysis was conducted only for location effects.

Exercise 12.1. Calculate the linearity check for the data in Table 12.4.

A normal plot of the estimated coefficients given in Table 12.4 are shown in Figure 12.5, from which it is evident that some two-factor interactions now approach the size of the main effects.

Additional Runs to Form a Central Composite Arrangement

To allow for the fitting of a second-order model, 12 additional runs were added, as shown in Table 12.5. These runs consisted of points placed at ±2 units along each of the four axes plus four additional center points. The estimated

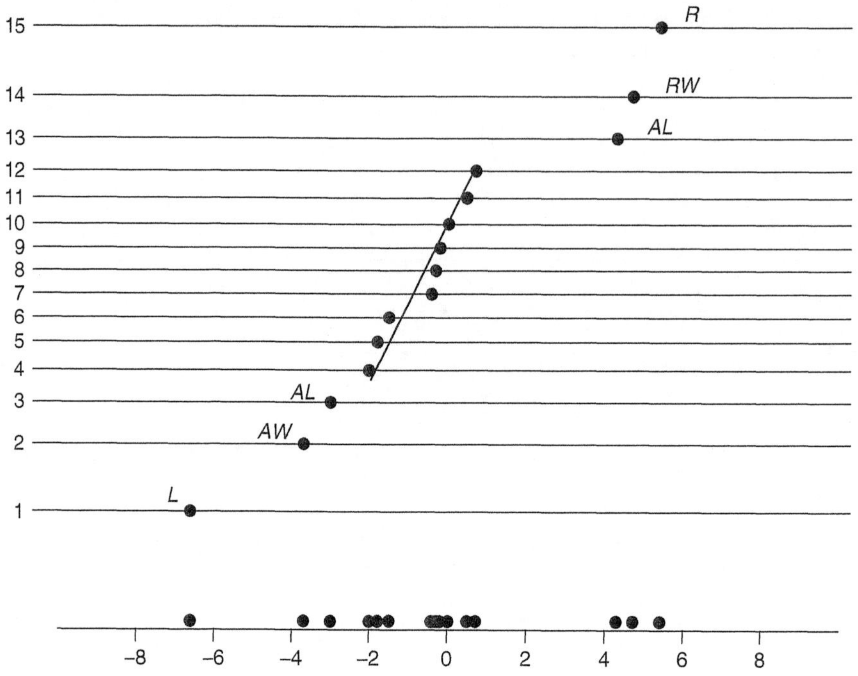

Figure 12.5. Normal plot of coefficients.

second-order model, allowing for possible mean differences between blocks, is given in Equation 12.2. The four linear coefficients are shown in the second line, the four quadratic coefficients on the third line, and the six two-factor interactions on the final lines. To the right are given the estimated standard errors of the coefficients for that line.

$$\begin{aligned}\hat{y} = {} & 372.06 \\ & - 0.08x_1 + 5.08x_2 + 0.25x_3 - 6.08x_4 \quad (0.64) \\ & - 2.04x_1^2 - 1.66x_2^2 - 2.54x_3^2 - 0.16x_4^2 \quad (0.61) \\ & - 2.88x_1x_2 - 3.75x_1x_3 + 4.38x_1x_4 \quad (0.78) \\ & + 4.63x_2x_3 - 1.50x_2x_4 - 2.13x_3x_4 \quad (0.78)\end{aligned} \qquad (12.2)$$

The ANOVA for location (average flight times) for the completed design is given in Table 12.6. There is no strong evidence of lack of fit and the residual mean square of 9.7 was used as the error variance. The overall F ratio (207.6/9.7) for the fitted second-degree equation is then over 20, exceeding its 5% significance level $F_{0.05,14,14} = 2.48$ by a factor of 8.6 more than complying with the requirement discussed in Chapter 10 that a factor of at least 4 is needed to ensure that the fitted equation is worthy of further interpretation. A similar analysis of the log variances produced little evidence for activity in dispersion.

Table 12.5. Helicopter Data: A Sequentially Constructed Central Composite Design

Factor	Symbol	−2	−1	0	+1	+2
Wing area (lw)	$x_1 = A$	11.20	11.80	12.40	13.00	13.60
Wing length ratio (l/w)	$x_2 = R$	1.98	2.25	2.52	2.78	3.04
Body width (in.)	$x_3 = W$	0.75	1.00	1.25	1.50	1.75
Body length (in.)	$x_4 = L$	1.00	1.50	2.00	2.50	3.00

Run Number	x_1 A	x_2 R	x_3 W	x_4 L	$\bar{y}$	100 log(s)
1	−1	−1	−1	−1	367	72
2	+1	−1	−1	−1	369	72
3	−1	+1	−1	−1	374	74
4	+1	+1	−1	−1	370	79
5	−1	−1	+1	−1	372	72
6	+1	−1	+1	−1	355	81
7	−1	+1	+1	−1	397	72
8	+1	+1	+1	−1	377	99
9	−1	−1	−1	+1	350	90
10	+1	−1	−1	+1	373	86
11	−1	+1	−1	+1	358	92
12	+1	+1	−1	+1	363	112
13	−1	−1	+1	+1	344	76
14	+1	−1	+1	+1	355	69
15	−1	+1	+1	+1	370	91
16	+1	+1	+1	+1	362	71
17	0	0	0	0	377	51
18	0	0	0	0	375	74
19	−2	0	0	0	361	111
20	2	0	0	0	364	93
21	0	−2	0	0	355	100
22	0	2	0	0	373	80
23	0	0	−2	0	361	71
24	0	0	2	0	360	98
25	0	0	0	−2	380	69
26	0	0	0	2	360	74
27	0	0	0	0	370	86
28	0	0	0	0	368	74
29	0	0	0	0	369	89
30	0	0	0	0	366	76

Canonical Analysis

Following the methodology described in Section 11.5 the canonical analysis of the fitted second-order model (Equation 12.2) goes as follows:

Table 12.6. Analysis of Variance of Average Flight Times for Completed Central Composite Design

Source	SS	df	MS	F
Blocks	16.8	1	16.8	
Regression	2906.5	14	207.6	21.34
Linear terms	1510.0	4	377.5	38.82
Interactions	1114.0	6	185.67	19.09
Square terms	282.5	4	70.6	7.26
Residual error	136.1	14	9.7	
Lack of fit	125.4	10	12.5	4.67
Pure error	10.8	4	2.7	
Total	3059.5	29		

Position of center S of canonical system	$x_{1s} = 0.86,\ x_{2s} = -0.33,\ x_{3s} = -0.84,\ x_{4s} = -0.12$ at which $\hat{y}_s = 371.4$
Shift from design center O to S	$\tilde{x}_1 = x_1 - 0.86,\ \tilde{x}_2 = x_2 + 0.33,\ \tilde{x}_3 = x_3 + 0.84,\ \tilde{x}_4 = x_4 + 0.12$
Rotation of axes	$X_1 = 0.39\tilde{x}_1 - 0.45\tilde{x}_2 + 0.80\tilde{x}_3 - 0.07\tilde{x}_4$ $X_2 = -0.76\tilde{x}_1 - 0.50\tilde{x}_2 + 0.12\tilde{x}_3 + 0.39\tilde{x}_4$ $X_3 = 0.52\tilde{x}_1 - 0.45\tilde{x}_2 - 0.45\tilde{x}_3 + 0.57\tilde{x}_4$ $X_4 = -0.04\tilde{x}_1 - 0.58\tilde{x}_2 - 0.37\tilde{x}_3 - 0.72\tilde{x}_4$
Canonical form	$\hat{y} = 371.4 - 4.66X_1^2 - 3.81X_2^2 + 3.27X_3^2 - 1.20X_4^2$

It had seemed likely to the experimenters that a maximum might now occur at S, which would mean that all four squared terms in the canonical equation would have had negative coefficients. However, the coefficient (+3.27) of X_3^2 was *positive*! Since its standard error would be roughly the same as that of the quadratic coefficients (about 0.6), the response surface almost certainly had a *minimum* at S in the direction represented by X_3. If this was so, it should be possible to move from the point S *in either direction* along the X_3 axis to *increase* flight times.

In terms of the centered $\tilde{x}$'s

$$X_3 = 0.52\tilde{x}_1 - 0.45\tilde{x}_2 - 0.45\tilde{x}_3 + 0.57\tilde{x}_4$$

Thus, beginning at S, one direction of ascent along the X_3 axis would be such that for each increase in $\tilde{x}_1$ of 0.52 units $\tilde{x}_2$ would be reduced by 0.45 units, $\tilde{x}_3$ reduced by 0.45 units, and $\tilde{x}_4$ increased by 0.57 units. To follow the opposite direction of ascent, you would make precisely the opposite changes. Helicopters were now designed and constructed for 16 points along this axis. Their dimensions, average flight times, and standard deviations are shown in Table 12.7. These tests confirm that you can indeed get longer flight times by proceeding in either direction along this axis. Figure 12.6 shows graphically the averages and standard deviations of the flight times for 16 helicopter designs constructed along this axis. As predicted, flight times show a central minimum with maxima in either direction. The standard deviations s remain reasonably constant except at the extremes, where instability causes rapid increase. Thus in *either direction* helicopter designs could be obtained with average flight times of over 400 centiseconds—almost twice that of the original prototype design. The designs for the two types of helicopters giving longest flight times are quite different in appearance (see Fig. 12.6*d*); one has a longer wing span and a shorter body while the other has a shorter wing span and a longer body. As always, such a surprising result can provide important ideas for further development. Obviously, experimentation could have continued but it was terminated at this point.

Of most importance is that this example may encourage the reader to experience process improvement and discovery by employing her imagination in studies of this kind. She can test her own ideas employing different starting points, varying different factors, and so forth.

In summary, this example shows you:

1. How you can experience the catalysis of the scientific method obtained by the use of statistical methods

Table 12.7. Experimental Data Employing Canonical Factor X_3

	Coded Values				Helicopter dimensions, in.				Responses, centisec	
Factor	A	R	W	L						
Coded Factor	x_1	x_2	x_3	x_4	W	L	w	l	Average	s
Coefficient	0.52	−0.45	−0.45	0.57						
$X_3 = 5.50$	3.73	−2.92	−3.34	2.95	2.91	5.03	0.42	3.48	332	12.7
$X_3 = 4.80$	3.37	−2.60	−3.02	2.55	2.82	5.12	0.50	3.28	373	5.8
$X_3 = 4.20$	3.05	−2.32	−2.75	2.20	2.74	5.19	0.56	3.10	395	5.9
$\mathbf{X_3 = 3.30}$	**2.59**	**−1.91**	**−2.35**	**1.69**	**2.64**	**5.29**	**0.66**	**2.85**	**402**	**7.5**
$X_3 = 2.67$	2.25	−1.62	−2.06	1.33	2.57	5.35	0.74	2.67	395	6.6
$X_3 = 1.86$	1.83	−1.25	−1.70	0.87	2.49	5.43	0.83	2.44	385	9.0
$X_3 = 0.70$	1.23	−0.71	−1.18	0.21	2.38	5.53	0.96	2.11	374	10.2
$X_3 = 0.30$	1.03	−0.53	−1.00	−0.02	2.34	5.56	1.00	1.99	372	7.6
$X_3 = 0.00$	0.87	−0.39	−0.86	−0.19	2.31	5.59	1.04	1.91	370	6.9
$X_3 = -0.70$	0.51	−0.07	−0.55	−0.59	2.25	5.64	1.11	1.71	376	6.3
$X_3 = -1.05$	0.32	0.09	−0.39	−0.79	2.22	5.66	1.15	1.61	379	8.4
$X_3 = -1.82$	−0.17	0.53	0.04	−1.33	2.15	5.72	1.26	1.34	387	9.0
$X_3 = -2.51$	−0.43	0.76	0.27	−1.62	2.11	5.75	1.32	1.19	406	5.4
$\mathbf{X_3 = -3.47}$	**−0.93**	**1.21**	**0.70**	**−2.17**	**2.04**	**5.81**	**1.43**	**0.92**	**416**	**6.2**
$X_3 = -3.70$	−1.05	1.31	0.81	−2.30	2.02	5.82	1.45	0.85	399	8.8
$X_3 = -4.22$	−1.32	1.55	1.04	−2.60	1.99	5.84	1.51	0.70	350	33.2

Note: The numbers in bold are "best" helicopters.

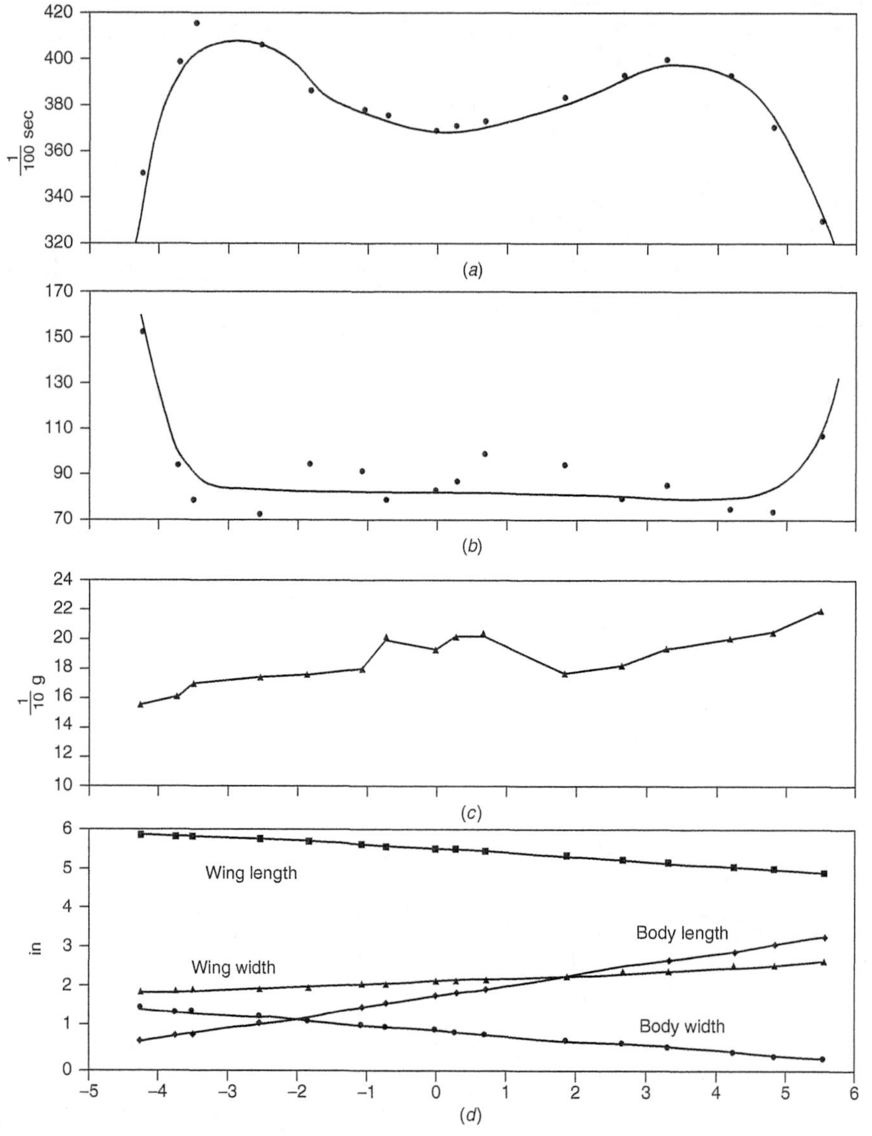

Figure 12.6. Characteristics of helicopters along axis X_3: (*a*) mean flight times; (*b*) 100 log s; (*c*) helicopter weight; (*d*) helicopter dimensions.

2. The use of fractional designs for *screening*
3. How you can follow an improvement trend with steepest ascent
4. Sequential assembly of a composite design by adding axial points and center points to a factorial

5. How canonical analysis can produce an unexpected surprise
6. How there can be more than one good answer to a problem

In this demonstration only a few of the almost limitless ideas for improving the helicopter were tested. The list of qualitative factors could be greatly extended and there are endless simple ways in which the general configuration of the design could be changed—split wings to give the helicopter four blades, other shapes of wings and body, and so on. Also, for example, it might have been better to use a PB_{12} design for the initial screening allowing more factors to be investigated, and it is easy to see that there might have been considerable room for improvement in other ways.

One thing is certain, had the experimenters been possessed of hindsight, they could have undoubtedly reached this answer more quickly.

12.2. SIMPLIFICATION OF A RESPONSE FUNCTION BY DATA TRANSFORMATION

The data shown in Table 12.8 are the "lifetimes" of test specimens of woolen thread subjected to repeated loadings under 27 different test conditions obtained from a 3^3 factorial design.* Four possible models for these data denoted by M_1, M_2, M_3, and M_4 are considered in what follows. The factor levels and coding for the 3^3 factorial are as follows:

Factors	Levels	Coding
l: length of specimen	250, 300, 350 mm	$x_1 = (l - 300)/50$
A: amplitude of loading cycle	8, 9, 10 mm	$x_2 = A - 9$
L: load	40, 45, 50 g	$x_3 = (L - 45)/5$

The Model M_1

The original authors had fitted the following second-degree equation by least squares:

$$\hat{y}_1 = 551 + 660x_1 - 536x_2 - 311x_3 + 239x_1^2 + 276x_2^2 - 48x_3^2 - 457x_1x_2 - 236x_1x_3 + 143x_2x_3$$

They found that all but one of the 10 estimated coefficients were statistically significant so that the model was thought to be adequate although difficult to interpret. This model will be called M_1. The estimated values $\hat{y}_1$ it produces for the 27 experimental conditions are shown in the sixth column of Table 12.8.

*These data are due to Barella and Sust published by Box and Cox (1964).

Table 12.8. Textile Experiment Data with Predicted Values for Four Different Models

Factor Levels				Models				
				Data	M_1	M_2	M_3	M_4
l	A	L	$G = A/l$	y	$\hat{y}_1$	$\hat{y}_2$	$\hat{y}_3$	$\hat{y}_4$
250	8	40	0.032	674	654	683	678	589
250	8	45	0.032	370	484	461	449	414
250	8	50	0.032	292	218	311	310	302
250	9	40	0.036	338	156	363	348	327
250	9	45	0.036	266	129	245	231	230
250	9	50	0.036	210	6	166	159	167
250	10	40	0.040	170	209	193	192	193
250	10	45	0.040	118	326	130	127	136
250	10	50	0.040	90	345	88	88	99
300	8	40	0.027	1414	1768	1569	1672	1467
300	8	45	0.027	1198	1362	1060	1107	1030
300	8	50	0.027	634	860	716	765	751
300	9	40	0.030	1022	813	835	859	814
300	9	45	0.030	620	551	564	569	572
300	9	50	0.030	438	192	381	393	417
300	10	40	0.033	442	410	444	474	481
300	10	45	0.033	332	290	300	313	338
300	10	50	0.033	220	74	203	217	246
350	8	40	0.023	3636	3359	3607	3586	3170
350	8	45	0.023	3184	2717	2436	2374	2226
350	8	50	0.023	2000	1980	1645	1641	1623
350	9	40	0.026	1568	1947	1919	1843	1759
350	9	45	0.026	1070	1449	1296	1220	1236
350	9	50	0.026	566	855	875	843	901
350	10	40	0.029	1140	1088	1021	1016	1039
350	10	45	0.029	884	733	690	672	730
350	10	50	0.029	360	281	466	465	532

Model M_2

These data cover a wide range, $y_{max}/y_{min} = 40.4$, and a natural question is whether by data transformation the model might be simplified and made more relevant and interpretable. Using the power transformation in the form

$$\frac{y^{\lambda} - 1}{\lambda} \qquad \text{when } \lambda \neq 0$$

which has the limit

$$\log y \qquad \text{when } \lambda = 0$$

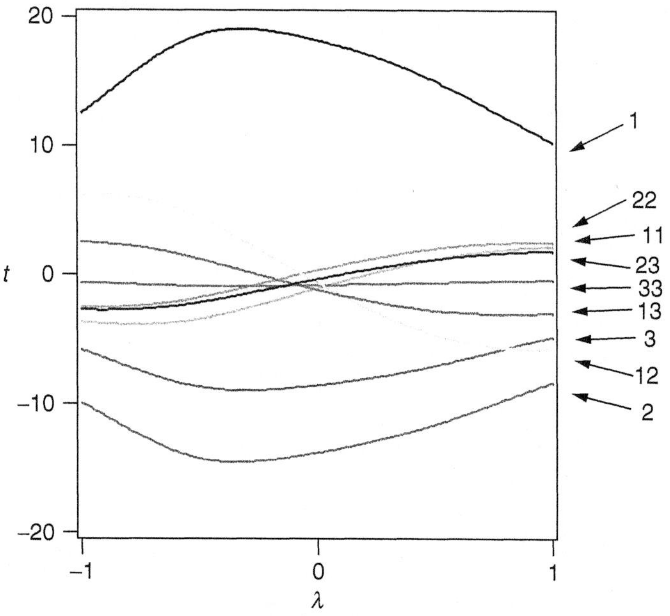

Figure 12.7. Lambda plot of t values.

a lambda plot is shown in Figure 12.7 for the t values of the coefficients in the fitted second-degree equation. Thus, for example, the line marked 2 is the plot of the t statistic for the coefficient of x_2 as λ is changed. Similarly that marked 22 is the plot for the coefficient of x_2^2. And so on. It is seen that when λ approaches zero all second-order terms essentially disappear while the absolute values of the linear terms 1, 2, and 3 become close to their maximum values. A simple first-order model in log y is thus suggested.

The value of this transformation is confirmed by the second λ plot in Figure 12.8. This shows the values of F obtained when a *first-order model* is fitted to the data where $F =$ (mean square for first order model)/(residual mean square). You will see that the F value for the logged data, when $\lambda = 0$, is about 10 times that of the untransformed data with $\lambda = 1$. Thus for this particular data, transformation is equivalent to making a 10-fold replication of the experiment! Setting $Y_2 = \log y$, the least squares fit for this model M_2 is*

$$\hat{Y}_2 = \underset{(0.02)}{2.75} + \underset{(0.02)}{0.36x_1} - \underset{(0.02)}{0.27x_2} - \underset{(0.02)}{0.17x_3}$$

where standard errors are given beneath each coefficient. For this model the predicted values $\hat{y}_2 = \text{antilog } \hat{Y}_2$ are shown in the seventh column of Table 12.8.

* logs are to base 10

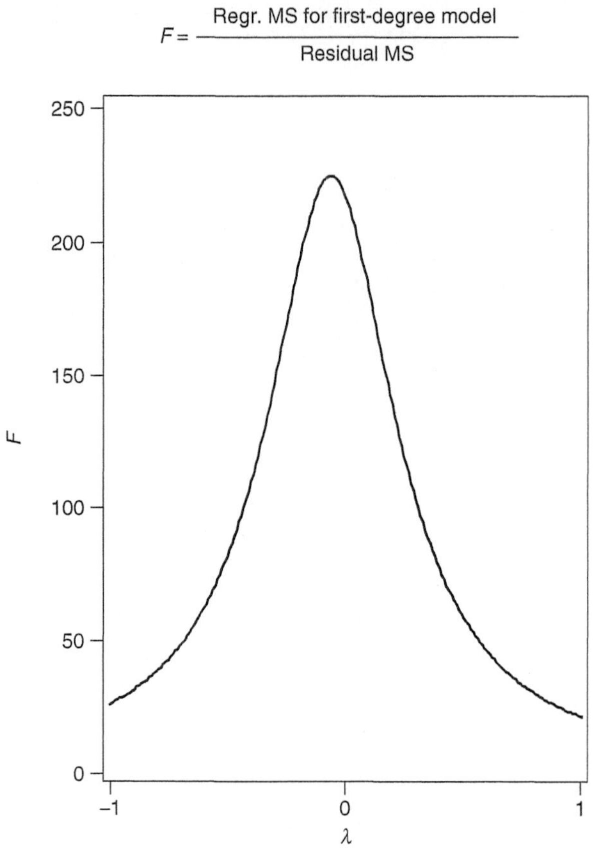

Figure 12.8. Lambda plot of F ratios.

Model M_3

Now in the sciences, and particularly in engineering, power relations are not uncommon. For this example such a model would be of the form

$$y \propto x_1^{\alpha} x_2^{\beta} x_3^{\gamma}$$

or $\log y = \text{const} + \alpha \log x_1 + \beta \log x_2 + \gamma \log x_3$. Fitting this model, M_3, to the data gives

$$\hat{Y}_3 = \text{const} + 4.95 \log l - 5.65 \log A - 3.5 \log L$$

Values of $\hat{y}_3 = \text{antilog}\ \hat{Y}_3$ are shown in the eighth column of Table 12.8.

Model M_4

The model M_3 is approximately $\hat{Y}_4 = \text{const} + 5 \log l - 5 \log A - 3 \log L$. Thus l/A is suggested as being possibly important in a redefinition of factors. In fact,

$A/l = G$, say, is the fractional amplitude of the loading cycle, whose relevance is suggested by dimensional analysis. Thus, a power function model M_4 with $y \propto G^{-5}L^{-3}$ was tried. The estimated values of $\hat{y}_4 = \text{antilog } \hat{Y}_4$ are shown in the ninth column of Table 12.8. You will see that this very simple model fits the data remarkably well.

A comparison of the models in their ability to estimate the response can be made using the fact that when n independent observations of constant variance σ^2 are fitted to a model linear in its p parameters, irrespective of the experimental design, the average variance of the n predictions $\hat{Y}$ is always $p\sigma^2/n$. The derivation is given in Appendix 12A. The mean square for each of the models, from which you can calculate the average variance of $\hat{y}$ for each model, is shown in the last column of Table 12.9. It is clear from the table that the much simpler models M_2, M_3, and M_4 provide considerably better estimates of the response than the more complicated quadratic model M_1. This is true even for model M_4, which is an imaginative attempt to postulate a more fundamental explanation of the data.

Relation between the Quadratic and Linear Models

Consider again the model M_2,

$$\hat{Y} = 2.75 + 0.36x_1 - 0.27x_2 - 0.17x_3$$

obtained by analyzing $Y = \log y$ instead of y. Figure 12.9*a* shows the appearance of the contours for this model in terms of the response $Y = \log y$. These contours are of course equally spaced planes. But the contours for the original response y shown in Figure 12.9*b* must also be a series of parallel contour planes. (Since if $\log y$ is constant on planes, then so must be y). But at first it might be thought, "How come we have planar contours when there are large quadratic and interaction effects in M_1?" However, the planes for M_1 would not be equally spaced because over this very large range $\log y$ is a highly nonlinear function of y. So there is a "canonical" variable X — a linear function of the factors x_1, x_2, x_3 — but a nonlinear function of y. It is *this* nonlinearity producing planar

Table 12.9. Comparing Models M_1, M_2, M_3, and M_4

	Residual Sum Squares $\times 10^{-3}$	Degrees of Freedom	Mean Square	p Parameters[a]	Average Variance of $\hat{y}$ = Mean Square $\times$ $(p/27)$
M_1 (quadratic, y)	1257	17	73.9	10	27.4
M_2 (linear, $\log y$)	1125	23	48.9	4	7.2
M_3 (log linear, $\log y$)	1169	23	50.8	4	7.5
M_4 (power)	1622	23	70.5	4	10.4

[a] The number of estimated parameters allocated to model M_4 is taken to be 4 because it is an approximation derived from model M_3.

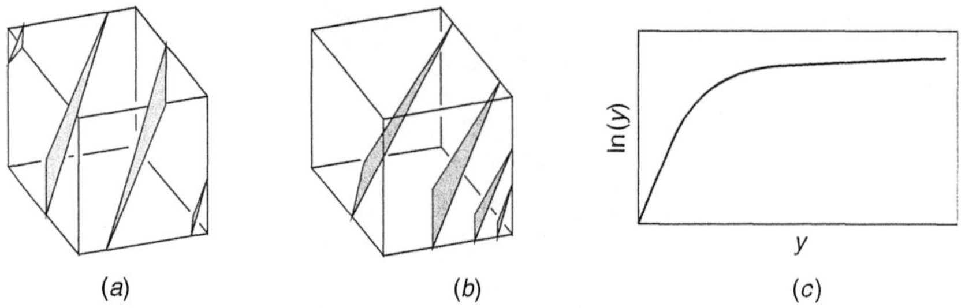

Figure 12.9. Contours for log y and y.

contours that are parallel but not equidistant that accounts for the necessity for all the quadratic and interaction terms that appear in model M_1. This example illustrates the point that one should not fall in love with a particular model.

Dropping Terms from the Model: Causation and Correlation

In the original analysis of the survival times of 27 wool test specimens given in Table 12.8 the fitted second-order model was

$$\hat{y} = 551 + 660x_1 - 536x_2 - 311x_3 + 239x_1^2 + 276x_2^2 - 48x_3^2 - 457x_1x_2 - 236x_1x_3 + 143x_2x_3$$

The standard errors of the coefficients were as follows:

$$\text{SE(linear terms)} = 64.1, \qquad \text{SE(quadratic terms)} = 111.0, \qquad \text{SE(cross-product terms)} = 78.4$$

Not all of these coefficients, were significantly different from zero at the 5% level. In similar circumstances some authors automatically omit nonsignificant terms from the model.

It does of course make sense to look for model simplification—to ask such questions as:

Is there evidence that a simpler model could be used (e.g., a linear model before or after suitable transformation)?

Is there evidence that certain factors are essentially inert (all the terms in which these factors appear are small)?

If desired, all such questions can be formally tested using the extra sum of square principle. But "cherry picking" of individual nonsignificant terms makes much less sense.

The dropping of a single term, say b_{33}, because it was "not significant" would be the same as replacing the optimal least squares estimate $b_{33} = -48$ by the estimate $b_{33} = 0$, which has nothing to recommend it, and the expression in which it appeared would not any longer be the least squares estimate of the model.

As was earlier explained, some confusion occurs because of the two uses to which "regression" equations can be put. In the present context they are being used to analyse experiments that have been designed to find out what *causes* what. A different use to which regression equations may be put is to find the subset of "x's" (from happenstance data) that *correlate* best with the response y. To do this, a number of "best subset" regression programs are available which drop or add individual terms. But a model obtained in this way makes no claim to represent causation, (although if the relationships between the variables remains the same it can be subsequently used to *forecast* y).

12.3. DETECTING AND EXPLOITING ACTIVE AND INACTIVE FACTOR SPACES FOR MULTIPLE-RESPONSE DATA

In the manufacture of a certain dyestuff three responses were of importance: strength y_1, hue y_2, and brightness y_3. Levels of these responses needed to be changed from time to time to meet the varying requirements of customers. It was important to find out how to manipulate the manufacturing process to obtain a product having the desired characteristics. Some operators believed they knew how to go about this, but their efforts frequently resulted in products very different from that desired. This was hardly surprising since they were attempting to produce desired levels for *three* different responses by adjusting the levels of *six* different process factors. These six adjustable factors were polysulfide index x_1, flux ratio x_2, moles of polysulfide x_3, reaction time x_4, amount of solvent x_5, and reaction temperature x_6. Variation in the measurements was known to be large.

The original experimenters* had heard of response surface methods and ran a one-shot experiment containing 80 runs in a complete 2^6 factorial design with 12 axial points and 4 center points. They proceeded to fit a second-order model in six factors for each of the three responses. (None of these fits were checked with the Box–Wetz criterion mentioned in Section 11.3 and their fitted surfaces were largely determined by noise). Not surprisingly they found these results impossible to interpret; the experiment was written off as an expensive failure and statistics was given, at least temporarily, a bad name. Later and more careful analysis showed that none of the second-order terms were detectably different from noise, which accounted for the strange second-order response functions that had been found. It was decided therefore to make a standard analysis of just the 64 runs that comprised the 2^6 factorial design. This data together with the estimated effects are shown in Appendix 12B. Normal plots of the 63 estimated effects for each of the three responses are shown in Figure 12.10 If you inspect

*This experiment was run by the Allied Chemicals Corporation. The portion of the data used here also appears in Box and Draper (1987).

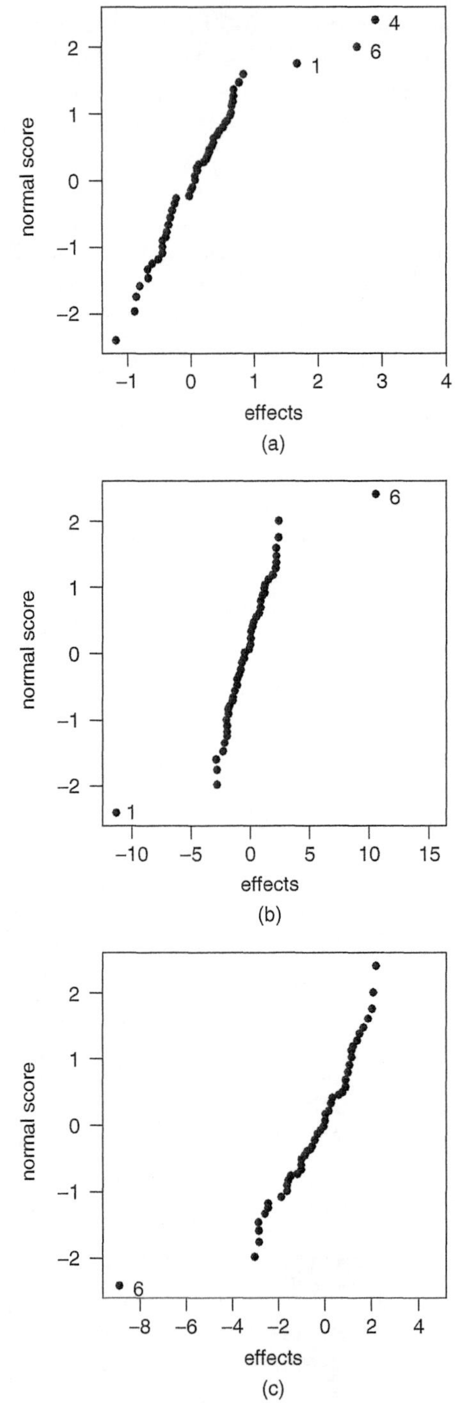

Figure 12.10. Normal plots, 3 responses, 64 estimates: (*a*) strength; (*b*) hue; (*c*) brightness.

these plots, you will see that three of the six factors, the reflux ratio x_2, moles of polysulfide x_3, and the amount of solvent x_5, showed no effects of any kind distinguishable from noise. They were thus regarded as locally inert. For the remaining three factors only main effects were detectable. Knowing the relations connecting the levels of these three active factors with the three responses would in principle allow you to calculate factor levels necessary to produce any given combination of the three responses. However, in this particular example further simplification occurred. It will be observed that the temperature x_6 affected all three responses, polysulfide index x_1 affected hue and strength, and reaction time x_4 affected only strength.

For these factors the three least squares fitted equations are as follows:

$$\text{For strength:} \quad \hat{y}_1 = 11.1 + 0.9x_1 + 1.5x_4 + 1.3x_6 \qquad (0.24) \quad (12.3a)$$

$$\text{For hue:} \quad \hat{y}_2 = 17.0 - 5.7x_1 \qquad + 5.4x_6 \qquad (0.74) \quad (12.3b)$$

$$\text{For brightness:} \quad \hat{y}_3 = 28.3 \qquad - 4.4x_6 \qquad (0.67) \quad (12.3c)$$

where the quantities shown in parentheses are the standard errors of the coefficients. Table 12.10 provides a convenient summary of the results.

Thus, provided the required factor combination could be run, a simple way for obtaining a dye of a given strength, hue, and brightness y_{10}, y_{20}, and y_{30} would be to first adjust x_6 to the value x_{60} for the required brightness y_{30}, then with $x_6 = x_{60}$ adjust x_1 to the value x_{10} to give y_{20}, and finally with $x_6 = x_{60}$ and $x_1 = x_{10}$ adjust x_4 to give y_{10}.

For illustration, suppose you wanted to find the operating conditions that gave a dye with brightness $y_{30} = 33$, hue $y_{20} = 15$, and strength y_1 as high as possible. From Equation 12.3c for $y_{30} = 33$ you need $x_{60} = -1.1$. From Equation 12.3b, given that $x_{60} = -1.1$, you need $x_{10} = -0.7$. After substitution you obtain from Equation 12.3a the equation for strength, $\hat{y}_1 = 8.9 + 1.5x_4$. Thus, to obtain the greatest strength, the reaction time x_4 must be increased. For example, if you set $x_4 = 2.0$, you should obtain a strength of about 12. However, since this is beyond the range of the experimental levels actually tested, you would need to check this by experiment.

Table 12.10. Summary of Results: Three Responses (Product Characteristics)

	Coefficients for Active Adjustable Factors			
Responses	Reaction Time, x_4	Polysulfide Index, x_1	Temperature, x_6 (°C)	Equation
y_1: Strength	1.5	0.9	1.3	12.3a
y_2: Hue		−5.7	5.4	12.3b
y_3: Brightness			−4.4	12.3c

Obviously, such calculations would be subject to considerable estimation error* and in any case apply only over the tested ranges of the factors. However, they provided valuable guidance. In particular, they showed the process operators the order in which the factors might be adjusted: temperature x_6 to attain desired brightness, polysulfide index to attain desired hue, and reaction time x_4 to attain desired strength.

Another very important finding from this experiment was that the remaining three process factors x_2, x_3, and x_5 (reflux rate, moles of polysulfide, and amount of solvent), were essentially inert over the tested ranges of the other factors. It was possible therefore to manipulate these to change other responses (e.g., cost, consistency, miscibility) without materially affecting the hue, brightness, and strength of the dye. You cannot, of course, expect that all problems will simplify like this one. However, it is always worth looking for simplification. (You can sometimes get lucky.)

Helping the Subject Matter Specialist

In many examples more than one response must be considered, and you would like to find conditions (combinations of factor levels) that, so far as possible, provide desirable values for all the responses. For example, suppose you had three responses y_1, y_2, and y_3 and four factors x_1, x_2, x_3, and x_4 and ideally you would like a product for which

(a) the level of y_1 was as high as possible;
(b) the level of y_2 was between two specific values, say y_{2L} and y_{2H}; and
(c) the level of y_3 was as low as possible.

In attempting to achieve, or approximate, these requirements, you should try to discover for each response which factors are important (active) in driving each response and which factors are unimportant (inert). With the help of graphics and suitable computer calculations. Information about the overlapping and nonoverlapping of the active and inactive factor spaces for the separate responses can often greatly simplify multiresponse problems, as is exemplified in the paint trial example in Chapter 6. In particular:

1. It may make it possible to meet all the desired requirements.
2. If not, the active and inactive factor information can show what might be available options for achieving some kind of acceptable compromise.
3. It may become clear that by using the present set of adjustable factors no useful compromise is possible. But the dimensional information can greatly help the subject matter specialist to conceive *new factor(s)* that might be introduced to resolve the difficulty.

* A confidence region for the solution of a set of linear equations with coefficients subject to error is given in Box and Hunter (1954).

12.4. EXPLORING CANONICAL FACTOR SPACES

So far active and inactive (inert) subspaces have been discussed whose the dimensions are those of the individual factors themselves. Thus in the dyestuff example, of the six factors studied three were inert—locally had no detectable effects on the outcome—and were thus available to bring additional responses to satisfactory levels. For some problems, however, although there are no inert subspaces in the original factors $x_1, x_2 \ldots$ themselves, analysis may reveal inert subspaces in the *canonical* factor $X_1, X_2 \ldots$. It may then be possible to use such inert *canonical* spaces to provide solutions for multiresponse problems.

A Two-Dimensional Subspace in Five Factors

The following example was from work originally carried out at the Dyestuffs Division of Imperial Chemical Industries. Five factors were studied and it was hoped that conditions could be found that produced satisfactory yields with *maximum throughput*. Preliminary application of the steepest ascent procedure had (as it turned out) brought the experimenters close to a nonstationary region. This also produced an estimated standard deviation of 1.2%.

The process occurred in two stages and the factors considered were the temperatures and times of reaction at the two stages and the concentration of one of the reactants at the first stage. Scientific considerations suggested that the times of reaction were best varied on a logarithmic scale, and it was known that the second reaction time needed to be varied over a wide range. Table 12.11 shows the levels used in the experiments to be described.

Coded factors expressed in terms of the natural units were

$$x_1 = \frac{T_1 - 112.5}{7.5}, \qquad x_2 = \frac{2(\log t_1 - \log 5)}{\log 2} + 1, \qquad x_3 = \frac{c - 70}{10}$$

$$x_4 = \frac{T_2 - 32.5}{7.5}, \qquad x_5 = \frac{2(\log t_2 - \log 1.25)}{\log 5} + 1$$

Table 12.11. Factor Levels: $2_{\rm V}^{5-1}$ Design

Reactor Stage	Factors	Factor Levels −1	+1
1	T_1: temperature (°C)	105	120
	t_1: time (hr)	2.5	5
	c: concentration	60	80
2	T_2: temperature (°C)	25	40
	t_2: time (hr)	0.25	1.25

A 2_{V}^{5-1} fractional factorial experiment provided the first 16 runs listed in Table 12.12. The estimated values for the 5 first-order coefficients $b_1, b_2, \ldots, b_5$ and the 10 two-factor interactions coefficients $b_{12}, b_{13}, \ldots, b_{45}$ were

$$\begin{array}{lll} b_1 = 3.194, & b_{12} = -1.869, & b_{24} = -0.169 \\ b_2 = 1.456, & b_{13} = 2.081, & b_{25} = -0.769 \\ b_3 = 0.981, & b_{14} = -0.506, & b_{34} = -4.019 \\ b_4 = 3.419, & b_{15} = -0.431, & b_{35} = -0.744 \\ b_5 = 1.419, & b_{23} = 0.743, & b_{45} = 0.144 \end{array}$$

and using a previous estimate of the standard deviation the estimated standard error for the coefficients was about 0.6.

It will be seen that, compared with some of the two-factor interactions, the first-order coefficients were not so large as to suggest that the present experimental

Table 12.12. Level of Experimental Factors and Observations Obtained

Trial Number	x_1	x_2	x_3	x_4	x_5	Yield
		2_{V}^{5-1} *Fractional Factorial*				
1	−1	−1	−1	−1	+1	49.8
2	+1	−1	−1	−1	−1	51.2
3	−1	+1	−1	−1	−1	50.4
4	+1	+1	−1	−1	+1	52.4
5	−1	−1	+1	−1	−1	49.2
6	+1	−1	+1	−1	+1	67.1
7	−1	+1	+1	−1	+1	59.6
8	+1	+1	+1	−1	−1	67.9
9	−1	−1	−1	+1	−1	59.3
10	+1	−1	−1	+1	+1	70.4
11	−1	+1	−1	+1	+1	69.6
12	+1	+1	−1	+1	−1	64.0
13	−1	−1	+1	+1	+1	53.1
14	+1	−1	+1	+1	−1	63.2
15	−1	+1	+1	+1	−1	58.4
16	+1	+1	+1	+1	+1	64.3
		Added Points to Form a Noncentral Composite Design				
17	+3	−1	−1	+1	+1	63.0
18	+1	−3	−1	+1	+1	63.8
19	+1	−1	−3	+1	+1	53.5
20	+1	−1	−1	+3	+1	66.8
21	+1	−1	−1	+1	+3	67.4

region was far removed from a near-stationary region or so small as to suggest that the region covered by the design was already centered there. In these circumstance it was appropriate to add further points forming a *noncentral* composite design of the type illustrated earlier that would allow the estimation of all coefficients up to second order. Points were therefore added to the fractional factorial design in the direction of increasing yield. It will be remembered that this form of augmentation has the effect of shifting the center of gravity of the design closer to the center of the near-stationary region of interest and thus reducing errors due to "lack of fit" of the postulated form of the equation.

In deciding where best to add the extra points it was noted that the highest yield (70.4%) was recorded at the point [+1, −1, −1, +1, +1] and this seemed a good point at which to augment the design. Five extra points having coordinates [+3, −1, −1, +1, −1], [+1, −3, −1, +1, +1], [+1, −1, −3, +1, +1], [+1, −1, −1, +3, +1], and [+1, −1, −1, +1, +3] were therefore added. These runs are numbered 17 to 21 in Table 12.12.

At this point in the investigation 21 coefficients were being estimated with only 21 experiments. This would not normally be regarded as good practice since no comparisons remain for tests of lack of fit. (However, as you will see, 11 confirmatory points were later added, making it possible to check the adequacy of the fit.) For this type of arrangement the least squares estimates of the first-order and two-factor interaction coefficients are unaffected by the new points, and estimates of b_0, the predicted yield at the new center of the design, and the quadratic effects b_{11}, b_{22}, b_{33}, b_{44}, and b_{55} were now calculated as follows:

$$b_0 = 68.173, \qquad b_{11} = -1.436, \qquad b_{22} = -1.348$$

$$b_{33} = -2.723, \qquad b_{44} = -2.261, \qquad b_{55} = -1.036$$

With estimates of all 21 coefficients in the five-factor second-order model we can carry out the canonical analysis as before. The coordinates of the center of the canonical system and the predicted yield are as follows:

$$x_{1S} = 1.2263, \qquad x_{2S} = -0.5611, \qquad x_{3S} = -0.0334$$

$$x_{4S} = 0.6916, \qquad x_{5S} = 0.6977, \qquad Y_S = 71.38$$

The canonical form of the model is

$$Y - 71.38 = -4.70X_1^2 - 2.58X_2^2 - 1.25X_3^2 - 0.48X_4^2 + 0.20X_5^2 \qquad (12.4)$$

The transformation whose rows on the latent vectors are given in Table 12.13. This gives the new X coordinates in terms of the old coordinates (or by transposition the old coordinates in terms of the new).

From Equation 12.4 it appears that two of the canonical factors, X_4 and X_5, play a minor role and that most of the activity could be ascribed to canonical factors X_1, X_2, and X_3. To determine whether these general findings could be

Table 12.13. Transformations Between Old x and New X Coordinates

$\tilde{x}$	X_1	X_2	X_3	X_4	X_5
$x_1 - 1.226$	−0.230	0.652	−0.225	0.426	−0.539
$x_2 + 0.561$	−0.129	0.599	0.249	−0.745	0.088
$x_3 + 0.033$	0.758	−0.034	0.252	−0.144	−0.583
$x_4 - 0.692$	0.595	0.412	−0.406	0.156	0.536
$x_5 - 0.698$	0.038	0.213	0.812	0.469	0.273
eigenvalue	−4.699	−2.580	−1.245	−0.483	+0.203

confirmed, 11 further experiments were carried out, one at S, the predicted center of the system, and the others in pairs along the axes X_1, X_2, X_3, X_4, and X_5. Each of these points was at a distance two units from the center. Thus, in terms of the *new* factors X_1, X_2, X_3, X_4, and X_5 the coordinates of the points were $[0, 0, 0, 0, 0]$, $[\pm 2, 0, 0, 0, 0]$, $[0, \pm 2, 0, 0, 0], \ldots, [0, 0, 0, 0, \pm 2]$.

The levels of the original factors corresponding to these points were calculated from the equations expressing the x in terms of the X's (Table 12.13) and are the coordinates of the experimental points 22 to 32 in Table 12.14. This table also shows the yield $\hat{y}$ predicted from the fitted second degree equation (in parentheses) and the values y actually found. These yields are in reasonably close agreement. Thus the additional experiments confirmed the overall features of the surface in a rather striking manner. In particular, it will be seen that the experimental points on either side of S along the X_1 axis are associated with the largest reductions in yield, approximately confirming the large coefficient of X_1^2 in the canonical equation. Somewhat smaller changes are associated with movement away from S along the X_2 and X_3 axes and these again are of the order

Table 12.14. Eleven Confirmatory Points Determined by Canonical Factors

Trial	X_1	X_2	X_3	X_4	X_5	x_1	x_2	x_3	x_4	x_5	Actual Yield, y	Predicted yield $\hat{y}$
22	0	0	0	0	0	1.23	−0.56	−0.03	0.69	0.70	72.3	(71.4)
23	2	0	0	0	0	0.77	−0.82	1.48	1.88	0.77	57.1	(52.6)
24	−2	0	0	0	0	1.69	−0.30	−1.55	−0.50	0.62	53.4	(52.6)
25	0	2	0	0	0	2.53	0.64	−0.10	1.51	1.12	62.3	(61.1)
26	0	−2	0	0	0	−0.08	−1.75	0.04	−0.13	0.27	61.3	(61.1)
27	0	0	2	0	0	0.78	−0.06	0.47	−0.12	2.32	64.8	(66.3)
28	0	0	−2	0	0	1.68	−1.06	−0.54	1.50	−0.93	63.4	(66.3)
29	0	0	0	2	0	2.08	−2.05	−0.32	1.00	1.63	72.5	(69.5)
39	0	0	0	−2	0	0.38	0.93	0.25	0.38	−0.24	72.0	(69.5)
31	0	0	0	0	2	0.15	−0.38	−1.20	1.76	1.24	70.4	(72.2)
32	0	0	0	0	−2	2.30	−0.74	1.13	−0.38	0.15	71.8	(72.2)

expected. Finally, movement along the X_4 and X_5 axes is associated with only minor changes in yield, thus confirming that there is an approximately stationary plane and we have a wide local choice of almost equally satisfactory conditions from which to choose. After including the information from these 11 extra points the canonical form of the second-degree equation became

$$Y - 73.38 = -4.46X_1^2 - 2.64X_2^2 - 1.80X_3^2 - 0.39X_4^2 - 0.004X_5^2$$

agreeing closely with what was found before. The transforming equations in Table 12.13 are affected very little and confirm that locally there is not much activity in the canonical subspace defined by X_4 and X_5. The estimated experimental error variance based on the residual sum of squares having 11 degrees of freedom was 1.51, and using this value, it appeared that the standard error

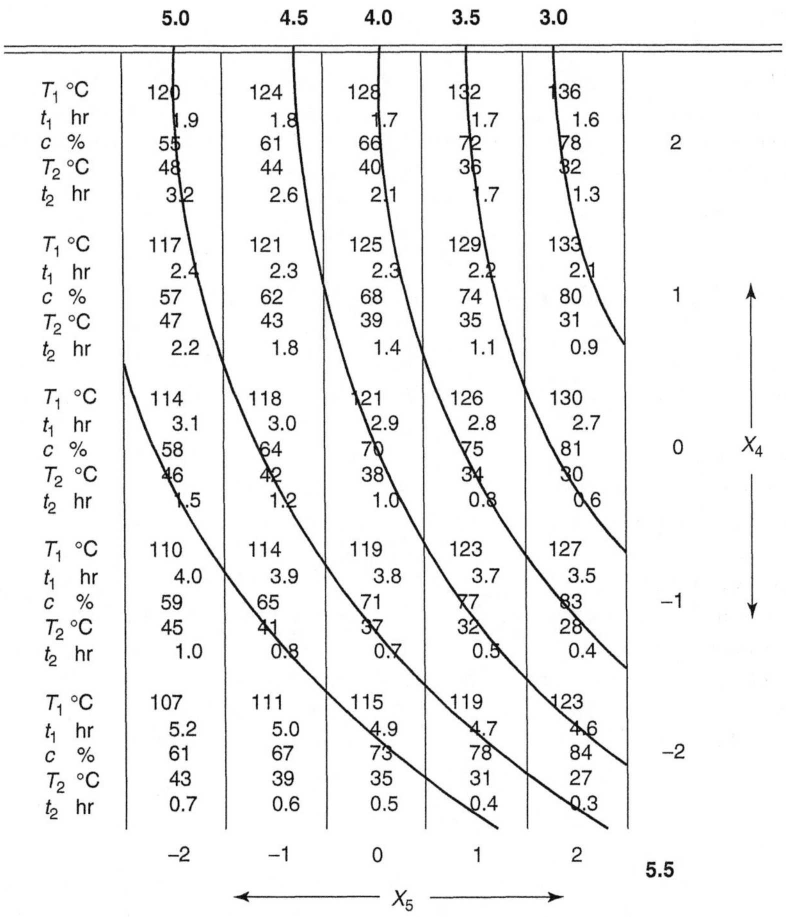

Figure 12.11. Sets of nearly alternative conditions on the plane of X_1 and X_2.

of the estimated coefficients was about 0.4. The sets of conditions which give approximately the same optimal yield are in the plane defined by X_4 and X_5 with $X_1 = 0$, $X_2 = 0$, $X_3 = 0$. These equations are readily transformed to equations in x_1, x_2, x_3, x_4, and x_5 and, into equations in the five natural factors.

We thus have two degrees of freedom in the choice of conditions in the inactive canonical subspace of two dimensions in the sense that, within the region, if we choose values for two of the factors we obtain three equations in the three unknowns, which can be solved to give appropriate levels for the remaining factors. To demonstrate the practical value of these findings, a plane of approximate alternative processes giving close to maximum yield in the range of interest is shown in Figure 12.11. The objective of increasing throughput was equivalent to minimizing the total reaction time (the sum of the reaction times for stage 1 and that of stage 2). Contours of the total reaction time are superimposed on the table of alternative processes in Figure 12.11. Over the range of the factors studied it will be seen that the conditions $T_1 = 136°\text{C}$, $t_1 = 1.6$ hours, $c = 78\%$, $T_2 = 32°\text{C}$, and $t_2 = 1.3$ hours give the smallest total reaction time of about 2.9 hours, potentially allowing a much greater throughout. What then was exploited here was a two-dimensional inert subspaces in the *canonical* variables. Also the figure suggested that, assuming the conditions checked out, the experimenters should explore even more extreme conditions implied by the trend of the contours—to apply steepest ascent to the canonical variables.

12.5. FROM EMPIRICISM TO MECHANISM

All models are approximations, and this applies to mechanistic models as well as to empirical models. However, when available, a mechanistic model often has advantages because it may provide a *physical understanding* of the system and greatly accelerate the process of problem solving and discovery. Also, the mechanistic model frequently requires fewer parameters and thus provides estimates of the fitted response with proportionately smaller variance.

Sometimes an empirical model can *suggest* a mechanism. This was the case for an example introduced in Table 11.1 that is now briefly reviewed. The objective was to find conditions of the temperature (T), the concentration of one of the reactants (c), and the time of reaction (t) that maximized an intermediate product y_2. The factor T was measured in degrees Celsius, factor c as a percent concentration, and factor t as the number of hours at temperature. The factors were coded as follows:

$$x_1 = \frac{T - 167}{5}, \qquad x_2 = \frac{c - 27.5}{2.5}, \qquad x_3 = \frac{t - 6.5}{1.5}$$

The first 15 runs consisted of a central composite design followed later by 4 verifying runs numbered 16, 17, 18, 19. The design and data, discussed earlier in Chapter 11 and listed in Table 11.1, are shown again in Table 12.15.

Table 12.15. (a) Central Composite Design plus Four Additional Exploratory Points, (b) Fitted Second-Order Model, (c) ANOVA, (d) Canonical Form

(a)

Run Number	x_1	x_2	x_3	y
1	−1	−1	−1	45.9
2	+1	−1	−1	60.6
3	−1	+1	−1	57.5
4	+1	+1	−1	58.6
5	−1	−1	+1	53.3
6	+1	−1	+1	58.0
7	−1	+1	+1	58.8
8	+1	+1	+1	52.4
9	−2	0	0	46.9
10	+2	0	0	55.4
11	0	−2	0	55.0
12	0	+2	0	57.5
13	0	0	−2	56.3
14	0	0	+2	58.9
15	0	0	0	56.9
16	2	−3	0	61.1
17	2	−3	0	62.9
18	−1.4	2.6	0.7	60.0
19	−1.4	2.6	0.7	60.6

(b)

$$\hat{y} = 58.78 + 1.90x_1 + 0.97x_2 + 1.06x_3 - 1.88x_1^2 - 0.69x_2^2 - 0.95x_3^2 - 2.71x_1x_2 - 2.17x_1x_3 - 1.24x_2x_3$$

(c)

Source	SS	df	MS	
2nd order model	371.4	9	41.3	$F = 12.5$
Residual	29.5	9	3.3	
Total	400.9	18		

(d)

$$\hat{y} - 59.1 = -3.4X_1^2 - 0.3X_2^2 + 0.2X_3^2$$

The ANOVA table produced a value for the "multiplied F" criterion of about 4, indicating that the model was worthy of further analysis. Using the residual mean square as an estimate of the error variance, the standard error of the linear and quadratic coefficients was about 0.4 and of the cross-product coefficient about 0.5. Analysis of the fitted equation showed the center of the system S to be close to that of the design.

In the canonical equation only the coefficient of X_1^2 was distinguishable from noise and the response was refitted* to give the model

$$\hat{y}_2 = 59.9 - 3.8X_1^2$$

where

$$X_1 = 0.75\tilde{x}_1 + 0.48\tilde{x}_2 + 0.46\tilde{x}_3$$

* An adequate approximation may be obtained by fitting the model $\hat{y}_2 = B_0 + BX_1^2$ by ordinary least squares. More exactly you can use nonlinear least squares, see Bates and Watts (1988).

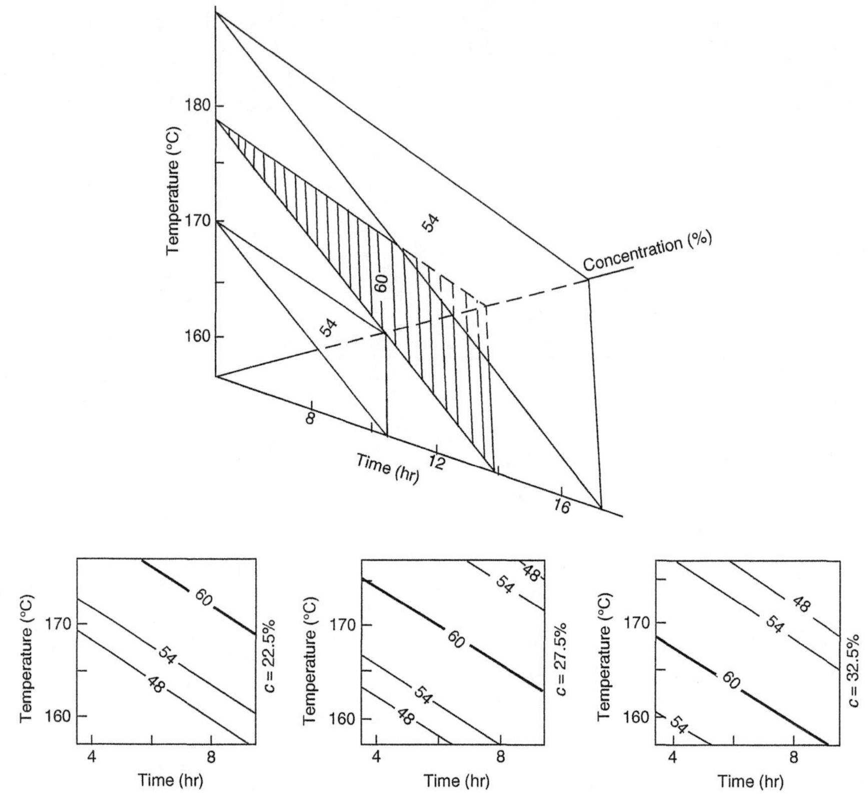

Figure 12.12. Contours of empirical yield surface with sections at three levels of concentration.

The contours of this empirical yield surface are shown in Figure 12.12*b* with sections at three different values of concentration in Figure 12.12*c*.

An approximate ANOVA is now as follows:*

Source	df	SS		df	SS	MS		
Full second-degree equation	9	371.4	Canonical form	4	357.4	89.35		F = 28.8
			Discrepancy	5	14.0	2.8	3.1	
Residual	9	29.5		9	29.5	3.3		
Total SS	18	400.9						

The mean square of 2.8 associated with the discrepancy between the canonical form and the full second-order model is slightly smaller than the residual mean

*The analysis assumes a model that is linear in its parameters. The full second-degree equation is linear in its parameters but the fitted canonical form is not. However, over the ranges of the factors explored here, the canonical model is approximately linear.

square of 3.3. Combining these mean squares produces an estimate of error of 3.1 and a value of $F = 28.8$ which is 8.7 times the 5% significance value. There is therefore no reason to believe that the directly fitted canonical form is an inadequate representation of the fitted response surface model.

The canonical model offered a wide choice of operating conditions that could produce approximately the same maximum yield. By choosing appropriate factor combinations that lie in the maximal plane, other responses such as cost and amount of impurity might be brought to more desirable levels. In particular, contours for other responses could be drawn on the stationary plane and the best compromise chosen.

A Mechanistic Model

When first obtained, this model was considered as somewhat of a curiosity. But a physical chemist, Philip Youle,* after cogitating over this sandwich-shaped contour system suggested a mechanism that might explain it. He postulated a model in which there were two consecutive reactions. Two chemical entities P and X, say, reacted together to form the *desired product* Q, which then reacted with more X to form R, an undesired product. Thus

$$\mathrm{P} + \mathrm{X} \overset{k_{1T}}{\rightarrow} \mathrm{Q} + \mathrm{X} \overset{k_{2T}}{\rightarrow} \mathrm{R}$$

where k_{1T} and k_{2T} are the rate constants for the two consecutive reactions and are both dependent on the temperature T. Now suppose c is the concentration of P. This is varied from one run to another and we will assume that X is always in sufficient excess that its concentration can be considered fixed in any given run. The law of mass action says that approximately the rate of reaction will be proportional to the molecular concentrations of the substances reacting. So if y_1, y_2, and y_3 are the concentrations of P, Q, and R, the proportions of the products will be determined by the three differential equations

$$\frac{dy_1}{dt} = k_{1T}cy_{1t} \qquad \frac{dy_2}{dt} = -k_{1T}cy_{1t} + k_{2T}cy_{2t} \qquad \frac{dy_3}{dt} = k_{2T}cy_{2t}$$

Now as the temperature increases the reaction rates increase.† Use was now made of the Arrhenius relationship between a rate constant k and the absolute temperature T, $k = k_0 e^{-\beta/T}$. For the two reaction rates then, you can write

$$k_{1T} = \rho\alpha e^{-\beta_1/T} \qquad \text{and} \qquad k_{2T} = \alpha e^{-\beta_2/T}$$

where the constants β_1 and β_2 are called the activation energies for the reactions and α and ρ are fixed constants. Youle postulated that a reason for the planar stationary ridge was that the constants β_1 and β_2 which determined the sensitivity of the reactions to temperature were approximately *equal.* With this

* See Box and Youle (1955)

† For instance a reaction rate might double for every 10°C increase in temperature.

assumption, $\beta_1 = \beta_2 = \beta$, say, the differential equations can be integrated to yield the approximate theoretical model of the system as*

$$
\begin{aligned}
-y_{1t} &= y_{10}\exp[-\rho\alpha \cdot f(T,c,t)] \\
y_{2t} &= y_{20}\alpha\left(\frac{\rho}{1-\rho}\right)\{\exp[-\rho\alpha \cdot f(T,c,t)] - \exp[-\alpha \cdot f(T,c,t)]\} \qquad (12.5) \\
y_{3t} &= y_{10} - y_{1t} - y_{2t}
\end{aligned}
$$

where $f(T, c, t) = \exp(-\beta/T + \ln c + \ln t)$.

For specific values of temperature T and concentration c a time plot for the yield y_2 is shown in Figure 12.13*a*. To determine the maximum yield of the desired product Q, reaction time t must be chosen so that y_2 is at its maximum.

Now since α, ρ, and β are fixed constants, y_1, y_2, and y_3 depend on the reaction conditions only via the single quantity $-\beta/T + \ln c + \ln t$. Any value of T, c, and t giving the same value to this quantity will produce identical values for y_{1t}, y_{2t}, and y_{3t}. Thus the formation of the desired intermediate product y_2 will always be represented by a graph like that of Figure 12.13*a* but the speed at which this occurs will be different for different choices of T and c. That is, the time scale will be different. The constants were estimated as†

$$\ln\alpha = 16.9 \pm 3.6, \qquad \beta = 10{,}091 \pm 1595 \qquad \text{with } \rho \approx 3.4$$

The contour plot for this mechanistic model, displayed in Figure 12.13*b*, closely resembles the empirical contour plot shown earlier.

An Analogy

Some understanding of the approximate mechanistic model can be gained from the following analogy. Imagine a billiard table on which a number of black and white balls are in continuous motion. Suppose that there is an excess of black balls and that whenever a white ball collides with a black ball the white ball disappears and a blue ball results and when in turn the blue ball collides with a black ball the blue ball disappears and a red ball results. In this analogy black and white balls correspond to the molecules of the starting materials P and X, blue balls to molecules of the desired product Q, and red balls to the unwanted product R. Starting off with a given number of black and white balls, we can see that as soon as the system is set into motion blue balls will begin to appear, but these in turn will give rise to an increasing number of red balls. Consequently,

*More generally, such equations will usually need to be integrated numerically; however, this is a simple task for the computer. Further, it is known that models of this kind, while providing reasonable approximations to the mechanism, oversimplify the detailed behavior of the system.

†It may be shown that the maximum value of y_2 is given by $\rho^{-1/(\rho-1)}$. Experimentally y_2 (max) was found to be about 0.6 (60%) whence $\rho \approx 3.4$.

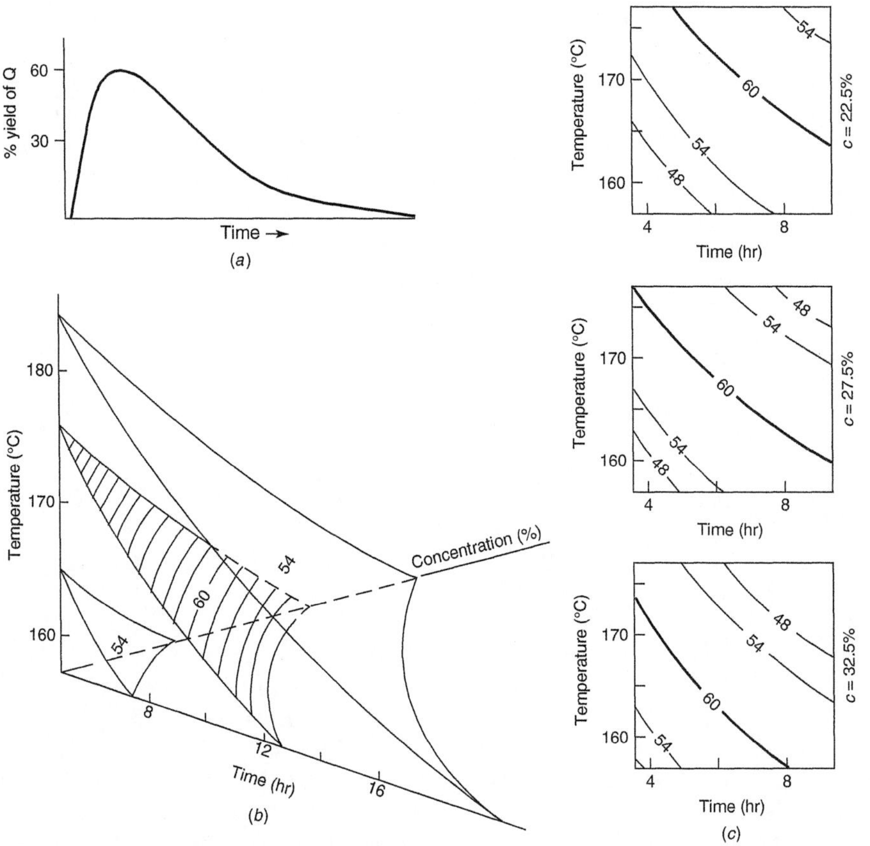

Figure 12.13. Graph and contours derived from a mechanistic model: (*a*) yield as a function of time with temperature and concentration fixed; (*b*) yield contours as a function of time, temperature, and concentration; (*c*) yield contours of three fixed levels of concentration.

in the course of time the number of blue balls representing the desired product will increase to a maximum and then fall off until, finally, only red balls and the excess of black balls remain.

Clearly, the proportions of the various sorts of balls on the table at a given instant will depend on the following:

the time that has elapsed since the start,
the relative speed with which the balls move, and
the relative number of black and white balls when the reaction is started.

These correspond to the experimental factors time of reaction t, temperature of reaction T, and concentration c.

Now we can increase the rate at which these events occur by increasing the probability of a collision occurring in any fixed time interval. One way to do this is to increase the initial proportion of white balls (like increasing c). But this will increase the probability of each kind of collision by the same factor. Changing temperature corresponds to changing the speed at which the balls move, and because of equal sensitivity to temperature change, the speed of both transitions will also be increased or decreased by the same factor. Thus, changing the value of T or c can only result in the speeding up or slowing down of an identical sequence of events. Thus what happens is represented by Figure 12.13*a* with only the time scale changed. In particular, the same maximum proportion of blue balls is produced.

A Mechanistic Canonical Factor

The mechanistic model highlights the important part played by the function

$$\frac{-\beta}{T} + \ln c + \ln t$$

This can be thought of as the "mechanistic canonical variable." If a reciprocal scale is used for absolute temperature and logarithmic scales for the time t and concentration c, the response is determined by a linear combination of these factors. Thus the ridge system appears as planes in the space of these factors. In Figure 12.14*a* the yield at a fixed concentration is shown with time now plotted on a logarithmic scale and absolute temperature on a reciprocal scale.* Thus to a very close approximation appropriate metrics would be T^{-1}, $\ln t$, and $\ln c$, where T^{-1} is the temperature in degrees Celsius. In particular, a closer fit might be expected from a second-degree empirical equation fitted to T^{-1}, $\ln t$, and $\ln c$. When this calculation is carried out, the residual mean square is further reduced from 3.3 to 2.3. More generally, empirical models employing log scales for such factors as time and overall concentration often produce closer fits.

So What?

The empirical model (Fig. 12.12) showed the existence of the stationary plane ridge. It also showed approximately which combinations of reaction time, temperature, and concentration would produce the same maximum yield. This allowed a choice to be made of the most convenient and least expensive manufacturing conditions. The fitted approximate mechanistic model (Fig. 12.13) showed essentially the same thing, so what was gained by study of this more theoretical model?

At this stage of the investigation the problem was that, no matter what combination of reaction conditions was chosen, on the maximal planar ridge the yield was only about 60%. Varying the temperature and concentration factors allowed you only to increase or decrease the speed at which you arrived at that maximum. What the mechanistic model did that the empirical model could not do was to

* Since the range of absolute temperatures involved is small compared to the average absolute temperature, $1/T$ is essentially a linear function of temperature.

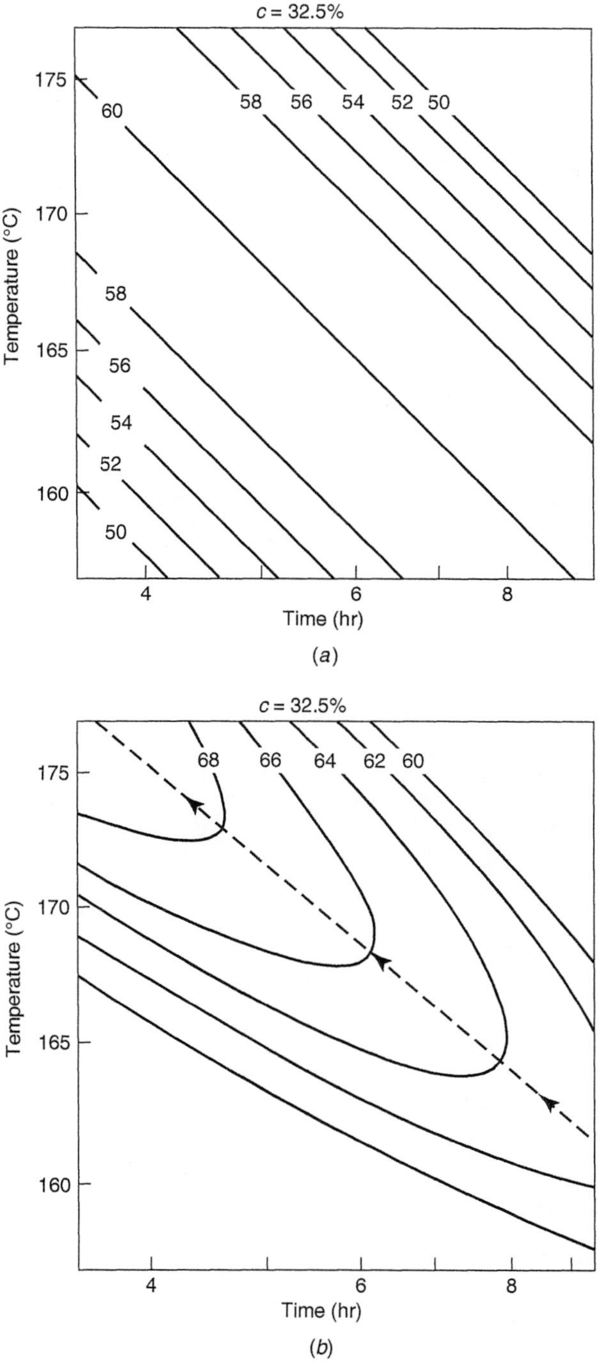

Figure 12.14. Contours of theoretical yield surfaces with time plotted on a log scale and temperature on a reciprocal scale.

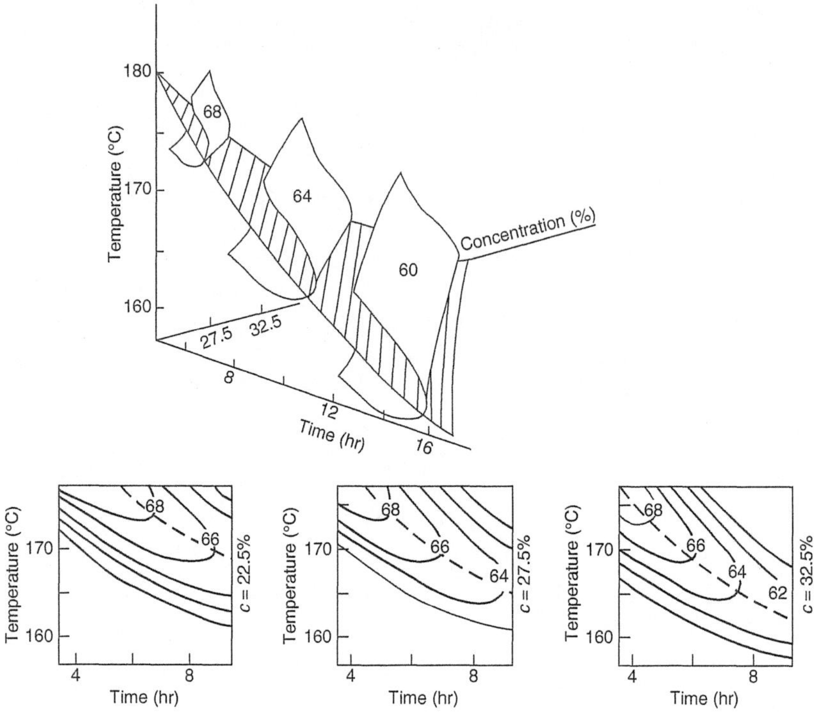

Figure 12.15. Contours of a yield surface for consecutive reactions with unequal sensitivities to temperature.

suggest *why* this was and what might be done about it. To increase the yield above 60%, some way was needed of increasing the speed of the first reaction relative to that of the second. It turned out this could be done by employing a different catalyst. A rising ridge was thus produced leading to yields above 60%. Mechanistic yield contours for an alternative catalyst are displayed in Figure 12.15 along with sections made at different concentrations. See also the effect of transformation of T and t in Figure 12.4*b*.

Frequently a mechanistic model (even an approximate one in the investigator's head) is a great help in deciding what to do next.

12.6. USES OF RSM

Response surface methods have been used in such widely different activities as storing bovine spermatozoa, bleaching cotton, removing radionuclides, growing lettuce, drying pulps, baking pie crusts, compounding truck tires, etching microchips, and the design of antenna arrays—the applications are almost endless.

APPENDIX 12A. AVERAGE VARIANCE OF $\hat{y}$

The average variance of the estimated responses at the Design Points is always $p\sigma^2/n$ whatever the experimental design.

One measure of the "goodness" of a design might be the average value of the variances of the estimated responses. What arrangement of experimental points will minimize this? Suppose that you set out any random scatter of n experimental points and used least squares to fit the data to any model linear in *p parameters*, then on IID assumptions the variance of the average of the fitted values $\bar{\hat{y}}$ would always be $p\sigma^2/n$ no matter where you put the points.

Examples of models linear in the parameters to which this theorem would apply are as follows:

$$
\begin{aligned}
&M_1 \quad y = \beta_0 + \beta_1 x_1 + \beta_2 x_2 + e \\
&M_2 \quad y = \beta_0 + \beta_1 x_1 + \beta_2 x_2 + \beta_{11} x_1{}^2 + \beta_{22} x_2{}^2 + \beta_{12} x_1 x_2 + e \\
&M_3 \quad y = \beta_0 + \beta_1 \log x_1 + \beta_2 \exp(-x_2) + \beta_3 \left(\sin \frac{x_3}{x_4}\right) + e
\end{aligned}
$$

If the values of the x's are assumed known, any such model containing p parameters may be written in the linear form

$$M\colon \quad y = \beta_0 + \beta_1 X_1 + \beta_2 X_2 + \cdots + \beta_p X_p + e$$

where X_1, X_2, ..., X_p are functions of the known x's. Thus, for the models above, for model M_1, $p = 3$; for M_2, $p = 6$; and for M_3, $p = 4$.

Derivation

The n relations associated with a particular model may be written in matrix form as

$$\mathbf{y} = \mathbf{X}\boldsymbol{\beta} + \mathbf{e}$$

where $\mathbf{y}$ is an $n \times 1$ vector of observations, $\mathbf{X}$ the $n \times p$ matrix of derivatives $\partial y/\partial \beta$, $\boldsymbol{\beta}$ a $p \times 1$ vector of unknown coefficients, and $\mathbf{e}$ an $n \times 1$ vector of errors assumed to be IID.

Under this assumption the y's have constant variance σ^2 and are independent so that $E(\mathbf{y} - \boldsymbol{\eta})(\mathbf{y} - \boldsymbol{\eta})^{\mathrm{T}} = \mathbf{I}_n\sigma^2$, where $\mathbf{I}_n$ is an $n \times n$ identity matrix. The least squares estimates of the parameters are then $\hat{\boldsymbol{\beta}} = (\mathbf{X}^{\mathrm{T}}\mathbf{X})^{-1}\mathbf{X}^{\mathrm{T}}\mathbf{y}$. Now $(\hat{\mathbf{y}} - \boldsymbol{\eta}) = \mathbf{X}(\hat{\boldsymbol{\beta}} - \boldsymbol{\beta}) = \mathbf{X}(\mathbf{X}^{\mathrm{T}}\mathbf{X})^{-1}\mathbf{X}^{\mathrm{T}}\mathbf{e}$. Thus the $n \times n$ covariance matrix for the $\hat{y}$'s is $V(\hat{\mathbf{y}}) = \mathbf{X}(\mathbf{X}^{\mathrm{T}}\mathbf{X})^{-1}\mathbf{X}^{\mathrm{T}}\mathbf{X}(\mathbf{X}^{\mathrm{T}}\mathbf{X})^{-1}\mathbf{X}^{\mathrm{T}}\sigma^2 = \mathbf{X}(\mathbf{X}^{\mathrm{T}}\mathbf{X})^{-1}\mathbf{X}^{\mathrm{T}}\sigma^2$. Now the average variance of the $\hat{y}$'s is tr $V(\hat{\mathbf{y}})/n$, the trace of $V(\hat{\mathbf{y}})$ divided by n. But for compatible matrices tr $\mathbf{AB}$ = tr $\mathbf{BA}$. If then we write $\underset{\sim}{\mathbf{A}} = \mathbf{X}(\mathbf{X}^{\mathrm{T}}\mathbf{X})^{-1}$, we find that tr $V(\hat{\mathbf{y}}) = \mathbf{X}^{\mathrm{T}}\mathbf{X}(\mathbf{X}^{\mathrm{T}}\mathbf{X})^{-1}\sigma^2 = \text{tr } \mathbf{I}_p\sigma^2$. But since $\mathbf{I}_p$ is a $p \times p$ identity matrix tr $\mathbf{I}_p = p$ and thus tr $V(\hat{\mathbf{y}}) = p\sigma^2$, whence the variance of the average $\bar{\hat{y}}$ is $V(\bar{\hat{y}}) = \dfrac{p\sigma^2}{n}$.

The remarkable thing is that this result applies to *any* linear model and any design haphazard or otherwise. So if this were the only criterion that had to be considered in designing an experiment, you might simply choose your experimental points to cover what you believed to be the region of interest. More specifically, you might chose them to obtain a satisfactory spread of information using the design information function.

APPENDIX 12 B.

Data and Estimated Effects* for Multiresponse Experiment in Section 12.3

x_1 Polysulfide index
x_2 Flux ratio
x_3 Moles of polysulfide
x_4 Reaction time
x_5 Solvent volume
x_6 Reaction temperature

y_1 Dye strength
y_2 Hue
y_3 Brightness

DATA									EFFECTS			
x_1	x_2	x_3	x_4	x_5	x_6	y_1	y_2	y_3		y_1	y_2	y_3
−1	−1	−1	−1	−1	−1	3.4	15	36	Av	11.12	16.95	28.28
1	−1	−1	−1	−1	−1	9.7	5	35	1	**1.75**	**−11.28**	0.38
−1	1	−1	−1	−1	−1	7.4	23	37	2	0.70	−2.72	1.50
1	1	−1	−1	−1	−1	10.6	8	34	12	0.89	0.72	0.94
−1	−1	1	−1	−1	−1	6.5	20	30	3	0.10	−1.91	−0.94
1	−1	1	−1	−1	−1	7.9	9	32	13	0.14	2.66	0.75
−1	1	1	−1	−1	−1	10.3	13	28	23	0.60	−1.91	0.25
1	1	1	−1	−1	−1	9.5	5	38	123	−0.65	−1.22	0.06
−1	−1	−1	1	−1	−1	14.3	23	40	4	**2.98**	−0.34	−3.00
1	−1	−1	1	−1	−1	10.5	1	32	14	0.15	−1.78	1.06
−1	1	−1	1	−1	−1	7.8	11	32	24	−0.86	0.16	0.81
1	1	−1	1	−1	−1	17.2	5	28	124	0.82	−1.41	0.00
−1	−1	1	1	−1	−1	9.4	15	34	34	0.40	−0.78	0.38
1	−1	1	1	−1	−1	12.1	8	26	134	0.40	0.03	−1.19
−1	1	1	1	−1	−1	9.5	15	30	234	−0.40	2.09	−2.56
1	1	1	1	−1	−1	15.8	1	28	1234	−0.22	0.28	−1.50
−1	−1	−1	−1	1	−1	8.3	22	40	5	−0.42	1.34	−1.44
1	−1	−1	−1	1	−1	8.0	8	30	15	−0.42	−0.97	2.13
−1	1	−1	−1	1	−1	7.9	16	35	25	−0.33	−1.91	1.25
1	1	−1	−1	1	−1	10.7	7	35	125	−0.27	0.16	−1.56
−1	−1	1	−1	1	−1	7.2	25	32	35	0.08	0.78	2.44
1	−1	1	−1	1	−1	7.2	5	35	135	0.07	−2.78	1.63

* The regression coefficients in Table 12.10 are one half of these values.

DATA									EFFECTS			
x_1	x_2	x_3	x_4	x_5	x_6	y_1	y_2	y_3		y_1	y_2	y_3
−1	1	1	−1	1	−1	7.9	17	36	235	0.12	0.16	−0.88
1	1	1	−1	1	−1	10.2	8	32	1235	0.65	2.47	−1.56
−1	−1	−1	1	1	−1	10.3	10	20	45	−0.32	1.09	1.13
1	−1	−1	1	1	−1	9.9	3	35	145	−0.33	0.03	1.19
−1	1	−1	1	1	−1	7.4	22	35	245	−0.33	−1.03	0.44
1	1	−1	1	1	−1	10.5	6	28	1245	−1.16	−1.97	−2.38
−1	−1	1	1	1	−1	9.6	24	27	345	0.72	1.16	0.38
1	−1	1	1	1	−1	15.1	4	36	1345	0.34	0.34	0.06
−1	1	1	1	1	−1	8.7	10	36	2345	−0.24	−1.59	−0.81
1	1	1	1	1	−1	12.1	5	35	12345	0.70	2.22	−0.50
−1	−1	−1	−1	−1	1	12.6	32	32	6	**2.69**	**10.84**	**−8.88**
1	−1	−1	−1	−1	1	10.5	10	34	16	−0.82	0.78	0.94
−1	1	−1	−1	−1	1	11.3	28	30	26	−0.18	−1.16	1.06
1	1	−1	−1	−1	1	10.6	18	24	126	−0.25	−1.09	1.75
−1	−1	1	−1	−1	1	8.1	22	30	36	−0.22	−1.84	0.13
1	−1	1	−1	−1	1	12.5	31	20	136	0.11	2.34	−0.94
−1	1	1	−1	−1	1	11.1	17	32	236	0.35	1.03	−0.69
1	1	1	−1	−1	1	12.9	16	25	1236	0.29	−2.16	−0.38
−1	−1	−1	1	−1	1	14.6	38	20	46	0.02	2.34	−0.31
1	−1	−1	1	−1	1	12.7	12	20	146	−0.56	−1.72	1.25
−1	1	−1	1	−1	1	10.8	34	22	246	0.30	0.22	1.00
1	1	−1	1	−1	1	17.1	19	35	1246	−0.32	−1.47	1.94
−1	−1	1	1	−1	1	13.6	12	26	346	0.17	−0.97	−0.94
1	−1	1	1	−1	1	14.6	14	15	1346	−0.77	−0.28	0.25
−1	1	1	1	−1	1	13.3	25	19	2346	−0.41	2.53	−2.88
1	1	1	1	−1	1	14.4	16	24	12346	0.74	1.09	−2.81
−1	−1	−1	−1	1	1	11.0	31	22	56	0.27	0.41	−1.88
1	−1	−1	−1	1	1	12.5	14	23	156	0.10	−0.53	0.94
−1	1	−1	−1	1	1	8.9	23	22	256	0.57	−2.22	−0.44
1	1	−1	−1	1	1	13.1	23	18	1256	0.02	−0.78	1.25
−1	−1	1	−1	1	1	7.6	28	20	356	−0.23	0.22	0.00
1	−1	1	−1	1	1	8.6	20	20	1356	−0.65	−1.47	2.19
−1	1	1	−1	1	1	11.8	18	20	2356	0.39	0.47	−0.06
1	1	1	−1	1	1	12.4	11	36	12356	0.50	−0.34	−0.50
−1	−1	−1	1	1	1	13.4	39	20	456	0.63	1.41	1.31
1	−1	−1	1	1	1	14.6	30	11	1456	−0.47	−0.53	−2.38
−1	1	−1	1	1	1	14.9	31	20	2456	0.03	−2.84	−1.63
1	1	−1	1	1	1	11.8	6	35	12456	0.47	−0.66	0.06
−1	−1	1	1	1	1	15.6	33	16	3456	−0.28	1.34	−0.94
1	−1	1	1	1	1	12.8	23	32	13456	0.71	0.16	−0.25
−1	1	1	1	1	1	13.5	31	20	23456	0.14	1.72	−0.38
1	1	1	1	1	1	15.8	11	20	123456	0.34	−0.84	−2.81

REFERENCES AND FURTHER READING

The technical literature concerning response surface methodology is vast. The following contains some 270 pertinent references:

Myers, R. H., Montgomery, D. C., Vining, G. G., Borror, C. M., and Kowalksi, S. M. (2004) Response surface methodology: A retrospective and literature survey, *J. Quality Technol.*, **36**, 53–78.

Textbooks devoted to the subject of response surface models, design, and analyses are:

Box, G. E. P., and Draper, N. R. (1987) *Empirical Model Building and Response Surfaces*, Wiley, New York.

Box, G. E. P., and Hunter, J. S. (1954) A confidence region for the solution for a set of simultaneous equations with an application to Experimental Design, *Biometrika*, **41** (pts. 1 and 2), 190–199.

Myers, R. H., and Montgomery, D. C. (2002) *Response Surface Methodology*, 2nd ed., Wiley, New York.

Response surface methodology was first developed and described in:

Box, G. E. P (1954) The exploration and exploitation of response surfaces: Some general considerations and examples, *Biometrics*, **10**, 16–60.

Box, G. E. P., and Coutie, G. A. (1956) Application of digital computers in the exploration of functional relationships, *Proc. Inst. Electrical Eng.*, **103**(1) 100–107.

Box, G. E. P., and Hunter, J. S. (1957) Multi-factor designs for exploring response surfaces, *A. Math. Statist.*, **28**, 195–241.

Box, G. E. P., and Wilson, K. B. (1951) On the experimental attainment of optimum conditions, *J. Roy. Statist. Soc. Ser. B*, **13**, 1.

Box, G. E. P., and Youle, P. V. (1955) The exploration and exploitation of response surfaces: An example of the link between the fitted surface and the basic mechanism of the system, *Biometrics*, **11**, 287–323.

Papers of particular interest are:

Alexsson, A.-K., and Carlson, R. (1995) Links between the theoretical rate law and the response surface. An example with a solvolysis reaction, *Acta Chem. Scand.*, **49**, 663–667.

Bartlett, M., and Kendall, D. G. (1946) The statistical analysis of variance-heterogeneity and the logarithmic transformation. *Supple. J. Roy. Stat. Soc.*, **8**, 128–138.

Barella, A., and Sust A. Unpublished report to the Technical Committee, International Wool Organization. See G. E. P. Box and D. R. Cox, (1964) An analysis of transformations, *J. Roy. Statist. Soc. Ser. B*, **26**, 211–252.

Box, G. E. P. (1982) Choice of response surface design and alphabetic optimality, *Utilitas Math.*, **21B**, 11–55.

Box, G. E. P. (1992) Sequential experimentation and sequential assembly of designs, *Quality Eng.*, **5**, 321–330.

Box, G. E. P., and Behnken, D. W. (1960) Some three level designs for the study of quantitative variables, *Technometrics*, **2**, 455–476.

Box, G. E. P., and Hunter, W. G. (1965) The experimental study of physical mechanisms, *Technometrics*, **7**, 23–42.

Box, G. E. P., and Liu, P. Y. T. (1999) Statistics as a catalyst to learning by scientific method, Part 1, An example, *J. Qual. Technol.*, **31**, 1–15.

Box, G. E. P., and Lucas, H. L. (1959) Design of experiments in non-linear situations, *Biometrika*, **46**, 77–90.

Box, G. E. P., and Wetz, J. (1973) *Criteria for Judging Adequacy of Estimation by Approximating Response Function*, Technical Report 9, Department of Statistics, University of Wisconsin, Madison.

Carlson, R. (2002) Canonical analysis of response surface models as tools for exploring organic reactions and processes, in *Proceedings of 1st International Symposiu: Optimising Organic Reaction Processes*, Oslo, Norway.

Hahn, G. J. (1984) Experimental design in a complex world, *Technometrics*, **26**, 19–31.

Hunter, W. G., and Mezaki, R. (1964) A model building technique for chemical engineering kinetics, *Am. Inst. Chem. Eng. J.*, **10**, 315–322.

Kim, K.-J., and Lin D. K. J. (2000) Dual response surface optimization: A fuzzy modeling approach, *J. Qual. Technol.*, **28**, 1–10.

Snee, R. D. (1985) Computer aided design of experiments — some practical experiences, *J. Quality*, **17**, 222–236.

QUESTION FOR CHAPTER 12

1. What is response surface methodology? Name three kinds of problems for which it can be useful.

2. Can you think of a problem, preferably in your own field, for which response surface methodology might be profitably used?

3. When the object is to find experimental conditions produces a high level of response, what techniques may be useful when the model that locally approximates the response function is a first-degree polynomial? A second-degree polynomial?

4. At what points in a response surface study could the method of least squares be appropriately used?

5. What checks might be employed to detect (a) bad values, (b) an inadequate model, and (c) the failure of assumptions?

6. When is it useful to have replicate points?

7. What is the path of steepest ascent and how is it calculated? Are there problems in which the path of steepest *descent* would be of more interest than a path of steepest *ascent*?

8. What is a composite design? How may it be built up sequentially? When would a non–central composite design be more appropriate than a central composite design?

9. Has response surface methodology been useful in your field? Has the one-factor-at-a-time approach been used? Comment.

10. What is the essential nature of canonical analysis of the fitted second-order model? Why is it helpful?

11. What checks would you employ to ensure that the fitted response equation was sufficiently well established to make further interpretation worthwhile?

12. Can you supply an example of a transition from an empirical to a theoretical model?

13. What are good clues to indicate the need for data transformation?

14. How can an inactive factor subspace prove useful?

15. What strategies can you suggest for handling multiple responses?

PROBLEMS FOR CHAPTER 12

1. This chapter discusses a number of uses for RSM: (a) to improve product design, (b) to provide experience in scientific investigation, (c) for possible simplification due to data transformation, (d) for detecting and exploiting active factor space with multiple-response problems, (e) for detecting and exploiting active canonical spaces for multiple-response problems, and (f) for movement from empirical to theoretical mechanism. Provide real or imagined examples for each kind of application. Do not use the illustrations in this chapter.

2. Criticize in detail the following statement: Response surface methodology is about fitting a second-degree equation to data using a central composite design.

3. The helicopter experiment in this chapter was intended to (a) allow all readers to gain experience in a self-directed iterative investigation, (b) demonstrate the method of steepest ascent, (c) demonstrate the use of the ANOVA to check the goodness of fit, (d) determine whether some function has been sufficiently well estimated to justify further analysis, (e) show how the path of discovery could be different for different experimenters, and (f) illustrate how there can be more than one satisfactory solution to a problem. Write a short paragraph dealing with each point.

4. If there were a number of models that might explain the data, as in the textile experiment, how might you decide empirically which model to employ? Would you also seek professional help? why?

5. The following data were obtained from a duplicated experiment where the response y was the breaking strength of a fiber and A, B, and C are the

coded levels of three ingredients in the fiber. Make a λ plot for the t values of the effects. What do you conclude?

A	−	+	−	+	−	+	−	+
B	−	−	+	+	−	−	+	+
C	−	−	−	−	+	+	+	+
	1.0	1.6	0.7	1.4	0.6	0.2	2.1	2.0
	0.8	1.6	0.6	1.5	0.8	0.5	2.2	1.8

6. The data given below were obtained when working on a lathe with a new alloy. From these data, which are in coded units, obtain the least squares estimates of β_0, β_1, and β_2 in the following model:

$$\eta = \beta_0 + \beta_1 x_1 + \beta_2 x_2$$

Criticize this analysis and carry out any further analysis that you think is appropriate.

Speed, x_1	**Feed,** x_2	**Tool Life,** y
−1	−1	1.3
0	−1	0.4
+1	−1	1.0
−1	0	1.2
0	0	1.0
+1	0	0.9
−1	+1	1.1
0	+1	1.0
+1	+1	0.8

7. The following data were collected by an experimenter using a composite design:

x_1	x_2	y
−1	−1	2
+1	−1	4
−1	+1	3
+1	+1	5
−2	0	1
+2	0	4
0	−2	1
0	+2	5
0	0	3

Assume that a second-degree equation is to be fitted to these data using the method of least squares.

(a) What is the matrix of independent variables?

(b) Set up the normal equations that would have to be solved to obtain the least squares estimates b_0, b_1, b_2, b_{12}, b_{11}, and b_{22}.

(c) Solve these normal equations.

(d) Criticize the analysis.

8. The following data were collected by a mechanical engineer studying the performance of an industrial process for the manufacture of cement blocks, where x_1 is travel, x_2 is vibration, and y is a measure of operating efficiency. All values are in coded units. High values of y are desired.

Travel, x_1	**Vibration,** x_2	**Efficiency,** y
−1	−1	74
+1	−1	74
−1	+1	72
+1	+1	73
−2	0	71
+2	0	72
0	−2	75
0	+2	71
0	0	73
0	0	75

Do the following:

(a) Fit a first-order model to these data using the method of least squares.

(b) Draw contour lines of the fitted response surface.

(c) Draw the direction of steepest ascent on this surface.

(d) Criticize your analysis.

9. Suppose that an experimenter uses the following design and wants to fit a second-degree polynomial model to the data using the method of least squares:

Run	x_1	x_2	x_3	**Run**	x_1	x_2	x_3
1	−1	−1	−1	9	−2	0	0
2	+1	−1	−1	10	+2	0	0
3	−1	+1	−1	11	0	−2	0
4	+1	+1	−1	12	0	+2	0
5	−1	−1	+1	13	0	0	−2
6	+1	−1	+1	14	0	0	+2
7	−1	+1	+1	15	0	0	0
8	+1	+1	+1	16	0	0	0

(a) Write the 16×10 matrix of independent variables $\mathbf{X}$.

(b) Is this a linear or nonlinear model?

(c) How many normal equations are there?

(d) What equation(s) would have to be solved to obtain the least squares estimate for β_{13}?

10. The object of an experiment, the results from which are set out below, was to determine the best settings for x_1 and x_2 to ensure a high yield and a low filtration time.

Trial	x_1	x_2	Yield (gs)	Field Color	Crystal Growth	Filtration Time (sec)
1	−	−	21.1	Blue	None	150
2	−	0	23.7	Blue	None	10
3	−	+	20.7	Red	None	8
4	0	−	21.1	Slightly red	None	35
5	0	0	24.1	Blue	Very slight	8
6	0	+	22.2	*Unobserved*	Slight	7
7	+	−	18.4	Slightly red	Slight	18
8	+	0	23.4	Red	Much	8
9	+	+	21.9	Very red	Much	10

Variable	Variable Level −	0	+
x_1 = condensation temperature (°C)	90	100	110
x_2 = amount of B (cm^3)	24.4	29.3	34.2

Source: W. J. Hill and W. R. Demler, *Ind. Eng. Chem.*, October 1970, 60–65.

(a) Analyze the data and draw contour diagrams for yield and filtration time.

(b) Find the settings of x_1 and x_2 that give (i) the highest predicted yield and (ii) the lowest predicted filtration time.

(c) Specify *one* set of conditions for x_1 and x_2 that will simultaneously give high yield and low filtration time. At this set of conditions what field color and how much crystal growth would you expect?

11. The following data are from a tire radial runout study. Low runout values are desirable.

	x_1	x_2	x_3	x_4	y
1	9.6	40	305	4.5	0.51
2	32.5	40	305	4.5	0.28
3	9.6	70	305	4.5	0.65
4	32.5	70	305	4.5	0.51
5	9.6	40	335	4.5	0.24
6	32.5	40	335	4.5	0.38
7	9.6	70	335	4.5	0.45
8	32.5	70	335	4.5	0.49
9	9.6	40	305	7.5	0.30
10	32.5	40	305	7.5	0.35
11	9.6	70	305	7.5	0.45
12	32.5	70	305	7.5	0.82
13	9.6	40	335	7.5	0.24
14	32.5	40	335	7.5	0.54
15	9.6	70	335	7.5	0.35
16	32.5	70	335	7.5	0.51
17	60.0	55	320	6.0	0.53
18	5.0	55	320	6.0	0.56
19	17.5	85	320	6.0	0.67
20	17.5	25	320	6.0	0.45
21	17.5	25	350	6.0	0.41
22	17.5	25	290	6.0	0.23
23	17.5	25	320	9.0	0.41
24	17.5	25	320	3.0	0.47
25	17.5	25	320	6.0	0.32

Here x_1 = postinflation time (minutes), x_2 = postinflation pressure (psi), x_3 = cure temperature (°F), x_4 = rim size (inches), and y = radial runout (mils).

(a) Make what you think is an appropriate analysis of the data.

(b) Make three separate two-dimensional contour plots of $\hat{y}$ versus x_1 and x_3 over the ranges $5 \leq x_1 \leq 60$ and $290 \leq x_3 \leq 350$, one for each of the following sets of conditions:
(i) $x_2 = 40$, $x_4 = 3$; (ii) $x_2 = 55$, $x_4 = 9$; (iii) $x_2 = 55$, $x_4 = 4.5$
Conditions (iii) correspond to the conditions used in the plant.

(c) Comment on this conclusion: It was surprising to observe that either very wide (9-inch) or very narrow (3-inch) rims could be used to reach low radial runout levels.

(d) Comment on any aspect of the design, the data, or the analysis which merits attention.
(*Source*: K. R. Williams, *Rubber Age*, August 1968, 65–71.)

12. The following data were collected by a chemist:

Trial	Variable: Temperature (°C)	Variable: Concentration (%)	Response: Yield (%)	Order in Which Experiments Were Performed
1	70	21	78.0	8
2	110	21	79.7	3
3	70	25	79.3	11
4	110	25	79.1	4
5	60	23	78.5	6
6	120	23	80.2	10
7	90	20	78.2	2
8	90	26	80.2	7
9	90	23	80.0	1
10	90	23	80.3	5
11	90	23	80.2	9

Using these data, do the following:

(a) Fit the model $\eta = \beta_0 + \beta_1 x_1 + \beta_2 x_2 + \beta_{11} x_1^2 + \beta_{22} x_2^2 + \beta_{12} x_1 x_2$ using the method of least squares (first code the data so that the levels of the variables are -1.5, -1.0, 0, $+1.0$, $+1.5$).

(b) Check the adequacy of fit of this model.

(c) Map the surface in the vicinity of the experiments.

(d) Suppose that the last three data values listed above (for trials 9, 10, and 11, which were the 1st, 5th, and 9th experiments performed) had been different, as follows:

Trial	Temperature (°C)	Concentration (%)	Yield (%)	Order in Which Experiments Were Performed
9	90	23	79.5	1
10	90	23	80.8	5
11	90	23	80.2	9

Would your answers for (*a*), (*b*), and (*c*) be different with these new data?

CHAPTER 13

Designing Robust Products and Processes: An Introduction

A product, process, or system that is *robust* is one that is designed to be insensitive to the varying conditions of its use in the real world. Thus, while it may be that the design is not "optimal" for any specific set of circumstances, it will give *good* performance over the range of conditions in which it is likely to be used.* Different problems in robust design are discussed:

1. ensuring insensitivity of performance to likely changes in environmental conditions and
2. ensuring insensitivity of performance to likely variations in the characteristics of components.

In this chapter, if there is a possibility of confusion between the *design* of a robust system and *designing* an experiment, an experimental layout such as a fractional factorial or response surface design, will be called an experimental *arrangement.*

13.1. ENVIRONMENTAL ROBUSTNESS

Robust Process Conditions for the Manufacture of a Medical Package†

The DRG Medical Packaging Company was anxious to design medical packages that would produce sterile seals over the wide range of process conditions employed by its customers. A cooperative study involving both the manufacturer and some volunteer customers was arranged as follows. Five possibly important

* The concept is closely related to that of sensitivity analysis.
† This study was conducted by engineers John Schuerman and Todd Algrim (1992).

product *design factors* (factors under the control of the manufacturer) were the paper supplier (PS), the solids content (SC), the cylinder type (CY), the oven temperature (OT), and the line speed (LS). Each design factor was to be tested at two levels to produce eight product designs in a 2^{5-2} fractional factorial arrangement. Two *environmental factors* not under control of the manufacturer, sealing temperature (ST) and dwell time (SD), were tested by the customer in a 3×2 factorial arrangement. The three levels for sealing temperature were denoted by $-$, 0, $+$ and the two levels of dwell time by $-$, $+$, as shown in Table 13.1. Thus the product was tested at six different environmental conditions.

The experimental program was conducted in the following manner. A number of packages were manufactured at each of the eight product design conditions. Batches of these eight types of product were then shipped to the cooperating customers. The customers subdivided the incoming batches into six subbatches and tested them at the six environmental conditions. This is therefore an example of the split-plot arrangement discussed in Chapter 9 in which the eight product designs were main plots and the six environmental conditions were subplots. Table 13.2 shows the actual layout of the experiment and the number of defective seals occurring for combinations of the design and environmental factors.

Table 13.1. Design Factors and Their Levels: Medical Packaging Example

Product Design Factors	−	+	Environmental Factors			
Paper supplier (PS)	UK	USA	Sealing temperature (ST)	−	0	+
Contents of solids (SC)	28%	30%	Dwell time (SD)	−		+
Cylinder (CY)	Fine	Coarse				
Oven temperature (OT)	180°C	215°C				
Line speed (LS)	Low	High				

Table 13.2. Layout of Experimental Arrangement Showing Number of Defective Seals for Each Available Combination of Design and Environmental Factors

Product	Design Factors						Environmental Factors					
						ST	−	0	+	−	0	+
Design	PS	SC	CY	OT	LS	SD	−	−	−	+	+	+
1	+	−	−	+	+		8	4	*	0	*	7
2	+	−	−	+	−		*	*	*	*	*	*
3	−	−	+	−	+		3	2	1	1	0	0
4	−	−	+	−	−		7	1	1	2	0	4
5	+	+	+	−	+		0	0	0	0	0	1
6	−	+	+	+	+		4	0	0	1	1	1
7	−	+	−	+	−		9	1	0	4	5	1
8	+	+	−	−	−		2	1	3	1	0	0

Note: The asterisks denote missing observations

The organization of an experimental arrangement of this kind can be difficult. In this case you will see that not only was the eight-run arrangement imperfectly carried out but there were also a number of missing observations, in particular a complete loss of information for the second product design. Nevertheless, this experiment was highly successful. It was easy to see from visual examination of the data in Table 13.2 that design 5 behaved almost perfectly under all the environmental conditions and moreover allowed a high line speed. This and consequent experiments produced a design for packaging material that would give sterile seals over the wide range of sealing process conditions used by customers.

Some History of Environmental Robustness

Experiments on environmental robustness go back at least to the beginning of the twentieth century. At that time the Guinness Company conducted field trials to find a variety of barley for brewing beer with properties that were insensitive to the many different soils, weather conditions, and farming techniques found in different parts of Ireland. At this early date those planning the trials did not have the benefit of statistical methods that were subsequently developed. Later, however, a carefully planned experiment of this kind, in which five different varieties (designs) of barley were tested under 12 (environmental) conditions (involving six different locations in the state of Minnesota on two successive years) were discussed by Fisher (1935) using data of Immer et al. (1934).

At about this time Fisher pointed out the importance of factorial experiments in such studies (1935, p. 98):

> [Extraneous factors] may be incorporated in experiments primarily designed to test other points with the real advantages, that if either general effects or interactions are detected, there will be so much knowledge gained at no expense to the other objects of the experiment and that, in any case, there will be no reason for rejecting the experimental results on the ground that the test was made in conditions different in one or other of these respects from those in which it is proposed to apply the results.

An early application in industry is due to Michaels (1964), who showed how the environmental robustness problem could be dealt with using split-plot designs. We owe to Taguchi (1986, 1991) our present awareness of the importance of such studies in achieving robust processes and products in industry. Vining and Myers (1990), Box and Jones (1992a, b), and Kim and Lin (1998) have shown how the concepts of robustness and response surface methods may be considered together.

Robust Design of an Industrial Detergent

An experiment of the kind described by Michaels will be used for illustration. The cleaning capacity of an industrial detergent was known to be affected by the temperature T, the hardness H of the wash water, and the concentration R of the detergent in the water. The manufacturer knew that in actual use by the customer the levels of T, H, and R could vary considerably. A study was conducted therefore to find out if manufacturing conditions could be found that would produce a product insensitive to likely changes in these environmental factors.

The investigating team believed this might be accomplished by making changes in some or all of four process factors, called here A, B, C, and D. Factors A and B were the amounts of two of the ingredients in the detergent itself and C and D were modifications in the manufacturing process. Thus A, B, C, and D were the design factors under the control of the manufacturer and T, H, and R were the environmental factors which in normal usage would not be under the manufacturer's control.

As shown in Table 13.3, the four design factors A, B, C, and D were tested using a 2_{IV}^{4-1} fractional factorial run in random order. Each of the eight resulting products were then tested under the four conditions of the environmental factors T, H, and R using a 2_{III}^{3-1} design with the minus and plus signs covering ranges likely to be met in routine use. Each product was tested by washing standard pieces of white cloth soiled in a standard manner. The results were assessed using a reflectometer.

It will be seen that this is a split-plot experiment with the design factors comprising the main plots and the environmental factors the subplots. A first rough appraisal of the data can be gained by simply looking at averages and ranges for each of the eight design conditions. The average supplies a measure of overall goodness of the product; the range supplies a rough measure of the design's robustness to changes in the environmental factors. You can see, for example, that product 4 yielded a high average of 90.8 but with a range of 8, possibly implying poor behavior at higher temperatures. Product 7, however, produced an even higher average of 92.0 with a much reduced range of 2, suggesting

Table 13.3. Study of Detergent Robustness

	Eight Products Produced by *Design* Factors A, B, C, and D					Four Washing Conditions Produced by *Environmental* Factors T, H, and R					
					T	−1	+1	−1	+1		
					H	−1	−1	+1	+1		
					R	+1	−1	−1	+1		
Product Number	A	B	C	D		i	ii	iii	iv	Average	Range
1	−1	−1	−1	−1		88	85	88	85	86.5	3
2	+1	−1	−1	+1		80	77	80	76	78.2	4
3	−1	+1	−1	+1		90	84	91	86	87.8	7
4	+1	+1	−1	−1		95	87	93	88	90.8	8
5	−1	−1	+1	+1		84	82	83	84	83.2	2
6	+1	−1	+1	−1		85	84	82	82	83.2	3
7	−1	+1	+1	−1		91	93	92	92	92.0	2
8	+1	+1	+1	+1		89	88	89	87	88.2	2

that this product design might produce a high mean response and also be more environmentally robust.

A More Careful Analysis

The main effects and interactions for the process factors and the environmental factors can be obtained by studying the complete set of observations in Table 13.3. If you temporarily ignore the split plotting, the 32 observations may be analyzed as if they came from a 2_{III}^{7-2} fractional factorial in the factors A, B, C, D, T, H, and R. (The generators for this design are $D = ABC$ and $R = TH$.) The 32 calculated effects so obtained may then be rearranged in the 8×4 configuration in Table 13.4. In this table the symbol I is used to indicate a mean value. For example, 86.25 was the grand average and −2.50, −0.25, and 0.25 were estimates of the *environmental* main effects; also −2.25, 6.88, 0.88, −3.75 were estimates of the *design* main effects. The row of estimates −0.50, −0.75, 0.00 were the interactions *TA, HA, RA* between the environmental factors and the design factor A, and so on.

We saw in Chapter 9 that an important characteristic of split-plot arrangements is that the subplot main effects and *all their interactions* with whole-plot effects are estimated with the same subplot error. Thus, the 24 effects to the right of the vertical line in Table 13.4 will have the same subplot error while the seven design (process) effects in the first column of the table will have the whole-plot error. Consequently, when data from this split plot arrangement are analyzed graphically, *two* normal plots are needed.

Figure 13.1a shows a normal plot for the seven estimated effects for the design factors. Figure 13.1b shows the plot of the 24 effects associated with the environmental factors: These are the 3 main effects and all 21 cross-product environment–design interaction effects. Notice that the slope of the "error line" for the subplot effects is much greater than that for the design factors, implying that the standard errors associated with the environmental factors are much less than those of the main-effect design factors. This was expected because changing

Table 13.4. Process and Environmental Estimated Effects

			Environmental effects		
		I	T	H	R
	I	86.25	**−2.50**	−0.25	+0.25
	A	−2.25	−0.50	−0.75	0.00
	B	**6.88**	−0.63	0.38	−0.13
	C	0.88	**2.13**	−0.38	−0.13
Design effects	D	−3.75	−0.25	0.50	0.00
	$AB + CD$	1.88	−0.38	0.13	0.38
	$AC + BD$	0.38	−0.13	−0.13	−0.13
	$AD + BC$	0.00	0.75	0.00	−0.75

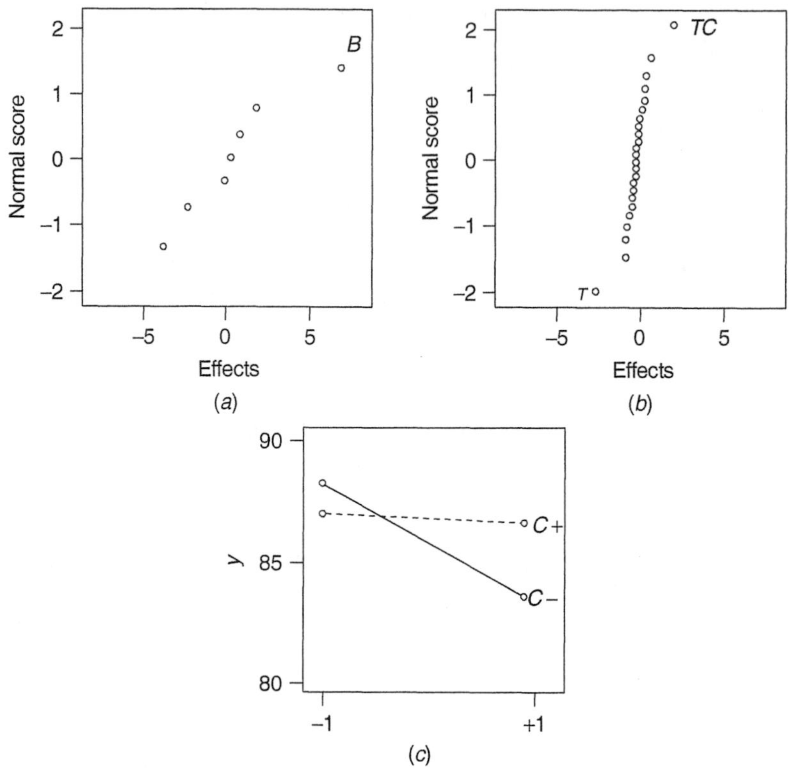

Figure 13.1. (*a*) Whole-plot design effects; (*b*) subplot environmental effects; (*c*) interaction plot.

the design (whole-plot) factors required resetting the design conditions whereas changing the environmental (subplot) factors only required testing the designed product under different washing conditions. The reduced subplot variation greatly helped in finding a product design that was environmentally insensitive.

Figure 13.1*a* suggests that an improved average response could be obtained with a design having process factor B at its plus level. The normal plot for the environmental effects (Fig. 13.1*b*) shows effects T and TC as distinguishable from noise. The nature of the TC interaction is made clear in Figure 13.1*c*.

An estimate of the subplot effect variance having nine degrees of freedom can be obtained by squaring and averaging the last nine estimates of the design–environment higher order interactions in the table: $(-0.38)^2 + (0.13)^2 + \cdots + (-0.75)^2$. Equivalently they can be obtained from a suitable ANOVA table. The standard error of an effect found in this way is 0.40 and in particular $T = -2.50 \pm 0.40$ and $TC = 2.12 \pm 0.40$. Thus the cleaning power of the product was sensitive to washing temperature, but this sensitivity could be greatly reduced by suitable manipulation of the design factor C. Specifically, Figure 13.1*c* suggests that, by switching to the upper level of C in the manufacturing process, sensitivity to temperature could be essentially eliminated.

More exactly, on appropriate assumptions discussed more fully in Appendix 13A, the estimated best level $\hat{C}$ for the process factor C is such that

$$-T = TC \times \hat{C}$$

that is

$$\hat{C} = \frac{-T}{TC} = \frac{2.50}{2.12} = +1.2 \tag{13.1}$$

Notice that with this kind of split-plot arrangement both the numerator and denominator of the ratio $-T/TC$ have the same smaller subplot error.

Exercise 13.1. Given the data in Table 13.3, confirm the estimated effects of Table 13.4 and show by an ANOVA that the within-plot variance $s^2 = 1.306$ and the standard error of an effect is 0.40.

Confidence Limits. The confidence limits for a ratio $\hat{C} = a/b$ may be obtained by a method due to Fieller (1940). In this particular application where the entries a and b in the numerator and denominator are uncorrelated random variables and have equal variance (σ_e^2, say) the argument goes as follows: Let γ be the "true" value of the *ratio a/b*. Then $a - b\gamma$ has mean zero and standard deviation $\sigma_e\sqrt{(1+\gamma^2)}$. In this example an estimate of σ_e is available, $s_e = 0.40$, having nine degrees of freedom. On normal theory therefore $a - b\gamma/s_e\sqrt{1+\gamma^2} = (2.5 - 2.12\gamma)/0.4\sqrt{1+\gamma^2}$ has a t distribution with nine degrees of freedom. For instance, the appropriate 95% limit for t with nine degrees of freedom is 2.26. Confidence limits for γ are thus given by the two roots of the quadratic equation $(2.5 - 2.12\gamma)^2 = 0.4^2 \times 2.26^2 \times (1+\gamma^2)$, that is, by $\gamma_- = 0.67$ and $\gamma_+ = 2.22$. These are the required confidence limits for $\hat{C}$. Notice that these confidence limits are rather wide, indicating that some confirmatory experiments should be conducted.

In general, there might be p environmental factors to desensitize and q design factors with which to do it. A general discussion of the mathematical aspects of this problem is given in Appendix 13A, but for immediate illustration suppose that there is one environmental factor to desensitize but two design factors having associated interactions. Then theoretically you would have an infinity of choices for the levels of two design factors that could produce desensitization. In practice, the magnitude of the interaction effects would determine how much "leverage" each design factor could exert on an environmental factor. Considerations of cost and convenience might then point to the use of one of the design factors or of a particular combination of both of them.

If there were two environmental factors to desensitize but only one design factor with associated environmental interactions, there would be two equations for one design condition and you would then need to consider some sort of compromise. In practice, graphical methods in combination with common sense will often provide a satisfactory solution. A fuller study of these issues can be found in Barrios (2005).

Looking for Useful Design Factors

You may find that one or more of the environmental factors cannot be desensitized with the design factors so far tested. It then becomes important to search for further design factors that might achieve this. Consultation with subject matter specialists may produce a number of possibilities. These can be screened using fractional experimental arrangements.

Split Plots and Main Plots: Choosing the Best Experimental Arrangement

A detergent experiment was conducted with $n = 8$ design conditions as whole plots and $m = 4$ environmental conditions as subplots. Call this arrangement (b). The types and numbers of operations with three other possible configurations (a), (c), and (d) are summarized in Table 13.5 and displayed in Table 13.6. The choice for any particular application will depend heavily on convenience and cost.

(a) In a fully randomized experimental arrangement, 32 runs would need to be made independently and each of the 32 resulting products tested individually.
(b) Only 8 whole-plot design runs would be needed to be made with random samples from each tested independently at all of the 4 environmental conditions as subplots. This arrangement uses fewer design runs but still requires 32 environmental tests.
(c) Here the 32 runs would be made in 4 complete and independent replications of the 8 design conditions. Each set of 8 designs would then be tested together as a whole plot at the 4 environmental conditions.
(d) In the so called *strip block* arrangement due to Yates (1935) only 1 run would be made at each of the 8 design conditions. Samples from each of these 8 runs would then be tested together at each of the 4 environmental conditions. Thus the arrangement would require only 8 design runs and only 4 environmental conditions. There would be one randomization for the design runs and one for the environmental runs. It is important to remember that in a strip block arrangement there will be *three* separate

Table 13.5. Four Possible Arrangements for the Detergent Example

	Example		General	
	Design Runs	Environmental Runs	Design Runs	Environmental Runs
(a) Fully randomized	32	32	$n \times m$	$n \times m$
(b) Split plot: designs are main plots	8	32	n	$n \times m$
(c) Split plot: environments are main plots	32	4	$n \times m$	m
(d) Strip block	8	4	n	m

Table 13.6. Alternative Blocking Arrangements

(a)	(b)		(c)		(d)			
32 Runs Randomized	Whole Design	Subplot Environmental	Whole Environmental	Subplot Design	Strip Block			
A B C D					*A*	*B*	*C*	*D*
1	1	A_1	*A*	1_A	1	1	1	1
2		B_1		2_A	2	2	2	2
3		C_1		3_A	3	3	3	3
4		D_1		4_A	4	4	4	4
5				5_A	5	5	5	5
6	2	A_2		6_A	6	6	6	6
7		B_2		7_A	7	7	7	7
8		C_2		8_A	8	8	8	8
		D_2						
	⋮	⋮	*B*	1_B				
				2_B				
				3_B				
				4_B				
	8	A_8		5_B				
		B_8		6_B				
		C_8		7_B				
		D_8		8_B				
	See Table 13.1		⋮	⋮				
			D	1_D				
				2_D				
				3_D				
				4_D				
				5_D				
				6_D				
				7_D				
				8_D				

types of errors associated with (i) the 8 design conditions, (ii) the 4 environmental conditions, and (iii) the interactions between the two.

Example: Strip Blocks Design

For the study of environmental robustness the strip block arrangement can be very economical in time and effort and in some instances the only feasible possibility. For example, suppose your company needs a new type of machine that performs a particular function and satisfies certain specifications. Each of n different suppliers has designed and produced a single prototype. Your company needs to know how sensitive is each machine's performance to each of m different environmental conditions. Since the machines are large and cumbersome, the environmental tests have to be made in a Biotron,* and because other researchers need to use this facility, the number of environmental runs must be minimal. The facility has a room large enough to accommodate simultaneously all the n prototypes. Thus in only m environmental runs of the Biotron each of the n designs can be simultaneously tested in a strip block arrangement. In drawing conclusions from the results, however, some caution is necessary. For example, if the above experiment was repeated with the *same* n prototypes, the size of the error would reflect only repeated measurements made on the single prototype design. It would not tell you anything about how that particular design might be subject to manufacturing variation.

Statistical Considerations in Choosing the Experimental Arrangement

It is important to be able to study the design–environmental interactions individually in any environmental robustness investigation, but in the present context the fully randomized arrangement (a) would usually be much too laborious and inefficient to merit consideration. When a cross-product configuration (b) or (c) is employed these design–environmental interactions will all be estimated with the subplot error, which is often relatively small. The split-plot arrangement (b) actually used for the detergent experiment and for the packaging experiment had the advantage that the effects of all the environmental main effects and interactions were all estimated with this same subplot error. Thus, for example, in Equation 13.1 the coefficients T and TC, which were needed to estimate the required desensitizing adjustment of C, were *both* estimated with the same small subplot error. This arrangement is even more desirable when there is more than one environmental effect to be minimized and/or more than one design factor to be adjusted. In such cases Equation 13.1 is replaced by a set of linear equations, and if arrangement (b) is used, then all the coefficients in these equations are estimated with the same small subplot error (see Appendix 13A).

Although the manner in which, for example, the detergent experiment was actually carried out would decide the appropriate analysis, the *calculated effects* that would be needed for any such analysis would always be those of Table 13.3. If the experiment had been conducted with split-plot arrangement (c), the 7 *design* effects and the 21 design–environment interactions would have the same subplot error and would appear on the same normal plot. If the strip block arrangement (d) had been employed, there would have been three separate sources of error to consider and three normal plots might be used in the analysis: one plot for the design effects, one for the environmental effects, and one for the design–environmental interactions. A detailed discussion of models, ANOVAs, and the statistical efficiency of each of the various arrangements can be found in Box and Jones (1992a).

* A Biotron is a facility containing large rooms within which various environments (light, temperature, humidity, etc.) can be produced and maintained for extensive periods of time.

Cost and Convenience Considerations in Choosing an Experimental Program

In considering which design configuration to use, usually cost and the difficulty of running the various possible experimental arrangements would be the deciding factors. For example, if we were designing a copying machine and if constructing alternative designs were expensive, then either arrangement (b) or arrangement (d) would almost certainly be used. Running the differently designed copiers at different environmental conditions (ambient temperature, humidity, type of paper, etc.) would usually be easy and inexpensive.

Sometimes the environmental factors are difficult or expensive to change. For example, an environmental factor might be the temperature of a furnace, which could not be easily manipulated (as in the split-plot example in Chapter 9). The favored arrangement would then be (c), in which the environmental factors are associated with main plots, or in some circumstances the strip block arrangement (d).

Notice that the important criterion is *not* the number of runs that are made but the number of separate operations and the cost and difficulty of making these operations. The objective is to maximize the amount of information produced for a given amount of effort.

Further Consultation with the Subject Matter Specialist

It is very important to check with the subject matter specialist how particular design and environmental effects and their interactions might have *arisen.* Good questions are: Would you have expected this? If yes, Why? If no, then what do you think might be going on? If your speculation is right, could the beneficial effects be amplified? Do these results enable you to conceive new ideas for further development of product?

It is important to discover which of the possible environmental factors need to be desensitized and to find *design* factors that could interact with these to achieve desensitization. The projective properties of the fractional arrangements make them particularly useful as screens to search for environmentally sensitive effects and for design factors that can neutralize them. When there is more than one environmental factor that cannot be neutralized with the design factors so far tested, it is important to seek expert help and make clear the nature of the difficulties. Close collaboration between the subject matter expert and the statistical practitioner almost always catalyzes the generation of new ideas.

13.2. ROBUSTNESS TO COMPONENT VARIATION

The problem of error transmission was discussed in Chapter 12. One example concerned the molecular weight of sulfur calculated from the formula

$$y = \frac{0.02x_3(100 - x_2)x_1^2}{100x_4(x_1 - x_2)} \tag{13.2}$$

where the variables x_1, x_2, x_3, and x_4 were all quantities subject to experimental error. Knowing the variances σ_1^2, σ_2^2, σ_3^2, and σ_4^2 of the x's, it was shown how the variance of y could be calculated to an adequate approximation from the transmission of the error formula

$$\sigma_y^2 = \theta_1^2\sigma_1^2 + \theta_2^2\sigma_2^2 + \theta_3^2\sigma_3^2 + \theta_4^2\sigma_4^2 \tag{13.3}$$

In that example the gradients $\theta_i = \partial y/\partial x_i$ were determined numerically and the total variance of y and the contribution from of each of the x's were as follows:

$$\sigma_y^2 = 27 + 28 + 178 + 204 \tag{13.4}$$

Notice how the transmission of the error formula identifies how much of the variance of y is accounted for by the variation is each of the x's. From Equation 13.4 you will recall that in this example most of the transmitted variation was due to x_3 and x_4 and only a small amount to x_1 and x_2. It was emphasized once again that, even though the variance of a particular component (say x_i) is small, it can still contribute a large portion of the variance of y if the gradient $\theta_i = \partial y/\partial x_i$ is sufficiently large. Conversely, large variation in a component (say x_j) may produce only a small contribution to the variance of y if $\theta_j = \partial y/\partial x_j$ is small.

Morrison (1957) provided an early demonstration of how the transmission of error could be used in engineering design. [See also Shewhart (1931).] Suppose a product contains an *assembly* of k components. Let y be a measure of the performance of the assembly which is a *known* function $y = f(x_1, x_2, \ldots, x_k)$ of measured characteristics of the k components $x_1, x_2, \ldots, x_k$. Suppose further that the variances of the relevant characteristics of these k components *also assumed known* are $\sigma_1^2, \sigma_2^2, \ldots, \sigma_k^2$ and that the designer can choose their nominal values $x_{10}, x_{20}, \ldots, x_{p0}$ with $p < k$. The variance contribution of each of the individual components to the variance of the assembly can then be calculated and used to help design the assembled product.

Morrison* illustrated his idea with the design of a product where the known functional relationship between the performance measure y and three component characteristics was

$$y = \frac{\pi}{4}(x_1^2 - x_2^2)x_3 \tag{13.5}$$

and the three nominal values x_{10}, x_{20}, and x_{30} were all at the designer's choice. Using 30 repeated measurements, *he estimated* the variances σ_1^2, σ_2^2, and σ_3^2 of each of the three components. He then determined the gradients $\theta_i = \partial y/\partial x_i$ by direct differentiation to obtain the error transmission formula

$$\sigma_y^2 = \left[\frac{\pi x_1 x_2}{2}\right]\sigma_1^2 + \left[\frac{\pi x_2 x_3}{2}\right]\sigma_2^2 + \left[\frac{\pi(x_1^2 - x_2^2)}{4}\right]\sigma_3^2 \tag{13.6}$$

The expression was evaluated by substituting the nominal values of the x's and the estimated variances. For his example he found

$$\hat{\sigma}_y^2 = 3.52 + 0.30 + 7.57 = 11.39$$

* The importance of Morrison's ideas for the design of auto mobiles was recently emphasized by R. Parry Jones (1999) vice president for design of the Ford Motor Company in his address to the Royal Academy of Engineering.

The largest contributor to the variation in performance of the measured assembly therefore was x_3. Also x_1 was of some importance but x_2 contributed little. Therefore, to reduce the value $\hat{\sigma}_y^2$, effort should be directed to reducing the variances σ_3^2 and σ_1^2.

Minimization of Variance Transmission

On the further assumption that the variances σ_1^2, σ_2^2, and σ_3^2 *remained constant* or could be derived for different nominal values x_{10}, x_{20}, and x_{30}, the values that produced *minimum* error transmission could be calculated. Equation 13.6 would now be treated as a function of the nominal values x_{10}, x_{20}, and x_{30} for which

$$\sigma_y^2 = F(x_{10}, x_{20}, x_{30}) \tag{13.7}$$

was to be minimized. This minimization is most easily done numerically using a standard computer program for minimization (or maximization). Notice, however, that to obtain Equation 13.7 and make such a calculation, you must *know* the function $y = f(x_1, x_2, x_3)$ and to a fair approximation the relative values of $\sigma_1^2, \sigma_2^2, \ldots, \sigma_k^2$. Also you need to know whether these component variances depend on the nominal values of the x's and, if so, in what way. These requirements were emphasized by Morrison, and their importance is best understood by considering a simple example.

Designing a Pendulum

Suppose you were designing a pendulum for inclusion in a clock mechanism. The period y of the swing would depend on the pendulum's length x so that manufacturing variations in x would be transmitted to the period y. Since you would need appropriate gears in the clock mechanism to achieve a constant speed of rotation of the clock hands, for any particular period y it is the *percentage* variation of y that is of interest. Thus you would need to minimize the coefficient of variation of y, or equivalently the standard deviation of $\log y$. For this reason in Figure 13.2 the period is shown on a log scale.

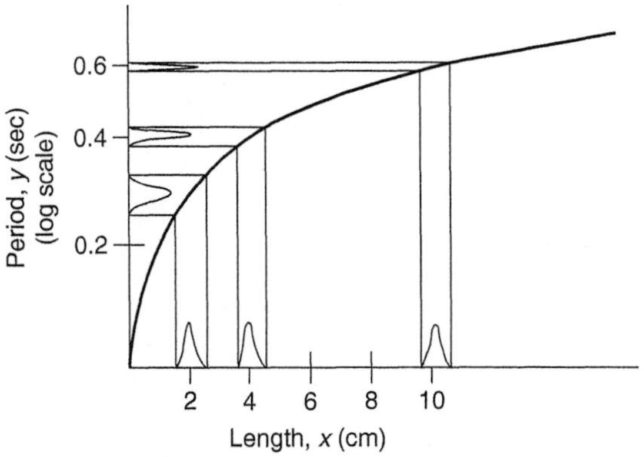

Figure 13.2. Period y vs. length x. (When the error variance of the length of the pendulum x is constant, the longest pendulum produces the smallest percentage error).

Suppose the relationship between the length x of the pendulum and its period y is *known* to an adequate approximation* from

$$y = 0.2\pi\sqrt{x} \tag{13.8}$$

and that the manufacturing variation in the length x of the pendulum has a constant standard deviation. Then, as illustrated in Figure 13.2, the precision of the clock would be maximized when the pendulum was made as long as possible.

But suppose you assumed instead that the length x was subject to a constant *percentage* error. Then it is easy to see that the error transmission would now be the same regardless of the length of the pendulum. Thus, as emphasized by Morrison, (1957) concrete knowledge about the variances of the x's (a single x in this case) is essential. Unfortunately, this precept has not been followed in many applications of this idea.

Parameter Design and Tolerance Design

For problems in product design Genichi Taguchi (1980, 1986, 1991) uses the error transmission idea extensively. He makes a distinction between what he calls parameter design and tolerance design. For parameter design he recommends using components that are inexpensive and finding nominal values that give smallest error transmission. If the solution obtained still gives too variable a product, he then employs the error transmission formula in parameter design to show when more expensive components with smaller tolerances are needed.

We believe that problems of this kind are best handled in the manner described earlier using the error transmission idea directly. The terminology as well as the methodology differ from those employed by Taguchi. The following example, due to Barker (1990, 1994) although extremely elementary, is sufficient to illustrate the difference.

An alternating current circuit is to be designed to maintain with minimum variance a constant resistance $R = 500$ ohms. From elementary physics†

$$R = \frac{10^{-6}}{2\pi f C} \tag{13.9}$$

where f is the frequency of the alternating current and C is the capacitance. Both f and C are known to vary from their nominal values. In addition to error transmitted by deviations in f and C there is a further source of variation caused by a final adjustment of R, independent of f and C, to the required value. Thus at final assembly $10^{-6}/2\pi f C + R_{\text{adj}} = 550$. In this example f and C are the design variables to be set and f, C, and R_{adj} the variables contributing to the noise.

Earlier in this book two methods have been described for estimating transmitted error: the analytical method used by Morrison and the numerical method illustrated in the molecular weight example. Barker used a third numerical method

* The formula is close to that of a "perfect" pendulum where y is measured in seconds and x in centimeters.

† A still simpler formula is obtained by taking logs on each side of the equation.

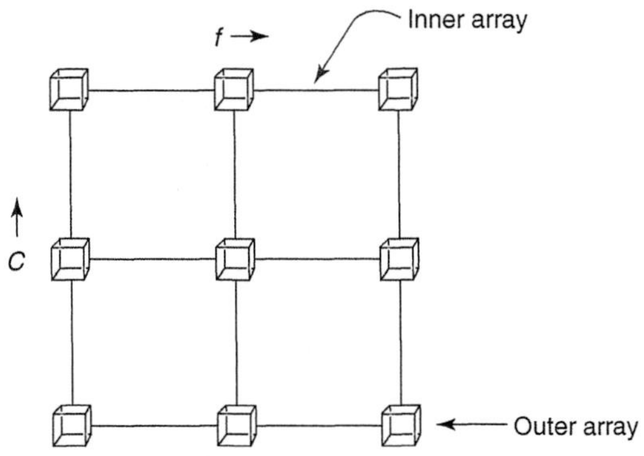

Figure 13.3. Inner and outer arrangement: a 3^2 arrangement with a 2^3 arrangement at the vertices.

due to Taguchi in which the calculated levels of each of the factor combinations are themselves varied over small ranges for each design configuration in what is called an inner and outer array. The concept is illustrated in Figure 13.3.

The tested levels of the design variables f and C intended to span the whole region of interest were as follows:

Variable	−	0	+
f	60	600	1200
C	5	25	50

The design levels (−, 0, +) form a 3^2 factorial arrangement called the *inner array*. At each of the nine design configurations eight circuits were constructed employing the following tolerances of the noise contributors:

Variable	Tolerance
f	±10%
C	±20%
R	±1%

Then for each of the nine inner array arrangements an *outer array* consisting of a 2^3 factorial arrangement was run with f, C, and R ranges spanning the tolerances. In split-plot terms the inner array identifies the main plots and the outer array the subplots. For example, for design configuration $f = 1200$ and $C = 50$ the outer array levels were $f = 1200 \pm 120$, $C = 50 \pm 10$, and $R = 500 \pm 5$. The complete inner and outer array configuration containing $3^2 \times 2^3 = 36$ points is illustrated in Figure 13.3.

Error transmission is then determined from the averages $\bar{y}$ and the estimates of variance s_y^2 calculated from the eight observations located at each of the nine 2^3 factorial inner array arrangements.

To determine the engineering design configuration that minimizes error transmission, Barker maximized Taguchi's "signal-to-noise" (S/N) ratio criteria: $H = -10\log_{10}(\overline{y}^2/s_y^2)$. The *marginal* averages of H calculated from the 3^2 inner array were as follows:

Variable	**−**	**0**	**+**
f	25.24	38.56	40.10
C	28.51	36.65	38.73

From these the author deduced that the highest level of the S/N ratio would be obtained by using the highest level of frequency and the highest level of capacitance.

In general, of course, overall maxima cannot be obtained from marginal maxima in this way. However, computer programs to maximize any function that can be calculated analytically or numerically are available.

The Anatomy of *H*

In this example the function maximized was the S/N criterion $H = -10\log_{10}(\overline{y}^2/s_y^2)$. Now $\overline{y}^2/s_y^2 = 1/V^2$, where V is the coefficient of variation. Thus maximizing $H = 20\log_{10} V$ is equivalent to minimizing the coefficient of variation, which in turn is equivalent to minimizing the standard deviation of $\log y$. For example, see Figure 13.4 and the derivation in Appendix 13B.

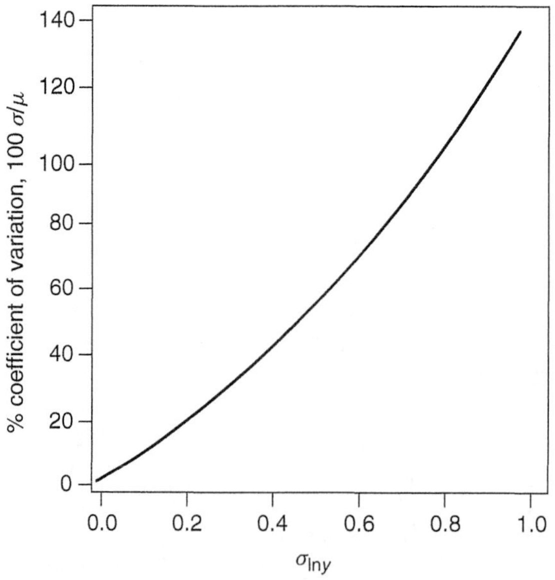

Figure 13.4. Plot of coefficient of variation versus $\sigma_{\log y}$.

Response Surface Ideas

A number of authors have pointed out that Robust design and response surface methods should be considered together to provide solutions more directly (for example Fung, 1986; Vining and Myers, 1990; Box and Jones, 1992a, b; Kim and Lin, 1998). Figure 13.5 shows the contours of $\sigma_{\log y}$ and hence of S/N for Barker's example. This elementary example can be used to illustrate cogent points that are clarified when the problem is looked at from the response surface view:

1. In this example, if f and C are constrained to be within the stated limits, Barker's conclusion is correct—the highest value of the S/N ratio of $H = 41.75$ within this region is obtained when f and C are both at their maxima. But the contour diagram of Figure 13.5 also shows that little would be lost by, for example, using the intermediate values $C = 27.5$, $f = 630$ where $H = 41.0$. This illustrates a more general consideration that it is rarely enough to calculate a point maximum (even when done correctly). You need to know what is happening in the neighborhood of that maximum. This is particularly true in cost analysis.
2. As in this example, the point of minimum error transmission is often found at a vertex or an edge of the "inner array." In practice, however, the ranges over which the design variables are changed would frequently be somewhat arbitrary and the designer would wish to explore what would happen if the present region were extended. For example, in a direction of less cost.
3. It is essential to consider whether the maximum is best approximated by a point or alternatively by a line or plane implying some kind of ridge system that might be explored and exploited to improve other attributes of the product design.

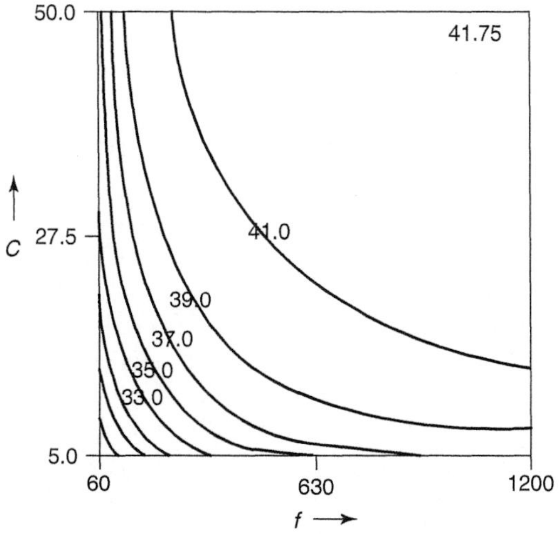

Figure 13.5. Contours of the S/N ratio and the (identical) contours of the standard deviation of $\log y$.

4. To carry through any analysis of this kind, you must *know*, at least to a reasonable approximation, the relative variances $\sigma_1^2, \sigma_2^2, \ldots, \sigma_k^2$ of the x's as well as the function $y = f(x_1, x_2, \ldots, x_k)$.

APPENDIX 13A. A MATHEMATICAL FORMULATION FOR ENVIRONMENTAL ROBUSTNESS

In a previous example it was found that the cleaning power of a detergent was highly sensitive to the temperature T of the wash water. Taking advantage of the fact that there was a large *TC* interaction effect, this sensitivity could be nullified by adjustment of the design variable C.

More generally, suppose that a change z in an environmental factor (in this case T) produces a change y in the effectiveness of the product (the cleaning capability of the detergent). Then approximately

$$y = z \times \frac{dy}{dz} \qquad (13A.1)$$

The equation emphasizes that even a large change z in the environmental factor will not produce a large change y in the measure of effectiveness if dy/dz is *or can be made* small. Thus dy/dz may be called the sensitivity factor* and will henceforth be denoted by S.

Now suppose for a prototype design the value of the sensitivity factor equals S and that there is a design factor x (process factor C in this example) which has a substantial interaction with x and is measured by the derivative $d\mathsf{S}/dx$. Then the sensitivity factor for some value of the design factor x is

$$\mathsf{S}_x = \mathsf{S} + (d\mathsf{S}/dx) \times x$$

So that the value $\hat{x}$ which will reduce the sensitivity to zero is such that

$$-\mathsf{S} = \frac{d\mathsf{S}}{dx} \times \hat{x} \qquad (13A.2)$$

In the example S is then the (main) effect of the environmental factor z (i.e., T) and the derivative $d\mathsf{S}/dx$ is the interaction effect between x and z (*TC*).

As the earlier example illustrated, since the main effect (T) and the interaction (*TC*) are both estimates, the value $\hat{x}$ may be subject to considerable error; also there will be a practical range within which the design factor x may be varied and that $\hat{x}$ may lie outside this range. However, even when the exact mathematical solution cannot be used, it may indicate a direction in which to go. This and associated questions are considered in more detail by Barrios (2005).

More generally, suppose that there are p environmental factors that need to be studied and thus p sensitivity factors $\mathsf{S}_1 = dy/dz_1$, $\mathsf{S}_2 = dy/dz_2$, ... Also

* Also called the robustness factor.

suppose there are q design factors $x_1, x_2 \ldots, x_q$ and that changes in one or more of these can produce changes in the 's. Then whether there is a unique mathematical solution or not will depend on whether $q < p, q = p$, or $q > p$. For illustration suppose $p = 1$ and $q = 2$, that is, there is one environmental factor and two design factors available to desensitize the response and the sensitivity factor $_1$ will be reduced to zero if

$$- \ _1 = \ _{11}\hat{x}_1 + \ _{12}\hat{x}_2 \qquad (13A.3)$$

where $_{ij} = d \ _i dx_j$. In this case, therefore, theoretically any values of the design variables x_1 and x_2 that satisfy this equation might be used to desensitize.

If $p = q$ we have two environmental factors to desensitize and two design factors to employ and to get the two sensitivity values to equal zero requires the solution of the two equations

$$- \ _1 = \ _{11}\hat{x}_1 + \ _{12}\hat{x}_2 \qquad - \ _2 = \ _{21}\hat{x}_1 + \ _{22}\hat{x}_2 \qquad (13A.4)$$

Again, both the limitations concerning the feasibility region for the manipulation of the design factors x_1 and x_2 and the errors of the estimates $\hat{x}_1$ and $\hat{x}_2$ must be borne in mind.

Finally, suppose $p = 2$ and $q = 1$, that is, there are two environmental factors to desensitize but only one design factor available to accomplish this end. Then

$$- \ _1 = \ _{11}\hat{x}_1 \quad \text{and} \quad - \ _2 = \ _{21}\hat{x}_1 \qquad (13A.5)$$

In this case it will be necessary to find some compromise value for $\hat{x}_1$.

Robustness and Optimality

Notice that none of these equations take into account the best *level* for the response. Suppose, for example that with the environmental variables fixed, the response is adequately represented by a second-degree equation in the design factors x_1 and x_2. Thus

$$\hat{y} = b_0 + b_1x_1 + b_2x_2 + b_{11}x_1^2 + b_{22}x_2^2 + b_{12}x_1x_2$$

The coordinates (x_1^0, x_2^0) of a maximum (minimum) in the immediate region of interest will then satisfy the equations

$$-b_1 = 2b_{11}x_1^0 + b_{12}x_2^0 \qquad - b_2 = b_{12}x_1^0 + 2b_{22}x_2^0 \qquad (13A.6)$$

But notice that the Equations 13A.6, which define "optimality," share no coefficients in common with those of Equations 13A.4, which define "robustness." Thus you could have an "optimal" solution that was highly nonrobust and a robust solution that was far from optimal (that would produce a product that was equally bad at all levels of the environmental factors). In practice, some kind of compromise is needed. This can be based on costs, the performance of a competitor's products, engineering feasibility, and the like. One approach which combines considerations of robustness and optimality and provides a locus of compromise between the two solutions is given in Box and Jones (1992, 1992b).

In practice the solutions to the linear equations will be subject to error because of uncertainties in the coefficients $b_1, b_{11}, b_{12}, \ldots$. Confidence regions for these solutions can be obtained by a method due to Box and Hunter (1954).

APPENDIX 13B. CHOICE OF CRITERIA

The Relation Between $\sigma_{\ln y}$, the Coefficient of Variation, and a Signal-to-Noise Ratio

As is well known, the coefficient of variation $\gamma = \sigma/\mu$ is, over extensive ranges, almost proportional to the standard deviation of $\ln y$. More specifically (see, e.g., Johnson and Kotz, 1970), if after a log transformation $Y = \ln y$ has mean μ_Y and *constant variance* σ_Y^2, then, exactly, if Y is normally distributed and approximately otherwise,

$$\mu = \exp(\mu_Y + \tfrac{1}{2}\sigma_Y^2) \tag{13B.1}$$

$$\sigma = \mu[\exp(\sigma_Y^2) - 1]^{1/2} \tag{13B.2}$$

Consider Equation 13B.2. You will see that σ, the standard deviation of y, is exactly proportional to its mean μ or, equivalently, that the coefficient of variation $\sigma/\mu = [\exp(\sigma_Y^2) - 1]^{1/2}$ is independent of μ. Moreover, σ/μ is the monotonic function of σ_Y. Thus, analysis in terms of σ/μ (or of μ/σ, the S/N, ratio) is essentially equivalent to the analysis of σ_Y — that is, of $\sigma_{\ln y}$.

Signal-to-Noise Ratios and Statistical Criteria

As was seen in earlier discussions, many familiar test statistics are S/N ratios in the sense that the term is commonly employed by communication engineers. Examples are the t statistic (or the normal deviate z when σ^2 is known):

$$t = \frac{\bar{y} - \eta}{\mathrm{SE}(\bar{y})} \qquad \text{or} \qquad t = \frac{(\bar{y}_1 - \bar{y}_2) - (\eta_1 - \eta_2)}{\mathrm{SE}(\bar{y}_1 - \bar{y}_2)}$$

or in general

$$t = \frac{\text{statistic} - E(\text{statistic})}{\sqrt{\mathrm{Var(statistic)}}}$$

In each case the numerator is a measure of the signal and the denominator a measure of the noise (appropriate to the statistic). The F statistics are also S/N ratios:

$$F = \frac{\text{mean square for treatments}}{\text{mean square for error}} \qquad \text{or} \qquad F = \frac{\text{mean square due to regression}}{\text{residual mean square}}$$

Notice that t and F are homogeneous functions of degree zero in the data and consequently are dimensionless and unaffected by changes in scale. For example, they are unchanged whether you measure y in grams or in pounds. Some confusion has arisen because Taguchi has called a number of his performance criteria "signal to noise ratios," which do not have this invariant property. According to his philosophy, the criteria should differ depending on the objective of the experimenter. Examples are as follows:

Objective	Criterion
Closeness to target	$\{SN_T\} = 10\log_{10}(\bar{y}^2/s^2)$
Response as large as possible	$\{SN_L\} = -10\log_{10}\left[\frac{1}{n}\sum_{i=1}^{n}\left(\frac{1}{y_i}\right)^2\right]$
Response as small as possible	$\{SN_S\} = -10\log_{10}\left[\frac{1}{n}\sum_{i=1}^{n}(y_i)^2\right]$

Such criteria pose a number of problems:

(a) Criteria, such as (SN_L) and (SN_S), are not S/N ratios; they are not dimensionless and linear transformation of data can change their information content. In particular the SN_L transformation blows up for any transformation that produces any data values close to zero.

(b) It is generally accepted that statistics should be efficient so that, on rational assumptions, they use all the information in the data. On reasonable distributional assumptions SN_L and SN_S are not statistically efficient and their use is equivalent to throwing some of the data away.

(c) It is unwise to be dogmatic in selecting criteria before an experiment is run. Criteria should be reconsidered after the data are available. This was illustrated in an example quoted earlier where the original objective was to maximize yield, but in the light of the data, the objective was changed to minimizing cost without changing yield.

REFERENCES AND FURTHER READING

Barker, T. B. (1990) *Engineering Quality by Design: Interpreting the Taguchi Approach*, Marcel Dekker, New York.

Barker, T. B. (1994) *Quality by Experimental Design*, 2nd ed., Marcel Dekker, New York.

Barrios, E. (2005) Topics on Engineering Statistics. Unpublished Ph.D Thesis. Department of Statistics, University of Wisconsin-Madison.

Box, G. E. P. (1988) Signal to noise ratios, performance criteria and transformations, *Technometrics*, **30**, 1–17.

Box, G. E. P., and Behnken, D. W. (1960) Some three level designs for the study of quantitative variables, *Technometrics*, **2**, 455–476.

Box, G. E. P., Bisgaard, S., and Fung, C. A. (1988) An explanation and critique of Taguchi's contributions to quality engineering, *Quality Reliability Eng. Int.*, **4** 123–131.

Box, G. E. P., and Fung, C. A. (1986) *Minimizing Transmitted Variation by Parameter Design*, Report 8, Center for Productivity and Quality Control, University of Wisconsin, Madison.

Box, G. E. P., and Hunter, J. S. (1954) A confidence region for the solution of a set of simultaneous equations with an application to experimental design, *Biometrika*, **41**, 190–198.

Box, G. E. P., and Hunter, W. G. (1965) The Experimental study of physical mechanisms, *Technometrics*, **7**, 23–42.

Box, G. E. P., and Jones, S. (1992a) Split plot designs for robust product experimentation, *J. Appl. Statist.*, **19**, 3–26.

Box, G. E. P., and Jones, S. (1992b) Designing products that are robust to the environment, *J. Total Quality Manag.*, **3**(3), 265–282.

Fieller, E. C. (1940–1941) The biological standardization of insulin. *Suppl. J. Roy. Statist. Soc.*, **7**, 1–64.

Fisher, R. A. (1935) The Design of Experiments, Edinburgh, Oliver & Boyd.

Fung, C. A. (1986) Statistical topics in off-line quality control, Ph.D thesis, Department of Statistics, University of Wisconsin-Madison.

Gunter, B. (1987) A perspective on Taguchi methods, *Quality Progress*, **June**, 44–52.

Hunter, J. S. (1985) Statistical design applied to product design, *J. Quality Technol.* **17**, 210–221.

Immer, F. R. (1934) Calculating linkage intensities from F_3 data. *Genetics*, **19**, 119–136.

Johnson, N. L., and Kotz, S. (1970). *Distributions in Statistics. Continuous Univariate Distributions I*, Mifflin, Boston.

Kim, K. J. and Lin, D. K. J. (1998). Dual response surface optimization: A fuzzy modeling approach, *J. Qual. Tech.*, **30** 1–10.

Michaels, S. E. (1964) The usefulness of experimental design, *Appl. Statist.*, **13**(3), 221–235.

Miller, A., and Wu, C. F. J. (1996) Parameter design for signal-response systems: A different look at Taguchi's dynamic parameter design, *Statist. Sci.*, **11**, 122–136.

Morrison, S. J. (1957) The study of variability in experimental design, *Appl. Statist.*, **6**(2), 133–138.

Myers, R. H., Khuri, A. I., and Vining, G. G. (1992) Response surface alternatives to the Taguchi robust parameter design approach, *Am. Statist.*, **468**, 131–139.

Nair, V. N. (Ed.) (1992) Taguchi's parameter design; a panel discussion, *Technometrics*, **34**, 127–161.

Parry-Jones, R. (1999) *Engineering for Corporate Success in the New Millennium*, Royal Academy of Engineering, Westminster, London.

Phadke, M. S. (1989) *Quality Engineering Using Robust Design*, Prentice-Hall, Englewood Cliffs, NJ.

Schuerman, J., and Algrim, T. (1992) Personal Communication.

Shewhart, W. A. (1931) *Economic Control of Quality of Product*, Van Nostrand, New York. Also reprinted (1981) American Society for Quality, Milwaukee.

Shoemaker, A. C., Tsui, K. L., and Wu, C. F. J. (1991) Economical experimentation methods for robust design, *Technometrics*, **33**, 415–427.

Taguchi, G. (1986) *Introduction to Quality Engineering*, UNIPUB, White Plains, NY.

Taguchi, G. (1991) *System of Experimental Design: Engineering Methods to Optimize Quality and Minimize Cost*, Quality Resources, White Plains, NY.

Taguchi, G., and Wu, Y. (1980) *Introduction to Off-Line Quality Control*, Nagoya, Japan, Central Japan Quality Control Organization; also American Supplier Institute, Dearborn, MI.

Vining, G. G., and Myers, R. H. (1990) Combining Taguchi and response surface philosophies, *J. Quality Technol.*, **28**, 38–45.

Yates, F. (1935) Complex experiments, *J. Roy. Stat. Soc. Ser. B.*, **2**, 181–223.

QUESTIONS FOR CHAPTER 13

1. What is meant by "an environmentally robust" product? How are the words "robust" and "environment" interpreted? Give two examples.

2. How can you run an experiment to discover an environmentally robust product?

3. What is a "design" factor and an "environmental" factor?

4. Why are design–environment interactions potentiality useful in designing environmentally robust products?

5. Why are split-plot designs often used in robust experimentation?

6. Suppose in a robustness experiment four different experimental arrangements are possible: (a) fully randomized, (b) split plot with the design factors as whole plots, (c) split plot with environmental factors as whole plots, and (d) a strip block arrangement. What are some circumstances in which each of these experimental arrangements might be preferred?

7. What is meant by "robustness to component variation"?

8. What would you need to know (exactly or approximately) in order to conduct a robustness to component variation study?

9. What are three different ways of determining the rate of change of a response with respect to a design factor?

10. What is meant by the coefficient of variation of the response y? How is it related to (a) the standard deviation of $\log y$ and (b) an S/N ratio?

11. How is it possible to design products that are robust to component variation but nevertheless give a response which is very far from optimal? Explain.

12. Why is it essential to have estimates of the component variances in order to estimate the variation transmitted to an assembly by the components?

13. One criterion proposed for use in robustness studies is the S/N ratio:

$$\text{SN} = -10\log_{10}(\bar{y}^2/s^2)$$

What relationship does this have to the coefficient of variation and to the variance of log y?

PROBLEMS FOR CHAPTER 13

1. Tests for environmental robustness are sometimes run only at the extreme conditions of the design factors. Thus is the example of Table 13.3 it might be decided that extreme conditions consisted of maximum hardness of the wash water with minimum temperature or vice versa. For this set of data, would your conclusions have been different from such a test? Why might such a strategy fail?

2. In the example of Table 13.2 suppose a full set of data for product designs 1 and 2 were available as follows: design 1: 8, 4, 3, 0, 4, 7; design 2: 7, 2, 1, 4, 5, 9. Analyze the completed set of data using so far as possible the techniques employed for the detergent study in Tables 13.3 and 13.4.

3. In an experiment of the kind shown in Table 13.3 it sometimes makes sense to work with a transformation of the response. For this example, (a) do you think a reciprocal transformation might be appropriate? Why? (b) Make an analysis with the metric $Y = 1/y$. Are your conclusions different from those reached in the chapter. Explain.

4. A response surface study showed an important characteristic of an assembly was related to two component characteristics x_1, x_2 by the equation

$$\hat{y} = 95.6 - 13.7x_1 + 10.7x_2 + 16.7x_1^2 + 18.2x_2^2 + 13.1x_1x_2$$

Given that manufacturing variation produces a standard deviation in the components of $\sigma_1 = 0.1$ and $\sigma_2 = 0.1$, what values of x_1 and x_2 will produce minimum error transmission with the smallest possible value of the variance of the response. Repeat your calculation with $\sigma_1 = 0.1$ and $\sigma_2 = 0.2$ what do you conclude?

5. Suppose the following data were obtained in an environmental robustness study concerning the manufacture of cakes which should be insensitive to the temperature T and the time t of baking. The cake mix employed different amounts of flour F, shortening S, and egg powder E. The factors where varied at two levels denoted by $-$ and $+$ in a factorial design with an intermediate value denoted by 0. The data are average scores of an hedonic index scaled from 1 to 7 as judged by a taste panel.

			T	0	−	+	−	+
F	S	E	t	0	−	−	+	+
0	0	0		6.7	3.4	5.4	4.1	3.8
−	−	−		3.1	1.1	5.7	6.4	1.3
+	−	−		3.2	3.8	4.9	4.3	2.1
−	+	−		5.3	3.7	5.1	6.7	2.9
+	+	−		4.1	4.5	6.4	5.8	5.2
−	−	+		5.9	4.2	6.8	6.5	3.3
+	−	+		6.9	5.0	6.0	5.9	5.7
−	+	+		3.0	3.1	6.3	6.4	3.0
+	+	+		4.5	3.9	5.5	5.0	5.4

(a) What are the design factors?

(b) What are the environmental factors?

(c) Calculate the average $\bar{y}$ and standard deviation s for each of the nine designs (recipes). What do you conclude from your analysis?

6. Suppose the analysis in problem 5 was carried out by preparing for each recipe enough cake mix to make five cakes to be baked together at the various environmental conditions. This is a split-plot design Which factors are associated with the design plots and which with subplots? How might the experimental program have been arranged differently? What might be the advantages and disadvantages of each arrangement?

Assume that the design factors are the whole plots and make a table of their main effects and two-factor interactions. How might this table be used to obtain the standard error of (a) an environmental main effect and (b) an environmental–design interaction?

7. It has been suggested that an experiment like that in problem 5 might be analyzed using the so-called quadratic loss for each recipe. For this experiment the quadratic loss is given by the expression $L = \sum(7 - y)^2$, which is a measure of how far a particular recipe falls short of the ideal value of 7. Calculate these losses for each of the nine recipes. How is the function L connected with the average $\bar{y}$ and the standard deviation s? Comment.

CHAPTER 14

Process Control, Forecasting, and Time Series: An Introduction

Two related statistical procedures are hypothesis (significance) testing and statistical estimation. Similarly there are two related problems of statistical process control. These are process *monitoring* and process *adjustment*. By process *monitoring** is meant *continual checking* of process operation to detect unusual data that can point to faults in the system. This activity is related to hypothesis testing. By contrast, process *adjustment* is used to *estimate* and *compensate* for drifts or other changes in the data.†

14.1. PROCESS MONITORING

In the 1930s the Bell Telephone and Western Electric companies needed to manufacture parts which, as nearly as possible, were identical. To ensure this, a program of continuous sampling and testing was instituted. Walter Shewhart, a member of the Bell research staff, knew that, to ensure survival, our eyes and brains have developed to be alert to the unexpected. To use this ability, he proposed that measurements be continuously displayed in sequence so that a comparison could be made between current data and data generated by a correctly operating process. A sequential record of this sort was called a *run chart*. Figure 14.1*a* shows such a run chart for a sequence of averages of four

* This system of monitoring is often called "statistical process control," but this term is avoided since process adjustment should also be part of statistical process control.

† Two frequently cited text in the chapter are *Time Series Analysis, Forecasting and Control, Third edition* by Box, G. E. P., Jenkins, G. M. and Reinsel, G. C. (1994) Prentice Hall, Englewood Cliffs, NJ. and *Statistical Control by Monitoring and Feedback Adjustment* by Box, G. E. P. and Luceño, A. (1997) Wiley, New York. In what follow these an referred to as **BJ&R** and **B&L** respectively.

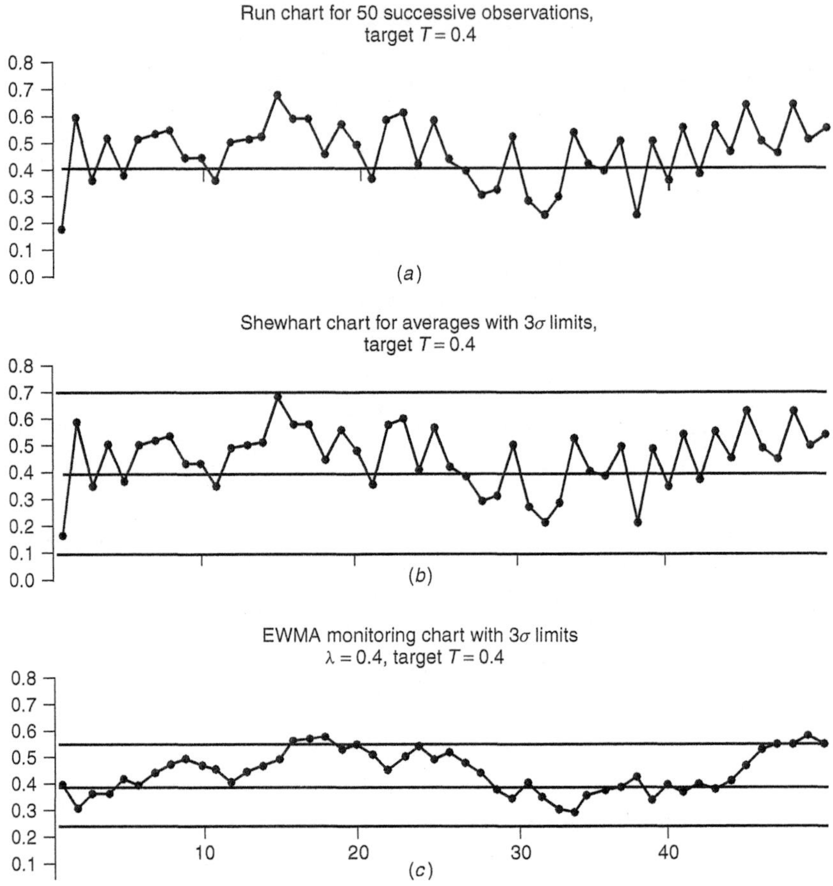

Figure 14.1. (*a*) Run chart for averages of four measurements, target $T = 0.4$. (*b*) Shewhart chart for the averages. (*c*) Shewhart chart for the EWMA of the averages, $\lambda = 0.4$.

measurements randomly recorded each hour. For simplicity an average of four observations taken at a particular time will be called an observation.

While a detectable deviation from the norm can point to the *existence* of a fault, its cause is not always easily recognized. However, when an assignable cause can be found, it may be possible to modify the system ensuring that the fault is corrected and cannot occur again. When followed as a regular policy, this procedure of continuous "debugging" (process monitoring) can gradually eliminate these faults. Consequently, continuous process improvement results due to reduced process variation and fewer process upsets.

Some process variation is bound to occur, but deciding what is an unacceptable deviation from the norm is not always easy. Nothing looks so unrandom as a series of random numbers and without proper checks you could start a wild goose chase for a nonexistent cause. Thus, to behave rationally, you must first ask the question, "In satisfactory routine production is this apparent deviation unusual or does it happen quite often?" What is needed is an appropriate reference set representing the behavior of the process when it is operating correctly. If what you at first thought was unusual behavior is something that has occurred frequently in past satisfactory operation, you should pay it little attention. On the other hand, if the distractive event has happened rarely or not at all in the past, you would have good reason to look for its cause.

When available, the direct source for a reference distribution is the history of past data recorded during periods of satisfactory performance. Comparison between the suspect event and previous experience requires no assumptions about independence and normality, and nowadays the computer can make a rapid search of past records. An example of the use of past data to obtain an appropriate reference distribution (in that case for the comparison of two means) was given in Chapter 3.

Shewhart said that you should try to keep your process in what he called a *state of statistical control.* By this he meant a state in which the variation of the data about the target value, called *common cause variability,* is stable about the target value.* For a reference distribution he assumed that there would be recognizable sequences of data taken when the process was running satisfactorily. For such data it was assumed that successive deviations from target would be NIID. To draw attention to "out-of-control" behavior, he proposed that limit lines be added to the run chart between which most of the data would be located when the process was operating correctly. Such "control limits" have customarily corresponded to deviations from target of "± three sigma" and "± two sigma," sometimes called action and warning limits. The "sigma" defining these limits is the standard deviation of the plotted points, usually averages of n measurements with $n = 4$ or 5.† In practice, exceeding a point outside action lines does not necessarily imply that immediate action should be taken—usually that would involve other considerations.

Figure 14.1*b* shows a Shewhart chart with appropriate action lines for the averages previously plotted on the run chart in Figure 14.1*a*. Shewhart also recommended that a second control chart be added to monitor the variability by plotting the ranges of successive samples.

The Shewhart quality control chart is a tool of great value and for decades has been the source of thousands of process improvements. It appears in a variety of forms: charts for averages, for percentages, for frequencies (counts), and so on.‡ The technical literature concerned with the construction and use of Shewhart control charts is vast. *Juran's Handbook of Quality* serves as an excellent

* Some books on statistical control speak of deviations from the long-run average or from the average of rational subgroups rather than deviations from a target value. We shall speak here of deviations from the target value (from specification) and not from some average value.

† Some authors recommend calculating the sigma of the plotted averages from the average variance of the data within each set of n observations, thus $\sigma_{\bar{y}} = \sigma/\sqrt{n}$. This is likely to underestimate the natural variability of the operating process.

‡ Shewhart charts have also been proposed that employ instead of the Normal distribution the hypergeometric, the double exponential, the truncated, folded normal distributions, and so on, (Juran and Godfrey, 1999).

general reference along with the official documents of ANSI/ASQ [The American National Standards Institute and the American Society for Quality (ANSI/ASQC), 1996].

Control by Looking at Recent Behavior

In the use of Shewhart charts many people believed that the single rule, one or more points outside the 3σ limits, was inadequate. This was because it did not take into account the *recent* behavior of the process. Additional rules proposed by Western Electric engineers postulated that a lack of statistical control would be indicated if any of the following occurred:

1. One point fell beyond the $\pm 3\sigma_{\bar{y}}$ limits.
2. Two of three *consecutive* points fell beyond the $\pm 2\sigma_{\bar{y}}$ limits.
3. Four out of five *consecutive* points were greater than $\pm \sigma_{\bar{y}}$.
4. Eight *consecutive* points occurred on the same side of the target value.

For a good review of additional rules that have been proposed see Nelson (1984).

If you look at Figure 14.1*b*, you will see that observations numbered 12 through 20 are nine successive values on the same side of the target value and activate rule 4, suggesting that the process was out of control in this interval. It must be understood that making use of four rules changes the probabilities of the testing system and will increase the possibility of wrongly asserting lack of statistical control.

A Forecasting Dilemma

The output of a process that is regularly monitored forms a time series. To better* understand process control systems you need to consider some elementary aspects of time series analysis and in particular the forecasting of such a series. For a start, consider a problem of sales forecasting.

Sam Sales is the sales manager and Pete Production the production manager for Grommets Incorporated. One day Sam bursts into Pete's office and says, "Guess what, Pete, we sold 50,000 more boxes of grommets last week than we forecast! Let's put up next week's forecast by 50,000." But Pete is more cautious. He knows that, on the one hand, if he produces too much he will be burdened with a big inventory but, on the other hand, if he produces too little he may not be able to meet higher sales demands should they develop. So Pete says, "Sam, we both know from experience that sales fluctuate up and down and are greatly influenced by chance occurrences. I always discount 60% of the difference between what we forecast and what we actually get. So in this case, let's discount 60% of the 50,000 extra boxes and increase the forecast by only 20,000 rather than 50,000."

At first sight Pete's forecasting rule seems too simple to be of much interest, but in fact it is more sophisticated than it looks. Suppose that last week (the

*A detailed account of what follows is given in B&L, pp. 105–154.

t−1th week say) you forecasted sales for week t as $\hat{y}_t$ (the caret, the "hat," on y indicates it is an estimate.) Now when the tth observation y_t actually occurred, it would differ somewhat from your forecast $\hat{y}_t$. Then, according to Pete's rule, to get the next forecast $\hat{y}_{t+1}$, you should discount 60% of the observed discrepancy $y_t - \hat{y}_t$ and adjust using only 40% of this discrepancy. Thus the amount $\hat{y}_{t+1} - \hat{y}_t$ by which you should adjust the forecast for next week would be

$$\hat{y}_{t+1} - \hat{y}_t = 0.4(y_t - \hat{y}_t)$$

which simplifies to

$$\hat{y}_{t+1} = 0.4y_t + 0.6\hat{y}_t \tag{14.1}$$

You will see that the new forecast is a *linear interpolation* between the new data value y_t and the old forecast $\hat{y}_t$.

Pete Had Rediscovered Exponential Forecasting!

Now if you had consistently used Pete's rule, in the previous week $t-1$ you would have made the forecast $\hat{y}_t = 0.4y_{t-1} + 0.6\hat{y}_{t-1}$ substituting this in the above equation you find

$$\hat{y}_{t+1} = 0.4y_t + 0.6(0.4y_{t-1} + 0.6\hat{y}_{t-1})$$

And in the week before that you would have made the forecast $\hat{y}_{t-1} = 0.4y_{t-2} + 0.6\hat{y}_{t-2}$, which on substitution would have given

$$\hat{y}_{t+1} = 0.4y_t + 0.6[0.4y_{t-1} + 0.6(0.4y_{t-2} + 0.6\hat{y}_{t-2})]$$

If you carry on this way and multiply out all the braces, you will eventually find that Pete's forecast for next week, $\hat{y}_{t+1}$, for the week $t+1$ is equivalent to calculating from the previous *observations* the quantity

$$\hat{y}_{t+1} = 0.4(y_t + 0.6y_{t-1} + 0.6^2y_{t-2} + 0.6^3y_{t-3} + \cdots)$$

that is,

$$\hat{y}_{t+1} = 0.400y_t + 0.240y_{t-1} + 0.144y_{t-2} + 0.0.086y_{t-3} + 0.052y_{t-4} + 0.031y_{t-5} + \cdots \tag{14.2}$$

14.2. THE EXPONENTIALLY WEIGHTED MOVING AVERAGE

Pete's rule is equivalent to representing the forecast $\hat{y}_{t+1}$ of y_{t+1} by a weighted average of past data with each of the weighting coefficients 0.400, 0.240, 0.144, ... being 0.6 times the previous one. (See Fig. 14.2.) The quantity 0.6 is therefore called the *smoothing constant* or *discount factor*, and Pete's rule is identical to

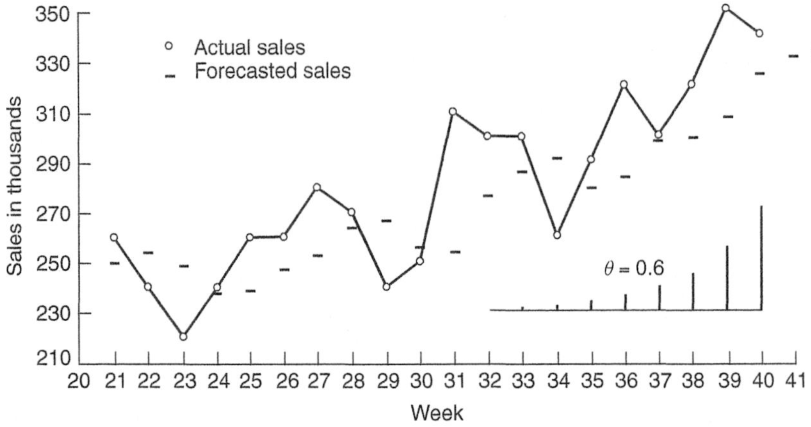

Figure 14.2. Grommet sales in thousands of boxes with forecast sales.

taking what is called an *exponentially weighted moving average* (EWMA) of past data with a discount factor of 0.6.

What is meant by "exponentially" weighted? Suppose you took, say, the ordinary *arithmetic* average of the last 10 observations as the forecast of the next observation. Then you would get

$$\hat{y}_{t+1} = \overline{y} = \frac{\sum_{u=-9}^{0} y_{t-u}}{10} = 0.1y_t + 0.1y_{t-1} + 0.1y_{t-2} + \cdots + 0.1y_{t-9}$$

In this average the weights are all equal to $\frac{1}{10}$ and sum to 1.0. But, obviously, to forecast what is happening at time $t+1$, observation y_{t-9} does not tell you as much as y_t. So it makes sense to systematically down-weight previous values as is done in the EWMA.

In general (see Equation 14.1) you can write Pete's rule as

$$\hat{y}_{t+1} = \lambda y_t + \theta \hat{y}_t \qquad \text{with} \qquad \lambda = 1 - \theta \tag{14.3}$$

where θ is the proportion to be discounted (the smoothing factor), in Pete's case 0.6, and $\lambda = 0.4$ is the proportion to be retained.* Since the new forecast is an interpolation between the forecast $\hat{y}_t$ and the new observed value y_t, the smaller is θ (the larger is λ), the more emphasis is placed on the recent data, as is illustrated in Figure 14.3. In the limit where $\theta = 0$ ($\lambda = 1.0$) only the last plotted observation is used to supply information, as with the Shewhart chart. At

*It is customary to use the symbol λ for the important quantity $1 - \theta$. We have used λ earlier in discussing power transformations, but we do not think that this duplication in notation should be troublesome.

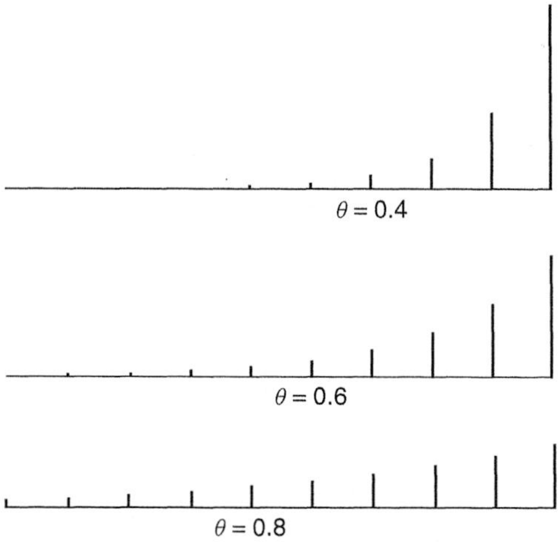

Figure 14.3. Weight factors for $\theta = 0.4, 0.6, 0.8$.

the other extreme, when θ gets close to 1.0 (λ close to zero), $\hat{y}_{t+1}$ approximates the arithmetic average of a long series of past data.

For illustration, the grommet sales y_t in thousands of boxes for a period of 20 weeks (from week 21 through week 40) are shown in Table 14.1 and plotted in Figure 14.2. Suppose you apply Pete's forecasting rule with $\theta = 0.6$ to these data. If this were all the data available, you could get the forecasting process started by using the first available value y_{21} as if it was the forecast $\hat{y}_{22}$ of y_{22} and proceeding from there. Small differences in the starting value will be quickly discounted and will have little effect on later forecasts. If in this way you apply Pete's rule successively to the whole series with $\theta = 0.6$ and $\lambda = 0.4$, you get, using Equation 14.3,

$$\hat{y}_{23} = 0.4(242) + 0.6(260) = 252.8 = 253,$$

$$\hat{y}_{24} = 0.4(225) + 0.6(253) = 241.8 = 242$$

and so on, leading eventually to the forecast $\hat{y}_{41} = 334$, as shown in Table 14.1 and Figure 14.2. Or equivalently, using the EWMA directly

$$\hat{y}_{t+1} = \lambda y_t + \theta \hat{y}_t = \lambda y_t + \lambda\theta y_{t-1} + \lambda\theta^2 y_{t-2} + \lambda\theta^3 y_{t-3} + \cdots \tag{14.4}$$

and you would have obtained the same result.

$$\hat{y}_{41} = 0.4(342) + 0.24(353) + 0.144(327) + 0.0864(305) + 0.05184(324)) + \cdots = 334$$

Clearly it is much easier to calculate the EWMA *successively* using Equation 14.3. The EWMA was first used by Holt (1957) and others in the 1950s, particularly for inventory control.

Table 14.1. Sales Forecasts and Forecast Errors with $\theta = 0.6$

Week, t	Observed, y_t	Forecast, $\hat{y}_t$	Error, $y_t - \hat{y}_t$
21	260		
22	242	260	−18
23	225	253	−28
24	241	242	−1
25	269	241	28
26	264	252	12
27	287	257	30
28	272	269	3
29	245	270	−25
30	250	260	−10
31	313	256	57
32	306	279	27
33	301	290	11
34	268	294	−26
35	299	284	15
36	324	290	34
37	305	304	1
38	327	304	23
39	353	313	40
40	342	329	13
41		334	

Sum of squared forecast errors: 12,126

What Value of θ Should You Choose?

Figure 14.3 shows the weight functions that would be obtained for $\theta = 0.4, 0.6, 0.8$. Choosing θ is evidently a question of finding a compromise between, on the one hand, achieving a fast reaction to change and, on the other hand, averaging out noise. The best values of θ will be different for different series, although the choice is not as critical as might be thought. If you look again at Table 14.1, you can see how θ can be estimated from the data. The table shows the original series y_t, the forecasts made one step previously $\hat{y}_t$, and the forecast errors $e_t = y_t - \hat{y}_t$. Also shown is the sum of squares of the errors of the forecast errors $S_\theta = S_{0.6} = 12,129$. You obtain the least squares estimate of θ, therefore, by computing S_θ from the same data for, say, $\theta = 0.0, 0.1, 0.2, \ldots, 0.9, 1.0$ and then plotting S_θ versus θ. You can then draw a smooth curve through the points and pick off a value that gives the smallest S_θ. However, you will not get a very precise estimate unless you have a fair amount of data. Thus the plot shown in Figure 14.4 is for *one hundred* values of the grommet series. (Only 20 values of this series were used in the forecasting example.) You will find that typically the sum of squares curve has a very high value when θ is close to 1 and descends very rapidly as

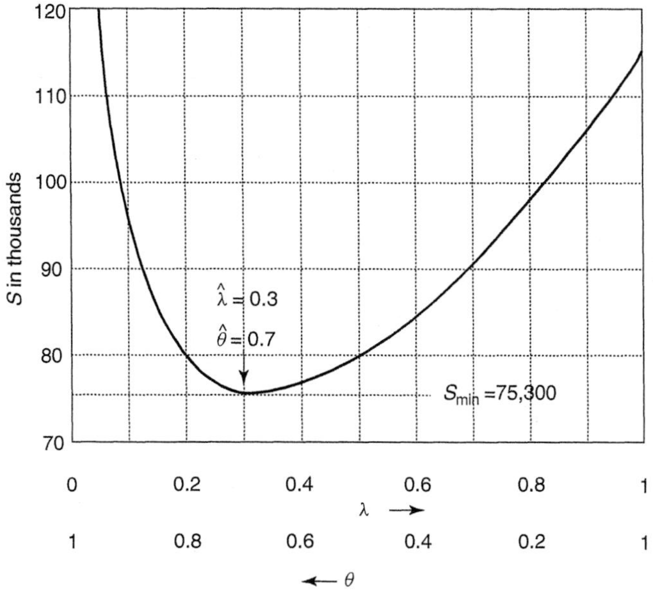

Figure 14.4. Plot of S_θ versus θ.

θ is decreased, becoming quite flat for the middle values of θ. This means that for values of θ in this "flat" region the EWMA method for forecasting is very "robust" to moderate changes in the smoothing constant. For many examples a value of $\theta = 0.6$ ($\lambda = 0.4$) seems to work quite well. In particular, you will see that Pete was quite close to guessing the right proportion to discount. The minimum value of the sum of squares in Figure 14.4 is at about $\theta = 0.7$ ($\lambda = 0.3$), where $S_{\min} = 75{,}300$, but as you see from the graph any value of θ as large as 0.8 or as small as 0.5 would have produced only a 7% increase in S_θ and so an increase of less than 3% in the forecast standard error. Notice also that from such a calculation you can obtain an estimate ($\hat{\sigma}_e = \sqrt{75{,}300/99} = 27.5$) of the standard deviation of σ_e of the optimal forecast errors.

Checking out a Forecasting System

Many different systems of forecasting have been used. Here is how you can assess the worth of any particular method whether based on data or on expert opinion. Suppose you had past data from some forecasting procedure XYZ—a record of the actual data and of the forecasts that had been made. From these data you could compute the series of one-step-ahead forecast errors. If now this error series turned out itself to be autocorrelated, this would mean it was to some extent forecastable. But any system for which the one-step-ahead errors themselves can be forecast can be improved upon. Thus in particular, if the one step ahead forecast errors appear to be white noise you can be happier with the forecasting method.

Use of the EWMA for Process Monitoring

In 1959 Roberts suggested that the EWMA be used for process monitoring. An appropriate chart is shown in Figure 14.1*c*. There are several things to say in favor of the EWMA:

1. It represents a weighting of recent past data that accords with common sense.
2. It is robust. Although there are more sensitive criteria for specific needs, the EWMA responds well to a number of common upsets and in particular to systematic trends and to changes in level.
3. It is easy to compute.

On NIID assumptions, for a process in a perfect state of control, the standard error of the EWMA to a good approximation is*

$$\sigma_{\text{EWMA}} = \sigma_y \sqrt{\frac{\lambda}{2-\lambda}} \tag{14.5}$$

For example, in the particular case when $\theta = 0.6$ and $\lambda = 0.4$, $\sigma_{\text{EWMA}} = \sigma_y \times 0.5$. So for this value of λ the action lines for $\hat{y}_t$ are half the distance from the target value as the corresponding Shewhart chart lines for y_t, as illustrated in Figure 14.1*c*.

Employing the same data as was used for the Shewhart chart in Figure 14.1*b*, an EWMA monitoring chart with 3σ limits for $\lambda = 0.4$ appears in Figure 14.1*c*. In this example you will see that EWMA chart is very sensitive to a shift in mean of modest size and signals a lack of control for observations 16, 17, and 18 which is in the same region as that signaled by the Western Electric rule 4.

14.3. THE CUSUM CHART

Another device (Page, 1954; Barnard, 1959) that has been successfully used for process monitoring, is the Cumulative sum chart. In its simplest form the "CuSum" Q where

$$Q = \sum_{i=1}^{t} (y_i - T) = (y_1 - T) + (y_2 - T) + (y_3 - T) + \cdots + (y_t - T)$$

is plotted against the time t. A change of mean in the series produces a change in *slope* of the CuSum. Details of extensive industrial application of these charts (particularly the textile industry) will be found for example in Kemp (1962) and Lucas (1982) and Ryan (1989).

Another valuable application of the CuSum chart is as a *diagnostic tool* to detect when slight changes in a process mean have occurred. Thus an example is given by Box and Luceño (1997) of a particular manufacture where it was

*More precisely, the standard error of the EMWA is $\sigma_{\text{EWMA}} = \sigma_y[\sqrt{1-(1-\lambda)^{2t}}]\sqrt{\lambda/(2-\lambda)}$. This quantity quickly approaches Equation 14.5 for practical values of λ and $t > 3$.

essential to maintain gas flow rate at a constant level. To achieve this very accurate flowmeters were used. However problems that were occurring suggested that something was amiss. A CuSum plotted over a period of eight months suggested that five small changes in flow rate, otherwise undetectable had occurred. The specific dates of the changes coincided almost exactly with times when the flowmeters were recalibrated. Modification of calibration procedure removed the problem.

How Should You Monitor Your Process?

In this chapter three different monitoring devices have been mentioned: the Shewhart chart, the EWMA chart, and the CuSum chart. Which you should use depends on what you are looking for—what you are afraid may happen.

The theory goes as follows: It is supposed that initially the process is in a state of statistical control, that is, a stationary state in which the deviations from target have mean zero and are normally, identically, and independently distributed—they are NIID. At some unknown point in time the process may depart from this state of control. You want to know with least possible delay when this change occurred and its possible nature.

Fisher's (1925) fundamental concept of efficient score statistics may be used to determine which function of the data (called in general the CuScore) is best to detect any specific type of deviation from the in-control, stationary state.

Here are two conclusions that follow from this approach (Box and Ramirez 1992).

If the feared deviation is a single spike, then $y_t - T$, the current deviation from the target, is the appropriate statistic and should be plotted and assessed with a Shewhart chart.

If the feared deviation is a change in level, then the CuSum, $\sum_{i=1}^{t}(y_i - T)$, is the appropriate statistic that should be plotted and assessed in the manner described in the above references.

Another possibility is for the process to change from a stationary to a nonstationary state. The CuScore criterion you come up with to detect this, will of course depend on how you define the non-stationary state.

An Important Nonstationary Model

For a process in a perfect* state of control the difference between data values one interval apart has the same variance as the difference between data at any other number of intervals apart. That is, $V(y_{t+1} - y_t) = V(y_{t+m} - y_t)$. Lack of control is indicated when the variance of the difference of data values m steps apart increases as m increases. If this rate of increase is approximately linear, it

*As we have said, a perfect state of control is a purely theoretical concept. In practice, as always, we deal with approximations.

was shown by Box and Kramer (1993) that the nonstationary model that must now describe the series is

$$y_t = y_{t-1} + a_{t-\theta} - a_{t-1} \tag{14.6}$$

This is called an integrated moving average (IMA) model. In this equation the a_t's are a white noise series (i.e., NIID with mean zero).

The transition from the stationary in-control state to the nonstationary state thus corresponds in Equation 14.6 to a change of the parameter θ from 1 to some value less than 1. It turns out that if e_t is the deviation $y_t - T$ from target and $\tilde{e}_t$ is an EWMA of post e's and so is a forecast from origin t−1 then the CuScore statistic is

$$\sum \tilde{e}_t e_t$$

The criteria asks at what point do the e_t's (which should be unforecastable white noise) becomes forecastable.

It is shown in Box and Ramirez (1992) how efficient CuScores can be obtained for other specific kinds of deviation from the state of control. Additional examples discussed are for the detection of process cycling and for a change in slope (e.g., a change in the rate of wear of a machine tool).

14.4. PROCESS ADJUSTMENT

Process monitoring detects unusual events and attempts to find their assignable causes. Continuous debugging of this kind can, over a period of time, greatly improve process operation. But not all disturbing influences can be easily found or eliminated. Diligent search may fail to discover the assignable cause for unexplained persistent deviations or drifts from target. For example, the characteristics of a feedstock such as crude oil, coal, or timber can have considerable natural variation, which can induce considerable variation and drifts in a product's characteristics. Thus, a cause may be assignable but cannot be removed. In such cases process *adjustment* is appropriate.

A Repeated Adjustment Control Scheme

For illustration, Table 14.2 lists 100 observations taken from an uncontrolled process in which the thickness of an extremely thin metallic deposit was measured at equally spaced intervals of time. Figure 14.5*a* shows a run plot of these uncontrolled data. It was desired to maintain the thickness of the deposit as close as possible to the target value $T = 80$. The process was very intricate and all attempts to stabilize it by monitoring methods had failed.

The thickness could be changed by adjusting the deposition rate, whose level at time t is denoted as X_t. For clarity, a system for *manual* adjustment is discussed here. Once understood, such systems can be computerized and automated. Control

Table 14.2. Metallic Film Series, $n = 100$, Target $T = 80$, Data Listed in Rows

80	92	100	61	93	85	92	86	77	82	85	102	93	90	94
75	75	72	76	75	93	94	83	82	82	71	82	78	71	81
88	80	88	85	76	75	88	86	89	85	89	100	100	106	92
117	100	100	106	109	91	112	127	96	127	96	90	107	103	104
97	108	127	110	90	121	109	120	109	134	108	117	137	123	108
128	110	114	101	100	115	124	120	122	123	130	109	111	98	116
109	113	97	127	114	111	130	92	115	120					

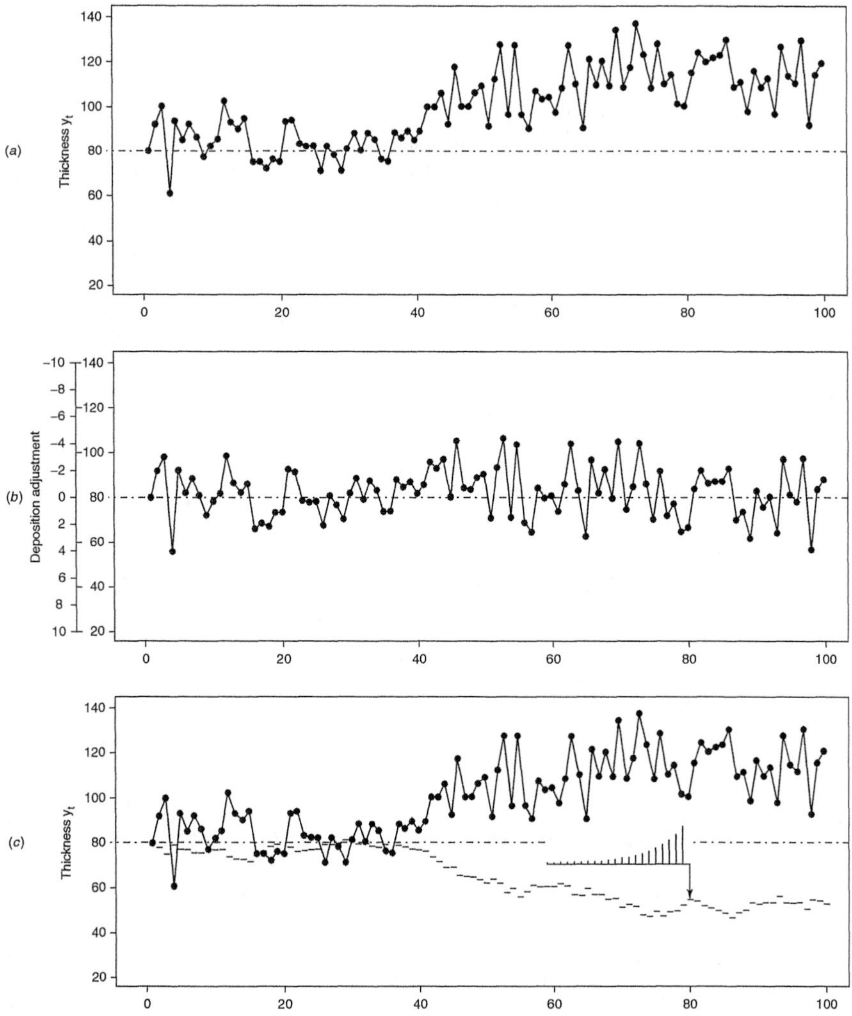

Figure 14.5. (*a*) The original uncontrolled process. (*b*) The controlled process after repeated adjustment. (*c*) How total adjustment is equivalent to using an EWMA with $\lambda = 0.2$ to predict the next value of the deviation from target and then adjusting to cancel it.

of the process could be obtained by repeated adjustment using the chart shown in Figure 14.5*b*. The operator plots the latest value of thickness on the chart and then reads off on the deposition adjustment scale the adjustment $X_t - X_{t-1}$ to be made in the deposition rate.

Figure 14.5*b* and Table 14.3 show the observations as they would be recorded after adjustment if such a scheme had been used. For example, the first recorded thickness, 80, is on target so no action is called for. The second value of 92 is 12 units above the target, calling for a reduction of −2 in the deposition rate, and so on. With the adjustment process engaged, the successive thickness values that would occur are displayed in Figure 14.5*b*.

It will be seen that the control equation being used here is

$$X_t - X_{t-1} = -\tfrac{1}{6}(y_t - T), \qquad T = 80 \tag{14.7}$$

Evidently repeated application of the adjustment determined by this equation can bring the process to a much more satisfactory state. The values seen by the operator, of course, are those of Table 14.3 plotted in Figure 14.5*b*. Two pieces of information were used in constructing the chart in Figure 14.5*b*:

1. The change in the deposition rate $X_t - X_{t-1}$ produced all of its effect on thickness within one time interval.
2. The "gain" in thickness caused by a one-unit change in X equals $g = 1.2$ units of thickness y.

A "*full*" adjustment would therefore have been obtained by setting

$$X_t - X_{t-1} = -\frac{1}{g}(y_t - T) = -\frac{1}{1.2}(y_t - 80) = -0.83\ (y_t - 80) \tag{14.8}$$

Endeavoring to compensate for the full deviation $y_t - 80$ in this way would correspond to what Deming (1986) called "tinkering" and, as he said, would increase the variance.

However in the *control equation* actually employed only a proportion λ of the full deviation would be compensated, so that

$$X_t - X_{t-1} = -\frac{\lambda}{g}(y_t - T) \tag{14.9}$$

Table 14.3. Metallic Film Series: Thickness Observed *After Repeated Adjustment*

80	92	98	56	92	82	79	83	73	80	83	99	87	83	86
66	69	67	74	74	93	92	79	78	78	68	81	77	70	82
88	79	87	83	74	74	88	84	87	82	85	95	93	97	80
105	84	83	89	90	70	93	106	71	103	68	64	84	79	80
73	85	104	83	62	96	81	92	79	104	74	84	104	86	70
91	71	77	64	66	83	92	86	87	87	93	69	73	61	82
75	80	64	97	81	78	97	56	83	88					

In this application λ was chosen to be $\lambda = 0.2$ so you would make a change of only one-fifth of the full adjustment. You thus obtain Equation 14.7, the actual control equation that would be used.

$$X_t - X_{t-1} = -\frac{0.2}{1.2}(y_t - 80) = -\frac{1}{6}(y_t - 80)$$

What was done exactly corresponds to Pete's rule to adjust for only a proportion (in this case 20%) of the full adjustment. As shown in Figure 14.5*c*, this form of control would be precisely equivalent to using an EWMA with $\lambda = 0.2$ to predict the next value $\hat{y}_{t+1} = \lambda y_t + \theta \hat{y}_t$ of the uncontrolled series and then adjusting X_t to cancel it. Notice that the plot labeled "Total effect of compensation" is the *total* compensation applied up to that time obtained by adding together all the individual repeated adjustments.

Nature of the Adjustment Process

Summing Equation 14.9 gives

$$X_t = X_0 - \frac{\lambda}{g}\sum_{i=1}^{t}(y_i - T) = X_0 - \frac{\lambda}{g}\sum_{i=1}^{t} e_i \tag{14.10}$$

Thus this adjustment is the discrete analogue of the control engineer's integral control action in which the integral is here replaced by a sum.

Exercise 14.1. Use the data of Table 14.2 to construct a manual adjustment chart using $\lambda = 0.4$.

Estimating λ from Past Data

This form of feedback control is insensitive (robust) to the precise choice of λ. Thus while in this example a good adjustment scheme was obtained using the value of $\lambda = 0.2$, it would have made little difference if you had used, say, $\lambda = 0.3$. However, it would have made a great deal of difference had you chosen $\lambda = 1.0$, thus tinkering with the process in an attempt to negate 100% of the observed deviation from target. Usually a value of λ between about 0.2 and 0.5 will give good control.

You can, if you wish, estimate λ from a past record of process deviations from target, as was done for forecasting in Figure 14.4. However, available records usually include periods in which adjustments of one kind or another *have* already been made. If adequate records of such past adjustments are available, the uncontrolled series can be approximately reconstructed by making appropriate allowance for these adjustments. If you knew, for example, that at some time $t = 53$ (say) an adjustment had been made that reduced the measured response by (say)1.5 units, you should adjust all subsequent observations in the series by adding this amount.

Reducing the Size of Needed Adjustments

Suppose the value of λ was exactly known. Then by using this value the output variance would be minimized. But there are situations where to achieve this the adjustment variance $V(X_t - X_{t-1})$ would be unsatisfactorily large. It may be shown that replacing λ by $\alpha\lambda$ where α was some positive fraction less than 1 could reduce the adjustment variance without causing much increase in the output variance. In practice substitution of an "ad hoc" value for λ between 0.2 and 0.4 usually gives good control.

Bounded Adjustment

Adjustment made after every observation is appropriate when the cost of adjustment is essentially zero. This happens, for example, when making a change consists simply of turning a valve or resetting a thermostat. In situations where process adjustment requires stopping a machine or taking some other action entailing cost, a scheme called *bounded adjustment* is used that calls for less *frequent* adjustment in exchange for a modest increase in process variation.

This employs an exponentially weighted moving average of past data to estimate the current deviation from target and calls for adjustment only when this estimate exceeds a specific limit. Thus a bounded adjustment chart is superficially like Robert's EWMA monitoring chart discussed earlier. In both cases an exponentially weighted moving average is plotted between two parallel limit lines. But the purpose and design of the bounded adjustment chart are very different from those of a monitoring chart. The limit lines for the monitoring chart are positioned to discover statistically significant events. For the bounded adjustment chart their positioning is such as will achieve the largest possible reduction in frequency of adjustment for an acceptably small increase in the standard deviation of the adjusted process. To do this, the limit lines for a bounded adjustment chart are typically much closer together than those for a monitoring chart.

For example, using the data from Table 14.2, the open circles in Figure 14.6 are for the process output after adjustment and the closed circles are forecasts $\hat{y}_t$ made one period previously using an EWMA with $\lambda = 0.2$. These forecasts are updated by the formula $\hat{y}_{t+1} = 0.2y_t + 0.8\hat{y}_t$ or, equivalently, $\hat{y}_{t+1} = \hat{y}_t + 0.2(y_t - \hat{y}_t)$. The chart in Figure 14.6(a) has boundary lines at $T \pm 8$ with the target value $T = 80$. When at time t the EWMA $\hat{y}_t$ falls below the lower limit 72 or above the upper limit 88, an adjustment $\hat{y}_t - 80$ is made to bring the forecast value on target. In this example you will see that the deposition rate, denoted by open circles, is held constant until time $t = 13$ when the *forecast* value $\hat{y}_{14} = 88.7$ falls outside the upper limit. Thus, at observation 13 the chart signals that a change is the deposition rate is required such as will change the thickness by $-(88.7 - 80.0) = -8.7$ units. The deposition rate remains unchanged until the next adjustment is needed. *After adjustment* the forecast becomes the target value, that is, $\hat{y}_{14} = 80$. The next observation is $y_{14} = 81.3$, and the next forecast error $y_t - \hat{y}_t = 81.3 - 80 = 1.3$. Thus adjustment does not upset the next EWMA calculation. After observation 20 the forecast value $\hat{y}_{20} = 71.1$ falls below the lower limit line, requiring a change in the deposition rate that will produce

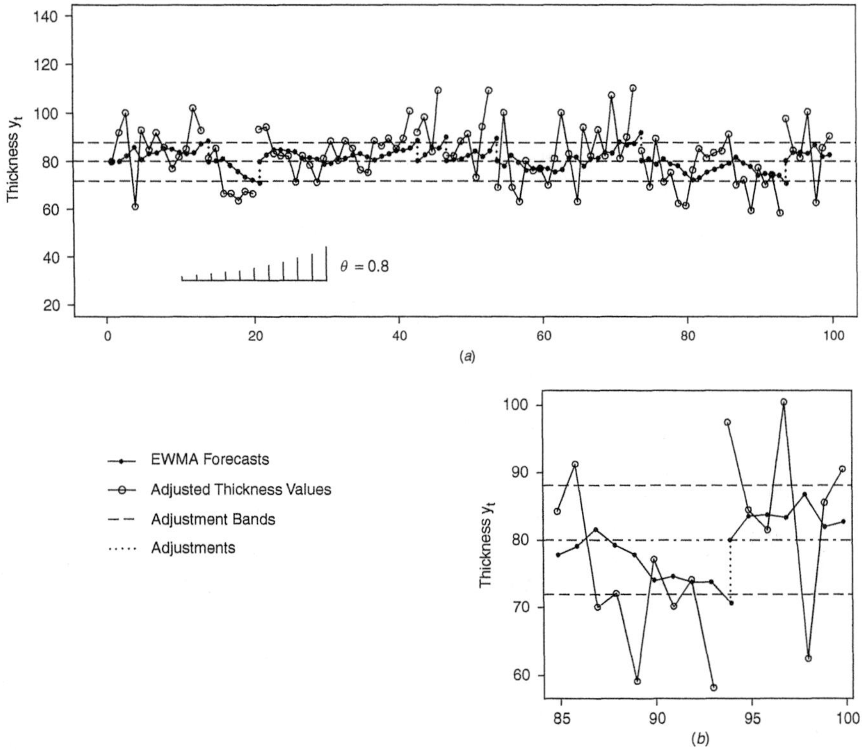

Figure 14.6. Bounded adjustment chart for metallic thickness data. (*a*) Open circles are for adjusted process output as in Figure 14.7*b*; Filled circles are EWMA's indicating where to change. (*b*) Blow-up of last 15 values of chart.

an increase of 8.9 in thickness. The inset in Figure 14.6(b) shows a magnified version of the adjustment process for the last 15 observations. Proceeding in this way for the whole series of 100 values, you find that the total number of needed changes is only 7. The empirical *average adjustment interval* (AAI) is thus $99/7 = 14$.

Figure 14.7*a* displays the 100 unadjusted thickness observations given previously in Table 14.2 and displayed earlier in Figure 14.5*a*. Figure 14.7*b* shows the output of the adjusted process as it would appear after bounded adjustment. Figure 14.7*c* illustrates how *bounded* adjustment produces a series of compensatory step changes. Each of these is initiated when the change of thickness from the target value of 80 estimated by the EWMA exceeds ± 8. Notice that the standard deviation of the controlled series in Figure 14.7*b* is only slightly larger than that of the series in Figure 14.5*b* obtained when adjustment was made after every observation. Limits such as 80 ± 8 will in general be denoted by $T \pm L$. One way to get a satisfactory value for L is to use some reconstructed past data (past data in which you have made allowance of known changes) and make "dry runs" for a few schemes with different values of L. Then choose one that in your particular circumstance balances the advantage of less frequent changes against

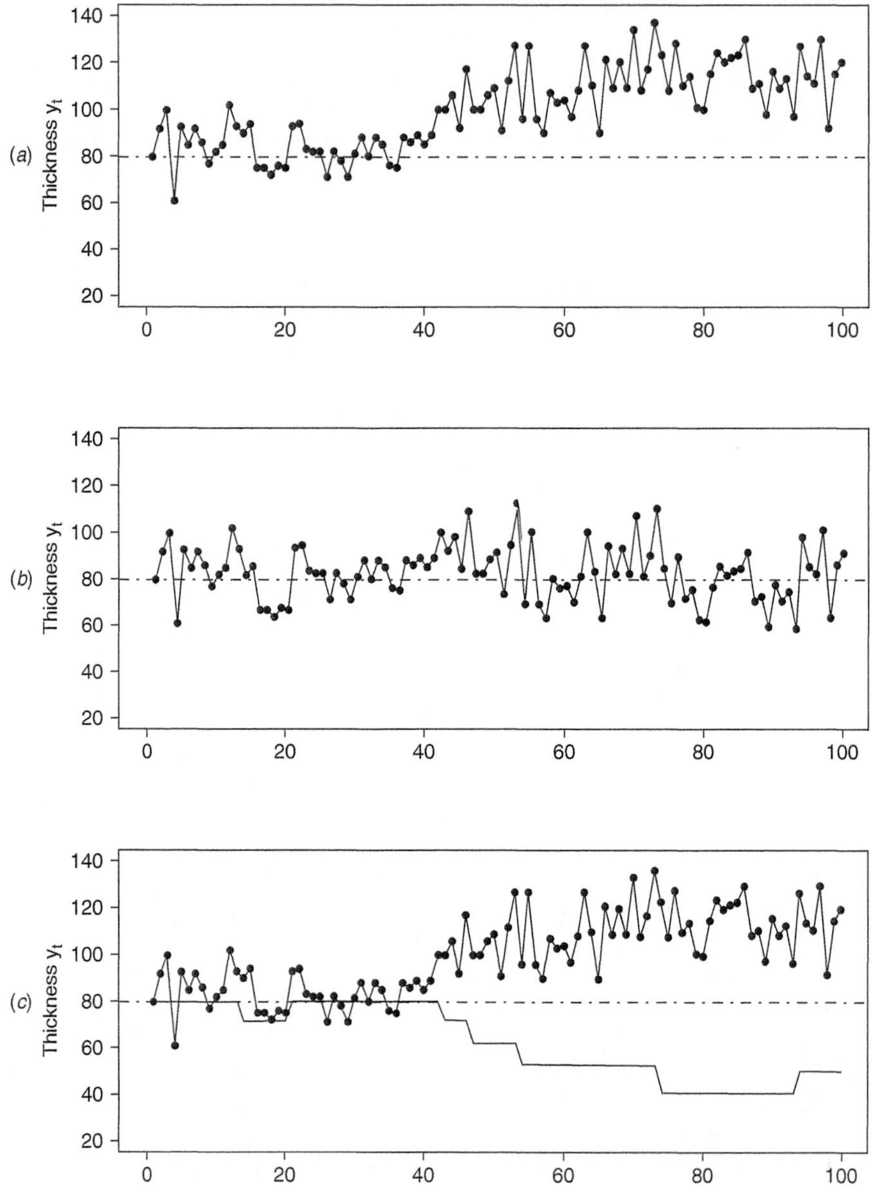

Figure 14.7. (*a*) Original uncontrolled process, (*b*) controlled process after bounded adjustment, and (*c*) compensatory step function adjustments.

the necessity of a somewhat larger standard deviation. Alternatively you can obtain an estimate of σ_e from a graph like Figure 14.4 and then use Table 14.4 to obtain L. Notice that σ_e is also the standard deviation for the output from an optimal "repeated adjustment" scheme.

Table 14.4. Theoretical Values of AAI and Percent Increase in Standard Deviation for Various Values of L/σ_e and λ

λ	L/σ_e	AAI	Percent Increase in Standard Deviation
0.1	0.5	32	2.4
	1.0	112	9.0
	1.5	243	18.0
	2.0	423	30.0
0.2	0.5	10	2.6
	1.0	32	9.0
	1.5	66	20.0
	2.0	112	32.0
0.3	0.5	5	2.6
	1.0	16	10.0
	1.5	32	20.0
	2.0	52	33.0
0.4	0.5	4	2.6
	1.0	10	10.0
	1.5	19	21.0
	2.0	32	34.0
0.5	0.5	3	2.5
	1.0	7	10.0
	1.5	13	21.0
	2.0	21	35.0

Table 14.4 shows the theoretical AAI for various values of L/σ_e and λ with the percent increases of the standard deviation. For example, suppose $\lambda = 0.3$ and that you were prepared to allow, in exchange for less frequent adjustment, an increase of 10% in the standard deviation. You will see from the table that you should set $L/\sigma_e = 1$ and that this would yield an AAI of about 16. Further tables and a fuller discussion of this mode of control can be found elsewhere. In particular in Box and Luceño (1997) which scheme you pick will depend on how difficult or expensive is the making of adjustments and how much leeway you have for slightly increasing the standard deviation.

Application of Process Adjustment in the Six-Sigma Initiative

The stated goal of the Six-Sigma initiative is to produce no more than 3.4 defects per million of individual items being sent to the customer.* On the assumption

* The Six-Sigma initiative, in addition to calling attention to basic statistical tools, emphasizes team work, organized problem solving, and quality costs. This discussion is concerned only with the statistical aspects of Six-Sigma.

that the outgoing items have a distribution that is normal and a fixed mean on target, this would require a specification/sigma ratio equal to 9.0. This is much less than the ratio of 12.0 implied by the term Six Sigma. However, the authors of this concept wisely assume that the process does not have a fixed mean but *undergoes drift*—specifically that the local mean may drift on either side of the target by about 1.5 standard deviations. Conventional wisdom would say that such a process drift must occur due to special causes and that these ought to be traced down and eliminated. But in practice such drift may be impossible to assign or to eliminate economically. It is refreshing to see the acknowledgment that *even when the best efforts are made* using the standard quality control methods of process monitoring the process mean can be expected to drift.

Using the EWMA To Adjust For Drift

A drifting process was simulated as follows. A series of 200 random normal deviates $a_t, a_{t-1}, a_{t-2}, \ldots$ were generated with mean zero and standard deviation $\sigma_a = 1$. The dotted lines in Figure 14.8 are for $\pm 3\sigma_a = 3$. A drift component was now added to produce constructed "data" y and the run plot in Figure 14.8*a* with

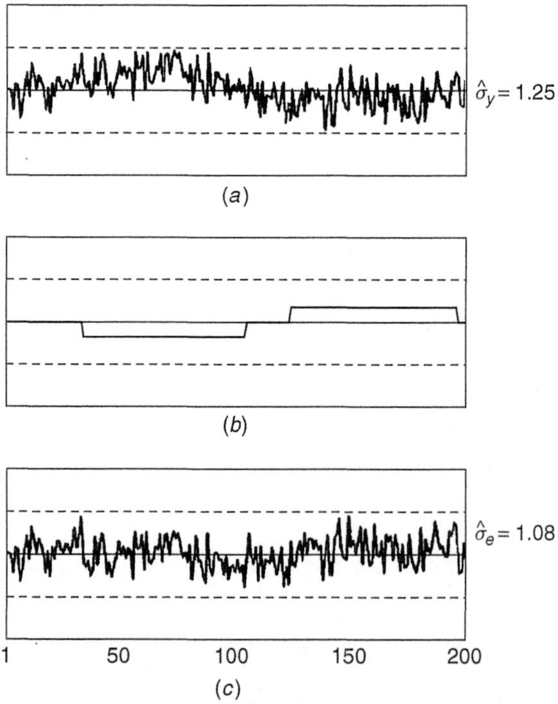

Figure 14.8. Performance of bounded adjustment scheme: (*a*) process subject to drift; (*b*) adjustments by bounded adjustment scheme; (*c*) controlled process output.

$\hat{\sigma}_y = 1.25$.* For a bounded adjustment scheme with $\lambda = 0.2$ Table 14.4 shows alternative possibilities for various choices of the $\pm L$ limits. The control scheme in Figure 14.8*b* is for $L = \pm 1$ with the table adjustment interval (AAI) of 32 and an output standard deviation of about $1.09\sigma_a$. This is in agreement with what you can see in Figure 14.8*b* obtained by actual calculation, where four adjustments are called for during this series of 200 observations. The empirical adjustment interval is thus 50. The empirical standard deviation of the data after control shown in Figure 14.8*c* is $\hat{\sigma}_e = 1.08\sigma_a$. This series is thus stabilized with only four needed adjustments.

14.5. A BRIEF LOOK AT SOME TIME SERIES MODELS AND APPLICATIONS

Modeling with White Noise

So far some practical problems in forecasting and adjustment control have been considered, but as is often the case, practice and theory have mutually supportive roles. It is rewarding to ask, therefore: What do these methods and in particular the central role of the EWMA tell you about models more generally and what further useful ideas and methods do these models suggest? Here only an elementary discussion can be attempted, but it is hoped that you may follow up on this brief sketch.

Reasons for the importance of the EWMA are as follows:

1. Exponential discounting of previous data agrees with common sense.
2. The EWMA is robust in the sense that for many series where it does not give an optimum forecast it can frequently supply a surprisingly good one.
3. Control with the EWMA produces the discrete analogue of the control engineer's "integral" action, which has found great practical application.
4. Pete's rationale of *making* an adjustment proportional to the observed discrepancy (closely related to Newton's law of cooling) is sensible.

This does not mean, of course, that all forecasting should be done with the EWMA. It does seem to imply, however, that the exponentially weighted moving average must take a central place in representing the level of an uncontrolled series, at a particular point in time just as the arithmetic average takes a central role in representing the location of the data for independent observations.

A sequence of IID random variables $a_t, a_{t-1}, a_{t-2}, \ldots$ each having mean zero is called a "white noise" time series. Sometimes it is also called a sequence of "random shocks." A characteristic of such a white noise series is that it is unforecastable; more precisely, given the mean η and standard deviation σ, it is impossible to predict anything about any value of the series by knowing other

* The mode of generation of this drift is discussed in Box and Luceño (2000). The amount of drift shown is considerably *less* than that which is allowed for in six-sigma specifications.

values in the same series. For example, knowing a_{t-3}, a_{t-2}, and a_{t-1} tells you nothing about the next value a_t. One very important way of representing serially dependent errors e_t, e_{t-1}, e_{t-2} is by relating them to a white noise series a_t, a_{t-1}, a_{t-2}. The simplest of such models expresses e_t as a function of the current and past shock so that

$$e_t = a_t - \theta a_{t-1}, \qquad -1 < \theta < 1 \tag{14.11}$$

The elements $e_t, e_{t-1}, \ldots$ will be *autocorrelated* (correlated with each other) because, for example, the model for e_{t-1} is

$$e_{t-1} = a_{t-1} - \theta a_{t-2} \tag{14.12}$$

so that Equations 14.11 and 14.12 both contain a_{t-1} and thus each value of the series $e_t, e_{t-1}, \ldots$ is correlated with its immediate neighbor. However, since

$$e_{t-2} = a_{t-2} - \theta a_{t-3} \tag{14.13}$$

e_t and e_{t-2} will be uncorrelated since Equations 14.11 and 14.13 have no terms in common. Thus the series has what is called a "lag 1" autocorrelation and has no correlation at other lags. The model of Equation 14.11 is called a first-order moving average, a MA(1) model.

If e_t were made to depend on a_t, a_{t-1}, and a_{t-2}, you would have a second-order moving average model having autocorrelations at lag 1, lag 2, and so on.

A different way of modeling dependence is by an autoregressive model. The first-order autoregressive model is $e_t = \phi e_{t-1} + a_t$, where $-1 < \phi < 1$. If e_t is made to depend on k previous values $e_{t-1}, e_{t-2}, \ldots, e_{t-k}$, you have a kth-order autoregressive series. In mixed autoregressive moving average models e_t depends both on previous e_t's and on a_t and previous a_t. In this elementary introduction to time series models it will be enough to consider and elaborate on this MA(1) model — the first order moving average.

Stationarity

The MA(1) model of Equation 14.11 describes a *stationary* series, that is, one that varies homogeneously about a fixed mean with a fixed standard deviation. Such a model was used in Chapter 3 to represent 210 consecutive values of yield from an industrial process. The data set out in Table 3B.1 and plotted in Figure 3.2 had an average value of 84.1 and a lag 1 estimated autocorrelation of $\hat{\rho} = -0.29$ corresponding to a model value of $\theta = 0.32$. No large autocorrelations were found at higher lags. An approximate model was therefore $\hat{y}_t = 84.1 + e_t$, where $e_t = a_t - 0.32a_{t-1}$, that is, $\hat{y}_{t+1} = 84.1 + a_{t+1} + 0.32a_t$. At time t the shock a_{t+1} has not yet happened and has an expected value of zero, so the one step ahead forecast made at time t is $\hat{y}_{t+1} = 84.1 + 0.32a_t$ with forecast error $y_{t+1} - \hat{y}_{t+1} = a_{t+1}$. Thus for a series of one-step-ahead forecasts of this kind the one-step-ahead forecast errors form a white noise series with no autocorrelation. As was mentioned earlier, this must be so, for if the one-step-ahead forecast errors were correlated, they would themselves be forecastable — the forecast errors could be forecast — and hence the forecasts themselves could not be optimal and the model would be inadequate.

Nonstationary Models

The stationary models mentioned above can represent time series that vary in a stable manner about a fixed mean. Such models are of limited practical use. Time series of interest occurring, for example, in economic, business, and technical applications are usually *nonstationary*. Two examples of nonstationary time series were the sequence of monthly sales of grommets listed in Table 14.1 and the uncontrolled series of thickness measurements in Table 14.2.

Now you have seen how the exponentially weighted average, the EWMA, was useful in both these time series. A sensible question is therefore, For what time series *model* does the EWMA provide an "optimal" forecast? We have seen that the updating formula for the EWMA forecast $\hat{y}_{t+1}$ may be defined in different ways, all equivalent in the sense that one can be obtained from the other; thus

$$\hat{y}_{t+1} = (1-\theta)(y_t + \theta y_{t-1} + \theta^2 y_{t-2} + \cdots) \tag{14.14}$$

$$\hat{y}_{t+1} = \hat{y}_t + (1-\theta)(y_t - \hat{y}_t) \tag{14.15}$$

$$\hat{y}_{t+1} = \lambda y_t + \theta \hat{y}_t \tag{14.16}$$

Now using any of these formulations to make a forecast (using any value of θ optimal or not) the one-step-ahead *forecast error* e_{t+1} will be the difference between the value y_{t+1} that actually occurred at time $t+1$ and the forecasted value $\hat{y}_{t+1}$, so that $\hat{y}_{t+1} = y_{t+1} - e_{t+1}$ and $\hat{y}_t = y_t - e_t$. Substituting these values for $\hat{y}_{t+1}$ and $\hat{y}_t$ in Equation 14.15, you arrive at a fourth equivalent model for the EWMA:

$$y_{t+1} - e_{t+1} = y_t - e_t + (1-\theta)e_t$$

or

$$y_{t+1} - y_t = e_{t+1} - \theta e_t \tag{14.17}$$

Now for *any* set of data $y_t, y_{t-1}, y_{t-2} \ldots$ and *any* value for θ, you can produce a series of forecast errors $e_t, e_{t-1}, e_{t-2}, \ldots$ by comparing the forecasts with what actually happened. Now it was argued previously that if the equation is to produce an optimal forecast then the one-step-ahead forecast errors — the e_t's–must be replaced by a_t's. Thus the EWMA with a discount factor θ provides an optimal forecast for a time series defined by

$$y_{t+1} - y_t = a_{t+1} - \theta a_t \tag{14.18}$$

where the a_t are IID. This is a time series in which the *first difference* $(y_{t+1} - y_t)$ is represented by a stationary moving average but the y_t's themselves are generated by $y_{t+1} = y_t + a_{t+1} - \theta a_t$, a nonstationary time series model called an *integrated first-order moving average* or an IMA model. This is the model for which the EWMA is an optimal forecast.

Why Is the IMA Times Series Model Important?

1. The identification of actually occurring industrial time series have frequently led to the IMA model or some close approximation.
2. It seems reasonable that for an *uncontrolled* series of equispaced data the variance of the *difference* between observations m steps apart should increase linearly with m. If this is so, then it can be shown that the appropriate model is the IMA.
3. It is the time series model that generates the EWMA forecast, whose merits have been discussed earlier.

14.6. USING A MODEL TO MAKE A FORECAST

Suppose you have time series data up to time t and you want to forecast y_{t+m}, the observation m steps ahead with $m \geq 1$. This forecast may be written $\hat{y}_t(m)$, where the quantity t is called the forecast *origin* and m the lead time. The general procedure for using a time series model to make a forecast is to write the model with the value y_{t+m} to be forecast on the left of the equal sign and the rest of the model on the right. Then for any element that has already happened, say y_{t-i} or a_{t-j} (with i and $j \geq 0$), you write the known value y_{t-i} or a_{t-j} itself. For any y_{t+j} or a_{t+h} (with j and $h \geq 0$) that has not already occurred, you substitute* the forecasted values $\hat{y}_t(j)$ and $\hat{a}_t(h) = 0$. For example, to make a one-step-ahead forecast from the IMA model of Equation 14.18, you write the model

$$y_{t+1} = y_t + a_{t+1} - \theta a_t$$

Now at the origin time t the terms on the right, except a_{t+1}, have already happened, and if you retain the simple notation $\hat{y}_{t+1} = \hat{y}_t(1)$ for the one-step-ahead forecast,

$$\hat{y}_{t+1} = y_t - \theta a_t \qquad \text{and} \qquad a_t = y_t - \hat{y}_t \tag{14.19}$$

so $\hat{y}_{t+1} = y_t - \theta(y_t - \hat{y}_t) = (1 - \theta)y_t + \theta\hat{y}_t = \lambda y_t + \theta\hat{y}_t$, which is Equation 14.16, defining the EWMA with smoothing constant θ. When the value y_{t+1} becomes available, the forecast error is $y_{t+1} - \hat{y}_{t+1} = a_{t+1}$. The standard deviation of the one-step-ahead forecast is thus σ_a, the standard deviation of the

*Technically you replace each term by its conditional expectation (mean value) at time t given the data up to time t are available (BJ&R).

shocks. Thus, as expected, the one-step-ahead forecast obtained from the IMA model is the EWMA.

But now that we have this model we can also obtain an optimal forecast for any lead time. Thus for lead time 2 the model is

$$y_{t+2} = y_{t+1} + a_{t+2} - \theta a_{t+1}$$

In this model, at origin t the value of y_{t+1} is unknown and so we substitute its one-step-ahead forecast $\hat{y}_t(1)$. Also the quantities a_{t+1} and a_{t+2} are future unknown shocks and are replaced by their mean value of zero, so that

$$\hat{y}_t(2) = \hat{y}_t(1)$$

For the IMA model the two-steps-ahead forecast from origin t is the same as the one-step-ahead forecast. In a similar way you can show that for the IMA model the m-steps-ahead forecast from origin t is the same for all values of m. The EWMA is thus a relevant measure of the current *local* average of the series at time t. For a nonstationary series the population mean η does not exist but the local average $\hat{y}_t$ can provide a useful measure of the *current* level. For example, the daily value of an ounce of gold as announced by the New York Stock Exchange has no mean value but if, as is likely, these daily values approximately follow an IMA model, the EWMA provides a local average for any given time origin.* Now, although future forecasts obtained for an IMA model from a fixed origin t remain the same, the forecast error does not. In fact the standard error of the forecast m steps ahead is

$$\sigma_{a,t+m} = \sigma_a\sqrt{1 + (m-1)\lambda^2}$$

So, as might be expected, the forecast becomes less and less relevant for longer lead times.

Estimating θ

Suppose you have a time series you believe could be represented by an IMA model. How can you estimate θ? You saw earlier (see Table 14.1 and Fig. 14.4) that the value of θ appropriate for an EWMA forecast could be estimated by making repeated calculations with different values of θ to minimize the sum of squares of the forecast errors. Precisely the same procedure is followed to estimate θ from the generating time series model. The model produces one-step-ahead forecasts and hence one-step-ahead forecast errors. The estimate of θ is the value which gives the minimum sum of squares of the forecast errors.

In these procedures there is always the question of how to get the estimation process started. How, for example, do you forecast $\hat{y}_t(1)$ when t = 1 and there

*However, the sad news is that, approximately, stock prices follow the IMA model with $\theta = 1$ (a model called the random walk) for which the best forecast for tomorrow's price is today's price.

are no previous values available? One fairly good approximation is provided by starting the estimation procedure at the second observation y_2 and treat the initial value of the series as if it were the required forecast. Then $\hat{y}_3 = \lambda y_2 + \theta y_1$. The error this introduces is rapidly discounted by the procedure.*

Other Applications of the IMA Model

Two further interesting applications of the IMA model are to *seasonal forecasting* and *intervention analysis*. Both subjects are too extensive to be described in any detail; the intention is to show what can be done and to give you an idea of how to do it. This will serve to show if the technique might help with your problem. If you think it might, you can study the more detailed referenced sources.

A Seasonal Model

The IMA models can be useful components in forecasting seasonal series. Suppose from some specific time origin you wished to make a forecast of, say, next January's average rainfall and you had extensive data of past average monthly rainfalls. Suppose at first that you used an EWMA *only* of past Januarys to forecast next January. This would be in accordance with an IMA model showing the yearly dependency,

$$y_t - y_{t-12} = u_t - \Theta u_{t-12} \tag{14.20}$$

with a 12-month interval and with u's representing the one-step- (12-month) ahead forecast errors. The symbol Θ is used to denote the *yearly* discount factor.

In the same way you could use past Februarys to forecast rainfall for the next February, past Marchs to forecast the next March, and so on, thus producing successive one-year-ahead forecasts with associated monthly errors $u_t, u_{t-1}, u_{t-2}, \ldots$. Now you would expect the *errors* calculated for January, February, March, and so on, to be themselves autocorrelated. If January had had an exceptionally high rainfall, then so might February. This *monthly* dependency is modeled by a second IMA,

$$u_t - u_{t-1} = a_t - \theta a_{t-1} \tag{14.21}$$

with the a_t's now representing white noise forecasting errors. If you substitute this expression for u_t in Equation 14.20, after some simplification you get the seasonal model

$$\begin{aligned} y_{t+m} = {} & y_{t+m-1} + y_{t+m-12} - y_{t+m-13} + a_{t+m} \\ & - \theta a_{t+m-1} - \Theta a_{t+m-12} + \theta\Theta a_{t+m-13} \end{aligned} \tag{14.22}$$

*A better approximation for the needed initial forecast values of a time series is to get them by "back-casting," that is, to turn the series around and treat the last value as the first and then back-cast the whole series until you produce the required starting values. (See BJ&R).

For example, the model for time $t+1$ obtained by substituting $m = 1$ in Equation 14.22 is

$$y_{t+1} = y_t + y_{t-11} - y_{t-12} + a_{t+1} - \theta a_t - \Theta a_{t-11} + \theta\Theta a_{t-12}$$

and the forecast $\hat{y}_t(1)$ made from the origin t with lead time $m = 1$ becomes

$$\hat{y}_t(1) = y_t + y_{t-11} - y_{t-12} - \theta a_t - \Theta a_{t-11} + \theta\Theta a_{t-12}$$

In general, the forecast is obtained, as before, by writing the model with the term to be forecast on the left of the equal sign. Then on the right those terms already known at time t remain as they are and appropriate forecasts replace the remaining terms.

Figure 14.9 shows the log monthly passenger sales for an 11-year period in the early days of transatlantic air travel. (The data are from series G in BJ& R.) The model in Equation 14.22 has come to be called the "airline model" because of its first application. You will see later how estimates of the needed parameters can be obtained from the data. For this example they were $\hat{\theta} = 0.4$ and $\hat{\Theta} = 0.6$ with $\hat{\sigma}_a = 0.037$. To illustrate the forecasting process, suppose you wanted a three months ahead forecast $\hat{y}_t(3)$ from origin t. Then, by setting $m = 3$ in Equation 14.22, you would write the model for y_{t+3} as

$$y_{t+3} = y_{t+2} + y_{t-9} - y_{t-10} + a_{t+3} - 0.4a_{t+2} - 0.6a_{t-9} + 0.24a_{t-10} \quad (14.23)$$

Then the forecast from origin t for lead 3, that is, $\hat{y}_t(3)$, is obtained by replacing unknown y's by their forecasts and unknown a's by zeroes in Equation 14.23 to give

$$\hat{y}_t(3) = \hat{y}_t(2) + y_{t-9} - y_{t-10} - 0.6a_{t-9} + 0.24a_{t-10}$$

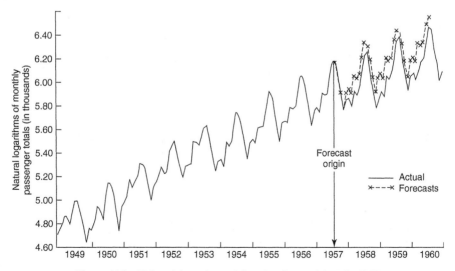

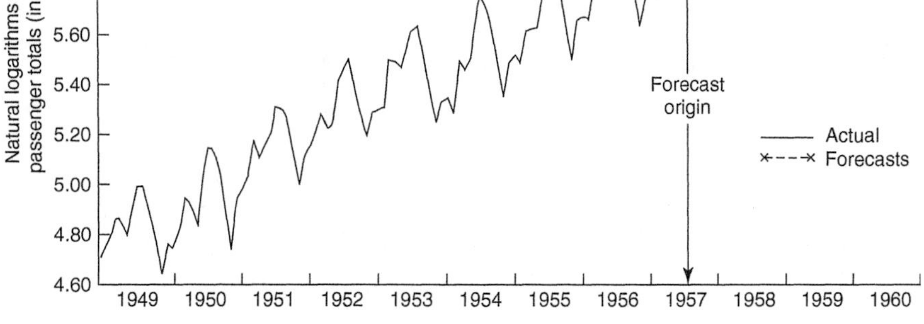

Figure 14.9. Airline ticket sales and forecasts from origin July 1957.

Figure 14.9 shows the forecasts made in this way from the arbitrarily selected origin (July 1957) for lead times up to 36 months. We see that the simple model containing only two parameters faithfully reproduces the seasonal patterns and supplies excellent forecasts for up to 6 months ahead. All predictions obtained from such a seasonal model are adaptive. When changes occur in the seasonal pattern, these will be appropriately projected into the forecast. Of course, a forecast for a long lead time, such as 36 months, must necessarily contain a large error. However, an initially remote forecast will be continually updated, and as the lead time shortens, greater accuracy will be obtained.

Estimating the Parameters

In this example, the estimates used for the monthly and yearly discount factors $\hat{\theta} = 0.4$ and $\hat{\Theta} = 0.6$ were obtained as follows. For any particular pair of values for θ and Θ you can calculate a set of one-step-ahead forecast errors. From this you can calculate their corresponding sum of squares. One way to proceed is illustrated in Figure 14.10, where a sum of squares grid in θ and Θ is shown and contours for the sum of squares surface are drawn. The coordinates for the lowest point on this surface provide least squares estimates of about $\hat{\theta} = 0.4$ and $\hat{\Theta} = 0.6$. An approximate 95% confidence region, obtained as in Chapter 10, is also shown. Computer programs for obtaining minima can rapidly perform such analyses.

Seasonal Weight Function

Shown in Figure 14.11 for the airline data is the weight function for the one-step-ahead seasonal forecast. For nonseasonal series the EWMA weight function makes sense since it gradually discounts past data. The weight function for the

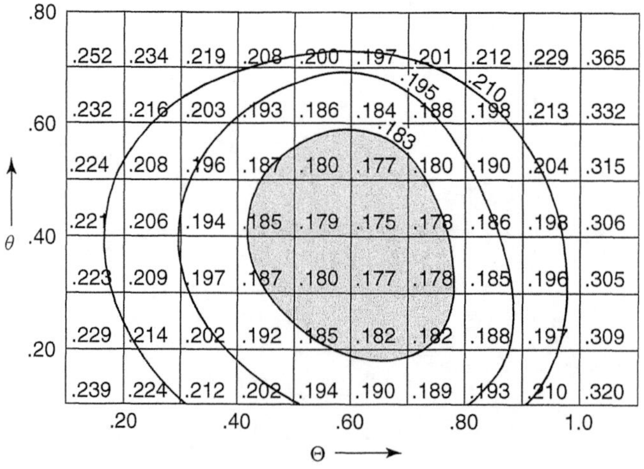

Figure 14.10. Estimating θ and Θ: the sum of squares surface $\sum(y_t - \hat{y}_t)^2$.

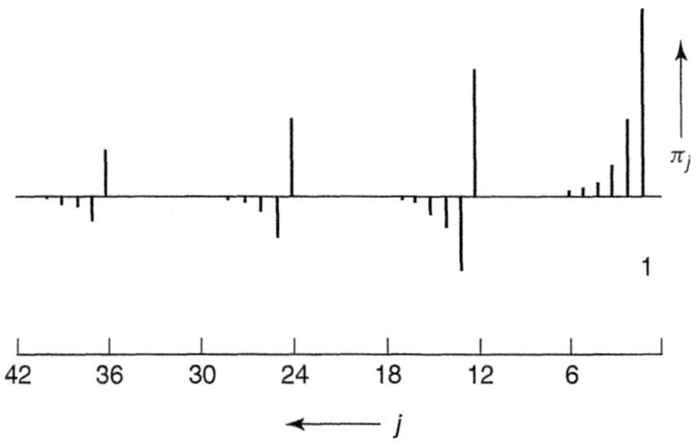

Figure 14.11. Weight function for seasonal forecasts.

seasonal model is a little more subtle. For illustration, suppose you used the seasonal model to predict monthly sales of a department store. Let us say it is November and from that origin you wished to make a one-step-ahead forecast for December (which, of course, has the distinction that it contains Christmas). Now look at Figure 14.11. The EWMA weights on the extreme right side of the figure might be applied with discount factor θ to November, October, September, and so on, of the current year and could alone produce a forecast for December. But this forecast would only be appropriate for a December that contained no Christmas! The remaining weights located to the left of Figure 14.11 correct for the sales in this exceptional month because they represent *differences* between what in previous years the naïve EWMA forecast made in November would have produced and what actually happened. This difference correction is itself down-weighted by a EWMA with discount factor Θ*. A somewhat different application of a seasonal model is illustrated in the next example.

14.7. INTERVENTION ANALYSIS: A LOS ANGELES AIR POLLUTION EXAMPLE

An indication of the level of smog in downtown Los Angeles is the oxidant concentration denoted by O_3. Monthly averages of O_3 for January 1955 to December 1965 are shown in Figure 14.12. In January 1960 a new law (Rule 63) was instituted that reduced the allowable proportion of reactive hydrocarbons in all gasoline sold in the state. Call this event an *intervention.* Did this intervention coincide with any difference in the O_3 concentration?

The ozone levels shown in Figure 14.12 vary a great deal; moreover, there are strong seasonal effects. Therefore, because of the obvious dependence between successive observations, it would have been totally misleading, for example, to apply a t test to the difference in averages before and after January 1960.

*This example uses the forecasting of Christmas sales for emphasis. The same weight function will of course apply to a one-step-ahead forecast made for any other month and the latter part of the weight function will make an appropriate correction if this is necessary.

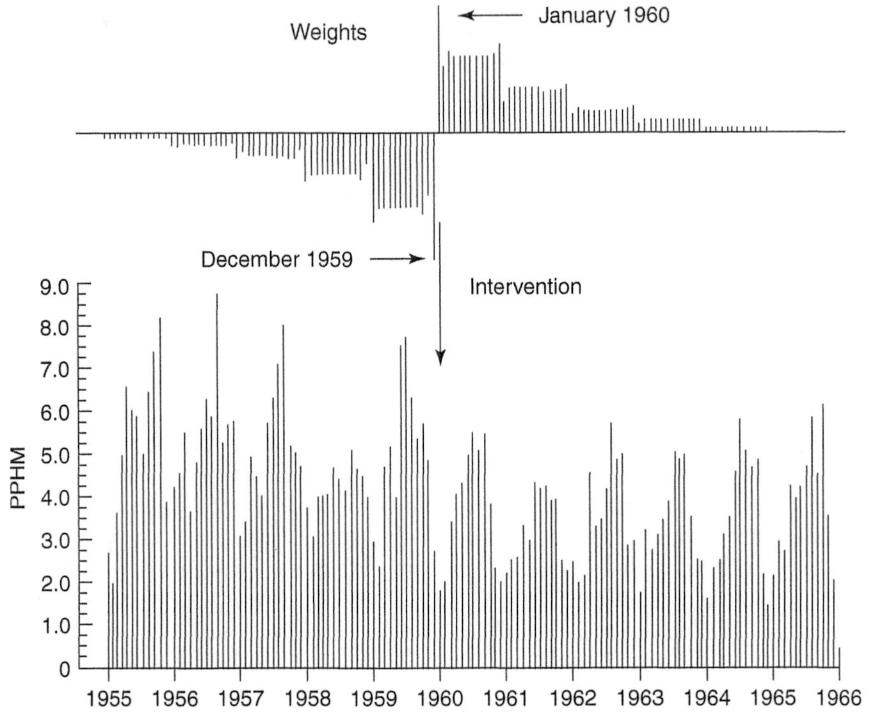

Figure 14.12. Monthly averages of O_3 in downtown Los Angeles in parts per hundred million (PPHM), January 1955 to December 1965, with weight function for determining the change in 1960.

Furthermore, no question of randomization can possibly arise since the ordering of the observations is fixed.

The problem can be tackled as follows. Denote the O_3 concentration at time t by y_t and then write

$$y_t = \beta x_t + z_t$$

where x_t is an indicator variable that takes the value zero before January 1960 and the value unity afterward and z_t is represented by a time series model that takes account of the local and seasonal dependence. With this arrangement β will represent the possible change in concentration of oxidant that occurred as a result of the intervention.

The model found appropriate for z_t was

$$z_t - z_{t-12} = a_t + 0.2a_{t-1} - 0.8a_{t-12} - 0.16a_{t-13}$$

With this model (see Box and Tiao, 1975) the least squares estimate of the change is

$$\hat{\beta} = -1.1 \pm 0.1$$

where ±0.1 is the standard error of $\hat{\beta}$. The analysis strongly suggests, therefore, that this intervention (or something else that occurred at the same time) produced an estimated reduction in the oxidant level of about 25%.

The weight function for $\hat{\beta}$ is shown at the top of Figure 14.12. Thus $\hat{\beta}$ is estimated by the difference of the two weighted averages before and after the event. Notice that the greatest weight is given to data immediately before and immediately after the intervention and the "steps" in the exponential discounting allow for averaging over the 12-month periods. Intervention methods of this kind can help the investigator find out what, if anything, happened when, for example, a new law was passed, a nuclear station started up, a teaching method changed, or free contraceptives were distributed. In this way it is possible to estimate the size of a possible change and its error. Such an analysis can establish whether a change has almost certainly occurred but the *reason* for such a change is necessarily open for debate.

Remember that investigations of this kind do not prove causality. Also such models cannot, of course, predict unexpected events like the September 11 attack on the World Trade Center, which certainly had a profound effect on airline travel. Notice, however, that if you had data for the relevant time period and needed to estimate the effect of such an event or to make a correction for it, you could obtain such an estimate by combining the two techniques of seasonal modeling and intervention analysis in the manner illustrated above.

REFERENCES AND FURTHER READING

Statistical Quality Control by Monitoring

A more detailed discussion of the topics introduced in this chapter will be found in:

Shewhart, W. A. (1931) *The Economic Control of Quality of Manufactured Product*, Van Nostrand, New York.

Juran, J. M., and Godfrey, A. B. (1999) *Juran's Quality Handbook*, 5th ed., McGraw-Hill, New York.

Deming, W. E. (1986) *Out of the Crisis*. MIT, Center of Advanced Engineering, Cambridge, MA.

ANSI/ASQC (1996) *B1-1996 Guide for Quality Control Charts, B2-1996 Control Chart Method of Analyzing Data, B3-1996 Control Chart method of Controlling Quality During Production*. American Society for Quality, Milwaukee, WI.

Montgomery, D. C. (2001) *Introduction to Statistical Quality Control*, 4th ed., Wiley, New York.

Nelson, L. S. (1984) The Shewhart control chart—tests for special causes, *J. Qual. Technol*, October, pp. 237–239.

Quesenberry, C. P. (1997) *SPC Methods for Quality Improvement*, Wiley, New York.

Woodall, W. H. (2000) Controversies and contradictions in statistical process control [with discussion], *J. Quality Technol.*, **32**, 341–378.

Woodall, W. H., and Montgomery, D. C. (1999) Research issues and ideas in statistical process control, *J. Quality Technol.*, **31**, 376–386.

EWMA Charts

Holt, C. C. (1957) *Forecasting Seasonals and Trends by Exponentially Weighted Moving Averages*, Carnegie Institute of Technology, Pittsburgh, PA.

Roberts, S. W. (1959) Control chart tests based on geometric moving averages, *Technometrics*, **1**, 239–250.

Hunter, J. S. (1986) The exponentially weighted moving average, *J. Quality Technol.*, **18**, 203–210.

CuSum and CuScore

Page, E. S. (1954) Continuous inspection schemes, *Biometrika*, **41**, 100–114.

Barnard, G. A. (1959) Control charts and stochastic processes, *J. Roy. Statist. Soc. Ser. B*, **21**, 239–271.

Kemp, K. W. (1962) The use of cumulative sums for sampling inspection schemes, *Appl. Statist.*, **11**, 16–31.

Lucas, J. M. (1982) Combined Shewhart-CuSum quality control schemes, *J. Quality Technol.*, **14**, 51–59.

Box, G. E. P., and Remirez, J. (1992) Cumulative score charts, *Quality Reliability Engineering International*, **8**, 17–27.

Process Adjustment

Box, G. E. P., and Luceño, A. (1997) *Statistical Control by Monitoring Feedback and Adjustment*, Wiley, New York.

Box, G. E. P., and Kramer, T. (1992) Statistical process control and automatic process control — a discussion, *Technometrics*, **34**(3), 251–267.

Box, G. E. P., and MacGregor, J. F. (1974) The analysis of closed loop dynamic-stochastic systems, *Technometrics*, **16**, 391–401.

Forecasting and Time Series and Intervention Analysis

Berthouex, P. M. (1996) *Time Series Models for Forecasting Wastewater Treatment Plant Performance*, CQPI Report 138, University of Wisconsin, Madison, WI.

Box, G. E. P., and Tiao, C. K. (1975) Intervention analysis with applications to economic and envirnmental problems, *J. Amer. Statist. Assoc.*, **70**, 70–79.

Box, G. E. P., Jenkins, G. M., and Reinsel, G. C. (1994), *Time Series Analysis, Forecasting and Control*, 3rd ed., Prentice-Hall, Englewood Cliffs, NJ.

QUESTIONS FOR CHAPTER 14

1. Can you explain the function of the two complementary aspects of statistical quality control: monitoring and adjustment?

2. What did Shewhart mean by (a) "state of statistical control" and (b) an "assignable cause"?
3. What is Pete's rule? Explain it from Pete's point of view.
4. (a) What is meant by process monitoring? (b) By process adjustment? (c) How could the EWMA be useful in both procedures?
5. Show a Shewhart chart. Can you describe how you would use it? Show a Robert's chart. Can you explain how you would use it? Describe the simultaneous use of these charts.
6. How would a chart for repeated adjustment look? How would you use it?
7. How would a bounded adjustment chart look? How might you use it?
8. What are some reasons for the importance of the EWMA in problems of control?
9. What is meant by a time series model? Give an example.
10. What is meant by (a) stationary time series and (b) nonstationary time series?
11. Why are nonstationary time series important for business and economics and problems of control.
12. For what time series is the EWMA the best estimator of future observations?
13. What is meant by (a) a seasonal model and (b) an intervention model?

PROBLEMS FOR CHAPTER 14

1. How could you set control limits for a Shewhart monitoring chart using past data (a) directly and (b) using specific assumptions?
2. What is the EWMA? How may it be used to make a forecast?
3. Give four different formulas for calculating the EWMA forecast for y_{t+1} given equally spaced data up to y_t. What is meant by the discount factor? What is λ ?
4. The number of model X30 clothes washers available in 19 successive weeks were

$$34, 35, 32, 30, 29, 26, 30, 30, 31, 25, 24, 27, 28, 26, 26, 24, 22, 24, 26$$

   Assuming it takes 3 weeks to get a fresh assignment of model X30 washers, construct an inventory control scheme whereby the chance that there are fewer than 20 washers in stock is about 5%.
5. Describe a process monitoring scheme using an EWMA with discount factor $\theta = 0.6$. Would you expect this scheme to be seriously affected if instead you set (a) $\theta = 0.5$, (b) $\theta = 0.7$. (c) $\theta = 1.0$, and (d) $\theta = 0$?

6. Describe how you would estimate a discount factor θ from past records. Illustrate using the data of problem 4. Do you think the data is adequate to make such an estimate.

7. Design a scheme for repeated adjustment assuming $\theta = 0.7$ for the series in Table 14.2.

8. **(a)** What is meant by "bounded adjustment"?

(b) When would bounded adjustment be useful?

(c) What is meant by the "average adjustment interval" (AAI)?

(d) Suppose you had a time series of process data. How could you *empirically* obtain a table showing the percent change in the standard deviation for various values of the AAI?

9. Discuss the statement, "A time series model is a recipe for transforming the data into white noise." Give examples. How might you detect deviant data?

10. To obtain useful estimates of time series parameters, moderately long series of data are necessary. Ignoring this fact, make calculations for the following series of 20 observations:

$$6, 4, 3, 4, 3, 3, 1, 2, 3, 5, 1, 1, -1, 3, 4, 7, 10, 13, 15, 14$$

Assume the model $y_t - y_{t-1} = a_t - \theta a_{t-1}$.

(a) Fit this time series model (find $\hat{\theta}$ and $\hat{\sigma}_a$).

(b) Show how you can get starting values by back-forecasting.

(c) Obtain forecasts for the next three values of the series, that is, for times $t = 21, 22, 23$

11. The following are quarterly totals of sales by a manufacturer of swimsuits in tens of thousands for the years 2000 to 2003:

	First Quarter	Second Quarter	Third Quarter	Fourth Quarter
2000	12	29	18	4
2001	15	27	21	5
2002	13	36	25	10
2003	17	41	32	11

What is a seasonal model that might be appropriate to represent such sales? Even though the above illustrative data are inadequate, perform a seasonal analysis with the data and illustrate how to (1) estimate the parameters of such a model, (2) use it for forecasting, and (3) obtain a confidence interval for a forecast.

CHAPTER 15

Evolutionary Process Operation

"A process can be routinely operated so as to produce not only product but also *information* on how to improve the process and the product."

Evolutionary process operation (EVOP) is a management tool for obtaining information from the full-scale process in its daily operation. In that respect it is like statistical process control (SPC). Also like SPC it is run largely by the plant personnel themselves. As you have seen earlier, SPC achieves improvement by continuously debugging the system—by finding and permanently removing causes of trouble. You owe to Murphy's law the many process improvements that can be obtained in this way. Murphy works hard to ensure that anything that can go wrong will go wrong, but with an adequate system of process monitoring, more and more of Murphy's tricks can be permanently stymied.

But the question remains: "Are you sure that the present process configuration (the settings of the factors and the manner in which the process is operated) is the best possible?" Since expert technical knowledge will have been used in designing the process, since many improvements may have been produced by special off-line and on-line experimentation, and because of process debugging by SPC, it would be easy to believe that the answer must be yes. But if you talk to people familiar with the operating system, in particular to the process operators and their supervisors, you may find that they are far from certain that the process is perfect and that they have a number of ideas for improvement. Many of these ideas would be worth trying if experimentation on the full scale was permissible.

Now in the day-to-day operation of any system some variation about the target value, called common-cause variation or system noise, is bound to occur. The problem is that an experimental process change whose effect is large

Statistics for Experimenters, Second Edition. By G. E. P. Box, J. S. Hunter, and W. G. Hunter

enough to be distinguished from this noise may turn out *not* to be an improvement and could produce undesirable consequences—and, in particular, unsatisfactory product. For this reason managers are usually very reluctant to allow departures from standard operating procedures. Thus the dilemma is that, while it is possible to conceive of potentially valuable process improvements, some of these are of the kind that cannot be realistically tested in small-scale experiments, and also are such that process changes large enough to test them might have undesirable consequences.*

Evolutionary operation resolves this dilemma by *repeated* application of a *small* process change. Because the iterated change is *small*, its effect can be virtually undetectable—it is buried in the noise. Nevertheless, however small the effect, continuous averaging over time can produce a detectable signal.†

An Example

The following data occurred in a test to determine whether the current operating temperature, which is called here process A, could be changed so as to produce a higher value for a measure of efficiency y. Modification B was an increase in temperature small enough that in a single trial it would be unlikely to have a noticeable effect. A schedule was followed whereby on the first day process A was run, on the second day process B was run, then back to A, and so on. The data shown in Table 15.1 are for 10 repetitions of pairs of runs.

The appropriate signal-to-noise ratio for judging the difference between the effects A and B is Student's t statistic $t = (\overline{y}_B - \overline{y}_A)/\text{SE}(\overline{y}_B - \overline{y}_A)$, where $\overline{y}_B - \overline{y}_A$ is the signal and $\text{SE}(\overline{y}_B - \overline{y}_A)$ is the relevant measure of noise. If d is the difference $B - A$ recorded in each cycle, then you will find that in such

Table 15.1. A simple EVOP Program to Test the Effect of a Small Increase in Temperature

Cycle	1	2	3	4	5	6	7	8	9	10
$\begin{bmatrix} B \\ A \end{bmatrix}$	$\begin{bmatrix} 72.3 \\ 72.0 \end{bmatrix}$	$\begin{bmatrix} 71.9 \\ 72.0 \end{bmatrix}$	$\begin{bmatrix} 72.4 \\ 72.1 \end{bmatrix}$	$\begin{bmatrix} 72.9 \\ 72.4 \end{bmatrix}$	$\begin{bmatrix} 72.9 \\ 72.1 \end{bmatrix}$	$\begin{bmatrix} 71.3 \\ 70.2 \end{bmatrix}$	$\begin{bmatrix} 71.7 \\ 71.4 \end{bmatrix}$	$\begin{bmatrix} 69.8 \\ 70.0 \end{bmatrix}$	$\begin{bmatrix} 73.7 \\ 73.1 \end{bmatrix}$	$\begin{bmatrix} 72.5 \\ 72.0 \end{bmatrix}$
$d = B - A$	0.3	−0.1	0.3	0.5	0.8	1.1	0.3	−0.2	0.6	0.5
$t = \dfrac{\sum d}{\sqrt{\sum d^2}}$	1.0	0.6	1.1	1.5	1.7	1.9	2.1	1.9	2.2	2.4

* *The bird dog line*: Where manufacturing is carried out with a number of simultaneously operating parallel lines, it is sometimes possible to dedicate just one line to experimentation. Improvements found can then be put to use on all the lines. However, this device can be used only for processes that employ this type of operation.

† To say that the effects of repeated small changes can be "buried in the noise" is somewhat misleading. Repeated change of an effective factor, though very small, must slightly increase the noise.

circumstances the standard t ratio reduces to the signal-to-noise ratio

$$t_\nu = \frac{\sum d}{\sqrt{\sum d^2}}$$

with degrees of freedom ν equal to the number of cycles. In the example in Table 15.1 it is clearly becoming very likely that B is better than A.*

A graphical representation of evolutionary process operation for this simple example is shown in Figure 15.1. It is supposed that the response yield y depends on temperature but (as is shown in this much exaggerated diagram) the standard process is not using the best temperature. A signal is injected into the temperature by running the process alternatively at a temperature slightly higher and slightly lower than the current normal temperature. Thus (unless the temperature gradient is zero) the output y consists of the concordant signal embedded in common-cause noise. Now suppose a *correlator signal* z is independently generated. This signal is the −, + sequence used in the statistical analysis synchronized with the input signal.† So for this example the computed effect

$$\sum y_1 z_1 = (-y_1 + y_2 - y_3 + y_4 - \cdots) = \sum d$$

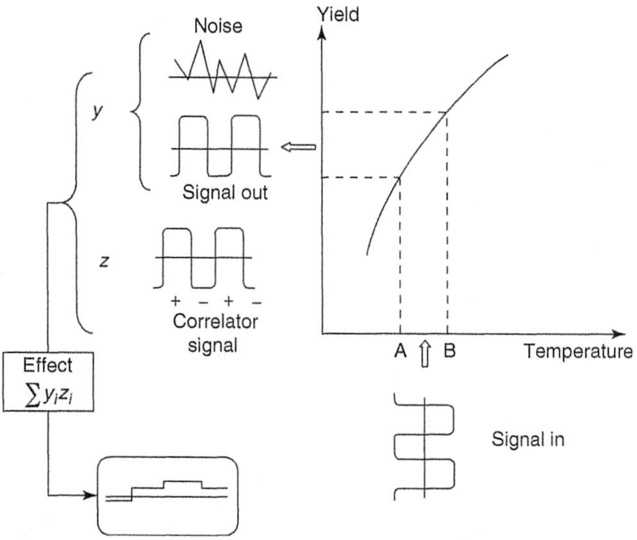

Figure 15.1. Graphical representation of evolutionary process operation: study of the effect on yield y of a small change in temperature.

*When t is used in this sequential manner, ordinary t tables do not give correct probabilities. However, using a sequential t test due to Barnard (1946) for the number of cycles greater than 6, t values of about 2.5 to 3.0 indicate probable change.

†This concept has also been used to produce a system of automatic adaptive process optimization (see e.g., Box G. and Chanmugam, 1962). In this application you may need to arrange that the corrector signal is slightly out of phase with the input signal to allow for the dynamics (inertia) of the system.

and $\sum d$ is the numerator of the t statistic (or equivalently of the signal-to-noise ratio) in which the transmitted signal is enhanced and the noise is reduced by averaging.

15.1. MORE THAN ONE FACTOR

Examples with More Than One Factor

Before discussing other examples, it is necessary to define the terms *run*, *cycle*, and *phase*:

> A run is a period of manufacture in which the process conditions are fixed.
>
> A cycle is a series of runs producing a simple experimental design.
>
> A phase consists of a number of cycles after which a decision to change is taken.

In the previous example there were 20 runs, 10 cycles of 2 runs each, and 1 phase.

In a manufacturing environment things must be kept simple. But it is usually practical to vary more than one factor. Often two or three factors can be varied in any given phase using a 2^2 or a 2^3 factorial design (sometimes with an added center point). An *information display* (where possible on a computer screen) shows conditions that are presently being tested and, estimates of the effects and their standard errors as results become available. This information display is continuously and expeditiously updated and is seen by for both the process superintendent and process operators. The process superintendent, with the help and advice of scientific and management colleagues, uses this information to lead the process to better operating conditions.

Evolutionary Process Operation on a Petrochemical Plant

In the following example the object was to *decrease* the cost per ton of product from a petrochemical plant. In the initial three phases shown here two factors were studied: the reflux ratio of a distillation column and the ratio of recycle flow to purge flow. The results at the end of the first phase obtained after five repetitions (cycles) are shown in Figure 15.2*a*. After five cycles it was decided that there was sufficient evidence to justify a move to the lower reflux ratio and a higher recycle–purge ratio. These were used in phase II shown in Figure 15.2*b*. Phase II terminated after five further cycles. The results confirmed that lower costs indeed resulted from these new conditions and suggested that still higher values of the recycle–purge ratio should be tried. This led to phase III in Figure 15.2*c*. After four more cycles, it was concluded that the lowest cost was obtained with the reflux ratio close to 6.3 and the recycle–purge ratio about 8.5. Throughout these three phases the process continued to produce satisfactory product, but by running the process in the evolutionary mode, the cost per ton of the product was reduced by almost 15%. At the end of these three phases new factors were

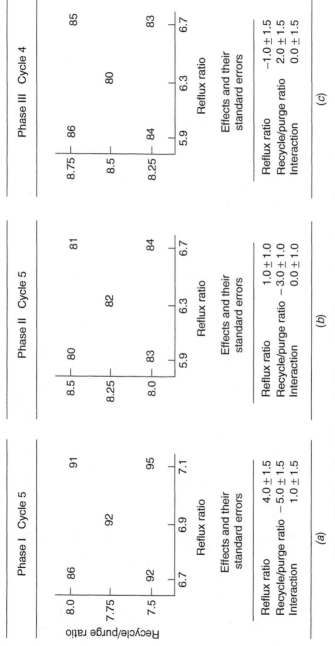

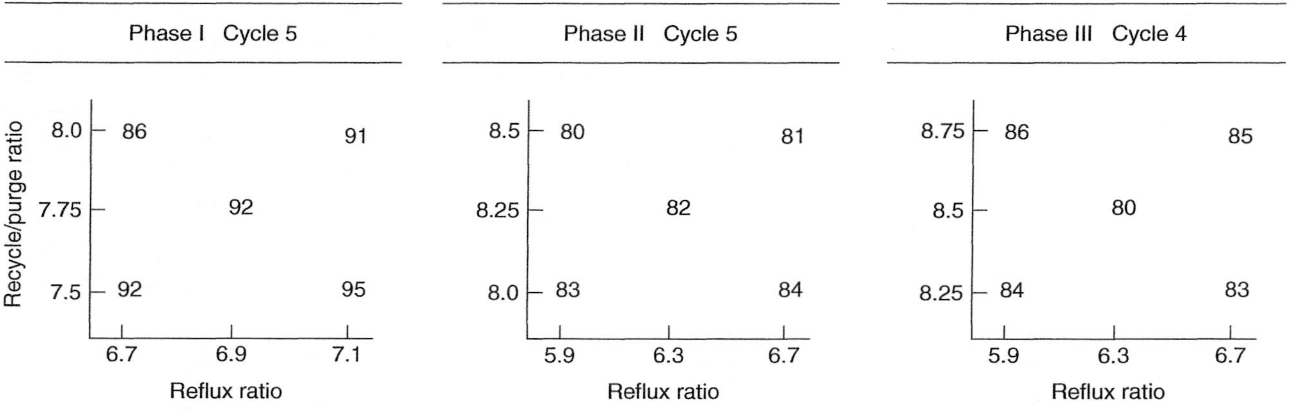

Effects and their standard errors

Reflux ratio	4.0 ± 1.5
Recycle/purge ratio	− 5.0 ± 1.5
Interaction	1.0 ± 1.5

(*a*)

Effects and their standard errors

Reflux ratio	1.0 ± 1.0
Recycle/purge ratio	− 3.0 ± 1.0
Interaction	0.0 ± 1.0

(*b*)

Effects and their standard errors

Reflux ratio	−1.0 ± 1.5
Recycle/purge ratio	2.0 ± 1.5
Interaction	0.0 ± 1.5

(*c*)

Figure 15.2. Petrochemical example. Information displays at the conclusions of three phases of EVOP.

introduced and the evolution of the process by small increments to better levels of performance was continued.

Exercise 15.1. Make scales on a single piece of graph paper covering the reflux ratio from 5.9 to 7.1 and the recycle–purge ratio from 7.5 to 8.75. Show the location and results from all three phases. In the light of hindsight, do you think the experimenters could have obtained their result faster by using steepest decent?

To be successful, it is essential that evolutionary operation be carefully organized and plant operators and managers responsible for these schemes receive proper training. Schedules for the current operation need to be prepared and made part of the computer display ensuring that the proper conditions are run in each cycle. It has been found that operators can become enthusiastic participants when they can see developing information on the display. They should be involved in the planning of the program, the generation of ideas, and the celebration of successes.

Evolutionary Process Operation in the Manufacture of a Polymer Latex

In this example the manufactured product was a polymer latex. The *original* objective was to determine conditions that, subject to a number of constraints, gave the *lowest* optical density. It was realized at the end of phase II that shorter addition times leading to increased throughput were possible. The objective was therefore broadened to *increasing plant throughput* subject to obtaining a satisfactory optical density of less than 40.

Initially, three process variables were studied: addition time (t), reactant temperature (T), and the stirring rate (S). Although it would have been possible to use all three factors simultaneously, it was decided for the sake of simplicity to run the factors in different combinations two at a time. Figure 15.3 displays five EVOP phases, each a 2^2 factorial, along with the recorded average optical density readings, estimated effects, and standard errors.

The standard operating conditions were $t = 3$ hours, $T = 120°\text{F}$, and $S = 88$ rpm. The course of events is summarized in Figure 15.3. In phase I, temperature was held at its standard condition and time and stirring rate were varied over what were fairly narrow limits. This was done because production feared that with wider ranges substandard product might be produced. The cycle times were short and 12 cycles were performed in this phase with no effects of any kind detected. This result, at first a disappointment, was actually quite helpful. It made those in charge of production more confident that changing these two factors over wider ranges would be unlikely to produce substandard material. As you see, wider ranges were therefore allowed in phase II, and after 18 cycles a stirring effect became apparent (1.6 ± 0.7), indicating that lower optical densities might be obtained at lower stirring rates. However, it was decided not to lower the stirring rate at this stage. The most important finding was the unexpected one that no deleterious effect resulted from reducing the addition time. A saving of 15 minutes in addition time would result in an 8% increase in throughput

Optical density reading

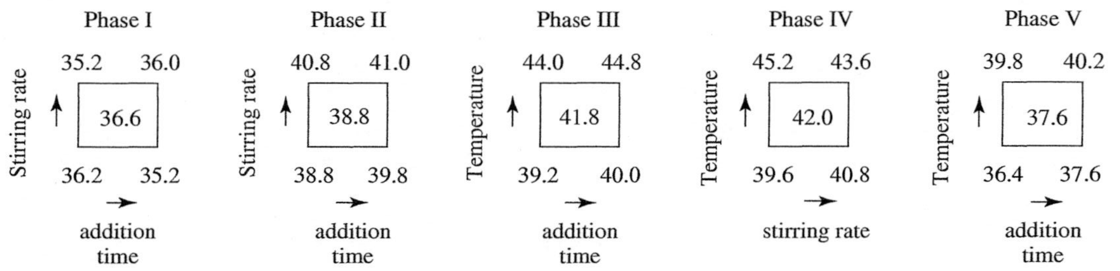

	Phase I			Phase II			Phase III			Phase IV			Phase V		
settings															
temperature		120			120		118	120	122	116	118	120	114	116	118
stirring rate	83	88	93	78	88	98		88		68	78	88		78	
addition time	2:50	3:00	3:10	2:35	3:00	3:25	2:30	2:45	3:00		2:45		2:15	2:30	2:45
effects with standards errors															
temperature		——			——			4.8 ± 1.2			4.2 ± 0.7			3.0 ± 0.4	
stirring rate		-0.4 ± 0.5			1.6 ± 0.7			——			-0.2 ± 0.7			——	
addition time		0.2 ± 0.5			0.6 ± 0.7			0.8 ± 1.2			——			0.8 ± 0.4	
interaction		0.6 ± 0.5			-0.4 ± 0.7			0.0 ± 1.2			-1.4 ± 0.7			-0.4 ± 0.4	

Figure 15.3. Optical density example. Information displays at the conclusions of five phases of EVOP.

and a considerable rise in profitability. To further investigate the possibilities of increased throughput, addition time and temperature were varied simultaneously in phase III. Continuing to contradict established belief, the results confirmed that increased throughput did not result in an inferior product and that optical density might be reduced by lowering the temperature. Phase IV used slightly lower temperatures while the stirring rate was varied about a lower value and addition time was held cautiously at the reduced value of 2 hours, 45 minutes.

In phase V both temperature and addition time were varied about lower values (116°F, 2 1/2 hours) without deleterious effect and the stirring rate was held at 78. The results indicated that temperature could be reduced even further.

Up to that point the net result of the evolutionary process operation in the manufacture of the product was that throughput was increased by a factor of 25%, stirring rate was reduced from 88 to 78, and temperature was reduced from 120 to 114°F. Notice, once again, that these experimenters exposed themselves to a *learning process*. All of these changes, but especially the first, resulted in economies which could certainly not have been foreseen at the outset.

15.2. MULTIPLE RESPONSES

In the above examples only one response was recorded. Usually it is desirable to consider a number of responses simultaneously. In the example that follows three responses were of importance:

(i) The cost of manufacturing a unit weight of the product. This was obtained by dividing the cost of running at the specified conditions by the observed weight yield at the conditions. It was desired to bring this cost to the smallest value possible, subject to certain restrictions listed in (ii) and (iii) below.

(ii) The percentage of a certain impurity. It was desired that this should not exceed 0.5%.

(iii) A measure of fluidity. It was preferred that this should lie between the limits 55 and 80.

In phase III of this scheme the information display after 16 cycles is shown in Figure 15.4. The display suggests a decrease in concentration would be expected to result in reduced cost, reduced impurity, and reduced fluidity. The effect on cost of a decrease in temperature is less certain but was more likely to be favorable than not and would almost certainly result in marked reductions in impurity and fluidity. These facts had to be considered, bearing in mind that a fluidity of less than 55 was undesirable, and although a further large reduction in impurity was welcome, it was not necessary in order to meet the specification. It was decided to explore the effects of reducing concentration alone. Phase III was terminated and in the next phase the three processes (13%, 126°F), (13.5%, 126°F), (14%, 126°F) were compared. The first of these gave an average cost of 32.1 with an impurity of 0.25 and a fluidity of 60.7 and was adopted as a base for further development.

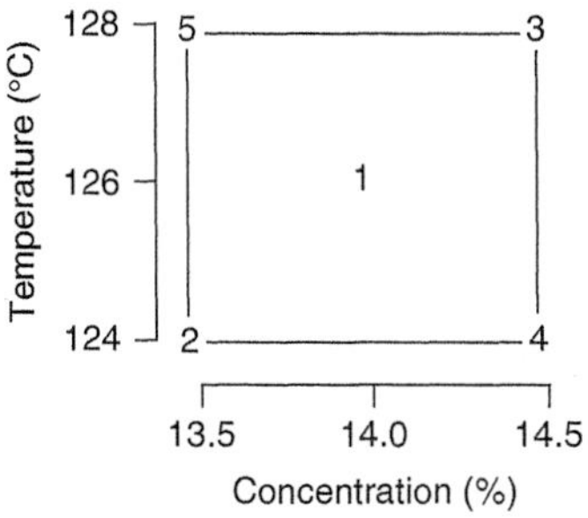

		Cost	Impurity (%)	Fluidity
Requirement		Minimum	Less than 0.50	Between 55 and 80
Running averages		32.6 33.9 32.8 32.3 33.4	0.29 0.35 0.27 0.17 0.19	73.2 76.2 71.3 60.2 67.6
95% error limits		± 0.7	± 0.03	± 1.1
Effects with 95% error limits	Concentration	1.2 ± 0.7	0.04 ± 0.03	5.2 ± 1.1
	Temperature	0.4 ± 0.7	0.14 ± 0.03	10.8 ± 1.1
	$C \times T$	0.1 ± 0.7	0.02 ± 0.03	−2.2 ± 1.1
	Change in mean	0.2 ± 0.6	−0.02 ± 0.03	−1.6 ± 1.0

Figure 15.4. Multiple responses information display after 16 cycles.

15.3. THE EVOLUTIONARY PROCESS OPERATION COMMITTEE

Selection of Variants

For many processes there is no lack of obvious candidates for factor testing and results from one phase raise questions for another. Successful innovations never previously considered for processes that had been running for a long time testify to the existence of valuable modifications *waiting to be conceived.* To make the evolutionary process effective, you must set up a situation in which useful ideas are likely to be continually forthcoming. The generation of such ideas is best induced by bringing together, perhaps every other month, a small group of people with complementary backgrounds to mull over results and help advise the process manager, who will be chairman of the committee. Candidates for such an EVOP committee would be engineers, chemists, and other technologists with intimate knowledge of the process and product as well as plant personnel. Different specialists might be consulted at various stages of the program. Rather more may be gotten from results, and more ambitious techniques adopted, if a person with statistical knowledge can be consulted. The intention should be to have complementary rather than common disciplines represented. Such a group can discuss the implications of current results and help decide on future phases of operation.

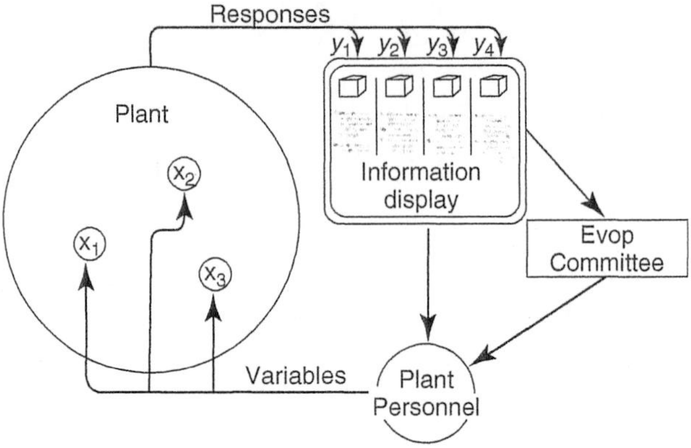

Figure 15.5. Evolutionary process operation by directed feedback.

With the establishment of this committee all the requirements for the efficient production in the evolutionary mode are satisfied and the "closed loop" illustrated in Figure 15.5 is obtained. You are thus provided with a practical method for continuous improvement of the full scale process which requires no special equipment and which can be applied to manufacturing processes with very little risk of producing bad product.

The advantages of randomization in experimental design have been frequently pointed out. With evolutionary operation the cycles are in effect small independent blocks providing experimental design data taken closely together in time. The advantages that randomization has in other experimental contexts are largely nullified in the EVOP context where the randomization of the trials within cycles can lead to confusion on the part of plant operators and is usually unnecessary.

REFERENCES AND FURTHER READING

Barnard, G. A. (1946). Sequential Test in Industrial Statistics Sup RSS, **8**, 1–26.

Box, G. E. P. (1957) Evolutionary operation: A method for increasing industrial productivity, *Appl. Statist.*, **6**, 81–101.

Box, G. E. P., and Draper, N. R. (1969) *Evolutionary Operation—A Statistical Method for Process Improvement*, Wiley, New York.

Box, G. E. P. (1994) Statistics and quality improvement, *J. Roy. Statist. Soc. Ser. A*, **157**(Pt. 2), 209–229.

Some applications of EVOP are described in:

Fogel, D. B. (1998) (Ed.) *Evolutionary Computation*, IEEE Press, New York.

Juran, J. M., and Godfrey, A. B. (1998) *Juran's Quality Handbook*, McGraw-Hill, New York.

Jenkins, G. M. (1969). A system study of a petrochemical plant, *J. Syst. Eng.*, **1**, 90.

Hunter, J. S. (2004) Improving an unstable process, *Quality Progress*, **Feb.**, 68–71.

Davies, O. L. (Ed) (1963) *Design and Analysis of Industrial Experiments*, Oliver & Boyd, London.

A discussion of Adaptive Optimization related to evolutionary process operation will be found in:

Box, G. E. P., and Chanmugam (1962) Adaptive optimization of continuous processes. *I and EC Fundamentals*, **1**, 2–16.

QUESTIONS FOR CHAPTER 15

1. Do you agree with the statement, "The philosophy of evolutionary process operation is profoundly different from that appropriate for small-scale experimentation in the lab or pilot plant"? How is it different?

2. What should an EVOP information display look like? What responses should be recorded? How are computers used in EVOP?

3. Can you think of possible changes to some system (real or imaginary) that could not be adequately tested on the small scale?

4. What is the role of the EVOP committee? Who might serve on this EVOP committee?

5. Deming said that when you transform an experimental phenomenon from one environment to another (e.g., you try to modify a full-scale process using data from experiments on the small scale) your expectations are based not on statistics but on a "leap of faith." Comment on this.

6. What can you do to make Deming's "leap of faith" less hazardous?

7. In running an EVOP scheme, what are appropriate activities for (a) process operators, (b) management personnel, and (c) the EVOP committee?

8. What is meant by an "Information Display"? How are computers used in an information display?

PROBLEMS FOR CHAPTER 15

1. Consider the statement, "While it is possible to conceive of potentially valuable full scale process changes, producing effects large enough to test such changes may have undesirable consequences."

(a) How does evolutionary process operation (EVOP) resolve this dilemma?

(b) What is meant by the signal-to-noise ratio?

(c) How is the signal-to-noise ratio magnified by the use of EVOP?

2. Consider two situations: In the first situation because of major process and product difficulties management has decided that for a short period, perhaps 2 weeks, experimentation on the full-scale process is permissible. They agree that during this period subgrade product may be produced and that highly

skilled personnel will be made available to design run and analyze experiments. In the second situation process improvement over the long term is desirable but management requires there should be little chance of producing substandard product and that only those personnel usually employed in operating the plant are available for the project.
If you were advising management, what questions might you ask, and what statistical strategies might you recommend in each of these situations?

3. What members of the operating team should see the information display? How will the display help plant personnel operate a process in the EVOP mode?

4. Suppose you were organizing the EVOP system shown in Figure 15.3. What training would you make available to (a) process operators and (b) process managers? What displays would you propose to ensure the correct running of the evolutionary process? What depth of knowledge of statistics do you think is desirable? For whom?

5. In discussing the value of EVOP how would you reply to the following comments:

(a) The process is finally operating satisfactorily and meeting specifications. Leave it alone.

(b) The EVOP technique gets information too slowly.

(c) The process is too trendy, too subject to upsets, for EVOP to work.

(d) Experimentation is the function of the research and development departments. We do not experiment with the process.

(e) Before EVOP can be run we would need to first establish that the process is in a state of statistical control.

6. What is meant by “a bird dog line”? Discuss when or how this idea might be used.

APPENDIX

Tables

Statistics for Experimenters, Second Edition. By G. E. P. Box, J. S. Hunter, and W. G. Hunter

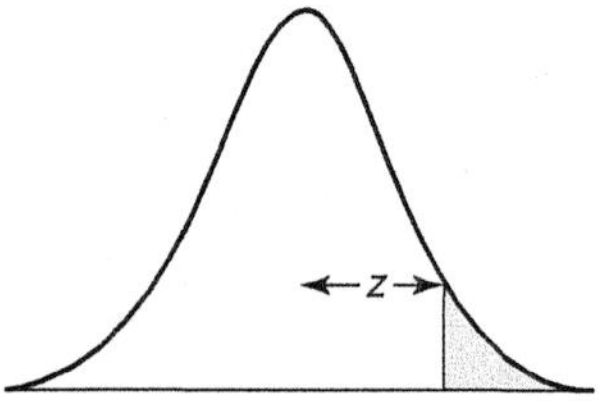

Table A. Tail Area of Unit Normal Distribution

z	0.00	0.01	0.02	0.03	0.04	0.05	0.06	0.07	0.08	0.09
0.0	0.5000	0.4960	0.4920	0.4880	0.4840	0.4801	0.4761	0.4721	0.4681	0.4641
0.1	0.4602	0.4562	0.4522	0.4483	0.4443	0.4404	0.4364	0.4325	0.4286	0.4247
0.2	0.4207	0.4168	0.4129	0.4090	0.4052	0.4013	0.3974	0.3936	0.3897	0.3859
0.3	0.3821	0.3783	0.3745	0.3707	0.3669	0.3632	0.3594	0.3557	0.3520	0.3483
0.4	0.3446	0.3409	0.3372	0.3336	0.3300	0.3264	0.3228	0.3192	0.3156	0.3121
0.5	0.3085	0.3050	0.3015	0.2981	0.2946	0.2912	0.2877	0.2843	0.2810	0.2776
0.6	0.2743	0.2709	0.2676	0.2643	0.2611	0.2578	0.2546	0.2514	0.2483	0.2451
0.7	0.2420	0.2389	0.2358	0.2327	0.2296	0.2266	0.2236	0.2206	0.2177	0.2148
0.8	0.2119	0.2090	0.2061	0.2033	0.2005	0.1977	0.1949	0.1922	0.1894	0.1867
0.9	0.1841	0.1814	0.1788	0.1762	0.1736	0.1711	0.1685	0.1660	0.1635	0.1611
1.0	0.1587	0.1562	0.1539	0.1515	0.1492	0.1469	0.1446	0.1423	0.1401	0.1379
1.1	0.1357	0.1335	0.1314	0.1292	0.1271	0.1251	0.1230	0.1210	0.1190	0.1170
1.2	0.1151	0.1131	0.1112	0.1093	0.1075	0.1056	0.1038	0.1020	0.1003	0.0985
1.3	0.0968	0.0951	0.0934	0.0918	0.0901	0.0885	0.0869	0.0853	0.0838	0.0823
1.4	0.0808	0.0793	0.0778	0.0764	0.0749	0.0735	0.0721	0.0708	0.0694	0.0681
1.5	0.0668	0.0655	0.0643	0.0630	0.0618	0.0606	0.0594	0.0582	0.0571	0.0559
1.6	0.0548	0.0537	0.0526	0.0516	0.0505	0.0495	0.0485	0.0475	0.0465	0.0455
1.7	0.0446	0.0436	0.0427	0.0418	0.0409	0.0401	0.0392	0.0384	0.0375	0.0367
1.8	0.0359	0.0351	0.0344	0.0336	0.0329	0.0322	0.0314	0.0307	0.0301	0.0294
1.9	0.0287	0.0281	0.0274	0.0268	0.0262	0.0256	0.0250	0.0244	0.0239	0.0233
2.0	0.0228	0.0222	0.0217	0.0212	0.0207	0.0202	0.0197	0.0192	0.0188	0.0183
2.1	0.0179	0.0174	0.0170	0.0166	0.0162	0.0158	0.0154	0.0150	0.0146	0.0143
2.2	0.0139	0.0136	0.0132	0.0129	0.0125	0.0122	0.0119	0.0116	0.0113	0.0110
2.3	0.0107	0.0104	0.0102	0.0099	0.0096	0.0094	0.0091	0.0089	0.0087	0.0084
2.4	0.0082	0.0080	0.0078	0.0075	0.0073	0.0071	0.0069	0.0068	0.0066	0.0064
2.5	0.0062	0.0060	0.0059	0.0057	0.0055	0.0054	0.0052	0.0051	0.0049	0.0048
2.6	0.0047	0.0045	0.0044	0.0043	0.0041	0.0040	0.0039	0.0038	0.0037	0.0036
2.7	0.0035	0.0034	0.0033	0.0032	0.0031	0.0030	0.0029	0.0028	0.0027	0.0026
2.8	0.0026	0.0025	0.0024	0.0023	0.0023	0.0022	0.0021	0.0021	0.0020	0.0019
2.9	0.0019	0.0018	0.0018	0.0017	0.0016	0.0016	0.0015	0.0015	0.0014	0.0014
3.0	0.0013	0.0013	0.0013	0.0012	0.0012	0.0011	0.0011	0.0011	0.0010	0.0010
3.1	0.0010	0.0009	0.0009	0.0009	0.0008	0.0008	0.0008	0.0008	0.0007	0.0007
3.2	0.0007	0.0007	0.0006	0.0006	0.0006	0.0006	0.0006	0.0005	0.0005	0.0005
3.3	0.0005	0.0005	0.0005	0.0004	0.0004	0.0004	0.0004	0.0004	0.0004	0.0003
3.4	0.0003	0.0003	0.0003	0.0003	0.0003	0.0003	0.0003	0.0003	0.0003	0.0002
3.5	0.0002	0.0002	0.0002	0.0002	0.0002	0.0002	0.0002	0.0002	0.0002	0.0002
3.6	0.0002	0.0002	0.0001	0.0001	0.0001	0.0001	0.0001	0.0001	0.0001	0.0001
3.7	0.0001	0.0001	0.0001	0.0001	0.0001	0.0001	0.0001	0.0001	0.0001	0.0001
3.8	0.0001	0.0001	0.0001	0.0001	0.0001	0.0001	0.0001	0.0001	0.0001	0.0001
3.9	0.0000	0.0000	0.0000	0.0000	0.0000	0.0000	0.0000	0.0000	0.0000	0.0000

Table B1. Probability Points of the *t* Distribution with ν Degrees of Freedom

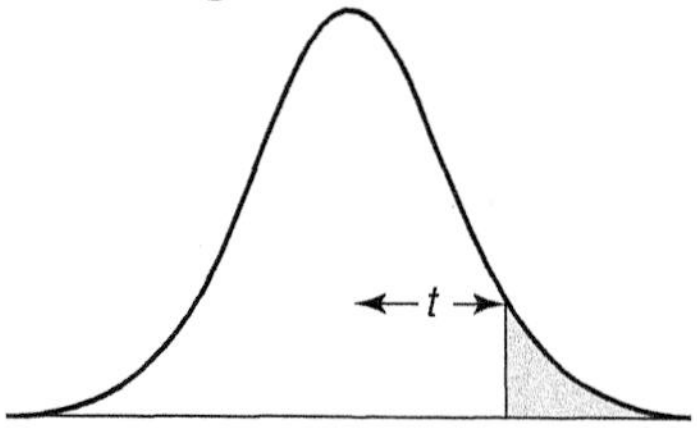

ν	Tail Area Probability									
	0.4	0.25	0.1	0.05	0.025	0.01	0.005	0.0025	0.001	0.0005
1	0.325	1.000	3.078	6.314	12.706	31.821	63.657	127.32	318.31	636.62
2	0.289	0.816	1.886	2.920	4.303	6.965	9.925	14.089	22.326	31.598
3	0.277	0.765	1.638	2.353	3.182	4.541	5.841	7.453	10.213	12.924
4	0.271	0.741	1.533	2.132	2.776	3.747	4.604	5.598	7.173	8.610
5	0.267	0.727	1.476	2.015	2.571	3.365	4.032	4.773	5.893	6.869
6	0.265	0.718	1.440	1.943	2.447	3.143	3.707	4.317	5.208	5.959
7	0.263	0.711	1.415	1.895	2.365	2.998	3.499	4.029	4.785	5.408
8	0.262	0.706	1.397	1.860	2.306	2.896	3.355	3.833	4.501	5.041
9	0.261	0.703	1.383	1.833	2.262	2.821	3.250	3.690	4.297	4.781
10	0.260	0.700	1.372	1.812	2.228	2.764	3.169	3.581	4.144	4.587
11	0.260	0.697	1.363	1.796	2.201	2.718	3.106	3.497	4.025	4.437
12	0.259	0.695	1.356	1.782	2.179	2.681	3.055	3.428	3.930	4.318
13	0.259	0.694	1.350	1.771	2.160	2.650	3.012	3.372	3.852	4.221
14	0.258	0.692	1.345	1.761	2.145	2.624	2.977	3.326	3.787	4.140
15	0.258	0.691	1.341	1.753	2.131	2.602	2.947	3.286	3.733	4.073
16	0.258	0.690	1.337	1.746	2.120	2.583	2.921	3.252	3.686	4.015
17	0.257	0.689	1.333	1.740	2.110	2.567	2.898	3.222	3.646	3.965
18	0.257	0.688	1.330	1.734	2.101	2.552	2.878	3.197	3.610	3.922
19	0.257	0.688	1.328	1.729	2.093	2.539	2.861	3.174	3.579	3.883
20	0.257	0.687	1.325	1.725	2.086	2.528	2.845	3.153	3.552	3.850
21	0.257	0.686	1.323	1.721	2.080	2.518	2.831	3.135	3.527	3.819
22	0.256	0.686	1.321	1.717	2.074	2.508	2.819	3.119	3.505	3.792
23	0.256	0.685	1.319	1.714	2.069	2.500	2.807	3.104	3.485	3.767
24	0.256	0.685	1.318	1.711	2.064	2.492	2.797	3.091	3.467	3.745
25	0.256	0.684	1.316	1.708	2.060	2.485	2.787	3.078	3.450	3.725
26	0.256	0.684	1.315	1.706	2.056	2.479	2.779	3.067	3.435	3.707
27	0.256	0.684	1.314	1.703	2.052	2.473	2.771	3.057	3.421	3.690
28	0.256	0.683	1.313	1.701	2.048	2.467	2.763	3.047	3.408	3.674
29	0.256	0.683	1.311	1.699	2.045	2.462	2.756	3.038	3.396	3.659
30	0.256	0.683	1.310	1.697	2.042	2.457	2.750	3.030	3.385	3.646
40	0.255	0.681	1.303	1.684	2.021	2.423	2.704	2.971	3.307	3.551
60	0.254	0.679	1.296	1.671	2.000	2.390	2.660	2.915	3.232	3.460
120	0.254	0.677	1.289	1.658	1.980	2.358	2.617	2.860	3.160	3.373
∞	0.253	0.674	1.282	1.645	1.960	2.326	2.576	2.807	3.090	3.291

Source: Taken with permission from E. S. Pearson and H. O. Hartley (Eds.) (1958), *Biometrika Tables for Statisticians*, Vol. 1, Cambridge University Press.

Parts of the table are also taken from Table III of Fisher, R.A., and Yates, F. (1963): *Statistical Tables for Biological, Agricultural and Medical Research*, published by Longman Group Ltd., London (previously published by Oliver and Boyd, Edinburgh), by permission of the authors and publishers.

Table B2. Ordinates of t Distribution with v Degrees of Freedom

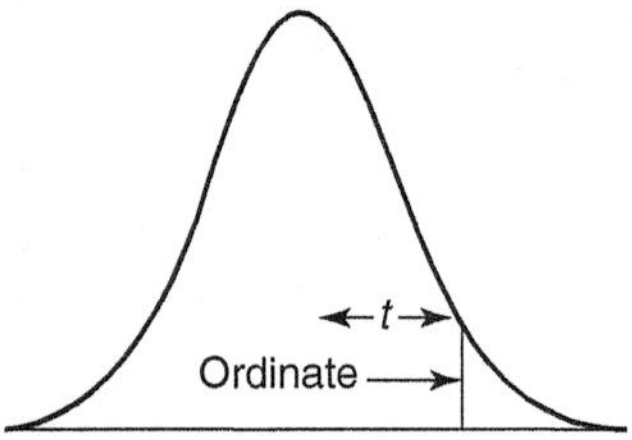

	Value of t												
v	0.00	0.25	0.50	0.75	1.00	1.25	1.50	1.75	2.00	2.25	2.50	2.75	3.00
1	0.318	0.300	0.255	0.204	0.159	0.124	0.098	0.078	0.064	0.053	0.044	0.037	0.032
2	0.354	0.338	0.296	0.244	0.193	0.149	0.114	0.088	0.068	0.053	0.042	0.034	0.027
3	0.368	0.353	0.313	0.261	0.207	0.159	0.120	0.090	0.068	0.051	0.039	0.030	0.023
4	0.375	0.361	0.322	0.270	0.215	0.164	0.123	0.091	0.066	0.049	0.036	0.026	0.020
5	0.380	0.366	0.328	0.276	0.220	0.168	0.125	0.091	0.065	0.047	0.033	0.024	0.017
6	0.383	0.369	0.332	0.280	0.223	0.170	0.126	0.090	0.064	0.045	0.032	0.022	0.016
7	0.385	0.372	0.335	0.283	0.226	0.172	0.126	0.090	0.063	0.044	0.030	0.021	0.014
8	0.387	0.373	0.337	0.285	0.228	0.173	0.127	0.090	0.062	0.043	0.029	0.019	0.013
9	0.388	0.375	0.338	0.287	0.229	0.174	0.127	0.090	0.062	0.042	0.028	0.018	0.012
10	0.389	0.376	0.340	0.288	0.230	0.175	0.127	0.090	0.061	0.041	0.027	0.018	0.011
11	0.390	0.377	0.341	0.289	0.231	0.176	0.128	0.089	0.061	0.040	0.026	0.017	0.011
12	0.391	0.378	0.342	0.290	0.232	0.176	0.128	0.089	0.060	0.040	0.026	0.016	0.010
13	0.391	0.378	0.343	0.291	0.233	0.177	0.128	0.089	0.060	0.039	0.025	0.016	0.010
14	0.392	0.379	0.343	0.292	0.234	0.177	0.128	0.089	0.060	0.039	0.025	0.015	0.010
15	0.392	0.380	0.344	0.292	0.234	0.177	0.128	0.089	0.059	0.038	0.024	0.015	0.009
16	0.393	0.380	0.344	0.293	0.235	0.178	0.128	0.089	0.059	0.038	0.024	0.015	0.009
17	0.393	0.380	0.345	0.293	0.235	0.178	0.128	0.089	0.059	0.038	0.024	0.014	0.009
18	0.393	0.381	0.345	0.294	0.235	0.178	0.129	0.088	0.059	0.037	0.023	0.014	0.008
19	0.394	0.381	0.346	0.294	0.236	0.179	0.129	0.088	0.058	0.037	0.023	0.014	0.008
20	0.394	0.381	0.346	0.294	0.236	0.179	0.129	0.088	0.058	0.037	0.023	0.014	0.008
22	0.394	0.382	0.346	0.295	0.237	0.179	0.129	0.088	0.058	0.036	0.022	0.013	0.008
24	0.395	0.382	0.347	0.296	0.237	0.179	0.129	0.088	0.057	0.036	0.022	0.013	0.007
26	0.395	0.383	0.347	0.296	0.237	0.180	0.129	0.088	0.057	0.036	0.022	0.013	0.007
28	0.395	0.383	0.348	0.296	0.238	0.180	0.129	0.088	0.057	0.036	0.021	0.012	0.007
30	0.396	0.383	0.348	0.297	0.238	0.180	0.129	0.088	0.057	0.035	0.021	0.012	0.007
35	0.396	0.384	0.348	0.297	0.239	0.180	0.129	0.088	0.056	0.035	0.021	0.012	0.006
40	0.396	0.384	0.349	0.298	0.239	0.181	0.129	0.087	0.056	0.035	0.020	0.011	0.006
45	0.397	0.384	0.349	0.298	0.239	0.181	0.129	0.087	0.056	0.034	0.020	0.011	0.006
50	0.397	0.385	0.350	0.298	0.240	0.181	0.129	0.087	0.056	0.034	0.020	0.011	0.006
∞	0.399	0.387	0.352	0.301	0.242	0.183	0.130	0.086	0.054	0.032	0.018	0.009	0.004

Table C. Probability Points of the χ^2 Distribution with ν Degrees of Freedom

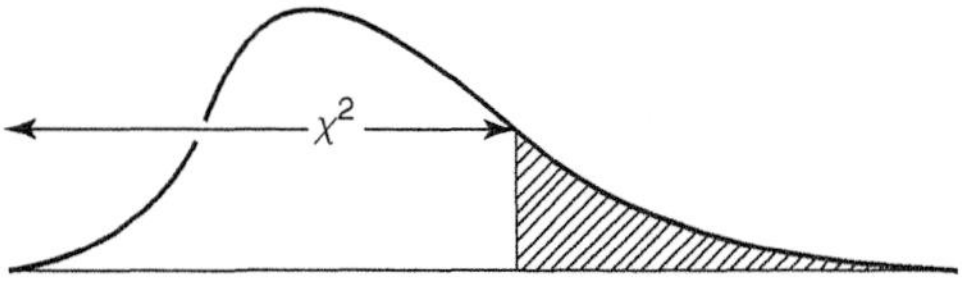

ν	Tail Area Probability													
	0.995	0.99	0.975	0.95	0.9	0.75	0.5	0.25	0.1	0.05	0.025	0.01	0.005	0.001
1	—	—	—	—	0.016	0.102	0.455	1.32	2.71	3.84	5.02	6.63	7.88	10.8
2	0.010	0.020	0.051	0.103	0.211	0.575	1.39	2.77	4.61	5.99	7.38	9.21	10.6	13.8
3	0.072	0.115	0.216	0.352	0.584	1.21	2.37	4.11	6.25	7.81	9.35	11.3	12.8	16.3
4	0.207	0.297	0.484	0.711	1.06	1.92	3.36	5.39	7.78	9.49	11.1	13.3	14.9	18.5
5	0.412	0.554	0.831	1.15	1.61	2.67	4.35	6.63	9.24	11.1	12.8	15.1	16.7	20.5
6	0.676	0.872	1.24	1.64	2.20	3.45	5.35	7.84	10.6	12.6	14.4	16.8	18.5	22.5
7	0.989	1.24	1.69	2.17	2.83	4.25	6.35	9.04	12.0	14.1	16.0	18.5	20.3	24.3
8	1.34	1.65	2.18	2.73	3.49	5.07	7.34	10.2	13.4	15.5	17.5	20.1	22.0	26.1
9	1.73	2.09	2.70	3.33	4.17	5.90	8.34	11.4	14.7	16.9	19.0	21.7	23.6	27.9
10	2.16	2.56	3.25	3.94	4.87	6.74	9.34	12.5	16.0	18.3	20.5	23.2	25.2	29.6
11	2.60	3.05	3.82	4.57	5.58	7.58	10.3	13.7	17.3	19.7	21.9	24.7	26.8	31.3
12	3.07	3.57	4.40	5.23	6.30	8.44	11.3	14.8	18.5	21.0	23.3	26.2	28.3	32.9
13	3.57	4.11	5.01	5.89	7.04	9.30	12.3	16.0	19.8	22.4	24.7	27.7	29.8	34.5
14	4.07	4.66	5.63	6.57	7.79	10.2	13.3	17.1	21.1	23.7	26.1	29.1	31.3	36.1
15	4.60	5.23	6.26	7.26	8.55	11.0	14.3	18.2	22.3	25.0	27.5	30.6	32.8	37.7
16	5.14	5.81	6.91	7.96	9.31	11.9	15.3	19.4	23.5	26.3	28.8	32.0	34.3	39.3
17	5.70	6.41	7.56	8.67	10.1	12.8	16.3	20.5	24.8	27.6	30.2	33.4	35.7	40.8
18	6.26	7.01	8.23	9.39	10.9	13.7	17.3	21.6	26.0	28.9	31.5	34.8	37.2	42.3
19	6.84	7.63	8.91	10.1	11.7	14.6	18.3	22.7	27.2	30.1	32.9	36.2	38.6	43.8
20	7.43	8.26	9.59	10.9	12.4	15.5	19.3	23.8	28.4	31.4	34.2	37.6	40.0	45.3
21	8.03	8.90	10.3	11.6	13.2	16.3	20.3	24.9	29.6	32.7	35.5	38.9	41.4	46.8
22	8.64	9.54	11.0	12.3	14.0	17.2	21.3	26.0	30.8	33.9	36.8	40.3	42.8	48.3
23	9.26	10.2	11.7	13.1	14.8	18.1	22.3	27.1	32.0	35.2	38.1	41.6	44.2	49.7
24	9.89	10.9	12.4	13.8	15.7	19.0	23.3	28.2	33.2	36.4	39.4	43.0	45.6	51.2
25	10.5	11.5	13.1	14.6	16.5	19.9	24.3	29.3	34.4	37.7	40.6	44.3	46.9	52.6
26	11.2	12.2	13.8	15.4	17.3	20.8	25.3	30.4	35.6	38.9	41.9	45.6	48.3	54.1
27	11.8	12.9	14.6	16.2	18.1	21.7	26.3	31.5	36.7	40.1	43.2	47.0	49.6	55.5
28	12.5	13.6	15.3	16.9	18.9	22.7	27.3	32.6	37.9	41.3	44.5	48.3	51.0	56.9
29	13.1	14.3	16.0	17.7	19.8	23.6	28.3	33.7	39.1	42.6	45.7	49.6	52.3	58.3
30	13.8	15.0	16.8	18.5	20.6	24.5	29.3	34.8	40.3	43.8	47.0	50.9	53.7	59.7

Source: Adapted from Table 8, Percentage points of the χ^2 distribution, in E. S. Pearson and H. O. Hartley (Eds.) (1966). *Biometrika Tables for Statisticians*, Vol. 1, 3rd ed., Cambridge University Press. Used by permission.

Table D. Percentage Points of the *F* Distribution: Upper 25% Points

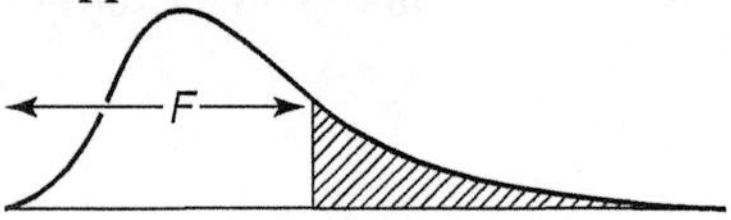

ν_2 \ ν_1	1	2	3	4	5	6	7	8	9	10	12	15	20	24	30	40	60	120	∞
1	5.83	7.50	8.20	8.58	8.82	8.98	9.10	9.19	9.26	9.32	9.41	9.49	9.58	9.63	9.67	9.71	9.76	9.80	9.85
2	2.57	3.00	3.15	3.23	3.28	3.31	3.34	3.35	3.37	3.38	3.39	3.41	3.43	3.43	3.44	3.45	3.46	3.47	3.48
3	2.02	2.28	2.36	2.39	2.41	2.42	2.43	2.44	2.44	2.44	2.45	2.46	2.46	2.46	2.47	2.47	2.47	2.47	2.47
4	1.81	2.00	2.05	2.06	2.07	2.08	2.08	2.08	2.08	2.08	2.08	2.08	2.08	2.08	2.08	2.08	2.08	2.08	2.08
5	1.69	1.85	1.88	1.89	1.89	1.89	1.89	1.89	1.89	1.89	1.89	1.89	1.88	1.88	1.88	1.88	1.87	1.87	1.87
6	1.62	1.76	1.78	1.79	1.79	1.78	1.78	1.78	1.77	1.77	1.77	1.76	1.76	1.75	1.75	1.75	1.74	1.74	1.74
7	1.57	1.70	1.72	1.72	1.71	1.71	1.70	1.70	1.69	1.69	1.68	1.68	1.67	1.67	1.66	1.66	1.65	1.65	1.65
8	1.54	1.66	1.67	1.66	1.66	1.65	1.64	1.64	1.63	1.63	1.62	1.62	1.61	1.60	1.60	1.59	1.59	1.58	1.58
9	1.51	1.62	1.63	1.63	1.62	1.61	1.60	1.60	1.59	1.59	1.58	1.57	1.56	1.56	1.55	1.54	1.54	1.53	1.53
10	1.49	1.60	1.60	1.59	1.59	1.58	1.57	1.56	1.56	1.55	1.54	1.53	1.52	1.52	1.51	1.51	1.50	1.49	1.48
11	1.47	1.58	1.58	1.57	1.56	1.55	1.54	1.53	1.53	1.52	1.51	1.50	1.49	1.49	1.48	1.47	1.47	1.46	1.45
12	1.46	1.56	1.56	1.55	1.54	1.53	1.52	1.51	1.51	1.50	1.49	1.48	1.47	1.46	1.45	1.45	1.44	1.43	1.42
13	1.45	1.55	1.55	1.53	1.52	1.51	1.50	1.49	1.49	1.48	1.47	1.46	1.45	1.44	1.43	1.42	1.42	1.41	1.40
14	1.44	1.53	1.53	1.52	1.51	1.50	1.49	1.48	1.47	1.46	1.45	1.44	1.43	1.42	1.41	1.41	1.40	1.39	1.38
15	1.43	1.52	1.52	1.51	1.49	1.48	1.47	1.46	1.46	1.45	1.44	1.43	1.41	1.41	1.40	1.39	1.38	1.37	1.36
16	1.42	1.51	1.51	1.50	1.48	1.47	1.46	1.45	1.44	1.44	1.43	1.41	1.40	1.39	1.38	1.37	1.36	1.35	1.34
17	1.42	1.51	1.50	1.49	1.47	1.46	1.45	1.44	1.43	1.43	1.41	1.40	1.39	1.38	1.37	1.36	1.35	1.34	1.33
18	1.41	1.50	1.49	1.48	1.46	1.45	1.44	1.43	1.42	1.42	1.40	1.39	1.38	1.37	1.36	1.35	1.34	1.33	1.32
19	1.41	1.49	1.49	1.47	1.46	1.44	1.43	1.42	1.41	1.41	1.40	1.38	1.37	1.36	1.35	1.34	1.33	1.32	1.30
20	1.40	1.49	1.48	1.47	1.45	1.44	1.43	1.42	1.41	1.40	1.39	1.37	1.36	1.35	1.34	1.33	1.32	1.31	1.29
21	1.40	1.48	1.48	1.46	1.44	1.43	1.42	1.41	1.40	1.39	1.38	1.37	1.35	1.34	1.33	1.32	1.31	1.30	1.28
22	1.40	1.48	1.47	1.45	1.44	1.42	1.41	1.40	1.39	1.39	1.37	1.36	1.34	1.33	1.32	1.31	1.30	1.29	1.28
23	1.39	1.47	1.47	1.45	1.43	1.42	1.41	1.40	1.39	1.38	1.37	1.35	1.34	1.33	1.32	1.31	1.30	1.28	1.27
24	1.39	1.47	1.46	1.44	1.43	1.41	1.40	1.39	1.38	1.38	1.36	1.35	1.33	1.32	1.31	1.30	1.29	1.28	1.26
25	1.39	1.47	1.46	1.44	1.42	1.41	1.40	1.39	1.38	1.37	1.36	1.34	1.33	1.32	1.31	1.29	1.28	1.27	1.25
26	1.38	1.46	1.45	1.44	1.42	1.41	1.39	1.38	1.37	1.37	1.35	1.34	1.32	1.31	1.30	1.29	1.28	1.26	1.25
27	1.38	1.46	1.45	1.43	1.42	1.40	1.39	1.38	1.37	1.36	1.35	1.33	1.32	1.31	1.30	1.28	1.27	1.26	1.24
28	1.38	1.46	1.45	1.43	1.41	1.40	1.39	1.38	1.37	1.36	1.34	1.33	1.31	1.30	1.29	1.28	1.27	1.25	1.24
29	1.38	1.45	1.45	1.43	1.41	1.40	1.38	1.37	1.36	1.35	1.34	1.32	1.31	1.30	1.29	1.27	1.26	1.25	1.23
30	1.38	1.45	1.44	1.42	1.41	1.39	1.38	1.37	1.36	1.35	1.34	1.32	1.30	1.29	1.28	1.27	1.26	1.24	1.23
40	1.36	1.44	1.42	1.40	1.39	1.37	1.36	1.35	1.34	1.33	1.31	1.30	1.28	1.26	1.25	1.24	1.22	1.21	1.19
60	1.35	1.42	1.41	1.38	1.37	1.35	1.33	1.32	1.31	1.30	1.29	1.27	1.25	1.24	1.22	1.21	1.19	1.17	1.15
120	1.34	1.40	1.39	1.37	1.35	1.33	1.31	1.30	1.29	1.28	1.26	1.24	1.22	1.21	1.19	1.18	1.16	1.13	1.10
∞	1.32	1.39	1.37	1.35	1.33	1.31	1.29	1.28	1.27	1.25	1.24	1.22	1.19	1.18	1.16	1.14	1.12	1.08	1.00

Table D. (*continued*), Percentage Points of the F Distribution: Upper 10% Points

ν_2 \ ν_1	1	2	3	4	5	6	7	8	9	10	12	15	20	24	30	40	60	120	∞
1	39.86	49.50	53.59	55.83	57.24	58.20	58.91	59.44	59.86	60.19	60.71	61.22	61.74	62.00	62.26	62.53	62.79	63.06	63.33
2	8.53	9.00	9.16	9.24	9.29	9.33	9.35	9.37	9.38	9.39	9.41	9.42	9.44	9.45	9.46	9.47	9.47	9.48	9.49
3	5.54	5.46	5.39	5.34	5.31	5.28	5.27	5.25	5.24	5.23	5.22	5.20	5.18	5.18	5.17	5.16	5.15	5.14	5.13
4	4.54	4.32	4.19	4.11	4.05	4.01	3.98	3.95	3.94	3.92	3.90	3.87	3.84	3.83	3.82	3.80	3.79	3.78	3.76
5	4.06	3.78	3.62	3.52	3.45	3.40	3.37	3.34	3.32	3.30	3.27	3.24	3.21	3.19	3.17	3.16	3.14	3.12	3.10
6	3.78	3.46	3.29	3.18	3.11	3.05	3.01	2.98	2.96	2.94	2.90	2.87	2.84	2.82	2.80	2.78	2.76	2.74	2.72
7	3.59	3.26	3.07	2.96	2.88	2.83	2.78	2.75	2.72	2.70	2.67	2.63	2.59	2.58	2.56	2.54	2.51	2.49	2.47
8	3.46	3.11	2.92	2.81	2.73	2.67	2.62	2.59	2.56	2.54	2.50	2.46	2.42	2.40	2.38	2.36	2.34	2.32	2.29
9	3.36	3.01	2.81	2.69	2.61	2.55	2.51	2.47	2.44	2.42	2.38	2.34	2.30	2.28	2.25	2.23	2.21	2.18	2.16
10	3.29	2.92	2.73	2.61	2.52	2.46	2.41	2.38	2.35	2.32	2.28	2.24	2.20	2.18	2.16	2.13	2.11	2.08	2.06
11	3.23	2.86	2.66	2.54	2.45	2.39	2.34	2.30	2.27	2.25	2.21	2.17	2.12	2.10	2.08	2.05	2.03	2.00	1.97
12	3.18	2.81	2.61	2.48	2.39	2.33	2.28	2.24	2.21	2.19	2.15	2.10	2.06	2.04	2.01	1.99	1.96	1.93	1.90
13	3.14	2.76	2.56	2.43	2.35	2.28	2.23	2.20	2.16	2.14	2.10	2.05	2.01	1.98	1.96	1.93	1.90	1.88	1.85
14	3.10	2.73	2.52	2.39	2.31	2.24	2.19	2.15	2.12	2.10	2.05	2.01	1.96	1.94	1.91	1.89	1.86	1.83	1.80
15	3.07	2.70	2.49	2.36	2.27	2.21	2.16	2.12	2.09	2.06	2.02	1.97	1.92	1.90	1.87	1.85	1.82	1.79	1.76
16	3.05	2.67	2.46	2.33	2.24	2.18	2.13	2.09	2.06	2.03	1.99	1.94	1.89	1.87	1.84	1.81	1.78	1.75	1.72
17	3.03	2.64	2.44	2.31	2.22	2.15	2.10	2.06	2.03	2.00	1.96	1.91	1.86	1.84	1.81	1.78	1.75	1.72	1.69
18	3.01	2.62	2.42	2.29	2.20	2.13	2.08	2.04	2.00	1.98	1.93	1.89	1.84	1.81	1.78	1.75	1.72	1.69	1.66
19	2.99	2.61	2.40	2.27	2.18	2.11	2.06	2.02	1.98	1.96	1.91	1.86	1.81	1.79	1.76	1.73	1.70	1.67	1.63
20	2.97	2.59	2.38	2.25	2.16	2.09	2.04	2.00	1.96	1.94	1.89	1.84	1.79	1.77	1.74	1.71	1.68	1.64	1.61
21	2.96	2.57	2.36	2.23	2.14	2.08	2.02	1.98	1.95	1.92	1.87	1.83	1.78	1.75	1.72	1.69	1.66	1.62	1.59
22	2.95	2.56	2.35	2.22	2.13	2.06	2.01	1.97	1.93	1.90	1.86	1.81	1.76	1.73	1.70	1.67	1.64	1.60	1.57
23	2.94	2.55	2.34	2.21	2.11	2.05	1.99	1.95	1.92	1.89	1.84	1.80	1.74	1.72	1.69	1.66	1.62	1.59	1.55
24	2.93	2.54	2.33	2.19	2.10	2.04	1.98	1.94	1.91	1.88	1.83	1.78	1.73	1.70	1.67	1.64	1.61	1.57	1.53
25	2.92	2.53	2.32	2.18	2.09	2.02	1.97	1.93	1.89	1.87	1.82	1.77	1.72	1.69	1.66	1.63	1.59	1.56	1.52
26	2.91	2.52	2.31	2.17	2.08	2.01	1.96	1.92	1.88	1.86	1.81	1.76	1.71	1.68	1.65	1.61	1.58	1.54	1.50
27	2.90	2.51	2.30	2.17	2.07	2.00	1.95	1.91	1.87	1.85	1.80	1.75	1.70	1.67	1.64	1.60	1.57	1.53	1.49
28	2.89	2.50	2.29	2.16	2.06	2.00	1.94	1.90	1.87	1.84	1.79	1.74	1.69	1.66	1.63	1.59	1.56	1.52	1.48
29	2.89	2.50	2.28	2.15	2.06	1.99	1.93	1.89	1.86	1.83	1.78	1.73	1.68	1.65	1.62	1.58	1.55	1.51	1.47
30	2.88	2.49	2.28	2.14	2.05	1.98	1.93	1.88	1.85	1.82	1.77	1.72	1.67	1.64	1.61	1.57	1.54	1.50	1.46
40	2.84	2.44	2.23	2.09	2.00	1.93	1.87	1.83	1.79	1.76	1.71	1.66	1.61	1.57	1.54	1.51	1.47	1.42	1.38
60	2.79	2.39	2.18	2.04	1.95	1.87	1.82	1.77	1.74	1.71	1.66	1.60	1.54	1.51	1.48	1.44	1.40	1.35	1.29
120	2.75	2.35	2.13	1.99	1.90	1.82	1.77	1.72	1.68	1.65	1.60	1.55	1.48	1.45	1.41	1.37	1.32	1.26	1.19
∞	2.71	2.30	2.08	1.94	1.85	1.77	1.72	1.67	1.63	1.60	1.55	1.49	1.42	1.38	1.34	1.30	1.24	1.17	1.00

Table D. (*continued*), Percentage Points of the F Distribution: Upper 5% Points

ν_2 \ ν_1	1	2	3	4	5	6	7	8	9	10	12	15	20	24	30	40	60	120	∞
1	161.4	199.5	215.7	224.6	230.2	234.0	236.8	238.9	240.5	241.9	243.9	245.9	248.0	249.1	250.1	251.1	252.2	253.3	254.3
2	18.51	19.00	19.16	19.25	19.30	19.33	19.35	19.37	19.38	19.40	19.41	19.43	19.45	19.45	19.46	19.47	19.48	19.49	19.50
3	10.13	9.55	9.28	9.12	9.01	8.94	8.89	8.85	8.81	8.79	8.74	8.70	8.66	8.64	8.62	8.59	8.57	8.55	8.53
4	7.71	6.94	6.59	6.39	6.26	6.16	6.09	6.04	6.00	5.96	5.91	5.86	5.80	5.77	5.75	5.72	5.69	5.66	5.63
5	6.61	5.79	5.41	5.19	5.05	4.95	4.88	4.82	4.77	4.74	4.68	4.62	4.56	4.53	4.50	4.46	4.43	4.40	4.36
6	5.99	5.14	4.76	4.53	4.39	4.28	4.21	4.15	4.10	4.06	4.00	3.94	3.87	3.84	3.81	3.77	3.74	3.70	3.67
7	5.59	4.74	4.35	4.12	3.97	3.87	3.79	3.73	3.68	3.64	3.57	3.51	3.44	3.41	3.38	3.34	3.30	3.27	3.23
8	5.32	4.46	4.07	3.84	3.69	3.58	3.50	3.44	3.39	3.35	3.28	3.22	3.15	3.12	3.08	3.04	3.01	2.97	2.93
9	5.12	4.26	3.86	3.63	3.48	3.37	3.29	3.23	3.18	3.14	3.07	3.01	2.94	2.90	2.86	2.83	2.79	2.75	2.71
10	4.96	4.10	3.71	3.48	3.33	3.22	3.14	3.07	3.02	2.98	2.91	2.85	2.77	2.74	2.70	2.66	2.62	2.58	2.54
11	4.84	3.98	3.59	3.36	3.20	3.09	3.01	2.95	2.90	2.85	2.79	2.72	2.65	2.61	2.57	2.53	2.49	2.45	2.40
12	4.75	3.89	3.49	3.26	3.11	3.00	2.91	2.85	2.80	2.75	2.69	2.62	2.54	2.51	2.47	2.43	2.38	2.34	2.30
13	4.67	3.81	3.41	3.18	3.03	2.92	2.83	2.77	2.71	2.67	2.60	2.53	2.46	2.42	2.38	2.34	2.30	2.25	2.21
14	4.60	3.74	3.34	3.11	2.96	2.85	2.76	2.70	2.65	2.60	2.53	2.46	2.39	2.35	2.31	2.27	2.22	2.18	2.13
15	4.54	3.68	3.29	3.06	2.90	2.79	2.71	2.64	2.59	2.54	2.48	2.40	2.33	2.29	2.25	2.20	2.16	2.11	2.07
16	4.49	3.63	3.24	3.01	2.85	2.74	2.66	2.59	2.54	2.49	2.42	2.35	2.28	2.24	2.19	2.15	2.11	2.06	2.01
17	4.45	3.59	3.20	2.96	2.81	2.70	2.61	2.55	2.49	2.45	2.38	2.31	2.23	2.19	2.15	2.10	2.06	2.01	1.96
18	4.41	3.55	3.16	2.93	2.77	2.66	2.58	2.51	2.46	2.41	2.34	2.27	2.19	2.15	2.11	2.06	2.02	1.97	1.92
19	4.38	3.52	3.13	2.90	2.74	2.63	2.54	2.48	2.42	2.38	2.31	2.23	2.16	2.11	2.07	2.03	1.98	1.93	1.88
20	4.35	3.49	3.10	2.87	2.71	2.60	2.51	2.45	2.39	2.35	2.28	2.20	2.12	2.08	2.04	1.99	1.95	1.90	1.84
21	4.32	3.47	3.07	2.84	2.68	2.57	2.49	2.42	2.37	2.32	2.25	2.18	2.10	2.05	2.01	1.96	1.92	1.87	1.81
22	4.30	3.44	3.05	2.82	2.66	2.55	2.46	2.40	2.34	2.30	2.23	2.15	2.07	2.03	1.98	1.94	1.89	1.84	1.78
23	4.28	3.42	3.03	2.80	2.64	2.53	2.44	2.37	2.32	2.27	2.20	2.13	2.05	2.01	1.96	1.91	1.86	1.81	1.76
24	4.26	3.40	3.01	2.78	2.62	2.51	2.42	2.36	2.30	2.25	2.18	2.11	2.03	1.98	1.94	1.89	1.84	1.79	1.73
25	4.24	3.39	2.99	2.76	2.60	2.49	2.40	2.34	2.28	2.24	2.16	2.09	2.01	1.96	1.92	1.87	1.82	1.77	1.71
26	4.23	3.37	2.98	2.74	2.59	2.47	2.39	2.32	2.27	2.22	2.15	2.07	1.99	1.95	1.90	1.85	1.80	1.75	1.69
27	4.21	3.35	2.96	2.73	2.57	2.46	2.37	2.31	2.25	2.20	2.13	2.06	1.97	1.93	1.88	1.84	1.79	1.73	1.67
28	4.20	3.34	2.95	2.71	2.56	2.45	2.36	2.29	2.24	2.19	2.12	2.04	1.96	1.91	1.87	1.82	1.77	1.71	1.65
29	4.18	3.33	2.93	2.70	2.55	2.43	2.35	2.28	2.22	2.18	2.10	2.03	1.94	1.90	1.85	1.81	1.75	1.70	1.64
30	4.17	3.32	2.92	2.69	2.53	2.42	2.33	2.27	2.21	2.16	2.09	2.01	1.93	1.89	1.84	1.79	1.74	1.68	1.62
40	4.08	3.23	2.84	2.61	2.45	2.34	2.25	2.18	2.12	2.08	2.00	1.92	1.84	1.79	1.74	1.69	1.64	1.58	1.51
60	4.00	3.15	2.76	2.53	2.37	2.25	2.17	2.10	2.04	1.99	1.92	1.84	1.75	1.70	1.65	1.59	1.53	1.47	1.39
120	3.92	3.07	2.68	2.45	2.29	2.17	2.09	2.02	1.96	1.91	1.83	1.75	1.66	1.61	1.55	1.50	1.43	1.35	1.25
∞	3.84	3.00	2.60	2.37	2.21	2.10	2.01	1.94	1.88	1.83	1.75	1.67	1.57	1.52	1.46	1.39	1.32	1.22	1.00

Table D. (*continued*), Percentage Points of the *F* Distribution: Upper 1% Points

ν_2 \ ν_1	1	2	3	4	5	6	7	8	9	10	12	15	20	24	30	40	60	120	∞
1	4052	4999.50	5403	5625	5764	5859	5928	5982	6022	6056	6106	6157	6209	6235	6261	6287	6313	6339	6366
2	98.50	99.00	99.17	99.25	99.30	99.33	99.36	99.37	99.39	99.40	99.42	99.43	99.45	99.46	99.47	99.47	99.48	99.49	99.50
3	34.12	30.82	29.46	28.71	28.24	27.91	27.67	27.49	27.35	27.23	27.05	26.87	26.69	26.60	26.50	26.41	26.32	26.22	26.13
4	21.20	18.00	16.69	15.98	15.52	15.21	14.98	14.80	14.66	14.55	14.37	14.20	14.02	13.93	13.84	13.75	13.65	13.56	13.46
5	16.26	13.27	12.06	11.39	10.97	10.67	10.46	10.29	10.16	10.05	9.89	9.72	9.55	9.47	9.38	9.29	9.20	9.11	9.02
6	13.75	10.92	9.78	9.15	8.75	8.47	8.26	8.10	7.98	7.87	7.72	7.56	7.40	7.31	7.23	7.14	7.06	6.97	6.88
7	12.25	9.55	8.45	7.85	7.46	7.19	6.99	6.84	6.72	6.62	6.47	6.31	6.16	6.07	5.99	5.91	5.82	5.74	5.65
8	11.26	8.65	7.59	7.01	6.63	6.37	6.18	6.03	5.91	5.81	5.67	5.52	5.36	5.28	5.20	5.12	5.03	4.95	4.86
9	10.56	8.02	6.99	6.42	6.06	5.80	5.61	5.47	5.35	5.26	5.11	4.96	4.81	4.73	4.65	4.57	4.48	4.40	4.31
10	10.04	7.56	6.55	5.99	5.64	5.39	5.20	5.06	4.94	4.85	4.71	4.56	4.41	4.33	4.25	4.17	4.08	4.00	3.91
11	9.65	7.21	6.22	5.67	5.32	5.07	4.89	4.74	4.63	4.54	4.40	4.25	4.10	4.02	3.94	3.86	3.78	3.69	3.60
12	9.33	6.93	5.95	5.41	5.06	4.82	4.64	4.50	4.39	4.30	4.16	4.01	3.86	3.78	3.70	3.62	3.54	3.45	3.36
13	9.07	6.70	5.74	5.21	4.86	4.62	4.44	4.30	4.19	4.10	3.96	3.82	3.66	3.59	3.51	3.43	3.34	3.25	3.17
14	8.86	6.51	5.56	5.04	4.69	4.46	4.28	4.14	4.03	3.94	3.80	3.66	3.51	3.43	3.35	3.27	3.18	3.09	3.00
15	8.68	6.36	5.42	4.89	4.56	4.32	4.14	4.00	3.89	3.80	3.67	3.52	3.37	3.29	3.21	3.13	3.05	2.96	2.87
16	8.53	6.23	5.29	4.77	4.44	4.20	4.03	3.89	3.78	3.69	3.55	3.41	3.26	3.18	3.10	3.02	2.93	2.84	2.75
17	8.40	6.11	5.18	4.67	4.34	4.10	3.93	3.79	3.68	3.59	3.46	3.31	3.16	3.08	3.00	2.92	2.83	2.75	2.65
18	8.29	6.01	5.09	4.58	4.25	4.01	3.84	3.71	3.60	3.51	3.37	3.23	3.08	3.00	2.92	2.84	2.75	2.66	2.57
19	8.18	5.93	5.01	4.50	4.17	3.94	3.77	3.63	3.52	3.43	3.30	3.15	3.00	2.92	2.84	2.76	2.67	2.58	2.49
20	8.10	5.85	4.94	4.43	4.10	3.87	3.70	3.56	3.46	3.37	3.23	3.09	2.94	2.86	2.78	2.69	2.61	2.52	2.42
21	8.02	5.78	4.87	4.37	4.04	3.81	3.64	3.51	3.40	3.31	3.17	3.03	2.88	2.80	2.72	2.64	2.55	2.46	2.36
22	7.95	5.72	4.82	4.31	3.99	3.76	3.59	3.45	3.35	3.26	3.12	2.98	2.83	2.75	2.67	2.58	2.50	2.40	2.31
23	7.88	5.66	4.76	4.26	3.94	3.71	3.54	3.41	3.30	3.21	3.07	2.93	2.78	2.70	2.62	2.54	2.45	2.35	2.26
24	7.82	5.61	4.72	4.22	3.90	3.67	3.50	3.36	3.26	3.17	3.03	2.89	2.74	2.66	2.58	2.49	2.40	2.31	2.21
25	7.77	5.57	4.68	4.18	3.85	3.63	3.46	3.32	3.22	3.13	2.99	2.85	2.70	2.62	2.54	2.45	2.36	2.27	2.17
26	7.72	5.53	4.64	4.14	3.82	3.59	3.42	3.29	3.18	3.09	2.96	2.81	2.66	2.58	2.50	2.42	2.33	2.23	2.13
27	7.68	5.49	4.60	4.11	3.78	3.56	3.39	3.26	3.15	3.06	2.93	2.78	2.63	2.55	2.47	2.38	2.29	2.20	2.10
28	7.64	5.45	4.57	4.07	3.75	3.53	3.36	3.23	3.12	3.03	2.90	2.75	2.60	2.52	2.44	2.35	2.26	2.17	2.06
29	7.60	5.42	4.54	4.04	3.73	3.50	3.33	3.20	3.09	3.00	2.87	2.73	2.57	2.49	2.41	2.33	2.23	2.14	2.03
30	7.56	5.39	4.51	4.02	3.70	3.47	3.30	3.17	3.07	2.98	2.84	2.70	2.55	2.47	2.39	2.30	2.21	2.11	2.01
40	7.31	5.18	4.31	3.83	3.51	3.29	3.12	2.99	2.89	2.80	2.66	2.52	2.37	2.29	2.20	2.11	2.02	1.92	1.80
60	7.08	4.98	4.13	3.65	3.34	3.12	2.95	2.82	2.72	2.63	2.50	2.35	2.20	2.12	2.03	1.94	1.84	1.73	1.60
120	6.85	4.79	3.95	3.48	3.17	2.96	2.79	2.66	2.56	2.47	2.34	2.19	2.03	1.95	1.86	1.76	1.66	1.53	1.38
∞	6.63	4.61	3.78	3.32	3.02	2.80	2.64	2.51	2.41	2.32	2.18	2.04	1.88	1.79	1.70	1.59	1.47	1.32	1.00

Table D. (*continued*), Percentage Points of the *F* Distribution: Upper 0.1% Points

ν_2 \ ν_1	1	2	3	4	5	6	7	8	9	10	12	15	20	24	30	40	60	120	∞
1	4053	5000	5404	5625	5764	5859	5929	5981	6023	6056	6107	6158	6209	6235	6261	6287	6313	6340	6366
2	998.5	999.0	999.2	999.2	999.3	999.3	999.4	999.4	999.4	999.4	999.4	999.4	999.4	999.5	999.5	999.5	999.5	999.5	999.5
3	167.0	148.5	141.1	137.1	134.6	132.8	131.6	130.6	129.9	129.2	128.3	127.4	126.4	125.9	125.4	125.0	124.5	124.0	123.5
4	74.14	61.25	56.18	53.44	51.71	50.53	49.66	49.00	48.47	48.05	47.41	46.76	46.10	45.77	45.43	45.09	44.75	44.40	44.05
5	47.18	37.12	33.20	31.09	29.75	28.84	28.16	27.64	27.24	26.92	26.42	25.91	25.39	25.14	24.87	24.60	24.33	24.06	23.79
6	35.51	27.00	23.70	21.92	20.81	20.03	19.46	19.03	18.69	18.41	17.99	17.56	17.12	16.89	16.67	16.44	16.21	15.99	15.75
7	29.25	21.69	18.77	17.19	16.21	15.52	15.02	14.63	14.33	14.08	13.71	13.32	12.93	12.73	12.53	12.33	12.12	11.91	11.70
8	25.42	18.49	15.83	14.39	13.49	12.86	12.40	12.04	11.77	11.54	11.19	10.84	10.48	10.30	10.11	9.92	9.73	9.53	9.33
9	22.86	16.39	13.90	12.56	11.71	11.13	10.70	10.37	10.11	9.89	9.57	9.24	8.90	8.72	8.55	8.37	8.19	8.00	7.81
10	21.04	14.91	12.55	11.28	10.48	9.92	9.52	9.20	8.96	8.75	8.45	8.13	7.80	7.64	7.47	7.30	7.12	6.94	6.76
11	19.69	13.81	11.56	10.35	9.58	9.05	8.66	8.35	8.12	7.92	7.63	7.32	7.01	6.85	6.68	6.52	6.35	6.17	6.00
12	18.64	12.97	10.80	9.63	8.89	8.38	8.00	7.71	7.48	7.29	7.00	6.71	6.40	6.25	6.09	5.93	5.76	5.59	5.42
13	17.81	12.31	10.21	9.07	8.35	7.86	7.49	7.21	6.98	6.80	6.52	6.23	5.93	5.78	5.63	5.47	5.30	5.14	4.97
14	17.14	11.78	9.73	8.62	7.92	7.43	7.08	6.80	6.58	6.40	6.13	5.85	5.56	5.41	5.25	5.10	4.94	4.77	4.60
15	16.59	11.34	9.34	8.25	7.57	7.09	6.74	6.47	6.26	6.08	5.81	5.54	5.25	5.10	4.95	4.80	4.64	4.47	4.31
16	16.12	10.97	9.00	7.94	7.27	6.81	6.46	6.19	5.98	5.81	5.55	5.27	4.99	4.85	4.70	4.54	4.39	4.23	4.06
17	15.72	10.66	8.73	7.68	7.02	6.56	6.22	5.96	5.75	5.58	5.32	5.05	4.78	4.63	4.48	4.33	4.18	4.02	3.85
18	15.38	10.39	8.49	7.46	6.81	6.35	6.02	5.76	5.56	5.39	5.13	4.87	4.59	4.45	4.30	4.15	4.00	3.84	3.67
19	15.08	10.16	8.28	7.26	6.62	6.18	5.85	5.59	5.39	5.22	4.97	4.70	4.43	4.29	4.14	3.99	3.84	3.68	3.51
20	14.82	9.95	8.10	7.10	6.46	6.02	5.69	5.44	5.24	5.08	4.82	4.56	4.29	4.15	4.00	3.86	3.70	3.54	3.38
21	14.59	9.77	7.94	6.95	6.32	5.88	5.56	5.31	5.11	4.95	4.70	4.44	4.17	4.03	3.88	3.74	3.58	3.42	3.26
22	14.38	9.61	7.80	6.81	6.19	5.76	5.44	5.19	4.99	4.83	4.58	4.33	4.06	3.92	3.78	3.63	3.48	3.32	3.15
23	14.19	9.47	7.67	6.69	6.08	5.65	5.33	5.09	4.89	4.73	4.48	4.23	3.96	3.82	3.68	3.53	3.38	3.22	3.05
24	14.03	9.34	7.55	6.59	5.98	5.55	5.23	4.99	4.80	4.64	4.39	4.14	3.87	3.74	3.59	3.45	3.29	3.14	2.97
25	13.88	9.22	7.45	6.49	5.88	5.46	5.15	4.91	4.71	4.56	4.31	4.06	3.79	3.66	3.52	3.37	3.22	3.06	2.89
26	13.74	9.12	7.36	6.41	5.80	5.38	5.07	4.83	4.64	4.48	4.24	3.99	3.72	3.59	3.44	3.30	3.15	2.99	2.82
27	13.61	9.02	7.27	6.33	5.73	5.31	5.00	4.76	4.57	4.41	4.17	3.92	3.66	3.52	3.38	3.23	3.08	2.92	2.75
28	13.50	8.93	7.19	6.25	5.66	5.24	4.93	4.69	4.50	4.35	4.11	3.86	3.60	3.46	3.32	3.18	3.02	2.86	2.69
29	13.39	8.85	7.12	6.19	5.59	5.18	4.87	4.64	4.45	4.29	4.05	3.80	3.54	3.41	3.27	3.12	2.97	2.81	2.64
30	13.29	8.77	7.05	6.12	5.53	5.12	4.82	4.58	4.39	4.24	4.00	3.75	3.49	3.36	3.22	3.07	2.92	2.76	2.59
40	12.61	8.25	6.60	5.70	5.13	4.73	4.44	4.21	4.02	3.87	3.64	3.40	3.15	3.01	2.87	2.73	2.57	2.41	2.23
60	11.97	7.76	6.17	5.31	4.76	4.37	4.09	3.87	3.69	3.54	3.31	3.08	2.83	2.69	2.55	2.41	2.25	2.08	1.89
120	11.38	7.32	5.79	4.95	4.42	4.04	3.77	3.55	3.38	3.24	3.02	2.78	2.53	2.40	2.26	2.11	1.95	1.76	1.54
∞	10.83	6.91	5.42	4.62	4.10	3.74	3.47	3.27	3.10	2.96	2.74	2.51	2.27	2.13	1.99	1.84	1.66	1.45	1.00

*Multiply these entries by 100.

Table E. Scales for Normal Plots

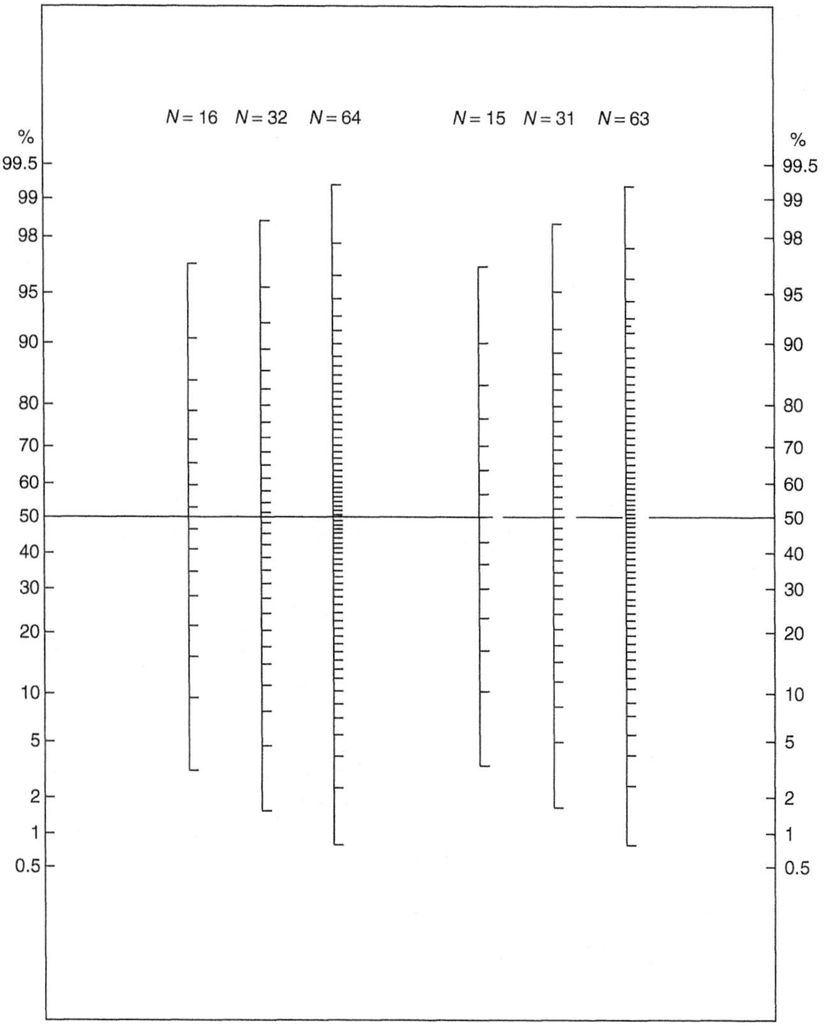

Table F. Confidence Limits for p in Binomial Sampling, Given a Sample Fraction $\hat{p} = y/n$: Confidence Coefficient = 0.95[a]

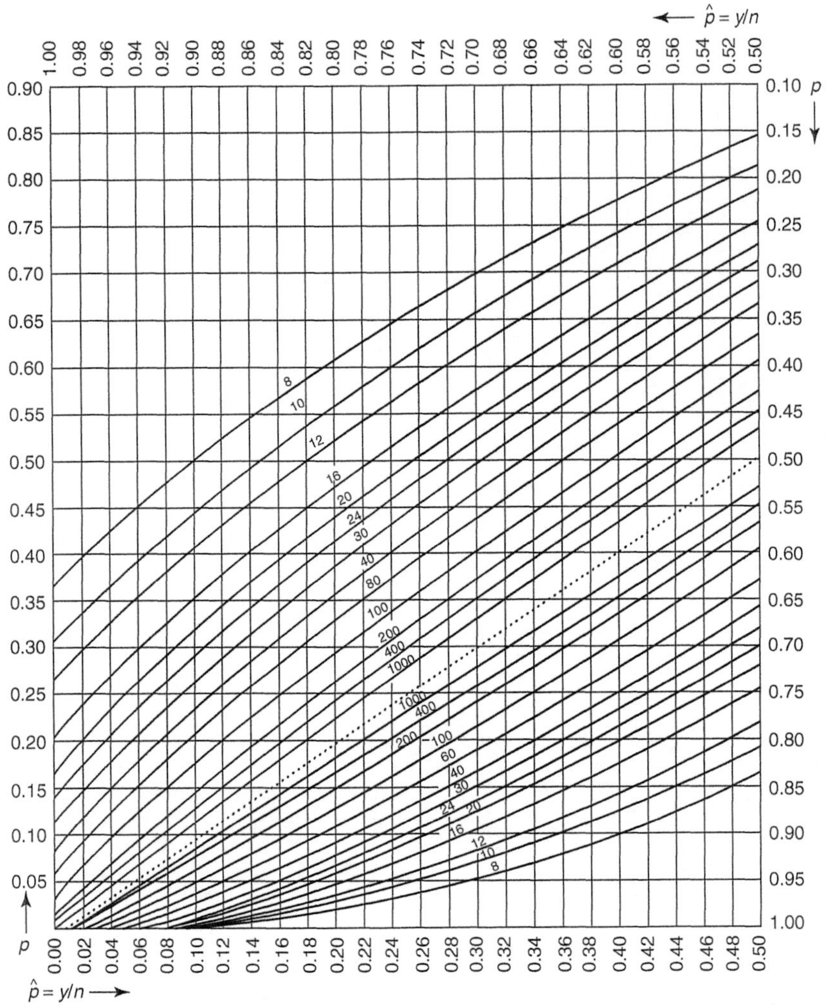

[a]The numbers printed along the curves indicate the sample size n. If, for a given value of the abscissa y/n, p_A and p_B are the ordinates read from (or interpolated between) the appropriate lower and upper curves, then $Pr(p_A \le p \le p_B) \le 0.95$.

Table F. (*continued*) **Confidence Coefficient = 0.99**

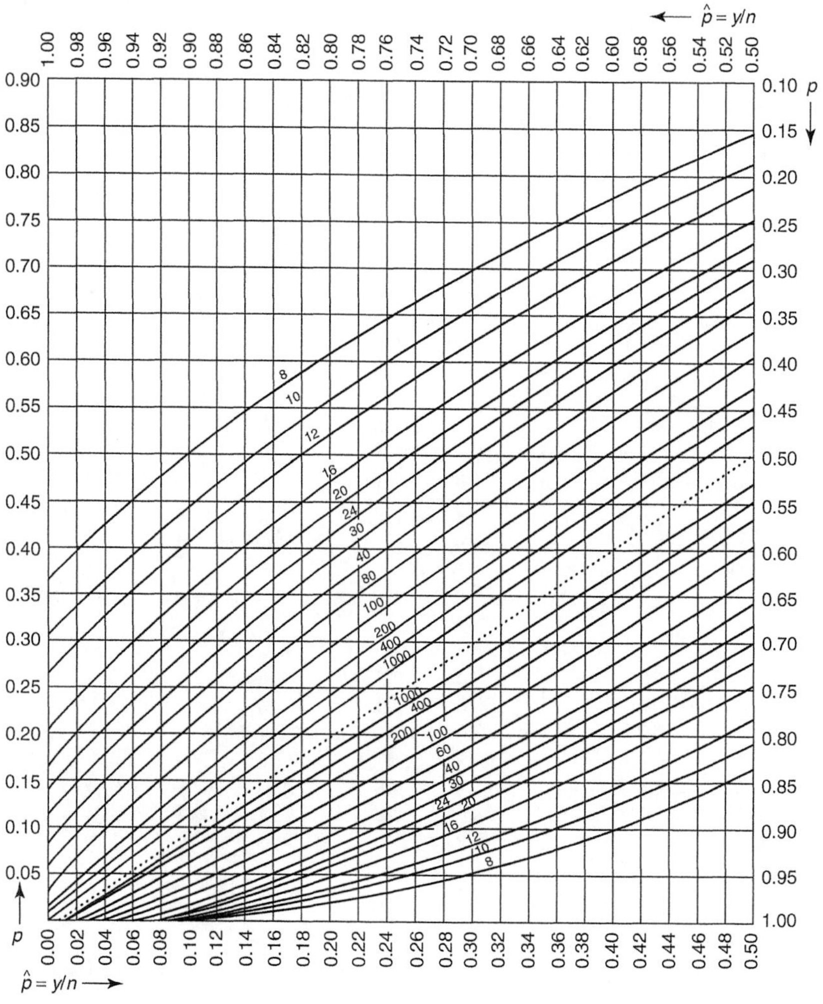

Table G. Confidence Limits for the Mean Value of a Poisson Variable

observed frequency	99% limit lower	99% limit upper	95% limit lower	95% limit upper	observed frequency	99% limit lower	99% limit upper	95% limit lower	95% limit upper
0	0.0	5.3	0.0	3.7					
1	0.0	7.4	0.1	5.6	26	14.7	42.2	17.0	38.0
2	0.1	9.3	0.2	7.2	27	15.4	43.5	17.8	39.2
3	0.3	11.0	0.6	8.8	28	16.2	44.8	18.6	40.4
4	0.6	12.6	1.0	10.2	29	17.0	46.0	19.4	41.6
5	1.0	14.1	1.6	11.7	30	17.7	47.2	20.2	42.8
6	1.5	15.6	2.2	13.1	31	18.5	48.4	21.0	44.0
7	2.0	17.1	2.8	14.4	32	19.3	49.6	21.8	45.1
8	2.5	18.5	3.4	15.8	33	20.0	50.8	22.7	46.3
9	3.1	20.0	4.0	17.1	34	20.8	52.1	23.5	47.5
10	3.7	21.3	4.7	18.4	35	21.6	53.3	24.3	48.7
11	4.3	22.6	5.4	19.7	36	22.4	54.5	25.1	49.8
12	4.9	24.0	6.2	21.0	37	23.2	55.7	26.0	51.0
13	5.5	25.4	6.9	22.3	38	24.0	56.9	26.8	52.2
14	6.2	26.7	7.7	23.5	39	24.8	58.1	27.7	53.3
15	6.8	28.1	8.4	24.8	40	25.6	59.3	28.6	54.5
16	7.5	29.4	9.4	26.0	41	26.4	60.5	29.4	55.6
17	8.2	30.7	9.9	27.2	42	27.2	61.7	30.3	56.8
18	8.9	32.0	10.7	28.4	43	28.0	62.9	31.1	57.9
19	9.6	33.3	11.5	29.6	44	28.8	64.1	32.0	59.0
20	10.3	34.6	12.2	30.8	45	29.6	65.3	32.8	60.2
21	11.0	35.9	13.0	32.0	46	30.4	66.5	33.6	61.3
22	11.8	37.2	13.8	33.2	47	31.2	67.7	34.5	62.5
23	12.5	38.4	14.6	34.4	48	32.0	68.9	35.3	63.6
24	13.2	39.7	15.4	35.6	49	32.8	70.1	36.1	64.8
25	14.0	41.0	16.2	36.8	50	33.6	71.3	37.0	65.9

Author Index

Statistics for Experimenters, Second Edition. By G. E. P. Box, J. S. Hunter, and W. G. Hunter

Subject Index

WILEY SERIES IN PROBABILITY AND STATISTICS
ESTABLISHED BY WALTER A. SHEWHART AND SAMUEL S. WILKS

Editors: *David J. Balding, Noel A. C. Cressie, Nicholas I. Fisher, Iain M. Johnstone, J. B. Kadane, Geert Molenberghs. Louise M. Ryan, David W. Scott, Adrian F. M. Smith, Jozef L. Teugels*
Editors Emeriti: *Vic Barnett, J. Stuart Hunter, David G. Kendall*

The ***Wiley Series in Probability and Statistics*** is well established and authoritative. It covers many topics of current research interest in both pure and applied statistics and probability theory. Written by leading statisticians and institutions, the titles span both state-of-the-art developments in the field and classical methods.

Reflecting the wide range of current research in statistics, the series encompasses applied, methodological and theoretical statistics, ranging from applications and new techniques made possible by advances in computerized practice to rigorous treatment of theoretical approaches.

This series provides essential and invaluable reading for all statisticians, whether in academia, industry, government, or research.

ABRAHAM and LEDOLTER · Statistical Methods for Forecasting
AGRESTI · Analysis of Ordinal Categorical Data
AGRESTI · An Introduction to Categorical Data Analysis
AGRESTI · Categorical Data Analysis, *Second Edition*
ALTMAN, GILL, and McDONALD · Numerical Issues in Statistical Computing for the Social Scientist
AMARATUNGA and CABRERA · Exploration and Analysis of DNA Microarray and Protein Array Data
ANDĚL · Mathematics of Chance
ANDERSON · An Introduction to Multivariate Statistical Analysis, *Third Edition*
* ANDERSON · The Statistical Analysis of Time Series
ANDERSON, AUQUIER, HAUCK, OAKES, VANDAELE, and WEISBERG · Statistical Methods for Comparative Studies
ANDERSON and LOYNES · The Teaching of Practical Statistics
ARMITAGE and DAVID (editors) · Advances in Biometry
ARNOLD, BALAKRISHNAN, and NAGARAJA · Records
* ARTHANARI and DODGE · Mathematical Programming in Statistics
* BAILEY · The Elements of Stochastic Processes with Applications to the Natural Sciences
BALAKRISHNAN and KOUTRAS · Runs and Scans with Applications
BARNETT · Comparative Statistical Inference, *Third Edition*
BARNETT and LEWIS · Outliers in Statistical Data, *Third Edition*
BARTOSZYNSKI and NIEWIADOMSKA-BUGAJ · Probability and Statistical Inference
BASILEVSKY · Statistical Factor Analysis and Related Methods: Theory and Applications
BASU and RIGDON · Statistical Methods for the Reliability of Repairable Systems
BATES and WATTS · Nonlinear Regression Analysis and Its Applications
BECHHOFER, SANTNER, and GOLDSMAN · Design and Analysis of Experiments for Statistical Selection, Screening, and Multiple Comparisons
BELSLEY · Conditioning Diagnostics: Collinearity and Weak Data in Regression

*Now available in a lower priced paperback edition in the Wiley Classics Library.
†Now available in a lower priced paperback edition in the Wiley–Interscience Paperback Series.

† BELSLEY, KUH, and WELSCH · Regression Diagnostics: Identifying Influential Data and Sources of Collinearity
BENDAT and PIERSOL · Random Data: Analysis and Measurement Procedures, *Third Edition*
BERRY, CHALONER, and GEWEKE · Bayesian Analysis in Statistics and Econometrics: Essays in Honor of Arnold Zellner
BERNARDO and SMITH · Bayesian Theory
BHAT and MILLER · Elements of Applied Stochastic Processes, *Third Edition*
BHATTACHARYA and WAYMIRE · Stochastic Processes with Applications
† BIEMER, GROVES, LYBERG, MATHIOWETZ, and SUDMAN · Measurement Errors in Surveys
BILLINGSLEY · Convergence of Probability Measures, *Second Edition*
BILLINGSLEY · Probability and Measure, *Third Edition*
BIRKES and DODGE · Alternative Methods of Regression
BLISCHKE AND MURTHY (editors) · Case Studies in Reliability and Maintenance
BLISCHKE AND MURTHY · Reliability: Modeling, Prediction, and Optimization
BLOOMFIELD · Fourier Analysis of Time Series: An Introduction, *Second Edition*
BOLLEN · Structural Equations with Latent Variables
BOROVKOV · Ergodicity and Stability of Stochastic Processes
BOULEAU · Numerical Methods for Stochastic Processes
BOX · Bayesian Inference in Statistical Analysis
BOX · R. A. Fisher, the Life of a Scientist
BOX and DRAPER · Empirical Model-Building and Response Surfaces
* BOX and DRAPER · Evolutionary Operation: A Statistical Method for Process Improvement
BOX, HUNTER, and HUNTER · Statistics for Experimenters: Design, Innovation, and Discovery, *Second Editon*
BOX and LUCEÑO · Statistical Control by Monitoring and Feedback Adjustment
BRANDIMARTE · Numerical Methods in Finance: A MATLAB-Based Introduction
BROWN and HOLLANDER · Statistics: A Biomedical Introduction
BRUNNER, DOMHOF, and LANGER · Nonparametric Analysis of Longitudinal Data in Factorial Experiments
BUCKLEW · Large Deviation Techniques in Decision, Simulation, and Estimation
CAIROLI and DALANG · Sequential Stochastic Optimization
CASTILLO, HADI, BALAKRISHNAN, and SARABIA · Extreme Value and Related Models with Applications in Engineering and Science
CHAN · Time Series: Applications to Finance
CHARALAMBIDES · Combinatorial Methods in Discrete Distributions
CHATTERJEE and HADI · Sensitivity Analysis in Linear Regression
CHATTERJEE and PRICE · Regression Analysis by Example, *Third Edition*
CHERNICK · Bootstrap Methods: A Practitioner's Guide
CHERNICK and FRIIS · Introductory Biostatistics for the Health Sciences
CHILÈS and DELFINER · Geostatistics: Modeling Spatial Uncertainty
CHOW and LIU · Design and Analysis of Clinical Trials: Concepts and Methodologies, *Second Edition*
CLARKE and DISNEY · Probability and Random Processes: A First Course with Applications, *Second Edition*
* COCHRAN and COX · Experimental Designs, *Second Edition*
CONGDON · Applied Bayesian Modelling
CONGDON · Bayesian Statistical Modelling
CONOVER · Practical Nonparametric Statistics, *Third Edition*
COOK · Regression Graphics
COOK and WEISBERG · Applied Regression Including Computing and Graphics

*Now available in a lower priced paperback edition in the Wiley Classics Library.
†Now available in a lower priced paperback edition in the Wiley–Interscience Paperback Series.

COOK and WEISBERG · An Introduction to Regression Graphics
CORNELL · Experiments with Mixtures, Designs, Models, and the Analysis of Mixture Data, *Third Edition*
COVER and THOMAS · Elements of Information Theory
COX · A Handbook of Introductory Statistical Methods
* COX · Planning of Experiments
CRESSIE · Statistics for Spatial Data, *Revised Edition*
CSÖRGŐ and HORVÁTH · Limit Theorems in Change Point Analysis
DANIEL · Applications of Statistics to Industrial Experimentation
DANIEL · Biostatistics: A Foundation for Analysis in the Health Sciences, *Eighth Edition*
* DANIEL · Fitting Equations to Data: Computer Analysis of Multifactor Data, *Second Edition*
DASU and JOHNSON · Exploratory Data Mining and Data Cleaning
DAVID and NAGARAJA · Order Statistics, *Third Edition*
* DEGROOT, FIENBERG, and KADANE · Statistics and the Law
DEL CASTILLO · Statistical Process Adjustment for Quality Control
DEMARIS · Regression with Social Data: Modeling Continuous and Limited Response Variables
DEMIDENKO · Mixed Models: Theory and Applications
DENISON, HOLMES, MALLICK and SMITH · Bayesian Methods for Nonlinear Classification and Regression
DETTE and STUDDEN · The Theory of Canonical Moments with Applications in Statistics, Probability, and Analysis
DEY and MUKERJEE · Fractional Factorial Plans
DILLON and GOLDSTEIN · Multivariate Analysis: Methods and Applications
DODGE · Alternative Methods of Regression
* DODGE and ROMIG · Sampling Inspection Tables, *Second Edition*
* DOOB · Stochastic Processes
DOWDY, WEARDEN, and CHILKO · Statistics for Research, *Third Edition*
DRAPER and SMITH · Applied Regression Analysis, *Third Edition*
DRYDEN and MARDIA · Statistical Shape Analysis
DUDEWICZ and MISHRA · Modern Mathematical Statistics
DUNN and CLARK · Basic Statistics: A Primer for the Biomedical Sciences, *Third Edition*
DUPUIS and ELLIS · A Weak Convergence Approach to the Theory of Large Deviations
* ELANDT-JOHNSON and JOHNSON · Survival Models and Data Analysis
ENDERS · Applied Econometric Time Series
ETHIER and KURTZ · Markov Processes: Characterization and Convergence
EVANS, HASTINGS, and PEACOCK · Statistical Distributions, *Third Edition*
FELLER · An Introduction to Probability Theory and Its Applications, Volume I, *Third Edition,* Revised; Volume II, *Second Edition*
FISHER and VAN BELLE · Biostatistics: A Methodology for the Health Sciences
FITZMAURICE, LAIRD, and WARE · Applied Longitudinal Analysis
* FLEISS · The Design and Analysis of Clinical Experiments
FLEISS · Statistical Methods for Rates and Proportions, *Third Edition*
FLEMING and HARRINGTON · Counting Processes and Survival Analysis
FULLER · Introduction to Statistical Time Series, *Second Edition*
FULLER · Measurement Error Models
GALLANT · Nonlinear Statistical Models
GHOSH, MUKHOPADHYAY, and SEN · Sequential Estimation
GIESBRECHT and GUMPERTZ · Planning, Construction, and Statistical Analysis of Comparative Experiments
GIFI · Nonlinear Multivariate Analysis

*Now available in a lower priced paperback edition in the Wiley Classics Library.
†Now available in a lower priced paperback edition in the Wiley–Interscience Paperback Series.

GIVENS and HOETING · Computational Statistics
GLASSERMAN and YAO · Monotone Structure in Discrete-Event Systems
GNANADESIKAN · Methods for Statistical Data Analysis of Multivariate Observations, *Second Edition*
GOLDSTEIN and LEWIS · Assessment: Problems, Development, and Statistical Issues
GREENWOOD and NIKULIN · A Guide to Chi-Squared Testing
GROSS and HARRIS · Fundamentals of Queueing Theory, *Third Edition*
† GROVES · Survey Errors and Survey Costs
* HAHN and SHAPIRO · Statistical Models in Engineering
HAHN and MEEKER · Statistical Intervals: A Guide for Practitioners
HALD · A History of Probability and Statistics and their Applications Before 1750
HALD · A History of Mathematical Statistics from 1750 to 1930
† HAMPEL · Robust Statistics: The Approach Based on Influence Functions
HANNAN and DEISTLER · The Statistical Theory of Linear Systems
HEIBERGER · Computation for the Analysis of Designed Experiments
HEDAYAT and SINHA · Design and Inference in Finite Population Sampling
HELLER · MACSYMA for Statisticians
HINKELMANN and KEMPTHORNE · Design and Analysis of Experiments, Volume 1: Introduction to Experimental Design
HINKELMANN and KEMPTHORNE · Design and Analysis of Experiments, Volume 2: Advanced Experimental Design
HOAGLIN, MOSTELLER, and TUKEY · Exploratory Approach to Analysis of Variance
HOAGLIN, MOSTELLER, and TUKEY · Exploring Data Tables, Trends and Shapes
* HOAGLIN, MOSTELLER, and TUKEY · Understanding Robust and Exploratory Data Analysis
HOCHBERG and TAMHANE · Multiple Comparison Procedures
HOCKING · Methods and Applications of Linear Models: Regression and the Analysis of Variance, *Second Edition*
HOEL · Introduction to Mathematical Statistics, *Fifth Edition*
HOGG and KLUGMAN · Loss Distributions
HOLLANDER and WOLFE · Nonparametric Statistical Methods, *Second Edition*
HOSMER and LEMESHOW · Applied Logistic Regression, *Second Edition*
HOSMER and LEMESHOW · Applied Survival Analysis: Regression Modeling of Time to Event Data
† HUBER · Robust Statistics
HUBERTY · Applied Discriminant Analysis
HUNT and KENNEDY · Financial Derivatives in Theory and Practice
HUSKOVA, BERAN, and DUPAC · Collected Works of Jaroslav Hajek—with Commentary
HUZURBAZAR · Flowgraph Models for Multistate Time-to-Event Data
IMAN and CONOVER · A Modern Approach to Statistics
† JACKSON · A User's Guide to Principle Components
JOHN · Statistical Methods in Engineering and Quality Assurance
JOHNSON · Multivariate Statistical Simulation
JOHNSON and BALAKRISHNAN · Advances in the Theory and Practice of Statistics: A Volume in Honor of Samuel Kotz
JOHNSON and BHATTACHARYYA · Statistics: Principles and Methods, *Fifth Edition*
JOHNSON and KOTZ · Distributions in Statistics
JOHNSON and KOTZ (editors) · Leading Personalities in Statistical Sciences: From the Seventeenth Century to the Present
JOHNSON, KOTZ, and BALAKRISHNAN · Continuous Univariate Distributions, Volume 1, *Second Edition*

*Now available in a lower priced paperback edition in the Wiley Classics Library.
†Now available in a lower priced paperback edition in the Wiley–Interscience Paperback Series.

JOHNSON, KOTZ, and BALAKRISHNAN · Continuous Univariate Distributions, Volume 2, *Second Edition*

JOHNSON, KOTZ, and BALAKRISHNAN · Discrete Multivariate Distributions

JOHNSON, KOTZ, and KEMP · Univariate Discrete Distributions, *Second Edition*

JUDGE, GRIFFITHS, HILL, LÜTKEPOHL, and LEE · The Theory and Practice of Econometrics, *Second Edition*

JUREČKOVÁ and SEN · Robust Statistical Procedures: Aymptotics and Interrelations

JUREK and MASON · Operator-Limit Distributions in Probability Theory

KADANE · Bayesian Methods and Ethics in a Clinical Trial Design

KADANE AND SCHUM · A Probabilistic Analysis of the Sacco and Vanzetti Evidence

KALBFLEISCH and PRENTICE · The Statistical Analysis of Failure Time Data, *Second Edition*

KASS and VOS · Geometrical Foundations of Asymptotic Inference

† KAUFMAN and ROUSSEEUW · Finding Groups in Data: An Introduction to Cluster Analysis

KEDEM and FOKIANOS · Regression Models for Time Series Analysis

KENDALL, BARDEN, CARNE, and LE · Shape and Shape Theory

KHURI · Advanced Calculus with Applications in Statistics, *Second Edition*

KHURI, MATHEW, and SINHA · Statistical Tests for Mixed Linear Models

* KISH · Statistical Design for Research

KLEIBER and KOTZ · Statistical Size Distributions in Economics and Actuarial Sciences

KLUGMAN, PANJER, and WILLMOT · Loss Models: From Data to Decisions, *Second Edition*

KLUGMAN, PANJER, and WILLMOT · Solutions Manual to Accompany Loss Models: From Data to Decisions, *Second Edition*

KOTZ, BALAKRISHNAN, and JOHNSON · Continuous Multivariate Distributions, Volume 1, *Second Edition*

KOTZ and JOHNSON (editors) · Encyclopedia of Statistical Sciences: Volumes 1 to 9 with Index

KOTZ and JOHNSON (editors) · Encyclopedia of Statistical Sciences: Supplement Volume

KOTZ, READ, and BANKS (editors) · Encyclopedia of Statistical Sciences: Update Volume 1

KOTZ, READ, and BANKS (editors) · Encyclopedia of Statistical Sciences: Update Volume 2

KOVALENKO, KUZNETZOV, and PEGG · Mathematical Theory of Reliability of Time-Dependent Systems with Practical Applications

LACHIN · Biostatistical Methods: The Assessment of Relative Risks

LAD · Operational Subjective Statistical Methods: A Mathematical, Philosophical, and Historical Introduction

LAMPERTI · Probability: A Survey of the Mathematical Theory, *Second Edition*

LANGE, RYAN, BILLARD, BRILLINGER, CONQUEST, and GREENHOUSE · Case Studies in Biometry

LARSON · Introduction to Probability Theory and Statistical Inference, *Third Edition*

LAWLESS · Statistical Models and Methods for Lifetime Data, *Second Edition*

LAWSON · Statistical Methods in Spatial Epidemiology

LE · Applied Categorical Data Analysis

LE · Applied Survival Analysis

LEE and WANG · Statistical Methods for Survival Data Analysis, *Third Edition*

LEPAGE and BILLARD · Exploring the Limits of Bootstrap

LEYLAND and GOLDSTEIN (editors) · Multilevel Modelling of Health Statistics

LIAO · Statistical Group Comparison

LINDVALL · Lectures on the Coupling Method

*Now available in a lower priced paperback edition in the Wiley Classics Library.

†Now available in a lower priced paperback edition in the Wiley–Interscience Paperback Series.

LINHART and ZUCCHINI · Model Selection
LITTLE and RUBIN · Statistical Analysis with Missing Data, *Second Edition*
LLOYD · The Statistical Analysis of Categorical Data
MAGNUS and NEUDECKER · Matrix Differential Calculus with Applications in Statistics and Econometrics, *Revised Edition*
MALLER and ZHOU · Survival Analysis with Long Term Survivors
MALLOWS · Design, Data, and Analysis by Some Friends of Cuthbert Daniel
MANN, SCHAFER, and SINGPURWALLA · Methods for Statistical Analysis of Reliability and Life Data
MANTON, WOODBURY, and TOLLEY · Statistical Applications Using Fuzzy Sets
MARCHETTE · Random Graphs for Statistical Pattern Recognition
MARDIA and JUPP · Directional Statistics
MASON, GUNST, and HESS · Statistical Design and Analysis of Experiments with Applications to Engineering and Science, *Second Edition*
McCULLOCH and SEARLE · Generalized, Linear, and Mixed Models
McFADDEN · Management of Data in Clinical Trials
* McLACHLAN · Discriminant Analysis and Statistical Pattern Recognition
McLACHLAN, DO, and AMBROISE · Analyzing Microarray Gene Expression Data
McLACHLAN and KRISHNAN · The EM Algorithm and Extensions
McLACHLAN and PEEL · Finite Mixture Models
McNEIL · Epidemiological Research Methods
MEEKER and ESCOBAR · Statistical Methods for Reliability Data
MEERSCHAERT and SCHEFFLER · Limit Distributions for Sums of Independent Random Vectors: Heavy Tails in Theory and Practice
MICKEY, DUNN, and CLARK · Applied Statistics: Analysis of Variance and Regression, *Third Edition*
* MILLER · Survival Analysis, *Second Edition*
MONTGOMERY, PECK, and VINING · Introduction to Linear Regression Analysis, *Third Edition*
MORGENTHALER and TUKEY · Configural Polysampling: A Route to Practical Robustness
MUIRHEAD · Aspects of Multivariate Statistical Theory
MULLER and STOYAN · Comparison Methods for Stochastic Models and Risks
MURRAY · X-STAT 2.0 Statistical Experimentation, Design Data Analysis, and Nonlinear Optimization
MURTHY, XIE, and JIANG · Weibull Models
MYERS and MONTGOMERY · Response Surface Methodology: Process and Product Optimization Using Designed Experiments, *Second Edition*
MYERS, MONTGOMERY, and VINING · Generalized Linear Models. With Applications in Engineering and the Sciences
† NELSON · Accelerated Testing, Statistical Models, Test Plans, and Data Analyses
† NELSON · Applied Life Data Analysis
NEWMAN · Biostatistical Methods in Epidemiology
OCHI · Applied Probability and Stochastic Processes in Engineering and Physical Sciences
OKABE, BOOTS, SUGIHARA, and CHIU · Spatial Tesselations: Concepts and Applications of Voronoi Diagrams, *Second Edition*
OLIVER and SMITH · Influence Diagrams, Belief Nets and Decision Analysis
PALTA · Quantitative Methods in Population Health: Extensions of Ordinary Regressions
PANKRATZ · Forecasting with Dynamic Regression Models
PANKRATZ · Forecasting with Univariate Box-Jenkins Models: Concepts and Cases
* PARZEN · Modern Probability Theory and Its Applications
PEÑA, TIAO, and TSAY · A Course in Time Series Analysis
PIANTADOSI · Clinical Trials: A Methodologic Perspective

*Now available in a lower priced paperback edition in the Wiley Classics Library.
†Now available in a lower priced paperback edition in the Wiley–Interscience Paperback Series.

PORT · Theoretical Probability for Applications
POURAHMADI · Foundations of Time Series Analysis and Prediction Theory
PRESS · Bayesian Statistics: Principles, Models, and Applications
PRESS · Subjective and Objective Bayesian Statistics, *Second Edition*
PRESS and TANUR · The Subjectivity of Scientists and the Bayesian Approach
PUKELSHEIM · Optimal Experimental Design
PURI, VILAPLANA, and WERTZ · New Perspectives in Theoretical and Applied Statistics
† PUTERMAN · Markov Decision Processes: Discrete Stochastic Dynamic Programming
QIU · Image Processing and Jump Regression Analysis
* RAO · Linear Statistical Inference and Its Applications, *Second Edition*
RAUSAND and HØYLAND · System Reliability Theory: Models, Statistical Methods, and Applications, *Second Edition*
RENCHER · Linear Models in Statistics
RENCHER · Methods of Multivariate Analysis, *Second Edition*
RENCHER · Multivariate Statistical Inference with Applications
* RIPLEY · Spatial Statistics
RIPLEY · Stochastic Simulation
ROBINSON · Practical Strategies for Experimenting
ROHATGI and SALEH · An Introduction to Probability and Statistics, *Second Edition*
ROLSKI, SCHMIDLI, SCHMIDT, and TEUGELS · Stochastic Processes for Insurance and Finance
ROSENBERGER and LACHIN · Randomization in Clinical Trials: Theory and Practice
ROSS · Introduction to Probability and Statistics for Engineers and Scientists
† ROUSSEEUW and LEROY · Robust Regression and Outlier Detection
* RUBIN · Multiple Imputation for Nonresponse in Surveys
RUBINSTEIN · Simulation and the Monte Carlo Method
RUBINSTEIN and MELAMED · Modern Simulation and Modeling
RYAN · Modern Regression Methods
RYAN · Statistical Methods for Quality Improvement, *Second Edition*
SALTELLI, CHAN, and SCOTT (editors) · Sensitivity Analysis
* SCHEFFE · The Analysis of Variance
SCHIMEK · Smoothing and Regression: Approaches, Computation, and Application
SCHOTT · Matrix Analysis for Statistics, *Second Edition*
SCHOUTENS · Levy Processes in Finance: Pricing Financial Derivatives
SCHUSS · Theory and Applications of Stochastic Differential Equations
SCOTT · Multivariate Density Estimation: Theory, Practice, and Visualization
* SEARLE · Linear Models
SEARLE · Linear Models for Unbalanced Data
SEARLE · Matrix Algebra Useful for Statistics
SEARLE, CASELLA, and McCULLOCH · Variance Components
SEARLE and WILLETT · Matrix Algebra for Applied Economics
SEBER and LEE · Linear Regression Analysis, *Second Edition*
† SEBER · Multivariate Observations
† SEBER and WILD · Nonlinear Regression
SENNOTT · Stochastic Dynamic Programming and the Control of Queueing Systems
* SERFLING · Approximation Theorems of Mathematical Statistics
SHAFER and VOVK · Probability and Finance: It's Only a Game!
SILVAPULLE and SEN · Constrained Statistical Inference: Inequality, Order, and Shape Restrictions
SMALL and McLEISH · Hilbert Space Methods in Probability and Statistical Inference
SRIVASTAVA · Methods of Multivariate Statistics
STAPLETON · Linear Statistical Models

*Now available in a lower priced paperback edition in the Wiley Classics Library.
†Now available in a lower priced paperback edition in the Wiley–Interscience Paperback Series.

STAUDTE and SHEATHER · Robust Estimation and Testing
STOYAN, KENDALL, and MECKE · Stochastic Geometry and Its Applications, *Second Edition*
STOYAN and STOYAN · Fractals, Random Shapes and Point Fields: Methods of Geometrical Statistics
STYAN · The Collected Papers of T. W. Anderson: 1943–1985
SUTTON, ABRAMS, JONES, SHELDON, and SONG · Methods for Meta-Analysis in Medical Research
TANAKA · Time Series Analysis: Nonstationary and Noninvertible Distribution Theory
THOMPSON · Empirical Model Building
THOMPSON · Sampling, *Second Edition*
THOMPSON · Simulation: A Modeler's Approach
THOMPSON and SEBER · Adaptive Sampling
THOMPSON, WILLIAMS, and FINDLAY · Models for Investors in Real World Markets
TIAO, BISGAARD, HILL, PEÑA, and STIGLER (editors) · Box on Quality and Discovery: with Design, Control, and Robustness
TIERNEY · LISP-STAT: An Object-Oriented Environment for Statistical Computing and Dynamic Graphics
TSAY · Analysis of Financial Time Series
UPTON and FINGLETON · Spatial Data Analysis by Example, Volume II: Categorical and Directional Data
VAN BELLE · Statistical Rules of Thumb
VAN BELLE, FISHER, HEAGERTY, and LUMLEY · Biostatistics: A Methodology for the Health Sciences, *Second Edition*
VESTRUP · The Theory of Measures and Integration
VIDAKOVIC · Statistical Modeling by Wavelets
VINOD and REAGLE · Preparing for the Worst: Incorporating Downside Risk in Stock Market Investments
WALLER and GOTWAY · Applied Spatial Statistics for Public Health Data
WEERAHANDI · Generalized Inference in Repeated Measures: Exact Methods in MANOVA and Mixed Models
WEISBERG · Applied Linear Regression, *Third Edition*
WELSH · Aspects of Statistical Inference
WESTFALL and YOUNG · Resampling-Based Multiple Testing: Examples and Methods for p-Value Adjustment
WHITTAKER · Graphical Models in Applied Multivariate Statistics
WINKER · Optimization Heuristics in Economics: Applications of Threshold Accepting
WONNACOTT and WONNACOTT · Econometrics, *Second Edition*
WOODING · Planning Pharmaceutical Clinical Trials: Basic Statistical Principles
WOODWORTH · Biostatistics: A Bayesian Introduction
WOOLSON and CLARKE · Statistical Methods for the Analysis of Biomedical Data, *Second Edition*
WU and HAMADA · Experiments: Planning, Analysis, and Parameter Design Optimization
YANG · The Construction Theory of Denumerable Markov Processes
* ZELLNER · An Introduction to Bayesian Inference in Econometrics
ZHOU, OBUCHOWSKI, and McCLISH · Statistical Methods in Diagnostic Medicine

*Now available in a lower priced paperback edition in the Wiley Classics Library.
†Now available in a lower priced paperback edition in the Wiley–Interscience Paperback Series.

Tue	1/19	1.1 - 2.6	ch.1) 1, 2, 3, 4	ch.2 1-3
Thu	1/21	2.7 - 2.14	5, 7, 11, 13, 14, 15	
Tue	1/26	3.1 - 3.4	1, 3, 5, 7, 9	
Thu	1/28	3.5 - 3.6	11, 15, 16, 18	
Tue	2/2	3.7 - 3.8, App 3A	21, 24, 26, 27	
Thu	2/4	4.1	1	
Tue	2/9	4.2 - 4.3	2, 3	
Thu	2/11	4.4 - 4.5, App 4A, 4D	4, 5	
Tue	2/16	Exam 1		
Thu	2/18	5.1 - 5.5	1, 3, 4, 5	

Designing an experiment is like gambling with the devil: Only a random strategy can defeat all his betting systems. (R. A. Fisher)

The most exciting phrase to hear in science, the one that heralds discoveries, is not "Eureka!" but "Now that's funny..." (Isaac Asimov)

Anyone who has never make a mistake has never tried anything new. (Albert Einstein)

Not everything that can be counted counts and not everything that counts can be counted. (Albert Einstein)

Compensatory **adjustment** *is essential when, even though we can correctly assign the cause of disturbances, we cannot otherwise eliminate them.*

The idea of a process in a perfect state of control contravenes the Second Law of Thermodynamics: Thus an exact state of control is unrealizable, and must be regarded as a purely theoretical, concept.

We should not be afraid of finding out something.

Seek computer programs that allow you to do the thinking.

When the ratio of the largest to the smallest observation y_{max}/y_{min}, *is large, you should question whether the data are being analyzed in the right metric (transformation).*

Original data should be presented in a way that will preserve the evidence in the original data. (Walter Shewhart)

You can see a lot by just looking. (Yogi Berra)

A computer should make both calculations and graphs. Both sorts of output should be studied; each will contribute to understanding. (F. J. Anscombe)

Murphy works hard to ensure that anything that can go wrong will go wrong. With an adequate system of process monitoring, therefore, more and more of the things that can go wrong will be corrected and more and more of Murphy's tricks will be permanently stymied.

There's never been a signal without noise.